Fabrizio Maria Maggi · Henrik Leopold ·
Flavia Santoro · Avigdor Gal
Editors

Business Process Management

24th International Conference, BPM 2026
Toronto, ON, Canada, September 27–October 2, 2026
Proceedings

Springer

Editors
Fabrizio Maria Maggi
Free University of Bozen-Bolzano
Bozen-Bolzano, Italy

Henrik Leopold
Kühne Logistics University
Hamburg, Hamburg, Germany

Flavia Santoro
Universidade do Estado do Rio de Janeiro
Rio de Janeiro, Rio de Janeiro, Brazil

Avigdor Gal
Israel Institute of Technology
Haifa, Israel

ISSN 0302-9743 ISSN 1611-3349 (electronic)
Lecture Notes in Computer Science
ISBN 978-3-032-37776-0 ISBN 978-3-032-37777-7 (eBook)
https://doi.org/10.1007/978-3-032-37777-7

This Springer imprint is published by the registered company Springer Nature Switzerland AG
The registered company address is: Gewerbestrasse 11, 6330 Cham, Switzerland

Preface

This volume contains the papers presented at BPM 2026: the 24th International Conference on Business Process Management, held during September 27– October 2, 2026 at York University in Toronto, Canada. For only the second time in its history, the conference returned to North America, bringing together researchers, practitioners, and industry leaders in a city defined by its diversity and energy. As the premier event in the field, the conference continued its tradition of bringing together academic and practitioner perspectives on the science and adoption of business processes. As in previous years, keynotes and presentations returned repeatedly to a central question: how to combine promising new methodologies and technologies, particularly agentic and generative AI, with responsible, process-aware practice.

In addition to the main conference, there was a multitude of affiliated events, including 9 workshops, four fora (Industry, Educators, Responsible BPM, and Process Technology), tutorials, a journal-first track, a demos and resources track, and a doctoral consortium.

BPM 2026 followed the history and philosophy of previous editions with respect to the three main research tracks reflecting the different communities of the conference series. Track I (Foundations, chaired by Fabrizio Maria Maggi) was dedicated to the underlying principles of BPM, including computational theories, formal methods, novel problems and languages, architectures, conceptual frameworks, taxonomies, techniques, and algorithms. Track II (Engineering, chaired by Henrik Leopold) dealt with engineering aspects of BPM research, covering tools and techniques for process modeling, the enactment of process models (including software infrastructure), and process monitoring and analytics (including process mining). Track III (Management, chaired by Flavia Maria Santoro) attracted papers focusing on the socio-technical, cognitive, and organizational aspects of BPM, as well as its managerial and governance dimensions, aiming to advance our understanding of how BPM concepts, models, methods, and information systems can transform organizations. Avigdor Gal served as the Consolidation Chair.

This year, the conference received a total of 169 submissions. 57 were submitted to Track I (Foundations), 73 to Track II (Engineering), and 39 to Track III (Management). The review process followed the established high-quality standards of the BPM conference series, with double-blind review maintained throughout. Each paper was reviewed by an average of three carefully selected members of the Program Committee of the respective track. An extensive discussion between the reviewers and a Senior Program Committee member followed, captured in a meta-review offering recommendations to the Track Chairs, who made the final acceptance decisions. This thorough process resulted in 11 accepted papers for Track I (19% acceptance rate), 21 for Track II (29%), and 10 for Track III (26%), leading to an overall acceptance rate of 25% and 42 accepted papers in total. Moreover, the review process resulted in the inclusion of 24 papers in the BPM Forum program, to be published in a separate volume of the Springer LNBIP series.

The program also included three invited keynote presentations covering contemporary topics in Business Process Management. Tom Davenport, from Babson College, spoke on "The Marriage of Process and AI," describing the obstacles organizations face in redesigning their processes to capture value from AI and the emerging role of process intelligence in that relationship. Opher Baron, from Rotman School of Management, University of Toronto and SiMLQ, presented on business process intelligence in congestion-driven systems, illustrating how process mining, queueing theory, simulation, and machine learning combine into AI-enabled digital twins for real-time operational decision-making. Emmanuelle Vaast, from McGill University's Desautels Faculty of Management, discussed how the growing availability of digital trace data reshapes the study of organizational processes, calling for a productive dialogue between computational and qualitative approaches to process research. In addition to the keynote abstracts, this volume also includes abstracts accompanying the tutorials.

The research papers in this volume show the wide variety of topics and methods that characterize the BPM community today. These proceedings report on academic insights obtained via various methodologies, including behavioural science (e.g., case studies), computational process science, and design science (e.g., artefact development). The topics presented and discussed at BPM 2026 span the discovery, prediction, and improvement of business processes: from event log extraction and generation to predictive and prescriptive process monitoring, causal process interventions, and object-centric and federated process mining. A recurring thread is the rise of agentic and LLM-based approaches to process modeling and automation, alongside the governance, quality, and fairness questions these raise as AI agents take on a more active role in business processes.

Open Science continued to be a major principle for the BPM conference, aiming at reproducibility and replicability of research results. Authors were explicitly requested to link one or more repositories with additional artefacts, such as data sets, prototypes, and interview protocols, to their papers. Papers that made artefacts available continue to be recognized with an "Artifacts Available" badge in these proceedings, following the ACM's artefact review and badging guidelines.

We also encouraged all participants to consider diversity, equity, and inclusion in their writing, reviews, presentations, and all other interactions related to the BPM conference. We are grateful for the work of the Diversity, Equity & Inclusion Chairs, Shazia Sadiq, Flavia Maria Santoro, and Araham Jesus Martinez Lagunas.

We would like to thank all authors, both regular and senior members of the Program Committees, and all external reviewers. Their diligent responses and effective collaboration made a rigorous, extensive, and timely review process possible and enabled the high-quality research reflected in this volume. We are equally grateful to the chairs and committees of the various workshops – led by Xixi Lu, Jorge Muñoz-Gama, and Daniel Beverungen as Workshop Management Chairs – the four fora, the Demonstration and Resources Track (chaired by Iris Beerepoot, Martin Kabierski, and Andrea Delgado), the Doctoral Consortium (chaired by Andrea Marrella, Jana-Rebecca Rehse, and Pnina Soffer), the Tutorial & Panel track (chaired by Daniel Amyot, Adela del Río-Ortega, and Thomas Grisold), and the Journal-First Track (chaired by Sarah Winkler, Sander

Leemans, and Carl Corea), all of whom did an outstanding job reviewing and selecting high-quality contributions across the program.

Weaving across all these tracks, Publicity Chairs Mar´ıa Salas Urbano and Jari Peeperkorn energetically mobilized the BPM research community, while Proceedings Chairs Johannes De Smedt and Alessandro Gianola provided outstanding support in preparing the conference and forum proceedings.

We acknowledge and are grateful to the sponsors of BPM 2026. Without their support such an event would not have been possible. We also appreciate the use of EasyChair for streamlining an intensive reviewing process, and CEUR Workshop Proceedings for supporting the publication of the workshop and forum proceedings.

Finally, we would like to express our special appreciation to Arik Senderovich and Ravi de Costa from York University as the General Chairs of BPM 2026. Together with the local organizing team – Jennifer Falzon, Melissa Zilberberg, Nathalia Wittingham, and Nazrin Vakilova – they created a flawless conference fostering inspiration, intellectual debate, and new friendships.

In conclusion, we hope that BPM 2026, a memorable week in Toronto, will remain as much a highlight for our fellow academic colleagues as these proceedings will provide a lasting resource for those dedicated to responsible Business Process Management.

September 2026

Fabrizio Maria Maggi
Henrik Leopold
Flavia Santoro
Avigdor Gal

Organization

Steering Committee

Jan Mendling (Chair)	Humboldt Universität zu Berlin, Germany
Avigdor Gal	Technion – Israel Institute of Technology, Israel
Chiara Ghidini	University of Trento, Italy
Manfred Reichert	Ulm University, Germany
Hajo A. Reijers	Utrecht University, the Netherlands
Stefanie Rinderle-Ma	Technical University of Munich, Germany
Adela del Río Ortega	University of Seville, Spain
Michael Rosemann	Queensland University of Technology, Australia
Flavia Santoro	Institute of Technology and Leadership (Inteli), Brazil
Shazia Sadiq	University of Queensland, Australia
Barbara Weber	University of St. Gallen, Switzerland
Matthias Weidlich	Humboldt Universität zu Berlin, Germany

Advisory Board

Wil van der Aalst	RWTH Aachen University, Germany
Mathias Weske	HPI, University of Potsdam, Germany
Marlon Dumas	University of Tartu, Estonia

Executive Committee

General Chairs

Arik Senderovich	York University, Canada
Ravi de Costa	York University, Canada

Main Conference Program Committee Chairs

Avigdor Gal	Technion - Israel Institute of Technology, Israel
Fabrizio Maria Maggi	Free University of Bozen-Bolzano, Italy
Henrik Leopold	Kühne Logistics University, Germany
Flavia Santoro	Institute of Technology and Leadership (Inteli), Brazil

Workshop Chairs

Xixi Lu	Utrecht University, The Netherlands
Jorge Muñoz-Gama	Pontificia Universidad Católica de Chile, Chile
Daniel Beverungen	Paderborn University, Germany

Demonstration and Resources Chairs

Iris Beerepoot	Utrecht University, The Netherlands
Martin Kabierski	University of Vienna, Austria
Andrea Delgado	Universidad de la República, Uruguay

Responsible BPM Forum Chairs

Marco Comuzzi	Ulsan National Institute of Science and Technology, South Korea
Stephan A. Fahrenkrog-Petersen	University of Liechtenstein, Liechtenstein
Amy Van Looy	Ghent University, Belgium

Process Technology Forum

Manfred Reichert	University of Ulm, Germany
Joerg Evermann	Memorial University of Newfoundland, Canada
Barbara Re	University of Camerino, Italy

Industry Forum Chairs

Najah El-Gharib	University of Ottawa, Canada
Julian Lebherz	SAP, Singapore
Fredrik Milani	University of Tartu, Estonia

Education Forum Chairs

Lucineia Heloisa Thom	Federal University of Rio Grande do Sul, Brazil
Karolin Winter	Eindhoven University of Technology, The Netherlands
Krzysztof Kluza	AGH University of Krakow, Poland

Tutorial Chairs

Daniel Amyot	University of Ottawa, Canada
Adela del Río-Ortega	University of Seville, Spain
Thomas Grisold	University of St. Gallen, Switzerland

Doctoral Consortium Chairs

Andrea Marrella	Sapienza University of Rome, Italy
Jana-Rebecca Rehse	University of Mannheim, Germany
Pnina Soffer	University of Haifa, Israel

Journal-First Track Chairs

Sarah Winkler	Free University of Bozen-Bolzano, Italy
Sander Leemans	RWTH Aachen University, Germany
Carl Corea	University of Koblenz, Germany

Publicity Chairs

María Salas Urbano	University of Seville, Spain
Jari Peeperkorn	KU Leuven, Belgium

Proceeding Chairs

Johannes De Smedt	KU Leuven, Belgium
Alessandro Gianola	Instituto Superior Técnico de Lisboa, Portugal

Diversity, Equity and Inclusion Chairs

Shazia Sadiq	University of Queensland, Australia
Flavia Santoro	Institute of Technology and Leadership (Inteli), Brazil
Araham Jesus Martinez Lagunas	Concordia University, Canada

Track IFoundations

Senior Program Committee

Johannes De Smedt	KU Leuven, Belgium
Claudio Di Ciccio	Utrecht University, Netherlands
Chiara Di Francescomarino	University of Trento, Italy
Thomas Hildebrandt	University of Copenhagen, Denmark
Sander J. J. Leemans	RWTH Aachen University, Germany
Xixi Lu	Utrecht University, Netherlands
Andrea Marrella	Sapienza University of Rome, Italy
Marco Montali	Free University of Bozen-Bolzano, Italy
Artem Polyvyanyy	University of Melbourne, Australia
Manfred Reichert	University of Ulm, Germany
Hagen Voelzer	University of St. Gallen, Switzerland
Matthias Weidlich	Humboldt-Universität zu Berlin, Germany
Mathias Weske	HPI, University of Potsdam, Germany
Wil van der Aalst	RWTH Aachen University, Germany
Jan Martijn van der Werf	Utrecht University, Netherlands

Program Committee

Simone Agostinelli	Universitas Mercatorum, Italy
Anti Alman	Free University of Bozen-Bolzano, Italy
Abel Armas Cervantes	University of Melbourne, Australia
Ahmed Awad	British University in Dubai, UAE
Mario Luca Bernardi	University of Sannio, Italy
Adam Burke	Queensland University of Technology, Australia
Izack Cohen	Bar-Ilan University, Israel
Marco Comuzzi	Ulsan National Institute of Science and Technology, South Korea
Patrick Delfmann	University of Koblenz, Germany
Jörg Desel	FernUniversität in Hagen, Germany
Ivan Donadello	Free University of Bozen-Bolzano, Italy
Rik Eshuis	Eindhoven University of Technology, Netherlands
Joerg Evermann	Memorial University of Newfoundland, Canada
Dirk Fahland	Eindhoven University of Technology, Netherlands
Peter Fettke	DFKI and Saarland University, Germany
Valeria Fionda	University of Calabria, Italy
Luciano García-Bañuelos	Tecnológico de Monterrey, Mexico
Laura Genga	Eindhoven University of Technology, Netherlands

Alessandro Gianola	University of Lisbon, Portugal
Georg Grossmann	University of South Australia, Australia
Maria Teresa Gómez López	University of Seville, Spain
Rick Hull	IBM, USA
Anna Kalenkova	University of Adelaide, Australia
Ekkart Kindler	Technical University of Denmark, Denmark
Akhil Kumar	Pennsylvania State University, USA
Francesco Leotta	Sapienza University of Rome, Italy
Hugo A. López	Technical University of Denmark, Denmark
Orlenys López-Pintado	University of Tartu, Estonia
Felix Mannhardt	Eindhoven University of Technology, Netherlands
Elisa Marengo	University of Turin, Italy
Chun Ouyang	Queensland University of Technology, Australia
Thomas Prinz	Friedrich Schiller University Jena, Germany
Barbara Re	University of Camerino, Italy
Daniel Ritter	SAP, Germany
Andrey Rivkin	Technical University of Denmark, Denmark
Massimiliano Ronzani	Fondazione Bruno Kessler, Italy
Lorenzo Rossi	University of Camerino, Italy
Tijs Slaats	University of Copenhagen, Denmark
Niek Tax	Meta, UK
Eric Verbeek	Eindhoven University of Technology, Netherlands
Sven Weinzierl	Friedrich-Alexander-Universität Erlangen-Nűrnberg, Germany
Sarah Winkler	Free University of Bozen-Bolzano, Italy

Track II: Engineering

Senior Program Committee

Boualem Benatallah	Dublin City University, Ireland
Andrea Burattin	Technical University of Denmark, Denmark
Marco Comuzzi	Ulsan National Institute of Science and Technology, South Korea
Jochen De Weerdt	KU Leuven, Belgium
Benoît Depaire	Hasselt University, Belgium
Remco Dijkman	Eindhoven University of Technology, Netherlands
Agnes Koschmider	University of Bayreuth, Germany
Niels Martin	Hasselt University, Belgium
Massimo Mecella	Sapienza University of Rome, Italy
Jan Mendling	Humboldt-Universität zu Berlin, Germany

Jorge Munoz-Gama	Pontificia Universidad Católica de Chile, Chile
Oscar Pastor	Universitat Politècnica de València, Spain
Luise Pufahl	Technical University of Munich, Germany
Manuel Resinas	University of Seville, Spain
Stefanie Rinderle-Ma	Technical University of Munich, Germany
Shazia Sadiq	University of Queensland, Australia
Marcos Sepúlveda	Pontificia Universidad Católica de Chile, Chile
Pnina Soffer	University of Haifa, Israel
Minseok Song	Pohang University of Science and Technology, South Korea
Barbara Weber	University of St. Gallen, Switzerland
Ingo Weber	Technical University of Munich, Germany
Moe Wynn	Queensland University of Technology, Australia
Francesca Zerbato	Eindhoven University of Technology, Netherlands
Massimiliano de Leoni	University of Padua, Italy
Boudewijn van Dongen	Eindhoven University of Technology, Netherlands
Han van der Aa	University of Vienna, Austria

Program Committee

Lars Ackermann-Igl	University of Bayreuth, Germany
Daniel Amyot	University of Ottawa, Canada
Robert Andrews	Queensland University of Technology, Australia
Michael Arias	University of Costa Rica, Costa Rica
Bernhard Axmann	Technische Hochschule Ingolstadt, Germany
Saimir Bala	Humboldt-Universität zu Berlin, Germany
Andrea Delgado	Universidad de la República, Uruguay
Claudia Diamantini	Università Politecnica delle Marche, Italy
Joerg Evermann	Memorial University of Newfoundland, Canada
Stephan Fahrenkrog-Petersen	University of Liechtenstein, Liechtenstein
Marcelo Fantinato	University of São Paulo, Brazil
Walid Fdhila	University of Vienna, Austria
Carlos Fernandez-Llatas	Universitat Politècnica de València, Spain
Walid Gaaloul	Télécom SudParis, France
Luciano García-Bañuelos	Tecnológico de Monterrey, Mexico
Laura Genga	Eindhoven University of Technology, Netherlands
José González Enríquez	University of Seville, Spain
Oscar González-Rojas	Universidad de los Andes, Colombia
Daniela Grigori	Paris-Dauphine University - PSL, France
Georg Grossmann	University of South Australia, Australia
Joscha Grüger	University of Trier, Germany
Lucinéia Heloisa Thom	Federal University of Rio Grande do Sul, Brazil

Andrés Jiménez Ramírez	University of Seville, Spain
Martin Kabierski	University of Vienna, Austria
Anna Kalenkova	University of Adelaide, Australia
Krzysztof Kluza	AGH University of Krakow, Poland
Julius Köpke	University of Klagenfurt, Austria
Manuel Lama Penin	University of Santiago de Compostela, Spain
Francesco Leotta	Sapienza University of Rome, Italy
Orlenys López-Pintado	University of Tartu, Estonia
Elisa Marengo	University of Turin, Italy
Sarajane Marques Peres	University of São Paulo, Brazil
Alfonso Marquez-Chamorro	University of Seville, Spain
Laura Maruster	University of Groningen, Netherlands
Renata Medeiros de Carvalho	Eindhoven University of Technology, Netherlands
Giovanni Meroni	Technical University of Denmark, Denmark
Andrea Morichetta	University of Camerino, Italy
Julian Neuberger	University of Bayreuth, Germany
Helen Paik	University of New South Wales, Australia
Cesare Pautasso	University of Lugano, Switzerland
Jari Peeperkorn	KU Leuven, Belgium
Pierluigi Plebani	Politecnico di Milano, Italy
Pascal Poizat	Université Paris Nanterre, France
Simon Poon	University of Sydney, Australia
Pascal Ravesteijn	Utrecht University of Applied Sciences, Netherlands
Barbara Re	University of Camerino, Italy
Kate Revoredo	Humboldt-Universität zu Berlin, Germany
Sareh Sadeghianasl	Queensland University of Technology, Australia
Stefan Schönig	Universität Regensburg, Germany
Natalia Sidorova	Eindhoven University of Technology, Netherlands
Vinicius Stein Dani	Utrecht University, Netherlands
Victoria Torres	Universitat Politècnica de València, Spain
Maxim Vidgof	WU Vienna, Austria
Karolin Winter	Eindhoven University of Technology, Netherlands
Roy Jing Yang	Queensland University of Technology, Australia
Nicola Zannone	Eindhoven University of Technology, Netherlands
Nick van Beest	CSIRO Data61, Australia
Sebastiaan van Zelst	Celonis, Germany

Track IIIManagement

Senior Program Committee

Fernanda Baião	Pontifical Catholic University of Rio de Janeiro, Brazil
Daniel Beverungen	Paderborn University, Germany
Paul Grefen	Eindhoven University of Technology, Netherlands
Thomas Grisold	University of St. Gallen, Switzerland
Mojca Indihar Štemberger	University of Ljubljana, Slovenia
Marta Indulska	University of Queensland, Australia
Christian Janiesch	TU Dortmund University, Germany
Mieke Jans	Hasselt University, Belgium
Ralf Plattfaut	University of Duisburg-Essen, Germany
Maximilian Röglinger	University of Bayreuth, Germany
Michael Rosemann	Queensland University of Technology, Australia
Peter Trkman	University of Ljubljana, Slovenia
Oktay Turetken	Eindhoven University of Technology, Netherlands
Amy Van Looy	Ghent University, Belgium
Irene Vanderfeesten	KU Leuven, Belgium

Program Committee

Adela del Río Ortega	University of Seville, Spain
Amine Abbad-Andaloussi	University of St. Gallen, Switzerland
Banu Aysolmaz	Eindhoven University of Technology, Netherlands
Christian Bartelheimer	Paderborn University, Germany
Markus Becker	University of Southern Denmark, Denmark
Iris Beerepoot	Utrecht University, Netherlands
Elisabetta Benevento	University of Pisa, Italy
Paolo Bottoni	Sapienza University of Rome, Italy
Edyta Brzychczy	AGH University of Krakow, Poland
Silvia Dallavalle	University of São Paulo, Brazil
Djordje Djurica	WU Vienna, Austria
Mahendrawathi Er	Institut Teknologi Sepuluh Nopember, Indonesia
Irene Bedilia Estrada Torres	University of Seville, Spain
Joklan Imelda Goni	Camelia Ghent University, Belgium
Thomas Haskamp	University of Münster, Germany
Michael Leyer	Philipps-Universität Marburg, Germany
Andrea Magalhaes	Dheka BPM Consulting, Brazil
Azumah Mamudu	Queensland University of Technology, Australia

Martin Matzner	FAU Erlangen-Nürnberg, Germany
Nijat Mehdiyev	DFKI, Germany
Patrick Mikalef	Norwegian University of Science and Technology, Norway
Fredrik Milani	University of Tartu, Estonia
Geert Poels	Ghent University, Belgium
Pascal Ravesteijn	Utrecht University of Applied Sciences, Netherlands
Jana-Rebecca Rehse	University of Mannheim, Germany
Dennis M. Riehle	Universität Koblenz, Germany
Ben Roelens	Open Universiteit, Netherlands
Estefanía Serral	KU Leuven, Belgium
Rehan Syed	Queensland University of Technology, Australia
Konstantinos Tsilionis	Eindhoven University of Technology, Netherlands
Sven Weinzierl	Friedrich-Alexander-Universität Erlangen-Nürnberg, Germany
Bastian Wurm	LMU Munich, Germany
Inge van de Weerd	Utrecht University, Netherlands

Additional Reviewers

Acitelli, Giacomo
Bahr, Charlotte
Banham, Adam
Brennig, Katharina
Camargo, Manuel
Delfmann, Patrick
El Haddad, Joyce
Grohs, Michael
Häge, Marie-Christin
Kalukapuge, Savandi
Käppel, Martin
Liessmann, Annina Lu, Yang
Mele, Alessandro
Min, Haeri
Valero-Ramon, Zoe
Wang, Weixin
Workman, Silas
Zilker, Sandra

Keynotes

The Marriage of Process and AI

Tom H. Davenport

Metropoulos Institute for Technology and Entrepreneurship, Babson College, Babson Park, MA, USA
tdavenport@babson.edu

Abstract. There are two unquestionable statements that are often made about AI and business processes:

1. AI enables new ways of doing work, i.e., new process and operating models [1];
2. For AI to provide value, it needs to be integrated into an organization's processes and workflows [5].

While both of these statements are widely believed and surely true, translating them into action and business value can be challenging. One impediment is that even after several decades of discussion that processes are an important way to view work, many organizations still are not process-focused [4]. They have not mapped their end-to-end process flows, they do not know the tasks that make up processes, they do not have clear process owners, and they do not apply new technology—AI or otherwise—in ways that make their processes more efficient and effective. In short, they are unlikely to see economic or customer value from their AI investments.

So how should organizations go about arranging an effective marriage between processes and AI? There are both prerequisites and specific analyses required. The prerequisites include:

- Identify and map important end-to-end processes, or at least a few that will be the focus for AI-enabled transformation;
- Identify the tasks or workflows within those processes. In the United States, which is not a process-oriented country, many people use the term *workflows* when referring to AI integration. However, workflows are narrower than processes and consist only of one or a few tasks;
- Assemble a team that understands both process and workflow redesign and how AI can transform work;
- Identify the process owners and stakeholders whose buy-in is needed to invest in and transform processes.

These sound like straightforward requirements, but as suggested above, it can take considerable time, human energy, and effective persuasion to make them happen. Many organizations are unwilling to emphasize processes (e.g., order management) over traditional business functions (e.g., sales, marketing, supply chain, and accounts receivable), or to designate owners of processes in addition to functional leaders. They may have employees who understand some aspects of process management, but these individuals are often more focused on "Lean Six Sigma" improvements to small processes rather than broad, end-to-end processes that can be dramatically transformed. Organizations may also have employees who understand AI. However, the two sets of capabilities are not often found within the same group or project team. Therefore, assembling these prerequisites is a major task in itself.

The specific analyses and activities needed to connect processes to AI in a production environment include the following:

- Prioritization of the processes that are most important to the company and most suited to transformation through AI. A large car manufacturer, for example, prioritized engineering, production, sales, and financial services processes for large-scale redesign, while other processes were designated for incremental improvements enabled by AI.
- Appointment of a project team with expertise in both AI and process redesign, incorporating inputs from process owners, stakeholders, and representatives of employees who work in the process.
- Measurement and documentation of the performance of the existing process. Today this can be accomplished relatively easily through the use of process mining technology [2]. Information gathered from process mining can also be used as contextual input for AI systems supporting the new process.
- Detailed design of the new process, with particular attention to how AI can enable and enhance key workflows.
- Determination of whether transactional systems supporting the existing process will be replaced, in whole or in part, by agentic AI capabilities offering greater intelligence and flexibility.
- Implementation of the required AI systems, production processes, employee upskilling initiatives, integration with existing technologies, and related capabilities.

These activities, together with the prerequisites that make them possible, comprise a major transformation for most organizations. They require substantial investment, executive support, and long-term commitment. If they involve the adoption of agentic AI, they also entail a significant degree of technological change throughout the transformation period.

As with business process reengineering, which in many companies eventually became associated with employee headcount reduction [3], such ambitious transformations will not always succeed. Nevertheless, this generation of technology-enabled change represents a major opportunity to improve organizational performance, and some companies are likely to realize substantial benefits from it.

References

1. Davenport, T.H., Hollweg, M., Jeavons, D.: How AI is helping companies redesign processes. Harvard Bus. Rev. (2023). https://hbr.org/2023/03/how-ai-is-helping-companies-redesign-processes
2. Davenport, T.H., Spanyi, A.: What process mining is and why companies should do it. Harvard Bus. Rev. (2019). https://hbr.org/2019/04/what-process-mining-is-and-why-companies-should-do-it
3. Kleiner, A.: Revisiting reengineering. Strategy Bus. (2000). https://www.strategy-business.com/article/19570
4. Majchrzak, A., Wang, Q.: Breaking the functional mind-set in process organizations. Harvard Bus. Rev. (1996). https://hbr.org/1996/09/breaking-the-functional-mind-set-in-process-organizations
5. McKinsey & Company. The state of AI in 2025, 5 November (2025). https://www.mckinsey.com/capabilities/quantumblack/our-insights/the-state-of-ai

From Digital Traces to Process Understanding: Rethinking Organizational Work and Change

Emmanuelle Vaast

Metropoulos Institute for Technology and Entrepreneurship, Desautels Faculty of Management of McGill University, Montreal, QC, Canada
emmanuelle.vaast@mcgill.ca

Abstract. Organizations generate detailed digital traces of decision-making, work, and coordination. This has led to huge opportunities to retrace and analyze organizational processes. At the same, this also raises a question for process research: how to understand organizational process when process is approached through data automatically captured by digital systems?

In this talk, I draw on scholarship on digital technologies, organizing, and theorizing from process to discuss how digital traces reshape the study of organizational processes. A key point is that trace data are not exact representations of organizational life. They are produced through particular technologies, infrastructures, and categorizations that shape how and which processes become visible and understandable.

This is importance since the growing availability of trace data has expanded the possibilities of studying process dynamics and oriented attention to forms of actions, interactions, sequences, and variation. Some aspects of organizational processes have become much more accessible. Analytical capabilities have tremendously expanded. Yet, other aspects of organizational processes remain difficult to apprehend through trace data alone. I thus suggest viewing trace data as partial yet consequential representations of organizing.

The keynote argues for a productive dialogue between computational and trace-based approaches on the other hand, and qualitative approaches of organizational process on the other. Such a dialogue is especially valuable to develop rich understanding of organizational processes and their dynamics in the digital age.

Contents

Keynotes

From Descriptive to Predictive Analytics in Healthcare Service Operations

A Strategic Analysis of AI-Driven Digital Twins for Emergency Department Capacity Management

Opher Baron[1,2,3(✉)], Shany Azaria[4], and Dmitry Krass[1,2]

[1] SiMLQ, Toronto, Canada
Opher.Baron@Rotman.Utoronto.Ca
[2] Rotman School of Management, University of Toronto, Toronto, Canada
[3] Hebrew University of Jerusalem, Jerusalem, Israel
[4] Tel Aviv University, Tel Aviv, Israel

Abstract. Many modern business processes operate under congestion: stochastic demand, shared and limited resources, complex routing, and tight service-level constraints. From healthcare and public services to logistics, financial operations, and large-scale customer support, these systems exhibit nonlinear behavior where small disruptions propagate quickly and performance deteriorates sharply. This paper examines how Business Process Management (BPM) can evolve to address congestion-driven environments by integrating process mining, queueing theory, simulation, and machine learning and artificial intelligence (ML&AI) into *digital twins*. We discuss how descriptive analytics must move beyond static dashboards toward data-driven process models that reveal how processes actually unfold in practice—capturing routing variability, rework loops, synchronization delays, and resource contention. Process mining provides structural visibility; queueing-aware modeling explains performance; ML&AI supports prediction under uncertainty. Building on these foundations, predictive and comparative analytics enable counterfactual "what-if" evaluation of staffing, routing, prioritization, and scheduling policies before implementation. Finally, prescriptive analytics embedded within real-time digital twins allow organizations to intervene proactively—anticipating congestion, reallocating capacity, and mitigating cascading delays. Drawing on industrial deployments through SiMLQ, we demonstrate how combining process intelligence with congestion-aware operational modeling transforms BPM from retrospective analysis into continuous, real-time decision support. The implications extend across service industries where variability and resource contention define performance, resilience, and competitiveness. Emergency departments are not failing because clinicians lack commitment; they are failing because the operating system that coordinates demand, staffing, beds, diagnostics, specialist consultation, and inpatient transfers remains largely reactive. Digital twins that are based on extensive descriptive analytics can transform fragmented event data into a safe, continuously updated environment for prediction, counterfactual evaluation, and prescriptive action. The managerial implication is blunt: modern businesses,

F. M. Maggi et al. (Eds.): BPM 2026, LNCS 16936, pp. 3–17, 2027.
https://doi.org/10.1007/978-3-032-37777-7_1

such as hospitals that rely only on after-the-fact statistics will remain trapped in decision blind spots. Businesses that build validated digital twins would move capacity management upstream, detect failure before it occurs, and redirect scarce resources to the highest-leverage bottlenecks in real time.

Keywords: emergency department crowding · digital twin · process mining · queueing theory · machine learning and artificial intelligence · simulation · patient flow · capacity management

1 The Global Crisis in Emergency Medicine Example

Modern systems and processes with different clockspeeds, from cloud computing to financial operations, large-scale customer support, manufacturing and logistics, healthcare, and public services, such as court systems, can all be thought of as *queueing networks*. In such systems stochastic demand, routing and service times create complexity in providing high service level. This paper examines how Business Process Management (BPM) can evolve to address congestion-driven environments by integrating process mining, queueing theory, simulation, machine learning and Artificial Intelligence (ML&AI) to create digital twins. We illustrate how descriptive analytics must move beyond static dashboards toward data-driven process models that reveal how processes actually evolve in practice—capturing routing variability, rework loops, synchronization delays, and resource contention. We illustrate these concepts using access to emergency care as an example, because of its direct impact on quality (and length) of life.

Emergency departments (EDs) are the open wound of modern health systems all around the world. They absorb unscheduled demand, function as a diagnostic front door, compensate for primary-care shortages, and serve as a holding area when inpatient beds are unavailable. From the perspective of a systems architect, the modern ED is far more than a clinical gateway; it can be conceptualized as a high-variability, high-acuity, multi-class queueing network. These systems operate under stochastic demand, shared and limited resources, and rigid precedence constraints. When downstream nodes, i.e., inpatient wards, are full or slow, the resulting "blocking" propagates upstream, manifesting as waiting-room crowding, ambulance offload delays, and the dangerous normalization of "hallway medicine" in the ED.

Recent Canadian data reinforce the scale of the problem. The Canadian Institute for Health Information [7,8] reported that in 2024–2025 half of ED patients spent four hours or more in care, and one in ten spent more than fourteen hours in the ED. One in ten admitted patients waited over two full days (48 h) in the ED (regardless of their initial acuity level). The same data describes a system in which approximately 12% of ED visits ended in admission, a small share of volume that dominates congestion because admitted patients require beds, transfers, and inpatient coordination rather than rapid discharge.

The human consequences of operational failure are no longer abstract. Public reporting from Alberta has described waiting-room and ED deaths amid overcrowding, lack of capacity, and long waits [6]. These episodes are not merely tragic clinical outliers; they are warning signals from a service system that cannot reliably distinguish between ordinary demand variability and imminent collapse. When patients with chest pain, sepsis risk, or possible aortic catastrophe wait for hours before assessment, the root cause is not only resource scarcity but also the absence of effective management.

In Quebec, the 2026 transition of dashboard responsibilities to Santé Québec led to a controversial reduction in publicly visible performance indicators, see [9,12]. For BPM, this highlights a critical governance flaw: descriptive dashboards that merely report yesterday's congestion expose failure without providing a mechanism for improvement. We argue for a paradigm shift: moving from reactive reporting to proactive, ML&AI driven digital twins that convert fragmented event data into a management science instrument for real-time anticipation and intervention.

We highlight two main themes: (i) the objective of descriptive analytics starting with process monitoring, often presented in dashboards, should encompass an improved understanding of the system. Such an understanding requires modeling, and (2) with ML&AI digital twins for queueing network process are relatively easy to build.

2 The Analytics Maturity Framework

Management Analytics Framework from Baron [5]: We extend the management analytics framework from Baron [5] the context of waits for healthcare organizations. The same ideas hold for other applications (with or without queueing). We emphasize the usage of ML&AI in supporting beneficial usage of management analytics in practice. For completeness, we copy the following definitions from Baron [5]:

- *Descriptive analytics* describes past performance of existing systems.
- *Predictive analytics* predicts future performance of systems.
- *Comparative analytics* compares performance of systems under different interventions.
- *Prescriptive analytics* prescribes interventions to improve future performance of systems.

In short, prescriptive analytics is the holy grail of analytics. Table 1 lists examples of managerial questions and decisions made based upon different management analytics components. The four-sphere framework in Fig. 1 extends Fig. 1 of Baron [5]; it separates analytics by time horizon and system state. This framework makes it explicit why descriptive dashboards, predictive models, causal comparisons, and optimization engines are complements rather than substitutes.

Table 1. Management Analytics – Spheres, Time, System, and Typical Decisions.

Sphere	Time / system state	Example	Typical ED decision
Descriptive	Past / existing	System monitoring: How long did patients wait?	Report Performance indicators
Predictive	Future / existing	System operation: How long would future arriving in the future wait?	Forecast crowding and boarding risk for the next two to six hours
Comparative	Past or future / alternative	System planning and what if analysis: What would change under alternative configurations?	Evaluate another nurse, faster consult request, or altered triage routing
Prescriptive	Future / alternative	System improvement: What should we do now and in the future?	Select staffing, escalation, and patient-flow policies to optimize performance

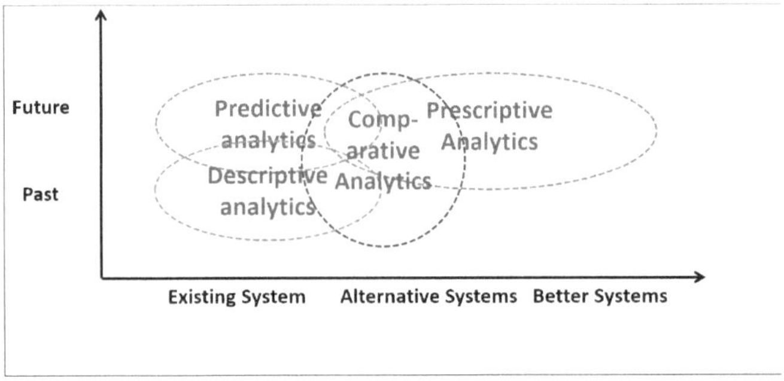

Fig. 1. Management Analytics.

The 5 Faces of a Process: To better conceptualize how descriptive analytics can help understanding the system, consider the 5 faces of a process depicted in Fig. 2: the process as (i) designed, by the process manager and its administrator, (ii) in reality, (iii) as seen by experts, such as employees and customers (thus there are many subfaces here) (iv) as captured in data, and (v) in its optimal, most effective, version. Obviously, these faces differ from each other. We think of management analytics as *using data to bring the real process towards the optimal one.*

System Understanding: We define a perfect understanding of a system as equivalent to being able to create *a digital simulator that responds as the system does*, i.e., can create both, the same data as is generated by the existing system, and the same data as would be generated by the real system after interventions. Having access to such a simulator in real time implies perfect comparative analytics capabilities. Then, all is left for prescriptive analytics, is searching among feasible interventions for the optimal ones. Such optimization is still challenging and time consuming, and optimization in a data driven context often requires

Fig. 2. The five faces of a process.

additional methods. An important part of this optimization is accurate and speedy performance evaluation, i.e., comparative analytics.

The focus of this paper is how to use ML&AI in descriptive analytics to support prescriptive one. While descriptive analytics is often understood as dashboards (no modeling), this view is too narrow. To identify interconnections within the "as is" system, modeling is often essential. The goal of descriptive analytics is to *understand* how the system operates, not just to describe its performance. This is where ML&AI can help—using descriptive analytics to understand, i.e., move towards comparative and prescriptive analytics.

3 How ML&AI Support Predictive, Comparative, and Prescriptive Analytics

We review the main usage of ML&AI in support for predictive, comparative, and prescriptive analytics today.

ML&AI in Predictive Analytics: Cheap, Accurate, and Timely. ML&AI tools have strategic importance to predictive analytics as they reduce the cost of prediction, improve its accuracy, and especially as more data is available, reduce the time to prediction. Agrawal, Gans, and Goldfarb [2] describe AI as prediction technology: it makes prediction cheaper, which changes the economics of decisions that depend on uncertainty.

In ED operations, cheaper prediction enables a shift from periodic (monthly or weekly) review to intra-day control. The hospital does not merely ask whether last period's performance target was missed; it asks "which patients are at risk now?," "what bottleneck will bind in three hours?".

As ML&AI can quickly provide an accurate answer to this and similar questions, managers now need answers to questions on interventions and their impact, such as "which action is most likely to prevent unsafe crowding in three hours?" This extra time allows managers to plan and optimize their response to system disruptions before standard of care is impacted.

ML&AI in Comparative Analytics: Challenges and Opportunities: Comparative analytics asks what would have happened under a credible alternative. This distinction is the difference between a performance report and a management science instrument, such as BPM. ED leaders do not merely need to know that the time to Provider Initial Assessment (PIA) target was missed; they need to know whether another nurse, a changed triage policy, faster consultation, or earlier bed-management escalation would have improved the outcome.

In general, comparative analytics has two forms. Demonstrating these in the context of an ED, we think of *Retrospective* comparative analytics: How long would the patient who arrived at 11:00 a.m. this morning have waited if another nurse had been on shift or if a consult had been requested 45 min earlier? And *Prospective* comparative analytics: How long a patient that will arrive at 11:00 a.m., tomorrow, wait if the current admitted-patient queue remains unresolved, or if an additional physician is redeployed from fast track to acute assessment?

The causal logic draws on a broader empirical tradition. The 2021 Nobel Prize in Economic Sciences recognized David Card, Joshua Angrist, and Guido Imbens. The first two for their empirical and methodological contributions that advanced the analysis of causal relationships, and the last one for contributions to empirical macroeconomics particularly through natural experiments [18]. The 2011 prize recognized Thomas Sargent and Christopher Sims for empirical research on cause and effect in the macroeconomy [17]. These examples matter for healthcare operations because many operational changes cannot ethically or practically be randomized in real time and thus causal analysis has to be used.

Data-driven analysis faces the challenge of causality since it needs to provide performance evaluation of different systems than the observed ones. A/B testing is the gold standard for comparative analytics because it directly measures the effect of an intervention under controlled conditions. In its simplest form, patients, clinicians, shifts, units, or sites would be randomly split between a "control" condition that preserves the current process and a "treatment" condition that introduces a change, such as a new triage protocol, staffing rule, routing policy, or discharge process. The random assignment isolates the causal effect of the intervention from background variation in demand, case mix, staffing, and seasonal conditions. For this reason, A/B testing is widely used in digital services, clinical trials, and operational improvement programs as the clearest way to answer comparative questions.

In EDs, direct A/B testing is often impossible or is severely constrained by ethics, patient-safety concerns, operational disruptions the complexity of isolating a certain system within a highly interconnected network. Moreover, hospitals should not learn by experimenting blindly on live patients. This is where ML&AI can help.

In the next Sections we discuss how ML&AI can support descriptive analytics in building a digital twin of a complex queueing network system. Digital twins extend the logic of A/B testing. Rather than experimenting on real patients, the organization can simulate the changes inside a calibrated model of the ED, comparing the current pathway against alternative ones before implementation. Thus, A/B testing remains the conceptual benchmark for comparative analytics, while ML&AI-driven digital twins provide a safer, faster, and more scalable way to conduct "what-if" experimentation in complex queueing networks.

ML&AI in Prescriptive Analytics: Point Vs System Solution: Current applications of ML and AI in prescriptive analytics are still dominated by automation of existing processes. Quoting Gates, Myhrvold, and Rinearson [11]: *"The first rule of any technology used in a business is that automation applied to an efficient operation will magnify the efficiency. The second is that automation applied to an inefficient operation will magnify the inefficiency."* Thus, one needs to be careful with the processes that are automated. Moreover, such implementations are point solutions rather than redesigning the system itself. The trick is to use ML&AI to support better processes that may not have been feasible without this technology.

In many hospitals, ML&AI is deployed as a point solution: a model predicts patient deterioration, flags imaging results, forecasts arrivals, or even recommends a scheduling adjustment within a narrow workflow. Such point solutions can be valuable, but they often optimize a local decision without changing the system architecture.

The distinction between point solution and a system one is central to the argument in Agrawal, Gans, and Goldfarb [3]: the deeper economic value of AI emerges not merely when prediction is inserted into current decisions, but when organizations *redesign* decision systems around better prediction capabilities.

In ED operations, moving beyond isolated automation toward system-level prescriptive analytics can be done by integrating prediction, simulation, queueing, and operational constraints into a digital twin that can evaluate the consequences of decisions on the performance indicators of the ED.

4 Descriptive Analytics for Understanding the Systems – Using Process Mining and Queueing Theory

ML&AI provide several tools that can help elevate descriptive analytics from reporting to understanding. Process mining and queueing theory bridge the gap between recorded reality, including performance indicators, and the operations model.

Descriptive Analytics Deep Dive: PIs That Matter: An important way to test the digital twin is to compare its performance, in- and out-of-sample with performance observed in practice. A digital twin raises the standard for the descriptive metrics – as it intends to capture the entire system's behaviour – not just "its mean performance." The ED must define the right operational variables, maintain consistent timestamp logic, and segment performance by clinically meaningful groups. Table 2 summarizes the core metrics used in many EDs around the world.

LOS is the most visible metric but not always the most actionable. A high LOS can arise from pre-triage waiting, diagnostic, consult, and disposition delay, or inpatient-bed boarding. Treating LOS as a single number is like treating a

Table 2. Performance Indicators in Emergency Department

Performance Indicator	Operational meaning	Significance?
Length of Stay (LOS)	Total time from ED arrival to departure from the ED	Measures total system burden and validates patient-pathway simulations
Left without Being Seen (LWBS)	Patients who depart before being seen by an appropriate clinician	Signals front-door failure and potential safety risk, especially when high-acuity patients leave
Time to Provider Initial Assessment (PIA)	Time from arrival or triage to physician initial assessment	Captures the first major bottleneck and often predicts LWBS risk
Time to Disposition (T2D)	Time to disposition decision: admit, discharge, transfer, or observation	Separates diagnostic/clinical decision delay from bed-exit delay
Time to Inpatient Bed (TIPB)	Time from admission decision to inpatient-bed transfer	Measures back-door blocking and ED boarding pressure
Time to Consult	Time from request for consult (often initial assessment) to arrival of the specialist	Identifies coordination delays that can dominate admitted-patient LOS

fever without diagnosing the infection. The digital twin can decompose LOS by pathway, daypart, and operational state. Most EDs measure LOS separately for admitted, non-admitted high acuity, and non-admitted low acuity patients. Similar distinctions exist for other performance measures recorded in Table 2.

LWBS rate is a key access quality measure. In standard queueing models, customers' balking or abandonment reflects low service valuation, and thus improve the efficiency of the system, see e.g., Hassin and Haviv [13]. In contrast, as diagnosis is a core ED service patients do not know their "valuation" ex ante since their true physical state can only be evaluated via assessment by the provider. Baker, Stevens and Brook [4] shows that within a week after LWBS, up to 11% of patients require hospitalization and nearly half of LWBS patients needed the ED services. Thus, LWBS signals a failure of the ED.

PIA is the front-door bottleneck. It is also a managerial lever because it can often be improved without adding beds: rapid assessment zones, physician-in-triage models, nurse-initiated protocols, or dynamic provider assignment can reduce PIA. The digital twin can test whether those changes improve total flow or merely move congestion downstream.

T2D is an important metric as it captures the efficacy of patient release procedures. For example, a high difference between LOS and T2D typically indicates issues of bed blocking within the hospital (for admitted patients) and alternative level of care (for both admitted and non-admitted patients). Thus, this measure highlights that the ED system is just a subsystem of the overall healthcare delivery system.

TIPB is the back-door bottleneck. Boarding admitted patients consumes ED capacity and reduces the department's ability to absorb incoming unscheduled demand. In queueing terms, boarding creates blocking: downstream inpatient capacity constraints propagate upstream to ambulances, waiting rooms, and clinical staff workload. The fact that admitted patients represent a minority of visits but a large share of congestion is precisely why pathway-specific modeling is essential. Moreover, adequate capturing of the impact of bed blocking within the hospital can support change management of the procedures taken outside the ED.

Time to Consult further highlight that the ED is not an independent system. In most modern EDs, admitted patients require a consult from the admitted ward to sign off on admission to the ward. Similarly, kids seen in an ED often require a pediatrician to see them even before being released home. Long consult wait times add much congestion to many EDs.

Given the high variability in EDs and that long waits may deteriorate patients' health, it is common to look at the 90th percentiles of these and other (e.g., ambulance offload time) performance measures. Still, their averages are often also very informative. Moreover, looking at these and other PIs by hour of the day, acuity and patient pathways, paves the way for a better understanding the inter-dependencies in the system.

When completing descriptive analytics to understand the system – it is important to consider the relevant performance measures and learn how these are balanced by the system. For example, it is intuitive that LWBS and PIA are correlated – if patients do not wait much before initial assessment – they have little chance to be LWBS. With this in mind – shorter PIA will increase the throughput of served patients.

In sum the top 3 performance indicators are mostly internal to the ED and its procedures and the bottom 3 ones remind that the ED is a part of a much better system. A digital twin for the entire healthcare delivery ecosystem is the next step in improving access to health.

Process Mining: Process mining extracts knowledge from event logs generated by information systems and uses that knowledge to discover, monitor, and understand the actual processes. The minimum event-log structure required for analyzing queueing networks is simple: "who" – a case identifier, "what" – an activity label, and "when" – a timestamp. While time stamps could include arrival, start of service, and end of service for each activity, in many cases only arrivals or departures from stations are recorded. In healthcare, the case is usually the patient encounter; activities include registration, triage, physician assessment, lab order, imaging, consult request, and disposition decision; timestamps establish the observed sequence [1,10,16]. Moreover, often there are many other relevant data, such as resources, i.e., beds, nurses, and doctors, and their availability; patients' characteristics, i.e., age, gender, and severity. The extra data supports better results of the ML&AI models, e.g., in understanding which patients follow specific pathways.

The directly-follows graph is the first operational translation of raw event data. It easily depicts if triage is followed by physician initial assessment in 65% of traces, and by waiting-room reassessment, lab-before-physician, or LWBS in the remainder. The difference between both pathways is not merely visualization – every discovered edge and node are a candidate delay, rework loop, bypass, or handoff failure.

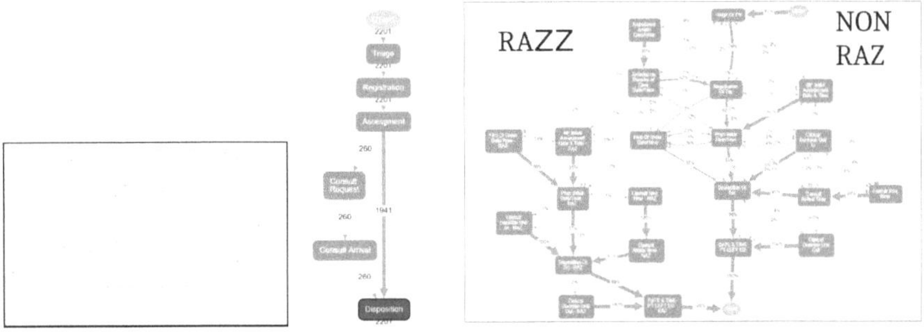

Fig. 3. Left: a spaghetti model, Center: a very simplistic ED process view, and Right: an informative ED process view with a rapid assessment zone (RAZ) and a main (non-RAZ) zone.

However, as depicted in Fig. 3, left, unfiltered process maps often look as a "spaghetti models": dense diagrams that faithfully represent complexity but fail to support managerial action. Too simplistic models, see Fig. 3, center, are not helpful either. The task is to filter without falsifying. High-frequency pathways, high-risk exceptions, and high-delay segments must remain visible, while rare noise and data-entry artifacts must not dominate the model. Figure 3, right, shows a useful simplified view of an ED that can support modeling.

Process mapping is part of the discovery process that reconstructs the real ED process from raw event logs. This process answers: What actually happens to patients, by acuity, mode of arrival, pathway, time of day, and disposition? This step often reveals that official protocols explain only part of the variation. Patients may move through diagnostics before physician assessment, wait for specialist consultation in undocumented holding states, or experience multiple cycles of order, wait, and reassessment.

The discovery process can expose differences between the designed and recorded workflows. Such conformance check compares actual flow with expected protocols. In the ED, conformance is not a rigid compliance exercise; many deviations are clinically justified. The value is to separate justified variation from preventable operational drift. For example, high-acuity patients should not normally experience long PIAs unless there is a clear clinical rationale or unavoidable capacity constraint.

Queueing Theory: Just depicting the process and supporting conformance checks delivers less than what descriptive analytics should. Queueing theory supplements process mining with the structural logic that captures the dependency of the system performance on its real time state. An ED is usefully conceptualized as a network of queues with stochastic arrivals, different patients' types, variable service times, finite servers, finite physical space, and downstream blocking. The same average arrival rate can produce radically different congestion depending on patient mix, variability, prioritization, pooling, and resource synchronization. Therefore, static staffing ratios routinely fail: they ignore the nonlinear relationship between utilization and delay. As utilization approaches capacity, waiting time increases exponentially, even before the nominal capacity limit is reached.

Queue mining that combines process mining and queueing theory allows us to gain much better understanding of the process by investigating the existing records. Moreover, with such an understanding of the process flow and internal dependencies, the analytics professional can start building a discrete-event simulation of the queueing network. Predictive analytics using ML&AI can calibrate arrival rates, service-time distributions, transition probabilities, and risk stratification. Queueing structure prevents the model from becoming a black-box curve fit, i.e., it maintains explainability. More importantly, changing the building blocks, such as the number of servers in a station, in the explainable simulation, i.e., provide comparative analytics, is straight-forward. Together, these methods form a defensible digital twin: data-driven enough to reflect reality, structured enough to simulate interventions, and transparent enough to support leadership decisions.

Note that a digital twin depends on this discipline because bad process discovery produces bad simulation logic. Thus, an important part of the process to create a digital twin is taking advantage of its explainability and support discussion with business domain experts, so that the resulting model is not missing adequately representing the system.

5 The SiMLQ Solution: Automated Digital Twins

SiMLQ (www.simlq.com) is a tool that combines simulation, ML&AI, and queueing solution for automated digital-twin generation. Traditional simulation projects often require months of manual process mapping, distribution fitting, stakeholder interviews, model coding, and validation. Automation changes the economics of simulation. Once a simulation model can be generated and refreshed orders of magnitude faster than traditional simulation models, it becomes a digital twin, a live operating asset, rather than a one-off improvement project.

The live-digital twin loop depicted in Fig. 4 has seven stages: data collection, creation of event logs, process map/dashboard (descriptive analytics), simulation and ML&AI (predictive analytics), what-if analysis (comparative analytics), and action selection, and action implementation (prescriptive analytics). The key is that the implemented actions impact the data collected, and thus feeds back to this 7-stage loop.

In contrast to traditional modeling, see Pidd [21], where the model is an exogenous representation of reality, the management analytics approach that leads to a digital twin model requires it to continuously interacts with reality and data. Thus, reality, data, and model impact each other in real time.

The benefits from live digital twin simulation models are tremendous. In contrast to traditional simulation models that are expensive and thus used to support big projects, the marginal cost of a digital twin is negligible. This low cost implies that a digital twin can support operational decisions. Common examples in ED would be when a patient needs to undergo both diagnostic imaging and lab work – the digital twin can direct this patient in a fashion that reduces the overall waits.

The design principle is closed-loop control. A conventional dashboard ends when it displays a metric. A digital twin begins there. It asks what the metric implies, whether the state is likely to deteriorate, what alternatives exist, and which action should be taken next. Thus, the digital twin should be viewed as an operating model rather than an analytics report.

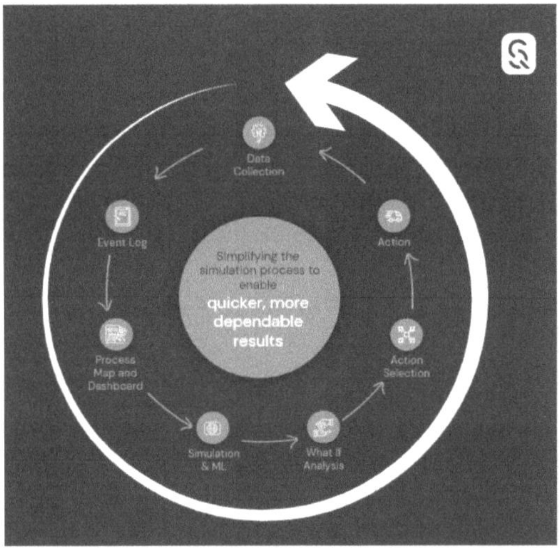

Fig. 4. Closed-loop digital twin operating model.

Recent healthcare digital-twin literature supports this direction. Reviews emphasize that healthcare digital twins combine monitoring, simulation, prediction, and optimization, while emergency-care-specific work highlights potential applications in capacity management, triage, and operational decision support. Simulation-based digital twins have also been used to evaluate emergency medical communication-center scenarios and broader healthcare workflow designs [14,15,19,20].

6 Conclusion

Implementing digital twins using the four components of management analytics is not without difficulties. Moreover, our discussion here hints at further opportunities to use digital twins generated with modern ML&AI as a system solution for EDs and other queueing network applications.

Risks, Limits, and Ethical Boundaries: The strongest argument for digital twins should not obscure their risks. The first risk is false precision. A digital twin can produce exact-looking numbers that depend on imperfect data, unstable demand, or undocumented clinical judgment. The solution is not to avoid modeling but to express uncertainty, validate forecasts, and communicate scenario ranges where appropriate.

The second risk is automation bias. Managers should not simply act in accordance with a recommendation just because it comes from ML&AI. A manager taking all decisions in accordance with "the algorithm," manages no more. Governance should require that recommendations be explainable: which bottleneck is binding, what intervention is proposed, what trade-off is expected, and what safety constraints apply.

The third risk is misaligned incentives. If a hospital optimizes only average LOS, it may neglect tail risk, high-acuity waiting, staff burnout, or equity. E.g., if physicians are focused solely on throughput maximization during their shifts, patients' waits may suffer. The objective must therefore include safety and fairness constraints. The goal is not to make the ED look good on paper; it is to improve real access and outcomes.

The fourth risk is privacy and cybersecurity. Digital twins depend on sensitive operational and clinical data. Implementation must follow applicable privacy law, role-based access, de-identification where feasible, audit logs, secure data pipelines, and clear data-retention rules. A hospital cannot build operational intelligence by weakening patient trust.

Toward the Optimal Healthcare System: The ED crisis is a service-operations failure with critical societal consequences. It is driven by stochastic demand, capacity and resource constraints, downstream boarding, diagnostic and consult dependencies. Currently decision systems often intervene only after the system deteriorates. The solution is not merely another dashboard. It is a progression from descriptive analytics to predictive, comparative, and prescriptive decision support.

Good descriptive analytics remains the foundation and should be pursued with modeling to support the next steps in the analytics journey. The ED needs a living model that can forecast crowding, test alternatives, and recommend action before patients experience avoidable harm. Many other queueing network systems can benefit from similar digital twin technology.

SiMLQ leads this strategic direction: a digital twin generated from operational data, calibrated by ML&AI, and activated through simulation. The premise of SiMLQ is not simply faster analysis (point solution) but a system change of the mode of ED management. In this mode leaders can see the system, anticipate failure, test interventions safely, and move toward the optimal hospital one operational decision at a time.

The transition from retrospective reporting to prescriptive digital twins marks a fundamental evolution in healthcare operations. By moving capacity management upstream, we transition from observing yesterday's PIs to anticipating and preventing tomorrow's failures.

The ultimate vision is an ecosystem of interconnected digital twins—linking the ED, the hospital-wide bed management system, and even the patient's digital twin. Such a network would synchronize clinical and operational capabilities across the entire continuum of care, including long-term care and primary physicians. By integrating these systems, we seek to transform the healthcare system from a collection of disconnected silos into a synchronized, anticipatory network capable of delivering optimal care in real time.

References

1. van der Aalst, W.M.P.: Process Mining: Data Science in Action. Springer, 2nd edn. (2016)
2. Agrawal, A., Gans, J., Goldfarb, A.: Prediction Machines: The Simple Economics of Artificial Intelligence. Harvard Business Review Press (2018)
3. Agrawal, A., Gans, J., Goldfarb, A.: Power and Prediction: The Disruptive Economics of Artificial Intelligence. Harvard Business Review Press (2022)
4. Baker, D.W., Stevens, C.D., Brook, R.H.: Patients who leave a public hospital emergency department without being seen by a physician. causes and consequences. J. Am. Med. Assoc. **266**(8), 1085–1090 (1991)
5. Baron, O.: Business analytics in service operations–lessons from healthcare operations. Nav. Res. Logist. **68**(5), 1–17 (2021)
6. Canadian Healthcare Technology: Doctors count 6 preventable deaths in Alberta EDs (2026). https://www.canhealth.com/2026/01/21/doctors-count-6-preventable-deaths-in-alberta-eds/. 21 Jan 2026
7. Canadian Institute for Health Information: Emergency department wait times in Canada: Longer ED stays reflect growing patient complexity and delays in admission (2025). https://www.cihi.ca/en/emergency-department-wait-times-in-canada-insights-from-a-health-system-perspective/longer-ed-stays-reflect-growing-patient-complexity-and-delays-in-admission. Accessed 3 July 2026
8. Canadian Institute for Health Information: NACRS emergency department visits and lengths of stay, 2024–2025 (2025). https://www.cihi.ca/en/nacrs-emergency-department-visits-and-lengths-of-stay. Accessed 3 July 2026
9. CanHealth: Quebec drops many ER stats from public dashboard (2026)
10. De Roock, E., Martin, N.: Process mining in healthcare: an updated perspective on the state of the art. J. Biomed. Inform. **127**, 103995 (2022)
11. Gates, B., Myhrvold, N., Rinearson, P.: The Road Ahead. Penguin Books, Revised edn. (1996)
12. Gouvernement du Québec: Data on the Québec health system and its services (2025)
13. Hassin, R., Haviv, M.: To queue or not to queue: equilibrium behavior in queueing systems. In: International Series in Operations Research & Management Science. Springer (2003)

14. Kuruppu Appuhamilage, G.D.K., Hussain, M., Zaman, M., Ali Khan, W.: A health digital twin framework for discrete event simulation based optimised critical care workflows. NPJ Digit. Med. **8**(1), 376 (2025)
15. Li, H., Zhang, J., Zhang, N., Zhu, B.: Advancing emergency care with digital twins. JMIR Aging **8**, e71777 (2025)
16. Munoz-Gama, J., et al.: Process mining for healthcare: characteristics and challenges. J. Biomed. Inform. **127**, 103994 (2022)
17. Prize, N.: The Sveriges Riksbank Prize in Economic Sciences in Memory of Alfred Nobel 2011 (2011)
18. Prize, N.: The Sveriges Riksbank Prize in Economic Sciences in Memory of Alfred Nobel 2021 (2021)
19. Olawade, D.B., et al.: The role of digital twin technology in modern emergency care. Int. J. Med. Inf., 106229 (2025)
20. Penverne, Y., et al.: A simulation based digital twin approach to assessing the organization of response to emergency calls. NPJ Digit. Med. **7**(1), 385 (2024)
21. Pidd, M.: Just modeling through: a rough guide to modeling. Interfaces **29**(2), 118–132 (1999)
22. Rabbi, F., Banik, D., Ibne Hossain, N.U., Sokolov, A.: Using process mining algorithms for process improvement in healthcare. Healthc. Anal., 100305 (2024)

Tutorials

In the Object-Centric Zoo: A Petri Net Perspective

Tutorial Paper

Alessandro Gianola[1], Ekkart Kindler[2], Marco Montali[3], Andrey Rivkin[2](✉), Jan Martijn E. M. van der Werf[4], and Sarah Winkler[3]

[1] INESC-ID/Instituto Superior Tecnico, Universidade de Lisboa, Lisboa, Portugal
alessandro.gianola@tecnico.ulisboa.pt
[2] Technical University of Denmark, 2800 Kongens Lyngby, Denmark
{ariv,ekki,ariv}@dtu.dk
[3] Free University of Bozen-Bolzano, Bolzano, Italy
{montali,winkler}@unibz.it
[4] Utrecht University, Utrecht, The Netherlands
j.m.e.m.vanderwerf@uu.nl

Abstract. In recent years, the object-centric paradigm has gained increasing momentum in process management and process mining. This development has led to the emergence of a diverse, yet still poorly systematized, set of modeling approaches, which we collectively refer to as the object-centric zoo. We call for a comparative overview of Petri net-based modelling approaches for object-centric processes, bringing froward an effort to provide structure into this zoo. By doing so, we aim to clarify the purpose and positioning of the various approaches. In particular, we offer practical guidance for selecting suitable modelling approaches and corresponding techniques for key tasks such as analysis, simulation, discovery, and conformance checking. By highlighting the capabilities and mutual relationships of the different approaches, it enables a coherent understanding of object-centric modeling approaches and their effective usage.

1 Introduction

Object-centric process management [9] and mining [2] study processes dealing with multiple, interrelated and co-evolving objects. This calls for a conceptual shift, moving beyond the traditional notion of case-centricity in event data, process models, and analysis and mining tasks alike.

Not surprisingly, several modelling approaches have been lately flourishing to reflect this conceptual shift, often as a by-product of novel object-centric discovery [1] and conformance checking techniques [5,6]. At the same, other approaches focused on studying correctness properties and their automated analysis [4]. This

F. M. Maggi et al. (Eds.): BPM 2026, LNCS 16936, pp. 21–25, 2027.
https://doi.org/10.1007/978-3-032-37777-7_2

rapid development has resulted in a *zoo* of approaches that, while often disconnected from one another, share common underlying principles. These include the need to handle object creation, deletion, and transfer, to manage object-to-object relationships (including one-to-many and many-to-many), and to support multi-object synchronisation.

We focus on the substantial part of the zoo that finds its roots in the domain of Petri nets, and set the following fourfold objective for its systematic characterisation:

1. identify key modelling features required for object-centricity;
2. map the identified features to existing approaches so to assess how well such features are supported and by which approaches;
3. understand under which assumptions the different modeling approaches operate and how they relate to each other;
4. highlight the suitability of different modelling approaches with respect to the intended purpose of use, such as analysis (e.g., conformance checking or verification), simulation, or discovery.

In our analysis, we examine the different proposed formalisms (see Table 1) and discuss purposes for which they are best suited.

Table 1. Comparison of Petri net-based object-centric process modelling languages along main object-centric modelling features. ✓ indicates full, direct support, ✗ no support, and ~ indirect or partial support.

	object creation	object removal	concurrent object flows	multi-object transfer	multi-object spawning	object relations	subset synchronisation	exact synchronisation	co-reference	data attributes	life-cyle per-type modularity	object reference
OC nets [1]	✓	✓	✓	✓	✓	✗	✗	✗	✗	✗	~	implicit
synchronous proclets [3]	✓	✓	✓	~	✓	~	✓	✓	✗	✗	✓	implicit
PNIDs/t-PNIDs [8,12]	✓	✓	✓	~	~	✓	~	✗	✓	✗	~	explicit
COA-nets [4]	✓	✓	✓	~	~	✓	~	~	✓	✓	✗	explicit
OPIDs [5]	✓	✓	✓	✓	✓	✓	✓	✗	✓	✗	✗	explicit
DOPIDs [6]	✓	✓	✓	✓	✓	✓	✓	✓	✓	✓	✗	explicit

2 Approach

Our idea for a systematic navigation of the object-centric zoo unfolds in three main phases, mirroring the four objectives outlined before.

The Whats and Hows of Object-Centricity. First, we introduce the object-centric paradigm for process modeling, mining, and analysis, drawing on recent literature, including the object-centric process management manifesto [9]. This requires to discuss core questions such as "*What is an object?*", and how such questions reflect in existing works on approaches for modeling, discovery and analysis of object-centric processes [7]. From this analysis, different core modeling features of Petri net-based object-centric modeling approaches consequently emerge [2,5], as summarized in the header of Table 1:

- creation and removal of objects;
- concurrent object flows, supporting the progression of single objects through (part of) the process (cf. *divergence* in [11]);
- spawning and transfer of unboundedly many objects in a single transition firing;
- object-to-object relations (that may evolve over time);
- subset and exact synchronisation, enabling a transition for an object o only if some or all objects related to o via some relations are in a certain state (cf. *convergence* in [11]);
- object co-reference, to explicitly retrieve, given an object, which other objects are related to it, and support corresponding fine-grained updates of relations;
- data attributes (and corresponding data-driven conditions);
- life-cycle per-type modularity, to obtain an explicit representation of the life-cycle of objects of a given type;
- whether objects are referred explicitly (i.e., with explicit binding mechanisms) or implicitly.

Notably, these features do not only pertain conceptual modelling of object-centric processes: their inclusion or exclusion impacts the complexity and feasibility of analysis and mining tasks.

Petri Net-Based Modelling Approaches and Their Relationships. The modelling dimensions recalled in the previous step provide a solid basis to map the different existing modelling appraoches.

We focus here on approaches that are based on Petri nets, come with formal syntax and semantics, and have been used for operational tasks related to analysis and/or mining (discovery and conformance checking in particular). Notably, a similar structure could be used to present and compare declarative modelling approaches as well.

Table 1 provides an overview of the covered formalisms, and how they relate to the elicited modelling features. Mapping the different approaches to the elicited dimensions enables a systematic understanding of both contemporary and traditional ("old-school") formalisms through the lens of object-centricity, contributing to a higher-level, bird-eye-view on object-centric models.

A further step is to establish mutual relationships between the modelling approaches, showing how they can be inter-translated, and how different modelling constructs can be (fully, or partially) reduced to each other. Small, illustrative examples showing "modelling gadgets" and their encoding in different formalisms can be used for this purpose, as done in the literature (see, e.g., [4–6, 10]).

The Purpose Metaphor: Linking Formalisms to Tasks. While providing a way to understand how various formal modelling approaches relate to one another is highly beneficial for navigating the object-centric landscape, we argue that this alone is not enough. Specifically, we introduce the notion of *purpose* as a guiding metaphor, enabling researchers and practitioners to more effectively identify which approaches suit best depending on:

- the analysis or mining tasks of interest (such as object-centric soundness checking, verification of temporal properties, automated discovery, conformance checking);
- the granularity of the task, e.g., how deeper the task goes in the evolution of objects, attributes, and relations;
- the type of event data available and with which assumptions they are interpreted, e.g., wether object-to-object relationships are explicitly recorded, and whether they can change over time or not.

For example, OC nets have been primarily introduced to discover object-centric processes from logs where events refer multiple objects, reusing existing case-centric techniques [1]. When applied to conformance checking, the resulting models are too under-specified to handle executions with dynamic object-to-object relations [5], while in the case where object-to-object relations do not change, they can be automatically enriched and suitably used for conformance checking as well [10].

3 Conclusions

The systematic navigation of the zoo of object-centric process modelling approaches is essential to understand their modelling capabilities and expressive power, as well as their suitability in relation to the analysis and mining tasks of interest.

Our proposal is to provide a structure for a systematic navigation by eliciting key modelling dimensions, by relating modelling approaches to these dimensions and to each other, and by linking them to the purpose of analysis/mining. This is an evolving effort, grounded in the literature and aiming at consolidating the fragmented proposals emerging in the community.

We believe that our systematic navigation of the landscape of object-centric modelling approaches, with the goal of understanding their relative expressive power and their linkage to purpose, can help researchers and practitioners alike.

References

1. van der Aalst, W.M.P., Berti, A.: Discovering object-centric Petri nets. Fundam. Informaticae **175**(1–4), 1–40 (2020)
2. Berti, A., Montali, M., van der Aalst, W.M.P.: Advancements and challenges in object-centric process mining: a systematic literature review. CoRR abs/2311.08795 (2023)
3. Fahland, D.: Describing behavior of processes with many-to-many interactions. In: Proceedings PETRI NETS (2019)
4. Ghilardi, S., Gianola, A., Montali, M., Rivkin, A.: Petri net-based object-centric processes with read-only data, vol. 107 (2022)
5. Gianola, A., Montali, M., Winkler, S.: Object-centric conformance alignments with synchronization. In: Proceedings 36th CAiSE. LNCS, vol. 14663, pp. 3–19 (2024)
6. Gianola, A., Montali, M., Winkler, S.: Object-centric processes with structured data and exact synchronization - formal modelling and conformance checking. In: Proceedings 37th CAiSE, vol. 15702, pp. 185–202. Springer (2025)
7. Kindler, E.: Objects, processes and the colours of the rainbow: towards a conceptual model for object-centric processes. In: Joint Workshop Proceedings of PNSE, ATAED, and PHOCON at PETRI NETS 2026 (2026). CEUR-WS.org
8. Polyvyanyy, A., van der Werf, J.M.E.M., Overbeek, S., Brouwers, R.: Information systems modeling: Language, verification, and tool support. In: Proceedings 31st CAiSE (2019)
9. Seidel, A., Weske, M., Montali, M., Rivkin, A.: Object-centric process management: A research manifesto. Inf. Syst., 141 (2026)
10. Seidel, A., et al.: To bind or not to bind? discovering stable relationships in object-centric processes. In: Pastor, O. (ed.) Proceedings of the 44th International Conference on Conceptual Modeling (ER 2025). LNCS, vol. 16189, pp. 223–241 (2025)
11. van der Aalst, W.M.P.: Object-centric process mining: dealing with divergence and convergence in event data. In: Software Engineering and Formal Methods, pp. 3–25 (2019)
12. van der Werf, J.M.E.M., Rivkin, A., Polyvyanyy, A., Montali, M.: Data and process resonance - identifier soundness for models of information systems. In: Proceedings PETRI NETS (2022)

Green Business Process Management: Approaches, Challenges, and Ways Forward

Finn Klessascheck[1,2(✉)] and Luise Pufahl[2,3]

[1] University of Münster, Münster, Germany
finn.klessascheck@uni-muenster.de
[2] Weizenbaum Institute, Berlin, Germany
[3] Technical University of Munich, Heilbronn, Germany
luise.pufahl@tum.de

Abstract. In light of increasing environmental concerns faced by organizations and society, the *Business Process Management* (BPM) community has developed the *Green BPM* sub-field. With a wide array of approaches and capabilities, environmental impacts of business processes can be assessed and reduced, allowing organizations to establish environmentally *sustainable* practices. While receiving significant research interest over the last fifteen years, Green BPM is still a marginal topic at the BPM conference, as well as in BPM practice. In this tutorial, we provide a general introduction into Green BPM, and present two alternative approaches for integrating notions of environmental sustainability into Green BPM. We discuss challenges associated with the "process lens" of Green BPM, both from a theoretical and practical angle. With our tutorial, we aim to create awareness for the rich stream of existing Green BPM research and ultimately for the potential for further relevant and valuable contributions to be made by BPM researchers.

1 Introduction

Green Business Process Management (Green BPM) has been developed in order to integrate increasing concerns for environmental impacts of organizations into the management of their business processes [21]. A rich body of conceptual and technical capabilities and approaches for Green BPM has emerged over the last fifteen years, offering support for the assessment and improvement of business processes for aspects of environmental *sustainability* [3]. Their overall purpose is to help organizations establish *sustainable business practices*, that is, practices that "exert minimal or no impact on the natural environment in terms of emissions and waste while using few resources, specifically non-renewable resources" [20] and thus can be considered *environmentally sustainable*.

A core position of Green BPM is that by assessing and intervening in business processes, such sustainable business practices can be established [19]. For this, Green BPM conceptualizes environmental sustainability, e.g. as expressed

F. M. Maggi et al. (Eds.): BPM 2026, LNCS 16936, pp. 26–30, 2027.
https://doi.org/10.1007/978-3-032-37777-7_3

in *environmental performance indicators*, as an extension of the "Devil's Quadrangle" of process performance dimensions [6] into the "Devil's Pentagon" [19], where environmental sustainability interacts with the *time*, *quality*, *cost*, and *flexibility* dimensions of business processes. Figure 1 illustrates this conceptualization and its interacting process performance dimensions.

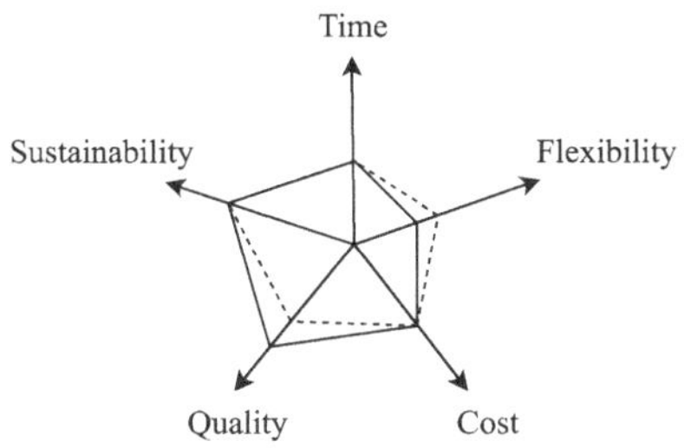

Fig. 1. The "Devil's Pentagon", adapted from [19].

This perspective has allowed extending traditional approaches and techniques across all *capability areas* of BPM [6] to integrate environmental sustainability in various ways: Green BPM now provides insights into *modelling*, *deploying*, *optimizing*, and *managing* business processes in view of environmental sustainability, as well as into *structural* and *cultural* aspects of organizations that are required [4].

2 Progress and Challenges

Recent developments in Green BPM include efforts to draw on *Life Cycle Assessment* [5,17] as a technique for environmental impact assessments to analyse business processes for their environmental impact *holistically*, that is, considering a wide range of environmental indicators traditionally left unconsidered [2,9,16]. Further, recognizing the *regulatory* relevancy of establishing and demonstrating sustainable business practices (for example, as formalized in the European Union's *Taxonomy for Sustainable Activities* [1,7]), efforts have been made to integrate regulatory frameworks that provide constraints and criteria for determining when business practices contribute to sustainability objectives into Green BPM [14,15]. Supporting these directions, increasing attention is being paid to questions of how (and in how far) data from process-aware information systems can be extracted into *event logs* to support Green BPM by data-driven techniques such as process mining [18,22].

However, despite continuing contributions to the field, as well as the topic's relevancy and timeliness, there is as of yet limited uptake of Green BPM concepts and approaches in practice; only in this year's edition of the BPM conference do we see a workshop dedicated to the Green BPM community. Therefore, in this tutorial we aim to provide an overview of existing approaches for Green BPM, describing and illustrating two of them in detail. We further discuss the issue of "conceptualizing" environmental sustainability on a process level and the various assumptions involved in doing so. Of particular concern are distinctions between approaching environmental sustainability from a *regulatory* and from an *indicator-based* angle, as well as between *weak* and *strong* notions of sustainability (cf. [10,12]) on a more fundamental level. The latter is especially important when considering the *ends* to which Green BPM serves as a *means* [11], and whether that end constitutes achieving 'actual' environmental sustainability. Crucially, Green BPM research has not yet systematically integrated a distinction between weak and strong sustainability, and issues and limitations arising thereof remain under-explored [13].

3 Ways Forward

Our key objective in this tutorial is to generate awareness for the rich stream of Green BPM research on the one hand, and to highlight the need for rigorous engagement with several ways of conceptualizing environmental sustainability on a process-level on the other [8]. We further aim to increase the awareness of how notions of weak and strong sustainability bring different, incompatible, assumptions regarding the *purpose* of sustainability, strategies for achieving it, and the underlying ethical stance [10–12], and how these come to bear on Green BPM research and practice. This involves critically assessing the *institutional context* and corresponding pressures [20] that shape the development and use of Green BPM approaches and techniques.

With these considerations, we hope to spark further research interest in this area and broaden the Green BPM community's reach, as well as to position Green BPM more strongly vis-à-vis industry, leading—in the long run—to increased interest *from practice* regarding what the Green BPM community has to offer. There is significant potential for impactful research on Green BPM, considering environmental sustainability from a *regulatory* and from a *normative* angle, and their integration into the technical and managerial dimensions of Green BPM [3].

Disclosure of Interests. The authors have no competing interests to declare that are relevant to the content of this article.

References

1. Alessi, L., Battiston, S., Melo, A.S., Roncoroni, A.: The EU sustainability taxonomy: a financial impact assessment. Pu1blications Office of the European Union (KJ-NA-29970-EN-N (online)) (2019). https://doi.org/10.2760/347810
2. Cortes-Cornax, M., Front, A., Oliveira, R., Roncancio, C.: A method for assessing environmental impacts of business processes. In: 2025 27th International Conference on Business Informatics (CBI), pp. 1–10 (2025). https://doi.org/10.1109/CBI68102.2025.00021

3. Couckuyt, D., Van Looy, A.: Green BPM as a business-oriented discipline: a systematic mapping study and research agenda. Sustainability **11**(15), 4200 (2019). https://doi.org/10.3390/su11154200
4. Couckuyt, D., Van Looy, A.: A systematic review of green business process management. Bus. Process. Manag. J. **26**(2), 421–446 (2019). https://doi.org/10.1108/BPMJ-03-2019-0106
5. de Bruijn, H., et al., (eds.): Handbook on Life Cycle Assessment: Operational Guide to the ISO Standards, Eco-Efficiency in Industry and Science, vol. 7. Springer Netherlands, Dordrecht (2002). https://doi.org/10.1007/0-306-48055-7
6. Dumas, M., La Rosa, M., Mendling, J., Reijers, H.A.: Fundamentals of Business Process Management. Springer, Berlin, Heidelberg (2018). https://doi.org/10.1007/978-3-662-56509-4
7. European Commission: EU taxonomy for sustainable activities - Finance - European Commission (2025). https://finance.ec.europa.eu/sustainable-finance/tools-and-standards/eu-taxonomy-sustainable-activities_en
8. Fritsch, A., Ullrich, M., Graves, N., Klessascheck, F., Nurkasanah, I.: Process Science for Sustainability: Research Gaps and Research Strategy, pp. 119–124. Gesellschaft für Informatik e.V., Heilbronn (2025). https://doi.org/10.18420/EMISA2025_17. EMISA 2025
9. Graves, N., Fritsch, A., Hensen, R., Koren, I., van der Aalst, W.M.P.: Object-centric process mining for semi-automated and multi-perspective sustainability analyses. In: 2025 11th International Conference on ICT for Sustainability (ICT4S), IEEE, Dublin, Ireland (2025)
10. Hahn, R.: Sustainability Management: Concepts, Instruments, and Stakeholders, 2nd edn. Pearson, Harlow (2025)
11. Hector, D.C., Christensen, C.B., Petrie, J.: Sustainability and sustainable development: philosophical distinctions and practical implications. Environ. Values **23**(1), 7–28 (2014). https://doi.org/10.3197/096327114X13851122268963
12. Heinrichs, H., Martens, P., Michelsen, G., Wiek, A. (eds.): Sustainability Science: An Introduction. Springer Netherlands, Dordrecht (2016). https://doi.org/10.1007/978-94-017-7242-6
13. Klessascheck, F.: Green BPM Solutions for Sustainable Business Processes. In: Reijers, H., et al., (eds.) Joint Proceedings of the Best Dissertation Award, Doctoral Consortium, and Demonstration & Resources Forum at BPM 2025. CEUR Workshop Proceedings, vol. 4032, pp. 16–24. CEUR-WS.org, Seville, Spain (2025). https://ceur-ws.org/Vol-4032/paper-04.pdf
14. Klessascheck, F., Bogatinovska, I., Andree, K., Pufahl, L.: A Conceptual Model and Supporting Approach for EU Taxonomy Alignment with Green BPM (2025). https://doi.org/10.2139/ssrn.5775427
15. Klessascheck, F., Fahrenkrog-Petersen, S.A., Mendling, J., Pufahl, L.: Unlocking sustainability compliance: characterizing the EU taxonomy for business process management. In: Borbinha, J., Prince Sales, T., Da Silva, M.M., Proper, H.A., Schnellmann, M. (eds.) Enterprise Design, Operations, and Computing, vol. 15409, pp. 339–359. Springer Nature Switzerland, Cham (2025). https://doi.org/10.1007/978-3-031-78338-8_18
16. Klessascheck, F., Weber, I., Pufahl, L.: SOPA: a framework for sustainability-oriented process analysis and re-design in business process management. IseB **23**(2), 489–537 (2025). https://doi.org/10.1007/s10257-024-00695-x
17. Klöpffer, W., Grahl, B.: Life Cycle Assessment (LCA): A Guide to Best Practice. Wiley-VCHVerlag GmbH & Co. KGaA, Weinheim (2014)

18. Schäfer, D., Klessascheck, F., Kampik, T., Pufahl, L.: Can we leverage process data from ERP systems for business process sustainability analyses? In: Delgado, A., Slaats, T. (eds.) Process Mining Workshops, pp. 764–777. Springer Nature Switzerland, Cham (2025). https://doi.org/10.1007/978-3-031-82225-4_56
19. Seidel, S., Recker, J., vom Brocke, J.: Green business process management. In: vom Brocke, J., Seidel, S., Recker, J. (eds.) Green Business Process Management, pp. 3–13. Springer, Berlin, Heidelberg (2012).https://doi.org/10.1007/978-3-642-27488-6_1
20. Seidel, S., Recker, J., vom Brocke, J.: Digital technology affordances for sustainable business practices. In: Cooper, V.A., Kranz, J.J., Mathew, S.K., Watson, R.T. (eds.) Research Handbook on Information Systems and the Environment, pp. 149–164. Edward Elgar Publishing (2023). https://doi.org/10.4337/9781802201864.00013
21. vom Brocke, J., Seidel, S., Recker, J. (eds.): Green Business Process Management: Towards the Sustainable Enterprise. Springer, Berlin, Heidelberg (2012). https://doi.org/10.1007/978-3-642-27488-6
22. Wolf, N., Wittges, H., Rinderle-Ma, S.: SuXXES: integrating sustainability data into process event logs. In: Fuentes, L., Plebani, P., Combi, C., Reijers, H. (eds.) Advanced Information Systems Engineering. pp. 213–229. Springer Nature Switzerland, Cham (2026). https://doi.org/10.1007/978-3-032-28110-4_12

Neuro-symbolic AI for BPM

A Tutorial on Predictive and Prescriptive Process Analytics

Gyunam Park[1(✉)] and Fabrizio De Santis[2]

[1] Eindhoven University of Technology, Eindhoven, Netherlands
g.park@tue.nl
[2] University of Parma, Parma, Italy
fabrizio.desantis@unipr.it

Abstract. Process analytics has become predominantly data-driven: deep sequence models such as LSTMs and Transformers now dominate predictive and prescriptive process monitoring and deliver strong average-case accuracy. These models are, however, opaque, require large amounts of data, and do not incorporate the process knowledge that organizations already hold in the form of regulations, declarative constraints, and reference process models. The symbolic side of Business Process Management (BPM) carries exactly this knowledge but is brittle on noisy, real-world event data. *Neuro-symbolic AI* (NeSy) aims to bridge these two traditions, and a growing body of BPM work has begun to use it. This tutorial gives the BPM community an accessible overview of the relationship between BPM and NeSy. It introduces Kautz's architectural taxonomy of NeSy and the *Learning from Logical Constraints* paradigm together with its main implementation, Logic Tensor Networks. It then traces how NeSy attaches to concrete BPM analytics tasks through a sequence of case studies spanning compliance-aware prediction, robust rule injection, healthcare outcome prediction, and prescriptive analytics. The tutorial highlights the *failure modes* that practitioners must avoid when injecting domain knowledge into deep process models, examines how to evaluate whether injected knowledge actually helps, and closes with open problems and a practical decision aid for when NeSy is worth adopting.

Keywords: Neuro-symbolic AI · Business process management · Predictive process monitoring · Prescriptive process analytics · Domain knowledge injection

1 Why BPM Meets Neuro-Symbolic AI

Modern process analytics rests on two intellectual traditions that rarely meet. The *data-driven* tradition learns predictors directly from event logs: given a running case, a neural model forecasts the next activity, the remaining time, the case outcome, or an entire suffix. The *knowledge-driven* tradition, which

F. M. Maggi et al. (Eds.): BPM 2026, LNCS 16936, pp. 31–36, 2027.
https://doi.org/10.1007/978-3-032-37777-7_4

has long been central to BPM, encodes what we already know about a process (DECLARE constraints, temporal-logic rules, service-level agreements, regulatory obligations, and reference process models) and reasons over it symbolically. Each tradition is strong exactly where the other is weak. Deep models generalize from data but ignore explicit knowledge and need many examples; symbolic models carry knowledge but degrade on the noisy, incomplete, and unlabeled data that real logs supply.

The cost of keeping the two apart is concrete and familiar to BPM practitioners. A purely neural outcome predictor can produce forecasts that violate constraints the organization *knows* must hold, such as predicting an activity that cannot legally precede another, or a decision that breaches a documented business rule. The problem is sharpest precisely where the stakes are highest and compliant behavior is rare. In rule-poor regimes a data-only model has almost nothing to imitate and collapses on rule adherence, even while its average accuracy looks acceptable.

Neuro-symbolic AI is the integration layer that lets a single model learn from data *and* respect knowledge. The field is young but maturing rapidly, and the BPM community has produced a distinct and growing cluster of NeSy work over the last two years. The aim of this tutorial is to make that work accessible to a general BPM audience: to explain what neuro-symbolic AI is, where it fits into the BPM analytics pipeline, what it provides, and when and how it fails. We assume familiarity with event logs and outcome prediction, and with the basics of deep learning, but *no* prior exposure to neuro-symbolic AI, fuzzy logic, or logic tensor networks [3,4].

2 Foundations of Neuro-Symbolic AI for BPM

We organize the landscape with Kautz's six-type architectural taxonomy of neuro-symbolic systems [3], which ranges from loose pipelines (a symbolic component feeding a neural one, or vice versa) to tight couplings in which logic is compiled into the network itself. Most current BPM work occupies the *Type-4* slot, in which symbolic knowledge is compiled into the loss function of a neural learner. A recurring caveat, which we emphasize throughout, is that Type-4 integration delivers *training-time soft* compliance (the network is *encouraged* to satisfy constraints) rather than the runtime *hard* guarantees that some safety-critical settings demand. Marra et al.'s survey [4] supplies a complementary vocabulary, casting neuro-symbolic systems along shared dimensions (how knowledge is represented, how it is combined with learning, how it is used at inference) that let participants compare BPM methods against the wider AI literature.

Learning from Logical Constraints (LfLC). The paradigm underlying most Type-4 BPM work treats knowledge injection as constrained optimization. Instead of training only against data, the model minimizes a combined objective

$$\min_{\theta} \; \mathcal{L}_{\text{data}}(\theta) \; + \; \lambda \cdot \mathcal{L}_{\text{logic}}(\theta),$$

where $\mathcal{L}_{\text{logic}}$ penalizes violations of domain rules and λ controls how strongly knowledge is enforced relative to fit. Building this objective takes three conceptual steps: predicates of interest (e.g. "this prefix is compliant", "lactate is high") become learnable neural functions; crisp logical connectives are relaxed into differentiable operators via fuzzy t-norms; and the resulting truth values are aggregated into a loss term that backpropagation can optimize.

Logic Tensor Networks (LTN). LTN [1] is the main implementation of LfLC and underlies the BPM case studies in this tutorial. Through its *Real Logic* formalism, first-order axioms are grounded onto tensors: constants and variables become real vectors, predicates become neural networks returning truth values in $[0, 1]$, and quantifiers become differentiable aggregations. This lets a practitioner write a domain rule as a first-order axiom and have it compiled into a training signal for a deep learning model. We give participants just enough Real Logic to read and write such axioms, without assuming any prior logic background.

3 Neuro-Symbolic AI Across Predictive and Prescriptive Process Analytics

The core of the tutorial walks through how neuro-symbolic AI attaches to concrete BPM analytics tasks. Rather than a single technique, the case studies each show a different point of contact between domain knowledge and a deep process model, and each surfaces a different lesson, including that knowledge injection can backfire.

Compliance-Aware Predictive Monitoring: Where Does Each Rule Attach? The first case study [7] addresses the most direct question a practitioner faces: given a rule and a predictive model, *where* should the rule be injected? It introduces a three-way injection scheme routed by the logical form of the rule: (A) *Feature Expansion*, augmenting the input with rule-derived signals; (B) *Output Refinement*, shaping the prediction head; and (C) *Parallel Constraints*, adding a dedicated constraint branch. Output Refinement gives the strongest empirical results, with F1 improvements of up to +18.4% on Sepsis and +10.1% on BPIC2012, and compliance that holds even when only 4.55% of training traces are compliant. This illustrates the data-efficiency argument for NeSy: knowledge substitutes for the labeled examples a pure-neural model lacks.

When Knowledge Hurts: Vacuous Satisfaction and Rule Pruning. The second case study [9] is the cautionary counterpart and is central to the tutorial's message. Naively adding a logic loss can make a model *worse* than using no knowledge at all: on Sepsis, an unguarded LTN drops F1 to 18.79%. The cause is *vacuous satisfaction*: the model learns to satisfy an implication by making its premise false rather than by learning the intended behavior, collapsing onto a degenerate solution that is logically "correct" yet useless. The remedy is a two-stage training procedure: a weighted pretraining phase that prioritizes data fit, followed by analytical rule pruning that gates each axiom by its satisfaction dynamics ($g_i = \bar{s}_i \cdot e^{-\lambda \operatorname{Var}(s_i)}$), retaining only rules that are both consistently satisfiable and informative, and a fine-tuning phase. This recovers performance to 75.82%. The takeaway is general: knowledge injection is not inherently beneficial; the mechanism that injects it must be designed with its failure modes in mind.

Outcome Prediction in Healthcare. The third case study [8] transfers the pattern to a high-stakes time-series setting by reading an ICU stay as a process: each patient stay is a trace of clinical events, and mortality is the case outcome. Domain rules are extracted automatically from SNOMED CT via AnyBURL, and predicates are *weakly anchored* at clinically meaningful thresholds (e.g. `HighLactate` initialized at > 4.0 mmol/L) so that learning starts from clinical knowledge but can adapt. The model offers *three-tier explainability*, at the level of concepts, rules, and individual patients, which makes it usable for real-time decision support and yields materially fewer false-positive alerts at a fixed recall.

Beyond Prediction: Prescriptive Analytics and Design Choices. The final part broadens from prediction to *prescription* and uses contrasts to teach design judgment. NeSyPA [6] brings neuro-symbolic reasoning to prescriptive process analytics (recommending interventions rather than forecasting outcomes) and deliberately steps *outside* the LfLC paradigm, which makes it a useful boundary case for when logic-as-loss is the wrong tool. Recent external work on suffix prediction [5] shows that logic-as-loss *can* enforce sequential, temporal constraints if fuzzy aggregation is replaced by differentiable automaton execution (a DeepDFA-style temporal-logic loss); this is a useful teaching point about why some constraints resist naive fuzzy injection. We close with a data-efficiency result from neuro-symbolic anomaly detection [2], where measurable gains arise from as few as ten conformant traces, reinforcing that knowledge is most valuable when data is scarce.

4 Open Problems and Takeaways

The case studies form a consistent picture and expose questions that remain open.

Where the Knowledge Comes From. Every case study assumes a usable set of rules, and their quality bounds what NeSy can achieve. In practice rules must be elicited or extracted, as in the automatic extraction from SNOMED CT via AnyBURL in the healthcare study. Scaling this step, detecting conflicts among rules, and weighting each rule by its confidence rather than treating all axioms as equally hard are largely unsolved.

The Soft-Compliance Gap. Type-4 integration buys compliance during training, not guarantees at inference. A model trained with a logic loss is more compliant on average, but nothing prevents it from violating a constraint on a particular case. Closing this gap by combining training-time logic losses with inference-time enforcement (constrained decoding, shielding, or post-hoc repair of predictions) is a pressing open problem for safety-critical BPM deployments.

Sequential Constraints and the Limits of Fuzzy Logic. The suffix-prediction contrast shows that fuzzy aggregation handles atemporal constraints well but struggles with the ordering and temporal-logic properties that are central to BPM. Differentiable automaton execution is one answer; how far fuzzy, LTN-style methods can be pushed before a different mechanism is required, and how to choose between them for a given constraint, is an open methodological question that the rule-pruning failure mode makes concrete.

Evaluating the Utility of Injected Knowledge. A subtler issue underlies every reported gain: injected knowledge has value only insofar as it offsets the data signal, so a sound evaluation must isolate what knowledge adds *beyond* what the data already provides. Two conditions can make that contribution illusory. First, if the constraint is already implicit in the training data, for instance when most traces are compliant, the network learns it from data alone and the logic term is redundant; the rule-poor regimes that recur in the case studies (only 4.55% compliant traces in Sepsis and 13.09% in BPIC2012) are precisely the conditions under which knowledge can demonstrably help. Second, if the test data does not exercise the constraint, no metric can register the knowledge's effect, whereas a test set skewed toward constraint-relevant cases can inflate it, so the test distribution decides whether utility is even observable.

A Decision Aid: Is NeSy Worth it for your Problem? The tutorial distills the case studies into five diagnostic questions practitioners can apply to their own work: (1) is labeled data abundant or scarce; (2) are reliable domain rules available; (3) what is the regulatory exposure of a wrong or non-compliant prediction; (4) how much model opacity can stakeholders tolerate; and (5) what is the deployment horizon. NeSy pays off most clearly when data is scarce, rules are available, and compliance or explainability is non-negotiable, exactly the regime of the compliance-aware and healthcare studies. When data is abundant and rules are thin, a well-tuned neural baseline is often the better engineering choice, and the rule-pruning study is a reminder that injecting knowledge carelessly can do real harm.

References

1. Badreddine, S., d'Avila Garcez, A.S., Serafini, L., Spranger, M.: Logic tensor networks. Artif. Intell. **303**, 103649 (2022)
2. Gaikwad, D., van der Aalst, W.M.P., Park, G.: Neuro-symbolic process anomaly detection. In: Fuentes, L., Plebani, P., Combi, C., Reijers, H.A. (eds.) CAiSE 2026. LNCS, vol. 16559, pp. 240–255. Springer (2026)

3. Kautz, H.A.: The Third AI Summer: AAAI Robert S. Engelmore Memorial Lecture. AI Mag. **43**(1), 105–125 (2022)
4. Marra, G., Dumancic, S., Manhaeve, R., Raedt, L.D.: From statistical relational to neurosymbolic artificial intelligence: a survey. Artif. Intell. **328**, 104062 (2024)
5. Mezini, A., Umili, E., Donadello, I., Maggi, F.M., Mancanelli, M., Patrizi, F.: Neuro-symbolic predictive process monitoring. CoRR abs/2509.00834 (2025)
6. Park, G.: Neuro-symbolic prescriptive process analytics: learning compliant and effective process recommendations through constraint distillation. In: Gal, A., Maggi, F.M., Leopold, H., Santoro, F. (eds.) BPM 2026, vol. 16936. Springer, LNCS (2026)
7. De Santis, F., Park, G., van der Aalst, W.M.P., Zanichelli, F.: Compliance-aware predictive process monitoring: A neuro-symbolic approach. In: Fuentes, L., Plebani, P., Combi, C., Reijers, H.A. (eds.) CAiSE 2026. LNCS, vol. 16559, pp. 223–239. Springer (2026)
8. De Santis, F., Park, G., Zanichelli, F.: Neuro-symbolic artificial intelligence for real-time sepsis mortality prediction. Eng. Appl. Artif. Intell. **177**, 114920 (2026)
9. De Santis, F., et al.: Neuro-symbolic learning for predictive process monitoring via two-stage logic tensor networks with rule pruning. In: Yiu, M.L. (ed.) PAKDD 2026. LNCS, vol. 16598, pp. 104–118. Springer (2026)

Track I: Foundations

Why Behavioral Representations Matter: A Framework for Comparing Their Expressiveness in Business Process Analysis

Kerstin Andree[1(✉)], Hannah Burke[2], Nick van Beest[2], Silvano Colombo Tosatto[2], and Luise Pufahl[1]

[1] School of Computation, Information and Technology, Technical University of Munich, Heilbronn, Germany
{kerstin.andree,luise.pufahl}@tum.de
[2] CSIRO, Brisbane, Australia
{hannah.burke,nick.vanbeest,silvano.colombotosatto}@csiro.au

Abstract. Behavioral representations are a fundamental abstraction mechanism in process mining, where an event log is abstracted into a set of behaviors in preparation for a subsequent analysis task, such as process discovery, log-delta analysis or concept drift. However, the choice of representation directly determines which behavioral distinctions remain observable in downstream analysis, yet no formal framework exists to compare representations with respect to this criterion. Consequently, the expressive capabilities and limitations of behavioral representations are undefined, impeding assessment of the quality and completeness of downstream analysis results. In this paper, we propose a framework for formally comparing the expressiveness and information preservation of behavioral representations. We introduce *losslessness* as a criterion for information preservation, *refinement* to capture when one representation subsumes the discriminative power of another, and *incomparability* when neither subsumes the other. We apply the framework to five representations commonly used in process mining. Our results show that representations differ not only in granularity but also in the behavioral dimensions they capture. In addition, commonly used representations are often orthogonal in their expressivity, showing the importance of selecting the most suitable representation for a particular task, as there is currently no single best representation.

Keywords: Downstream Analysis · Process Mining · Behavioral Representation · Representation Comparison · Log Abstraction

F. M. Maggi et al. (Eds.): BPM 2026, LNCS 16936, pp. 39–55, 2027.
https://doi.org/10.1007/978-3-032-37777-7_5

1 Introduction

Behavioral representations are a fundamental abstraction mechanism in business process management and, in particular, in process mining [2]. Rather than operating directly on raw event logs, process mining techniques typically abstract process executions into *behavioral abstractions of event logs*, capturing relations between activities. Such representations are used across a wide range of disciplines, including process discovery [1,10], log delta analysis [7], and concept drift localization [17]. As shown in Fig. 1, the selected behavioral representation strongly influences the quality and expressiveness of the analysis result [2]. As a consequence, representational limitations may render certain behavioral phenomena entirely invisible to downstream analysis techniques. This becomes particularly important when downstream analysis techniques depend on identifying behavioral differences between two event logs. If the chosen representation cannot distinguish certain behavioral aspects, such differences in the event logs will be entirely invisible to the downstream analysis task.

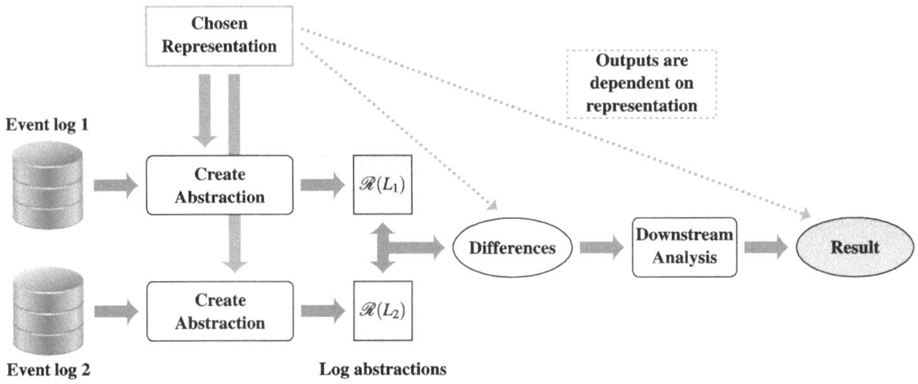

Fig. 1. Diagram of how behavioral representations are used to extract differences in event logs for downstream analysis, and how the choice of representation affects the resulting analysis output.

To illustrate this, consider two event logs L_1, L_2 of two purchase approval processes. In L_1, every purchase request (a) had to be followed by a managerial approval (b). L_2, however, integrates automatic approvals for low-value requests, thus, making b optional, meaning that b can still occasionally follow a but is no longer mandatory.

Using a directly-follows representation, the relation (a, b) remains observable whenever both occur, so the behavioral difference is not reflected and cannot be localized. In contrast, a declarative representation, e.g., Declare, captures the difference explicitly: for L_1, a succession constraint enforces that each request must eventually be followed by an approval, and for L_2, a precedence constraint reflects that b is no longer mandatory. This makes the behavioral difference directly observable.

This example illustrates the fact that some representations lose behavioral information when comparing two logs, and the ability to accurately represent process behavior depends on how explicitly a behavioral representation captures activity relationships. Depending on the requirements of the respective use case, this information loss is fundamental regarding the quality of analysis results.

To select suitable representations for a given analysis task, we need to understand how representations differ in terms of expressiveness, limitations, and behavioral distinctions they preserve. Existing work compares representations mostly in a descriptive manner based on process models [2], or studies which process behavior can be recovered from a given representation [10]. However, these approaches remain representation-specific and do not provide a general, context-independent framework for formally comparing behavioral representations. In particular, existing work lacks formal notions for determining whether one representation subsumes another, or which behavioral differences become unobservable under a given abstraction.

To investigate differences in representational expressiveness, we study behavioral comparison between event logs as a general setting in which process behavior is abstracted and compared using the same behavioral representation. This setting makes it possible to analyze which behavioral distinctions remain observable under a given representation and which information is lost. Based on these observations, explicit behavioral aspects captured by a representation can be identified. Moreover, behavioral comparison forms the basis of downstream tasks, such as log delta analysis and concept drift localization (see Fig. 1), making it a natural setting for studying representational limitations.

Based on this setting, we propose a general framework for comparing behavioral representations and introduce three fundamental criteria: *losslessness* indicates whether a representation can distinguish behavioral aspects and, thus, does not lose any information; *refinement* as a relation to capture when representations subsume others in terms of their expressiveness; and *incomparability* meaning that both representations cover behavioral aspects that the other does not include. The proposed framework is applied to five commonly used behavioral representations in process mining to showcase its application. For these representations, we conduct a comparative study and present detailed results. Nevertheless, the framework can also be applied to any other representation.

The remainder of this paper is organized as follows: Sect. 2 reviews related work while Sect. 3 introduces the behavioral representations used for the comparative study. Subsequently, Sect. 4 presents the formalism of the proposed comparison framework. The comparative study is explained in Sect. 5, while Sect. 6 discusses the results and concludes the paper.

2 Related Work

Behavioral abstractions are a common mechanism in process mining that are typically used to simplify comparison between heterogeneous sources of information or to increase efficiency of process discovery algorithms, process similarity analysis or process drift detection. Common abstractions are, e.g., "directly follows" [2] or "behavioral profiles" [16].

Process discovery aims to obtain a process model that captures the common behavior of an unknown process as observed in an event log [1]. The level of abstraction used for process discovery is a trade-off between computational efficiency and accurate discovery of more complex behaviors. Leemans and Fahland [10] discuss information preservation of process abstractions with a focus on recovering behavioral aspects for process discovery. While the authors make a contribution in investigating the transition from log abstractions to model abstractions that are used to discover models such that they more accurately preserve complex behavior, their approach is restricted to block structured processes. In addition to examining information preservation, the framework presented in this paper provides a more fine-grained analysis of behavioral abstractions by introducing the concepts of losslessness, refinement and incomparability.

Process similarity aims to identify behavioral differences between process executions. Similarity analysis has been studied for comparisons between two event logs, two process models, or between an event log and a process model. For log–log comparison, van Beest et al. [7] propose an approach based on event structures to capture behavioral differences between logs by identifying behavior that is present in one log but not in the other. Process model similarity approaches compare two models, typically assuming that both models are expressed using the same modeling notation [4,18]. Schützenmeier et al. [19] present an approach that compares Declare with imperatively modeled process models based on state automata. The comparison between event logs and process models is studied in the field of conformance checking [8], where techniques such as alignment-based methods and token-based replay measure how well a log conforms to a given model. Overall, these approaches demonstrate that process similarity measurement is an important task across multiple downstream analysis scenarios, yet they do not explicitly account for the inherent representational bias introduced by different modeling formalisms.

Furthermore, several studies discuss the limitations and properties of behavioral representations used in downstream analysis techniques. In [1], an overview is presented of the differences and limitations of the α relations, whereas [2] discusses the differences of directly-follows relationships to other types of abstractions, such as transition systems and cardinality constraint models. [3] extends this discussion by pointing out the difficulties of directly-follows relations to distinguish loops and concurrency. Furthermore, Weidlich et al. [21] compare behavioral profiles with causal footprints, analyzing their computational efficiency and their relation to trace equivalence for determining behavioral equivalence between process models. While these works analyze specific properties or limitations of behavioral representations, they only focus on applications such as process discovery or model comparison. In [13], a method is presented to detect process drift for unpredictable processes, in which they mention a very brief comparison between some different log abstractions before settling on α-

relations. However, no details on the criteria for the comparison are reported. Overall, each of these works lack a general framework for systematically comparing heterogeneous behavioral representations across different analysis contexts.

3 Background

3.1 Event Logs

An *event log* is a collection of traces, where each *trace* consists of a sequence of timestamped *events* recorded during the execution of a single instance of a business process. Each event in an event log refers to an activity executed at a particular time as part of a particular process. Given a universe of activities captured as events A, an event log is defined as follows.

Definition 1 (Trace, Event log). *A* trace τ *is a finite sequence of* n *events* $\langle a_1, \ldots, a_n \rangle$*, with* $a_i \in A$ *and* $i \in [1..n]$*. An* event log L *is a finite multiset of traces over* A*.*

3.2 Behavioral Representations

This section briefly introduces the behavioral representations commonly used in downstream analysis techniques, which we use to demonstrate the applicability of the proposed framework (see Sect. 5). For completeness, we include the undefined constraint $\perp$ for all representations that do not explicitly define an absence of a relationship.

Directly-Follows Relationships. The directly-follows representation [1] (DF) captures whether an activity is immediately followed by another activity in at least one trace of the log. Formally, for an event log L, the representation contains two types of relations:

- $a_i > a_j$ with $(a_i, a_j) \in A \times A$ if a_i directly precedes a_j in some trace of L, and
- $a_i \perp a_j$ with $(a_i, a_j) \in A \times A$ if a_i never directly precedes a_j in all traces of L.

Behavioral Profile. Behavioral profiles [20] (BP) are based on the weak order relation $a_i \succ a_j$ iff $\exists \sigma = \langle n_1, n_2, ..., n_m \rangle \in L$ with $k, l \in \{1, ..., m\}$ so that $k < l \leq m \wedge n_k = a_i \wedge n_l = a_j$. For an event log L, the behavioral profile representation assigns to each activity pair a_i, a_j exactly one of four mutually exclusive relations:

- Strict order ($a_i \rightsquigarrow a_j$): $a_i \succ a_j$ and $a_j \not\succ a_i$,
- Reverse strict order ($a_j \rightsquigarrow a_i$): $a_j \succ a_i$ and $a_i \not\succ a_j$,
- Exclusiveness ($a_i + a_j$): $a_i \not\succ a_j$ and $a_j \not\succ a_i$,
- Interleaving ($a_i \parallel a_j$): $a_i \succ a_j$ and $a_j \succ a_i$,
- Relationship absence ($a_i \perp a_j$).

α^+-Relations. The α^+ representation [5] classifies the behavioral relationship between two activities based on directly-follows observations and short-loop detection. Let L be a loop-complete event log over activity set A, i.e., $L \in \mathcal{P}(A^*)$, and let $a_i, a_j \in A$. Then, α+-relations are defined as follows:

- The directly-follows relation $a_i >_L a_j$ holds if there exists a trace $\sigma = a_1, a_2, \ldots, a_m \in L$ such that $j = i + 1$ and $i \in \{1, .., m-1\}$,
- $a \triangle b$ holds if there exists a trace $\sigma = a_1, a_2, \ldots, a_m \in L$ and an index $i \in \{1, \ldots, m-2\}$ such that $a_i = a_{i+2} = a$ and $j = i + 1$,
- $a_i \diamond a_j$ holds if and only if $a_i \triangle a_j$ and $a_j \triangle a_i$,
- A causal relation $a_i \rightarrow a_j$ holds if $a_i > a_j$ and neither $a_j > a_i$ nor a short loop $a_i \triangle a_j$ exists,
- Concurrency $a_i \parallel a_j$ holds if both $a_i > a_j$ and $a_j > a_i$ occur while no short loop is present,
- $a_i \# a_j$ denotes the absence of directly-follows relations in both directions.

Declare Constraints. Declare [15] was first introduced as a system to support declarative process models that also provides the possibility to generate constraint templates to define a modeling language. The ConDec language, for example, was created using Declare [14] and provides the set of constraints that is mostly used when talking about "Declare process models". Thus, we refer to the ConDec language explained by [11] when speaking about Declare constraints.

Declare constraints can be grouped into four categories: *existence constraints* (e.g., absence(), exactly(), init()) are unary relations indicating starting activities or how often an activity occurs per case; *choice constraints* choice() and ex_choice() are n-ary constraints indicating inclusive and exclusive OR relations; *binary constraints*, such as response(), precedence(), and succession(), define relations between two different activities and also include alternating behavior; and *negated binary constraints*, e.g., neg_response(), that exclude certain behavior from the process. For this paper, we consider the full set of Declare constraints.

Activity Relationships Notation. For two activities a_i, a_j, an activity relationship is defined as a tuple $(a_i, a_j) = (d_{temp}, d_{exist})$ with d_{temp} indicating the temporal dependency, i.e., ordering, between the activities, and d_{exist} describing their existential dependency [6]. For temporal dependency, we differentiate between

- Direct ($a_i <_d a_j$) temporal dependency denoting a_i directly precedes a_j in every trace in L,
- Eventual ($a_i < a_j$) temporal dependency denoting an eventual ordering that might include direct temporal orderings for some traces in L, and
- Independence ($a - b$), i.e., any ordering is allowed once both activities occur together in a trace.

For existential dependency, seven types are distinguished following Boolean Algebra operators. Note that each operator holds for all traces considered.

- $a_i \Rightarrow a_j$: Forward existential implication, i.e., every activity occurrence of a_i implies an occurrence of a_j,

- $\Leftarrow$: Backward existential implication, i.e., every activity occurrence of a_j implies an occurrence of a_i,
- $\Leftrightarrow$: Whenever a_i occurs in a trace, a_j occurs too, and vice versa,
- $\not\Leftrightarrow$: In every trace either a_i or a_j occurs,
- $\bar{\wedge}$: a_i and a_j cannot occur together,
- $\vee$: At least one of a_i or a_j occurs, and
- $-$: None of the above applies.

Relationships between multiple instances of the same activity a_i are represented as (a_i, a_i). If multiple instances of a_i do not exist, we refer to $(a_i, a_i) = \bot$. Combining these two types of dependencies leads to the set of constraints provided by the activity relationships representation $\mathcal{R}_{AR}$.

4 Formalization

This section formalizes behavioral representations, the behavioral difference between two event logs, and introduces *losslessness*, *refinement* and *incomparability* as measures of how finely a representation distinguishes behavior.

Behavioral representations provide a set of constraints that are assigned to an activity set $a^+ \in \mathcal{P}(A)$, with A being the set of activities extracted from an event log L, and $\mathcal{P}(A)$ the powerset of A. This leads to the following formalization:

Definition 2 (Behavioral Representation). *A behavioral representation is a function*

$$\mathcal{R} : \mathcal{L}(A) \to (\mathcal{P}(A) \to C_{\mathcal{R}}),$$

that assigns to each event log $L \in \mathcal{L}(A)$ a function $\mathcal{R}(L) : \mathcal{P}(A) \to C_{\mathcal{R}}$, where $\mathcal{P}(A)$ is the powerset of A, and $C_{\mathcal{R}}$ is the set of behavioral constraints expressible in representation $\mathcal{R}$.

That is, for each set $a^+ = \{a_1, a_2, ...a_n\} \in \mathcal{P}(A)$, the value $\mathcal{R}(L)(a^+)$ specifies the set of behavioral constraints describing the given activities' relation in L. Note that activity sets a^+ are assigned to $\bot$ if there are no constraints that hold for $a_i \in a^+$.

For example, for the log $L = \{\langle a, b, c\rangle\}$, the directly-follows representations assigns the following constraints to sets of activities: $\{a, b\} \to a > b$ and $\{b, c\} \to b > c$. For all other $a^+ \in \mathcal{P}(A)$, it is true that $a^+ \to \bot$. Note that the ordering of activities is set by the constraint defined by the representation, not through a^+.

To assess the expressiveness of a behavioral representation, we examine which behavioral aspects can be preserved when comparing the behavior of two event logs. By analyzing the behavioral differences between logs, we understand which behavioral aspects can be distinguished by a representation. Therefore, we define the behavioral difference as follows:

Definition 3 (Behavioral Difference). *Given two event logs L_1 and L_2, and a behavioral representation $\mathcal{R}$, the* behavioral difference *between the logs, considering $\mathcal{R}$, is defined as follows:*

$$\oplus_{\mathcal{R}} (L_1, L_2) = \mathcal{R}(L_1) \oplus \mathcal{R}(L_2) = \{a^+ | \mathcal{R}(L_1)(a^+) \neq \mathcal{R}(L_2)(a^+)\}$$

That is, the set of activities that are assigned constraints that differ when comparing two logs L_1, L_2 under $\mathcal{R}$.

Consider, for example, two event logs $L_1 = \{\langle a, b\rangle, \langle b, c\rangle\}$ and $L_2 = \{\langle a\rangle, \langle a, b\rangle, \langle b, c\rangle\}$. The behavior between the two logs is different, as b is optional in L_2, resulting in a behavioral difference of $\{\{a, b\}\}$. The relationship between a and b is different.

The *quality* of the output depends on how well those constraints pinpoint the actual structural modification, which is captured by the criterion *representation resolution* $\rho_{\mathcal{R}} = |C_{\mathcal{R}}|$. It captures the number of distinct behavioral constraints a representation comprises. However, it does not capture *which* configurations are distinguished and is, thus, only considered as a lightweight, practical property of behavioral representations. Two representations with the same ρ may be sensitive to different types of behavioral changes. To formally compare representations, we need a structural relation that captures when one representation subsumes the discriminative power of another.

Definition 4 (Losslessness). *Representation $\mathcal{R}_1$ is lossless to $\mathcal{R}_2$, written $\mathcal{R}_1 \sqsupseteq \mathcal{R}_2$, if $\forall L_1, L_2 \in \mathcal{L}(A) : \mathcal{R}_1(L_1) \oplus \mathcal{R}_1(L_2) \supseteq \mathcal{R}_2(L_1) \oplus \mathcal{R}_2(L_2)$.*

Intuitively, a representation is lossless with respect to another representation if every behavioral distinction observable under the latter remains observable under the former. That is, every behavioral distinction made by $\mathcal{R}_2$ is also made by $\mathcal{R}_1$ as illustrated in Fig. 2. Every constraint in $\mathcal{R}_1$ can be mapped to a constraint in $\mathcal{R}_2$, indicating a surjective function. If such a function does not exist, $\mathcal{R}_1$ is not lossless to $\mathcal{R}_2$ as illustrated on the left-hand side of Fig. 2.

Lemma 1 (Derivability Characterization). *$\mathcal{R}_1 \sqsupseteq \mathcal{R}_2$ iff there exists a surjective[1] function φ such that $\forall L \in \mathcal{L}(A), a^+ \in \mathcal{P}(A) : \varphi(\mathcal{R}_1, L, a^+) = \mathcal{R}_2(L)(a^+)$.*

Proof. ($\Rightarrow$) We define a function φ as: $\forall L \in \mathcal{L}(A), a^+$, $\varphi(\mathcal{R}_1(L), a^+) = \mathcal{R}_2(L)(a^+)$. In this, φ can be defined such that it finds the constraints of $\mathcal{R}_1(L)$ for any relevant permutation of a^+ and thereby deduce $\mathcal{R}_2(L)(a^+)$.

To show that φ is a well-defined function, suppose for L_1, L_2 we have $\mathcal{R}_1(L_1)(a^+) = \mathcal{R}_1(L_2)(a^+)$. If $\mathcal{R}_2(L_1)(a^+) \neq \mathcal{R}_2(L_2)(a^+)$ holds, then $(a^+) \in \mathcal{R}_2(L_1) \oplus \mathcal{R}_2(L_2)$ but $(a^+) \notin \mathcal{R}_1(L_1) \oplus \mathcal{R}_1(L_2)$, contradicting $\mathcal{R}_1 \sqsupseteq \mathcal{R}_2$. Hence φ is well-defined. Now we show surjectivity by contradiction. Assume φ is not surjective. Then there exists an $\mathcal{R}_2(L)(a^+)$ for which we do not find a corresponding $\mathcal{R}_1(L)(a^+)$, which contradicts the subset relation of the losslessness condition.

[1] This includes bijective and surjective-only.

($\Leftarrow$) Suppose such a surjective function φ exists. Then, it is true that $\forall L_1, L_2 : \mathcal{R}_2(L_1)(a^+) \neq \mathcal{R}_2(L_2)(a^+) \implies \varphi(\mathcal{R}_1(L_1)(a^+)) \neq \varphi(\mathcal{R}_1(L_2)(a^+))$ which is exactly the losslessness condition.

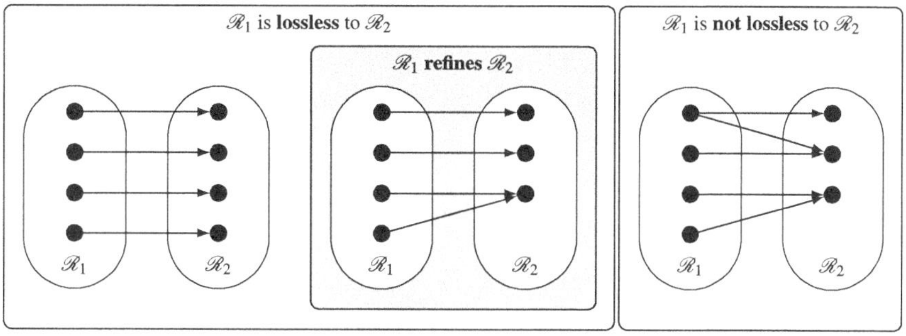

Fig. 2. Losslessness and refinement.

If φ is surjective-only, multiple behavioral constraints in $\mathcal{R}_1$ collapse into the same behavioral constraint in $\mathcal{R}_2$ (shown in the middle of Fig. 2), meaning $\mathcal{R}_1$ makes strictly finer distinctions. In this case it is true that $\mathcal{R}_2 \not\sqsupseteq \mathcal{R}_1$. We use this observation to define resolution dominance, i.e., *refinement*.

Definition 5 (Refinement). *Representation* $\mathcal{R}_1$ refines $\mathcal{R}_2$ *in representation resolution, written* $\mathcal{R}_1 \sqsupset \mathcal{R}_2$ *iff* $\mathcal{R}_1 \sqsupseteq \mathcal{R}_2 \wedge \mathcal{R}_2 \not\sqsupseteq \mathcal{R}_1$.

Based on these definitions, we can formulate the following proposition regarding the existence of blind spots under refinement.

Proposition 1 (Existence of Blind Spots under Refinement). *If* $\mathcal{R}_1 \sqsupset \mathcal{R}_2$*, then there exist sublogs* L_1, L_2 *such that for an activity set* (a^+) *it is true that* $\mathcal{R}_1(L_1)(a^+) \neq \mathcal{R}_1(L_2)(a^+)$ *but* $\mathcal{R}_2(L_1)(a^+) = \mathcal{R}_2(L_2)(a^+)$*. As such, the behavioral difference* $\oplus_{\mathcal{R}_2}(L_1, L_2) \subset \oplus_{\mathcal{R}_1}(L_1, L_2)$*.*

Proof. Since $\mathcal{R}_1 \sqsupset \mathcal{R}_2$, we have $\mathcal{R}_2 \not\sqsupseteq \mathcal{R}_1$. By negation of Definition 4 for R_2, R_1, it is true that $\exists L_p, L_q, a^+ \in \mathcal{L}(A) : a^+ \in \mathcal{R}_1(L_p) \oplus \mathcal{R}_1(L_q) \wedge a^+ \notin \mathcal{R}_2(L_p) \oplus \mathcal{R}_2(L_q)$ which means that the two logs L_p, L_q are identical under $\mathcal{R}_2$ for (a^+) but not under $\mathcal{R}_1$: $\mathcal{R}_2(L_p)(a^+) = \mathcal{R}_2(L_q)(a^+) \wedge \mathcal{R}_1(L_p)(a^+) \neq \mathcal{R}_1(L_q)(a^+)$.

Setting $L_p = L_1$ and $L_q = L_2$, the activity set a^+ that is part of $\oplus_{\mathcal{R}_1}(L_1, L_2)$ is not part of $\oplus_{\mathcal{R}_2}(L_1, L_2)$, thus proving the claim.

□

If $\mathcal{R}_1 \not\sqsupseteq \mathcal{R}_2$ and $\mathcal{R}_2 \not\sqsupseteq \mathcal{R}_1$ holds, the representations R_1, R_2 are *incomparable*, meaning one can distinguish certain changes, while the other cannot and vice versa.

Definition 6 (Incomparability). *Two representations* $\mathcal{R}_1$ *and* $\mathcal{R}_2$ *are incomparable, written* $\mathcal{R}_1 \bowtie \mathcal{R}_2$*, if neither refines the other:* $\mathcal{R}_1 \not\sqsupseteq \mathcal{R}_2 \wedge \mathcal{R}_2 \not\sqsupseteq \mathcal{R}_1$.

5 Comparative Analysis of Behavioral Representations

Given the formalism introduced in Sect. 4, we are able to compare different behavioral representations with regard to their representation resolution $\rho_{\mathcal{R}}(a^+)$ and expressiveness in terms of incomparability and refinement. Table 1 provides an overview of the results.

Table 1. Results of the systematic analysis of behavioral representations comparing representation resolution $\rho_{\mathcal{R}}$, refinement, and incomparability of different representations R (DF: directly-follows, BP: behavioral profiles, AR: activity relationships).

$\mathcal{R}$	**DF**	**BP**	α^+	**AR**	**Declare**
$\rho_{\mathcal{R}}$	2	3	6	22	29
DF	-	$\bowtie$	$\sqsubset$	$\sqsubset$	$\sqsubset$
BP		-	$\sqsubset$	$\sqsubset$	$\sqsubset$
α^+			-	$\bowtie$	$\bowtie$
AR				-	$\bowtie$
Declare					-

Representation resolution is the number of different behavioral constraints that can be assigned to an activity set a^+. We illustrate the computation of the resolution by using the example of directly-follows relations. All other results shown in Table 1 were derived similarly.

For a set of activities $a^+ = \{a_1, a_2\}$, the directly-follows representation indicates that either a_1 is never directly followed by a_2, or there exists a directly-follows relation in at least one trace. Behavioral differences can only be identified when a change affects the existence of an adjacency relation between two activities, while other changes in their behavioral relationship remain indistinguishable. Thus, $\rho_{\mathcal{R}_{DF}} = |\{\perp, \{a_1 > a_2\}\}| = 2$.

To determine the relation between two representations, we analyzed each pair with respect to losslessness, refinement, and incomparability. For a given pair $\mathcal{R}_1, \mathcal{R}_2$, Definition 4 states that $\mathcal{R}_1$ is lossless to $\mathcal{R}_2$ if all relation changes detected in $\mathcal{R}_1$ can be detected in $\mathcal{R}_2$. As such, we identified the pairs where $\mathcal{R}_1$ is *not* lossless to $\mathcal{R}_2$ by finding counterexamples of two logs that can be distinguished by $\mathcal{R}_2$ but not by $\mathcal{R}_1$. These counterexamples are given in Table 2. For pairs that *do* have losslessness, we show this based on Lemma 1, which states that losslessness holds if there exists a surjective mapping function from constraints in $\mathcal{R}_1$ to constraints in $\mathcal{R}_2$. As such, we identified the relations where this function exists, and provided this function in Table 3. Finally, once the losslessness condition had been evaluated in each direction for every pair, we were then able to identify whether each pair has refinement or is incomparable.

For example, consider the representations $\mathcal{R}_{DF}$ and $\mathcal{R}_{BP}$. It holds true that neither representation is lossless to the other, because we can find two sets of logs that cannot be distinguished by one representation (see Table 2). Following Definition 6, the representations are considered incomparable. For $\mathcal{R}_{\alpha^+}$ and $\mathcal{R}_{DF}$, we can find a surjective function mapping from directly-follows constraints to constraints included in the α^+ relations (see Table 3), proving that $\mathcal{R}_{\alpha^+}$ is lossless to $\mathcal{R}_{DF}$. This function is surjective-only, thus indicating refinement. Blind spots can be identified as shown in Table 2 since $\mathcal{R}_{DF}$ is not lossless to $\mathcal{R}_{\alpha^+}$. The reader is referred to the schema explained above on how to read the tables to interpret the results shown in Table 1.

The analysis reveals two important observations. First, behavioral representations differ significantly in their representation resolution, which determines how many behavioral configurations between activities can be distinguished. Higher resolution enables more precise identification of behavioral changes. Second, even representations with higher resolution do not necessarily dominate others in terms of behavioral expressiveness. Several representations are incomparable, meaning that each representation captures behavioral aspects that the other cannot represent.

As shown in Table 1, directly-follows relations have the lowest resolution, distinguishing only whether a_2 directly follows a_1. Thus, behavioral comparison is limited to adjacency changes. Behavioral profiles increase resolution by capturing concurrency and choice patterns, further extended α^+ relations which include loops. Declare constraints and activity relationships provide a much higher resolution by capturing richer semantics such as temporal and existential dependencies, enabling finer-grained analysis of behavioral difference.

However, a higher resolution alone does not determine representational power, as lower-resolution representations may capture different behavioral aspects. This aspect is illustrated by Table 4 summarizing the behavioral aspects that are covered by the representations, respectively. The aspects shown are derived from the conducted comparative analysis and constraint types captured by the different representations. We formalized each aspect in a general manner to ensure comparability across representations. Note that the coverage of explicit existential dependencies means whether the representation includes constraints that do not indicate temporal ordering.

Although directly-follows relations are suitable for detecting behavioral differences resulting from immediate sequencing of activities, they cannot detect many higher-level ones. For example, a directly-follows relation between two activities in both direction can indicate loop behavior and concurrency and is, thus, not explicit on the exact behavioral difference. Behavioral profiles focus on global ordering relations, enabling the detection of behavioral differences affecting concurrency or exclusiveness between activities. Compared to activity relationships and Declare, behavioral profiles only cover the exclusive choice constraint, whereas the other two representations can distinguish inclusive, exclusive and nand behavior. Interestingly, only activity relationships cover temporal ordering in inclusive choice structures, but Declare defines choice patterns as an n-ary constraint. The α^+ relations extend this perspective by incorporating loop patterns derived from directly-follows relations that remain indistinguishable in

Table 2. Detailed counter-examples for incomparability and refinement.

$\mathcal{R}_1 \not\sqsupseteq \mathcal{R}_2$	**Logs**	**Derived Relationships**
$\mathcal{R}_{BP} \not\sqsupseteq \mathcal{R}_{DF}$	$L_1 : \{\langle a,b,c\rangle\}$ $L_2 : \{\langle a,b,c\rangle, \langle a,c\rangle\}$	$R_{BP}(L_1) = R_{BP}(L_2) = \{a \rightsquigarrow b,\ b \rightsquigarrow c,\ a \rightsquigarrow c\}$ $R_{DF}(L_1) = \{a > b, b > c, a \perp c\} \neq R_{DF}(L_2) = \{a > b, b > c, a > c\}$
$\mathcal{R}_{DF} \not\sqsupseteq \mathcal{R}_{BP}$	$L_1 : \{\langle a,b,c\rangle, \langle c,b,a\rangle\}$ $L_2 : \{\langle a,b,a\rangle, \langle c,b,c\rangle\}$	$R_{BP}(L_1) = \{a \parallel b, a \parallel c, c \parallel b\} \neq R_{BP}(L_2) = \{a \parallel b, a + c, c \parallel b\}$ $R_{DF}(L_1) = R_{DF}(L_2) = \{a > b,\ b > c,\ c > b,\ b > a\}$
$\mathcal{R}_{Declare} \not\sqsupseteq \mathcal{R}_{\alpha+}$	$L_1 : \{\langle a,b\rangle, \langle a\rangle, \langle b\rangle\}$ $L_2 : \{\langle b,a\rangle, \langle a\rangle, \langle b\rangle\}$	$R_{Declare}(L_1) = \{choice(1 \text{ of } 2,\ [a,b])\} = R_{Declare}(L_2) = \{choice(1 \text{ of } 2,\ [a,b])\}$ $R_{\alpha+}(L_1) = \{a > b\} \neq R_{\alpha+}(L_2) = \{b > a\}$
$\mathcal{R}_{\alpha+} \not\sqsupseteq \mathcal{R}_{Declare}$	$L_3 : \{\langle a,b\rangle, \langle a,c\rangle\}$ $L_4 : \{\langle a,b\rangle, \langle a,c\rangle, \langle c\rangle\}$	$R_{Declare}(L_3) =$ {precedence(a,b), precedence(a,c), ex_choice(1 of 2,[b, c])} $\neq$ $R_{Declare}(L_4) =$ {precedence(a,b), precedence(a,c), not_coexistence([b, c])} $R_{\alpha+}(L_3) = R_{\alpha+}(L_4) = \{a < b, a < c\}$
$\mathcal{R}_{AR} \not\sqsupseteq \mathcal{R}_{\alpha+}$	$L_1 : \{\langle a,b,b,a\rangle, \langle b,c,b\rangle\}$ $L_2 : \{\langle a,b,a,b\rangle, \langle b,c,b\rangle\}$	$R_{AR}(L_1) = \{(a,b) = (b,c) = (-, \Leftrightarrow), (a,a) = (b,b) = (<, \Leftarrow), (a,c) = (-, \not\Leftrightarrow)\}$ $= R_{AR}(L_2)$ $R_{\alpha+}(L_1) = \{a \parallel b, a \# c, b \triangle c\} \neq R_{BP}(L_4) = \{a \diamond b, a \# c, b \triangle c\}$
$\mathcal{R}_{\alpha+} \not\sqsupseteq \mathcal{R}_{AR}$	$L_1 : \{\langle a,b\rangle, \langle b,a\rangle\}$ $L_2 : \{\langle a,b\rangle, \langle b,a\rangle, \langle a\rangle, \langle b\rangle\}$	$R_{\alpha+}(L_1) = \{a \parallel b\} = R_{\alpha+}(L_2) = \{a \parallel b\}$ $R_{AR}(L_1) = \{(-, \Leftrightarrow)\} \neq R_{AR}(L_2) = \{(-, \vee)\}$
$\mathcal{R}_{AR} \not\sqsupseteq \mathcal{R}_{Declare}$	$L_1 : \langle b,c,a,c,a,c,b\rangle$ $L_2 : \{\langle a,c,b,c,a,c,b\rangle\}$	$R_{AR}(L_1) = \{(a,b) = (-, \Leftrightarrow), (a,a) = (b,b) = (<, \Leftarrow), (a,c) = (c,b) = (-, \Leftrightarrow)\}$ $= R_{AR}(L_2)$ $R_{Declare}(L_1) =$ {alt_precedence(a, b), alt_precedence(c, b), alt_succession(a, c)} $\neq R_{Declare}(L_2) =$ {alt_succession(a, b), alt_precedence(c, b), alt_response(a, c)}
$\mathcal{R}_{Declare} \not\sqsupseteq \mathcal{R}_{AR}$	$L_3 : \{\langle a,b\rangle, \langle a\rangle, \langle b\rangle\}$ $L_4 : \{\langle b,a\rangle, \langle a\rangle, \langle b\rangle\}$	$R_{Declare}(L_3) =$ {choice(1 of 2, [a, b])} $= R_{Declare}(L_4)$ $R_{AR}(L_3) = \{(a,b) = (<_d, \vee)\} \neq R_{AR}(L_4) = \{(a,b) = (>_d, \vee)\}$
$\mathcal{R}_{DF} \not\sqsupseteq \mathcal{R}_{\alpha+}$	$L_1 : \{\langle a,b,c\rangle, \langle a,c,b\rangle\}$ $L_2 : \{\langle a,b,c\rangle, \langle a,c,b,c\rangle\}$	$R_{DF}(L_1) = R_{DF}(L_2) = \{a > b, b > c, a > c, c > b\}$ $R_{\alpha+}(L_1) = \{a \rightarrow b, c \parallel b, a \rightarrow c\} \neq R_{\alpha+}(L_2) = \{a \rightarrow b, a \rightarrow c, c \triangle b\}$
$\mathcal{R}_{DF} \not\sqsupseteq \mathcal{R}_{Declare}$	$L_1 : \{\langle a,b\rangle\}$ $L_2 : \{\langle a,b\rangle, \langle a\rangle\}$	$R_{DF}(L_1) = R_{DF}(L_2) = \{a > b\}$ $R_{Declare}(L_1) =$ {chain_succession(a, b) } $\neq R_{Declare}(L_2) =$ {chain_precedence(a, b)}
$\mathcal{R}_{DF} \not\sqsupseteq \mathcal{R}_{AR}$	$L_1 : \langle a,b\rangle$ $L_2 : \{\langle a,b\rangle, \langle a\rangle\}$	$R_{DF}(L_1) = R_{DF}(L_2) = \{a > b\}$ $R_{AR}(L_1) = \{(a,b) = (<_d, \Leftrightarrow)\} \neq R_{AR}(L_2) = \{(a,b) = (<_d, \Leftarrow)\}$
$\mathcal{R}_{BP} \not\sqsupseteq \mathcal{R}_{Declare}$	$L_1 : \langle a,b,b,a\rangle$ $L_2 : \langle b,a,a,b\rangle$	$R_{BP}(L_1) = R_{BP}(L_2) = \{a \parallel b\}$ $R_{Declare}(L_1) =$ {alt_response(a, b)} $\neq R_{Declare}(L_2) =$ { alt_precedence(a, b)}
$\mathcal{R}_{BP} \not\sqsupseteq \mathcal{R}_{AR}$	$L_1 : \{\langle a,b\rangle, \langle b,a\rangle\}$ $L_2 : \{\langle a,b\rangle, \langle b,a\rangle, \langle a\rangle, \langle b\rangle\}$	$R_{BP}(L_1) = R_{BP}(L_2) = \{a \parallel b\}$ $R_{AR}(L_1) = \{(a,b) = (-, \Leftrightarrow)\} \neq R_{AR}(L_2) = \{(a,b) = (-, \vee)\}$

Table 3. Existence of a surjective function to show refinement.

Relation	Function
$\mathcal{R}_{\alpha^+} \sqsupset \mathcal{R}_{DF}$	$\varphi(v) = \begin{cases} a_i > a_j & v \in \{\triangle, \diamond, >, \rightarrow, \parallel\}, \\ \perp & v = \# \end{cases}$
$\mathcal{R}_{Declare} \sqsupset \mathcal{R}_{DF}$	$\varphi(v) = \begin{cases} \perp & \text{if } v \in \{\text{neg\_chain\_response, neg\_chain\_precedence, neg\_chain\_succession, neg\_response, neg\_precedence, neg\_succession, resp\_absence}\} \\ a_i > a_j & \text{otherwise} \end{cases}$
$\mathcal{R}_{\alpha^+} \sqsupset \mathcal{R}_{BP}$	$\varphi(v) = \begin{cases} a_i \rightsquigarrow a_j & v \in \{\rightarrow\}, \\ a_i + a_j & v \in \{\#\}, \\ a_i \parallel a_j & v \in \{\parallel\}, \\ \perp & \text{otherwise} \end{cases}$
$\mathcal{R}_{Declare} \sqsupset \mathcal{R}_{BP}$	$\varphi(v) = \begin{cases} a_i \parallel a_j & v \in \{\text{coexistence, resp\_existence}\}, \\ a_i + a_j & v \in \{\text{not\_coexistence, resp\_absence, neg\_succession}\}, \\ a_i \rightsquigarrow a_j & v \in \{\text{response, precedence, succession, chain\_response, chain\_precedence, chain\_succession, neg\_chain\_succession, neg\_chain\_precedence, neg\_chain\_response}\}, \\ \perp & \text{otherwise} \end{cases}$
$\mathcal{R}_{AR} \sqsupset \mathcal{R}_{DF}$	$\varphi((d_{temp}, d_{exist})) = \begin{cases} a_i > a_j & d_{temp} \in \{<_d, <\}, \\ \perp & \text{otherwise} \end{cases}$
$\mathcal{R}_{AR} \sqsupset \mathcal{R}_{BP}$	$\varphi((d_{temp}, d_{exist})) = \begin{cases} a_i \rightsquigarrow a_j & d_{temp} \notin \{-\}, \\ a_i + a_j & (d_{temp}, d_{exist}) \in \{(-, \Leftrightarrow), (-, \overline{\wedge})\}, \\ a_i \parallel a_j & d_{temp} \in \{-\}, \\ \perp & \text{otherwise} \end{cases}$

behavioral profiles. Declare constraints provide a constraint-based representation of process behavior. They capture various dependencies such as response, precedence, succession, and alternating patterns, enabling detailed comparison of differences in event logs. However, it does not capture pure temporal dependencies with existential independence but is the only representation providing an explicit constraint for alternating behavior.

The behavioral aspect of looping behavior includes self-loops and short loops. While α^+ introduces an explicit constraint for short loops, activity relationships and Declare cover this aspect by a set of constraints without having an explicit operator. For an event log $L : \{\langle a, b, a \rangle\}$, the following constraints describe the contained loop: exactly$(2, a)$, exactly$(1, b)$, alt_response(a, b) for Declare and $(a, a) = (<, \Leftarrow), (a, b) = (-, \Leftrightarrow), (b, b) = \perp$ for activity relationships.

Explicit existential dependencies that do not entail a temporal ordering between activities are included in Declare and activity relationships, with the latter having a focus on distinguishing temporal and existential dependency.

Overall, Table 4 shows, that representations do not only differ in their coverage of behavioral aspects but also in terms of scope. Declare and the activity relationships notation, for instance, define constraints that are valid for all traces of an event log, whereas directly-follows, behavioral profiles, and α^+ relations only need at least one trace for a constraint to hold.

Table 4. Behavioral aspects that are covered by different representations (DF: directly-follows, BP: behavioral profiles, AR: activity relationships). Behaviors that can be explicitly represented are indicated by ✓, while those that can be represented only implicitly, through a combination of constraints, are indicated by (✓).

Behavioral Aspect	DF	BP	α^+	AR	Declare
Directly-follows adjacency					
$\exists \sigma \in L : a_i, a_j \in \sigma \wedge j = i + 1$	✓		✓	✓	✓
$\forall \sigma \in L : a_i, a_j \in \sigma \rightarrow j = i + 1$				✓	✓
Eventually-follows ordering					
$\exists \sigma \in L : a_i, a_j \in \sigma \wedge i < j$		✓		✓	✓
$\forall \sigma \in L : a_i, a_j \in \sigma \rightarrow i < j$				✓	✓
Concurrency / parallel execution					
$\exists \sigma_1, \sigma_2 \in L : a_i, a_j \in \sigma_1, \sigma_2 \wedge i_{\sigma_1} < j_{\sigma_1} \wedge j_{\sigma_2} < i_{\sigma_2}$		✓	✓	✓	✓
Choice structures (inclusive, exclusive, nand)					
$\forall \sigma \in L : a_i \in \sigma \vee a_j \in \sigma$				✓	✓
$\forall \sigma \in L : (a_i \in \sigma \wedge a_j \notin \sigma) \vee (a_i \notin \sigma \wedge a_j \in \sigma)$		✓		✓	✓
$\forall \sigma \in L : (a_i \in \sigma \wedge a_j \notin \sigma) \vee (a_i \notin \sigma \wedge a_j \in \sigma) \vee (a_i, a_j \notin \sigma)$				✓	✓
Ordering in choice structures					
$\forall \sigma \in L : a_i \vee a_j \in \sigma \wedge (a_i, a_j \in \sigma \rightarrow i < j)$				✓	
Loop behavior					
$\exists \sigma \in L : a_i, a_j, a_k \in \sigma \wedge a_i = a_k \neq a_j \wedge j = i + 1 \wedge k = i + 2$			✓	(✓)	(✓)
$\exists \sigma \in L : a_i, a_j \in \sigma \wedge a_i = a_j \wedge j = i + 1$				✓	
Alternating behavior					
$\forall \sigma \in L : a_i, a_j \in \sigma \wedge a_i = a_j \rightarrow j \neq i + 1 \wedge \exists a_k \neq (a_i \vee a_j) : i < k < j$					✓
Co-occurrence patterns					
$\forall \sigma \in L : a_i, a_j \in \sigma$				✓	✓
Temporal ordering with existential independence					
$\exists \sigma_1, \sigma_2, \sigma_3, \sigma_4 : a_i, a_j \notin \sigma_1 \wedge (a_i \in \sigma_2 \wedge a_j \notin \sigma_2) \wedge$ $(a_i \notin \sigma_3 \wedge a_j \in \sigma_3) \wedge (a_i, a_j \in \sigma_4) \wedge (a_i, a_j \in \sigma_4 \rightarrow i < j)$	✓	✓	✓	✓	
Explicit existential dependencies					
$\exists c = (a_i, a_j) \in \mathcal{R} : c = \text{occurrence}(a_i, a_j) \wedge c \not\models (i < j) \wedge c \not\models (j < i)$				✓	✓

The refinement analysis further shows that some representations refine others. For example, Declare constraints and activity relationships refine behavioral profiles and directly-follows relations. This means that these representations preserve all behavioral distinctions of the less expressive representation while introducing additional distinctions. However, the analysis also reveals several cases of incomparability, where neither representation refines the other, such as, for instance, Declare constraints and activity relationships. Overall, the results highlight that no single representation dominates all others with respect to determining behavioral differences between two event logs. Instead, different representations capture complementary aspects of process behavior.

6 Discussion & Conclusion

Behavioral representations are fundamental for process mining techniques and a necessity for model abstractions for subsequent analysis. In this paper, we present a framework that allows to compare two representations in terms of losslessness and refinement and apply it to commonly used representations in process mining.

The proposed framework shifts the perspective on downstream analysis techniques, such as concept drift localization or log delta analysis, from a purely algorithmic problem towards a representational observability problem. Our results show that the ability to distinguish behavioral aspects is fundamentally bounded by the expressive capabilities of the underlying behavioral representation. Consequently, downstream analysis techniques cannot be evaluated independently from representation choice.

The systematic analysis presented in Sect. 5 provides formal insights into these capabilities of representations to preserve and distinguish behavioral differences between event logs. The results demonstrate that behavioral representations differ not only in their representation resolution, but also in the behavioral aspects they preserve explicitly. Thus, the quality and interpretability of downstream analysis tasks strongly depend on the representational assumptions encoded in the selected abstraction. Depending on the specific requirements and expectations, e.g., the granularity, of the desired result, the representation should be chosen accordingly. For instance, if the analysis focuses primarily on adjacency relations and local ordering behavior, directly-follows relations may provide a sufficiently lightweight abstraction without requiring computationally more intensive discovery procedures such as Declare or activity relationship extraction. In contrast, analyses requiring richer distinctions between temporal, existential, or alternating behavior benefit from more expressive representations.

The overview in Table 4, therefore, allows practitioners to better reason the choice of a representation for applying process mining techniques and algorithms, as well as to investigate potential representational limitations in existing ones. This is especially useful when evaluating the output quality of concept drift localization algorithms, e.g., [9,12], or log delta analysis [7], which directly depends on the expressiveness of the underlying behavioral representation. Similarly, process discovery algorithms can only represent those behavioral aspects that can be captured by the chosen log abstraction, i.e., behavioral representation [2].

Interestingly, our results also show that commonly used representations are often incomparable; they do not necessarily subsume the behavioral distinctions captured by other representations. Instead, different representations preserve complementary behavioral aspects. This observation has important implications for process mining research. Improving downstream analysis techniques alone cannot overcome behavioral information that has already been lost during abstraction. Consequently, the expressive capabilities and assumptions of behavioral representations should be considered explicitly when designing, evaluating, and comparing process mining approaches. Nevertheless, future work can make use of the results to extend existing representations so that more behavioral aspects are captured. Due to its general concept, the proposed framework can be applied to other representations to further extend the comparative overview of representations.

Note that we study commonly used representations in process mining and base our comparison and analysis on the original sets of constraints that define each representation. Some techniques also include additional information, such as frequencies, for example, in directly-follows graphs, where edges can be weighted by how often a relation occurs. We do not explicitly include such frequency information in our comparison because it adds an extra feature that makes representations no longer directly comparable to those that do not support frequencies. Furthermore, extending all representations under comparison with equivalent frequency attributes would not alter the conclusions of this paper, as this property could then be uniformly subsumed by all representations. It is also important to note that frequencies are not only about how often a relation between two activities occurs, but can also refer to how often individual activities occur, as in Declare with its existence constraints. This additional property of Declare makes the representation incomparable to those that cover the same set of binary constraints but lack the information of activity frequencies. Nevertheless, our framework supports a more detailed investigation into frequencies.

Finally, the presented analysis focuses on the theoretical expressive power of behavioral representations rather than the computational effort of their discovery from event logs. This provides a foundation that can be complemented by future empirical studies investigating how factors such as extraction methods and parameters influence downstream analysis results in practice.

Acknowledgments. We thank our colleague Leon Bein for his input on untangling formalisms.

References

1. van der Aalst, W.M.P.: Process Discovery: An Introduction, pp. 163–194. Springer, Heidelberg (2016)
2. van der Aalst, W.M.P.: Process discovery from event data: Relating models and logs through abstractions. WIREs Data Min. Knowl. Discovery **8**(3), e1244 (2018)
3. van der Aalst, W.M.P.: A practitioner's guide to process mining: limitations of the directly-follows graph. Procedia Comput. Sci. **164**, 321–328 (2019)
4. Alves de Medeiros, A.K., van der Aalst, W.M.P., Weijters, A.: Quantifying process equivalence based on observed behavior. Data Knowl. Eng. **64**(1), 55–74 (2008)
5. Alves De Medeiros, A.K., van Dongen, B.F., van der Aalst, W.M.P., Weijters, A.: Process mining for ubiquitous mobile systems: an overview and a concrete algorithm. In: International Workshop on Ubiquitous Mobile Information and Collaboration Systems, pp. 151–165. Springer (2004)
6. Andree, K., Bano, D., Weske, M.: A closer look at activity relationships to improve business process redesign. Softw. Syst. Model. **24**, 69–86 (2025)
7. van Beest, N.R.T.P., Dumas, M., García-Bañuelos, L., La Rosa, M.: Log delta analysis: Interpretable differencing of business process event logs. In: Int. Conf. on Business Process Management, pp. 386–405. Springer (2015)

8. Dunzer, S., Stierle, M., Matzner, M., Baier, S.: Conformance checking: a state-of-the-art literature review. In: Proceedings of the 11th International Conference on Subject-Oriented Business Process Management, pp. 1–10 (2019)
9. Kraus, A., van der Aa, H.: Machine learning-based detection of concept drift in business processes. Process Sci. **2**(1), 5 (2025)
10. Leemans, S.J.J., Fahland, D.: Information-preserving abstractions of event data in process mining. Knowl. Inf. Syst. **62**(3), 1143–1197 (2020). https://doi.org/10.1007/S10115-019-01376-9. https://doi.org/10.1007/s10115-019-01376-9
11. Montali, M.: The condec language. In: Specification and Verification of Declarative Open Interaction Models: A Logic-Based Approach, pp. 47–75. Springer (2010)
12. Ostovar, A., Maaradji, A., La Rosa, M., ter Hofstede, A.H.M.: Characterizing drift from event streams of business processes. In: Dubois, E., Pohl, K. (eds.) Int. Conf. on Advanced Information Systems Engineering. LNCS, vol. 10253, pp. 210–228. Springer (2017)
13. Ostovar, A., Maaradji, A., La Rosa, M., ter Hofstede, A.H., van Dongen, B.F.: Detecting drift from event streams of unpredictable business processes. In: International Conference on Conceptual Modeling, pp. 330–346. Springer (2016)
14. Pesic, M., van der Aalst, W.M.P.: A declarative approach for flexible business processes management. In: Int. Conf. on Business Process Management. pp. 169–180. Springer (2006)
15. Pesic, M., Schonenberg, H., van der Aalst, W.M.P.: Declare: full support for loosely-structured processes. In: 11th IEEE international enterprise distributed object computing conference (EDOC 2007), p. 287. IEEE (2007)
16. Polyvyanyy, A., Armas-Cervantes, A., Dumas, M., García-Bañuelos, L.: On the expressive power of behavioral profiles. Formal Aspects Comput. **28**(4), 597–613 (2016)
17. Sato, D.M.V., Freitas, S.C.D., Barddal, J.P., Scalabrin, E.E.: A survey on concept drift in process mining. CoRR abs/2112.02000 (2021)
18. Schoknecht, A., Thaler, T., Fettke, P., Oberweis, A., Laue, R.: Similarity of business process models-a state-of-the-art analysis. ACM Comput. Surv. (CSUR) **50**(4), 1–33 (2017)
19. Schützenmeier, N., Jablonski, S., Käppel, M., Ackermann, L.: Comparing the expressiveness of imperative and declarative process models. In: Babkin, E., Barjis, J., Malyzhenkov, P., Merunka, V., Molhanec, M. (eds.) Model-Driven Organizational and Business Agility, pp. 16–31. Springer, Cham (2023)
20. Smirnov, S., Weidlich, M., Mendling, J.: Business process model abstraction based on synthesis from well-structured behavioral profiles. Int. J. Cooperative Inform. Syst. **21**(01), 55–83 (2012)
21. Weidlich, M., Mendling, J., Weske, M.: Efficient consistency measurement based on behavioral profiles of process models. IEEE Trans. Softw. Eng. **37**(3), 410–429 (2011)

Online Resource-Aware Process Configuration with Uncertain Arrivals and Stochastic Resource Consumption

Janik Bürgermeister(✉), Felix Schumann, Martin Bichler, and Stefanie Rinderle-Ma

TUM School of Computation, Information and Technology, Technical University of Munich, Garching, Germany
{janik.buergermeister,felix.schumann,m.bichler,stefanie.rinderle-ma}@tum.de

Abstract. Business processes are often executed in multiple variants, which affect operational costs and the consumption of shared resource capacities. We study an *online* planning problem in which process instances arrive sequentially and must be configured irrevocably upon arrival under horizon-wide resource capacity constraints. For example, in a hospital, an unknown number of patients arrive sequentially during a shift. The patient's diagnosis consumes capacity of available resources and the resources then chosen for allocation, in turn, affect the resulting control flow configuration for this patient. We formulate the problem as an online integer program (IP) with a fixed configuration space and model uncertainty through (i) a random number of arrivals and (ii) stochastic, positive resource consumptions revealed upon arrival; costs are modeled as a weighted sum of consumed workload, reflecting processing time, and thus are also stochastic. We assess rolling-horizon and resource-pricing (primal-dual) approaches and compare them with simple heuristics and the optimal ex post solution. We instantiate the framework for a hospital diagnostics data set. Using distribution estimates based on this data, we evaluate online decision policies under robustness perturbations. The findings show that optimization-based policies that account for remaining capacities and expected demand substantially improve throughput and resource utilization under uncertain arrivals.

Keywords: Process configuration · Online optimization · Stochastic modeling · Resource allocation

1 Introduction

Business processes are often executed with limited shared resources, such as staff or equipment. In many domains, a single process can be executed in multiple *configurations*, for example, by assigning tasks to different resources, or in terms of process variants by selecting alternative control flows [26]. These configuration choices affect both operational cost and resource consumption. Traditional business process management (BPM) typically considers either configuration [20] or

F. M. Maggi et al. (Eds.): BPM 2026, LNCS 16936, pp. 56–73, 2027.
https://doi.org/10.1007/978-3-032-37777-7_6

resource allocation [10,16] in isolation or under offline planning assumptions [22]. However, in many operational settings, configuration decisions must be taken repeatedly as new cases arrive and compete for shared and scarce resources, while future arrivals and workloads remain uncertain. Due to this uncertainty, configuration and allocation decisions must balance current cost with the preservation of resources for future arrivals. Decisions that minimize immediate cost may exhaust scarce resources too early and increase costs later in the planning horizon.

Figure 1 visualizes the studied problem in the context of a hospital, where each process instance corresponds to the diagnosis of a single patient. Patients arrive sequentially throughout the day, but the total number of future patients and their conditions are unknown in advance. A *configuration* corresponds to an admissible assignment of tasks (`report`, `read`, `approve`) to physician qualification profiles. We distinguish between three qualification profiles: high (H, Chief physician), mid (M, physician), and low (L, intern). Highly qualified physicians perform tasks more quickly, but they are scarce resources. Further, the control flow and resource perspective are coupled, i.e., if the report is done by H, the subsequent tasks, read and approval, are removed from the process instance (see *Next Instance t4*). Upon arrival, instances are configured and tasks are allocated to resources. Thus, instances consume different amounts of resources (H, M, L) depending on their configuration. An instance can also be rejected. This serves as an abstraction for real-world strategies to handle overflow, such as outsourcing an instance, adding expensive overtime capacity, or moving a process instance to the next planning horizon. Rejecting an instance, therefore, saves resource capacity now but incurs additional costs later. From the perspective of the optimization problem, the ability to avoid using capacity by rejecting an instance ensures the existence of a feasible solution.

In this paper, we study the problem of *online resource-aware process configuration* under uncertainty. Each chosen configuration induces workloads on shared resources that accumulate over the planning horizon. We model this problem as an online integer program where each arriving instance must be assigned a configuration from a finite configuration space with the corresponding resource allocation. Configuring & allocating an instance on arrival is aligned with existing work on resource allocation and scheduling [15,22]. The configurations differ in cost, resource consumption, and control flow. Decisions are coupled through horizon-wide capacity constraints. To address this problem, we connect resource-aware process configuration with recent advances in online integer optimization.

In particular, this paper makes the following contributions. First, we introduce an online optimization model for *resource-aware runtime process configuration*. In contrast to traditional configurable process models, where variant choices are often analyzed offline or per case, our model captures the fact that configuration decisions for one case influence future choices for other cases through shared resource constraints. Second, we adapt and analyze online decision policies for this setting, including rolling-horizon optimization and primal-dual methods that dynamically update prices for using resource capacity. This provides a formal

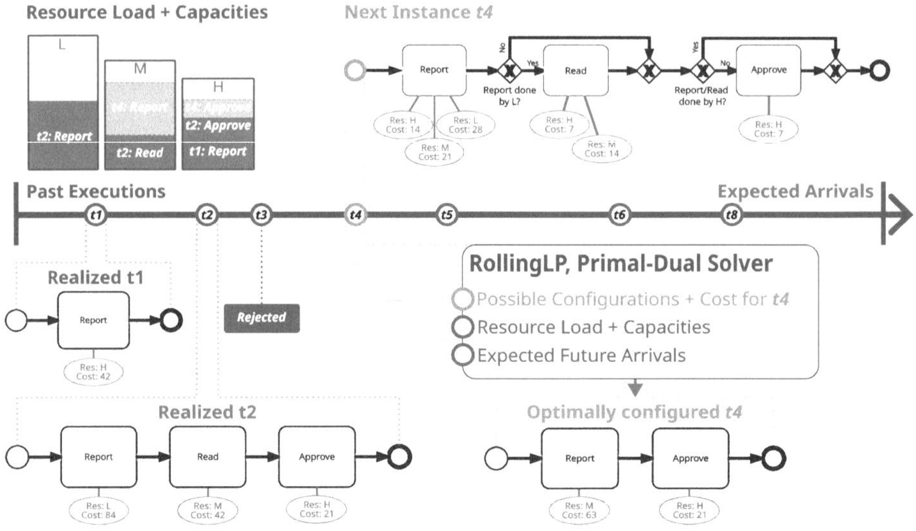

Fig. 1. Illustration of the online problem. Realized resource consumptions are depicted in purple, additions by an arriving instance in orange. (Color figure online)

algorithmic perspective on enactment-time process configuration under uncertainty. Third, we demonstrate how the proposed model can be instantiated in a hospital process setting based on a real-world log of resource requirements and patient-specific characteristics. This shows how configurable process models can be operationalized in a dynamic, data-informed runtime environment. Though the paper employs a hospital dataset, the provided methods also fit online resource allocation problems in other domains, e.g., manufacturing or logistics. Here, the choice of rejecting an instance naturally reflects the outsourcing of an instance to a supplier for a certain cost. Fourth, our computational study identifies when more anticipatory online methods yield substantial benefits over myopic assignment policies, thereby generating BPM-relevant insights on the value of coordination across simultaneously active process instances.

The remainder of the paper is organized as follows. Section 2 reviews related literature. Section 3 introduces the online integer programming model. Section 4 presents the online decision policies. Section 5 describes the hospital case study and the empirical evaluation. Section 6 discusses implications on BPM. Section 7 concludes the paper.

2 Related Work

This work relates to configurable and resource-aware BPM, as well as stochastic, and online resource allocation. Both areas have studied the coordination of configuration and allocation decisions over shared resources and under uncertainty.

Process Configuration and Resource-Aware BPM: Structural variability in business processes has been widely studied in the BPM literature through configurable and variant-rich process models. Such models capture multiple variants whose control flow structure is determined through configuration decisions [7,12]. These approaches focus primarily on representing and managing structural variability of the control flow perspective. Resource-aware BPM extends this classical work by explicitly considering the allocation of tasks to resources such as staff or equipment. Workflow resource patterns [18,19] and subsequent work on resource allocation and workload management [2] formalize how tasks are assigned during execution. These approaches take local allocation decisions at runtime instead of horizon-wide coupling across many concurrently executed instances. Recent approaches in the BPM field apply predictive process monitoring [17], or deep reinforcement learning [16] to improve such allocation decisions under uncertainty. To predict waiting times during the execution of running cases [23] employs queue mining. [14] uses information gathered by queue mining for offline scheduling. The focus of our work is on online decision-making for process configuration in combination with resource allocation where queue mining can serve as an additional input for decision making. Resource-Augmented Process Structure Trees (RA-PSTs), model how resource assignments can modify process structure through task insertion, deletion, or replacement [22]. In [21], the joint configuration and scheduling of multiple process instances under temporal constraints, without uncertainty, is studied. Uncertainty-aware scheduling is addressed on a hospital use case in [15]. In contrast, our work abstracts from detailed scheduling and focuses on the higher-level problem of selecting capacity-feasible configurations for sequentially arriving instances under uncertainty.

Online and Stochastic Resource Allocation: Our problem is related to stochastic packing and online linear and integer programming (IP) for online resource allocation. In these problems, items arrive sequentially, decisions must be made irrevocably, and shared capacity constraints couple decisions across the planning horizon. Seminal work by Feldman et al. [6] and Agrawal et al. [3] develops near-optimal algorithms based on prices for consuming capacity under stochastic arrivals. Such price-based approaches are applied in revenue management systems [24,25]. Our setting differs from classical online packing problems as every arriving process instance has multiple feasible configurations that induce different resource consumption patterns, whereas most online packing formulations assume a single item per arrival. Our decisions correspond to higher-dimensional configuration choices for a BPM process model rather than simple one-dimensional accept–reject decisions. We show that our problem can be modeled as a generalized assignment problem. While the offline version, with all information present, is NP-hard, there are LP relaxations that yield a relatively tight approximation. For the online version with uncertain future arrivals, several recent papers derive performance bounds in the random-arrival model [4,9]. However, this theoretical work assumes that the exact number of arrivals is known. We relax this assumption and require only distributional knowledge.

Despite substantial progress in configurable process modeling and resource-aware BPM, less attention has been paid to the online selection of process variants during enactment when decisions are coupled across cases through shared capacity constraints. Existing approaches often either abstract from resource scarcity, treat cases independently, or optimize allocation decisions conditional on a fixed process structure. Our paper addresses *runtime process configuration* in a stochastic multi-case environment by proposing a formal online model.

3 Modeling Online Configuration and Allocation

The following online IP model represents a variant of the generalized assignment problem (GAP) for a fixed planning horizon. We index process instances by their order of arrival t, and denote the total number of arrivals by T. For each new process instance, exactly one feasible configuration $i \in [n]$ is selected. We encode this selection with one-hot vectors $x_t \in \{0,1\}^n$, where $e_i = (0,\ldots,0,1,0,\ldots,0)^\top$ represents selecting the i-th feasible configuration. Each configuration allocates resources $r \in [m]$ with capacities $b_r \geq 0$, and the resource consumption of the selected configuration is given by $A_t x_t$. The matrix $A_t \in \mathbb{R}^{m \times n}$ describes how many resources an instance t consumes for a given configuration assignment. An example is given in Table 1a where each row of the matrix describes one possible process configuration over the resources. The vector $c_t \in \mathbb{R}^n$ represents the costs of an allocation. Throughout this paper, the costs are equal to the resource capacity consumed by an allocation. Generally, the costs can also be set independently of the resource consumption. The goal of an allocation policy is to minimize cost over all process instances arriving in the horizon, while adhering to the capacity constraints

$$\min_{x \in \mathcal{X}} \sum_{t=1}^{T} c_t^\top x_t \quad \text{s.t.} \quad \sum_{t=1}^{T} A_t x_t \leq b, \tag{GAP}$$

where the feasible region $\mathcal{X}$ is given by

$$\mathcal{X} = \prod_{t=1}^{T} \mathcal{X}_t, \quad \mathcal{X}_t = \left\{ \sum_{i=1}^{n} x_{t,i} = 1, \quad x_{t,i} \in \{0,1\}, \forall i \in [n] \right\}.$$

The inequality constraints $\sum_{t=1}^{T} A_t x_t \leq b$ couple decisions across the entire time horizon and represent shared capacity constraints.

In the example shown in Fig. 1, resources correspond to physician qualification levels $R = \{H, M, L\}$ (chief physician, physician, intern). We aggregate multiple resources of a given type into a common capacity $b = (b_H, b_M, b_L)$; e.g., b_L represents the combined available work time of 10 different interns. Each arriving patient t must be assigned one configuration $a \in \mathcal{A} = \{H, HM, HL, HML, R\}$, where each configuration induces a workload vector across the resources. For example, selecting configuration HM assigns the task `report` of the diagnosis process to a physician and the task `approval` to a chief physician. The task `read` must not be executed, since a resource of type L is not involved in the `Report` task (see Fig. 1, $t4$). Besides the configurations shown in Fig. 1, the configuration assignment R models the rejection of an instance and has no resource consumption. This modeling choice ensures the existence of a feasible solution; i.e., an instance is rejected if no configuration except R can satisfy the capacity constraints. Configuration R represents an expensive configuration inducing the highest cost (see Assumption 4). The capacity constraint $\sum_{t=1}^{T} A_t x_t \leq b$ ensures that the total workload assigned to each qualification level over the shift does not exceed the staff capacity. Intuitively, each configuration decision for a process instance consumes part of the available resource capacity, and these consumptions accumulate across all process instances in the horizon.

Information Structure. In the online version, the capacity vector b is known at the start of the horizon, i.e., the beginning of a shift (cf. Resource Capacities Fig. 1). Information is revealed sequentially: for each instance t, the realized costs c_t and capacity consumption coefficients A_t become known at arrival. Based on this revealed information, a configuration for t is chosen via x_t.

Assumptions on Problem Setting. We impose some mild assumptions to ensure that the model captures realistic and meaningful settings that are also reflected in the data used for evaluation.

In the online optimization literature, there are two predominant arrival models: the adversarial model and the random-order model. Adversarial arrival models make no assumptions on the arrival process, treating inputs as arbitrary or worst-case sequences and evaluating algorithms through worst-case performance guarantees. This perspective is overly pessimistic. In line with simulation and resource allocation approaches in BPM, the arrival distribution can be estimated from historical data. We adopt the random-order model and assume i.i.d. arrivals from a known distribution F to determine the resource consumption of arriving instances [1,16].

Assumption 1 (i.i.d Arrivals). $(A_t, c_t) \overset{i.i.d}{\sim} F$.

A central regularity assumption in stochastic packing and online integer programming is that individual arrivals have a limited impact on global capacity.

Assumption 2 (Bounded Impact). $\exists \alpha \in (0,1)$ *such that* $\forall x \in \mathcal{X}_t$,

$$(A_t x)_r \;\leq\; \alpha\, b_r \quad \forall t, \forall r.$$

If α is small, then Assumption 2 rules out dominant arrivals that could consume a large fraction of resources in a single step. This aligns with the hospital example, where no single patient monopolizes an entire shift of staff.

The regime of interest is $\alpha > 1/T$. When $\alpha \leq 1/T$, capacity is effectively divisible at the level of individual arrivals, and simple greedy policies that always select the cheapest feasible configuration are optimal. Our focus is therefore on intermediate scarcity regimes where no single arrival dominates the system.

Prior works [4,9] assume that the number of decisions to be made T is fixed and known. In contrast, we assume distributional knowledge.

Assumption 3 (Random Number of Arrivals). $T \sim G$.

Assumption 3 allows us to study a more realistic setting, as the number of arriving cases is typically unknown, e.g., the arrival of emergency cases.

4 Algorithms

We present online decision policies for solving the previously introduced resource-aware process configuration problem. These policies select process configurations for each arriving instance based on the current system state, such as remaining resource capacities and expected future demand (cf. Fig. 1 *Past Executions*). This allows configuration choices, typically treated at the design phase of the BPM lifecycle, to be integrated with online optimization techniques that impact the enactment phase during process execution.

We consider a set of policies that reflect different levels of sophistication. First, we introduce simple heuristic baselines (Greedy and Conservative) that represent natural operational strategies. Second, we adapt two approaches from the online optimization literature: a rolling-horizon method that repeatedly solves a planning problem with estimated future demand, and a price-based primal–dual policy that maintains adaptive resource prices.

The Greedy baseline ignores future arrivals and selects the currently cheapest feasible configuration under the remaining capacity for each instance t. Let $\bar{b}$ be the remaining capacity. Greedy chooses

$$x_t \in \arg\min_{x \in \mathcal{X}_t} c_t^\top x \quad \text{s.t.} \quad A_t x \leq \bar{b}.$$

The Conservative baseline aims to preserve scarce resources by choosing a capacity-sparing configuration, even if this increases immediate cost. Operationally, it selects the non-reject option with the *largest* immediate cost, or the reject option if no such option exists. This is a simple proxy for choosing a configuration that consumes less of the bottleneck resources.

$$x_t \in \arg\max_{x \in \mathcal{X}_t \setminus \{e_R\}} c_t^\top x \quad \text{s.t.} \quad A_t x \leq \bar{b}.$$

The RollingLP (Algorithm 1) is a rolling horizon approach that repeatedly optimizes over the remaining number of expected arrivals. After observing the resource consumption at the current point in time and the costs for the current instance t, it generates samples for future instances that serve as estimates of the remaining demand. Using this estimated demand, the corresponding GAP is solved (see line 5 in Algorithm 1), but only the immediate decision x_t is implemented. Decisions for future steps are discarded for now and will be recalculated in the next iteration. This allows the algorithm to continuously adapt to unfolding events while accounting for future arrivals.

The ROLLINGLP solves the LP relaxation of the IP model described in Sect. 3, which removes the integrality constraints of the GAP and thus can produce fractional solutions. In this case, we order the resulting actions by decreasing LP value and select the first one that satisfies the capacity constraints. This might yield slightly worse results, since the rounded solution may not always coincide with the optimal integer solution. However, in practice, the LP relaxation is relatively tight, so the solution quality does not degrade, while the runtime substantially decreases.

Algorithm 1 RollingLP

Initialize remaining capacity $\bar{b} \leftarrow b$ and expected number of arrivals $\hat{T} = \mathbb{E}[T]$
for $t = 1$ to T **do**
  Observe c_t, A_t
  Sample $(c_\tau, A_\tau) \sim F$ for $\tau \in \{t+1, \ldots, \hat{T}\}$
  Solve the LP of:

$$\min_{x_{\geq t}, x_\tau \in \mathcal{X}_\tau} \sum_{\tau=t}^{\hat{T}} c_\tau^\top x_\tau \quad \text{s.t.} \quad \sum_{\tau=t}^{\hat{T}} A_\tau x_\tau \leq \bar{b},$$

  If x_t is fractional, round to a feasible integer solution
  Implement x_t, discard $x_{>t}$
  Update remaining capacity $\bar{b} \leftarrow \bar{b} - A_t x_t$
  Update expected number of arrivals $\hat{T} = \lfloor \mathbb{E}[T|T > t] \rfloor$
end for

Besides ROLLINGLP, we consider PRIMALDUAL (Algorithm 2), which leverages dynamic resource pricing to balance demand and capacity. This approach is referred to as *primal–dual*, because the resource prices are derived from the *dual problem* of the underlying GAP. In contrast to ROLLINGLP, not the GAP over the full set of expected arrivals is solved, but the price aware one-step problem is solved, which reduces the problem size. By adding λ_t to the original resource costs of the current instance, the prices for the current instance are adapted to the system state and the expected future arrivals. The initial dual variables $\lambda_t \in \mathbb{R}_+^m$, interpreted as resource prices, are estimated from historical data. Solving the price-aware problem for the current resource prices yields the next decision. Subsequently, dual prices are updated based on remaining capacity and expected future arrivals. For example, for an expensive process instance a configuration is chosen that consumes much of the capacity of a certain resource. This resource's price increases in future allocations. For a less expensive instance, the updated prices for all resources might decrease, as the price update considers average capacities over the full horizon.

Primal–dual algorithms admit strong theoretical guarantees. In the case of accept–reject decisions and a known number of arrivals, the regret bounds of [13] imply an expected competitive ratio $ECR = 1 + O\left(\frac{\log T \log\log T}{T}\right) \xrightarrow{T \to \infty} 1$, since

Algorithm 2 PrimalDual

1: Initialize remaining capacity $\bar{b} \leftarrow b$ and expected number of arrivals $\hat{T} = \mathbb{E}[T]$
2: Initialize dual variables λ
3: **for** $t = 1$ to T **do**
4: Observe A_t, c_t
5: Solve the price aware one-step problem

$$x_t \in \arg\min_{x \in \mathcal{X}_t} \quad c_t^\top x + \lambda^\top A_t x \quad \text{s.t.} \quad A_t x \leq \bar{b}$$

6: Implement x_t
7: Update capacity $\bar{b} \leftarrow \bar{b} - A_t x_t$
8: Update dual variables $\lambda_{t+1} \leftarrow \max\{\lambda_t + \eta_t (A_t x_t - b/\hat{T}), 0\}$
9: Update expected number of arrivals $\hat{T} = \lfloor \mathbb{E}[T|T > t] \rfloor$
10: **end for**

the optimal objective value grows at least linearly with the number of instances in our problem. Hence, the primal-dual policy is asymptotically optimal in this case. A formal definition of the (expected) competitive ratio is provided in Sect. 5.1

5 Evaluation

We evaluate the approach based on a data set abstracted from a real-world clinical event log[1]. Note that the approach can be applied to event logs from any application domain as long as information about resources and distributional information on instance arrival can be derived from it.

The data set spans slightly more than one year and contains approximately 67,000 cases. We consider only the subset of weekdays in our evaluation, with a mean volume of about 255 cases per day. Arrivals are time-of-day dependent, given a three shift model with segments 06:00–14:00, 14:00–22:00, and 22:00–06:00. Consequently, we choose a separate planning horizon for each shift. In practice, methods from business process simulation can be employed to derive expected arrivals and consumption of process configurations directly from an event log in a data-driven manner. This has no influence on the performance of the presented algorithms. The corresponding process model for the hospital example is shown in Fig. 1. The process variants and their resource consumption for each qualification profile are summarized in Table 1a), Table 1b) shows default cost initialization.

Assumption 4 summarizes the restrictions on costs and workload reductions across skill tiers in the experiment, which state that (1) higher-qualified doctors are faster, and (2) lower-tier workers reduce the workload of higher-tier workers. The hierarchy-reflecting cost matrix can be adapted to any given use-case.

[1] Due to confidentiality restrictions, the dataset cannot be made publicly available. The source code for the experiments and algorithm implementations is publicly available at https://github.com/janik-buergermeister/OnlineOptBPM/.

Table 1. Cost structures across treatment variants.

Variant	Cost	H	M	L
H	c_H	c_H	0	0
HM	$c_{H,M} + c_M$	$c_{H,M}$	c_M	0
HL	$c_{H,L} + c_L$	$c_{H,L}$	0	c_L
HML	$c_{H,M} + c_{M,L} + c_L$	$c_{H,M}$	$c_{M,L}$	c_L
R	c_R	0	0	0

(Columns H, M, L form A.)

(a) General cost breakdown by variant

Variant	Cost	H	M	L
H	1	1	0	0
HM	2	0.5	1.5	0
HL	2.75	0.75	0	2
HML	3.25	0.5	1	2
R	10	0	0	0

(b) Baseline costs satisfying (A4)

Assumption 4 (Cost ordering and workload reduction).

1. Cost ordering. $c_H < c_{H,M} + c_M < c_{H,L} + c_L < c_{H,M} + c_{M,L} + c_L < c_R$.
2. Workload reduction. $c_H > c_{H,M} > c_{H,L}; c_M > c_{M,L}$.

To instantiate the GAP for the hospital example, we need to define c_t and A_t. We do so by modeling the task difficulty via a random variable Y_t and obtain the objective and capacity constraints as

$$c_t = Y_t A \mathbb{1} + (Y_t + 1) c_R e_R, \quad A_t = Y_t A^\top.$$

A is defined in Table 1a), $\mathbb{1}$ denotes a vector containing all ones of suitable size, and $e_R = (0, \ldots, 0, 1)^\top$ corresponds to the rejection decision. The penalty term $(Y_t + 1)c_R$ ensures that rejection is the most costly decision. For any no reject configuration a, $c_{t,a}$ is the sum of resource consumption for this configuration.

5.1 Experimental Setup and Parameter Estimation

As no algorithm can be guaranteed to achieve a bounded (expected) competitive ratio (ECR) for an uncertain number of arrivals [5], we computationally study the performance of the proposed algorithms in our experimental setup.

Performance metrics. We evaluate online algorithms by comparing their performance to an offline benchmark with full knowledge of the realization of the online information. Let $\omega = (T, (c_t, A_t)_{t=1}^T)$ denote a realization of the number of arrivals T and the realized per-step coefficients (c_t, A_t). We denote the corresponding clairvoyant (offline) optimal solution as $x^\star(\omega)$ and its objective value by $\mathrm{OPT}(\omega) = c^\top x^\star(\omega)$. Let π denote an online policy that produces a feasible decision vector $x^\pi(\omega)$. The (expected) competitive ratio is then measured by

$$\mathrm{CR}(\pi, \omega) = \frac{c^\top x^\pi(\omega)}{\mathrm{OPT}(\omega)}, \quad \mathrm{ECR}(\pi) = \mathbb{E}\left[\mathrm{CR}(\pi, \omega)\right],$$

where values closer to 1 indicate better performance. We also evaluate the (expected) resource utilization

$$U(\pi, \omega) = \frac{\sum_t A_t x_t^\pi(\omega)}{b}, \quad EU(\pi) = \mathbb{E}[U(\pi, \omega)].$$

An additional metric of interest is the fraction of rejected process instances. The (expected) rejection rate is given by

$$RR(\pi,\omega) = \frac{\sum_t \mathbb{1}\{x^{\pi}_{t,R}(\omega) = 1\}}{T}, \quad ERR(\pi) = \mathbb{E}[RR(\pi,\omega)].$$

Distribution Estimation. To determine suitable distributions for the number of arrivals T during one shift and the task difficulty Y_t, we fit several candidate distributions to the data using maximum likelihood estimation (MLE). Model comparison is performed using the Akaike Information Criterion (AIC) and the Bayesian Information Criterion (BIC), which balance goodness of fit with model complexity. The distribution with the lowest information criterion value is selected as the best-fitting model. Based on this, T is modeled using a negative binomial distribution, while we use a Weibull distribution for task difficulty:

$$T \sim \text{NegBin}(r,p), \quad Y_t \sim \text{Weibull}(k,\lambda).$$

Beyond the statistical goodness-of-fit, the selected distributions are consistent with the dataset's structural properties. The negative binomial distribution reflects arrivals in which the variance exceeds the mean. The Weibull distribution captures right-skewness of the task durations, with most cases being completed relatively quickly, while a smaller number requires longer processing times.

Random Number of Arrivals Model $T \sim G$ (Negative Binomial). As a first-order summary, the night shift has the smallest volume (roughly ≈ 30 cases) while the two other shifts have substantially higher volumes (roughly ≈ 90 cases each). To account for different expected numbers of arrivals, we evaluate our policies over a grid of $\mu_T = \mathbb{E}[T] \in \{30, 60, 90, 120\}$. For a given target mean μ_T, we estimate the dispersion $\hat{r} \approx 37$ from the dataset and we adjust

$$p(\mu_T) = \frac{\hat{r}}{\hat{r} + \mu_T},$$

across the grid of expected arrivals. This construction yields variances that scale consistently with the mean (smaller variance for smaller segments).

Task-Difficulty Model $Y_t \sim F$ (Weibull). The mean duration of the full process across all model variants is 4.5h (median 3h). From our parametric estimation, we obtain a shape parameter of $k = 0.9$. As we normalize $c_H = 1$, we use $\lambda = 2$ as the scale parameter, for the full duration of the diagnosis.

Resource Capacities b. We scale resource capacities b by the expected number of arrivals μ_T, ensuring that capacities increase as μ_T increases. Moreover, more resources are available at lower qualification levels. Specifically, for each chief physician, there are two normal physicians and four interns,

$$b(\mu_T) = (b_H, b_M, b_L)(\mu_T) = \beta\mu_T(1, 2, 4).$$

We investigate different capacity regimes characterized by $\beta \in \{1, 1.5, \ldots 5\}$. We note that the cheapest variant (H) does not have sufficient capacity to handle all the processes for low to medium values of β, while it does so for high values of β. While the first case ($\beta = 1$) characterizes a situation with resource scarcity, the last case ($\beta = 5$) describes a setting with abundant capacity.

5.2 Results

Our model relies on distributions for the number of arrivals T and the task difficulty Y_t, estimated from data. This allows us to calibrate the model to a realistic hospital setting, but the performance may be sensitive to model misspecification. *Robustness.* To assess the robustness of our results, we conducted additional experiments under alternative distributions. Specifically, we considered different parameterizations of the same parametric families and alternative distributions with comparable means and variances. Across all tested variants, the qualitative behavior of the evaluated policies remained unchanged: RollingLP and PrimalDual consistently outperformed simple heuristics in capacity-constrained regimes, while the relative ranking of the heuristics depends on the level of resource scarcity. This suggests that the conclusions are robust across distributions. Due to space constraints, we omit a detailed presentation of the additional experiments and only report the results in detail for $T \sim \text{NegBin}(37, 0.29)$ and $Y_t \sim \text{Weibull}(0.9, 2)$, which corresponds to the distributions estimated from our real-world dataset.

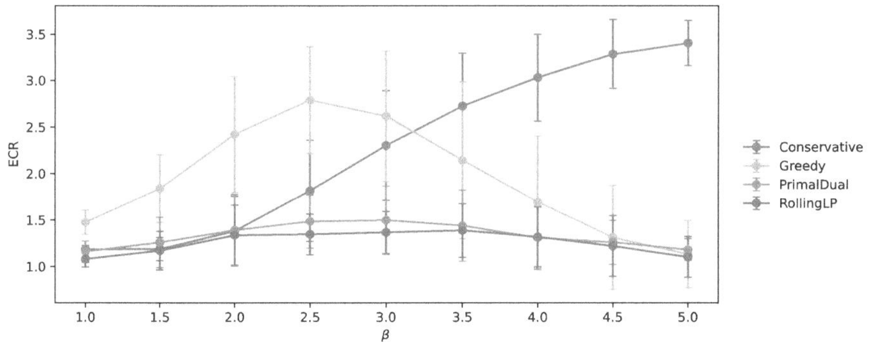

Fig. 2. Performance comparison (ECR) across resource levels β.

Figure 2 summarizes the performance of the evaluated policies across resource scarcity levels β for a fixed expected number of arrivals ($\mu_T = 90$). Performance is measured by the mean ECR relative to the clairvoyant offline IP across 50 instances, with error bars indicating the standard deviation.
Impact of Resource Scarcity. Resource availability strongly influences the relative performance of the simple heuristics, while the optimization-based approaches RollingLP and PrimalDual consistently achieve low ECR values.

In the low-capacity regime (small β), Conservative achieves solution quality close to the more advanced methods RollingLP and PrimalDual. By prioritizing capacity-sparing configurations, it preserves scarce resources and reduces the risk of infeasible states later in the horizon. The GREEDY policy performs significantly worse because it prioritizes immediate costs and tends to exhaust critical resources early. In the high-capacity regime (large β), the relative ordering of the heuristics changes. Since the risk of premature resource depletion vanishes, selecting the cheapest configuration becomes advantageous. GREEDY now outperforms CONSERVATIVE.

Decision Patterns. Figure 3 reports the frequency of selected configurations in both capacity regimes. In the low-capacity setting, the CONSERVATIVE policy almost exclusively selects the configuration HML, which distributes workload across all resources and thus minimizes the risk of exhausting any single capacity. Choosing HML incurs high cost compared to other non-reject configurations. GREEDY predominantly selects the cheapest configuration H. While this minimizes immediate cost, the workload concentrates on the constrained resource.

The more advanced policies, PRIMALDUAL and ROLLINGLP, adopt a more balanced strategy by selecting a mix of configurations, primarily H, HM and HL. This allows them to exploit cheap configurations when possible while preserving flexibility for future arrivals. When capacity is abundant, policies shift toward lower-cost configurations because resource constraints are rarely binding.

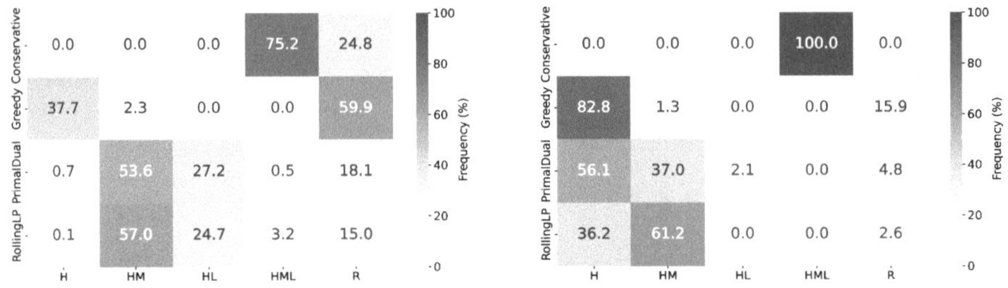

(a) Low-capacity setting. ($\beta = 1.5$) (b) High-capacity setting. ($\beta = 3.5$)

Fig. 3. Comparison of decision frequency under different capacity settings.

Rejection Rate. The last columns of Fig. 3 show the rejection frequencies. In the low-capacity regime, rejection becomes a frequent outcome for Greedy. In contrast, the Conservative policy spreads workload across all resources, which delays capacity exhaustion and results in substantially fewer rejections. PRIMALDUAL and ROLLINGLP, achieve even lower rejection rates by dynamically balancing immediate cost with remaining capacity and expected future demand.

For high capacities, rejection rates decrease as capacity constraints become less binding. Nevertheless, differences remain: GREEDY exhibits a noticeable rejection rate due to its systematic reliance on the cheapest configuration, whereas the optimization-based policies maintain marginal rejection levels. Overall, accounting for future arrivals and capacity scarcity improves cost performance and substantially reduces rejection rate. CONSERVATIVE does not have the lowest rejection rate, because RollingLP and PrimalDual can reject instances with high resource consumption even when sufficient capacity is available. By incurring the highest cost, rejecting a case has the highest impact on ECR. Lowering the cost gap between the rejection and the other configurations makes the decision for a rejection more tolerable in terms of ECR, setting a higher rejection cost makes the optimization-based policies RollingLP and PrimalDual less prone to choose rejections.

Resource Utilization. The resulting utilization is shown in Fig. 4. In the low-capacity regime, Conservative achieves nearly full utilization by consistently selecting capacity-balanced configurations. In contrast, Greedy quickly saturates the high-tier resource while leaving the remaining resources largely unused. The optimization-based policies achieve more balanced utilization patterns across resources, which explains their improved overall performance.

In the high-capacity regime, utilization levels are generally lower and the differences between policies become smaller. Nevertheless, PrimalDual and RollingLP still maintain slightly more balanced resource usage.

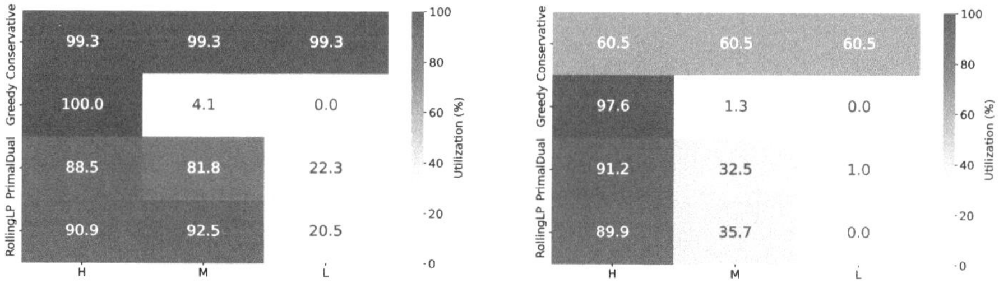

(a) Low-capacity setting. ($\beta = 1.5$) (b) High-capacity setting. ($\beta = 3.5$)

Fig. 4. Comparison of resource utilization Ax/b under different capacity settings.

Effect of the expected number of arrivals. The benefit of the optimization-based policies increases for larger μ_T. When resources are scarce, their ECR decreases as the expected number of arrivals μ_T increases. This behavior is natural, as larger sample sizes allow better approximations. When capacity is abundant, the performance becomes largely insensitive to the expected number of arrivals, with both policies achieving stable ECR values around 1.4.

Runtime. Table 2 reports the mean runtime per decision. RollingLP incurs the highest computational cost due to the repeated solution of large LPs, whereas PrimalDual requires only solutions to much smaller problems, since future instance arrivals are accounted for through the dual prices. Despite this difference, the absolute runtime remains small for our exemplary evaluation dataset: solving the LP relaxation requires approximately 0.2 seconds per decision step in our implementation. For larger allocation sets, RollingLP scales less well due to its per-step computational cost increasing with the expected time horizon, whereas PrimalDual, Greedy, and Conservative are independent of horizon length. All approaches incur increasing computational costs as the number of configurations increases. However, this scaling is linear in the number of configurations, and thus the resulting LPs for the PRIMALDUAL approach can still be computed efficiently for large scale instances in practice.

Table 2. Runtime comparison of policies.

Policy	RollingLP	PrimalDual	Greedy	Conservative
Runtime per Step (s)	0.231	0.018	0.001	0.001

Summary. ROLLINGLP provides the best overall performance, but requires solving a large linear program at every step. PRIMALDUAL achieves comparable solution quality while requiring substantially less computational effort, at the cost of tuning the step-size parameters η_t for the dual updates.

6 Implications for BPM and Extensions

Arrival Distributions. The current evaluation assumes that task characteristics are independent and identically distributed throughout the planning horizon. In practice, arrival patterns may exhibit non-stationary behavior due to time-of-day effects. Extending the model accordingly would provide a more accurate representation of realistic workloads and could be integrated naturally into the framework. Future work can further leverage more sophisticated algorithms from business process simulation, such as those explored in [8,11], to model complex arrival distributions that capture correlation and non-stationarity. Prediction approaches based on queue mining, as presented in [14,23], could also be combined with the online decision-making algorithms presented in this work. Introducing the queue lengths into the allocation presents an interesting extension.
Scalability and Algorithmic Tuning. While ROLLINGLP yields strong empirical performance, it requires solving a large optimization problem at each step. Larger configuration spaces or larger time horizons increase computational cost. Conversely, PRIMALDUAL scales better for larger configuration spaces, but requires careful tuning of step-size parameters. In our experiments, we used a constant step size of $\eta_t = 0.5$, selected via hyperparameter grid search across all experiments. Using adaptive step-sizes would improve its robustness in practice.
Constraint Violation Metrics. In real-world systems, temporary capacity violations may be tolerated (e.g., through overtime). Incorporating penalty terms or soft capacity constraints allows the framework to capture operational flexibility and trade off solution quality and feasibility. Common metrics in this setting are

$$\mathbb{E}\big[\|(A(\omega)x^\pi(\omega) - b)_+\|\big], \qquad \mathbb{P}\big(A(\omega)x^\pi(\omega) \le b\big) \ge 1 - \delta,$$

which quantify expected constraint violation and high-probability feasibility.

Integrated Offline–Online Optimization. A natural extension to our single-period problem is a multi-period setting in which rejected instances are reconsidered in subsequent periods. A decision maker would face a set of known instances at the start of each planning period, while new instances are expected to arrive. This leads to an integrated offline–online problem: present instances constitute an offline allocation problem, while arrivals are handled under uncertainty.

Managerial Implications. From a managerial perspective, the optimization-based algorithms achieve dynamic configuration and allocation of process instances. In real-world settings, workers pull reports manually, realizing a local optimization of workload without considering global optimization. The proposed algorithms allocate instances directly to workers, thereby balancing global workload capacity and achieving global cost optimization within an organization.

7 Conclusion

We studied the problem of online process configuration under shared resource constraints and uncertain arrivals. Motivated by hospital diagnostics, we formulated the decision problem as an online integer program in which process instances arrive sequentially and must be configured irrevocably while respecting horizon-wide resource capacities. The model captures two common sources of uncertainty: stochastic task characteristics and an unknown number of arrivals.

We connected our problem with methods from online optimization. A rolling-horizon optimization method that repeatedly solves a planning problem, and a price-based primal–dual policy that maintains adaptive resource prices. These methods were compared against heuristics that serve as operational baselines.

Using a hospital diagnostics case-study calibrated from real-world data, we evaluated the policies across different number of arrivals and capacity regimes. The results show that both rolling-horizon and primal–dual methods substantially improve solution quality. While the former achieves the best overall performance, the latter offers competitive performance with lower computational overhead. From a practical perspective, the computational effort of all approaches remains small, implying a viable support for real-time operational decisions.

Our work demonstrates how ideas from online optimization can be integrated into resource-aware BPM and that configuration decisions can benefit from online optimization techniques. The proposed framework provides a principled way to coordinate configuration decisions under uncertainty while respecting shared capacity constraints. Several directions for future research remain. In particular, extending the model to non-stationary arrival processes and developing learning-based approaches that adapt to distributional changes are promising avenues. Beyond healthcare, applying the framework to other domains with configurable processes and shared resources may further demonstrate its practical relevance.

Acknowledgments. This project has received funding from the European Research Council (ERC) under the European Union's Horizon Europe research and innovation programme (grant agreement No 101198689).

This project was funded by the Deutsche Forschungsgemeinschaft (DFG, German Research Foundation) - GRK 2201/2 - Project Number 277991500 and BI 1057/9.

Disclosure of Interests.. The authors have no competing interests to declare that are relevant to the content of this article.

References

1. van der Aalst, W.M.P.: Business process simulation survival guide. In: Handbook on Business Process Management, pp. 337–370 (2015)
2. van der Aalst, W.M.: Process Mining: Data Science in Action. Springer (2016)
3. Agrawal, S., Wang, Z., Ye, Y.: A dynamic near-optimal algorithm for online linear programming. Oper. Res. **62**(4), 876–890 (2014)
4. Albers, S., Khan, A., Ladewig, L.: Improved Online Algorithms for Knapsack and GAP in the Random Order Model. Algorithmica **83**(6), 1750–1785 (2021). Jun
5. Devanur, N.R., Hayes, T.P.: The adwords problem: online keyword matching with budgeted bidders under random permutations. In: Electronic Commerce, pp. 71–78 (2009)
6. Feldman, J., Henzinger, M., Korula, N., Mirrokni, V., Stein, C.: Online stochastic packing applied to display ad allocation. ACM Trans. Algorithms **8**(1), 1–22 (2012)
7. Hallerbach, A., Bauer, T., Reichert, M.: Contextualized process variants. Int. J. Cooperative Inf. Syst. **19**(03n04), 211–251 (2010)
8. Kirchdorfer, L., Özdemir, K., Kusenic, S., van der Aa, H., Stuckenschmidt, H.: A divide-and-conquer approach for modeling arrival times in business process simulation. In: Business Process Management. pp. 325–342 (2025)
9. Klimm, M., Knaack, M.: Generalized assignment and knapsack problems in the random-order model. In: Integer Programming and Combinatorial Optimization, pp. 341–354. Springer Nature Switzerland (2025)
10. Kunkler, M., Rinderle-Ma, S.: Online resource allocation to process tasks under uncertain resource availabilities. In: Process Mining, pp. 137–144 (2024)
11. Kunkler, M., Rinderle-Ma, S.: Probabilistic learning of processing and waiting times: a scalable approach for process event log data. Process Sci. **3**(1), 3 (2026)
12. La Rosa, M., Dumas, M., ter Hofstede, A.H., Mendling, J.: Configurable process models. Inf. Syst. **36**(2), 313–340 (2011)
13. Li, X., Ye, Y.: Online linear programming: Dual convergence, new algorithms, and regret bounds. Oper. Res. **70**(5), 2948–2966 (2022)
14. Mandelbaum, A., Momčilović, P., Trichakis, N., Kadish, S., Leib, R., Bunnell, C.A.: Data-driven appointment-scheduling under uncertainty: The case of an infusion unit in a cancer center. Manage. Sci. **66**(1), 243–270 (2020). Jan
15. Meneghello, F., Senderovich, A., Ronzani, M., Francescomarino, C.D., Ghidini, C.: Proactive data-driven scheduling of business processes. In: International Joint Conference on Artificial Intelligence, pp. 8572–8581 (2025)
16. Middelhuis, J., Bianco, R.L., Sherzer, E., Bukhsh, Z., Adan, I., Dijkman, R.: Learning policies for resource allocation in business processes. Inf. Syst. **128**, 102492 (2025)
17. Park, G., Song, M.: Prediction-based resource allocation using lstm and minimum cost and maximum flow algorithm. In: International Conference on Process Mining, pp. 121–128 (2019)

18. Russell, N., van der Aalst, W.M., ter Hofstede, A.H., Edmond, D.: Workflow resource patterns. BPM Center Report (2005)
19. Russell, N., van der Aalst, W.M., ter Hofstede, A.H., Edmond, D.: Workflow resource patterns: Identification, representation and tool support. Advanced Information Systems Engineering, pp. 216–232 (2006)
20. Santos, E., Pimentel, J., Castro, J., Finkelstein, A.: On the dynamic configuration of business process models. In: Enterprise. Business-Process and Information Systems Modeling, pp. 331–346. Springer, Berlin Heidelberg (2012)
21. Schumann, F., van der Heijden, G., Rinderle-Ma, S.: Instance configuration and scheduling based on the resource-augmented process structure tree. In: Business Process Management, vol. 564, pp. 110–127. Springer (2026)
22. Schumann, F., Rinderle-Ma, S.: Optimizing resource-driven process configuration through genetic algorithms. In: Business Process Management, pp. 3–20 (2024)
23. Senderovich, A., Weidlich, M., Gal, A., Mandelbaum, A.: Queue mining for delay prediction in multi-class service processes. Inf. Syst. **53**, 278–295 (2015)
24. Talluri, K., van Ryzin, G.: An analysis of bid-price controls for network revenue management. Manage. Sci. **44**(11), 1577–1593 (1998)
25. Talluri, K.T., Van Ryzin, G.J.: The Theory and Practice of Revenue Management. Springer (2004)
26. Taymouri, F., Rosa, M.L., Dumas, M., Maggi, F.M.: Business process variant analysis: survey and classification. Knowl.-Based Syst. **211**, 106557 (2021)

DIFF-ERO: A Conformance-Aware Loss for Deep Learning in Process Mining

Johannes De Smedt[1(✉)], Jari Peeperkorn[1], Artem Polyvyanyy[2], and Jochen De Weerdt[1]

[1] KU Leuven, Leuven, Belgium
{johannes.desmedt,jari.peeperkorn, jochen.deweerdt}@kuleuven.be
[2] The University of Melbourne, Parkville, VIC 3010, Australia
artem.polyvyanyy@unimelb.edu.au

Abstract. Deep learning has driven many recent advances in process analytics, especially for predictive and prescriptive monitoring. However, standard objectives such as cross-entropy optimize local next-step likelihoods and only implicitly capture control-flow structure. As a result, models can achieve high token-level accuracy while permitting imprecise global behaviour. We introduce DIFF-ERO, a conformance-aware loss function for deep learning models on process data. DIFF-ERO is a differentiable formulation of entropy-based stochastic conformance that incorporates control-flow information during training. Our approach constructs batch-level stochastic transition matrices with soft edge memberships, allowing structural precision and recall signals to directly inform backpropagation. The loss is model-agnostic and can be applied whenever the final representation parametrizes stochastic transitions. We instantiate DIFF-ERO in transformer encoderdecoder pipelines for next-activity prediction and use it jointly with cross-entropy to analyse its theoretical components with respect to convergence. Across benchmarks comparing other loss functions and targets, DIFF-ERO shows improved predictive performance where structure matters most while maintaining parity elsewhere. At the same time, the learned stochastic automaton converges towards the structural ground truth, indicating that the network internalizes process model structure.

Keywords: process mining · predictive process monitoring · deep learning · conformance-aware training · differentiable loss function

1 Introduction

Deep neural models, especially recurrent and transformer-based architectures, have advanced predictive and prescriptive process monitoring. Since the early use of LSTMs [25], researchers have applied deep learning to outcome prediction, next-activity and suffix prediction, and remaining-time estimation. Graph neural networks and transformers have further broadened the toolbox [12,29]. In

F. M. Maggi et al. (Eds.): BPM 2026, LNCS 16936, pp. 74–92, 2027.
https://doi.org/10.1007/978-3-032-37777-7_7

prescriptive settings, LSTMs enable deep reinforcement learning to optimize process performance [7], and even process discovery can be cast as a deep learning task [23].

Despite this progress, most approaches solely optimize standard objectives such as cross-entropy or squared error. Cross-entropy is effective for local prediction targets, yet it is precision-focused, susceptible to class imbalance [26], and unable to capture structural information such as correlations in the output space. This has led researchers in image and text classification to formulate alternative losses [11,18]. In event logs, cross-entropy loss does not explicitly represent process structures such as loops, long-distance dependencies, concurrency, and choice. As a result, models can achieve high next-step accuracy while permitting imprecise global behaviour. Deep learning models may then fail to generalize to unseen, but correct, control-flow behaviour [20,21].

Recent efforts encode control flow into features or tensors [8], or impose control-flow constraints on generated outputs via offline conformance calculations during inference [2,24]. However, these strategies do not inform the learning signal during gradient calculations and can yield high next-event accuracy without guarantees of process model conformance. We therefore seek a training objective that makes control-flow conformance part of the signal received during learning, encouraging structural fidelity in addition to local accuracy.

We address this gap with a loss function specifically designed for deep learning in process analytics. Building on entropic relevance, a stochastic conformance metric, we develop a DIFFerentiable EntRopic Process LOss `DIFF-ERO` that brings control-flow conformance information directly into training. `DIFF-ERO` complements the recall-oriented behaviour encouraged by cross-entropy with a precision-oriented signal rooted in process conformance. We demonstrate the objective in a multi-modal transformer encoderdecoder that decodes an automaton representation of traces and compares it against a ground-truth control-flow model.

Our contributions are as follows:

- **Differentiable entropic process loss.** A neural-compatible formulation of entropic conformance with soft edge memberships in a stochastic automaton and a batch-level fallback term, yielding a fully differentiable conformance-aware loss.
- **Batchable conformance supervision.** A scheme to construct ground-truth stochastic transition information from (mini-)batches or the full training log to supervise structural fit during backpropagation.
- **Theoretical analysis.** The proposed loss functions are shown to converge to desirable properties.
- **Integration with predictive tasks.** A multi-task setup that combines `DIFF-ERO` with cross-entropy for transformer-based next-activity prediction, enabling both local accuracy and global structural fidelity.
- **Empirical study.** An evaluation on benchmark transformer architectures for next-activity prediction that shows convergence in terms of learned process model structure and identifies where `DIFF-ERO` improves predictive accuracy.

The paper is structured as follows. Section 2 introduces background and motivation. Section 3 presents the preliminaries, and Sect. 4 introduces the `DIFF-ERO` loss and its differentiable formulation. Section 5 describes the experimental setup and results, Sect. 6 discusses limitations, and Sect. 7 concludes.

2 Background and Motivation

Conformance checking quantifies how well a process model explains an event log [3]. Early works, such as token-based replay, evaluate fitness and precision by replaying traces on a Petri net and counting produced and remaining tokens. Alignment-based conformance finds the lowest-cost sequences of synchronous and model/log moves so that each trace fits the model, typically solved with search or ILP. Behavioural precision and recall measure, respectively, how much model behaviour appears in the log and how much observed behaviour is allowed by the model. Recent conformance checking approaches shifted towards comparing *distributions* or the stochasticity of behaviour. Earth Mover's Distance (EMD) measures the cost of transforming one probability distribution over traces into another, but becomes expensive over large trace spaces and long prefixes [14]. The entropic relevance (ER) family models behaviour with a stochastic automaton and uses entropy to quantify bits needed to replay a trace, with a regularization that discourages overly permissive models [1]. Standard formulations of these metrics, however, are not directly differentiable as they rely on discrete calculations.

Deep learning approaches have surfaced for various predictive and prescriptive process analytics tasks. Neural sequence models such as Long Short-term Memory networks (LSTMs) and Transformers are widely used for next-activity prediction, remaining time, and suffix prediction. These black box models offer no guarantees for capturing control flow information directly [20]. Hence, recent approaches have shifted towards injecting conformance information into these models as additional input or auxiliary signals, for example, by providing a process model alongside the prefix [5,15], or via fuzzy/tensor encodings of control-flow constraints [8,27]. These strategies enrich representations at the input or provide an offline-calculated conformance metric [2], but do not by themselves define a training objective based on conformance.

Beyond process mining, losses tailored to evaluation criteria for sequential data include minimum-risk and calibration objectives in NLP [10,32], and Soft-DTW for time series [6]. These are differentiable and align gradients with task semantics, but they do not account for process-model conformance. Additionally, contrastive learning and learning-to-rank losses have emerged in the context of models working over sequences such as text [4,30] to improve robustness and generalization capabilities of predictive models. E.g., [30] breaks up next token tasks in transformer networks into a double loss component to provide contrastive/negative samples to the loss function. In the further neuro-symbolic learning domain, various recent works have proposed differentiable loss functions to inform the training about, e.g., temporal relations [17,31].

Given that conformance checking techniques such as token replay, alignments, or sequence constraints are discrete in nature, they are not compatible with the continuous, differentiable backpropagation-based learning of neural networks. The neuro-symbolic approaches which offer the potential to input conformance-based or constraint information into a deep learning model either use offline pre-calculated conformance results [2,24], or use discrete approximations using Gumbel-softmax [17,27] which typically rely on expensive sampling procedures to achieve these approximations [19]. Furthermore, these approaches typically require either prior domain knowledge [9,27] or a dedicated discovery step except for [17,24].

In the next sections, we introduce a natively differentiable objective derived from the natively stochastic, continuous domain of entropic relevance that connects these lines of work by bringing process-model conformance directly into neural training.

3 Preliminaries

We recall basic concepts on event logs, entropic relevance, and predictive model training.

3.1 Event Logs

Let $\mathcal{U}_A$ and $\mathcal{U}_C$ be the universes of activities and case identifiers, respectively. An *event log* E is a finite set of events. An *event* $e \in E$ is a triple (a, t, c), where $a \in A \subseteq \mathcal{U}_A$ is the activity that triggered the event, $t \in \mathbb{R}$ is the timestamp at which the event was triggered or recorded, and $c \in C \subseteq \mathcal{U}_C$ is the identifier of the case in which activity a was executed (e.g., the case of handling a specific loan application). We assume that no two events share the same timestamp. By e^a, e^t, and e^c, we denote the activity, timestamp, and the case identifier of event e, respectively. A *trace* of case c is denoted by σ_c, and is the sequence $\langle e_1, \ldots, e_n \rangle$, $n \in \mathbb{N}$, of all events in E with case identifier c ordered by their timestamps, i.e., $\forall\, i \in [1\mathinner{..}n] : e_i^c = c$, $\forall\, i, j \in [1\mathinner{..}n], i < j : e_i^t < e_j^t$, and $\forall e \in E : (e^c = c) \implies (\exists\, i \in [1\mathinner{..}n] : e_i = e)$. For ease of notation, we also use $\sigma \in E$ to denote a trace whose events are contained in E.

For instance, consider event log $E = \{e_1, e_2, e_3, e_4, e_5, e_6\}$ over activities $\mathcal{A} = \{\texttt{start}, \texttt{review}, \texttt{approve}, \texttt{complete}\}$ with events as follows:

$$\begin{aligned} e_1 &= (\texttt{start},\ 1.00,\ c_1), & e_2 &= (\texttt{review},\ 1.05,\ c_1),\\ e_3 &= (\texttt{approve},\ 1.10,\ c_1), & e_4 &= (\texttt{complete},\ 1.15,\ c_1),\\ e_5 &= (\texttt{start},\ 2.00,\ c_2), & e_6 &= (\texttt{complete},\ 2.05,\ c_2). \end{aligned}$$

This event log describes two traces:

$$\sigma_{c_1} = \langle e_1, e_2, e_3, e_4 \rangle, \qquad \sigma_{c_2} = \langle e_5, e_6 \rangle.$$

Throughout this paper, we assume that the first and last events of each trace are triggered by special `start` and `complete` activities, respectively.

3.2 Entropic Relevance

Entropic relevance is a stochastic conformance checking metric that quantifies how well a stochastic process model, that is, a process model that describes the probabilities of the decisions in the process, discovered from a given event log, explains the distribution of traces the system that generated the event log executes [1,22]. The entropic relevance of a stochastic process model is the average number of bits used to compress a trace from the event log using the probabilities of traces described by the model. Fewer bits indicate that the model better captures the trace probabilities of the system.

A Stochastic Deterministic Finite Automaton (SDFA) $\mathcal{A}$ is a tuple $(S, A, \delta, \delta_p, s_0)$, where S is a finite set of states, A is a set of activities, $\delta : S \times A \rightarrow S$ is the transition function annotated with probabilities by function $\delta_p : S \times A \rightarrow [0, 1]$, such that for each state $s \in S$ it holds that $\sum_{a \in A} \delta_p(s, a) \leq 1$, and s_0 is the initial state.

An SDFA $\mathcal{A}$ induces a stochastic language $L_{\mathcal{A}}$. In this language, the probability of a trace is defined as the product of the probabilities of the transitions along the unique execution in the automaton induced by the trace, multiplied by the termination probability of the state reached after replaying the trace. The termination probability of a state $s \in S$ is given by $1 - \sum_{a \in A} \delta_p(s, a)$.

Consider SDFA $\mathcal{A}$ in Fig. 1. It holds that $L_{\mathcal{A}}(\sigma_{c_1}) = \Pr(\sigma_{c_1} \mid \mathcal{A}) = 1.0 \times 0.5 \times 1.0 \times 1.0 \times 1.0 = 0.5$ and $L_{\mathcal{A}}(\sigma_{c_2}) = \Pr(\sigma_{c_2} \mid \mathcal{A}) = 1.0 \times 0.5 \times 1.0 = 0.5$.

By $\Pr(\sigma \mid E)$, we denote the relative likelihood of trace σ in event log E , that is, the empirical probability of observing trace σ among all traces in E. By $E_{\mathcal{A}}$, we denote the sublog of E obtained by retaining all traces σ such that $L_{\mathcal{A}}(\sigma) > 0$, that is, all the traces in E that are described in $\mathcal{A}$.

Then, by $\rho(E, \mathcal{A})$, we denote the probability that a trace in E is described in $\mathcal{A}$:

$$\rho(E, \mathcal{A}) = \sum_{\sigma \in E_{\mathcal{A}}} \Pr(\sigma \mid E). \tag{1}$$

For example, $\rho(E, \mathcal{A}) = \Pr(\sigma_{c_1} \mid E) + \Pr(\sigma_{c_2} \mid \mathrm{E}) = \frac{1}{2} + \frac{1}{2} = 1$, where E and $\mathcal{A}$ are the example event log and automaton in Sect. 3.1 and Fig. 1, respectively.

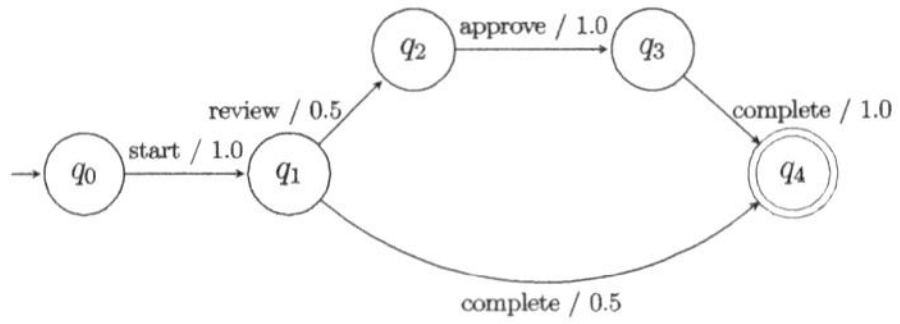

Fig. 1. An example SDFA.

By $\|E\|$, we denote the total number of traces recorded in E. Then, *entropic relevance* of SDFA $\mathcal{A}$ to E is denoted by $ER(E, \mathcal{A})$ and defined below:

$$ER(E, \mathcal{A}) = H_0(\rho(E, \mathcal{A})) + \frac{1}{\|E\|} \sum_{\sigma \in E} cost_{bits}(\sigma, E, \mathcal{A}). \tag{2}$$

$$cost_{bits}(\sigma, E, \mathcal{A}) = \begin{cases} -\log_2 L_{\mathcal{A}}(\sigma) & \text{if } L_{\mathcal{A}}(\sigma) > 0\,, \\ \text{bits}(\sigma, E, \mathcal{A}) & \text{otherwise}\,. \end{cases} \tag{3}$$

$$H_0(p) = -p\ \log_2(p) - (1-p)\ \log_2(1-p)\,, \quad p \in [0,1]\,. \tag{4}$$

Function $\text{bits}(\sigma, E, \mathcal{A})$ implements a background procedure that penalizes traces not compatible with the model. For example, the uniform background coding model is calculated as $(1 + |\sigma|)\log_2(1 + |A|)$, see [22] for details.

For the running example, it holds that $\rho(E, \mathcal{A}) = 1$, resulting in $H_0(\rho(E, \mathcal{A})) = 0$; taking $H_0(0) = H_0(1) = 0$ by definition. In addition, it holds that:

$$cost_{bits}(\sigma_{c_1}, E, \mathcal{A}) = cost_{bits}(\sigma_{c_2}, E, \mathcal{A}) = -\log_2 0.5 = 1 \text{ bit}.$$

Hence, it holds that $ER(E,\ \text{A}) = 0 + \frac{1}{2}(1+1) = 1$ bit. For an example non-fitting trace $\sigma_{nf} = \langle e_1, e_2, e_6 \rangle$, with $|A| = 4$, the penalty-term is engaged as below:

$$\text{bits}(\sigma_{nf}, E,\ \text{A}) = (1 + |\sigma_{\text{nf}}|)\log_2(1 + |A|) = 4\ \log_2\ 5 \approx 9.29 \text{ bits}.$$

If this trace is included in E to obtain event log E', then it holds that $ER(E', \mathcal{A}) \approx 0.92 + \frac{1}{3}(1 + 1 + 9.29) \approx 4.68$ bits.

3.3 Predictive Model Training

Predictive models, like next-activity or remaining-time prediction models, typically rely on prefixes of events [12]. The prefix of event e of length l is the prefix of length $l \in \mathbb{N}$ of trace σ_{e^c}. We define the suffix of event e analogously as the suffix of length l of the trace defined by the case of event e.

Event prefixes and suffixes are often truncated and/or padded to ensure a consistent length for neural network models that use fixed input and output dimensions. Therefore, by $\text{pref}(e, l)$ and $\text{suff}(e, l)$ we denote functions that return padded prefixes and suffixes, respectively, of event e of length l. Table 1 shows padded prefixes and suffixes of length three for events from the example event log in Sect. 3.1; the padding symbol is '$\perp$'.

Table 1. Prefixes and suffixes ($l = 3$) for the example event log from Sect. 3.1.

Event	Prefix	$\text{pref}(e_i, 3)$	Suffix	$\text{suff}(e_i, 3)$
e_1	$\langle\rangle$	$\langle\perp,\perp,\perp\rangle$	$\langle e_2, e_3, e_4\rangle$	$\langle e_2, e_3, e_4\rangle$
e_2	$\langle e_1\rangle$	$\langle\perp,\perp,e_1\rangle$	$\langle e_3, e_4\rangle$	$\langle e_3, e_4, \perp\rangle$
e_3	$\langle e_1, e_2\rangle$	$\langle\perp,e_1,e_2\rangle$	$\langle e_4\rangle$	$\langle e_4, \perp, \perp\rangle$
e_4	$\langle e_1, e_2, e_3\rangle$	$\langle e_1, e_2, e_3\rangle$	$\langle\rangle$	$\langle\perp,\perp,\perp\rangle$
e_5	$\langle\rangle$	$\langle\perp,\perp,\perp\rangle$	$\langle e_6\rangle$	$\langle e_6, \perp, \perp\rangle$
e_6	$\langle e_5\rangle$	$\langle\perp,\perp,e_5\rangle$	$\langle\rangle$	$\langle\perp,\perp,\perp\rangle$

4 DIFF-ERO: Differentiable Entropic Process Loss

Sect. 2 outlined various conformance checking and process model featurisation approaches used in deep learning. On the one hand, the use of discrete operations in conformance checking, such as aligning traces and models through ILP or $\in$-operators, impedes differentiable optimisation. On the other hand, the use of Declare constraints [2,17,24] or other behavioural profiles would again require a discrete one-hot encoding inside the loss function. [17] provide a differentiable DeepDFA with approximations using Gumbel-softmax. There is a trade-off between one the one hand having very expressive temporal imputation and tractability on the other hand as differentiable approximations relying on expensive sampling for discrete information which also do not estimate the joint distribution of several one-hot encoded variables as they do not natively account for correlations [19] , and multivariate solutions such as Gumbel-Sinkhorn can become expensive for larger feature sets [16].

We aim to develop a natively differentiable, neural network-compatible loss function that approximates the continuous distributions of stochastic process models, in analogy to how cross-entropy minimizes the Kullback-Leibler divergence in classification. We achieve this by first adapting the ideas of ER to the batch structure required for neural training and then making all terms differentiable so that the loss is compatible with backpropagation.

4.1 Batch Calculations

Entropic relevance is calculated from the whole event log E. However, in a neural network, the loss is typically calculated over batches (e.g., prefixes of events in E with their corresponding targets), and gradients are backpropagated to inform tasks such as next-activity or outcome prediction. We therefore adapt the ER calculations to obtain, for each batch b, a probability matrix PM_b that plays the role of a ground-truth SDFA for that batch and share it across all samples or prefixes in b (when calculating the loss for backpropagation). This ground-truth SDFA can then be compared with an estimated SDFA produced by the model.

To construct PM_b we use directly-follows relations $DF_\sigma : A \times A \rightarrow \mathbb{N}$. For a sequence σ, i.e., $DF_\sigma(a_1, a_2) = |\{(e_1, e_2) \mid e_1^t < e_2^t, \nexists e_1^t < e_3^t < e_2^t \wedge e_1^a = a_1 \wedge e_2^a = a_2, \forall e_1, e_2, e_3 \in \sigma\}|$.[1] We can obtain DF_σ for a set of sequences, e.g., all suffixes from all events in a batch, or even all suffixes from all events in a training log, by considering a subset of events $b \subseteq E_{train} \subset E$. E.g., for a batch b of events we can obtain the prefixes $\sigma_{pre} = \{\text{pref}(e_i, l) \mid e_i \in b\}$ and their corresponding suffixes $\sigma_{suf} = \{\text{suff}(e_i, l) \mid e_i \in b\}$ for a chosen l so we can calculate $DF_{\sigma_{suf}}(a_1, a_2) = \Sigma_{\sigma \in \sigma_{suf}} DF_\sigma(a_1, a_2)$. The use of directly-follows relations allows us to obtain a ground-truth probability matrix $PM_b \in [0,1]^{|A| \times |A|}$ with its elements $p_{a_1,a_2} = \frac{DF_b(a_1,a_2)}{\Sigma_{a_3} DF_b(a_1,a_3)+\epsilon}$ for batch b. Note that we include ϵ to avoid numerical instability in tensor calculations.

---

[1] Note that there are other representations from which an SDFA could be derived, e.g., stochastic Petri nets [13].

4.2 Differentiable Calculations

As explained in Sect. 2, conformance-based metrics such as ER rely on some form of replay. In the case of ER this is expressed in the form of $\sigma \in L_{\mathcal{A}}$ per Eq. 3. This discrete check impedes the formula from being used in a differentiable setup such as the backpropagation in neural networks.

The first change towards a differentiable ER is to introduce a soft membership function over the edges of the original SDFA's transition probabilities δ_p. In a neural network, we obtain such a tensor from the output of the final hidden layer $h_O \in \mathbb{R}^{|B|\times|A|\times|A|}$ produced by, e.g., an LSTM or Transformer architecture ingesting prefix sequences, to derive O from a softmax layer returning probabilities for the full transition matrix $O = softmax(h_O)$ with $softmax(x_i) = \frac{e^{x_i}}{\Sigma_j e^{x_j}}$. We refer to $O_b \in \mathbb{R}^{|A|\times|A|}$ as the batch-level representation of δ_p with a (near) 0 probability indicating the arc not being present in δ. O_b can be used to calculate a batch-specific tensor L_{PM_b} representing an output SDFA of the model to calculate the final difference with the batch-level ground-truth PM_b. O_b is globally normalized due to the softmax operator. An SDFA requires row-level normalization to ensure that for each state the outgoing probabilities sum to one. Therefore, we apply an additional row-wise normalization L_{PM_b} to obtain a valid conditional distribution. L_{PM_b} is calculated by

$$L_{PM_b}(a_1, a_2) = \frac{O_b[a_1, a_2]}{\Sigma_{a_3 \in A} O_b[a_1, a_3] + \epsilon} \in [0,1]^{|A|\times|A|} \tag{5}$$

to transform the edge memberships into probabilities which sum to 1.

A second challenge is the *bits*-term from Eq. 3 being calculated based on the size of t and A. We propose a smoothed negative log-likelihood derived from PM_b, which signifies a form of fallback regularisation based on the ground truth. If an edge is incorrectly predicted to (not) be part of the model, this term penalizes the neural network for such errors. Other possible alternatives for this term include using the Gumbel-softmax trick to approximate the actual discrete values of $|t|$ and $|A|$, which might prove computationally expensive [19], or differentiable placeholder terms learned with the model, which will mostly facilitate gradient calculations and have a similar purpose to our solution. We obtain the fallback using standard cross-entropy to approximate $\rho(E, \mathcal{A})$ on batch-level as $\rho(E_b, PM_b) = \Sigma_{t \in E_b} Pr(t|E_b)$ by calculating $-\log_2(PM_b)$.

Finally, we can rewrite Eq. 3 as the `DIFF-ERO` equation to measure the loss between ground-truth PM_b and trained representation O_b as

$$\mathcal{L}_{DIFF-ERO}(E, PM_b, O_b) = O_b \cdot (-\log_2 L_{PM_b}) + (1 - O_b)\cdot(-\log_2 PM_b) \tag{6}$$

or in element-wise form

$$\mathcal{L}_{DIFF-ERO}(E, PM_b, O_b) = \Sigma_{a_i \in A} \Sigma_{a_j \in A} [O_b[a_i, a_j] \cdot (-\log_2 L_{PM_b}(a_i, a_j)) + (1 - O_b[a_i, a_j]) \cdot (-\log_2 PM_b(a_i, a_j))] \tag{7}$$

When edges are predicted to be present in O_b and approach 1, they are penalized by $-log_2 L_{PM_b}$ representing the predicted distribution and when they are not, they are penalized by $-log_2 PM_b$ representing the ground truth. The loss computes the weighted cross-entropy between the predicted distribution L_{PM_b} and the ground-truth edge distribution PM_b with O_b as a soft gate. The left-hand term captures the *entropy* part of the equation which aims to have a compact approximation of the ground-truth SDFA PM_b using as few bits as possible. The right-hand term captures the *relevance* part of the equation, punishing for any mismatches in SDFA structure. They inform the learning process about both recall and precision, contrary to, e.g., standard cross-entropy loss which is often used to optimise next-activity or suffix prediction with a focus on precision only [26].

4.3 `DIFF-ERO`-Driven Topologies

The loss function can be used in various ways which require different training procedures. We opt for a fully transformer-based architecture as they tend to return state-of-the-art performance in many tasks [12,29], however, our loss function is compatible with any neural network-based approach returning a hidden state, including LSTMs. We start from a single sequence $\mathbf{x} = \langle x_1, \ldots, x_l \rangle$ where each item is a one-hot encoded vector with length $|A|$. They are embedded with dimension d and fed to a positional encoding layer which is processed by a multi-head self-attention block which returns an output $\mathbf{H}_{enc} \in \mathbb{R}^{|B| \times max_{l_i} \times d}$ where max_{l_i} is the longest input sequence. We obtain the prefixes for all events in the training log by using pref() with $l = max_{l_i} - 1$ set to the longest observed trace minus 1 and use padding according to Table 1. We do the same for every event using suff() with the same l and padding to obtain the $DF_{\sigma_{suff}}$ and PM_b as described in Sect. 4.1 at the output. In other words, we obtain a teacher forcing setup that is conformance-informed by the output's real control flow model to inform the modelling of the prefixes.

We can use the hidden state of the encoder to obtain O by using a linear mapping layer over the mean-pooled version $\overline{h}$ of $\overline{\mathbf{H}}_{enc} \in \mathbb{R}^{|B| \times d}$

$$z = \overline{h}\mathbf{W}^T + b$$

which is then passed on to a softmax layer to obtain

$$softmax(z) \in [0,1]^{|B| \times P}$$

where $P = |A|^2$. Finally, we can get the soft tensor by re-structuring $softmax(z)$:

$$O \in [0,1]^{|B| \times |A| \times |A|}$$

Auto-encoding. Firstly, `DIFF-ERO` can be used as an autoencoder to perform stochastic process discovery. By learning batched prefixes with their corresponding PM_b values as outputs for the `DIFF-ERO` function, the bottleneck captures a representation which can be used, for example, for clustering or model comparison. However, given that such an unsupervised outcome is hard to verify for its usefulness, we currently did not include an experimental evaluation for this setup.

Next Activity Prediction. Secondly, DIFF-ERO can be used in a multitask setup to inform next-activity and suffix prediction. In this setup, the additional loss will inform and regularize the training process.

Then, the next-activity prediction can be informed by a joint loss function

$$\mathcal{L}_{na}^{DIFF-ERO} = \mathcal{L}_{na} + \lambda \cdot \mathcal{L}_{DIFF-ERO}$$

with $\mathcal{L}_{na} = \frac{1}{|B|} \Sigma_{b=1}^{|B|} - \log(p_b(y_b))$ the batch cross-entropy loss for a vector of $|A|$ probabilities which can be used for next-activity predictions and λ controlling the contribution of the DIFF-ERO term.

The setup is illustrated in Fig. 2.

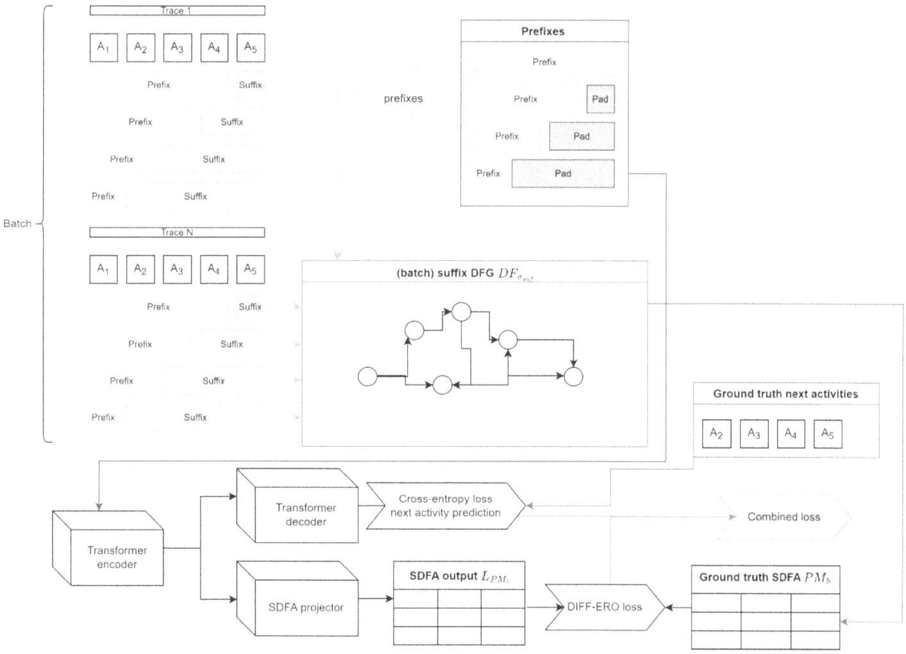

Fig. 2. Architectural overview of DIFF-ERO.

4.4 Theoretical Properties

Below, we sketch the theoretical properties of DIFF-ERO with regard to its convergence under certain distributions and the original event log-based ER.

Convergence. We investigate the gradients to understand how O_b informs the backpropagation in neural networks using $\mathcal{L}_{DIFF-ERO}(E, PM_b, O_b)$, or $\mathcal{L}$ for notational ease.

Firstly, we note that

$$\frac{\partial L_{PM_b}(a_i, a_k)}{\partial O_b[a_i, a_j]} = \frac{\mathbf{1}[a_k = a_j] - L_{PM_b}(a_i, a_k)}{\Sigma_{a_3} O_b[a_i, a_3] + \epsilon}$$

for all a_k sharing the row with a_i which is analogous to the Jacobian of the softmax function. Assuming that PM_b is independent of O_b and hence $\frac{\partial PM_b}{\partial O_b} = 0$ and knowing that $\frac{\partial}{\partial L_{PM_b}}[-log_2 L_{PM_b}] = -\frac{1}{\ln\ 2} \cdot \frac{1}{L_{PM_b}}$ results in the derivative

$$\begin{aligned} &\frac{\partial}{\partial O_b[a_i, a_j]} O_b[a_i, a_j] \cdot (-\log_2 L_{PM_b}(a_i, a_j)) = \\ &- \Sigma_{a_k \in A} O_b[a_i, a_k] \cdot \frac{1}{\ln\ 2 \cdot L_{PM_b}(a_i, a_k)} \cdot \frac{\mathbf{1}[a_k = a_j] - L_{PM_b}(a_i, a_k)}{\Sigma_{a_3} O_b[a_i, a_3] + \epsilon} \end{aligned} \tag{8}$$

Calculating the full derivative over

$$\begin{aligned} &\frac{\partial}{\partial O_b[a_i, a_j]} \Big[O_b[a_i, a_j] \cdot (-\log_2 L_{PM_b}(a_i, a_j)) + (1 - O_b[a_i, a_j]) \cdot (-\log_2 PM_b(a_i, a_j)) \Big] \\ &= -\log_2 L_{PM_b}(a_i, a_j) + \log_2 PM_b(a_i, a_j) \quad = \log_2 \frac{PM_b(a_i, a_j)}{L_{PM_b}(a_i, a_j)} \end{aligned} \tag{9}$$

and using Eqs. 5 and 8 finally results in

$$\begin{aligned} &\frac{\partial \mathcal{L}_{DIFF-ERO}}{\partial O_b[a_i, a_j]} = \\ &\log_2 \frac{PM_b(a_i, a_j)}{L_{PM_b}(a_i, a_j)} - \frac{1}{\ln\ 2 \cdot (\Sigma_{a_k} O_b[a_i, a_k] + \epsilon)} \cdot (\frac{O_b[a_i, a_j]}{L_{PM_b}(a_i, a_j)} - \Sigma_{a_k} O_b[a_i, a_k]) \end{aligned} \tag{10}$$

This captures three elements indicating the convergence and properties of the loss function. Firstly, $log_2 \frac{PM_b}{L_{PM_b}}$ will align the trained L_{PM_b} with the ground truth transition system in PM_b by returning a positive gradient requiring further updates in the weights when $L_{PMb} < PM_b$ and by approaching 0 when $L_{PM_b} \simeq PM_b$. Secondly, $O_b[a_i, a_j]$ is present as a soft gate; it reaches 0 when transitions are to be excluded thereby reducing the impact of the second term and stopping any gradients from flowing towards updating information concerning this transition. Finally, ϵ only appears in the denominator and avoids numerical instability by avoiding division by 0 and exploding magnitude.

Consistency with ER. Secondly, as `DIFF-ERO` is a soft edge membership interpretation of ER, the consistency with ER becomes clear by comparing the corresponding terms of ER and `DIFF-ERO`. $cost_{bits}(\sigma, E, \mathcal{A}) = -log_2\ L_{\mathcal{A}}(\sigma)$ (Eq. 3) in ER corresponds with $-\Sigma_{(a_i, a_j) \in \sigma}\ log_2\ L_{PM}(a_i, a_j) = -log_2 L_A(\sigma)$ in `DIFF-ERO`

(Eq. 7). Hence, if the output of a parametrized model under θ, O_θ, converges to PM_E being the SDFA of the full log, then, given that ϵ converges to 0, DIFF-ERO converges to $L_{DIFF-ERO}(E, PM_E, O_\theta) = \frac{1}{|E|}\Sigma_{\sigma \in E}(-\log_2 L_{\mathcal{A}}(\sigma)) + C(E) = ER(E, \mathcal{A}) + C(E)$. Given a batch size $|b|$ equal to the event log size $|E|$ and $\epsilon = 0$, DIFF-ERO converges towards $ER(E, A)$ with an additive constant independent of the model's parameters.

5 Evaluation

We have implemented the loss function in PyTorch[2]. We next describe the experimental setup, present the next-activity prediction results, and discuss limitations.

5.1 Experimental Setup

We use six widely-used BPI Challenge event logs from of 2012, 2017, 2019, and three logs from the 2020 edition[3]. Detailed statistics are provided in Table 2.

Table 2. Overview of the event log characteristics before and after preprocessing.

	BPI12		BPI17		BPI19		BPI20PTC		BPI20RFP		BPI20TPD	
Preproc.	*No*	*Yes*	*No*	*Yes*	*No*	*Yes*	*No*	*Yes*	*No*	*Yes*	*No*	*Yes*
$\|A\|$	24	26	26	28	42	38	29	31	19	20	51	53
$\overline{\|\sigma\|}$	20.04	12.86	38.16	24.99	6.34	7.10	8.69	10.48	5.34	7.13	12.25	13.21
median($\|\sigma\|$)	11	8	35	23	5	7	8	10	5	7	11	12
$\|C\|$	13,087	12,595	31,509	30,189	251,734	239,734	2,099	2,030	6,886	6,579	7,065	6,714
max($\|\sigma\|$)	175	35	180	41	990	10	21	15	20	10	90	23

To streamline the predictive task, we retain traces within the 95% trace length distribution and apply out-of-time correction to prevent temporal leakage [28]. This results in an 80/20% train-test split without temporal overlap. The prefix input-length l is set to the maximum observed training prefix length. Transformer dimensionality (encoder, decoder, and PM_b projection layer) is varied across 16, 32, and 64, while the number of heads and layers is fixed at 4 and 2 following [12]. For DIFF-ERO, we evaluate local batch-level PM_b tensors instead of training-wide PM tensors to study the effect of regularisation and varying SDFA granularity. Since batch size influences the discovered variants, we vary it between 16, 32, and 128, similarly to imputing different discovered models [17].

[2] github.com/JohannesDeSmedt/differentiable-er and doi.org/10.5281/zenodo.20841070, which also contains additional results including epoch time, additional statistical analysis, and full hyperparameter results.

[3] BPI12,BPI17,BPI19,BPI20.

Our focus, however, is on the effect of the loss function rather than SDFA discovery quality.

Finally, we vary λ controlling the contribution of the entropic process loss. Each configuration is evaluated over 10 random seeds for statistical analysis.

For next-activity prediction, we report the weighted precision and F1-score as the harmonic mean of recall and precision of predicting the next activity in the trace. The weighing corrects the macro-average for class imbalance, which is appropriate given the skewed activity frequency distribution. Finally, we also visualise the loss over training to assess whether the model is actually informed by the loss during training, and acts on it during backpropagation, i.e., whether the values for $\mathcal{L}_{DIFF-ERO}$ decrease.

We compare `DIFF-ERO` to other loss and target baselines.

1. Base model (`base`): we start from a baseline encoder-decoder transformer-based model using cross-entropy for training because of its strong performance for next-activity prediction [12]. Note that all models below can also be used directly with other neural network architectures like LSTMs.
2. Base model with SDFA (`SDFA`): to verify whether adding the extra information of the ground truth SDFA can already inform the training of the network without dedicated structural loss, we add the SDFA to the output of the model and backpropagate using standard cross-entropy loss. This can happen both in a batch form and a global form as explained in Sect. 4.1.
3. Base model with rank loss (`rank_loss`): in order to validate whether altering the loss in the form of another information retrieval baseline can improve the performance, we employ the rank loss [4] which performs negative sampling to understand whether a learning-to-rank baseline already can improve a cross-entropy optimisation while still not directly incorporating a process structural point of view. The rank loss is $L_b = \frac{1}{|\mathcal{N}_b|} \Sigma_{a_j \in \mathcal{N}_b} \log(1+e^{logit(a_j)-logit(y_b)}$ where $\mathcal{N}_b$ is the number of negative samples and $logit()$ is the logit scores of these samples.
4. Base model with `DIFF-ERO` (`DIFF-ERO`): finally, we use the base model with the full `DIFF-ERO` loss to leverage the SDFA information as an additional target in both local/batch and global form.

5.2 Results

The results for next-activity prediction (with optimal hyperparameters), averaged over random seeds, are shown in Table 3. `DIFF-ERO`-based models report the best or among the best results in terms of F1-score and are comparable in terms of precision, suggesting stronger recall. This further underlines that it steers the model in additional ways compared to standard cross-entropy-based models. Still, the strong performance of the `SDFA`-based models, with the exception of BPI12, indicates that adding the additional data at the output of the model is a major driver of this effect. Rank loss does not seem to improve training and rather reports worse results compared to the baseline. Based on training time

Table 3. Results of the next activity prediction tests with best performances in boldface.

	F1				Precision				Time (seconds per epoch)			
Dataset	Base	Rank loss	SDFA	DIFF-ERO	Base	Rank loss	SDFA	DIFF-ERO	Base	Rank loss	SDFA	DIFF-ERO
BPI12	0.786	0.716	0.715	**0.826**	0.878	0.83	0.863	**0.882**	19.447	49.41	25.528	19.709
BPI17	0.823	0.795	0.83	**0.865**	0.895	0.883	0.899	**0.901**	136.267	250.709	144.111	144.7
BPI19	0.722	0.735	0.769	**0.79**	0.918	0.914	0.918	0.916	85.574	293.725	113.466	88.273
BPI20PTC	0.855	0.818	0.887	**0.89**	0.939	0.936	**0.955**	0.944	2.323	7.17	2.922	2.408
BPI20RFP	0.948	0.948	0.949	0.949	0.99	0.989	0.99	0.99	4.343	15.415	5.474	4.558
BPI20TPD	0.808	0.737	0.854	**0.862**	0.906	0.884	**0.926**	**0.924**	11.46	32.429	17.736	12.13

per epoch, it is clear that DIFF-ERO adds only a modest overhead compared to rank loss or incorporating SDFAs using cross-entropy.

A rank-based Friedman test is used to compare the significance in the results of each model over the different datasets where a consistently lower rank indicates a better performing model. The Nemenyi test is a post-hoc test which reports which pair-wise comparisons are significantly different. The average rank of each technique (lower is better) and the corresponding Nemenyi outcome is visualised in Fig. 3a .

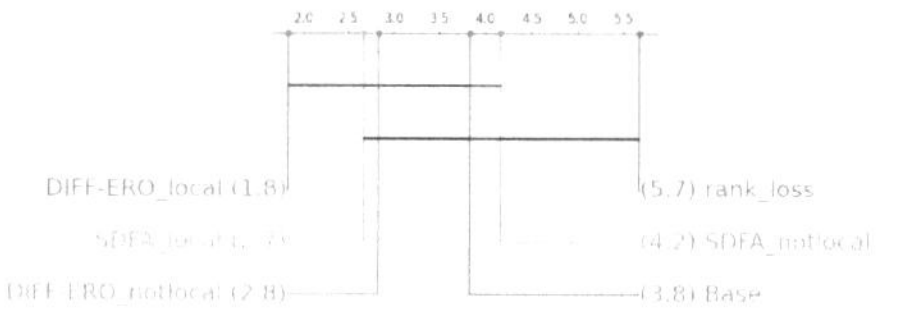

(a) Nemenyi post-hoc test of NAP results.

	(2)	(3)	(4)	(5)	(6)
(1) Base	0.432	0.940	0.889	1.000	0.534
(2) DIFF-ERO$_{\text{local}}$		0.940	0.972	0.257	0.005
(3) DIFF-ERO$_{\text{notlocal}}$			1.000	0.820	0.092
(4) SDFA$_{\text{local}}$				0.734	0.061
(5) SDFA$_{\text{notlocal}}$					0.734

(b) Pairwise p-values between methods.

Fig. 3. Nemenyi post-hoc results and corresponding pairwise p-values of the F1-score.

DIFF-ERO does not outperform the SDFA and base models at a significance level of 95% due to the relatively small number of datasets and relatively strong performance of all models for BPI20PTC and BPI20RFP. While DIFF-ERO does not statistically outperform the baselines on pure predictive accuracy due to the use of a limit number of datasets, it achieves the highest overall rank while guaranteeing structural convergence (as seen in Fig. 4) while incurring only modest additional computational cost. This demonstrates that the network internalizes global process constraints without sacrificing local predictive power.

A closer examination shows that non-local models, typically with higher λ values and batch sizes, achieve the best performance. Unlike standard cross-entropy, DIFF-ERO uses batch size as a structural regularizer: larger batches provide a more complete representation of the ground-truth SDFA, increasing the emphasis on structural recall through the left-hand term of Eq. 7. Table 3 further shows that DIFF-ERO is less sensitive to the global-local setup than SDFA, where global SDFAs generally perform worse. In this sense, the DIFF-ERO construction appears more robust. Finally, Fig. 4 presents the min-max scaled cross-entropy loss and DIFF-ERO over training epochs for the best-performing configurations on the BPI12/17/19 logs. All models show a clear and rapidly converging decrease in DIFF-ERO alongside the standard cross-entropy loss, particularly for the larger BPI17 and BPI19 logs. This indicates that the model successfully propagates the structural signal from the SDFA through its hidden layers during training.

Table 4. Detailed F1-score results with non-local (top 4 rows for standard batch size 32) and local models with different batch sizes. Higher values are in red, lower values in blue. Boldface indicates best performance.

$\|b\|$	lambda	BPI12	BPI17	BPI19	BPI20PTC	BPI20RFP	BPI20TPD
32	0	0.769	0.800	0.690	0.812	0.948	0.779
	0.1	0.778	0.801	0.685	0.806	0.948	0.784
	0.2	0.783	0.804	0.700	0.795	0.947	0.812
	0.5	0.788	0.819	0.692	0.709	0.943	0.795
16	0.1	0.773	0.790	0.688	0.814	0.947	0.793
	0.2	0.777	0.799	**0.702**	0.788	0.946	0.801
	0.5	0.785	0.811	0.679	0.699	0.943	0.802
32	0.1	0.769	0.811	0.686	0.820	0.947	0.780
	0.2	0.774	**0.821**	0.680	0.794	0.947	0.813
	0.5	0.773	0.813	0.686	0.708	0.944	0.801
64	0.1	0.752	0.808	0.695	**0.822**	**0.948**	0.789
	0.2	0.770	0.813	0.679	0.816	0.946	0.792
	0.5	**0.788**	0.818	0.688	0.733	0.944	0.811
128	0.1	0.783	0.808	0.698	0.801	0.947	0.797
	0.2	0.766	0.801	0.662	0.790	0.947	0.804
	0.5	0.775	0.812	0.671	0.715	0.943	**0.814**

6 Discussion and Limitations

DIFF-ERO introduces a differentiable mechanism to incorporate conformance information directly into neural training, enabling models to learn structures that more closely reflect process behaviour. Currently, the approach is operationalized as a regularization for next activity prediction with similar aims like the rank-based baselines. Contrary to prior approaches incorporating conformance information [9,17], it offers a trade-off between inputting a coarser, continuous stochastic automaton to incorporate conformance information versus a more fine-grained constraint-based model or procedural model [9,27] which require approximations to obtain differentiability over discrete domains. Imputation currently happens through batches of the training set, however, similar to

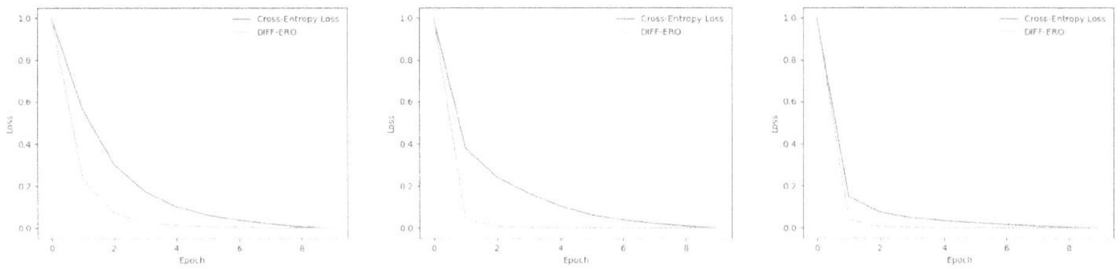

(a) BPI12 (b) BPI17 (c) BPI19

Fig. 4. Evolution of training loss of cross-entropy and DIFF-ERO for next activity prediction.

[17,24,27], an additional discovery step could further lead DIFF-ERO to be used to input specific information in the form of a ground truth PM_b external to the (training) data. Table 4 illustrates the impact of changing the model used in the loss function.

Our experiments show significant improvements in next-activity prediction. These findings suggest that the proposed loss function may also be useful beyond prediction tasks, particularly in settings where adherence to process structure is essential. However, we have not yet evaluated additional predictive tasks that could benefit from the structural information. Most notably, it would be interesting to use DIFF-ERO in the context of suffix prediction, where the state information infused into the training process could improve the generation of next activities that adhere to a process model.

At the same time, several limitations accompany these contributions. The composite loss introduces computational overhead that remains modest for small and medium alphabets. Still, it may increase memory usage and training time for logs with large activity sets or highly imbalanced distributions. The weighting parameter λ offers flexibility but adds an additional hyperparameter, possibly requiring tuning. Moreover, the current formulation captures only pairwise directly-follows relations, limiting its ability to express higher-order or concurrent behaviour such as loops, parallelism, and so on, without further extension. The results in Table 4 and Fig. 4 also suggest that a further study into the relation between convergence, batch size, and even learning rate could further clarify the impact of DIFF-ERO. Secondly, the grouping within batches was not explored in detail. Clustering-based approaches to group prefixes could potentially further leverage the power of DIFF-ERO. Finally, the experiments focus exclusively on predictive models that rely on control-flow information, excluding payload or contextual data. Although the inclusion of such data can improve predictive performance [29], it was intentionally omitted here, given the nature of the experiments where the focus lies on demonstrating its potential use for learning control-flow behaviour.

7 Conclusion

This paper introduced a process model-informed loss function for deep learning on event logs that exposes structural process model features to backpropagation. The entropy-based process loss `DIFF-ERO` proposed is fully differentiable, can be computed at batch level, and is compatible with architectures that parametrise stochastic transitions. We instantiated `DIFF-ERO` in a transformer encoder-decoder architecture for next-activity prediction. Our empirical results show that `DIFF-ERO` significantly improves next-activity prediction accuracy when compared to a cross-entropy baseline. In both tasks, the learned representations converge towards the structural ground truth, leading to models whose internal stochastic automata better reflect the underlying process. This suggests that `DIFF-ERO` supports structurally faithful representation learning, even when predictive gains are limited. For future work, we envision applying `DIFF-ERO` in other deep learning contexts, including auto-encoding for unsupervised tasks and prescriptive settings in which predictive models are used to generate generalizable recommendations to optimise process performance.

Acknowledgments. This work was supported in part by the Research Foundation Flanders (FWO) under Project 1294325N as well as grant number G039923N, and Internal Funds KU Leuven under grant number C14/23/031.

References

1. Alkhammash, H., Polyvyanyy, A., Moffat, A., García-Bañuelos, L.: Entropic relevance: a mechanism for measuring stochastic process models discovered from event data. Inf. Syst. **107**, 101922 (2022)
2. Buliga, A., Di Francescomarino, C., Ghidini, C., Donadello, I., Maggi, F.M.: Guiding the generation of counterfactual explanations through temporal background knowledge for predictive process monitoring. Data Min. Knowl. Disc. **39**(5), 1–49 (2025)
3. Carmona, J., Van Dongen, B., Solti, A., Weidlich, M.: Conformance checking. Switzerland: Springer. **56**, 12 (2018)
4. Chen, W., Liu, T.Y., Lan, Y., Ma, Z.M., Li, H.: Ranking measures and loss functions in learning to rank. Adv. Neural Inf. Process. Syst. **22** (2009)
5. Chiorrini, A., Diamantini, C., Mircoli, A., Potena, D.: Exploiting instance graphs and graph neural networks for next activity prediction. In: International Conference on Process Mining, pp. 115–126. Springer (2021)
6. Cuturi, M., Blondel, M.: Soft-dtw: a differentiable loss function for time-series. In: International Conference on Machine Learning, pp. 894–903. PMLR (2017)
7. De Moor, J., Weytjens, H., De Smedt, J.: Procause: generating counterfactual outcomes to evaluate prescriptive process monitoring methods. In: 2025 7th International Conference on Process Mining (ICPM), pp. 1–8. IEEE (2025)
8. Donadello, I., Felli, P., Innes, C., Maggi, F.M., Montali, M.: Conformance checking of fuzzy logs against declarative temporal specifications. In: International Conference on Business Process Management, pp. 39–56. Springer (2024)

9. Donadello, I., Ko, J., Maggi, F.M., Mendling, J., Riva, F., Weidlich, M.: Knowledge-driven modulation of neural networks with attention mechanism for next activity prediction. arXiv preprint arXiv:2312.08847 (2023)
10. Edunov, S., Ott, M., Auli, M., Grangier, D., Ranzato, M.: Classical structured prediction losses for sequence to sequence learning. arXiv preprint arXiv:1711.04956 (2017)
11. Graber, C., Meshi, O., Schwing, A.: Deep structured prediction with nonlinear output transformations. Adv. Neural Inf. Process. Syst. **31** (2018)
12. Käppel, M., Ackermann, L., Jablonski, S., Härtl, S.: Attention please: what transformer models really learn for process prediction. In: International Conference on Business Process Management, pp. 203–220. Springer (2024)
13. Leemans, S.J., Li, T., Montali, M., Polyvyanyy, A.: Stochastic process discovery: can it be done optimally? In: International Conference on Advanced Information Systems Engineering, pp. 36–52. Springer (2024)
14. Leemans, S.J., Syring, A.F., van der Aalst, W.M.: Earth movers' stochastic conformance checking. In: International Conference on Business Process Management, pp. 127–143. Springer (2019)
15. Lischka, A., Rauch, S., Stritzel, O.: Directly follows graphs go predictive process monitoring with graph neural networks. arXiv preprint arXiv:2503.03197 (2025)
16. Mena, G., Belanger, D., Linderman, S., Snoek, J.: Learning latent permutations with gumbel-sinkhorn networks. In: ICLR (2018)
17. Mezini, A., Umili, E., Donadello, I., Maggi, F.M., Mancanelli, M., Patrizi, F.: Neuro-symbolic predictive process monitoring. Available at SSRN 5420992
18. Mnih, V., Larochelle, H., Hinton, G.E.: Conditional restricted boltzmann machines for structured output prediction. arXiv preprint arXiv:1202.3748 (2012)
19. Paulus, M., Choi, D., Tarlow, D., Krause, A., Maddison, C.J.: Gradient estimation with stochastic softmax tricks. Adv. Neural. Inf. Process. Syst. **33**, 5691–5704 (2020)
20. Peeperkorn, J., vanden Broucke, S., De Weerdt, J.: Can recurrent neural networks learn process model structure? J. Intell. Inf. Syst. **61**(1), 27–51 (2023)
21. Peeperkorn, J., vanden Broucke, S., De Weerdt, J.: Validation set sampling strategies for predictive process monitoring. Inf. Syst. **121**, 102330 (2024)
22. Polyvyanyy, A., Moffat, A., García-Bañuelos, L.: An entropic relevance measure for stochastic conformance checking in process mining. In: 2020 2nd International Conference on Process Mining (ICPM), pp. 97–104. IEEE (2020)
23. Sommers, D., Menkovski, V., Fahland, D.: Process discovery using graph neural networks. In: 2021 3rd International Conference on Process Mining (ICPM), pp. 40–47. IEEE (2021)
24. Stevens, A., Ouyang, C., De Smedt, J., Moreira, C.: Generating feasible and plausible counterfactual explanations for outcome prediction of business processes. IEEE Trans. Serv. Comput. (2025)
25. Tax, N., Verenich, I., La Rosa, M., Dumas, M.: Predictive business process monitoring with LSTM neural networks. In: CAiSE, pp. 477–492. Springer (2017)
26. Terven, J., Cordova-Esparza, D.M., Romero-González, J.A., Ramírez-Pedraza, A., Chávez-Urbiola, E.: A comprehensive survey of loss functions and metrics in deep learning. Artif. Intell. Rev. **58**(7), 195 (2025)
27. Umili, E., Capobianco, R., De Giacomo, G.: Grounding LTLF specifications in image sequences. In: Proceedings of the International Conference on Principles of Knowledge Representation and Reasoning, vol. 19, pp. 668–678 (2023)

28. Weytjens, H., De Weerdt, J.: Creating unbiased public benchmark datasets with data leakage prevention for predictive process monitoring. In: International Conference on Business Process Management, pp. 18–29. Springer (2021)
29. Wuyts, B., Vanden Broucke, S., De Weerdt, J.: Sutran: an encoder-decoder transformer for full-context-aware suffix prediction of business processes. In: 2024 6th International Conference on Process Mining (ICPM), pp. 17–24. IEEE (2024)
30. Xie, X., et al.: Contrastive learning for sequential recommendation. In: 2022 IEEE 38th International Conference on Data Engineering (ICDE), pp. 1259–1273. IEEE (2022)
31. Xu, Z., Rawat, Y., Wong, Y., Kankanhalli, M.S., Shah, M.: Don't pour cereal into coffee: differentiable temporal logic for temporal action segmentation. Adv. Neural. Inf. Process. Syst. **35**, 14890–14903 (2022)
32. Zhao, Y., Khalman, M., Joshi, R., Narayan, S., Saleh, M., Liu, P.J.: Calibrating sequence likelihood improves conditional language generation. arXiv preprint arXiv:2210.00045 (2022)

Validity Criteria for Explaining Variation in Process Complexity

Lennart Ebert[1(✉)], Kate Revoredo[1], Jan Mendling[1,2,3], and Benoît Depaire[4]

[1] Humboldt University of Berlin, Berlin, Germany
{lennart.ebert,kate.revoredo,jan.mendling}@hu-berlin.de
[2] Vienna University of Economics and Business, Vienna, Austria
[3] Weizenbaum Institute, Berlin, Germany
[4] Digital Future Lab, Hasselt University, Hasselt, Belgium
benoit.depaire@uhasselt.be

Abstract. Business process complexity is frequently used to assess behavioral change. It is measured repeatedly over consecutive trace windows, resulting in a time series of complexity scores. When analyzing this data, it is a problem that complexity measures can fluctuate even when the underlying process remains stable. So far, business process management research lacks a formal observation model to distinguish process change from measurement-induced variation. In this paper, we build upon signal processing theory to develop an observation model for process complexity. The model decomposes observed complexity scores into a process complexity signal and measurement noise component. The measurement noise component consists of systematic measurement bias and stochastic measurement error. We derive three validity criteria to analyze signal-and-noise characteristics of complexity measures: (1) reliability, reflecting sensitivity to stochastic measurement error when process and observation setting do not change, (2) robustness, reflecting sensitivity to observation setting change, and (3) responsiveness, reflecting sensitivity to process change. We operationalize these criteria and apply them in a large-scale simulation study. The results reveal systematic differences in the signal-and-noise behavior of complexity measures. Our work contributes a measurement-theoretic foundation for interpreting changes in process complexity and for validating complexity measures.

Keywords: Business Process Complexity · Process Mining · Signal Processing Theory · Measurement Validity

1 Introduction

Process complexity captures the variety of process execution sequences based on digital trace data [5,13]. Empirical studies have shown that complexity affects process efficiency [2] and the performance of process discovery algorithms [22]. Furthermore, complexity time series have been used to assess process change

F. M. Maggi et al. (Eds.): BPM 2026, LNCS 16936, pp. 93–110, 2027.
https://doi.org/10.1007/978-3-032-37777-7_8

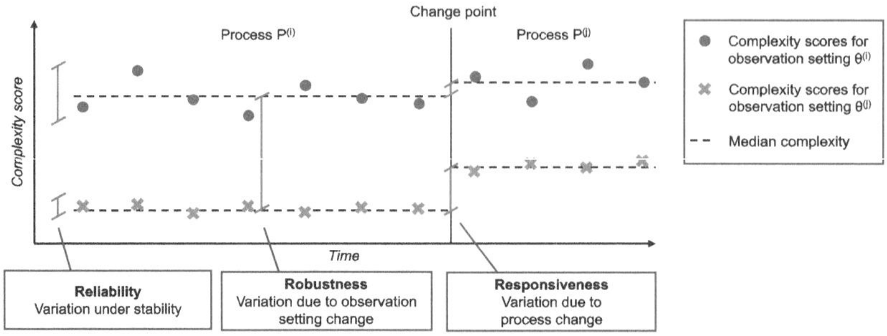

Fig. 1. Illustration of the three validity criteria considering one change in the process (change from $P^{(i)}$ to $P^{(j)}$) and two observation settings ($\theta^{(i)}$ and $\theta^{(j)}$).

and behavioral patterns [14,21]. Process complexity is operationalized through process complexity measures that map sets of process traces to numerical complexity scores. These measures can serve as valid indicators of process change only to the extent that (1) they reflect changes in the underlying process and (2) variation induced by their measurement process can be distinguished from variation due to true process change.

The complexity measurement process involves a sequence of steps, such as event log windowing and measure calculation, each of which may introduce measurement-induced variability. For instance, consider that an event log is split into trace windows, for each of which a complexity score is calculated [21]. Suppose windows are set to contain 100 traces, while a particular activity occurs on average only once every 200 traces. Even when the underlying process remains unchanged, this activity will appear in some windows but not in others. Consequently, complexity measures based on activity counts may fluctuate due to windowing effects alone. Currently, we lack concepts that enable distinguishing variation due to true process change and measurement-induced variation. This ambiguity leads to under-specified analysis methods for process complexity.

To address this research problem, we develop a process complexity observation model based on concepts from signal processing theory [20]. More specifically, we model observed complexity scores as consisting of a signal component and measurement-induced noise. Based on this observation model, we derive three fundamental validity criteria as illustrated in Fig. 1. *Reliability* captures variability of complexity scores under process and observation setting stability. *Robustness* quantifies sensitivity to observation setting changes. *Responsiveness* quantifies sensitivity to process change. We evaluate these criteria for established complexity measures in a large-scale simulation study. Our work makes three main contributions: (i) a formal observation model for process complexity grounded in signal processing theory, (ii) the derivation and formalization of reliability, robustness, and responsiveness as validity criteria for complexity measures, and (iii) an operationalization of these validity criteria in a controlled simulation setting, delivering concrete results and demonstrating the criteria's effectiveness. These contributions provide a measurement-theoretic foundation for interpreting variation in process complexity and for validating complexity measures as indicators of process change.

The remainder of this paper is organized as follows. Section 2 introduces key concepts from process mining and process complexity. Section 3 reviews related work on process complexity and process change. Section 4 presents the proposed process complexity observation model and the validity criteria. Section 5 applies the proposed validity criteria to uncover signal-and-noise characteristics of process complexity measures. Section 6 concludes the paper.

2 Preliminaries

In this section, we introduce the core concepts underlying our observation model. We build on standard process mining terminology [18] and formalize the notions of trace windows, windowing function and complexity measures.

Definition 1 (Event, trace, event log). *Let $\mathcal{U}_{event}$ be the universe of events, $\mathcal{U}_{case}$ be the universe of cases, $\mathcal{U}_{act}$ be the universe of activity labels, and $\mathcal{U}_{time}$ be the universe of timestamps.*

An event *e is a tuple $e = (c, a, t) \in \mathcal{U}_{case} \times \mathcal{U}_{act} \times \mathcal{U}_{time}$ that represents the execution of an activity (a) in a process instance (c) at some point in time (t).*

A trace $\sigma = \langle e_1, \ldots, e_n \rangle$ is a non-empty sequence of events, *where $e_i = (c_i, a_i, t_i)$ for all $i \in \{1, \ldots, n\}$, such that $\forall i \in \{1, \ldots, n-1\} : t_i \leq t_{i+1} \wedge c_i = c_{i+1}$.*

An event log L is a set of traces.

Definition 2 (Trace window, sequence of trace windows). *Let L be an event log and let $\langle \sigma_1, \ldots, \sigma_{|L|} \rangle$ denotes a sequence of all traces in L ordered by the timestamp of the first event of each trace. If two or more traces have the same timestamp for their first event, their relative order in the sequence is determined randomly. A trace window is a contiguous subsequence of this sequence, i.e., $w = \langle \sigma_i, \sigma_{i+1}, \ldots, \sigma_j \rangle$ for some $1 \leq i \leq j \leq |L|$. We denote a sequence of trace windows as $W = \langle w_1, \ldots, w_m \rangle$.*

Definition 3 (Windowing function). *A windowing function $win_{\theta_{win}}$ maps an event log L to a finite sequence of trace windows $win_{\theta_{win}}(L) = W = \langle w_1, \ldots, w_m \rangle$. The parameter θ_{win} specifies the windowing setting.*

Studies of complexity over time typically segment event logs into fixed-sized sublogs (e.g., [3,21]). In such a case, the windowing setting defines the size of the windows, which can be time-based (e.g., 1-month windows) or trace-based (e.g., 200-trace windows) size [16]. Furthermore, a step size may define the time or trace offset between consecutive windows [21]. The configuration parameter (θ_{win}) captures the required settings for the windowing function.

Definition 4 (Complexity Measure). *A process complexity measure C induces a function $comp_C$ that maps a trace window w to a real-valued complexity score $comp_C(w) \in \mathbb{R}$, where $comp_C(w) \geq 0$.*

In process complexity time series analysis, process complexity measures compute real-valued scores for trace windows. Most commonly, complexity measures represent variation in the control-flow perspective [18], although complexity measures have also been defined on other perspectives, e.g., the data perspective [15]. Control-flow complexity measures can be categorized into four dimensions [22]: (i) *Size measures* represent counts (e.g., the number of distinct activities) (ii) *Variation measures* capture the degree of unique behavior (e.g., the estimated number of acyclic paths) (iii) *Distance measures* reflect how distant traces are from each other (e.g., average edit distance between traces) (iv) *Graph entropy measures* capture length, size, variation, and distance by building a graph of the traces and calculating its entropy [2].

3 Related Work

The connection between process complexity and process change spans two research streams: (i) process complexity and (ii) process drift detection.

In the first stream, process complexity captures process variation, calculated from observed process traces. The areas of routine dynamics (e.g., [13]) and business process management (e.g., [5]) study process complexity. Simulation-based research in routine dynamics indicates that digitized processes show so-called bursts of complexity [14]. Their findings are supported by case studies that observe process complexity bursts at known process change points [3]. To create complexity time series, routine dynamics studies apply a windowing function with overlapping consecutive windows. Studies in business process management that explore the effect of process complexity consider the entire event log as part of a single trace window. These studies found that process complexity is a factor that affects process efficiency [2] and process mining performance [22].

In the second stream, process drift detection identifies process change points, i.e., whether and when a process change occurred [10,16]. Most approaches split event logs into trace windows, extracting control-flow attributes from each window [11,16]. These attributes are then compared using statistical hypothesis testing, which distinguishes between insignificant differences (random fluctuation) and significant ones. Despite performing statistical tests, concept drift detection techniques remain sensitive to configuration parameter choice (e.g., p-value threshold) [1]. Furthermore, recent research at the intersection of process complexity and drift detection indicates that process change points align with increases in complexity scores [21]. If complexity scores reflect process change points, they could potentially serve as direct indicators of process change. However, it remains unclear which types of process change are captured by complexity measures and which measures enable reliable detection of such changes.

Overall, both research streams analyze changes in observed process characteristics, yet neither explicitly separates genuine change of the underlying process from variability introduced by the measurement process.

4 Signal-and-Noise Perspective on Process Complexity

In this section, we develop a signal-and-noise perspective on process complexity in three steps. First, we identify the *observation pipeline* (previously termed the measurement process) that formalizes the steps for making complexity observations from an underlying process. Second, we leverage the observation pipeline to introduce an *observation model* which specifies signal and noise in the context of process complexity. Third, we derive *validity criteria* for studying signal-and-noise characteristics under the observation model, based on analytical scenarios.

4.1 Observation Pipeline

Case studies of process change compute process complexity from windows of observed traces, while the underlying business process remains a latent construct. We refer to the sequence of steps that transforms this latent construct to complexity observations as the *observation pipeline*. It consists of three steps: (i) trace observation, (ii) trace windowing, and (iii) process complexity calculation.

Business processes are generative systems [7] from which complexity observations are obtained through the observation pipeline. As a first step of this pipeline, process executions are recorded in an event log. The recorded trace observations represent only sampled behavior, as they may be an incomplete representation of a process's control flow [7,19]. Event logs may also contain recording errors that result from incorrect logging [8]. The next step, trace windowing, may further reduce the completeness of trace samples, as windows are subsets of event logs. Finally, complexity scores are calculated for each window (Definition 4). Definition 5 formalizes this observation pipeline.

Definition 5 (Observation pipeline). *Let P denote a process, C a complexity measure, and $\theta = (\theta_{sample}, \theta_{win})$ an observation setting. Let $\mathcal{L}$, $\mathcal{W}$, and $\mathcal{X}_C$ be random variables over event logs, trace windows, and complexity scores, respectively. The complexity observation pipeline consists of three steps:*

$$P \xrightarrow{sample_{\theta_{sample}}} \mathcal{L} \xrightarrow{win_{\theta_{win}}} \mathcal{W} \xrightarrow{comp_C} \mathcal{X}_C$$

where

1. *$sample_{\theta_{sample}}(P) \sim \mathcal{L}$ defines a random variable over event logs by stochastically sampling from process P under sampling setting θ_{sample}. θ_{sample} may include settings such as activity definitions or, in simulation, the frequency of recording errors.*
2. *$win_{\theta_{win}}(L) \sim \mathcal{W}$ for $L \sim \mathcal{L}$ defines a random variable over trace windows by applying a windowing procedure to realizations L of $\mathcal{L}$.*
3. *$comp_C(w) \sim \mathcal{X}_C$ for $w \sim \mathcal{W}$ defines a real-valued random variable by applying a complexity function to realizations w of $\mathcal{W}$.*

In practice, the process may change, but for simplicity, the observation pipeline describes the generation of windows conditional on a single given process P. Changes to P are captured by comparing the results of two separate observational pipelines, one for P before the change (pre-change) and another for after the change (post-change).

The observation pipeline is conceptually analogous to a digital communication system [20]. In such systems, a source signal is transmitted through a channel and observed as a received signal that may contain noise. Similarly, process complexity is not directly observable but is inferred through the observation pipeline, yielding noisy complexity scores.

4.2 Observation Model

Signal processing theory provides an observation model in which at every time t, a source signal (s) is transmitted through a channel, adding noise (n) to the signal, and resulting in a received signal r that contains both signal and noise [20]. For discrete-time signals, this observation model is defined as $r(t) = s(t) + n(t)$.

Transferring this perspective to process complexity, we model observed complexity scores as the sum of the complexity signal and a noise term. This measurement noise term may contain both systematic deviation (i.e., systematic measurement bias) as well as stochastic variance (i.e., stochastic measurement error). This model aligns with estimation theory in signal processing [9], since its goal is to estimate the process complexity signal.

This notion of measurement noise differs from the concept of event log noise [8]. Event log noise may arise from recording errors, infrequent behavior, deviations from process specification, and external influences [8]. In contrast, the measurement noise term comprises only those factors that distort the measured complexity of the underlying generative process. Hence, it includes recording errors, as well as distortions arising from (1) incomplete sampling of process traces and (2) trace windowing and complexity computation. Infrequent behavior and deviations from process specification are not considered noise under this view, as they constitute characteristics of the observed process behavior rather than artifacts of the measurement process.

Definition 6 (Observation model). *Let P be a process and $\theta = (\theta_{sample}, \theta_{win})$ an observation setting. The observation model is*

$$\mathcal{X}_C = s_C(P) + n_C(P,\theta) \quad \textit{with} \quad n_C(P,\theta) = B_C(P,\theta) + \varepsilon_C(P,\theta) \quad \textit{where}$$

- $\mathcal{X}_C$ *denotes the random variable over complexity scores induced by the observation pipeline (Definition 5) under (P,θ) for complexity measure C.*
- $s_C(P)$ *denotes the* process complexity signal *of process P according to measure C. It is a deterministic quantity given P and C.*
- $n_C(P,\theta)$ *denotes the* measurement noise *in complexity observations from process P according to measure C and observation setting θ. It is a random variable due to the stochasticity of the observation pipeline.*

- $B_C(P, \theta)$ *denotes the* systematic measurement bias *induced by the observation pipeline under process P and observation setting θ according to measure C. It is deterministic given P, θ and C.*
- $\varepsilon_C(P, \theta)$ *denotes* stochastic measurement error *induced by the observation pipeline under* (P, θ) *according to measure C. It is modeled as a random variable with* $\mathbb{E}[\varepsilon_C \mid P, \theta] = 0$.

The three components of the observation model decompose complexity scores ($comp_C(w)$). The term $s_C(P)$ denotes the complexity of process P according to measure C, only depending on the process and not on the observation setting. In the real world, this quantity is typically unknown, although in simulation, the process complexity signal may be known. The term $B_C(P, \theta)$ captures systematic measurement bias introduced by the observation pipeline, such as bias from rare behavior [7]. It depends on both the process P and the observation setting θ because the magnitude of such bias may vary with properties of the process and with the settings of the observation pipeline. Finally, $\varepsilon_C(P, \theta)$ represents stochastic measurement error induced by the random realization of the observation pipeline. It is modeled as a zero-mean random variable and captures variability due to observation pipeline stochasticity. Based on this observation model, we develop the following problem statement.

Problem 1 (Interpretation of observed complexity variations). Let P be a process and let $\mathcal{X}_C$ denote the random variable of observed complexity values induced by the observation pipeline under (P, θ) for complexity measure C. In empirical settings, one observes finite realizations of $\mathcal{X}_C$. According to the observation model, variation in these observations may arise from: (i) changes in process complexity signal ($s_C(P)$) induced by changes in the underlying process, or (ii) changes in measurement noise ($n_C(P, \theta)$) induced by the observation pipeline. Hence, observed variation in complexity values is not uniquely attributable to process change. The problem is to derive validity criteria that characterize to what extent a complexity measure C allows distinguishing between process change and measurement-induced variation.

4.3 Validity Criteria

To study variation in complexity measures, we consider analytical scenarios defined by whether the underlying process or the observation setting changes between two observations.

Definition 7 (Analytical scenarios). *Let* $\mathcal{X}_C^{(i)}$ *and* $\mathcal{X}_C^{(j)}$ *denote random variables of complexity scores induced by the observation pipeline (Definition 5) under observation settings* $(P^{(i)}, \theta^{(i)})$ *and* $(P^{(j)}, \theta^{(j)})$*, where* $P^{(j)}$ *reflects changes to* $P^{(i)}$*. Depending on whether the process and observation setting remain constant or change, four analytical scenarios arise:*

1. *Observation setting and process constant. The process and observation setting do not change* ($P^{(i)} = P^{(j)}$, $\theta^{(i)} = \theta^{(j)}$)*. Variation in complexity scores is due to stochastic measurement error.*

Table 1. Signal-and-noise interpretation of validity criteria

Criterion	Analytical scenario	Ratio
Reliability	Observation setting and process constant	Systematic components (s_C and B_C) vs. stochastic measurement error (ε_C)
Robustness	Observation setting change	Systematic components (s_C and B_C) vs. shifts in measurement bias B_C
Responsiveness	Process change	Shifts in systematic components (s_C and B_C) vs. stochastic measurement error (ε_C)

2. *Observation setting change. The process remains constant but the observation setting changes ($P^{(i)} = P^{(j)}$, $\theta^{(i)} \neq \theta^{(j)}$). Systematic measurement bias and stochastic measurement error may cause variation in observed complexity scores.*
3. *Process change. The observation setting remains constant but the underlying process changes ($P^{(i)} \neq P^{(j)}$, $\theta^{(i)} = \theta^{(j)}$). Hence, the process complexity signal may change, but because measurement noise depends on the process and the observation setting, also systematic measurement bias and stochastic measurement error may be affected.*
4. *Observation setting and process change. Both the process and the observation setting change ($P^{(i)} \neq P^{(j)}$, $\theta^{(i)} \neq \theta^{(j)}$). All components of the observation model may induce variation in observed complexity scores.*

We derive validity criteria that characterize the behavior of complexity measures under these analytical scenarios. The *observation setting and process change* scenario does not introduce new types of effects beyond those present in *process change* and *observation setting change*, but rather combines them, making the sources of variation harder to disentangle. Therefore, it is not considered further. The *observation setting and process constant* scenario allows assessing the stochastic measurement error of a measure. The *observation setting change* scenario reflects the measure's sensitivity to changes in the observation pipeline settings. The *process change* scenario captures the strength of response to process change. Accordingly, measures are valid to the extent that they exhibit (i) low stochastic measurement error when the observation setting and process remain constant, (ii) low sensitivity to observation setting change, and (iii) a strong response to process change. On this basis, we derive three validity criteria that quantify how complexity measures behave in response to analytical scenarios one to three in Definition 7. *Reliability* captures stochastic measurement error under observation setting and process constant. *Robustness* measures sensitivity to observation setting change, and *responsiveness* reflects the ability to detect process change. Table 1 summarizes these criteria.

Reliability captures the stability of a measurement instrument under repeated observations of an invariant construct [17], where variability reflects random measurement error [12]. In our setting, fixing (P, θ) corresponds to observing an invariant construct, and repeated realizations of $\mathcal{X}_C \mid (P, \theta)$ correspond to repeated measurements. Consequently, variability in $\mathcal{X}_C$ directly reflects stochastic measurement error induced by the observation pipeline.

Definition 8 (Reliability). *Let C be a complexity measure, P a process, and θ an observation setting. Let $\mathcal{X}_C$ denote the random variable of complexity scores induced by the observation pipeline under (P, θ) for complexity measure C. The reliability of C under (P, θ) is defined as*

$$Rel_C(P, \theta) := \frac{\mathbb{E}[\mathcal{X}_C \mid P, \theta]}{\sqrt{Var[\mathcal{X}_C \mid P, \theta]}}.$$

Reliability is defined as the inverse of the coefficient of variation (CV). The CV yields a unitless representation of variation that enables comparison across different complexity measures. Definition 8 computes the inverse of CV so that high values are desirable and low values are undesirable.

Under the observation model (Definition 6), $s_C(P)$ and $B_C(P, \theta)$ are deterministic given (P, θ), while all stochasticity is captured by $\varepsilon_C(P, \theta)$. Hence,

$$\mathbb{E}[\mathcal{X}_C \mid P, \theta] = s_C(P) + B_C(P, \theta). \tag{1}$$

and

$$Var[\mathcal{X}_C \mid P, \theta] = Var[\varepsilon_C(P, \theta) \mid P, \theta], \tag{2}$$

Reliability thus captures the relative magnitude of systematic components (process complexity signal and systematic measurement bias) relative to stochastic measurement error.

Systematic differences in measurements under invariant constructs reflect construct-irrelevant measurement error [12,17]. In our setting, fixing P while varying θ corresponds to repeated measurements of an invariant construct under different observation conditions. Since $s_C(P)$ is invariant with respect to θ, any systematic differences in $\mathbb{E}[\mathcal{X}_C \mid P, \theta]$ arise from changes in $B_C(P, \theta)$ and thus reflect construct-irrelevant shifts in measurement bias.

Definition 9 (Robustness). *Let C be a complexity measure, P a process and $\theta^{(i)}, \theta^{(j)}$ two observation settings. Let $\mathcal{X}_C^{(i)}$ and $\mathcal{X}_C^{(j)}$ denote the random variables of complexity scores induced by the stochastic observation pipeline under $(P, \theta^{(i)})$ and $(P, \theta^{(j)})$, respectively. The robustness of C under $(P, \theta^{(i)}, \theta^{(j)})$ is defined as*

$$Rob_C(P, \theta^{(i)}, \theta^{(j)}) := \frac{\mathbb{E}\left[\mathcal{X}_C^{(i)} \mid P, \theta^{(i)}\right]}{\left|\mathbb{E}\left[\mathcal{X}_C^{(i)} \mid P, \theta^{(i)}\right] - \mathbb{E}\left[\mathcal{X}_C^{(j)} \mid P, \theta^{(j)}\right]\right|}.$$

Robustness is defined as the stability of expected complexity under observation setting change. It is high if variations in θ induce only small changes in expected complexity. Under the observation model (Definition 6), $\mathbb{E}[\varepsilon_C \mid P, \theta] = 0$ and $s_C(P)$ is invariant with respect to θ, yielding

$$Rob_C(P, \theta^{(i)}, \theta^{(j)}) = \frac{s_C(P) + B_C(P, \theta^{(i)})}{\left|B_C(P, \theta^{(i)}) - B_C(P, \theta^{(j)})\right|}. \tag{3}$$

Analogous to reliability (i.e., a ratio of systematic components to stochastic error), robustness can be interpreted as a ratio of systematic components (process complexity signal and systematic measurement bias) to shifts in measurement bias across observation settings. High robustness indicates that observed complexity is stable under observation setting change, whereas low robustness indicates that changes in the observation setting induce substantial shifts in measurement bias.

Responsiveness assesses systematic shifts in response to process change relative to stochastic measurement error. In signal detection theory, this corresponds to the sensitivity index d', which quantifies the detectability of a signal in the presence of noise [20]. The sensitivity index can be interpreted as a signal-to-noise ratio.

Definition 10 (Responsiveness). *Let C be a complexity measure, $P^{(i)}, P^{(j)}$ be two processes where $P^{(j)}$ reflects change of $P^{(i)}$, and θ an observation setting. Let $\mathcal{X}_C^{(i)}$ and $\mathcal{X}_C^{(j)}$ denote the random variables of complexity scores induced by the observation pipeline under $(P^{(i)}, \theta)$ and $(P^{(j)}, \theta)$, respectively. The responsiveness of C with respect to process $P^{(i)}$ and $P^{(j)}$ is defined as*

$$Resp_C(P^{(i)}, P^{(j)}, \theta) := \frac{\left|\mathbb{E}[\mathcal{X}_C^{(j)} \mid P^{(j)}, \theta] - \mathbb{E}[\mathcal{X}_C^{(i)} \mid P^{(i)}, \theta]\right|}{\sqrt{\frac{Var[\mathcal{X}_C^{(i)} \mid P^{(i)}, \theta] + Var[\mathcal{X}_C^{(j)} \mid P^{(j)}, \theta]}{2}}}.$$

Definition 10 corresponds to the process change scenario in Definition 7, where the observation setting remains fixed while the process changes. Using the decomposition established in Eq. 1 and Eq. 2, responsiveness can be rewritten as

$$Resp_C(P^{(i)}, P^{(j)}, \theta) = \frac{\left|\left(s_C(P^{(j)}) - s_C(P^{(i)})\right) + \left(B_C(P^{(j)}, \theta) - B_C(P^{(i)}, \theta)\right)\right|}{\sqrt{\frac{Var[\varepsilon_C(P^{(i)}, \theta)] + Var[\varepsilon_C(P^{(j)}, \theta)]}{2}}}. \tag{4}$$

Responsiveness quantifies the relative magnitude of systematic change (shifts in process complexity signal and shifts in measurement bias) relative to stochastic measurement error. The inclusion of shifts in measurement bias in the numerator reflects that process change may also affect the behavior of the observation pipeline. High responsiveness indicates that process-induced changes dominate stochastic measurement error, whereas low responsiveness indicates that such changes are difficult to distinguish from stochastic measurement error.

5 Evaluation

In this section, we evaluate the signal-and-noise behavior of established control-flow complexity measures by applying the proposed validity criteria. The evaluation demonstrates the effectiveness of the criteria in characterizing these behaviors. We begin the evaluation by introducing the design of the large-scale simulation study, then we report results from computing the criteria and, finally, discuss the findings.

5.1 Experimental Design

We perform experiments using a factorial simulation design that systematically varies process characteristics and observation settings. For each parameter combination, we generate event logs with one process change, sample windows from each log and compute complexity scores per log window. We simulate a scenario with a single process change rather than multiple changes to simplify the experimental setting. The complexity scores are used to empirically estimate the three validity criteria. In the following, we provide details on the experimental parameters, simulation procedure and operationalization of validity criteria.

Table 2. Summary of experimental parameters

Category	Parameter	Value
Process	Base complexity	Middle, complex
	Change magnitude	10%, 20%, 30%
	Change type	Activity insertion, activity deletion, activity replacement, resequentialization, operator replacement, mixed
Observation setting	Log noise level	0%, 1%, 2%
	Window size	50, 100, 150, 200 traces

Experimental Parameters. The process parameters in Table 2 define the structure of $P^{(i)}$ and its change to $P^{(j)}$. *Base complexity* describes the complexity of the generated process models, following the parameters described in [10]. E.g., a middle-complexity process model has 14–20 activities while a complex process model has 20–30 activities. The *change magnitude* determines the proportion of the model affected during change, while the *change type* constrains the allowed modifications (e.g., activity insertion, activity deletion). The *mixed* setting allows combinations of all change types and resembles realistic process evolution. The observation setting parameters in Table 2 correspond to configurations (θ) of the observation pipeline. The *recording error level* specifies the proportion of traces affected by recording errors (component θ_{sample}), implemented following [8]. The *window size* defines the number of traces per window (component θ_{win}). Windowing is performed by sampling with replacement, drawing 200 windows from each part of the log without overlap across the drift boundary. The parameter values are chosen in accordance with prior research. Values for *base complexity* and *change magnitude* were applied in [10], *recording error level* values in [8] and *window sizes* in [14]. *Change types* are all those available in the CDLG simulator [4].

Simulation Procedure. Each log contains 10,000 traces, where the first 5,000 traces are generated from process $P^{(i)}$ and the remaining 5,000 traces from the changed process $P^{(j)}$. Log generation (process play-out and noise injection) and window sampling are controlled by random seeds. To account for stochasticity, each configuration is repeated for 30 seeds. Complexity scores are computed for all parameter combinations, seeds, and sampled windows. The simulation is implemented in Python using `pm4py`[1], Vidgof et al. [21]'s process complexity library for entropy-based measures, PTandLogGenerator for generating process models [6], and an extension of the CDLG event log generator for simulating process drifts [4][2].

Operationalization of Validity Criteria. The validity criteria are computed based on empirical realizations of complexity scores obtained from the simulation. Reliability is estimated as the ratio of the empirical mean to the empirical standard deviation of complexity scores within the pre-change part of the log. Robustness is operationalized as the sensitivity of complexity values to changes in the observation setting relative to a fixed baseline configuration. For each process and seed, we estimate expected complexity values by the empirical mean and compute deviations between a baseline setting and alternative settings. Specifically, we compare (i) a baseline noise level (0% noise) to higher noise levels at fixed window size, and (ii) a baseline window size (50) to larger window sizes at fixed noise level. These deviations are computed within the same process configuration and seed. The reported robustness value for each measure is the median of all resulting deviations across seeds, process configurations, and setting variations. Responsiveness is operationalized as the difference in observed pre-change and post-change mean complexity values relative to their pooled standard deviation under a fixed observation setting. Reported responsiveness values are medians across seeds and process configurations. To reflect realistic process evolution, we report responsiveness values for the mixed change type if not further broken down.

---

1 https://processintelligence.solutions/pm4py.

2 Source code, interim data and final results shared publicly at https://doi.org/10.5281/zenodo.19119437.

5.2 Evaluation Results

Table 3 summarizes the empirical validity scores across all criteria. Three main insights can be identified. First, measures differ substantially in their reliability, indicating large stochastic measurement error. Second, sensitivity to observation settings varies even more strongly, with some measures being highly robust while others are strongly affected by the observation pipeline. Third, measures generally respond to process change, but their sensitivity is limited, suggesting that detecting process change from complexity alone is inherently difficult.

Table 3. Medians of reliability, robustness, and responsiveness across experiment settings.

Dimension	Measure	Label	Rel_C	Rob_C	$Resp_C$
Size	# Events	A1	19.23	1.01	0.89
	# Distinct activities	A2	20.74	46.26	1.03
	# Distinct traces	A3	11.16	2.30	<u>0.26</u>
	# Distinct directly-follows relations	A4	8.99	6.04	0.69
	Min. trace length	A5	10.38	26.00	0.91
	Avg. trace length	A6	19.23	**2245.35**	0.89
	Max. trace length	A7	5.52	9.40	0.36
Variation	Percentage of distinct traces	B1	11.16	7.20	<u>0.26</u>
	Avg. distinct activities per trace	B2	41.63	1043.43	**1.29**
	Structure	B3	23.59	30.21	0.49
	Estimated # acyclic paths	B4	<u>1.48</u>	0.93	0.31
	# Ties in paths to goal	B5	7.53	2.14	0.52
	Lempel-Ziv complexity	B6	17.91	1.60	0.71
Distance	Avg. affinity	C1	80.30	459.56	0.68
	Deviation from random	C2	**125.74**	555.03	1.12
	Avg. edit distance	C3	12.37	276.53	0.39
Graph entropy	Sequence entropy	D1	11.16	<u>0.88</u>	0.67
	Normalized sequence entropy	D2	32.54	28.73	0.36
	Variant entropy	D3	6.45	1.48	0.54
	Normalized variant entropy	D4	34.83	15.75	0.38

Formatting: **Bold** = highest per column, <u>underlined</u> = lowest per column.

Reliability varies strongly across measures (see Table 3, column Rel_C). For example, the distance measure *deviation from random* achieves a reliability of 125.74, meaning that the expected value is more than two orders of magnitude larger than its standard deviation. Hence, observations of this measure are highly stable when process and observation setting are not changing. In contrast, *estimated number of acyclic paths* exhibits very low reliability. Its values are expected

Table 4. Responsiveness per process change type.

Dimension	Label	Del.	Ins.	Repl.	Reseq.	Op. repl.	Mix
Size	A1	1.26	0.98	0.12	0.18	0.82	0.89
	A2	1.30	1.31	**0.14**	0.15	0.15	1.03
	A3	0.46	0.26	0.12	0.21	0.61	0.26
	A4	1.11	0.84	0.10	0.23	0.42	0.69
	A5	1.06	1.00	0.13	0.21	0.43	0.91
	A6	1.26	0.98	0.12	0.18	0.82	0.89
	A7	0.56	0.33	0.13	0.18	0.42	0.36
Variation	B1	0.46	0.26	0.12	0.21	0.61	0.26
	B2	**1.51**	**1.42**	0.13	0.21	0.75	**1.29**
	B3	0.61	0.56	0.11	0.21	0.43	0.49
	B4	0.40	0.38	0.11	0.19	0.32	0.31
	B5	0.92	0.59	0.12	0.20	0.67	0.52
	B6	1.06	0.76	0.12	0.23	0.71	0.71
Distance	C1	1.05	0.95	0.11	0.18	0.80	0.68
	C2	1.35	1.10	0.12	**0.24**	**1.14**	1.12
	C3	0.85	0.39	0.14	0.21	0.91	0.39
Graph entropy	D1	1.01	0.76	0.11	0.22	0.83	0.67
	D2	0.54	0.31	0.11	0.23	0.85	0.36
	D3	0.86	0.60	0.12	0.24	0.72	0.54
	D4	0.56	0.37	0.12	0.23	0.65	0.38

Measure labels: See Table 3. *Formatting:* **Bold** = highest per column, underlined = lowest per column. *Abbreviations:* Del.= Activity deletion, Ins.=Activity insertion, Repl.=Activity replacement, Reseq.=Resequentialization, Op. repl.=Operator replacement, Mix=Mixed.

to fluctuate solely from stochasticity in the observation pipeline. The low reliability for this measure can be explained due to its exponential calculation logic. *Estimated number of acyclic paths* is computed by the exponential formula $10^{0.08*(1+A4-A2)}$ [14]. Variations in measures *number of distinct directly-follows relations* (A4) and *number of distinct activities* (A2) are hence amplified.

Robustness shows even stronger dispersion than reliability. Measures based on aggregation, such as *average trace length* and *average number of distinct activities per trace*, are largely invariant to changes in window size and recording errors. This indicates that averaging absorbs rare behavior and noise-induced variation. In contrast, entropy-based measures are more sensitive to observation settings. For instance, *sequence entropy* exhibits low robustness due to its dependence on window size, whereas *normalized sequence entropy* mitigates this effect by explicitly adjusting for it. This highlights that normalization can substantially reduce measurement-induced bias.

Responsiveness is generally low across all measures. For reference, a value of $Resp_C = 1$ implies that process change shifts the mean by one standard deviation, which still results in considerable overlap between pre- and post-change distributions. The highest observed responsiveness (1.29 for *average distinct activities per trace*), therefore, indicates only moderate detectability of process change. Measures such as *percentage of distinct traces* and *number of distinct traces* show substantially lower responsiveness, suggesting that structural changes are difficult to distinguish from stochastic variation.

Responsiveness further depends on the type of process change (Table 4). Measures are most responsive to activity insertion, deletion, and operator replacement, which directly affect control-flow structure. In contrast, activity replacement and resequentialization are poorly captured. This is expected, as control-flow measures abstract from activity labels and are therefore insensitive to label substitutions, while changes in execution order do not necessarily affect aggregate structural properties.

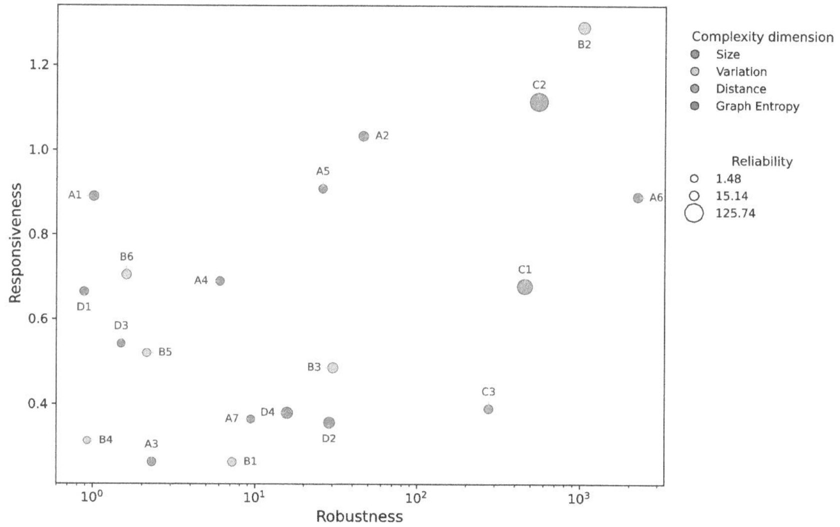

Fig. 2. Scatter plot of robustness (x-axis; log), responsiveness (y-axis) and reliability (dot size) for established control-flow complexity measures observed in experiment. See Table 3 for measure labels.

Joint Perspective. Figure 2 summarizes the joint behavior across all three validity criteria. Measures in the top-right region combine high robustness and responsiveness, while dot size reflects reliability. The visualization shows that a small subset of measures (notably *deviation from random*, *average distinct activities per trace*, and *average trace length*) outperforms others within their respective dimensions. These measures combine high stability, low configuration sensitivity, and comparatively strong responsiveness, making them particularly suitable for complexity time series analysis.

5.3 Discussion

The results imply that variation in complexity scores cannot be interpreted independently of the signal-and-noise characteristics of the employed measures. Even in absence of change, measures may exhibit substantial stochastic measurement error or systematic shifts induced by the measurement process. Furthermore, the sensitivity by which process change is captured is generally low and varies significantly between measures. Conclusions about process change and stability depend on the properties of the selected measure, and ignoring these properties risks conflating process change with measurement-induced variation.

The proposed validity criteria make these characteristics explicit. They allow assessing whether a measure is reliable, robust to configuration choices, and responsive to structural change. Consequently, they support validating complexity measures with respect to their use in process change detection and enable more defensible interpretations of complexity time series. Overall, the validation of complexity measures is a necessary condition for meaningful empirical analysis.

The proposed signal-and-noise perspective suggests several directions for future research. First, the generally low responsiveness of existing measures motivates the development of alternative measures with higher responsiveness, which may be promising candidates for complexity-based process mining drift detection. Second, the current evaluation is based on simulation settings, which enable controlled analysis but do not capture real-world complexity. Future work could therefore apply the proposed criteria in empirical settings using event logs with known process changes. Third, the operationalization of robustness focuses on sensitivity to recording errors and window-size effects, which are major contributors to measurement-induced variability, but additional sources of variation, such as activity definitions, remain to be explored. Finally, the proposed measurement perspective may be generalized beyond control-flow complexity to encompass further process mining perspectives.

6 Conclusion

In this paper, we established validity criteria for explaining variation in process complexity by introducing an observation pipeline and a corresponding observation model that decomposes observed complexity into structural signal and measurement-induced noise. Based on distinct analytical scenarios, we developed the validity criteria of reliability, robustness, and responsiveness, each capturing a unique signal-and-noise characteristic. We evaluated these criteria in a large-scale simulation study and showed that commonly used complexity measures differ substantially in their signal-and-noise behavior: some exhibit low reliability, weak robustness, and limited responsiveness, while alternative measures perform consistently better across all criteria. However, even the best-performing measures show only moderate responsiveness to change. The observation pipeline and derived validity criteria provide a measurement-theoretic foundation for interpreting variations in process complexity. Furthermore, our evaluation of existing measures contributes concrete reference points for the selection of measures in future complexity time series analysis. Therefore, changes in complexity scores should not be interpreted as evidence of process change without considering the signal-and-noise characteristics of the underlying complexity measure. Overall, we show the importance of differentiating between measurement-induced variation and variation arising from changes in the underlying construct.

Acknowledgments. The research of the authors was supported by the Einstein Foundation Berlin under grant EPP-2019-524, by the German Federal Ministry of Research, Technology and Space under grant 16DII143, and by Deutsche Forschungsgemeinschaft under grants 496119880 (VisualMine), 531115272 (ProImpact), and SFB 1404/2 (FONDA). Kate Revoredo is funded by the Berliner Chancengleichheitsprogramm (BCP) as part of the DiGiTal Graduate Program.

Disclosure of Interests. The authors have no competing interests to declare that are relevant to the content of this article.

References

1. Adams, J.N., Pitsch, C., Brockhoff, T., van der Aalst, W.M.P.: An experimental evaluation of process concept drift detection. Proc. VLDB Endow. **16**(8), 1856–1869 (2023)
2. Augusto, A., Mendling, J., Vidgof, M., Wurm, B.: The connection between process complexity of event sequences and models discovered by process mining. Inf. Sci. **598**, 196–215 (2022)
3. Franzoi, S., Hartl, S., Grisold, T., van der Aa, H., Mendling, J., vom Brocke, J.: Explaining process dynamics: a process mining context taxonomy for sense-making. Process Sci. **2**(1) (2025)
4. Grimm, J., Kraus, A., van der Aa, H.: CDLG: A tool for the generation of event logs with concept drifts. In: BPM 2022 (PhD/Demos). CEUR Workshop Proceedings, vol. 3216, pp. 92–96. CEUR-WS.org (2022)
5. Günther, C.: Process Mining in Flexible Environments. Phd thesis, Technische Universiteit Eindhoven, Eindhoven (2009)
6. Jouck, T., Depaire, B.: Ptandloggenerator: A generator for artificial event data. In: BPM 2016 (Demos). CEUR Workshop Proceedings, pp. 23–27. CEUR-WS.org (2016)
7. Kabierski, M., Richter, M., Weidlich, M.: Quantifying and relating the completeness and diversity of process representations using species estimation. Inf. Syst. **130**, 102512 (2025)
8. Karunaratne, A., Polyvyanyy, A., Moffat, A.: The effects of log noise in process mining. IEEE Access **13**, 198540–198563 (2025)
9. Kay, S.M.: Fundamentals of Statistical Signal Processing. 1: Estimation Theory. Prentice Hall PTR. NJ, 20th edn. (2013)

10. Kraus, A., van der Aa, H.: Comprehensive characterization of concept drifts in process mining. Inf. Syst. **135**, 102584 (2026)
11. Maaradji, A., Dumas, M., Rosa, M.L., Ostovar, A.: Detecting sudden and gradual drifts in business processes from execution traces. IEEE Trans. Knowl. Data Eng. **29**(10), 2140–2154 (2017)
12. O'Leary-Kelly, S.W., Vokurka, J.R.: The empirical assessment of construct validity. J. Oper. Manag. **16**(4), 387–405 (1998)
13. Pentland, B.T.: Conceptualizing and measuring variety in the execution of organizational work processes. Manage. Sci. **49**(7), 857–870 (2003)
14. Pentland, B.T., Liu, P., Kremser, W., Hærem, T.: The dynamics of drift in digitized processes. MIS Q. **44**(1), 19–47 (2020)
15. Revina, A., Aksu, Ü.: An approach for analyzing business process execution complexity based on textual data and event log. Inf. Syst. **114**, 102184 (2023)
16. Sato, D.M.V., de Freitas, S.C., Barddal, J.P., Scalabrin, E.E.: A Survey on Concept Drift in Process Mining. ACM Comput. Surv. **54**(9) (2021)
17. Straub, D.W.: Validating instruments in MIS research. MIS Q. **13**(2), 147–169 (1989)
18. van der Aalst, W.M.P.: Process Mining: Data Science in Action, 2nd edn. Springer, Cham (2016)
19. van der Aalst, W., et al.: Process Mining Manifesto. In: Daniel, F., Barkaoui, K., Dustdar, S. (eds.) BPM 2011. LNBIP, vol. 99, pp. 169–194. Springer, Heidelberg (2012). https://doi.org/10.1007/978-3-642-28108-2_19
20. Van Trees, H.L.: Detection, Estimation, and Linear Modulation Theory. No. 1 in Detection, Estimation, and Modulation Theory, Wiley-Interscience (2002)
21. Vidgof, M.: First insights into the impact of concept drift on process complexity. In: Gdowska, K., Gómez-López, M.T., Rehse, JR. (eds.) Business Process Management Workshops 2024. LNBIP, vol. 534, pp. 332–337. Springer, Cham (2024). https://doi.org/10.1007/978-3-031-78666-2_25
22. Vidgof, M., Wurm, B., Mendling, J.: The impact of process complexity on process performance: A study using event log data. In: BPM. LNCS, vol. 14159, pp. 413–429. Springer, Cham (2023). https://doi.org/10.1007/978-3-031-41620-0_24

Trust No Pool: Formal Security Analysis of BPMN Collaborations

Martin Farkas(✉) and Imre Kocsis

Department of Artificial Intelligence and Systems Engineering, Budapest University of Technology and Economics, Budapest, Hungary
martin.farkas@edu.bme.hu, kocsis.imre@vik.bme.hu

Abstract. Cross-organizational collaborations face inherent trust challenges: any participant may want to deviate from the agreed-upon process model, trying to deceive others about the shared state. Traditional verification approaches over collaborations – e.g., over BPMN collaboration models – verify correctness properties under the assumption that all participants faithfully execute their roles. They do not consider what a strategically dishonest participant can achieve, and to our knowledge, no formal framework exists for doing so. We introduce potentially adversarial participants for BPMN collaborations as a first-class concept, analyze them as security protocols, and demonstrate that this framing is both formally precise and directly actionable. We utilize a structural correspondence between BPMN's token-based collaboration semantics and the multiset rewriting formalism of the Tamarin security protocol verifier. This correspondence allows us to define a formal, epistemically grounded threat model and to encode a core subset of collaboration semantics as Tamarin theories, enabling the formal verification of collaboration security: the automated discovery of successful attack strategies or proving of their absence. We introduce a composable attacker model whose elements reflect the underlying platform's security guarantees – from loosely coupled integrations through smart contract orchestration to Verifiable Credential-based protocols. We demonstrate the approach on an example, examining how the attack surface shifts across the three execution platforms and revealing platform-specific attacks that arise from the interplay of business logic and platform semantics. These attacks are undetectable by standard correctness verification and sophisticated enough to warrant systematic analysis in business-critical settings.

Keywords: BPMN collaborations · formal verification · security protocol analysis · adversarial participants · multiset rewriting · epistemic security modeling

F. M. Maggi et al. (Eds.): BPM 2026, LNCS 16936, pp. 111–127, 2027.
https://doi.org/10.1007/978-3-032-37777-7_9

1 Introduction

Consider a four-party procurement collaboration (Fig. 1). The Procurement Authority (A) obtains regulatory clearance from the Regulatory Authority (R), then at an exclusive gateway selects exactly one of two structurally symmetric suppliers (B or C) to fulfill an order. The topology is hub-and-spoke: A mediates all cross-pool communication. Standard soundness analysis would confirm that the model is correct: every trace terminates, no deadlocks arise, no activities are dead.

Now suppose A is dishonest and awards both suppliers simultaneously. One of the multiple realistic incentives is hedging with asymmetric cost transfer: A recognizes a delivery risk and secretly awards both, so whichever delivers first or better gets accepted; the other gets rejected on a technicality. A eliminates its supply risk, but the "losing" supplier has already mobilized, based on the false belief of an exclusive award. The cost of A's hedge is borne entirely by the deceived party.

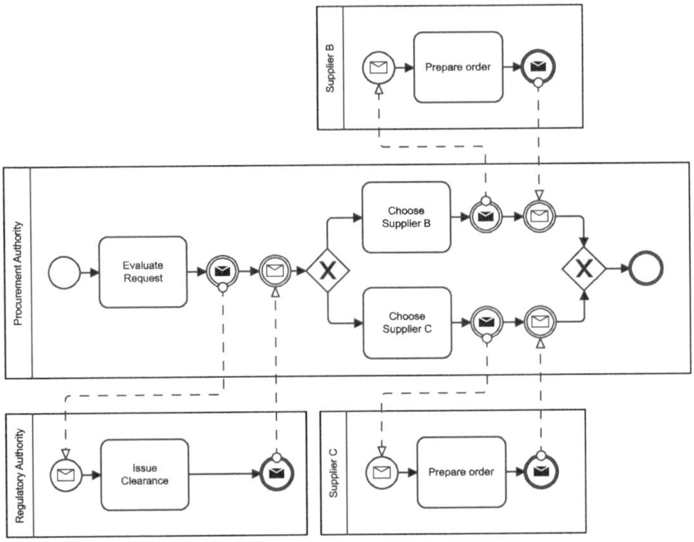

Fig. 1. Running example: a four-party procurement collaboration.

Can A succeed? It depends on the execution platform approach used to realize the collaboration specification. In the classic integration-by-loosely-coupled-APIs collaboration-realization model, nothing prevents attack success: each supplier receives an authentic message and has no evidence that the other supplier was also selected. Under smart contract-based orchestration, the on-chain state enforces the necessary mutual exclusion – but every participant, including competing suppliers, may be able to read every other's order details on-chain. In a protocol where orchestration is based on the participants exchanging prior execution trace as a cryptographically linked series of activity attestations [12] (using standard Verifiable Credentials – VCs), cryptographic commitments can prevent most such attacks; yet, as we will show, a timing window in the credential revocation mechanism may open a "split brain" vulnerability with respect to B and C.

Thus, the same collaboration logic admits three qualitatively different security postures at the application level, depending on the business logic *as well as* the execution platform.

Existing formal verification approaches for BPMN specifically, and to a lesser extent, business collaborations in general seem to be oblivious to such concerns. Soundness analysis, Petri net model checking, and tools such as BProVe [8] and LoLA [20] verify correctness properties (deadlock freedom, proper termination, liveness) under the assumption that all participants faithfully execute their pools. Security annotation frameworks such as SecBPMN [19] and PE-BPMN [18] allow designers to express confidentiality and integrity requirements on process elements, but neither define what a dishonest participant *can do* given a specific execution platform; nor verify whether the platform's mechanisms prevent the attack.

Security has many aspects. The vulnerabilities we illustrated above fall in the class of a system not meeting its *epistemic* requirements: what participants are supposed to *know* and *believe* through rational inference and exchanging knowledge and beliefs, even in the presence of malicious participants. We argue that under an assumption of imperfect trust among participants, epistemic security properties should be a key concern for business-critical collaborations and we need to become able to analyze them in a systematic, efficient and repeatable way.

Our key insight is that, to that end, we need to treat collaborations as security protocols: the participants are principals, the platform is the trusted third party (or lack thereof), and the potential attacker goals derive from the guarantees that BPMN's security-implicit execution model takes for granted. In this paper, we make the following novel contributions.

1. We derive an epistemic threat model for BPMN collaborations and specific attacker goals from the gap between what BPMN's security-implicit execution model assumes and what execution platforms actually enforce.
2. We establish a structural correspondence between BPMN token passing semantics and multiset rewriting, enabling a systematic encoding of collaborations as security protocol theories for the security protocol analysis tool Tamarin [16].
3. We define a composable attacker model, for which platform enforcement profiles parameterize adversary capabilities and platform constraints, and provide Tamarin instantiations for three deployment platforms.
4. We report on the verification of our example across the three platforms, confirming that from a composite security point of view, none of the three platforms is "best", and discover a timing-dependent state bifurcation attack under non-atomic credential revocation for the protocol published in [12].

All artifacts are available at https://doi.org/10.5281/zenodo.20273971.

2 An Epistemic Model of BPMN Collaboration Security

In this paper, we work with a heavily simplified subset of well-formed BPMN: we admit pools (participants), simple tasks and sequence flows, start and end events; and intermediate catch/throw events with message flows. Regarding gateways, we only admit exclusive ones. This subset is sufficient for the purposes of this paper; future work will extend the set of supported elements.

Let $\mathcal{C} = (P, E, F, M)$ be a BPMN collaboration where $P = \{P_1, \ldots, P_n\}$ is the set of pools (participants), E is the set of flow elements, F is the flow relation, and M is the set of message flows. A *trace* τ is a sequence of task and gateway activations and message throw/catch events during collaboration execution. We write $\varphi@\#i$ to denote that event φ occurs at (logical) time point $\#i$ in a trace. An *exclusive gateway* $g \in E$ with outgoing branches $\{b_1, \ldots, b_k\}$ requires exactly one b_j to be taken per execution instance. We write pid for a unique process instance identifier.

2.1 Epistemic States

The BPMN 2.0 standard [1] implicitly assumes an epistemic model in which participants truthfully execute their own pool, send messages that faithfully represent their process state, and those messages cannot be intercepted or tampered with. Under this assumption, every *belief* a participant derives from a received message is also *knowledge*, because no participant would ever deceive the others.

Once adversarial behavior is admitted, the epistemic distinction between knowledge and belief becomes load-bearing. A signed message on an authenticated channel yields K_i(message is from P_j) but only B_i(P_j followed the prescribed path). The distinction is not a property of the message content, but of the delivery mechanism: channel authentication provides unforgeable evidence of sender identity, but no evidence of the sender's prior execution history. Under the VC-based orchestration model [12], a receiver can additionally verify non-revocation of execution-relevant message payload statements through querying a so-called "Verifiable Data Registry" (VDR) service, obtaining temporal knowledge $K_i^t(\varphi)$: i knows φ held at time t (if the statement is reliable), but this may change after the check.

Knowledge (K), belief (B), and temporal knowledge (K^t) are our three epistemic building blocks; the definitions below make them precise for our setting. Intuitively, participant i *knows* φ if φ holds in every execution consistent with what i has observed; i merely *believes* φ if its observations are consistent with φ but do not rule out executions where φ fails. In the running example, B *knows* its award came from A (channel authentication is unforgeable evidence of the sender) but only *believes* it is the sole awardee: B cannot distinguish the world in which A awarded only B from the one in which A also awarded C. This is the indistinguishability the Kripke structure below makes precise.

Definition 1 (Epistemic execution model). *An epistemic execution model for collaboration* $\mathcal{C}$ *is a Kripke structure* $\mathcal{M} = (W, \sim_i, V)_{i \in P \setminus P_{adv}}$ *where:* W *is the* *set of possible execution traces (worlds), including both honest and adversarial traces;* $\sim_i \subseteq W \times W$ *is participant* i*'s indistinguishability relation* ($w \sim_i w'$ *iff* i*'s observations in* w *and* w' *are identical); and* V *assigns truth values to propositions about collaboration state at each world. Then,* $K_i(\varphi)$ *holds in world* w *iff* φ *holds in every* w' *such that* $w \sim_i w'$. *We write* $B_i(\varphi)$ *when* i*'s local observations are consistent with* φ *holding; this is strictly weaker than* $K_i(\varphi)$, *which requires* φ *in every indistinguishable world.*

The BPMN standard's implicit assumption is that W contains only honest traces: no adversarial worlds are admitted, so every $w \in W$ already satisfies each guarantee φ, and $B_i(\varphi) \Rightarrow K_i(\varphi)$ holds trivially even where a participant cannot tell two honest worlds apart. Our model relaxes this by admitting adversarial traces into W, creating worlds that honest participants cannot distinguish, opening the gap between belief and knowledge that attacks can exploit. The question is: for a given execution platform, which of these implicit beliefs are actually knowledge, and which remain mere belief, exploitable by a strategically dishonest participant?

Definition 2 (Secure and Insecure Execution States). *Given epistemic execution model* $\mathcal{M}$ *and security properties* $\Phi = \{\varphi_1, \ldots, \varphi_k\}$, *a world* $w \in W$ *is secure iff for every honest participant* i *and every* $\varphi_j \in \Phi$*: if* $B_i(\varphi_j)$ *holds in* w, *then* φ_j *holds in* w. *A world is insecure if some honest participant holds a false security-relevant belief.*

The attacker seeks to reach an insecure world; the platform mechanisms constrain which worlds are reachable. Under the standard's implicit model (only honest worlds), soundness and security coincide; once adversarial traces enter W, the two separate: a sound collaboration can contain insecure worlds, and a secure collaboration can fail soundness criteria irrelevant to cross-organizational security.

2.2 The Case for Tamarin

Analyzing for this insecurity notion requires a framework that's aware of BPMN's token-based execution semantics and one that models adversarial behavior as a first-class primitive at the same time. Process verification tools such as LoLA [20] and BProVe [8] assume faithful execution and cannot express adversarial behavior.

A known structural correspondence provides the bridge to existing security protocol verification tooling. Cervesato and Scedrov [7] showed that place and transition nets embed directly into multiset rewrite systems, where places correspond to fact symbols, tokens to linear fact instances, transitions to rewrite rules, and markings to multiset states. A rule fires by consuming the facts on its left-hand side and producing those on its right, just as a Petri net transition consumes and produces tokens; the rule syntax is given in Sect. 3. The Tamarin tool [16] extends multiset rewriting with structured terms, fresh value generation, a built-in Dolev-Yao [10] adversary that controls the network (it can read, drop, and inject messages, but cannot break cryptography), and a trace logic in which the security properties of Definition 2 can be expressed directly. Each Tamarin trace corresponds to a world $w \in W$; and the prover's backward search exhaustively explores the space of adversarial strategies, determining whether any trace reaches an insecure world. Thus, we proceed with modeling security properties, platform profiles and attacker capabilities by providing epistemic definitions and the equivalent Tamarin constructs.

2.3 Security Properties

Through a semistructured analysis, we identified seven attacker goals that are relevant to the epistemic collaboration security setting (see Table 1). Further work will explore possible extensions of this model and a grounded taxonomization. We model each property as the constructive negation of an execution guarantee φ: the attacker seeks an insecure world (Definition 2) where $B_i(\varphi) \wedge \neg\varphi$ for some honest participant i.

The epistemically foundational goal is state bifurcation (G1): honest participants accept mutually exclusive process continuations, violating common knowledge $C_G(\varphi)$ (every member of group G knows φ, knows that all others know it, and so on). In the procurement example, mutual exclusion requires $C_{\{B,C\}}$("exactly one supplier selected"); when this fails, the Procurement Authority awards both suppliers simultaneously, each believing it is the sole awardee while neither knows what the other received ($\neg C_{\{B,C\}}(b_j)$). The remaining goals arise by construction from Definition 2; Table 1 summarizes all seven.

G1–G4, G6a, and G7 are safety properties with finite violating prefixes [3]; for each, the honest participant necessarily adopts the corresponding belief upon message receipt, so a violating trace witnesses both $\neg\varphi$ and $B_i(\varphi)$, reducing verification to trace analysis. G6b requires observational equivalence rather than a trace property. Whether an attacker can realize these goals depends on the execution platform's evidence mechanisms: the boundary between K and B.

G5 and G6b fall outside finite-trace safety. Selective withholding (G5) is a liveness property: the adversary can always refuse to send or act, and no protocol can force a participant to take a step (FLP impossibility), so no platform precludes it and we do not verify it per profile. Structural secrecy (G6b) concerns whether an observer can tell which exclusive branch was taken from its own observable events, without reading any payload; deciding it requires observational equivalence between two executions (diff-equivalence), not a property of a single trace.

2.4 Platform Profiles

We formalize the platform's role through an *attacker configuration* $\mathcal{A} = (P_{\text{adv}}, \Pi)$: $P_{\text{adv}} \subset P$ is the set of adversary-controlled pools ($P_{\text{adv}} \neq P$); $\Pi = (\mathcal{R}^+, \mathcal{R}^-)$ is a *platform enforcement profile* where $\mathcal{R}^+$ replaces the honest rules of pools in P_{adv} and $\mathcal{R}^-$ captures platform constraints (delivery rules,

Table 1. Attacker goals G1–G7. Each goal negates a BPMN execution guarantee; the epistemic violation instantiates $B_i(\varphi) \wedge \neg\varphi$ from Definition 2.

Goal	Name	Epistemic violation	Trace formulation
G1	State bifurcation	$\neg C_{\{P_k,P_l\}}(b_j)$: honest participants hold inconsistent gateway beliefs	$\exists$ Received events downstream of distinct XOR branches in same pid
G2	Repudiation	Attribution belief unjustified: no matching send event	Receive attributed to P_i; no Send by P_i in trace
G3	History falsification	Compliance belief false: sender skipped prerequisites	Send by P_i without preceding Act event
G4	Unauth. progression	Origin belief false: no send event exists	Honest advance on message with no Send
G5	Selective withholding	Uniformly B by FLP impossibility; outside safety verification scope	Liveness property; no platform can force a participant to act
G6a	Value secrecy	$K_{\text{adv}}(v)$: adversary learns unaddressed message content	P_{adv} obtains value from flow where $P_{\text{adv}} \notin \text{recv}(M)$
G6b	Structural secrecy	Requires observational equivalence; analytically violated under all profiles where the adversary's observable events differ between branches	Diff-equivalence, not a trace property
G7	Instance confusion	Instance binding false: artifact from different instance accepted	Message from pid_1 accepted in pid_2

trace-level restrictions) that bind regardless of participant honesty. The adversarial theory is:

$$\mathcal{T}_{\mathcal{A}} = \left(\mathcal{T}(\mathcal{C}) \setminus \bigcup_{P_i \in P_{\text{adv}}} R_{P_i}\right) \cup \mathcal{R}^{+} \cup \mathcal{R}^{-}$$

Two diagnostic profiles validate the methodology. Π_0 (all pools honest, secure channels) confirms encoding correctness: all lemmas pass. Π_{DY} (full Dolev-Yao adversary, trivial $\mathcal{R}^{-}$) establishes a lower bound: all safety lemmas fail uniformly, confirming that platform constraints are necessary for every guarantee. Table 2 summarizes the three substantive profiles. Full Tamarin rules for all profiles are provided in the verification artifact.

Table 2. Platform enforcement profiles at a glance. $\mathcal{R}^+$ specifies adversary capabilities; $\mathcal{R}^-$ specifies platform-enforced constraints.

Profile	$\mathcal{R}^+$ (adversary)	$\mathcal{R}^-$ (platform)	Channel	$B_i(\varphi) \Rightarrow K_i(\varphi)$?
Π_{LC}	Any message under own identity; no model compliance; independent sends	Authenticated channel; sender signature only	Authenticated	Identity: K. All else: B.
Π_{SC}	Submit transactions; choose valid transitions; observe all state	Contract validates model; linear state (atomicity)	Contract-mediated	Safety (G1–G4, G7): K. Confidentiality: **lost**.
Π_{VC}	Own VCs; choose timing; selective disclosure	Credential chain + NiTC non-revocation; VDR revocations	Credential-mediated	Identity: K. History: B. Gateway: K^t / B.

The last column records which implicit beliefs each platform turns into knowledge: "Identity: K" means a receiver knows who sent a message, while "All else: B" leaves every other guarantee as belief, and so open to attack. In effect, it states which of BPMN's assumed guarantees the platform backs with evidence.

Π_{LC}: Loose Coupling. Participants exchange messages over authenticated point-to-point channels with no shared execution logic. The adversary's $\mathcal{R}^+$ consumes no linear precondition, so nothing enforces mutual exclusion: the adversary can send to any subset of recipients independently. The authenticated channel ($\mathcal{R}^-$) guarantees sender identity is K (G2, G4 hold); gateway exclusivity (G1), history compliance (G3), and instance binding (G7) remain B. Here $\mathcal{R}^+$ is a single rule that consumes no token and emits a MsgOut to a chosen recipient under the adversary's authenticated identity; with no shared precondition to deplete, it can fire once per recipient, modeling an award sent to both B and C.

Π_{SC}: Smart Contract Orchestration. Smart contract orchestration [21] deploys collaboration logic on-chain: a contract validates every state transition against the BPMN model, consuming a linear state fact to enforce atomicity. The adversary's remaining freedom is choosing which valid transition to take and observing all on-chain state. All safety properties (G1–G4, G7) hold; confidentiality (G6a, G6b) fails, since the adversary observes the full execution trace. We model the transparent case (e.g. permissionless or fully replicated ledgers), in which every participant observes all on-chain state; this is the worst case for confidentiality. Permissioned ledgers with channel- or collection-level access control (e.g. private data collections) can restrict who observes which payloads, mitigating the G6 loss; the safety guarantees (G1–G4, G7) derive from the shared validated state and are unaffected.

Π_{VC}: VC-Based Self-Orchestration. This profile encodes the Verifiable Credential based [15] collaboration protocol of [12]. Before execution, each recipient issues one non-invocation token credential (NiTC) per message flow, registered as unrevoked in a Verifiable Data Registry (VDR). The adversary consumes no linear state at the sender side. The protocol ($\mathcal{R}^-$) proceeds in five steps:

1. **Declaration.** The sender submits a presentation bundling the message payload and the NiTC for the target flow.
2. **Verification.** The receiver checks that the NiTC is unrevoked, confirming the flow has not been previously invoked.
3. **Bifurcation check.** For each exclusive alternative, the receiver verifies that the alternative's NiTC is also unrevoked.
4. **Commitment.** The receiver revokes the current NiTC by consuming the linear unrevoked fact, irreversibly committing to this path.
5. **Acceptance.** The receiver accepts the message; the protocol issues an updated credential attesting to the completed transmission.

The VDR stores only the revocation status of each NiTC (one unrevoked flag per message flow), not the message payloads. A receiver's evidence of exclusivity is thus the non-revocation of the competing branch's NiTC, not A's send: at the bifurcation check (step 3), B inspects C's NiTC and C inspects B's, each confirming that the alternative branch has not committed rather than re-examining A's declaration.

The critical variable is VDR atomicity, which determines whether G1 evidence constitutes K or K^t. Under atomic VDR, the bifurcation check at step 3 requires that no revocation of alternative NiTCs exists at *any* time point; under non-atomic VDR, only that none exists *before* the check. This single temporal conjunct is crucial: the timing gap between steps 3 and 4 creates a TOCTOU (Time-of-Check to Time-of-Use) vulnerability under non-atomic VDR, analyzed later.

3 Theory Encoding and Verification

Figure 2 illustrates the overall workflow encoding, theory composition and verification. A Tamarin theory $\mathcal{T}$ is a set of rewriting rules $[l_1, \ldots, l_m]{-}[a_1, \ldots, a_k]{\rightarrow}[r_1, \ldots, r_n]$, where each l_i, a_j, r_i is a *fact* (a predicate applied to terms). Premises l_i are consumed from a global multiset state; conclusions r_i are added to it; action facts a_j are recorded on the trace without affecting the state. Facts are *linear* by default—each instance is consumed exactly once—modeling resources such as tokens. Facts prefixed with ! are *persistent* and remain available indefinitely. Security properties are first-order formulas universally quantified over the resulting action trace. For properties not expressible as trace formulas—such as whether an adversary can distinguish which exclusive branch was taken—Tamarin provides a *diff-equivalence* mode that checks observational equivalence between two protocol variants.

The structural correspondence of Sect. 2.2 provides the formal basis: the encoding is a (currently) manual, pattern-based translation of a core BPMN subset (sequence flows, exclusive gateways, message throw/catch events, process

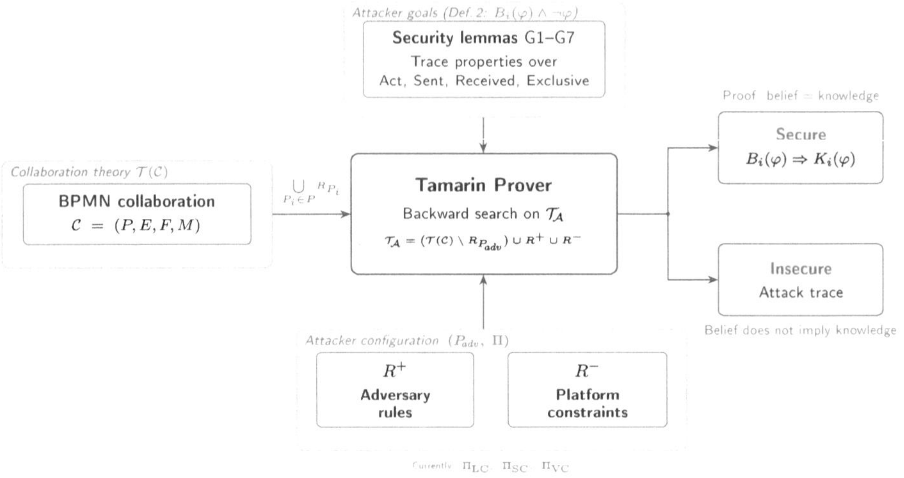

Fig. 2. From BPMN collaboration to adversarial Tamarin theory. A BPMN collaboration model is encoded as MSR rules (one rule set per pool, coupled by message facts). The attacker configuration selects an adversary pool and a platform enforcement profile $\Pi = (\mathcal{R}^+, \mathcal{R}^-)$: $\mathcal{R}^+$ replaces the adversary pool's honest rules, $\mathcal{R}^-$ constrains delivery. The configured theory is passed to Tamarin together with the security lemma under scrutiny. Tamarin returns either a proof (*Secure*) or a concrete attack trace (*Insecure*).

instantiation) into Tamarin's multiset rewriting (MSR) formalism. The Tamarin theory of a collaboration $\mathcal{C} = (P, E, F, M)$ is

$$\mathcal{T}(\mathcal{C}) = \bigcup_{P_i \in P} R_{P_i}$$

where R_{P_i} is the rewriting rule set encoding pool P_i's execution. Pools interact exclusively through message facts: MsgOut produced by throw rules is transformed into MsgIn by the platform's delivery mechanism model.

We write $L \xrightarrow{A} R$ for a rewriting rule consuming facts L, emitting action facts A, and producing facts R. Three representative patterns illustrate the translation:

$$\begin{aligned}
\textit{Seq. flow:}\quad & \mathrm{St}(\mathrm{pid}, P_i, e_1) \xrightarrow{\mathrm{Act}(\mathrm{pid}, P_i, P_i, e_1)} \mathrm{St}(\mathrm{pid}, P_i, e_2)\\
\textit{Msg. throw } (P_i \to P_j)\textit{:}\quad & \mathrm{St}(\mathrm{pid}, P_i, e_{\mathrm{th}}) \xrightarrow{\mathrm{Act,\ Sent}} \mathrm{St}(\mathrm{pid}, P_i, e_{\mathrm{next}}),\ \mathrm{MsgOut}(\mathrm{pid}, P_i, P_j, m)\\
\textit{Excl. gateway:}\quad & \mathrm{St}(\mathrm{pid}, P_i, g) \xrightarrow{\mathrm{Act}(\mathrm{pid}, P_i, g, b_j)} \mathrm{St}(\mathrm{pid}, P_i, b_j)
\end{aligned}$$

Each rule can be read as a token firing: the premise St fact is the token on a sequence flow, the rule fires the corresponding element, and the conclusion places the token on the outgoing flow. The message-throw rule additionally emits a MsgOut fact onto the message flow, which the platform's delivery model turns into a MsgIn at the receiver.

The remaining constructs (message catch, process instantiation) follow the same pattern; complete rule sets are provided in the artifact repository. Since every St fact is tagged with P_i, rules in distinct pools consume disjoint state facts, and the theory decomposes along pool boundaries.

The initialization rule emits two classes of static model annotations derived from $\mathcal{C}$ by static graph analysis. Exclusive(m_k, m_l) marks payloads downstream of distinct branches of an exclusive gateway as mutually exclusive, enabling the G1 lemma; by construction, it is emitted only for distinct messages ($m_k \neq m_l$). Secret(v) marks fresh values on message flows whose intended recipient is not in P_{adv}, enabling the G6a lemma. Both are platform-independent properties of $\mathcal{C}$; the same annotations and lemmas are checked against every profile, so cross-profile comparison tests enforcement, not annotation.

The security lemmas quantify over two classes of action facts: *runtime facts* Act (activity execution), Sent (message dispatch), and Received (message acceptance); and *static annotations* Exclusive and Secret. The translation preserves token consumption (linearity of St), gateway mutual exclusion (k rules sharing one linear precondition), message causality (MsgOut/MsgIn ordering), and instance isolation (fresh pid generation); each follows from MSR semantics.

Each goal from Table 1 is expressed as a Tamarin lemma universally negating the corresponding violation pattern. The safety lemmas (G1–G4, G7) and G6a are trace properties verified via backward search; G6b requires diff-equivalence (Sect. 2.3). The full lemmas are provided in the verification artifact.

For a given attacker configuration $\mathcal{A} = (P_{\text{adv}}, \Pi)$, automated verification yields either a proof or a concrete counterexample trace.

4 Verification Results

We instantiate the framework on the motivating example and verify the security lemmas under each platform enforcement profile. The Tamarin theories and reproduction scripts for this paper are archived at https://doi.org/10.5281/zenodo.20273971.

4.1 Setup and Results

The adversary configuration is $\mathcal{A} = (\{A\}, \Pi)$ for each profile $\Pi \in \{\Pi_{\text{LC}}, \Pi_{\text{SC}}, \Pi_{\text{VC}}\}$. The Procurement Authority controls the main exclusive gateway, producing the sharpest cross-profile differentiation and the scenario where the TOCTOU result emerges. For each analysis, A's honest rules are replaced by the profile's adversary rules (Sect. 2.4) and the corresponding delivery rules instantiated. All seven goals are evaluated under every profile (Table 3): G1–G4, G6a, and G7 by Tamarin, G5 and G6b analytically; the comparison thus covers the full goal set, not bifurcation alone.

Diagnostic profiles confirm the encoding: under Π_0 all properties hold; under Π_{DY} all fail. Table 3 presents both the epistemic predictions from the profile analysis (Sect. 2.4) and the Tamarin verification outcomes.

The Π_{LC} and Π_{SC} columns confirm the epistemic analysis of Sect. 2.4. Under loose coupling, channel authentication suffices for non-repudiation (G2) and

Table 3. Verification results ($P_{\text{adv}} = \{A\}$). ✓ = Tamarin proof; × = Tamarin counterexample. [a] = analytical argument;—= subsumed (G6b under Π_{SC}: full state visibility makes structural secrecy vacuous); † = Sect. 4.2.

Goal	Property	Π_{LC}	Π_{SC}	Π_{VC} (atomic)	Π_{VC} (non-at.)
G1	No-bifurcation	×	✓	✓	×[†]
G2	Non-repudiation	✓	✓	✓	✓
G3	Faithful history	×	✓	✓	×
G4	Auth. progression	✓	✓	✓	✓
G5	Process progress	B[a]	B[a]	B[a]	B[a]
G6a	Value secrecy	×	×	✓	✓
G6b	Structural secrecy	×[a]	—	×[a]	×[a]
G7	Instance isolation	×	✓	✓	✓

sender existence (G4) but provides no exclusion proof (G1), execution trace enforcement (G3), or instance binding (G7). Under smart contract orchestration, atomic on-chain state enforces all safety properties; the public ledger destroys both value and structural secrecy (G6a, G6b). Π_{SC} is the only profile where all safety properties hold, and the only one where confidentiality completely fails. G5 is uniformly B across all profiles: no platform can force a participant to act, an instance of FLP impossibility in the BPMN domain.

The Π_{VC} columns reveal the section's central finding: the atomic and non-atomic variants produce identical results *except* for G1 and G3. The G1 discrepancy, from ✓ under atomic VDR to ×[†] under non-atomic, is the TOCTOU attack analyzed next.

4.2 The TOCTOU Attack Under Non-atomic VDR

The VC-based approach does not constitute a single enforcement profile. Whether the VDR processes revocations atomically or with propagation delay produces two distinct configurations that share identical $\mathcal{R}^+$ and differ only in the temporal semantics of a single restriction in $\mathcal{R}^-$. This difference flips two goals: G1 (no-bifurcation) and G3 (faithful history) move from ✓ to ×.

Under atomic VDR, the bifurcation check at step 3 of the presentation submission requires that no *Revoked*(nitc_id) event exists at *any* time point: the check sees the complete VDR state. Tamarin proves the no-bifurcation lemma. The NiTC mechanism establishes negative knowledge, $K_B(\neg$"C's path was taken"$)$, via the non-revocation proof; this knowledge is genuine because the evidence reflects the global state.

Under non-atomic VDR, the check requires only that no revocation exists *before* the check ($\#k < \#i$). Revocations concurrent with or after the check are invisible. This single temporal weakening opens a timing window between step 3 (check) and step 4 (commitment) that Tamarin exploits as a **TOCTOU attack:**

1. Adversarial A initiates the protocol (step 1) with both B and C simultaneously.
2. B reaches the bifurcation check (step 3): C's NiTC has not been revoked, because C's run has not yet reached commitment (step 4).
3. C reaches the bifurcation check (step 3): B's NiTC has not been revoked, because B's run has not yet reached step 4.
4. Both checks pass. Both proceed to commitment (step 4) and acceptance (step 5).
5. Bifurcation succeeds: B and C each believe they are the sole awardee.

The encoding follows check-before-commit ordering (step 3 before step 4), the strongest variant a protocol designer could implement. The attack persists, confirming the vulnerability is inherent to non-atomic VDR semantics, not a protocol design choice.

Epistemic Reading. $K_B^{t_{\text{check}}}(\neg\text{bifurcated}) \wedge \neg K_B^{t_{\text{commit}}}(\neg\text{bifurcated})$. At check time, B's non-revocation proof was valid: genuine cryptographic evidence that no bifurcation had occurred. By commit time, C's concurrent run has not yet made its commitment visible, rendering the evidence stale. Knowledge degrades from K to B across the temporal gap, the K^t phenomenon from Sect. 2.1 made concrete. VDR atomicity is necessary and sufficient for bifurcation prevention under Π_{VC}: the infrastructure assumption, not the protocol logic, determines whether the receiver's evidence constitutes knowledge or belief.

4.3 Synthesis

The verification confirms the epistemic predictions of Sect. 2.4 and adds one finding the profile analysis anticipated but could not mechanically verify: the TOCTOU vulnerability under non-atomic VDR. No profile dominates: Π_{SC} satisfies G1–G4 and G7 but fails G6a; Π_{VC} preserves G6a but introduces timing-dependent evidence (K^t); Π_{LC} satisfies only G2 and G4. No profile achieves K for both history verification (G3) and confidentiality (G6) simultaneously; this is a structural tension the framework surfaces but cannot resolve. The $K/B/K^t$ distinction recasts the TOCTOU result as a precise degradation pattern: evidence that was genuine at observation ($K_B^{t_{\text{check}}}$) becomes stale belief by commit time, wherever platform evidence has propagation delay (e.g., transaction reordering under Π_{SC}). Security depends on the conjunction of business logic, platform guarantees, and infrastructure: changing one parameter (VDR atomicity) flips two goals (G1, G3) while the rest stay fixed, so analyzing any single dimension in isolation is insufficient.

5 Discussion

Our analysis on a minimalistic BPMN subset constitutes a *lower bound* on the attack surfaces: richer constructs expand the set of adversarial strategies (e.g., timer events introduce timing manipulation; parallel forks introduce ordering manipulation), so identified vulnerabilities persist in any superset. Goals G1–G7 are not claimed to be exhaustive for all of BPMN; they capture the primary attack classes for the supported subset. The derivation methodology of Sect. 2.3 extends naturally as constructs are added, with each new construct yielding execution guarantees whose negation defines new attacker goals.

The single-adversary assumption ($|P_{\text{adv}}| = 1$) is standard in security protocol verification; the framework supports collusion by replacing multiple pools' honest rules with coordinated adversary rules, but the combinatorial growth ($\binom{n}{k}$ configurations) makes exhaustive analysis questionable and is deferred to future work. The BPMN-to-Tamarin encoding is currently manual, but highly automatable. Tamarin termination is not guaranteed in general; for the running example, all models terminated within an encouraging ~42 s on a modern 12 core x86-64 CPU with 96 GB of RAM. The Π_{SC} safety results are conditional on honest consensus, fair transaction ordering, and finalization permanence; extending Π_{SC}'s $\mathcal{R}^+$ to model a miner-adversary with reordering capability remains future work.

Three properties are inherent boundaries of process-level security analysis, not shortcomings of the framework. Selective withholding (G5) is achievable on every platform (Table 1). Information extraction (G6) is trivially achievable under Π_{SC} because the public ledger destroys both value and structural confidentiality; this is an inherent property of transparent execution. History falsification (G3) for internal activities remains at B under both Π_{LC} and Π_{VC}: even Verifiable Credentials prove only that a claim was *issued*, not that the underlying activity was *executed*. These bounds are themselves results of the analysis, delimiting what platform enforcement can and cannot guarantee.

6 Related Work

6.1 BPMN Security Modeling and Formal Verification

BPMN security extensions [14,18,19], and the threat representation of Meland and Gjære [17], let designers annotate models with security and privacy requirements, but assume an external or unspecified attacker and do not model what a dishonest participant controlling their own pool can achieve. Formal verification assumes the complementary extreme: van der Aalst's soundness [2], BProVe [8], and LoLA [20] verify correctness properties under an implicit, total honest-execution assumption with no adversary of any kind. Our approach is complementary: soundness verification corresponds to Π_0, the baseline where all participants follow the model, and the central thesis is that soundness $\neq$ security, since a sound collaboration can be insecure (Sect. 2.1). The structural correspondence between Petri nets and multiset rewriting (Cervesato and Scedrov [7]) connects both lines of work at the formal level; Tamarin extends verification from correctness to security by introducing an adversary, not by replacing the underlying formalism.

6.2 Security Protocol Verification

The Tamarin prover [16] and ProVerif [5] are the principal contemporary tools for security protocol verification, applied to TLS, OAuth, Kerberos, electronic voting, and 5G. In all existing applications, the verified protocol is a communication protocol between principals; the connection to business process models, where the protocol structure derives from a BPMN collaboration rather than being hand-specified, has not been made. We bridge this gap: the structural correspondence (Sect. 2.2) provides the formal basis, and the encoding procedure realizes it as a Tamarin theory. Here BPMN is more than a syntactic front-end: the attacker goals (G1–G7) and the per-pool rule decomposition both derive from BPMN's token-passing semantics, so we analyze the models BPM practitioners actually build, and the encoding is meant to integrate with existing BPM tooling rather than to re-specify each protocol by hand.

6.3 Choreographies and Epistemic Reasoning

Multiparty session types [13] and choreography conformance [4] classify protocol deviations syntactically (type mismatch, missing message, extra message), not whether a deviation *succeeds* given platform enforcement. Our goals G1–G7 provide the missing semantic layer: they classify what the adversary achieves, not which rule is broken, and the profile formalism makes the platform-dependence of deviation success explicit and verifiable. Fagin, Halpern, Moses, and Vardi [11] establish epistemic foundations for distributed systems; BAN logic [6] pioneered epistemic reasoning about authentication protocols. Neither tradition has been connected to business process collaborations; our epistemic execution model (Sect. 2.1) bridges this gap. The connection to Byzantine fault tolerance is direct, since BFT equivocation is precisely G1, but BFT's fault model cannot express strategic, goal-directed deception, nor does it address confidentiality or accountability [9].

7 Conclusion

This paper proposed that multiparty collaborations with adversarial participants are best analyzed as security protocols, formalized through an epistemic execution model that derives attacker goals from BPMN's implicit knowledge assumptions and encodes them as Tamarin theories via the structural correspondence between token semantics and multiset rewriting. The profile formalism parameterizes this analysis: each execution platform maps to a pair $(\mathcal{R}^+, \mathcal{R}^-)$ that determines both the adversary's capabilities and the platform's enforcement guarantees.

Verification across three profiles confirmed that platform capabilities are structurally incomparable and revealed a timing-dependent bifurcation attack for one of the platforms considered. The attack exemplifies temporal knowledge degradation ($K^t \rightarrow B$), a general pattern wherever platform evidence has a propagation delay. Platform choice is a security design decision whose trade-offs the profile formalism makes formally precise.

The BPM community has mature tools for correctness verification. Security protocol verification, sharing multiset rewriting foundations with Petri net analysis, is a natural and powerful complement for epistemic security analysis.

Immediate extensions to our work include an automated BPMN-to-Tamarin translator and enrichment of the supported BPMN subset with parallel gateways, timer events, and data-aware processes. Multi-adversary configurations ($|P_{\text{adv}}| > 1$) will enable analysis of collusion strategies, where coordinated adversaries combine individually achievable attack goals to produce new vulnerabilities.

Acknowledgments. The work of Martin Farkas supported by the Doctoral Excellence Fellowship Programme (DCEP) is funded by the National Research Development and Innovation Fund of the Ministry of Culture and Innovation and the Budapest University of Technology and Economics.

Disclosure of Interests. The authors have no competing interests to declare.

References

1. Business Process Model and Notation. 2.0.2 (2014)
2. Aalst, W.M.P.: Verification of workflow nets. In: Azéma, P., Balbo, G. (eds.) ICATPN 1997. LNCS, vol. 1248, pp. 407–426. Springer, Heidelberg (1997). https://doi.org/10.1007/3-540-63139-9_48
3. Alpern, B., Schneider, F.B.: Defining liveness. Inf. Process. Lett. **21**(4), 181–185 (1985)
4. Barbanera, F., Lanese, I., Tuosto, E.: Choreography Automata. In: Bliudze, S., Bocchi, L. (eds.) COORDINATION 2020. LNCS, vol. 12134, pp. 86–106. Springer, Cham (2020). https://doi.org/10.1007/978-3-030-50029-0_6
5. Blanchet, B.: Modeling and verifying security protocols with the applied pi calculus and ProVerif. Found. Trends® Priv. Secur. **1**(1-2), 1–135 (2016)
6. Burrows, M., Abadi, M., Needham, R.M.: A logic of authentication. Proceedings of the Royal Society of London. A. Math. Phys. Sci. **426**(1871), 233–271 (1989)
7. Cervesato, I., Scedrov, A.: Relating state-based and process-based concurrency through linear logic (full-version). Inf. Comput. **207**(10), 1044–1077 (2009)
8. Corradini, F., Fornari, F., Polini, A., Re, B., Tiezzi, F., Vandin, A.: A formal approach for the analysis of BPMN collaboration models. J. Syst. Softw. **180**, 111007 (2021)
9. van Ditmarsch, H., Fruzsa, K., Kuznets, R., Schmid, U.: A Logic for Repair and State Recovery in Byzantine Fault-Tolerant Multi-agent Systems (2024)
10. Dolev, D., Yao, A.: On the security of public key protocols. IEEE Trans. Inf. Theory **29**(2), 198–208 (1983)
11. Fagin, R., Halpern, J.Y., Moses, Y., Vardi, M.: Reasoning About Knowledge, The MIT Press (1995)
12. Farkas, M., Péter, B.Z., Kocsis, I.: A Self-orchestration Model for Business Collaborations with Verifiable Process History Credentials. In: Senderovich, A., Cabanillas, C., Vanderfeesten, I., A. Reijers, H. (eds.) Business Process Management

Forum 2025, pp. 58–74. Springer, Cham (2026). https://doi.org/10.1007/978-3-032-02929-4_4
13. Honda, K., Yoshida, N., Carbone, M.: Multiparty Asynchronous Session Types. J. ACM **63**(1), 1–67 (2016)
14. Köpke, J., Meroni, G., Salnitri, M.: Designing secure business processes for blockchains with SecBPMN2BC. Futur. Gener. Comput. Syst. **141**, 382–398 (2023)
15. Sporny, M., Chadwick, D., Longley, D.: Verifiable credentials data model v2.0 (2024)
16. Meier, S., Schmidt, B., Cremers, C., Basin, D.: The TAMARIN Prover for the Symbolic Analysis of Security Protocols. In: Sharygina, N., Veith, H. (eds.) CAV 2013. LNCS, vol. 8044, pp. 696–701. Springer, Heidelberg (2013). https://doi.org/10.1007/978-3-642-39799-8_48
17. Meland, P.H., Gjære, E.A.: Representing Threats in BPMN 2.0. In: 2012 Seventh International Conference on Availability, Reliability and Security, pp. 542–550. IEEE, Prague, Czech Republic (2012)
18. Pullonen, P., Tom, J., Matulevičius, R., Toots, A.: Privacy-enhanced BPMN: enabling data privacy analysis in business processes models. Softw. Syst. Model. **18**(6), 3235–3264 (2019)
19. Salnitri, M., Dalpiaz, F., Giorgini, P.: Designing secure business processes with SecBPMN. Softw. Syst. Model. **16**(3), 737–757 (2017)
20. Schmidt, K.: LoLA A Low Level Analyser. In: Nielsen, M., Simpson, D. (eds.) ICATPN 2000. LNCS, vol. 1825, pp. 465–474. Springer, Heidelberg (2000). https://doi.org/10.1007/3-540-44988-4_27
21. Weber, I., Xu, X., Riveret, R., Governatori, G., Ponomarev, A., Mendling, J.: Untrusted Business Process Monitoring and Execution Using Blockchain. In: La Rosa, M., Loos, P., Pastor, O. (eds.) BPM 2016. LNCS, vol. 9850, pp. 329–347. Springer, Cham (2016). https://doi.org/10.1007/978-3-319-45348-4_19

Customizable Declarative Process Specification Scenarios via Automated Planning

Alessandra Giovannetti[1(✉)], Giacomo Acitelli[1], Angelo Casciani[1], Claudio Di Ciccio[2], and Andrea Marrella[1]

[1] Sapienza University of Rome, Rome, Italy
{a.giovannetti,acitelli,casciani,marrella}@diag.uniroma1.it
[2] Utrecht University, Utrecht, The Netherlands
c.diciccio@uu.nl

Abstract. Declarative specifications define the behaviour of processes by means of temporal rules exerting dependencies between the occurrence of activities. Those rules can constitute the system requirements and normative boundaries within which the workflow is framed. The multitude of rules and their interrelations, though, make it reportedly difficult for users to linearly figure out unfoldings that depict the admitted behaviour in the form of explicative scenarios. While simulation engines can produce example traces of customizable length and in varying numerosity, these state-of-the-art tools adopt random approaches that do not permit the selection of a subset of runs exhibiting characteristic traits. In other words, simulations could produce traces that show common behavior, leaving out significant parts of the possible execution paths, and thus substantially hindering a proper validation of specifications. To mitigate this issue, we propose a principled solution that leverages automated planning to produce execution scenarios depicting the specified behavior according to desirable execution costs, diversifying the enacted behavior, or matching a given subsequence of steps. We demonstrate its efficiency and efficacy through experiments demonstrating its scalability and superiority in terms of behavioral coverage with respect to the state-of-the-art simulation tool for declarative process specifications.

Keywords: Automated planning · Simulation · Process science · Declare

F. M. Maggi et al. (Eds.): BPM 2026, LNCS 16936, pp. 128–147, 2027.
https://doi.org/10.1007/978-3-032-37777-7_10

1 Introduction

Declarative process specifications express conditions for the execution of activities based on given conditions. An example of such rules is PRECEDENCE (Send payment request, Fill in payment information), dictating that if payment information is filled in, then a payment request must have been sent at an earlier stage of the process. These specifications thus enable the definition of process behavior by means of rules that dictate what can, or cannot, be done, without explaining how [37]. Therefore, they are naturally apt for setting the boundaries to a process behavior [17], akin to the requirement specifications that are of interest to reactive systems in software engineering (SE) [27, Chap. 2] and more specifically specifications from a functionality standpoint [40]. Taking inspiration from the SE literature, then, process specifications can be considered as a means to elicit, or *discover* in more Business Process Management (BPM) words [20, Chap. 5], the behavior to be ultimately enacted in a process-aware information system based on domain expertise [1,2,36].

However, the information modeling lifecycle in SE [23] and the quality assurance framework in BPM [20] remind us that the specification should not only be sound, which can be verified through automated techniques, but also validated, i.e., properly representing the desiderata of the domain experts and stakeholders once turned from a universe of discourse to formalized behavioral rules. For validation, declarative specifications alone are reportedly not suitable, because of their intrinsic visual complexity and the demanded cognitive effort to bear in mind multiple rules at once [21,26]. As a more effective approach, research work in BPM has proposed to let the domain experts validate and iteratively refine a specification through exemplary runs that it would accept [38,42,43].

To this end, research in SE resorts to *scenarios* [39], namely step-by-step executions derived from the provided specifications. Such scenarios are, among others, at the core of a well-known methodology for the specification of reactive systems releasing the user from handling formal representations, the play-in/play-out approach [28].

BPM research has provided noticeable contributions in the simulation of declarative process specifications [9,14,19], mainly with the aim of generating event logs to be later processed via process mining. The outputs are thus designed to incorporate a diverse array of possible traces, broad enough to cover an extensive spectrum of execution paths and put mining systems to the test in their capacity to rediscover the original models (the quest for "perfect logs" [9]). Scenarios, instead, serve a different purpose: to exemplify a few runs that are representative of the typical expected process behaviour.

In this paper, we propose an approach based on automated planning [24] tailored for the generation of declarative process specification scenarios. Figure 1 provides a graphical sketch of our investigation, pertaining to the encoding of specification constraints into a planning problem, and the generation of scenarios. To support scenario customization, our solution allows the user to optionally specify the patterns the scenarios should match (e.g., the occurrence of certain tasks or a sequence thereof). Also, we let the user indicate maximum execution costs for scenarios, their ranking, and numerosity. We experimentally demonstrate that our solution is applicable to real-world specifications and that the scenario generation problem remains tractable with inputs of growing size.

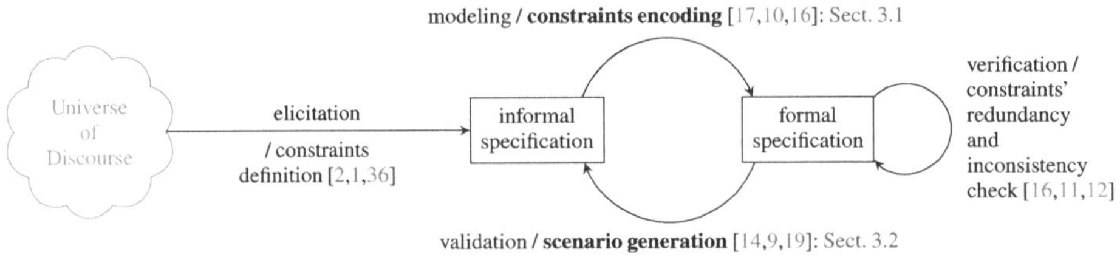

Fig. 1. The process specification cycle, reinterpreted from the *information modeling lifecycle* [23]. The phases considered in this paper are highlighted in bold.

The remainder of the paper is organized as follows. Section 2 outlines the core notions underpinning our work and provides a running example scenario from the public administration domain. Section 3 presents our automata-based manipulation technique and shows how to reduce it to a planning problem. In Sect. 4, we evaluate the scalability of our technique. In Sect. 5, we discuss related work. Finally, in Sect. 6 we conclude the paper, pinpoint the technique's limitations, and outline future work.

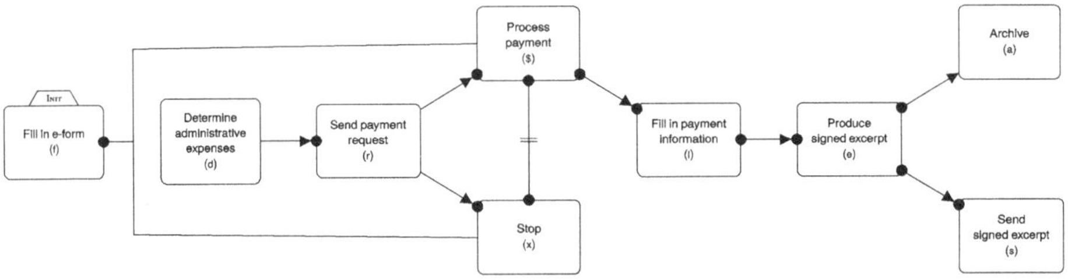

Fig. 2. A DECLARE specification, based on the case study of Schunselaar et al. [37].

2 Running Example and Background

Figure 2 illustrates the running example and motivating scenario for our paper. It is inspired by the real-world specification introduced by Schunselaar et al. for their work on configurable process behavior [37]. The represented process excerpt pertains to the management of requests from the civil registration in a municipality. In particular, it denotes a *compliance model*, namely a set of rules stemming from expert knowledge, and made configurable to apply those in different contexts. The process behavior is there described with DECLARE, a declarative specification language for the representation of (temporally extended) constraints that must be satisfied throughout the process execution [17]. Table 1 illustrate some of those typical constraints. A DECLARE specification consists of a set of activities involved in a process and a collection of constraints defined over such activities. The orderings of activities are implicitly specified by these constraints, and anything that does not violate them is possible during execution.

Table 1. DECLARE constraints involved in the running example of Fig. 2

Notation	Constraint	Explanation
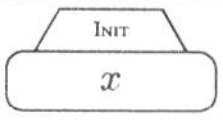	INIT(x)	x is the *first* to occur in a trace
	RESPONDEDEXISTENCE(x,y)	If x occurs in the trace, then y occurs as well
	RESPONSE(x,y)	If x occurs, then y eventually occurs after x in the trace
	PRECEDENCE(x,y)	y occurs only if preceded by x in the trace
	SUCCESSION(x,y)	x occurs if and only if it is followed by y
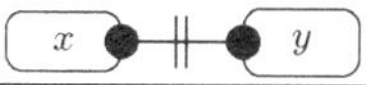	NOTCOEXISTENCE(x,y)	x and y cannot both occur in the same trace

The case study in Fig. 2 describes a municipal process beginning with the filling of a form (INIT (f)). Other necessary passages are the payment ($) or the stop (x) of the procedure. We express this aspect with a so-called branched constraint [15]: RESPONDEDEXISTENCE (f, {$, x}), which means that if f occurs, $ *or* x should occur in the same trace. The NOTCOEXISTENCE ($, x) constraint additionally imposes that f and x are mutually exclusive, though. To take place, they require the sending of a payment request (PRECEDENCE (r, $), PRECEDENCE (r, x)) and, in turn, the payment request necessitates that administrative expenses are previously determined (PRECEDENCE (d, r)). If the payment is processed, then payment information should be filled in afterwards (SUCCESSION ($, i)), and then a signed transaction excerpt produced (SUCCESSION (i, e)). Finally, the signed excerpt should later be archived (RESPONSE (e, a)) and sent via mail (SUCCESSION (e, s)).

Schunselaar et al. [37] presented an approach to automatically derive those models out of configurable specifications, in order to apply those models, upon adaptation, in different contexts (municipalities). Later research contributions showed that we can verify that those rules are not contradictory or redundant [11, 12, 16]. However, how can domain experts verify whether the rules adhere to the expected behaviour, even though they are satisfiable? Next, we describe the background information that our approach leverages to generate possible scenarios depicting the allowed unfoldings.

2.1 Representing Declarative Constraints with DECLARE

DECLARE constraints are template instantiations that define parameterized classes of properties. For example, the RESPONSE (a, b) template specifies that when a occurs, then b should eventually occur after a in a process execution.

Every DECLARE constraint has a formal semantics based on Linear Temporal Logic over finite traces (LTL_f) [13], unambiguously characterising the notion of conforming process execution to a DECLARE specification. Since LTL_f is interpreted on finite execution traces, every LTL_f formula (hence every DECLARE constraint) can be captured as a *Deterministic Finite State Automaton* (DFA) recognizing all and only its conforming traces [13]. Formally, a DFA is a finite-state machine that accepts or rejects a given trace, by running it through a state sequence uniquely determined by the sequence of activities included in the trace [3]. Deterministic refers to the uniqueness of the computation run. In the following, we call *constraint automaton* a DFA used to encode a DECLARE constraint.

Definition 1 (Constraint Automaton). *A constraint automaton is a DFA* $\mathcal{A} = (\Sigma, Q^{\mathcal{A}}, q_0^{\mathcal{A}}, \delta^{\mathcal{A}}, F^{\mathcal{A}})$, *where:*

- Σ *is the* input alphabet*;*
- $Q^{\mathcal{A}}$ *is the (non-empty) finite set of* automaton states*;*
- $q_0^{\mathcal{A}} \in Q^{\mathcal{A}}$ *is the* initial state*;*
- $\delta^{\mathcal{A}} : Q^{\mathcal{A}} \times \Sigma \to Q^{\mathcal{A}}$ *is the* transition function*;*
- $F^{\mathcal{A}} \subseteq Q^{\mathcal{A}}$ *is the set of* final states.

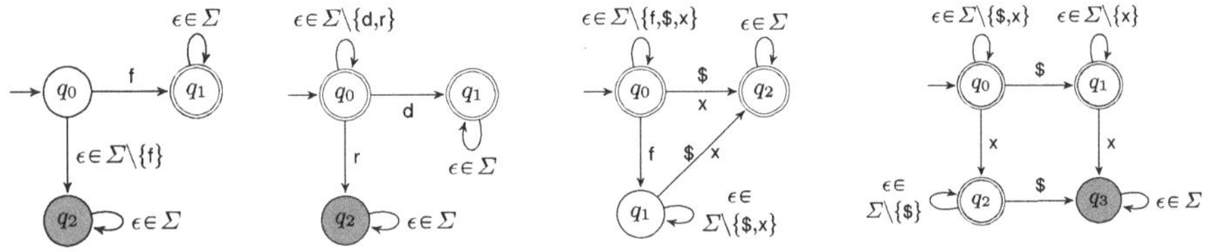

(a) INIT(f) (b) PREC.(d,r) (c) RES.EX(f,{$,x}) (d) NOTCOEX.($,x)

Fig. 3. Example DFAs of DECLARE constraints.

Figure 3 depicts four DFAs for the encoding of INIT (f), PRECEDENCE (d, r), RESPONDEDEXISTENCE (f, {$, x}) and NOTCOEXISTENCE ($, x), which are part of the DECLARE specification introduced in the running example. In the remainder of the paper, in DFA diagrams, we adopt the convention that double circles denote final states, while gray circles represent dead-end states. A DFA $\mathcal{A}$ reads in input sequences of symbols ("*string*") of its input alphabet. It starts in its initial state $q_0^{\mathcal{A}}$ and updates the state after having read each symbol via the state transition function $\delta^{\mathcal{A}}$. We say that a DFA $\mathcal{A}$ *accepts* a string π ($\pi \models \mathcal{A}$) if after reading it is in one of its accepting states (i.e., a state in $F^{\mathcal{A}}$), and otherwise we say that it *rejects* that string. The set of strings accepted by $\mathcal{A}$ is called the *language* of $\mathcal{A}$, i.e., $\mathcal{L}(\mathcal{A})$.

2.2 Automated Planning

Automated Planning (AP) is a research field focused on developing model-based autonomous solutions in Artificial Intelligence (AI) [24]. Planning systems (i.e., *planners*) are problem-solving algorithms that operate on explicit representations of states and actions. PDDL [22] is the standard Planning Domain Definition Language; it allows us to formulate a *planning model* with the description of the initial state of the world, the desired goal condition, and the planning domain.

A *planning domain* is built from a set of *propositions* describing the state of the world (i.e., the set of propositions that are true) and a set of *actions* Ω that can be executed. An *action schema* $a \in \Omega$ defines the list of *input parameters* for a, the *preconditions* under which a can be executed, and the *effects* of a on the state of the world. Both preconditions and effects are stated in terms of propositions in the planning domain, represented as *boolean predicates* and *numeric fluents*. While predicates represent facts about *domain objects* (i.e., things in the world that interest us) that can be true or false, fluents are number-valued functions that specify how the action increases the total cost of the plan being computed. The execution of actions can change the values of propositions in the planning domain, thereby transitioning the planning domain to a new state.

There exist several forms of AP in the literature (e.g., to capture durative and nondeterministic actions, etc.). In this paper, we focus on *classical planning* techniques characterized by *fully observable* and *deterministic* domains. To encode a classical planning problem, we use PDDL 2.1 [22], which is considered the most significant evolution over the original version of the language (called STRIPS) and is supported by the majority of state-of-the-art planners.[1] A solution to a planning problem is a sequence of operators—a *plan*—whose execution transforms the initial state into a state satisfying the goal by optimizing a pre-specified metric, generally related to the minimization of the plan's total cost.

From a computational perspective, classical planning is a PSPACE-complete problem [29]: finding a plan requires polynomial memory, although the computation time may still require exponential time in the worst case. Nonetheless, recent years have seen the emergence of classical planners with efficient heuristics that scale to large and challenging problems across several Computer Science domains [35]. In this paper, we generate multiple candidate execution scenarios for a given declarative behaviour using three classical planners that produce distinct plans under different criteria: Top-K [33], which finds the k best plans by cost (without excluding lower-cost alternatives); Top-Quality [32], which returns all plans within a cost threshold; and Diverse [31], which generates maximally distinct plans (in terms of reordering of the plan's activities).

[1] http://icaps-conference.org/index.php/main/competitions. Accessed: 1/6/2026.

3 Planning-Based Approach

Our approach targets to automatically generate execution scenarios ensuring compliance with (i) a declarative process specification and (ii) a set of user-defined requirements. We express these requirements as *preference constraints* on activity occurrence and their partial ordering, allowing users to customize the behavioural characteristics of the generated scenarios.

In this section, we first formalize the scenario generation problem and discuss how it can be solved using a DFA-based manipulation technique (Sect. 3.1). Then, we demonstrate how this technique reduces to a planning problem (Sect. 3.2).

3.1 Scenario Generation Problem

Let π be an *execution scenario*, i.e., a finite sequence of activities $\pi = \langle e_1, \dots, e_n \rangle$ over an alphabet Σ. We consider the following bipartite input for it:

- A declarative specification $D = \{d_1, \dots, d_k\}$, consisting of a set of k DECLARE constraints, each specifying a behavioural requirement over the activities in Σ (e.g., RESPONSE, PRECEDENCE, INIT, etc.).
- A set $P = \{p_1, \dots, p_m\}$ of m *preference constraints*, each expressing a user-defined requirement to constrain process execution. Specifically, building on the definition of relevant sequence patterns by Gusfield in [25], we consider five types of preference constraints: *prefix*, requiring the trace to start with a given sub-sequence of activities; *suffix*, requiring the trace to end with a given sub-sequence; *pattern*, requiring the trace to contain a given sub-sequence of activities; *pattern-gap*, requiring the trace to contain a given sub-sequence of activities not necessarily as consecutive elements, i.e., with possible activities between them; additionally, *min-len*, indicating the minimum trace length.

Notice that pattern (or pattern-gap) constraints with sub-sequences of length 1 (e.g., PATTERN(e)) can be used to instruct the generator to include a given activity in the scenarios and, as a consequence, indicate that related constraints (e.g., RESPONSE (e, a)) must be triggered rather than only satisfied in a so-called *vacuous* manner [17] (that is, simply because e does not occur).

Each constraint in $D \cup P$ is translated into a corresponding DFA. Specifically, for each DECLARE constraint $d_i \in D$, we derive a *constraint automaton* $\mathcal{D}_i = (\Sigma, Q^{\mathcal{D}_i}, q_0^{\mathcal{D}_i}, \delta^{\mathcal{D}_i}, F^{\mathcal{D}_i})$ following standard DFA constructions for DECLARE semantics, as shown in Sect. 2. Similarly, for each preference constraint $p_j \in P$, we construct a *preference automaton* $\mathcal{P}_j = (\Sigma, Q^{\mathcal{P}_j}, q_0^{\mathcal{P}_j}, \delta^{\mathcal{P}_j}, F^{\mathcal{P}_j})$, defined as a DFA that characterizes a specific regular language by restricting the set of accepted traces according to conditions on their prefix, suffix, inclusion of a given sub-sequence, or minimum length.

We illustrate the DFA construction with two examples built over our case study in Sect. 2. Figure 4(a) shows the DFA encoding the DECLARE template RESPONSE (e, a), which requires that every occurrence of activity e

(Produce signed excerpt) is eventually followed by a (Archive). The automaton starts in the accepting state q_0 (the constraint is always satisfied if e never occurs); upon executing e, it moves to the non-accepting state q_1, from which it returns to q_0 only upon executing a. Any other activity leaves the automaton in its current state. Figure 4(b) shows the DFA encoding the PREFIX (f, d) preference constraint, which requires that the trace starts with the sub-sequence ⟨f, d⟩. The automaton advances through states q_0, q_1, and q_2 upon executing f (Fill in e-form) and then d (Determine administrative expenses) in this order; any deviation from this prefix leads to a non-accepting sink state q_3 from which no recovery is possible.

We refer to the *process frame* $\mathcal{F}$ as the collection of DFAs representing the DECLARE and preference constraints. We formally enunciate the scenario generation problem as follows.

Definition 2 (Scenario Generation). *Let* $\mathcal{F} = \{\mathcal{D}_1, ..., \mathcal{D}_k, \mathcal{P}_1, ..., \mathcal{P}_m\}$ *be a process frame that includes* k *constraint automata* $\mathcal{D}_1, ..., \mathcal{D}_k$ *and* m *preference automata* $\mathcal{P}_1, ..., \mathcal{P}_m$*. The generation of a scenario* π *over* $\mathcal{F}$ *consists of computing a sequence* $\rho = \langle e_1 \cdots e_n \rangle \in \Sigma^*$ *such that, for each* $z = 0, \ldots, n-1$*, there exist:*

- *(in each* $\mathcal{D}_i$*,* $1 \leq i \leq k$*) a transition* $(q_z^{\mathcal{D}_i}, e_{z+1}, q_{z+1}^{\mathcal{D}_i}) \in \delta^{\mathcal{D}_i}$*, denoted as* $q_z^{\mathcal{D}_i} \xrightarrow{e_{z+1}} q_{z+1}^{\mathcal{D}_i}$*, such that* $q_0^{\mathcal{D}_i} \xrightarrow{e_1} q_1^{\mathcal{D}_i} \cdots q_{n-1}^{\mathcal{D}_i} \xrightarrow{e_n} q_n^{\mathcal{D}_i}$ *, where* $q_0^{\mathcal{D}_i}$ *is the initial state of* $\mathcal{D}_i$ *and* $q_n^{\mathcal{D}_i} \in F^{\mathcal{D}_i}$*.*
- *(in each* $\mathcal{P}_j$*,* $1 \leq j \leq m$*) a transition* $(q_z^{\mathcal{P}_j}, e_{z+1}, q_{z+1}^{\mathcal{P}_j}) \in \delta^{\mathcal{P}_j}$*, denoted as* $q_z^{\mathcal{P}_j} \xrightarrow{e_{z+1}} q_{z+1}^{\mathcal{P}_j}$*, such that* $q_0^{\mathcal{P}_j} \xrightarrow{e_1} q_1^{\mathcal{P}_j} \cdots q_{n-1}^{\mathcal{P}_j} \xrightarrow{e_n} q_n^{\mathcal{P}_j}$ *, where* $q_0^{\mathcal{P}_j}$ *is the initial state of* $\mathcal{P}_j$ *and* $q_n^{\mathcal{P}_j} \in F^{\mathcal{P}_j}$*.*

We say that $\mathcal{F}$ accepts *a scenario* π *(*$\pi \models \mathcal{F}$*) if there exists an execution* ρ *on the declarative automata* $\mathcal{D}_1, ..., \mathcal{D}_k$ *and on the preference automata* $\mathcal{P}_1, ..., \mathcal{P}_m$ *such that the last state of all these automata is final at the end of the execution.*

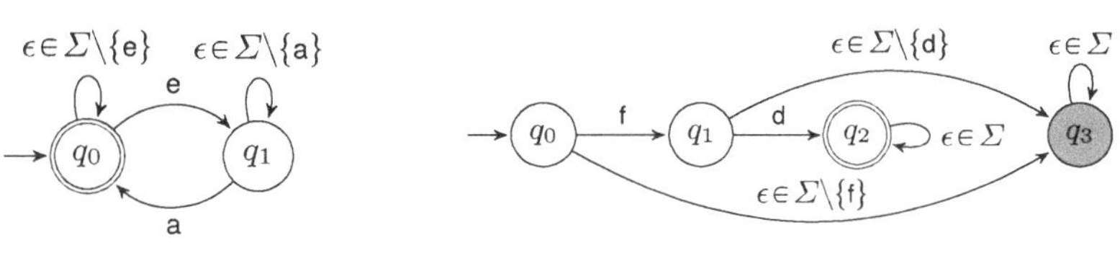

(a) RESPONSE(e,a) (b) PREFIX(f,d)

Fig. 4. DFA encoding of RESPONSE (e, a) and PREFIX (f, d).

Therefore, a scenario π satisfies all declarative and preference constraints defined in a process frame $\mathcal{F}$ if and only if it is *accepted* by every DFA in $\mathcal{F}$; i.e., $\pi \in \mathcal{L}(\mathcal{D}_i)$ and $\pi \in \mathcal{L}(\mathcal{P}_j)$ for each $\mathcal{D}_i$ and $\mathcal{P}_j$ in $\mathcal{F}$. For example, considering the process frame specified in Fig. 4, any scenario that begins with the subsequence ⟨f, d⟩ and in which activity a occurs after the execution of e can be considered a valid scenario with respect to the frame.

3.2 Scenario Generation as Planning

In this section, based on the formalization presented above, we show how we can reduce the scenario generation problem to a deterministic planning problem. The idea behind the reduction is to model the activities from the alphabet Σ as planning actions whose execution triggers state changes for each $\mathcal{D}_i$ and $\mathcal{P}_j$ included in a process frame $\mathcal{F}$, according to their respective transition function.

Planning actions are executable based on the current state of the DFAs: an action e can be executed only if the current states of each $\mathcal{D}_i$ and $\mathcal{P}_j$ have an outgoing transition labelled with the related process activity e. The problem then consists in finding a deterministic plan that takes such DFAs from their initial state to a final state, where actions corresponding to process activities are assigned a cost depending on a predefined cost function $\mathcal{C}$.

Inspired by common practice in conformance checking [6], we associate costs with activities. Here, costs represent penalties incurred when less preferred activities are included during scenario generation. For example, considering a process frame that includes the DFAs derived from the DECLARE specification in Fig. 2 and the Prefix constraint in Fig. 4(b), some valid scenarios may include the x (Stop) activity, which corresponds to stopping the payment procedure (e.g., $\pi_1 = \langle \mathsf{f}, \mathsf{d}, \mathsf{r}, \mathsf{x} \rangle$), while others follow the standard payment flow (e.g., $\pi_2 = \langle \mathsf{f}, \mathsf{d}, \mathsf{r}, \mathsf{e}, \$, \mathsf{i}, \mathsf{e}, \mathsf{a}, \mathsf{s} \rangle$). If interrupting the procedure is considered an undesirable yet still valid outcome, assigning a significantly higher cost to x than to the other activities (which may retain a standard unit cost) leads a cost-aware planning technique such as TOP-K [33] to prioritize the generation of scenarios that do not include the Stop activity (such as π_1) over those that do (such as π_2). We note that designing custom cost functions to weight individual activities is beyond the scope of this work, as it is inherently domain-specific and requires case-by-case customization. In contrast, our approach is intentionally domain-agnostic, focusing on a rigorous formulation of the scenario generation problem with broad applicability across diverse contexts. Interested readers can consult the work [5] for an in-depth exploration of this topic.

The obtained plan is a representation of an *execution scenario* that solves the generation problem. We provide the details of this reduction below.

A *deterministic planning domain with action costs* over a set of propositions Φ is a tuple $\mathcal{N} = (S, \Omega, cost, \tau)$, where:

- $S \subseteq 2^{\Phi}$ is the finite set of *domain states*, with each state $s = \{\phi_1, \ldots, \phi_n\}$ representing the propositional interpretation that assigns *true* to a proposition ϕ if and only if $\phi \in s$;
- Ω is the finite set of *domain actions*;
- $cost : \Omega \mapsto \mathbb{N}_0$ is a *cost function*, assigning a non-negative cost to each planning action;
- $\tau : S \times \Omega \mapsto S$ is the domain *transition function*.

A *planning problem* is a tuple $\mathcal{R} = (\mathcal{N}, s_0, G)$, where $s_0 \in S$ is the *initial state* of the problem and G is a propositional formula over Φ representing the *goal* condition. A *plan* for $\mathcal{R}$ is a finite sequence of actions $\pi = e_1 \cdots e_n \in \Omega^*$ whose execution from s_0 leads to a state s_n such that $s_n \models G$. A solution π of $\mathcal{R}$ is said to be *optimal* if, for all other solutions π', it is the case that $cost(\pi) \leq cost(\pi')$.

We encode the scenario generation problem as a planning problem as follows. Let $\mathcal{F} = \{\mathcal{D}_1, ..., \mathcal{D}_k, \mathcal{P}_1, ..., \mathcal{P}_m\}$ be a process frame that includes k constraint automata $\mathcal{D}_1, ..., \mathcal{D}_k$ and m preference automata $\mathcal{P}_1, ..., \mathcal{P}_m$. The current state of each DFA is modelled through suitable propositions. We define the planning domain $\mathcal{N} = (S, \Omega, cost, \tau)$ over the set of propositions $\Phi = Q^{\mathcal{D}_1} \cup \cdots \cup Q^{\mathcal{D}_k} \cup Q^{\mathcal{P}_1} \cup \cdots \cup Q^{\mathcal{P}_m}$ (with all state sets assumed pairwise disjoint), where:

1. $S \subseteq \{\{\bigcup_{i=1,\dots,k} q^{\mathcal{D}_i}\} \cup \{\bigcup_{j=1,\dots,m} q^{\mathcal{P}_j}\} \mid q^{\mathcal{D}_i} \in Q^{\mathcal{D}_i}, q^{\mathcal{P}_j} \in Q^{\mathcal{P}_j}\}$ is the set of domain states, each containing exactly one state per DFA in $\mathcal{F}$;
2. $\Omega = \Sigma$, i.e., each activity $e \in \Sigma$ corresponds to a planning action, modeling the synchronous advancement of all DFAs upon executing e;
3. For all $e \in \Sigma$, $cost(e) = \mathcal{C}(e)$, i.e., it depends on the specific cost function $\mathcal{C}$;
4. For $e \in \Omega$, $q_a^{\mathcal{D}_i}, q_b^{\mathcal{D}_i} \in Q^{\mathcal{D}_i}$, $q_a^{\mathcal{P}_j}, q_b^{\mathcal{P}_j} \in Q^{\mathcal{P}_j}$, we define the domain transition function as $\tau(\{\bigcup_{i=1,\dots,k} q_a^{\mathcal{D}_i}\} \cup \{\bigcup_{j=1,\dots,m} q_a^{\mathcal{P}_j}\}, e) = \{\bigcup_{i=1,\dots,k} q_b^{\mathcal{D}_i}\} \cup \{\bigcup_{j=1,\dots,m} q_b^{\mathcal{P}_j}\}$ if:
   (a) $(q_a^{\mathcal{D}_i}, e, q_b^{\mathcal{D}_i}) \in \delta^{\mathcal{D}_i}$, for each $\mathcal{D}_i \in \mathcal{F}$;
   (b) $(q_a^{\mathcal{P}_j}, e, q_b^{\mathcal{P}_j}) \in \delta^{\mathcal{P}_j}$, for each $\mathcal{P}_j \in \mathcal{F}$.

We call $\mathcal{N}$ the *scenario generation planning domain* of $\mathcal{F}$. Since all automata are DFAs, the transition function τ is deterministic. Moreover, since $\Omega = \Sigma$, every plan for $\mathcal{N}$ directly corresponds to a scenario executed over Σ.

We then define the planning problem $\mathcal{R} = (\mathcal{N}, s_0, G)$, where:

- $s_0 = \{q_0^{\mathcal{D}_1}, \dots, q_0^{\mathcal{D}_m}, \dots, q_0^{\mathcal{P}_1}, \dots, q_0^{\mathcal{P}_k}\}$ is the initial state, corresponding to the initial state of each DFA in $\mathcal{F}$;
- $G = (\bigvee_{q^i \in F^{\mathcal{D}_i}} q^i) \wedge (\bigvee_{q^j \in F^{\mathcal{P}_j}} q^j)$ is the goal condition, requiring every DFA to be simultaneously in an accepting state.

We call $\mathcal{R}$ the *scenario generation planning problem* of $\mathcal{F}$. In planning terms, finding a *plan* $\pi = e_1 \cdots e_n \in \Omega^*$ for $\mathcal{R}$ amounts to navigating the synchronous product of all DFAs from their joint initial state to a joint accepting state.

One of the major advantages of our planning-based formulation is that it naturally supports the generation of multiple and diverse scenarios by adopting different planning strategies. Depending on the specific generation objectives, various planners and parameter settings can be employed to control the scenario generation process, as discussed in the next section.

4 Evaluation

We have developed a planning-based tool in Python that implements the technique described in Sect. 3. The tool is available for reproducibility alongside the testbeds at: https://doi.org/10.5281/zenodo.20842315. Based on a set of DECLARE and preference constraints in input, and following the translation approach of Sect. 3.2, it converts such constraints into DFAs and generates the planning tasks in PDDL. To generate multiple scenarios, our tool leverages the FORBID-ITERATIVE classical planner,which can produce multiple plans that vary depending on the underlying planning algorithm. In particular, we employ three well-known planning algorithms: TOP-K [33], DIVERSE [31], and TOP-QUALITY [32].

4.1 Experiments with Synthetic Specifications

To assess the scalability of our technique, we conducted two experiments with synthetic input specifications of increasing complexity. For both experiments, we used a unit-cost function for all planning actions, as varying action costs are unlikely to significantly affect performance. Conversely, differentiated costs could improve efficiency by reducing equivalent-cost alternatives and adding structure to the planner's search space. Under this assumption, cost-optimal plans coincide with shortest plans. Tests were run on a 12th-gen. Intel Core i7 with 8 GB RAM.

In the first experiment, we evaluated the performance of our technique against scenario inputs of increasing size (see Sect. 3.1), namely specification and preference constraints. Different types of selected constraints were systematically added to assess their impact on the overall performance of the scenario generation task. In the second experiment, we evaluated the scalability of our technique by using a One-Factor-at-a-Time (OFAT) design. Specifically, we varied one input parameter at a time (i.e., the size of the activity alphabet, the number of constraints, or the number of generated scenarios) while keeping the other two fixed. This controlled setup allowed us to assess each parameter's impact independently.

Scalability w.r.t. Scenario Input Size: Setup. We considered two activity alphabets, with sizes 10 and 20, respectively. For each alphabet, we incrementally built specifications with 3, 5, 7, and 9 constraints: the 3-constraint specification contains only DECLARE constraints; the 5-, 7-, and 9-constraint specifications contain 4, 5, and 6 DECLARE constraints and 1, 2, and 3 preference constraints, respectively. Constraints were added in this order: ATLEASTONE (t0), RESPONSE (t1, t2), NOTCOEXISTENCE (t3, t4), CHAINPRECEDENCE (t5, t4), CHOICE(t5,t6), and NOTSUCCESSION (t7, t8) as DECLARE constraints, and PATTERN-GAP(t4,t6), PATTERN(t7,t8), and MIN-LEN(8) as preference constraints. For TOP-K and DIVERSE, we generated 10, 25, and 50 plans (i.e., scenarios) per test. Note that TOP-K returns cost-optimal plans, while DIVERSE returns mutually distinct plans. For TOP-QUALITY, we generated 20 plans per test with cost bounds of 10, 15, and 20.

Table 2. First-experiment results on synthetic specifications

	10 activities				20 activities			
	3 constraints	5 constraints	7 constraints	9 constraints	3 constraints	5 constraints	7 constraints	9 constraints
Grounding time	0.0017 s	0.0017 s	0.0028 s	0.0183 s	0.0021 s	0.0022 s	0.0028 s	0.0227 s
Planner	**Average planning time [s]**							
diverse-10	0.25	0.27	0.29	0.4	0.22	0.3	0.31	0.4
diverse-25	0.18	0.31	0.3	0.4	0.19	0.28	0.31	0.34
diverse-50	0.17	0.3	0.32	0.45	0.16	0.31	0.31	0.33
topk-10	0.2	0.21	$4.52 \cdot 10^{-2}$	0.41	0.19	0.24	$6.56 \cdot 10^{-2}$	0.45
topk-25	0.17	0.31	0.17	0.44	0.17	0.29	0.26	0.51
topk-50	0.15	0.32	0.25	0.61	0.16	0.31	0.27	0.71
topq-10	0.28	0.37	0.45	0.42	0.32	0.34	0.48	0.5
topq-15	0.32	0.42	0.48	0.49	0.35	0.39	0.53	0.6
topq-20	0.27	0.37	0.55	0.57	0.33	0.47	0.64	0.74

Table 3. OFAT experiment results on synthetic specifications

	15 constraints - 15 activities					15 constraints - 15 plans					15 activities - 15 plans				
	10 plans	15 plans	20 plans	25 plans	30 plans	10 act.	15 act.	20 act.	25 act.	30 act.	10 const.	15 const.	20 const.	25 const.	30 const.
Grounding time	0.0412 s					0.0699 s	0.0733 s	0.0822 s	0.0931 s	0.1067 s	0.0234 s	0.0567 s	0.1685 s	0.5865 s	1.9058 s
Planner	Average planning time [s]														
Diverse	0.31	0.31	0.32	0.32	0.32	0.25	0.26	0.28	0.3	0.3	0.3	0.39	1.01	1.73	3.97
Top-k	0.41	0.42	0.43	0.44	0.45	0.52	0.53	0.59	0.6	0.68	0.42	0.54	3.98	5.85	16.31
Top-q	0.37	0.38	0.39	0.4	0.42	0.3	0.31	0.32	0.32	0.32	0.39	0.5	10.54	15.38	68.21

Scalability Experiment w.r.t. Scenario Input Size: Results. Table 2 reports the results of the first experiment. For each test, we measured the *grounding time*, i.e., the preprocessing time needed to instantiate the PDDL specification into an explicit planning model that can be processed by the planner, and the average *planning time* required by the planner to explore that model and generate a scenario. The results suggest that for all three planning algorithms, increasing the number of constraints leads to a moderate increase in both grounding and planning time. However, this increase remains manageable across all configurations, indicating that the proposed approach scales well with respect to the number of constraints and can potentially support compliance specifications of considerable size. Moreover, the results also indicate that performance degradation depends on the number of constraints rather than their type, with no significant differences observed across the specific DECLARE and preference constraints considered.

A further observation concerns the number of plans/scenarios to be generated. The results show that this parameter does not significantly affect performance: the average time per plan remains largely constant across configurations, with no clear correlation between the number of requested plans and the time required to generate them. For instance, examining the results in Table 2 related to the configurations with an alphabet size of 20 activities, for both TOP-K and DIVERSE the per-plan generation time for 10, 25, and 50 plans does not vary substantially, confirming that the computational cost scales linearly with the number of plans rather than increasing exponentially. A similar behaviour is observed for TOP-QUALITY: the average time per plan remains largely stable as the cost bound increases from 10 to 15 and 20, suggesting that relaxing the quality threshold does not introduce significant computational overhead.

Finally, increasing the size of the alphabet also affects both grounding and planning time. However, the approach scales well along this dimension, with the observed increase in execution time remaining modest. Notably, the number of constraints has a stronger impact on performance than the size of the activity alphabet, suggesting that the complexity of the compliance specification is a more influential factor than the number of activities.

OFAT Experiment: Setup. We adopted a controlled setup in which two parameters of the input configuration (e.g., number of constraints and activity alphabet size) were fixed at 15 (e.g., 15 constraints and 15 activities), while the third (e.g., number of generated plans) was varied over the values 10, 15, 20, 25, and 30. This yielded three experimental configurations, each isolating the effect of one parameter of interest. This allowed us to independently assess the impact of each parameter, namely (i) the number of generated plans, (ii) the size of the activity alphabet, and (iii) the number of constraints, on the performance of the three tested planning algorithms, while keeping the remaining parameters fixed.

Table 4. Experimental results for the real-world specifications

	Municipality A	Municipality B	Municipality C	Fracture treatment	University admission	Surgery process
Grounding time	0.0067 s	0.0236 s	0.0235 s	0.0040 s	0.1380 s	0.0142 s
Planner			Average planning time [s]			
topk-10	0.22	0.48	0.38	0.31	0.93	0.14
topk-25	0.18	0.41	0.19	0.31	0.58	$6.93 \cdot 10^{-2}$
topk-50	0.19	0.47	0.19	0.34	0.51	$4.94 \cdot 10^{-2}$
topq-10	0.64	0.48	0.44	0.29	0.78	0.24
topq-15	0.44	0.5	0.49	0.3	0.86	0.35
topq-20	0.43	0.57	0.54	0.32	0.87	0.42
Constraints	10	11	12	7	15	21

OFAT Experiment: Results. Table 3 shows the results of the second experiments. For each configuration, we report the grounding time and the average planning time per plan. Each experiment was repeated five times, and the reported time is the average across runs. Note that, in the first configuration, grounding time remains constant across different numbers of plans, as it depends only on the fixed number of activities and constraints. In the other two configurations, it increases with the number of activities or constraints, reflecting the growing complexity of the planning task. Regarding the average planning time, the results in Fig. 5 support the following observations. (I) In the first configuration (15 constraints, 15 activities), the average planning time increases almost linearly with the number of generated plans, with DIVERSE being faster than TOP-K and TOP-QUALITY, which exhibit similar trends. (II) In the second configuration (15 constraints, 15 plans), planning times grow moderately when the number of activities increases, with DIVERSE emerging as the fastest planner. These results can be explained by the fact that a larger alphabet directly increases the number of transitions in the DFAs encoding the constraints, thereby expanding the state space that the planner must explore. (III) The third configuration (15 activities, 15 plans) further confirms the findings of the first experiment, namely that the number of constraints is the most influential factor affecting planning time. While variations in the number of activities or plans result in only modest changes, increasing the number of constraints causes a substantial increase in computation time across all three planning algorithms.

It is worth noting that among the three planning algorithms, DIVERSE demonstrates the best overall performance. This can be attributed to the fact that, unlike TOP-K and TOP-QUALITY, which balance both plan diversity and quality, DIVERSE focuses solely on generating mutually distinct plans without optimizing for cost. As a result, its search is less constrained and thus more efficient.

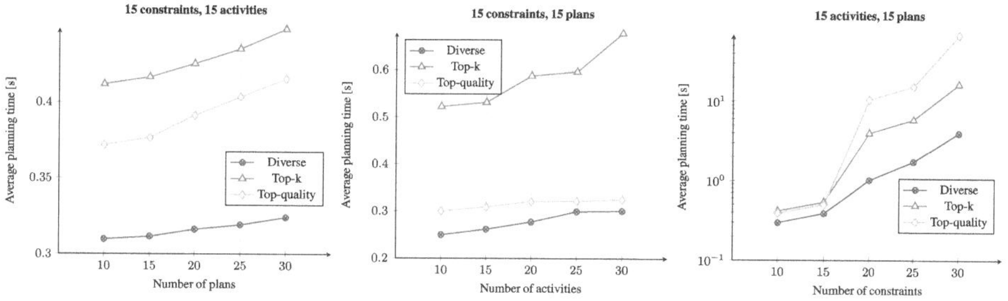

Fig. 5. Average planning time of the OFAT experiment.

4.2 Experiments with Real-World Specifications

To further demonstrate the applicability of our approach, we tested it on four real-world declarative process specifications from the process science literature: *Municipality* [37], *Fracture Treatment* [14], *University Admission* [17], and *Surgery Process* [41]. Specifically, Municipality [37] is an extract of a process for requesting an excerpt from the civil registration. We considered three variants of this specification, namely *MunicipalityA*, *MunicipalityB*, and *MunicipalityC*, which differ in the number and type of constraints. Fracture Treatment [14] describes the clinical process involved in the management of bone fractures. University Admission [17] specifies the process of admitting students to a university program. Finally, Surgery Process [41] describes the pre-, intra-, and post-operative procedures for colorectal surgery.

For these real-world scenarios, we evaluated the scenario generation using TOP-K and TOP-QUALITY, as these cases call for scenarios that balance both diversity and quality. DIVERSE, while effective in terms of performance as shown in Sect. 4.1, does not account for plan costs and therefore may produce plans that, although distinct, do not reflect meaningful differences in process behavior. Moreover, in the considered real-world scenarios, the ordering of activities is a key dimension of diversity: plans that involve the same set of activities but differ in their execution order should be regarded as genuinely distinct. This is particularly evident in the Surgery Process case, where the sequencing of activities is clinically significant. For example, performing a medical treatment before a surgical intervention, or vice versa, can lead to substantially different clinical outcomes.

Table 4 summarizes our findings with the real-world use cases. The results confirm that our approach delivers scalable performance on real process specifications, with execution times in the order of a tenth of a second or less, despite the modest hardware infrastructure.

A comparative example. Beyond scalability, a key objective of the real-world evaluation is to demonstrate the practical utility of our approach in generating meaningful and realistic process scenarios. To this end, we compare our approach with MINERful [30], a well-known tool for declarative process discovery that also supports trace generation. MINERful allows the user to configure the minimum and maximum length of generated traces, without support for preference constraints or ranking criteria. Such a limitation can lead to the generation of traces that are semantically trivial or unrealistic, though formally compliant with the specified constraints. Next, we provide an exemplificative illustration of this statement, leaving a full-scale comparative evaluation of the effectiveness of planning-based scenario generations against randomized approaches for future work.

As an example, we take into account the municipality use case of Sect. 2. We make MINERful generate 100 traces (its default setting) of length ranging between 5 (the minimum number of activity occurrences required by the specification) and 20 (the uppermost value for the TOP-QUALITY parameter in our tests). The traces generated by MINERful contain repeated occurrences of the same activities, resulting in unrealistic scenarios. In contrast, our approach generates scenarios that are both compliant and avoid repetitions of activities that are vacuously satisfied by the DECLARE constraints. As an example, the first five scenarios produced by our tool have a significantly more contained length between 4 and 10 activities, 8.4 on average, compared to the 10 to 20 activities of MINERful, 16.2 on average. They simulate plausible runs as expected. As an example, one of the top-5 generated scenarios is the following: $\pi = \langle$Fill in e-form, Determine administrative expenses, Send payment request, Fill in payment information, Process payment, Produce signed excerpt, Archive, Send excerpt$\rangle$.

This scenario reflects a realistic execution of the municipality process, with activities occurring in a semantically coherent order and without unnecessary repetitions. In contrast, the first two traces generated by our run of MINERful, with and without the knock-out occurrence of x respectively, are ⟨f, a, a, a, f, a, d, a, d, f, d, $, x⟩ and ⟨f, d, a, f, a, a, d, a, $, i, i, d, f, p, i, i, i, p, s, a⟩ using the abbreviations introduced in Sect. 2. The reader, irked by the need to filter out the unnecessary repetitions of tasks or navigating through so long sequences out of which few significant steps are considered (especially compared to the preceding scenario), can intuitively experience the feeling that motivated our research. To give a quantitative account of this aspect, we can consider the coverage of the generated runs for the constraints. Intuitively, coverage [8] measures the ratio of events triggering a constraint, like the occurrence of e for RESPONSE(e, a). Excluding INIT(f), which is trivially triggered by every initial event in a trace, the coverage of the constraints is 0.180 in the top-5 scenarios our tool generates (each constraint is triggered by 18 % of the events), whereas it halves to 0.092 with the 100 traces generated by MINERful (0.067 if we take the first five).

5 Related Work

Zugal et al. [42,43] empirically demonstrated the effectiveness that a test-driven approach has on the modeling and maintainability of DECLARE specifications, whereby test cases are exemplary traces against which a process specification is checked (akin to what we name scenarios in this paper). Slaats et al. [38] theorize the use of short, open-world-like test cases for the iterative modeling of DCR Graphs. These valuable contributions strengthen the argument motivating our work, as we aim to yield automated generators of such test cases.

Several approaches in the literature leverage computational logic and formal models to generate event logs from declarative process specifications. A seminal work in the area [14] simulates DECLARE models by translating constraints into regular expressions and DFAs, and generates logs compliant with a given model by navigating the product of automata derived from individual constraints. Building upon logic-based foundations, other authors have explored different paradigms. Chesani et al. [9] propose using abductive logic programming to generate both positive and negative traces (i.e., those violating the model) for DECLARE models. Similarly, Answer Set Programming (ASP) was adopted as a highly flexible technique for synthetic log generation [10]. Donadello et al. [19] explore the purpose-guided log generation via ASP, demonstrating high trace diversification, and Maggi et al. [34] propose a translation of DECLARE models augmented with data into Alloy programs.

While these declarative and procedural methods offer flexibility, they often face limitations when attempting to generate interpretable scenarios for model validation. State-of-the-art simulation engines typically adopt random approaches to trace generation. Consequently, they tend to produce traces that show common behavior while leaving out significant parts of the possible execution paths. On the other hand, solver-based approaches (i.e., ASP or SAT) heavily rely on internal search heuristics. Indeed, as noted by Donadello et al. [18], the problem of computing diverse solutions for combinatorial problems is intrinsically difficult: solvers often yield a relatively small number of solutions and require approximation techniques to operate within reasonable time bounds. Hence, these limitations might hinder the selection of a curated subset of runs exhibiting characteristic traits, which is central for the proper validation of complex declarative specifications. To mitigate these limitations, our technique formally encodes the scenario generation problem as an automated planning task. Unlike random simulation approaches or solver-dependent searches, our solution enables the customization and ranking of the generated scenarios. By navigating the synchronous product of DFAs through classical planning techniques encoded in PDDL 2.1, our approach allows users to tailor scenario generation to specific needs.

6 Discussion and Concluding Remarks

In this paper, we presented an AI planning-based approach to generate representative scenarios for declarative process specifications. To cater for succinct yet characterizing exemplary runs, we leverage three deterministic planning algorithms, namely TOP-K, TOP-QUALITY, and DIVERSE, where actions carry associated costs and valid plans must meet specific criteria. These algorithms facilitate the recommendation of alternative unfoldings of process behaviour within the conforming space defined by the process frame, guided by controlled criteria that shape both the quality and diversity of the generated plans.

A more general advantage of adopting planning for scenario generation is that it provides solutions that are correct, transparent, and scalable (three key requirements for trustworthy systems). Specifically, planning leverages mature reasoning mechanisms developed over decades of research to guarantee solution correctness [24]; it supports explainability through verification techniques such as Model Checking [4], enabling the analysis of relevant plan properties (for instance, explaining why a specific action was selected); despite its worst-case computational complexity (see Sect. 2.2), advances in domain-independent heuristics have made most practical problems, particularly in BPM and process mining, tractable and scalable in practice [35], as confirmed by our empirical evaluation. It is worth noting that our approach is applicable to the full DECLARE repertoire of templates and beyond (see, e.g., the preference constraints). In fact, our approach can handle any constraint expressible via DFAs, and DFAs have a strictly higher expressiveness than LTL_f [13], on which DECLARE is based. Therefore, its applicability includes and extends beyond DECLARE.

Our approach paves the path for future amelioration. First and foremost, its applicability in the field calls for an investigation with users into the most effective framings for scenario generation, in terms of preference automata and activity costs. To this end, the integration of a dedicated specification language for log creation is highly promising [18]. Also, we aim to design specific heuristics that make it applicable for process specifications discovered from event logs, which often consist of a larger number of constraints. In that regard, it becomes compelling to make use of process mining quality measures of the mined constraints, like confidence and support, to guide the generation of plans. Our reliance on classical planners is tailored to the generation of linear execution traces. While sound for scenario generation, this choice outlines a clear boundary if the execution context shifts to a runtime environment driven by uncontrollable exogenous events (e.g., unexpected customer cancellations). This would require coupling the approach with more expressive knowledge representation formalisms to natively manage non-deterministic control flows and exogenous events [7]. We also plan to study how the scenarios we generate can be effectively used for validation and refinement of process specifications, thus closing the cycle presented in Fig. 1. Inspired by the play-out approach of Harel and Marelly [28], we foresee as a highly interesting research avenue the inclusion of user interactions to guide the generation of scenarios. Although originally conceived for live sequence charts (LSCs), the authors advocate the adoption of their approach with "languages other than LSCs, such as appropriate variants of temporal logic," on which DECLARE is indeed based. Finally, we aim to develop a comparative framework to empirically analyze the properties of scenarios generated by our planning-based approach against those produced by randomized simulation techniques, in order to better assess the utility of our approach with respect to existing techniques in the literature.

Acknowledgement. The work of A. Marrella has been supported by the Sapienza projects CAI4PPI and FOND-AIBPM, and the EU Horizon Europe Research and Innovation Programme under Grant 101299128 (euroFMX). A. Casciani conducted this work while enrolled in the National Doctorate in AI run by Sapienza (CUP B53C23003500006, Mission 4 NRRP Generic Research Scholarship, Component 1).

References

1. van der Aa, H., Balder, K.J., Maggi, F.M., Nolte, A.: Say it in your own words: defining declarative process models using speech recognition. In: BPM (Forum), pp. 51–67 (2020)
2. van der Aa, H., Di Ciccio, C., Leopold, H., Reijers, H.A.: Extracting declarative process models from natural language. In: CAiSE, pp. 365–382 (2019)
3. Agostinelli, S., Chiariello, F., Maggi, F.M., Marrella, A., Patrizi, F.: Process mining meets model learning: Discovering deterministic finite state automata from event logs for business process analysis. Inf. Syst. **114** (2023)
4. Baier, C., Katoen, J.P.: Principles of Model Checking. MIT Press (2008)
5. Boltenhagen, M., Chatain, T., Carmona, J.: A discounted cost function for fast alignments of business processes. In: BPM'21, pp. 252–269 (2021)
6. Carmona, J., van Dongen, B., Weidlich, M.: Conformance checking: foundations, milestones and challenges. In: Process Mining Handbook, pp. 155–190 (2022)
7. Casciani, A., Agostinelli, S., Lespérance, Y., Marrella, A., Sardiña, S.: Formal semantics for knowledge representation and automated reasoning in BPMN process models. Inf. Syst. **140** (2026)
8. Cecconi, A., De Giacomo, G., Di Ciccio, C., Maggi, F.M., Mendling, J.: Measuring the interestingness of temporal logic behavioral specifications in process mining. Inf. Syst. **107** (2022)
9. Chesani, F., et al.: Towards the generation of the "perfect" log using abductive logic programming. In: CILC, pp, 179–192 (2019)
10. Chiariello, F., Maggi, F.M., Patrizi, F.: ASP-based declarative process mining, pp. 5539–5547. AAAI (2022)

11. Corea, C., Delfmann, P.: Quasi-inconsistency in declarative process models. In: BPM Forum, pp. 20–35 (2019)
12. Corea, C., Kuhlmann, I., Thimm, M., Grant, J.: Paraconsistent reasoning for inconsistency measurement in declarative process specifications. Inf. Syst. **122** (2024)
13. De Giacomo, G., Vardi, M.Y.: Linear temporal logic and linear dynamic logic on finite traces. In: IJCAI, pp, 854–860 (2013)
14. Di Ciccio, C., Bernardi, C., Cimitile, M.L., Maggi, M., F.M.: Generating event logs through the simulation of Declare models. In: EOMAS, pp, 20–36 (2015)
15. Di Ciccio, C., Maggi, F.M., Mendling, J.: Efficient discovery of target-branched declare constraints. Inf. Syst. **56**, 258–283 (2016)
16. Di Ciccio, C., Maggi, F.M., Montali, M., Mendling, J.: Resolving inconsistencies and redundancies in declarative process models. Inf. Syst. **64**, 425–446 (2017)
17. Di Ciccio, C., Montali, M.: Declarative process specifications: reasoning, discovery, monitoring. In: van der Aalst, W.M.P., Carmona, J. (eds.) Process Mining Handbook. LNBIP, vol. 448, pp. 108–152. Springer, Cham (2022). https://doi.org/10.1007/978-3-031-08848-3_4
18. Donadello, I., Maggi, F.M., Patrizi, F., Tessaris, S., Zorzi, M.: Flexible event log generation using answer set programming. Inf. Syst. **139**, 102714 (2026)
19. Donadello, I., Maggi, F.M., Riva, F., Singh, M.: ASP-based log generation with purposes in Declare4Py. In: ICPM Doctoral Consortium/Demo (2023)
20. Dumas, M., La Rosa, M., Mendling, J., Reijers, H.A.: Fundamentals of Business Process Management, 2nd edn. Springer, Heidelberg (2018). https://doi.org/10.1007/978-3-662-56509-4
21. Figl, K., Di Ciccio, C., Reijers, H.A.: Do declarative process models help to reduce cognitive biases related to business rules? In: ER, pp. 119–133 (2020)
22. Fox, M., Long, D.: PDDL2.1: an extension to PDDL for expressing temporal planning domains. JAIR **20** (2003)
23. Frederiks, P.J.M., van der Weide, T.P.: Information modeling: the process and the required competencies of its participants. Data Knowl. Eng. **58**(1), 4–20 (2006)
24. Geffner, H., Bonet, B.: A Concise Introduction to Models and Methods for Automated Planning. Synthesis Lectures on AI and Machine Learning. Springer, Cham (2013). https://doi.org/10.1007/978-3-031-01564-9
25. Gusfield, D.: Algorithms on Strings, Trees, and Sequences - Computer Science and Computational Biology. Cambridge University Press (1997)
26. Haisjackl, C., Barba, I., Zugal, S., et al.: Understanding declare models: strategies, pitfalls, empirical results. Softw. Syst. Model. **15**(2), 325–352 (2016)
27. Harel, D., Marelly, R.: Come, Let's Play: Scenario-Based Programming Using LSCs and the Play-Engine. Springer, Heidelberg (2003). https://doi.org/10.1007/978-3-642-19029-2
28. Harel, D., Marelly, R.: Specifying and executing behavioral requirements: the play-in/play-out approach. Softw. Syst. Model. **2**, 82–107 (2003)
29. Helmert, M.: New complexity results for classical planning benchmarks. In: 16th International Conference on Automated Planning and Scheduling, ICAPS '06, pp. 52–62 (2006)
30. Iacometta, C., Di Ciccio, C.: Declarative process mining with minerful, reloaded. In: BPM (Demos / Resources Forum), pp. 328–335 (2025)
31. Katz, M., Sohrabi, S.: Reshaping diverse planning. In: AAAI 2020 (2020)
32. Katz, M., Sohrabi, S., Udrea, O.: Top-quality planning: finding practically useful sets of best plans. In: AAAI 2020, pp. 9900–9907 (2020)
33. Katz, M., Sohrabi, S., Udrea, O., Winterer, D.: A novel iterative approach to top-k planning. In: ICAPS 2018, pp. 132–140 (2018)

34. Maggi, F.M., Marrella, A., Patrizi, F., Skydanienko, V.: Data-aware declarative process mining with SAT. ACM Trans. Intell. Syst. Tech. **14**(4), 75:1–75:26 (2023)
35. Marrella, A.: Automated planning for business process management. J. Data Semant. **8**(2) (2019)
36. Park, G., Kofferath, J., Cho, M.: Extracting declarative constraints for process modeling from natural language descriptions with large language models. Decis. Support Syst. **203**, 114627 (2026)
37. Schunselaar, D.M.M., Maggi, F.M., Sidorova, N., van der Aalst, W.M.P.: Configurable declare: designing customisable flexible process models. In: CoopIS (2012)
38. Slaats, T., Debois, S., Hildebrandt, T.T.: Open to change: a theory for iterative test-driven modelling. In: BPM, pp. 31–47 (2018)
39. Uchitel, S., Kramer, J., Magee, J.: Implied scenario detection in the presence of behaviour constraints. VISS@ETAPS **65**, 65–84 (2002)
40. Uchitel, S., Kramer, J., Magee, J.: Synthesis of behavioral models from scenarios. IEEE Trans. Softw. Eng. **29**(2), 99–115 (2003)
41. Vespa, M., Bellodi, E., Chesani, F.: Probabilistic compliance in declarative process mining. In: PMAI@ECAI24, pp. 11–22 (2024)
42. Zugal, S., Pinggera, J., Weber, B.: Creating declarative process models using test driven modeling suite. In: CAiSE Forum (Selected Papers), pp. 16–32 (2011)
43. Zugal, S., Pinggera, J., Weber, B.: The impact of testcases on the maintainability of declarative process models. In: BMMDS/EMMSAD, pp. 163–177 (2011)

Towards an OR-Join Semantics that Meets Expectations

Andreas V. Hense[1(✉)] and Wil M. P. van der Aalst[2]

[1] Bonn-Rhein-Sieg University oAS, Grantham-Allee 20, 53757 Sankt Augustin, Germany
andreas.hense@h-brs.de
[2] RWTH Aachen University, Ahornstr. 55, 52074 Aachen, Germany
wvdaalst@pads.rwth-aachen.de

Abstract. The OR-join is an essential but semantically complex construct in process modeling that has challenged process modelers and analysts for decades. Because it possesses *non-local* semantics – an OR-join may need to wait for tokens to arrive from "further away" – making it difficult to determine when it is enabled or not. The BPMN specification provides "official" semantics, but despite its complexity, these semantics lead to "surprising" behavior in certain scenarios. This paper investigates the complexities of the BPMN OR-join specification and explores whether alternative semantics better align with the *expectations* of modeling experts.

To assess different OR-join policies, a *questionnaire* featuring 14 example models was distributed among a large group of process experts. Analysis of the responses shows a significant *lack of consensus* among experts regarding the expected behavior of OR-joins. None of the experts gave answers fully consistent with the official BPMN standard. We propose a new policy that aligns more closely with expert expectations and is also able to determine ambiguous situations.

Keywords: Business Process Modeling · BPMN · OR-Join Semantics

1 Introduction

The OR-Join [23] is an essential and valuable construct in business process modeling that has been used in Event-Driven Process Chains [12,24] and other modeling languages for over thirty years. The OR-join is a point in a process model where two or more alternative branches executed earlier are merged. It waits until all active incoming branches that can still reach the join have completed, and only then does it allow the process to continue.

In the early days, when process models were used only to describe business processes, the OR-Join was regarded as a straightforward construct, like the AND-Join and the XOR-Join. However, when the focus shifted from process modeling to business process automation and advanced analytics (e.g., conformance checking), it was discovered that some of the synchronization constructs

F. M. Maggi et al. (Eds.): BPM 2026, LNCS 16936, pp. 148–165, 2027.
https://doi.org/10.1007/978-3-032-37777-7_11

of Event-Driven Process Chains were problematic [1,13,18,22] and since then, many authors have worked on the semantics of the OR-Join. For example, [1] was the first paper to identify the vicious cycle paradox, i.e., whatever general rule is applied, it is always wrong. The problem with the OR-Join is that it has a non-local semantics [16] because it may have to wait for tokens to arrive from "further away" before being activated. In [1], it was shown that a cyclic definition does not need to have a fixpoint, proving that earlier definitions were incorrect. Despite these semantical challenges, the concept is widely used and the OR-Join is part of the BPMN specification, where it is called a merging inclusive gateway. The BPMN OR-Join semantics [19] originates from [26].

Although the early approaches for defining an executable semantics for the OR-Join are over twenty years old [15,28] the problem does not seem to be completely solved. When we look at the support for the OR-Join in the BPMN tool Camunda, for example, we can see that it is supported as a graphical symbol, but that neither the token simulation nor the automation component supports the OR-Join.[1]

Further motivation for this investigation comes from industry projects in which we used the workflow management system YAWL [11], which is based on workflow patterns [23] and was developed around the same time as BPMN. The OR-Join has received considerable attention during the design and implementation of YAWL and uses a state-space-based approach [3,28]. Nevertheless, in one industry project, we unexpectedly encountered a major performance issue: some OR-join computations took up to twenty minutes, making complex models with state-space-based OR-joins impractical to use. This illustrates that efficiency problems of state-space-based approaches to OR-Join activation [7] are not only of academic interest but also relevant in practice.

Given these challenges, it is no surprise the OR-Join resurfaces regularly. For example, in a recent paper [6], the authors present some example workflow specifications and call the BPMN semantics "surprising" and "anomalous".

The remainder of this paper is structured as follows. We define a core BPMN language that is the basis for formalizing the semantics of the main OR-join approaches proposed in the literature. We formalize the *state-space-based* semantics, the *BPMN 2.0* semantics, and the *post-dominance* semantics. We then try to *repair* the BPMN 2.0 semantics in a naïve way and discuss why this approach fails. Then we present our new *total order semantics* as an improvement of the post-dominance semantics. Our total order semantics can be configured to return unknown and thus avoids making arbitrary choices when OR-joins depend on each other. We do *not* claim that this is the best solution. However, the total order semantics exposes the problems, while trying to stay close to the expectations of the modeler. All five semantics are then evaluated in a survey for modeling experts. The survey includes 14 example situations inspired by the process models we identified in our literature review.

[1] Only recently, Camunda added OR-join support. Until version 8.5, Camunda only supported inclusive split gateways, i.e., no OR-joins, only OR-splits.

2 Representing Core BPMN Models

The *Business Process Modeling Notation* (BPMN) emerged as the de facto industry standard. Although our findings are not limited to BPMN models and apply to any notation involving OR-joins, we use a subset of BPMN to explain the different OR-join policies. Note that the OR-join problems already existed before the uptake of BPMN as a standard, see, for example, the "vicious cycle" paradox in the context of event-driven process chains [1].

The current OMG document describing the standard has 532 pages [19]. Depending on the counting methodology, the total number of distinct BPMN symbols can range from about 50 to over 100. Most of these are rarely used and not relevant to the OR-join semantics of BPMN. As shown in [17], typically only a small subset of BPMN is used. Therefore, we provide an abstraction that we call *core BPMN model* that only considers tasks, three different types of gateways (XOR, AND, and OR), and start and end events. We abstract from swimlanes, data, and other elements, focusing solely on control flow. Moreover, we adapt terminology to be more precise.

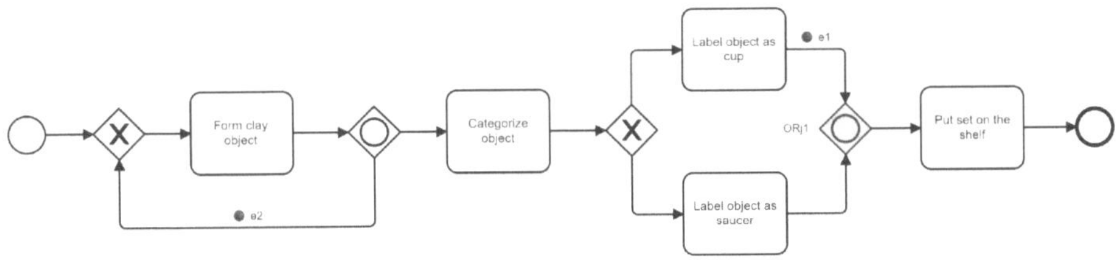

Fig. 1. A core BPMN model. According to the BPMN semantics, the highlighted OR-join is activated by token $e1$ and will not wait for token $e2$.

Figure 1 depicts such a core BPMN model with a highlighted OR-join and a concrete situation denoted by tokens *e1* and *e2*. The informed reader will have noted that this model lacks the soundness property and exhibits several modeling errors. We have deliberately chosen this example because many of the interesting OR-join models we have found in the literature are also unsound. The intention of the modeler was to create pairs of matching cups and saucers at the OR-join *ORj1*. However, according to the official BPMN semantics, the OR-join would be activated by token $e1$ and would not wait for token $e2$, although the token $e2$ enables the occurrence sequence involving activities "Form clay object", "Categorize object", and "Label object as saucer". This would put a token on the other input arc of the OR-join. The example already illustrates the challenge of deciding whether an OR-join is enabled or not. Later, we present the official BPMN 2.0 semantics and various alternative semantics.

As mentioned earlier, we abstract from irrelevant details described in the 532-page standard [19]. Instead, we focus on the things most relevant for control flow and the OR-join. Moreover, for clarity, we streamline language and deviate from the official BPMN terms where appropriate. We also assume a unique start and end event.

Definition 1 (Core BPMN Model). *A core BPMN model is a tuple (N, E) with nodes $N = T \cup \{se, ee\} \cup G_{XOR} \cup G_{AND} \cup G_{OR}$ and edges $E \subseteq N \times N$ such that:*

- *T is the set of tasks,*
- *se is the unique start event,*
- *ee is the unique end event,*
- *G_{XOR} is the set of XOR gateways,*
- *G_{AND} is the set of AND gateways,*
- *G_{OR} is the set of OR gateways, and*
- *the following constraints need to be satisfied:*
  - *each node is on a path from the start event se to the end event ee,*
  - *the start event has no incoming edges and the end event has no outgoing edges, and*
  - *each task has one incoming edge and one outgoing edge.*

Given a core BPMN model (N, E), we introduce the following notations. For $n \in N$: $\bullet n = \{(n_1, n_2) \in E \mid n_2 = n\}$ and $n\bullet = \{(n_1, n_2) \in E \mid n_1 = n\}$. Hence, $|\bullet n| = |n\bullet| = 1$ for $n \in T$, $\bullet se = \emptyset$, and $ee\bullet = \emptyset$. For $e = (n_1, n_2) \in E$: $\pi_1(e) = n_1$ and $\pi_2(e) = n_2$. A gateway $n \in G_{XOR} \cup G_{AND} \cup G_{OR}$ is a *split gateway* if $|n\bullet| \geq 2$. A gateway $n \in G_{XOR} \cup G_{AND} \cup G_{OR}$ is a *join gateway* if $|\bullet n| \geq 2$. Since we are particularly interested in the join behavior of the OR-join, we define $G_{OR}^{join} = \{n \in G_{OR} \mid |\bullet n| \geq 2\}$ to denote OR-joins with *multiple* incoming edges. Next to having an explicit notation for non-trivial OR joins (i.e., G_{OR}^{join}), it is helpful to distinguish two types of paths in a BPMN graph.

Definition 2 (Node and Edge Paths). *Let (N, E) be a core BPMN model. A node path of (N, E) is a sequence $\langle n_1, n_2, \ldots, n_k \rangle \in N^*$ such that $k \geq 2$ and $(n_i, n_{i+1}) \in E$ for any $1 \leq i < k$. An edge path of (N, E) is a sequence $\langle e_1, e_2, \ldots, e_k \rangle \in E^*$ such that $k \geq 1$ and $\pi_2(e_i) = \pi_1(e_{i+1})$ for any $1 \leq i < k$.*

For a given set A, $\mathbb{B}(A)$ is the set of all multisets over A. In a multiset, the same element can appear multiple times, but the order is not important. Formally, $\mathbb{B}(A) = A \to \mathbb{N}$, i.e., each element of A is mapped onto a natural number. If $a \in A$ and $B \in \mathbb{B}(A)$, then $B(a)$ is the number of occurrences of a in multiset B. $B_1 \uplus B_2$ is the union of two multisets and $B_1 \setminus B_2$ is the difference of two multisets. $[a_1, a_2, \ldots a_n]$ is the multiset consisting of the elements a_1, a_2, ... a_n. Note that the order does not matter, and the same elements can appear multiple times. For example, $[a, b, a, b, c] = [a, a, b, b, c] = [a^2, b^2, c]$. Sets are treated as multisets when needed, e.g., $[a, b, c, a, b, c] \uplus \{b, c, d\} = [a^2, b^3, c^3, d]$.

Definition 3 (State). *Let (N, E) be a core BPMN model. $S \in \mathbb{B}(E)$ is a state of (N, E). $S \in \mathbb{B}(E)$ is the initial state if $S = [e \in E \mid \pi_1(e) = se]$. $S \in \mathbb{B}(E)$ is a final state if for all $e \in S$: $\pi_2(e) = ee$.*

An element $e \in S$ is called a *token* or *marker*. In the initial state, all outgoing edges of the start event are marked (i.e., carry a token). In the final state, only ingoing edges of the end event are marked (i.e., carry a token). Tokens are depicted as dots next to an edge. For an edge $e \in E$, $S(e)$ is the number of tokens on the edge. Gateways and tasks can be *enabled*. For tasks and XOR/AND gateways, enabling is deterministic and well-defined. Tasks and XOR gateways are enabled if at least one of the incoming edges is marked. AND gateways are enabled if all of the incoming edges are marked. *For now, we leave the precise semantics of OR gateways open and define different alternatives in Sect.* 3. However, a minimal requirement is that at least one of the incoming edges is marked.

Definition 4 (Enabling). *Let (N, E) be a core BPMN model and $S \in \mathbb{B}(E)$ a state of the model. Tasks and gateways can be enabled in S using the following enabling rules.*

- $n \in T \cup G_{XOR}$ *is enabled if there exists an $e \in \bullet n$ such that $e \in S$ (at least one incoming edge is marked).*
- $n \in G_{AND}$ *is enabled if for all $e \in \bullet n$: $e \in S$ (all incoming edges are marked).*
- $n \in G_{OR}$ *may be enabled if there exists an $e \in \bullet n$ such that $e \in S$ (only partially defined).*

An enabled node can *occur* (i.e., it happens/is executed). We assume a fixed semantics for all nodes (including OR gateways). Tasks and XOR gateways remove one token from one of the incoming edges and produce one token for one of the outgoing edges when they occur. An AND gateway removes one token from each of the incoming edges and produces one token for each of the outgoing edges when it occurs. The OR gateway removes one token from each of the incoming edges *that is marked* when it occurs. If an incoming edge is not marked, then it was not needed for enabling and no token is removed. The OR gateway produces *at most* one token for each of the outgoing edges. However, at least one outgoing edge receives a token.

Definition 5 (Occurrence). *Let (N, E) be a core BPMN model, $S \in \mathbb{B}(E)$ a state of (N, E), and $n \in N$ a task or gateway enabled in S. S' is the state after n occurred (notation $S \rightarrow_n S'$).*

- *If $n \in T \cup G_{XOR}$, then there are edges $e_1 \in \bullet n$ and $e_2 \in n\bullet$ such that $e_1 \in S$ and $S' = (S \setminus [e_1]) \uplus [e_2]$.*
- *If $n \in G_{AND}$, then $S' = (S \setminus \bullet n) \uplus n\bullet$.*
- *If $n \in G_{OR}$, then $E_1 = \{e \in S \mid e \in \bullet n\}$, $E_2 \subseteq n\bullet$, $E_2 \neq \emptyset$, and $S' = (S \setminus E_1) \uplus E_2$.*

Note that, based on this definition, the join behavior of the OR-gateway is *deterministic* (apart from deciding whether it is enabled or not). The split behavior is not (a non-empty subset of outgoing arcs is selected). An *occurrence sequence* of a core BPMN model (N, E) is a sequence of states $\langle S_1, S_2, \ldots S_k \rangle$ such that for any $1 \leq i < k$: $S_i \rightarrow_n S_{i+1}$. The model's behavior is defined by all possible occurrence sequences that start in the initial state.

Note that we *fully defined the semantics* of core BPMN models *except* for the enabling of non-trivial OR joins in G_{OR}^{join}. Next, we consider different types of enabling policies and evaluate their effects.

3 Characterizing the Main OR-Join Policies

To determine whether an OR-Join is enabled or not, one can either use a state-space-based or graph-based approach [26]. A state-space-based approach analyzes the possible future states of the model. A graph-based approach considers only tokens on edges and paths to the OR-Join, thereby reducing computational complexity. The following subsections describe one *state-space-based OR-join policy*, followed by four *graph-based OR-join policies.*

3.1 The State-Space-Based Approach

Before describing the OR-join semantics based on the OMG standard, we first present the OR-join policy that captures the intuitive meaning of the OR-join semantics best. Think of the OR-join as a "bus driver" who waits for the passengers who can still reach the bus stop. This means that at least one incoming edge is marked, and it is not possible to mark any additional unmarked incoming edges in the future. Seemingly, the most precise semantics of the OR-Join is the state-space-based approach given by Definition 5. When evaluating if a certain OR-join n is enabled in a particular state of the core BPMN model, we calculate all possible occurrence sequences from this state, and we treat all other OR-joins according to the partially defined enabling of Definition 4. To determine whether n is enabled, we temporarily assume that the enabling of all OR-joins is the same as an XOR-join, i.e., one marked input edge suffices.

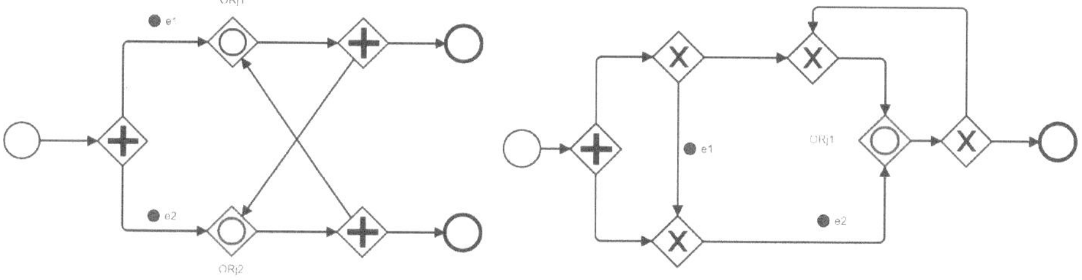

(a) *DI10* (b) *DI13*

Fig. 2. Two situations: (a) the vicious cycle, originally defined in the context of EPCs [1,26], and (b) a situation where the post-dominance semantics would wait for $e1$ [6].

Definition 6 (Enabling OR Join (State-space)). *Let (N, E) be a core BPMN model and $S \in \mathbb{B}(E)$ a state of the model. An OR-join $n \in G_{OR}^{join}$ is enabled if*

- *there exists an $e \in \bullet n$ such that $e \in S$*
- *for all occurrence sequences $\langle S_1, S_2, \ldots, S_k \rangle$ with $S_1 = S$ not involving OR-join n (i.e., for all $1 \leq i < k$: $S_i \not\to_n S_{i+1}$), there is no state S_j for all $1 \leq j \leq k$ with $set(S(\bullet n)) \subsetneq set(S_j(\bullet n))$, where set is the function that transforms a multiset into an ordinary set.*[2]

This approach calculates all possible future states, approximating the behavior of other OR-joins and looking for tokens that can arrive in the future at incoming edges of the considered OR-join that do not have a token yet. These semantics are close to the OR-join semantics called "optimistic" in the context of YAWL [27,28]. This can also lead to deadlocks when there are several OR-Joins waiting for each other, like the vicious cycle of Fig. 2a.[3]

3.2 Where BPMN Does Not Behave As Expected

The semantics of the OR-Join as defined in the specification of BPMN goes back to [26] and is as follows [19]: *"The inclusive gateway is activated if*

- *at least one incoming sequence flow has at least one token and*
- *for every directed path formed by sequence flow that*
  1. *starts with a sequence flow f of the diagram that has a token,*
  2. *ends with an incoming sequence flow of the inclusive gateway that has no token*
  3. *does not visit the inclusive gateway*
- *there exists a directed path formed by sequence flow that*
  1. *starts with f,*
  2. *ends with an incoming sequence flow of the inclusive gateway that has a token*
  3. *does not visit the inclusive gateway"*

We have cited the BPMN semantics verbatim here and will now translate it to our terminology of core BPMN models introduced in Sect. 2.

Definition 7 (Enabling OR Join (BPMN)). *Let (N, E) be a core BPMN model and $S \in \mathbb{B}(E)$ a state of the model. An OR-join $n \in G_{OR}^{join}$ is enabled if*

- *there exists an $e \in \bullet n$ such that $e \in S$*
- *for every edge path $\langle e_1, e_2, \ldots, e_k \rangle$ with*
  1. *$e_1 \in S$*

[2] Note that we defined occurrence sequences modulo an OR-join enabling policy. Here, we assume that any OR-join (but not n) is enabled the moment one of the input edges has a token (see Definitions 4 and 5).

[3] If a BPMN diagram has more than one end event, as in this example, we can transform it to a BPMN core model by inserting a new OR-join in front of a new unique end event and by directing all sequence flows to the original end events to this new OR-join. The original end events are thus eliminated.

   2. $\pi_2(e_k) = n$ *and* $e_k \notin S$
   3. *and for all* $1 \leq i < k : \pi_2(e_i) \neq n$
- *there exists an edge path* $\langle f_1, f_2, \ldots, f_l \rangle$ *with*
   1. $f_1 = e_1$
   2. $\pi_2(f_l) = n$ *and* $f_l \in S$
   3. *and for all* $1 \leq j < l : \pi_2(f_j) \neq n$

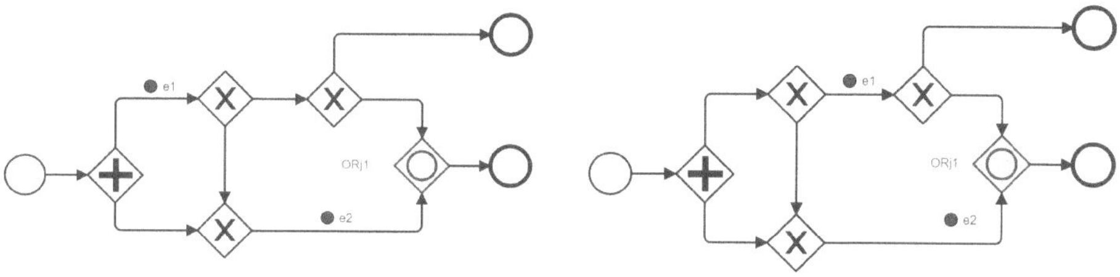

(a) *DI03* (b) *DI03* after making the first choice

Fig. 3. The BPMN semantics enables the OR-Join in situation (a), but after the first choice (b) all policies agree that the OR-join needs to wait for $e1$ [6].

The first two requirements of the OR-Join semantics seem to roughly align with what we expect. However, the motivation for the last requirement is not clear at first glance and leads to surprising behavior, as noted by [6]. For example, in Fig. 3a, we have a token on e_2 at the OR-Join. Although there is a path from the token on e_1 to the empty incoming arc of the OR-Join, we do not have to wait for it because there is also a path from e_1 to the incoming arc e_2 that has a token. Therefore, according to the BPMN definition above, the OR-Join is activated.

In Fig. 3b, we see the next situation after the token has passed the first exclusive gateway and is now before the second exclusive gateway. Now the OR-Join is not activated anymore. To activate the OR-Join in Fig. 3a is against expectations because it may very well be that the token takes the path to the empty arc of the OR-Join and we couldn't agree more with the authors of [6] who call this behavior "surprising" and "anomalous".

3.3 A Naïve Variant of the BPMN Semantics

We have seen that the BPMN semantics are complex, and especially the motivation for the third bullet point is not immediately clear. We have also seen that it produced surprising behavior in simple situations. Therefore, we now modify the BPMN semantics by "deleting" the third bullet point and obtain the following OR-join policy.

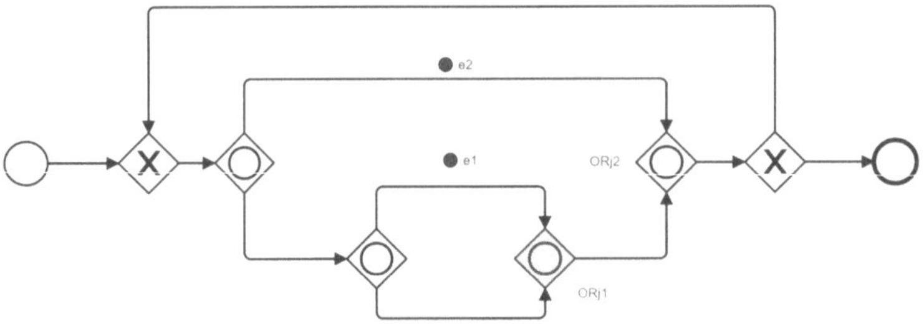

Fig. 4. *DI09* The vast majority of experts consider OR-join *ORj1* to be enabled, but multiple semantics block *ORj1* waiting for the other OR-join [26].

Definition 8 (Enabling OR Join (Naïve)). *Let* (N, E) *be a core BPMN model and* $S \in \mathbb{B}(E)$ *a state of the model. An OR-join* $n \in G_{OR}^{join}$ *is enabled if*

- *there exists an* $e \in \bullet n$ *such that* $e \in S$
- *there exists no edge path* $\langle e_1, e_2, \ldots, e_k \rangle$ *with*
  1. $e_1 \in S$
  2. $\pi_2(e_k) = n$ *and* $e_k \notin S$
  3. *and for all* $1 \leq i < k : \pi_2(e_i) \neq n$

This policy is much simpler and solves the problem of Fig. 3a, namely, it waits for token $e1$ to arrive. However, this policy fails dramatically with the cyclic example of Fig. 4. It would wait with the activation of $ORj1$ until the token $e2$ arrives. A deadlock would be the consequence. This represents a severe flaw of this variant, because the example of Fig. 4 is well-structured and has the soundness property. Therefore, this policy is definitely *not* an improvement of the official BPMN semantics.

3.4 Ordering the OR-Joins with Post-Dominance

The problem we had with the simplified semantics applied to the example of Fig. 4 stems from the fact that we were unable to differentiate between the two OR-Joins $ORj1$ and $ORj2$. For the human eye, it is clear that $ORj1$ precedes $ORj2$. Therefore, $ORj1$ should be activated, and $ORj2$ should wait. [20] has introduced the concept of post-dominance that starts from the end node and solves the problem for the example of Fig. 4.

Definition 9 (Post-Dominance). *Let* (N, E) *be a core BPMN model. A node* $n \in N$ post-dominates *node* $m \in N$ *if for all node paths* $\langle n_1, n_2, \ldots, n_k \rangle \in N^*$ *with* $n_1 = m$ *and* $n_k = ee$ *there exists an* $1 \leq i \leq k$ *such that* : $n_i = n$.

Based on post-dominance, the semantics of the OR-Join can be defined as in [20].

Definition 10 (Enabling OR Join (Post-Dominance)). *Let* (N, E) *be a core BPMN model and* $S \in \mathbb{B}(E)$ *a state of the model. An OR-join* $n \in G_{OR}^{join}$ *is enabled if*

- *there exists an* $e \in \bullet n$ *such that* $e \in S$
- *for each node path* $\langle n_1, n_2, \ldots, n_k \rangle$ *with* $S(n_1, n_2) \geq 1$, $n_k = n$, *there exists an* $1 < j < k$ *such that* $n_j \in G_{OR}^{join}$ *and* n_j *post-dominates* n *or* $n_j = n$.

This OR-join policy solves the problem for the example of Fig. 4, but it fails for the example of Fig. 2b. In this case, *ORj1* will wait for token e_1 to arrive, although we would expect it to be activated immediately. The specification of Fig. 2b does not have the soundness property: it lacks proper completion.

3.5 Total Order Semantics

Figure 2b is the example where the post-dominance semantics differ from the expectation of the experts. Therefore, we improve this OR-join policy by considering only node paths that lead to an unmarked incoming arc of the OR-join. Furthermore, we extend the partial order of post-dominance to a total order to deal with situations like the vicious cycle of Fig. 2a. The basic idea is either to break symmetry and assume a total order on OR-joins, or to detect the problem and return "unknown". Unlike all other OR-join policies, we *avoid making an arbitrary choice.*

Definition 11 (Total Order Semantics). *Let* (N, E) *be a core BPMN model with a reflexive total ordering relation* $\preceq_{OR} \subseteq G_{OR}^{join} \times G_{OR}^{join}$ *on OR join gateways, and* $S \in \mathbb{B}(E)$ *a state of the model. An OR-join* $n \in G_{OR}^{join}$ *is enabled if*

- *there exists an* $e \in \bullet n$ *such that* $e \in S$
- *there is no node path* $\langle n_1, n_2, \ldots, n_k \rangle$ *such that* $S(n_1, n_2) \geq 1$, $S(n_{k-1}, n_k) = 0$, $n_k = n$, *and* $\{n_j \in G_{OR} \mid 1 < j < k \ \wedge \ n \preceq_{OR} n_j\} = \emptyset$.

Note that the definition is parameterized by a reflexive total ordering relation $\preceq_{OR} \subseteq G_{OR}^{join} \times G_{OR}^{join}$. This may be a refinement of the partial order defined by post-dominance. Note that any partial order or preorder can be turned into a reflexive total order. Such a total order is needed to resolve paradoxes, such as the vicious cycle in Fig. 2a. Alternatively, we may decide to return "unknown" if the order defined by post-dominance leaves OR-joins unordered.

The first requirement in Definition 11 is as before. The second requirement can also be phrased as follows: for all node paths $\langle n_1, n_2, \ldots, n_k \rangle$ such that $S(n_1, n_2) \geq 1$, $S(n_{k-1}, n_k) = 0$, and $n_k = n$: $\{n_j \in G_{OR} \mid 1 < j < k \ \wedge \ n \preceq_{OR} n_j\} \neq \emptyset$. Moreover, the last part can be rewritten as: there is a $1 < j < k$ such that $n_j \in G_{OR}$ and $n \preceq_{OR} n_j$ (i.e., n itself or a "later" OR-join is on the path to n). The paths considered in the second requirement are of length $k \geq 3$ because $S(n_1, n_2) \geq 1$ and $S(n_{k-1}, n_k) = 0$. Also, n_1 could be an OR-join gateway or even n itself. n_1 has already occurred because $S(n_1, n_2) \geq 1$ and is not relevant for the enabling of n.

The total order semantics are simple and purely syntactical. It is the only policy addressing the vicious cycle paradox. The total order semantics are parameterized by the ordering relation $\preceq_{OR}$. In the evaluation, we assume that this is a refinement of the partial order defined by post-dominance. However, if there are multiple refinements leading to different answers, we return "unknown" or "undefined" in the evaluation. This corresponds to a human process analyst stating that she does not know or cannot decide. Of the models considered in the paper, this only applies to the vicious cycle example in Fig. 2a.

Definition 12 (Well-Ordered BPMN Model). *Let (N, E) be a core BPMN model. (N, E) is well-ordered if for any $n, m \in G_{OR}^{join}$ on a common node path (i.e., connected): $n = m$, n post-dominates m, or m post-dominates n.*

In a *well-ordered* BPMN model, two connected OR-joins must have a clear ordering relation: one is last when moving toward the end of the process. If the OR-joins are not connected (i.e., not on a common node path), then one cannot wait for the other, and the ordering is irrelevant. In the evaluation, we return "unknown" for the total order policy if the BPMN is *not* well-ordered. *DI10* shows a BPMN model that is not well-ordered. All other models shown thus far are well-ordered.

4 Evaluating the Expected OR-Join Behavior

Literature Search for Example Workflows. Startled by the surprising behavior of the example specifications in [6], we conducted a literature search consisting of four phases:

1. Backward search starting with [6]: we have examined all publications that addressed OR-Join semantics and included process models with OR-Joins.
2. Forward search using Google Scholar: from the publications found in the first phase, we have looked at all new publications where the title indicated some relationship to the OR-Join topic.
3. Literature for the terms "OR-Join" and "inclusive gateway" using Google Scholar and DBLP.
4. Relevant examples of workflow specifications containing OR-Joins were found here: [2,4–6,8–10,14,20,21,25,26,28,29].

We have found four examples with OR-Joins and YAWL cancellation regions: Fig. 10 in [28], the visa process in [29], Fig. 6c in [2], and Fig. 2 in [25]. Most of them could be translated into BPMN without problems using sub-processes and interrupting boundary events. While cancellation regions in general can affect the question of whether a process model has deadlocks, the concrete example diagrams did not add any new aspects to our discussion here.

Creating Example Situations in BPMN. We have translated all remaining specifications to Core BPMN and have removed all elements that are irrelevant for OR-join activation. The removed elements include unnecessary tasks and uncritical structures of corresponding AND- or XOR-splits and joins. From each workflow specification, we have created one or more example situations – 39 in total. An example *situation* consists of a core BPMN diagram in a certain state and a single highlighted OR-Join.

Eliminating Indisputable Examples. We have investigated 39 example situations according to the five OR-join policies discussed in this paper. Among these were 8 situations where the different policies disagreed on the question"is the OR-Join activated" – the disputable examples – and 31 situations where the policies agreed – the indisputable examples. These came from earlier publications: [4,6,8,9,20,21,23,26,28,29].

Conducting the Survey. From the 39 example situations, we picked all 8 disputable ones and added 6 indisputable ones for control purposes. The resulting 14 situations were arranged in random order and used in the questionnaire.[4] The survey was conducted between 05.11.2025 and 30.11.2025 with *SoSci Survey* (https://www.soscisurvey.de). The invitations to the survey were sent by e-mail to 82 experts in the field of business process modeling. These experts also included the authors of OR-Join papers, where we were able to identify current e-mail addresses. 37 of these experts completed the questionnaire. 97% of the participants were familiar with BPMN, 81% with Petri Nets, 62% with Workflow Nets, 59% with UML Activity Diagrams, 51% with Event-Driven Process Chains, and 51% with YAWL.

5 Evaluation of the Survey Results

There are 14 situations labeled *DI01*, *DI02*, ..., *DI14* answered by 37 participants, labeled *P16*, *P18*, ..., *P70*. The participants are not numbered consecutively, e.g., *P17* does not exist. These identifiers correspond to invitations. The goal was to ensure anonymity and prevent a single person from filling out the questionnaire multiple times. Next to the participants, also the five *OR-join policies* described in Sect. 3 were used to answer the questions. The following labels were used: *State-space* refers to the semantics described in Definition 6, *BPMN* refers to the semantics described in Definition 7, *Naïve* refers to the semantics described in Definition 8, *Post Dom* refers to the semantics described in Definition 10, and *Total Order* refers to the improved semantics described in Sect. 3.5 (returning "unknown" if the model is not well-ordered). This implies that for each of the 14 situations, there are $37 + 5 = 42$ answers (37 participants and 5 policies). Possible answers are *yes*, *no*, or *unk* (unknown).

Table 1. The entropy values of the answers given for each situation. Higher values correspond to a lack of consensus.

DI01	DI02	DI03	DI04	DI05	DI06	DI07	DI08	DI09	DI10	DI11	DI12	DI13	DI14
1.09	1.12	1.00	1.30	0.81	1.40	1.09	1.09	0.88	1.25	1.04	1.14	1.05	1.14

[4] To save space, we do not show all diagrams here. All 14 BPMN situations can be found in the supplementary material https://doi.org/10.5281/zenodo.18511692.

Table 1 shows the entropy of the answers given for each situation. Since there are three possible values, the maximal entropy is $log_2(3) \approx 1.58$. The higher the entropy, the more diverse the answers are. *DI06* was the question having the highest entropy (entropy of 1.4 considering all 42 answers, distributed as follows: 55% *no*, 31% *yes*, 14% *unk*) followed by *DI04* and *DI10* having entropy values of 1.30 and 1.25, respectively. These were apparently the most "controversial" situations. For situation *DI04*, 17 participants answered *yes*, 17 participants answered *no*, and three answered *unk*. Also, the OR-join policies disagreed (three *no* and two *yes*). *DI05* and *DI09* were the situations with the lowest entropy, indicating that these were situations with the highest level of consensus. However, full agreement would mean an entropy of 0. The relatively high entropy values in Table 1 illustrate that consensus is limited. To compute correlations, we need to convert the categorical answers into numerical values. We map *yes* to 1, *no* to -1, and *unk* to 0. The idea is that the "distance" between *yes* and *no* is larger than the 'distance" between *unk* and *yes* or *no*. Table 2 shows the pairwise correlation of a small subset of participants and policies using these values. By definition, the diagonal values are 1.0. A correlation of 1 means a perfect positive linear relationship between two variables. A negative correlation means that as one variable increases, the other variable tends to decrease (and vice versa). For example, the correlation between participants *P43* and *P49* is -0.73. This means that when participant *P43* says yes (value 1.0), participant *P49* most likely says

Table 2. Pairwise correlation of participants and policies. Values close to 1 indicate that two participants or policies agree on most or all situations. Negative values indicate disagreement. See https://doi.org/10.5281/zenodo.18494165 for the full table and more detailed analysis results.

	P16	P18	P19	P20	...	P66	P68	P70	State-space	BPMN 2.0	Naïve	Post Dom	Total Order
P16	1.00	0.21	0.28	0.06	...	−0.10	-0.14	0.58	−0.03	0.15	−0.14	1.00	0.79
P18	0.21	1.00	0.17	−0.19	...	−0.47	0.11	0.21	−0.12	−0.28	0.11	0.21	0.29
P19	0.28	0.17	1.00	0.65	...	0.47	0.68	0.28	0.12	0.78	0.68	0.28	0.62
P20	0.06	−0.19	0.65	1.00	...	0.41	0.44	0.06	0.19	0.44	0.44	0.06	0.25
⋮	⋮	⋮	⋮	⋮	⋱	⋮	⋮	⋮	⋮	⋮	⋮	⋮	⋮
P66	−0.10	−0.47	0.47	0.41	...	1.00	0.32	0.25	0.56	0.60	0.32	−0.10	0.02
P68	−0.14	0.11	0.68	0.44	...	0.32	1.00	−0.14	0.37	0.53	1.00	−0.14	0.42
P70	0.58	0.21	0.28	0.06	...	0.25	−0.14	1.00	-0.03	0.15	-0.14	0.58	0.40
State-space	-0.33	-0.19	0.19	0.30	...	0.73	0.44	0.06	1.00	0.06	0.44	−0.33	−0.10
BPMN 2.0	0.15	−0.28	0.78	0.44	...	0.60	0.53	0.15	0.06	1.00	0.53	0.15	0.40
Naïve	−0.14	0.11	0.68	0.44	...	0.32	1.00	-0.14	0.44	0.53	1.00	−0.14	0.42
Post Dom	1.00	0.21	0.28	0.06	...	−0.10	−0.14	0.58	−0.33	0.15	−0.14	1.00	0.79
Total Order	0.79	0.29	0.62	0.25	...	0.02	0.42	0.40	−0.10	0.40	0.42	0.79	1.00

no (value -1.0) or vice versa. Participants *P43* and *P49* agree only on situation *DI02*; for all other situations, they give different answers.

Six participants have a pairwise correlation of 1.0: *P19*, *P21*, *P29*, *P34*, *P46*, and *P47*. This means that they agreed on all 14 situations. The answers of these six participants were closest to the *BPMN* semantics described in Definition 7 (correlation 0.78).

Only two participants gave answers fully consistent with one of the policies. Participant *P16* and the *Post Dom* policy (i.e., the semantics described in Definition 10) agreed on all 14 situations. This is evidenced by the correlation value 1.0. Participant *P68* and the *Naïve* policy (i.e., the semantics described in Definition 8) agreed on all 14 situations. This is also evidenced by the correlation value 1.0.

The average correlation between the *Total Order* policy (Definition 3.5) and all participants is 0.40. This is the highest value. The average correlation between the *State-space* policy (Definition 6) and all participants is the lowest: 0.12.

Table 3. Among the evaluated policies, *Total Order* is closest to the participants' responses, exhibiting the smallest distance. The *State-space* policy has the greatest distance from the participants' answers.

	Sum Manhattan Distance	Average Manhattan Distance	Sum Euclidean Distance	Average Euclidean Distance
Total Order	320	8.65	139.64	3.77
BPMN 2.0	331	8.95	141.92	3.84
Naïve	343	9.27	142.42	3.85
Post Dom	349	9.43	147.47	3.99
State-space	409	11.05	162.59	4.39

Each participant can be represented by a vector of length 14, e.g., *P16* corresponds to $\boldsymbol{v}_{P16} = (-1, -1, -1, 1, -1, 1, -1, -1, 1, -1, -1, -1, -1, -1)$. Policies can be presented in the same manner, e.g., *Total Order* corresponds to $\boldsymbol{v}_{TO} = (-1, -1, -1, 1, -1, 1, -1, -1, 1, 0, -1, -1, 1, -1)$. Hence, we can compute the *distance between a participant and a policy*. For example, the Manhattan distance between $\boldsymbol{v}_{P16}$ and $\boldsymbol{v}_{TO}$ is 3 and the Euclidean distance is 2.24. The sum of all Manhattan distances between $\boldsymbol{v}_{TO}$ and all 37 participants is 320, i.e., 8.65 on average. Table 3 shows the distances for all five policies. Independent of the chosen distance measure, the *Total Order* policy aligns most closely with the participants' responses, whereas the *State-space* policy deviates the most. This matches the findings using average correlation.

Next to the results in Tables 1, 2, and 3, we also applied more advanced methods such as computing mutual information and Theil's uncertainty coefficients (see https://doi.org/10.5281/zenodo.18494165 for more analysis results).

These also support the findings that we have seen before. The *Total Order* policy aligns best with the answers given by participants. The *State-space* policy diverges most from the answers given by the participants. However, differences are rather small, and there is no consensus among participants. A few additional situations could have changed the results. Also, factors such as the layout influence participants' answers.

Table 4. Comparing policies and participants per situation.

	DIO1	DIO2	DIO3	DIO4	DIO5	DIO6	DIO7	DIO8	DIO9	DIO10	DIO11	DIO12	DIO13	DIO14
Total Order	*no*	*no*	*no*	*yes*	*no*	*yes*	*no*	*no*	*yes*	*unk*	*no*	*no*	*yes*	*no*
BPMN 2.0	*no*	*no*	*yes*	*no*	*no*	*no*	*no*	*no*	*yes*	*no*	*no*	*no*	*yes*	*no*
Nave	*no*	*no*	*no*	*no*	*no*	*no*	*no*	*no*	*no*	*no*	*no*	*no*	*yes*	*no*
Post Dom	*no*	*no*	*no*	*yes*	*no*	*yes*	*no*	*no*	*yes*	*no*	*no*	*no*	*no*	*no*
State-space	*no*	*yes*	*no*	*no*	*no*	*no*	*no*	*yes*	*no*	*no*	*yes*	*no*	*yes*	*no*
yes	9	11	6	17	4	10	5	14	**31**	6	12	7	**27**	7
no	**25**	**24**	**28**	17	**30**	**21**	**26**	**22**	3	**24**	**24**	**25**	7	**25**
unk	3	2	3	3	3	6	6	1	3	7	1	5	3	5

After the detailed statistical analysis of the results, we analyze where participants deviate from the policies defined before. Table 4 shows the results of the survey together with the "answers" given by the five policies. We mention a few of the more qualitative findings. Situations *DI01*, *DI12*, and *DI14* are isomorphic, but using very different layouts. This can be confusing, even for experts, as indicated by the fact that inconsistent answers are given.

The handling of situation *DI03* by the BPMN 2.0 semantics is different from the expectations of the majority of experts. The BPMN semantics does not wait for token *e1*. For situation *DI04*, opinions are mixed with an equal split between yes and no. For situation *DI10*, which contains a vicious cycle, only the *Total Order* policy returns an unknown. This is also the example with the highest number of participants choosing unknown. However, all other policies, as well as the majority of participants, answered no (despite the paradox) (Fig. 5).

6 Conclusion

Motivated by the surprising results of the official BPMN 2.0 semantics and our own experience with OR-joins in industry projects, we initiated this study to systematically collect all OR-join examples and existing OR-join specifications in the literature. After analyzing these, we created a survey among process modeling experts. We invited 82 experts and received 37 complete responses, which we compared with five representative OR-join policies.

The results show a lack of consensus among experts and that some OR-join policies (including the official BPMN 2.0 standard) yield counterintuitive and surprising results. The idea of ordering the OR-joins in the *Post Dom* semantics

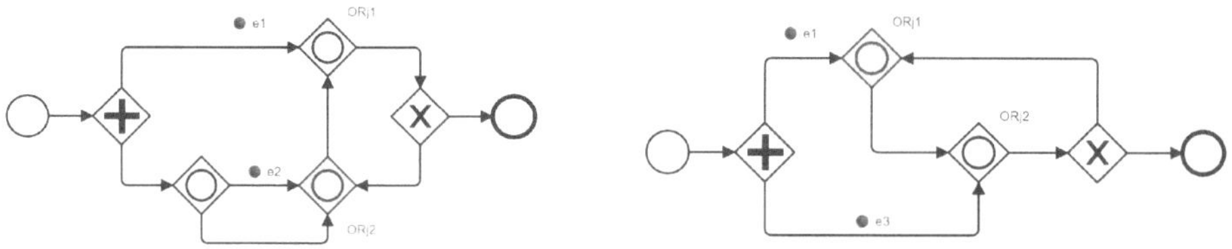

(a) *DI04* (b) *DI06*

Fig. 5. For situation (a), opinions are mixed equally among participants and policies [6], and for (b), it is surprising that the majority of participants say no [20]. In both cases, enabling the OR-join seems more logical given the ordering.

solved many problems in situations with more than one OR-join and circular structures. However, the *Post Dom* has problems with unsound process models and cannot handle paradoxes such as the vicious cycle.

Therefore, we proposed the *Total Order* policy, which can handle such situations and, if desired, resolve them by either highlighting the problem or deliberately breaking symmetry when a paradox emerges. The policy also allows for an efficient implementation. Moreover, a detailed statistical analysis of the results shows that the *Total Order* policy is closest to the experts' expectations.

This paper's title suggests that we are working toward a new OR-join semantics that meets experts' expectations. Although we have shown that our semantics comes closest to experts' expectations, we do not claim that our *Total Order* policy is optimal. We encourage modelers to rethink their BPMN model when it is not well-ordered.

Our questionnaire focused on BPMN process models and used BPMN syntax throughout. While our intention was to capture the genuine expectations of experts, we cannot exclude the possibility that many participants were already familiar with the official BPMN 2.0 semantics and therefore answered accordingly. We hope that, in the future, as experts become more familiar with our total-order semantics, their expectations may gradually shift in favor of this new approach.

Acknowledgment. We thank Arthur ter Hofstede for his support during the initial phase of this research and Michael Adams for proposing the original idea. We are also grateful to all colleagues from the BPM community who took the time to respond to our extensive survey.

References

1. van der Aalst, W.M.P., Desel, J., Kindler, E.: On the Semantics of EPCs: A Vicious Circle. In: Nüttgens, M., Rump, F.J. (eds.) EPK 2002 - Geschäftsprozessmanagement mit Ereignisgesteuerten Prozessketten, pp. 71–79 (2002)

2. van der Aalst, W.M.P., ter Hofstede, A.H.M.: YAWL: yet another workflow language. Inf. Syst. **30**(4), 245–275 (2005)
3. Adams, M., Hense, A.V., ter Hofstede, A.H.M.: YAWL: an open source business process management system from science for science. SoftwareX **12**, 100576 (2020)
4. Börger, E., Sörensen, O., Thalheim, B.: On defining the behavior of OR-joins in business process models. J. Univers. Comput. Sci. **15**(1), 3–32 (2009)
5. Christiansen, D.R., Carbone, M., Hildebrandt, T.T.: Formal Semantics and Implementation of BPMN 2.0 Inclusive Gateways. In: WS-FM 2010. LNCS, vol. 6551, pp. 146–160., Hoboken, NJ, USA (2010)
6. Corradini, F., Muzi, C., Re, B., Rossi, L., Tiezzi, F.: BPMN 2.0 OR-join semantics: Global and local characterisation. Inf. Syst. **105**, 101934 (2022)
7. Cuntz, N., Kindler, E.: On the semantics of EPCs - efficient calculation and simulation. Enterp. Model. Inf. Syst. Archit. Int. J. Concept. Model. **1**(1), 14–24 (2005)
8. Dumas, M., Großkopf, A., Hettel, T., Wynn, M.T.: Semantics of standard process models with OR-Joins. In: CoopIS 2007, LNCS, vol. 4803, pp. 41–58., Vilamoura, Portugal (2007)
9. Favre, C., Völzer, H.: The difficulty of replacing an inclusive OR-Join. In: Barros, A., Gal, A., Kindler, E. (eds.) BPM 2012, LNCS, vol. 7481, pp. 156–171., Tallinn, Estonia (2012)
10. Gfeller, B., Völzer, H., Wilmsmann, G.: Faster OR-Join Enactment for BPMN 2.0. In: Dijkman, R.M., Hofstetter, J., Koehler, J. (eds.) BPMN 2011, Lucerne, Switzerland. LNBIP, vol. 95, pp. 31–43 (2011)
11. ter Hofstede, A.H.M., van der Aalst, W.M.P., Adams, M., Russell, N. (eds.): Modern Business Process Automation - YAWL and its Support Environment. Springer, Cham (2010)
12. Keller, G., Nüttgens, M., Scheer, A.: Semantische Prozessmodellierung auf der Grundlage ereignisgesteuerter Prozessketten (EPK). Veröffentlichungen des Instituts für Wirtschaftsinformatik, Iwi (1992)
13. Kindler, E.: On the semantics of EPCs: resolving the vicious circle. Data Knowl. Eng. **56**(1), 23–40 (2006)
14. Kubovy, J., Küng, J.: Refinement of BPMN 2.0 Inclusive and Complex Gateway Activation Concept Towards Process Engine. In: ERP Future 2013 Conference, vol. 8, pp. 55–62., Vienna, Austria. LNISO (2013)
15. Mendling, J., van der Aalst, W.M.P.: Formalization and Verification of EPCs with OR-Joins Based on State and Context. In: CAiSE 2007, Trondheim, Norway. LNCS, vol. 4495, pp. 439–453 (2007)
16. Mendling, J., van Dongen, B.F., van der Aalst, W.M.P.: Getting rid of the OR-Join in business process models. In: EDOC 2007, pp. 3–14. Annapolis, MD, USA (2007)
17. Muehlen, M., Recker, J.: How much language is enough? Theoretical and practical use of the business process modeling notation. In: CAiSE'08). LNCS, vol. 5074, pp. 465–479 (2008)
18. Nüttgens, M., Rump, F.J.: Syntax und Semantik Ereignisgesteuerter Prozessketten (EPK). In: Promise 2002, Potsdam. LNI, vol. P-21, pp. 64–77 (2002)
19. Object Management Group: Business Process Model and Notation (BPMN) (2013). version 2.0.2
20. Prinz, T.M., Amme, W.: A complete and the most liberal semantics for converging OR gateways in sound processes. Complex Syst. Inf. Model. Q. **4**, 32–49 (2015)
21. Prinz, T.M., Choi, Y., Ha, N.L.: Soundness Unknotted: Efficient Algorithm for Process Models with Inclusive Gateways by loosening Loops. SSRN (2023)
22. Rittgen, P.: Quo vadis EPK in ARIS? - Ansätze zu syntaktischen Erweiterungen und einer formalen Semantik. Wirtschaftsinf. **42**(1), 27–35 (2000)

23. Russell, N., van der Aalst, W.M.P., ter Hofstede, A.H.M.: Workflow Patterns: The Definitive Guide. MIT Press (2016)
24. Scheer, A.: ARIS Toolset: A Software Product is Born. Inf. Syst. **19**(8), 607–624 (1994)
25. Verbeek, H.M.W., van der Aalst, W.M.P., ter Hofstede, A.H.M.: Verifying workflows with cancellation regions and OR-joins: an approach based on relaxed soundness and invariants. Comput. J. **50**(3), 294–314 (2007)
26. Völzer, H.: A New Semantics for the Inclusive Converging Gateway in Safe Processes. In: BPM 2010. LNCS, vol. 6336, pp. 294–309. Springer, Hoboken, NJ, USA (2010)
27. Wynn, M.T., van der Aalst, W.M.P., ter Hofstede, A.H.M., Edmond, D.: Verifying workflows with cancellation regions and OR-Joins: an approach based on reset nets and reachability analysis. In: BPM 2006, Vienna, Austria. LNCS, vol. 4102, pp. 389–394 (2006)
28. Wynn, M.T., Edmond, D., van der Aalst, W.M.P., ter Hofstede, A.H.M.: Achieving a general, formal and decidable approach to the or-join in workflow using reset nets. In: ICATPN 2005, Miami, USA. LNCS, vol. 3536, pp. 423–443 (2005)
29. Wynn, M.T., Verbeek, H.M.W., van der Aalst, W.M.P., ter Hofstede, A.H.M., Edmond, D.: Business process verification - finally a reality! Bus. Process. Manag. J. **15**(1), 74–92 (2009)

Exact Language Computation for Stochastic Process Trees via Discrete-Time Markov Chain Construction

András Horváth[1(✉)], Paolo Ballarini[2], and Pierre Cry[2]

[1] Università di Torino, Torino, Italy
horvath@di.unito.it
[2] CentraleSupélec, Université Paris Saclay, Gif-sur-Yvette, France
{paolo.ballarini,pierre.cry}@centralesupelec.fr

Abstract. Stochastic process discovery requires efficient procedures to compute the stochastic language emitted by a model. A natural approach consists in unfolding the underlying state-transition graph; however, for Stochastic Workflow Nets (SWN), this is severely hampered by silent transitions, which hinder scalability and may prevent termination in the presence of silent loops. Stochastic Process Trees (SPTs) were introduced as a more compact alternative, yet so far they have lacked semantics enabling exact, rather than simulation-based, language computation. We address this gap by defining a Markovian semantics for SPTs in terms of Labeled Absorbing Discrete-Time Markov Chains (LADTMCs), enabling exact language computation via DTMC unfolding while entirely bypassing the silent-loop issue. The resulting framework drastically reduces computational complexity and allows optimization engines to scale to large real-world event logs.

Keywords: Stochastic Process Trees · Discrete-Time Markov Chains · Stochastic Languages

1 Introduction

Business processes exhibit variability, with executions differing in activity ordering, timing, and frequency. While *logical* process discovery returns a model whose language is a plain set of traces, leaving frequent and rare behaviors indistinguishable, *stochastic* process discovery returns a stochastic model M, typically a Labeled Generalized Stochastic Petri Net (LGSPN) [12] or a Stochastic Workflow Net (SWN) [7], whose stochastic language $\mathcal{L}_M$ is a probability distribution over traces. Model quality is then captured by *stochastic conformance*: the similarity between $\mathcal{L}_M$ and the empirical trace distribution of the log, quantified via measures such as Earth Mover's Distance [14] or Entropic Relevance [2].

Measuring stochastic conformance requires computing $\mathcal{L}_M(\sigma)$ for each log trace σ, which is what makes stochastic process discovery computationally hard:

F. M. Maggi et al. (Eds.): BPM 2026, LNCS 16936, pp. 166–183, 2027.
https://doi.org/10.1007/978-3-032-37777-7_12

loops induce an infinite language support, while concurrency causes a combinatorial proliferation of execution paths over which per-trace probabilities must be summed. The issue is exacerbated by widely used discovery algorithms such as the Inductive Miner (IM) [9] and its infrequent variant (IMf) [10], whose Petri-net output encodes loops and concurrency through a proliferation of *silent transitions* that inflate execution paths and, in the worst case, give rise to non-terminating silent loops. Practical approaches therefore frequently fall back on Monte-Carlo simulation [6] to *approximate* $\mathcal{L}_M$ rather than computing it exactly.

The ultimate goal of identifying the *optimal* instance of a stochastic process model can then be cast as an optimization problem: find the parameters that maximizes stochastic conformance against a target log. The cost of this optimization is dominated by the dimensionality of the parameter space and the per-iteration cost of evaluating $\mathcal{L}_M$ over the log traces, both of which are aggravated by the silent-transition proliferation characteristic of WN-based models.

Concretely, the per-iteration evaluation reduces to the following computational problem: given a stochastic model M and a trace σ, compute the exact probability $\mathcal{L}_M(\sigma)$ that M emits σ (as opposed to estimating $\mathcal{L}_M(\sigma)$ by Monte-Carlo simulation). Even when M's language is infinite, which happens as soon as M contains a loop, $\mathcal{L}_M(\sigma)$ remains well-defined and computable for any single σ, and in particular for every trace in the support $Supp(E)$ of an event log E; the resulting values are then normalized over $Supp(E)$ for conformance evaluation. This is what we refer to as *exact language computation.*

Ourcontribution. We address both factors (the dimensionality of the parameter space and the exact evaluation $\mathcal{L}_M$) by working with the *Stochastic Process Tree* (SPT) formalism, recently introduced in [6] as a more compact alternative to stochastic Petri nets. In [6], the language of a SPT was determined through Monte-Carlo simulation that *approximates* $\mathcal{L}_Q(\sigma)$ for a given SPT Q. In this paper we fill this gap by proposing a construction of a LADTMC whose stochastic language is proven to be identical to that of a given SPT, and which is silent-transition-free by construction. This enables *exact* computation of $\mathcal{L}_Q(\sigma)$ for any trace σ in the support of an event log E, with two complementary benefits: optimization operates directly in the more compact SPT parameter space, and simulation-based approximation is replaced by exact computation, allowing optimization engines to scale to large real-world event logs.

In Sect. 2 we introduce preliminaries. Section 3 provides the rules to construct a LADTMC corresponding to a given SPT and proves that the two formalisms emit the same stochastic language. Section 4 outlines the experimental results. Concluding remarks are given in Sect. 5.

1.1 Related Work

Several contributions have addressed stochastic model discovery from event logs, most targeting WNs via an *indirect* approach: a non-stochastic workflow model is first extracted, then converted into a stochastic one by associating *weight* parameters to events. Burke *et al.* [4] introduced six computationally efficient weight estimators combining log and model statistics, though at the cost of conformance. In a follow-up work [5], the same authors proposed a framework for directly discovering an *untimed* LGSPN from a log via mapping of probabilistic process trees (PPTs).

In [3], an approach is introduced that optimizes the Earth Mover's Stochastic Conformance score of a discovered stochastic WN via subgradient ascent: the stochastic language of the net is first derived from its structure, then subgradient optimization is performed on a conformance loss function, with the resulting subgradient propagated to update the Petri net weights.

In [11], a weight optimization scheme for WN models is introduced by extracting symbolic expressions for the probability of each log trace as a function of the WN weights, which are then fed to an optimization engine targeting either the unit Earth Mover's Stochastic Conformance (uEMSC) [14] or inverted entropic relevance (ER^{-1}) [2]. The approach is demonstrated on underfitting WN models produced by IMf [10].

Two complementary exact approaches to computing the stochastic language of a Petri-net-based stochastic process model have recently been proposed. The first [7], adopted as the SWN-side baseline in this paper, computes trace probabilities by breadth-first *unfolding* of the reachability graph, producing a DAG of symbolic functions parametrized by transition weights; memoization amortizes the setup cost across all traces and optimization iterations. The second [12] builds, for each target trace σ, a silenced deterministic finite automaton $\bar{A}_\sigma$ accepting σ modulo silent-transition insertions, takes the cross product with the reachability graph, and solves a linear system encoding absorption probabilities on the resulting Markov chain. The two approaches have complementary strengths: the latter absorbs silent loops analytically but builds a fresh product chain and linear system per trace, while the former amortizes one setup phase across all traces and weight vectors but inflates with silent-path multiplicity and may fail to terminate in the presence of silent loops.

The LADTMC construction we propose in Sect. 3 sidesteps both difficulties by eliminating silent transitions by construction. The LADTMC of a SPT Q contains no silent transitions, since silent leaves contribute no states and the operator rules propagate this property through composition. The unfolding procedure of [7] can then be applied to the LADTMC unchanged. Since the model has no silent transitions to enumerate and no silent loops to threaten termination, this yields the substantial speed-ups reported in Sect. 4.

Finally, note that PPTs introduced in [5] are inherently semantically different from the SPTs we consider in this work, as, by definition, their semantics is given in terms of a mapping to equivalent LGSPN models and, as a consequence, they suffer from silent transitions proliferation.

2 Preliminaries

Alphabets, traces, languages. We denote Σ an *alphabet* of process activities and use letters to represent its elements (e.g., $\Sigma = \{a, b, c, d\}$) and $\tau \notin \Sigma$ to represent the *silent action*. A *trace* $\sigma = \langle a_1, a_2, ..., a_n \rangle \in \Sigma^\star$ is a finite sequence of activities with $\sigma[i]$ $(1 \le i \le n)$ being its ith activity, $|\sigma|$ the length of σ and $\epsilon = \langle\rangle \in \Sigma^*$ the empty trace. A set of traces is called a *language*. For any two traces σ_1, σ_2, we denote $\sigma_1.\sigma_2 = \langle \sigma_1[1], \dots, \sigma_1[|\sigma_1|], \sigma_2[1], \dots, \sigma_2[|\sigma_2|] \rangle$ the concatenation of σ_2 to σ_1 (e.g., with $\sigma_1 = \langle a, b, b, a, c \rangle$ and $\sigma_2 = \langle e, f \rangle$ then $\sigma_1.\sigma_2 = \langle a, b, b, a, c, e, f \rangle$).

Event log and stochastic language. An event log E is a multi-set of traces. Given a trace $\sigma \in E$, we denote by $f(E, \sigma)$ its multiplicity (i.e., its frequency in E). A stochastic language (SL) over the traces built on an alphabet Σ is a function $\mathcal{L} : \Sigma^* \to [0, 1]$ that provides the probability of the traces such that $\sum_{\sigma \in \Sigma^*} \mathcal{L}(\sigma) = 1$. The SL of a log E is obtained by computing the relative frequency of the traces as $p(E, \sigma) = f(E, \sigma) / \sum_{\omega \in Supp(E)} f(E, \omega)$ where $Supp(E)$ denotes the set of unique traces in E, i.e., its support. For simplicity we sometime denote a SL as $\mathcal{L} = \{\sigma_1^{p_1}, \dots, \sigma_n^{p_n}\}$ where σ_i is the ith trace and $p_i = \mathcal{L}(\sigma_i)$ its probability.

Shuffling. Given traces $\sigma_1, ..., \sigma_n$, their shuffling can be described by a *shuffling sequence* $sh = [i_1, i_2, \dots, i_{|\sigma|}]$, with $1 \le i_j \le n$, $1 \le j \le |\sigma| = |\sigma_1| + \dots + |\sigma_n|$, where entry $sh[k] = i_k$ indicates that the kth element of σ is drawn from trace σ_{i_k}, preserving the order of activities within each trace. E.g., if $\sigma_1 = \langle a, a, b \rangle$, $\sigma_2 = \langle c, d \rangle$ and $sh = [1, 1, 2, 2, 1]$, then shuffling of σ_1 and σ_2 according to sh yields $\sigma = \langle a, a, c, d, b \rangle$. To obtain *stochastic shuffling*, every involved trace $\sigma_i, 1 \le i \le n$, is associated with a parameter p_i used to calculate the probability that the next activity in the resulting trace σ is drawn from σ_i. Naturally, the k-th entry in σ can only come from traces that have not yet been *exhausted* at position k. To account for this, $\mathcal{C}(sh, k)$ denotes the set of trace indices corresponding to those traces that have not yet been fully consumed in σ prior to position k. For the previous example, we have $\mathcal{C}(sh, 1) = \mathcal{C}(sh, 2) = \mathcal{C}(sh, 3) = \mathcal{C}(sh, 4) = \{1, 2\}$ and $\mathcal{C}(sh, 5) = \{1\}$. Then, the probability that the stochastic shuffling of traces $\sigma_1, \dots, \sigma_n$ with parameters $p_1, \dots, p_n$ yields σ is

$$P(Sh(\sigma_1, \dots, \sigma_n, p_1, \dots, p_n) = \sigma) = \sum_{sh \in \mathcal{S}(\sigma_1, \dots, \sigma_n, \sigma)} \prod_{k=1}^{|\sigma|} \frac{p_{sh[k]}}{\sum_{i \in \mathcal{C}(sh, k)} p_i}, \quad (1)$$

where $\mathcal{S}(\sigma_1, \dots, \sigma_n, \sigma)$ denotes the set of all shufflings of $\sigma_1, \dots, \sigma_n$ that produce σ. With the above two traces σ_1, σ_2, and $p_1 = \frac{1}{3}$, $p_2 = \frac{2}{3}$ we have, among others,

$$P(Sh(\sigma_1, \sigma_2, p_1, p_2) = \langle a, a, c, d, b \rangle) = \frac{1}{3} \cdot \frac{1}{3} \cdot \frac{2}{3} \cdot \frac{2}{3} \cdot 1 = \frac{4}{81},$$
$$P(Sh(\sigma_1, \sigma_2, p_1, p_2) = \langle c, d, a, a, b \rangle) = \frac{2}{3} \cdot \frac{2}{3} \cdot 1 \cdot 1 \cdot 1 = \frac{4}{9}.$$

A convenient family of models for trace generation is labeled absorbing discrete-time Markov chains (LADTMCs).

Definition 1 (LADTMC). *A LADTMC has a set of* transient *states and one (implicit) absorbing state, and it is given by a tuple* $\mathcal{D} = (S, \pi_0, P, T)$ *where*

- *S is the set of transient states,*

- $\pi_0 : S \rightarrow [0,1]$ *gives the initial probabilities of the transient states such that*

$$\sum_{s \in S} \pi_0(s) \leq 1 ,$$

with $\pi_\epsilon = 1 - \sum_{s \in S} \pi_0(s)$ *being the probability of generating the empty trace (which can also be seen as the initial probability of the absorbing state),*
- $P : S \times S \times \Sigma \rightarrow [0,1]$ *provides the transition probabilities among transient states, that is,* $P(s, s', l)$ *is the probability of the transition from state* s *to state* s' *producing the label* l, *such that* $\forall s \in S$ *we have*

$$\sum_{s' \in S} \sum_{l \in \Sigma} P(s, s', l) \leq 1 .$$

- $T : S \times \Sigma \rightarrow [0,1]$ *provides the transition probabilities to the absorbing state, that is,* $T(s, l)$ *is the probability of the transition from transient state* s *to the absorbing state producing the label* l, *such that* $\forall s \in S$ *we have*

$$\sum_{s' \in S} \sum_{l \in \Sigma} P(s, s', l) + \sum_{l \in \Sigma} T(s, l) = 1 .$$

The process terminates, that is, the generation of a trace is completed, as soon as the absorbing state is reached.

Given a LADTMC $\mathcal{D} = (S, \pi_0, P, T)$, a sequence of its transient states $\omega = \langle s_0, \ldots, s_{n-1} \rangle$ and a trace $\sigma = \langle a_1, \ldots, a_n \rangle$, the probability that $\mathcal{D}$ arrives to the absorbing state along ω and produces σ is obtained and denoted as

$$G(\omega, \sigma) = \pi_0(s_0) \cdot \left(\prod_{i=1}^{n-1} P(s_{i-1}, s_i, a_i) \right) \cdot T(s_{n-1}, a_n) .$$

By noting that a given trace can be generated along different state sequences, we obtain the SL of a LADTMC as provided in the following definition.

Definition 2 (SL of a LADTMC). *The stochastic language of a LADTMC* $\mathcal{D} = (S, \pi_0, P, T)$ *is a function* $\mathcal{L}_\mathcal{D} : \Sigma^* \rightarrow [0,1]$ *defined as*

$$\mathcal{L}_\mathcal{D}(\sigma) = \begin{cases} 1 - \sum_{s \in S} \pi_0(s) & \text{if } \sigma = \epsilon \\ \sum_{\langle s_0, \ldots, s_{n-1} \rangle \in S^n} G(\langle s_0, \ldots, s_{n-1} \rangle, \sigma) & \text{if } \sigma = \langle a_1, \ldots, a_n \rangle . \end{cases} \quad (2)$$

Figure 1 shows an example LADTMC (top) and the corresponding stochastic language (bottom). Note that Definition 2 is not directly applicable for the computation of an SL because it requires considering every possible state sequence.

Finally, we recall the *stochastic process tree* (SPT) formalism as defined in [6], where process tree [1] operators *choice* ($\times$), *parallel* ($\wedge$) and *loop* ($\circlearrowleft$) have been extended with probabilistic parameters. First, we provide the syntax.

Definition 3 (SPT syntax). *With an alphabet* Σ *and SPT operators* $\{\rightarrow, \times, \wedge, \circlearrowleft\}$, *the set of SPTs over* Σ *is recursively defined as:*

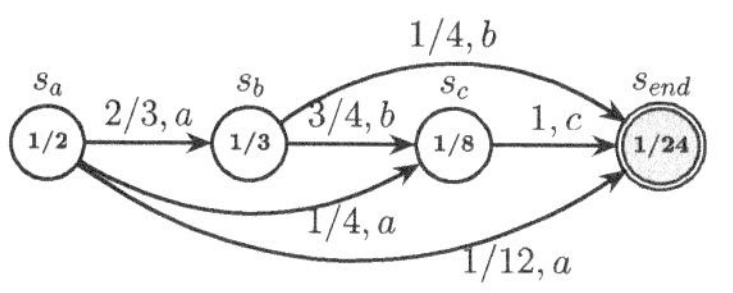

$\Sigma = \{a, b, c\}$

$$P = \begin{bmatrix} 0 & (2/3, a) & (1/4, a) \\ 0 & 0 & (3/4, b) \\ 0 & 0 & 0 \end{bmatrix}$$

$$T = \begin{bmatrix} (1/12, a) \\ (1/4, b) \\ (1, c) \end{bmatrix}$$

$\pi_0 = [1/2, 1/3, 1/8]$

$(\pi_\epsilon = 1 - (1/2 + 1/3 + 1/8) = 1/24)$

$\mathcal{L}_\mathcal{D} = \{\epsilon^{1/24}, a^{1/24}, ac^{1/8}, ab^{1/12}, abc^{1/4}, b^{1/12}, bc^{1/4}, c^{1/8}\}$

Fig. 1. Example of a 4-state LADTMC with its transient probabilities (P), absorption probabilities (T), and initial probabilities (π_0); transient states are depicted in white single stranded circles, while the absorbing state is violet and double stranded.

- *if* $a \in \Sigma \cup \{\tau\}$ *then* $Q = a$ *is a SPT,*
- *if* $Q_1, ..., Q_n$ *with* $n \geq 2$ *are SPTs then* $\rightarrow (Q_1, Q_2, ..., Q_n)$ *is a SPT,*
- *if* $Q_1, ..., Q_n$ *with* $n \geq 2$ *are SPTs,* $0 \leq p_i \leq 1, i = 1, 2, ...n$ *with* $\sum_{i=1}^{n} p_i = 1$*, and* $\oplus \in \{\times, \wedge\}$ *then* $\oplus(Q_1, ..., Q_n, p_1, ..., p_n)$ *is a SPT,*
- *if* Q_1 *and* Q_2 *are SPTs and* $0 \leq p < 1$ *then* $\circlearrowleft (Q_1, Q_2, p)$ *is a SPT.*

The semantics of SPTs can be summarized as follows. The sequence operator is kept non-stochastic, that is, $\rightarrow (Q_1, Q_2, \ldots, Q_n)$ corresponds to the subsequent executions of the involved SPTs. Choice, $Q = \times(Q_1, \ldots, Q_n, p_1, \ldots, p_n)$, simply selects one operand to execute, namely SPT Q_i with probability p_i. The parallel operator, $Q = \wedge(Q_1, \ldots, Q_n, p_1, \ldots, p_n)$, chooses a trace from every involved SPT and then shuffles them in a stochastic manner, as defined in Eq. (1). Regarding the loop operator, $Q = \circlearrowleft (Q_1, Q_2, p)$, after every execution of Q_1, trace generation terminates with probability $1 - p$ while, with probability p, Q_2 is executed before Q_1 is re-executed. The probability that Q_1 is executed m times (and Q_2 $m - 1$ times) is therefore $p^{m-1}(1 - p)$, meaning the number of executions of Q_1 follows the geometric distribution. A detailed explanation with several simple examples is given in [6]. The formal semantics follows.

Definition 4 (Semantics of SPTs.). *The stochastic language* $\mathcal{L}_Q$ *issued by a SPT* Q *with alphabet* Σ *is recursively defined as follows, in function of the stochastic languages* $\mathcal{L}_{Q_i}$ *of the corresponding sub-trees* Q_i*:*

- *if* $Q = a$ *with* $a \in \Sigma$ *then* $\mathcal{L}_Q(\langle a \rangle) = 1$*,*
- *if* $Q = \tau$ *then* $\mathcal{L}_Q(\epsilon) = 1$*,*
- ***(sequence:)*** *if* $Q = \rightarrow (Q_1, Q_2, ..., Q_n)$ *then*

$$\mathcal{L}_Q(\sigma) = \sum_{\sigma_1 \cdot \sigma_2 \cdot \cdots \cdot \sigma_n = \sigma} \prod_{i=1}^{n} \mathcal{L}_{Q_i}(\sigma_i), \tag{3}$$

- ***(choice:)*** *if* $Q = \times(Q_1, ..., Q_n, p_1, ..., p_n)$ *then*

$$\mathcal{L}_Q(\sigma) = \sum_{i=1}^{n} p_i \cdot \mathcal{L}_{Q_i}(\sigma), \tag{4}$$

- ***(parallel:)*** *if* $Q = \wedge(Q_1, ..., Q_n, p_1, ..., p_n)$ *then*

$$\mathcal{L}_Q(\sigma) = \sum_{\sigma_1, \sigma_2, \ldots, \sigma_n} \left(P(Sh(\sigma_1, \ldots, \sigma_n, p_1, \ldots, p_n) = \sigma) \prod_{i=1}^{n} \mathcal{L}_{Q_i}(\sigma_i) \right), \quad (5)$$

- ***(loop:)*** *if* $Q = \circlearrowleft (Q_1, Q_2, p)$ *then*

$$\mathcal{L}_Q(\sigma) = \sum_{\substack{m=1,2,3,\ldots \\ \sigma_1.\sigma'_1.\sigma_2.\sigma'_2. \ \ldots \ .\sigma'_{m-1}.\sigma_m = \sigma}} \left(p^{m-1}(1-p) \prod_{i=1}^{m} \mathcal{L}_{Q_1}(\sigma_i) \prod_{i=1}^{m-1} \mathcal{L}_{Q_2}(\sigma'_i) \right). \quad (6)$$

3 Markov Chain Semantics for SPTs

We show that the stochastic process underlying a SPT is a LADTMC. The base case, that is, a SPT consisting of a single leaf, follows directly from the definition. For composite SPTs, we proceed inductively: given a SPT Q constructed via any of the operators $\{\rightarrow, \times, \wedge, \circlearrowleft\}$ and LADTMCs corresponding to its sub-SPTs, we provide rules to construct a LADTMC corresponding to Q. The resulting LADTMC then allows us to compute the probability of a trace σ in Q, by applying an unfolding procedure [7].

First, in Sect. 3.1, we provide the mapping rules for obtaining a LADTMC from a SPT, illustrating the process with examples. Then, in Sect. 3.2, Theorem 1 shows that the SL of the LADTMC obtained through the mapping rules is equal to that of the corresponding SPT.

3.1 Mapping of SPT to LADTMC

Given a SPT Q, the LADTMC $\mathcal{D}$ underlying Q can be constructed as follows:
Action-leaf: if $Q = a$, with $a \in \Sigma$, then $\mathcal{D} = (S, \pi_0, P, T)$ is defined as follows.
State space: A single transient state $S = \{s\}$.
Initial probabilities: $\pi_0(s) = 1$.
Transition probabilities: A single transition to the absorbing state, $T(s, a) = 1$.
Silent-leaf: if $Q = \tau$ then in $\mathcal{D} = (S, \pi_0, P, T)$ we have the following.
State space: no transient states are present in $\mathcal{D}$, that is, $S = \emptyset$.
Initial probabilities: terminating directly in the absorbing state with probability 1 (this follows from $1 - \sum_{s \in S} \pi_0(s) = 1$).
Transition probabilities: no transitions are present in $\mathcal{D}$.

The above two cases are illustrated graphically in Fig. 2.

To deal inductively with the SPT operators $\{\rightarrow, \times, \wedge, \circlearrowleft\}$, the SPTs involved in an operation will be denoted by $\{Q_i\}_{i=1}^{n}$ and their corresponding LADTMCs by $\mathcal{D}_i = (S_i, \pi_{0,i}, P_i, T_i)$.

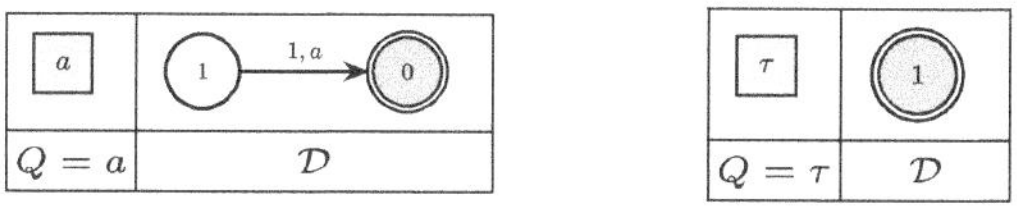

Fig. 2. Leaves SPTs and corresponding LADTMCs.

Choice: if $Q = \times(Q_1, ..., Q_n, p_1, \ldots, p_n)$ then $\mathcal{D} = (S, \pi_0, P, T)$ is as follows.
State space: $S = \bigcup_{i=1}^{n} S_i$.
Initial Probabilities: they reflect the choice among the operands by

$$\pi_0(s) = p_i \pi_{0,i}(s), \ \forall i, 1 \leq i \leq n, \ \forall s \in S_i \ . \tag{7}$$

Transition Probabilities: the transitions among the transient states and to the absorbing state are simply copied from the operands as, $\forall l \in \Sigma$,

$$P(s, s', l) = \begin{cases} P_i(s, s', l) & \forall i, 1 \leq i \leq n, \ \forall s, s' \in S_i \\ 0 & \text{otherwise} \ . \end{cases} \tag{8}$$

$$T(s, l) = \begin{cases} T_i(s, l) & \forall i, 1 \leq i \leq n, \ \forall s \in S_i \\ 0 & \text{otherwise} \ . \end{cases} \tag{9}$$

Examples are shown in Fig. 3

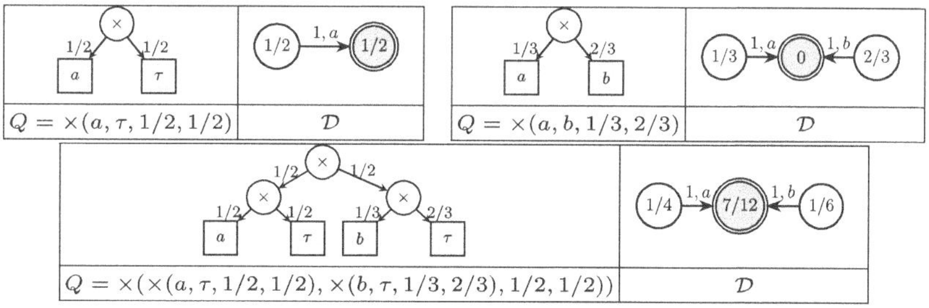

Fig. 3. Choice SPTs and corresponding LADTMCs.

Sequence: if $Q = \rightarrow (Q_1, Q_2)$ then $\mathcal{D} = (S, \pi_0, P, T)$ is as follows.

State Space: $S = S_1 \bigcup S_2$.

Initial Probabilities: those of the states in S_1 are simply copied from $\mathcal{D}_1$ while states in S_2 has non-zero inital probabilities only if Q_1 can produce the empty trace; this is reflected by

$$\pi_0(s) = \begin{cases} \pi_{0,1}(s) & \forall s \in S_1 \\ \pi_{\epsilon,1} \cdot \pi_{0,2}(s) & \forall s \in S_2 \end{cases} \tag{10}$$

where $\pi_{\epsilon,i} = 1 - \sum_{s \in S_i} \pi_{0,i}(s)$, that is, the probability that $\mathcal{D}_i$ produces the empty trace. It follows that, as expected, direct absorption in $\mathcal{D}$ is with probability $\pi_\epsilon = 1 - \sum_{s \in S} \pi_0(s) = \pi_{\epsilon,1} \cdot \pi_{\epsilon,2}$.

Transition Probabilities: those inside S_1 and inside S_2 are simply copied from $\mathcal{D}_1$ and $\mathcal{D}_2$ while the ones leading from a transient state in S_1 to the absorbing state in $\mathcal{D}_1$ are redirected to transient states in S_2 according to the corresponding initial probabilities of $\mathcal{D}_2$; we get

$$P(s, s', l) = \begin{cases} P_i(s, s', l) & \forall i = 1, 2,\ \forall s, s' \in S_i \\ T_1(s, l) \cdot \pi_{0,2}(s') & \forall s \in S_1, \forall s' \in S_2 \\ 0 & \text{otherwise}\ , \end{cases} \tag{11}$$

whereas transition probabilities to the absorbing state from states in S_2 is simply copied from $\mathcal{D}_2$ while from states in S_1 they are calculated using the probability of direct absorption in $\mathcal{D}_2$ as

$$T(s, l) = \begin{cases} T_1(s, l) \cdot \pi_{\epsilon,2} & \forall s \in S_1 \\ T_2(s, l) & \forall s \in S_2\ . \end{cases} \tag{12}$$

The n-ary sequence operator is managed recursively as

$$\rightarrow (Q_1, \ldots, Q_n) = \rightarrow (\rightarrow (Q_1, \ldots, Q_{n-1}), Q_n)\ , \tag{13}$$

which follows directly from Definition 3. An example is shown in Fig. 4 (top).

Parallel: if $Q = \wedge(Q_1, \ldots, Q_n, p_1, \ldots, p_n)$ then $\mathcal{D}$ is defined as follows.

State Space: $S = \times_{i=1}^{n}(S_i \cup \{s_{i,end}\}) \setminus \{(s_{1,end}, \ldots, s_{n,end})\}$ where states $s_{i,end}$ allow us to represent that component i terminated and $\{(s_{1,end}, \ldots, s_{n,end})\}$ is removed because it is not a transient state.

Initial Probabilities: given a state $s = (s_1, \ldots, s_n) \in S$, the contribution of $\pi_{0,i}$ to its initial probability depends on $\mathcal{D}_i$ producing the empty trace or starting in a transient state of $\mathcal{D}_i$:

$$\pi_0(s) = \pi_0((s_1, \ldots, s_n)) = \prod_{\substack{i=1,\ldots,n \\ \text{s.t. } s_i = s_{i,\text{end}}}} \pi_{\epsilon,i} \quad \cdot \prod_{\substack{i=1,\ldots,n \\ \text{s.t. } s_i \neq s_{i,\text{end}}}} \pi_{0,i}(s_i)\ , \tag{14}$$

which, as expected, leads to the probability that $\mathcal{D}$ produces the empty trace being $\pi_\epsilon = \prod_{i=1}^{n} \pi_{\epsilon,i}$.

Transition Probabilities: first, we introduce $\mathcal{A}(s) = \{i \mid 1 \leq i \leq n, s_i \neq s_{i,end}\}$ to indicate the set of components not yet in their absorbing state. Following the semantics of the parallel operator, for $i \in \mathcal{A}(s)$, the probability that a transition in state s is due to component $\mathcal{D}_i$ is $p_i / \sum_{j \in \mathcal{A}(s)} p_j$, which we denote by $w(s, i)$.

Among the transient states the transition probabilities are

$$P(s,s',l)=\begin{cases}\sum_{i\in\mathcal{A}(s)} w(s,i)P_i(s_i,s_i,l) & \text{if } s=s'\\ w(s,i)P_i(s_i,s_i',l) & \text{if } \exists i: s_i\neq s_i' \wedge s_i\neq s_{i,end}\wedge \\ & s_i'\neq s_{i,end}\wedge \bigwedge_{j\neq i}(s_j=s_j')\\ w(s,i)T_i(s_i,l) & \text{if } \exists i: s_i\neq s_{i,end}\wedge s_i'=s_{i,end}\wedge \\ & \bigwedge_{j\neq i}(s_j=s_j')\wedge \exists j: s_j\neq s_{j,end}\\ 0 & \text{otherwise}\end{cases} \tag{15}$$

where the first case corresponds to self-loops producing the label l in any of the components active in s, the second to a transition among transient states in exactly one component, and the third to termination in exactly one component (this can happen only if at least another component is in a transient state). Termination can happen only if exactly one component is still active. We get

$$T(s,l)=\begin{cases}T_i(s_i,l) & \text{if } \mathcal{A}(s)=\{i\}\\ 0 & \text{otherwise}\end{cases} \tag{16}$$

Note that the weights $p_1,\ldots,p_n$ play no role here, since only one component can be active at termination.

An examples is shown in Fig. 4 (center).

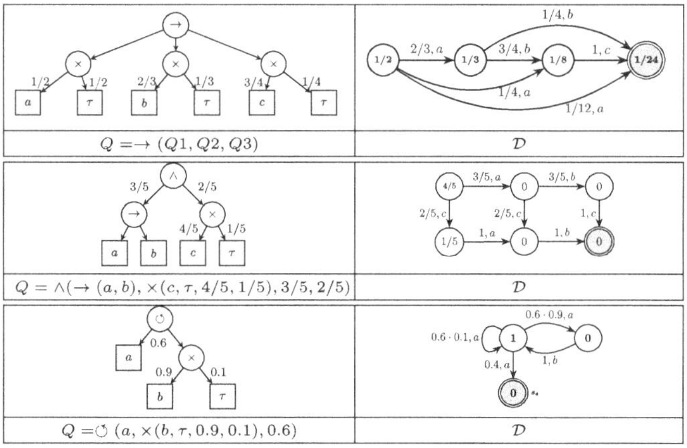

Fig. 4. A sequence (top), a parallel (center) and a loop (bottom) SPT and corresponding LADTMC.

Loop: if $Q=\circlearrowleft(Q_1,Q_2,p)$ then $\mathcal{D}=(S,\pi_0,P,T)$ is defined as follows.

State Space: $S=S_1\cup S_2$.

Initial Probabilities: if both $\mathcal{D}_1$ and $\mathcal{D}_2$ produce the empty trace with non-zero probability, then the model can exhibit a silent loop. We denote the probability of such an event by $\alpha = \pi_{\epsilon,1} \cdot p \cdot \pi_{\epsilon,2}$. To consider any number of consecutive silent loops we introduce also $\alpha^\star = \sum_{i=0}^{\infty} \alpha^i = 1/(1-\alpha)$. The process starts in a state $s \in S_1$ if it is chosen after any number of silent loops. For a state $s \in S_2$ to be the initial state, $\mathcal{D}_1$ must produce the empty trace at least once and the process must loop back, possibly preceded by any number of silent loops. We get

$$\pi_0(s) = \begin{cases} \alpha^\star \cdot \pi_{0,1}(s) & \text{if } s \in S_1 \\ \pi_{\epsilon,1} \cdot p \cdot \alpha^\star \cdot \pi_{0,2}(s) & \text{if } s \in S_2 \end{cases} \tag{17}$$

The model $\mathcal{D}$ produces the empty trace if, after any number of silent loops, $\mathcal{D}_1$ produces the empty trace and then exits the loop. This happens with probability $\pi_\epsilon = \alpha^\star \pi_{\epsilon,1}(1-p)$. It is easy to verify that this is in accordance with Eq. (17), that is, $\pi_\epsilon = 1 - \sum_{s \in S} \pi_0(s) = \alpha^\star \pi_{\epsilon,1}(1-p)$. (Note that if silent loops are not possible, that is, $\alpha = 0$, then $\alpha^\star = 1$.)

Transition Probabilities: among transient states they are based on similar considerations:

$$P(s,s',l) = \begin{cases} P_1(s,s',l) + T_1(s,l) \cdot p \cdot \pi_{\epsilon,2} \cdot \alpha^\star \cdot \pi_{0,1}(s') & \text{if } s, s' \in S_1 \\ P_2(s,s',l) + T_2(s,l) \cdot \alpha^\star \cdot \pi_{\epsilon,1} \cdot p \cdot \pi_{0,2}(s') & \text{if } s, s' \in S_2 \\ T_2(s,l) \cdot \alpha^\star \cdot \pi_{0,1}(s') & \text{if } s \in S_2, s' \in S_1 \\ T_1(s,l) \cdot \alpha^\star \cdot p \cdot \pi_{0,2}(s') & \text{if } s \in S_1, s' \in S_2 \end{cases} \tag{18}$$

among which let us explain two. A transition from a state $s \in S_1$ to a state $s' \in S_1$ can happen directly inside $\mathcal{D}_1$ (with probability $P_1(s,s',l)$) or by the following sequence of events: absorption in $\mathcal{D}_1$ ($T_1(s,l)$), looping back to $\mathcal{D}_2$ (p), generating the empty trace in $\mathcal{D}_2$ ($\pi_{\epsilon,2}$), performing any number of silent loops ($\alpha^\star$) and then choosing s' as initial state in $\mathcal{D}_1$ ($\pi_{0,1}(s')$). A transition from $s \in S_1$ to $s' \in S_2$ happens if $\mathcal{D}_1$ terminates from s (with probability $T_1(s,l)$) followed by any number of silent loops ($\alpha^\star$), looping back once more (p) and choosing s' as initial state in $\mathcal{D}_2$ ($\pi_{0,2}(s')$).
Termination from a state $s \in S_1$ happens if $\mathcal{D}_1$ terminates followed by any number of silents loops and not looping back. From a state $s \in S_2$ the process terminates if $\mathcal{D}_2$ terminates, followed by any number of silent loops, then $\mathcal{D}_1$ produces the empty trace and the process does not loop back. We get

$$T(s,l) = \begin{cases} T_1(s,l) \cdot \alpha^\star \cdot (1-p) & \forall s \in S_1 \\ T_2(s,l) \cdot \alpha^\star \cdot \pi_{\epsilon,1} \cdot (1-p) & \forall s \in S_2 \end{cases} \tag{19}$$

An example is given in Fig. 4 (bottom).

Figure 5 illustrates the construction on two more complex SPTs. The parallel example $Q = \wedge(Q_1, Q_2, 1/3, 2/3)$ (first row), with $Q_1 = \circlearrowleft (a, \tau, 3/5)$ and $Q_2 = \circlearrowleft (b, \tau, 3/4)$ whose component LADTMCs $\mathcal{D}_1$ and $\mathcal{D}_2$ are shown in the mid column, yields the three transient states of the cross-product $(\{s_a, s_{1,end}\} \times \{s_b, s_{2,end}\}) \setminus \{(s_{1,end}, s_{2,end})\}$. Two points are worth noting: component transitions are rescaled by the parent parameters (1/3, 2/3) only while both components are active, becoming unscaled once one is absorbed; and the two differently labeled self-loops at (s_a, s_b) show why transition probabilities must be label-dependent (matrices P and T in Def. 1).

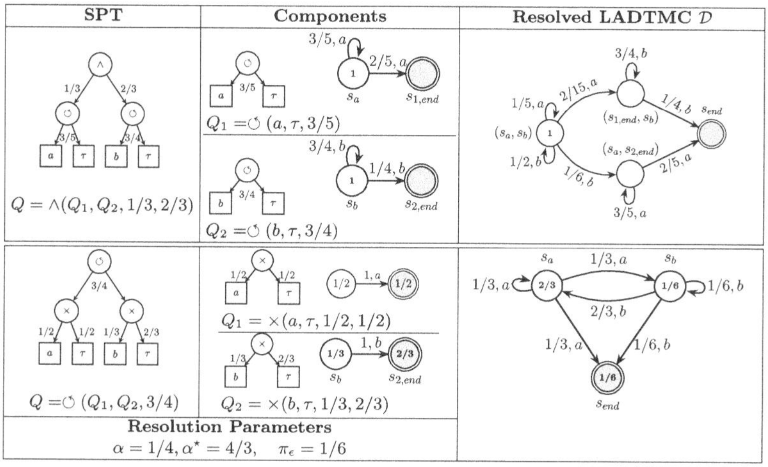

Fig. 5. Examples of LADTMC construction for given SPTs.

The loop example $Q = ⟲ (Q_1, Q_2, 3/4)$ (second row of Fig. 5), with $Q_1 = \times(a, \tau, 1/2, 1/2)$ and $Q_2 = \times(b, \tau, 1/3, 2/3)$ (component LADTMCs in the mid column), resolves to two transient states $\{s_a, s_b\}$. Here the silent-loop factors $\alpha = \pi_{\epsilon,1} \cdot p \cdot \pi_{\epsilon,2} = 1/4$ and $\alpha^\star = 1/(1-\alpha) = 4/3$ (derived from the looping probability $p{=}3/4$) rescale the local transition probabilities (e.g. 1/3 from s_a to s_b), capturing the infinite geometric series of the stochastic loop language.

3.2 Language Equivalence

Theorem 1 (Language equivalence). *For a SPT Q and $\mathcal{D} = (S, \pi_0, P, T)$, the corresponding LADTMC obtained with the constructions rules in Sect. 3.1, the stochastic languages $\mathcal{L}_Q$ and $\mathcal{L}_\mathcal{D}$ are equivalent.*

Proof. We proceed by structural induction on Q.
Base case. If Q contains a single leaf (labeled by an action or silent), the SLs $\mathcal{L}_Q$ and $\mathcal{L}_\mathcal{D}$ are trivially equivalent.
Inductive step. Assume, as inductive hypothesis (IH), the theorem holds for all sub-SPTs Q_i with corresponding LADTMCs $\mathcal{D}_i = (S_i, \pi_{0,i}, P_i, T_i)$, $i = 1, \ldots, n$.
• **Choice:** $Q = \times(Q_1, \ldots, Q_n, p_1, \ldots, p_n)$. Recall $S = \bigcup_{i=1}^n S_i$, $\pi_0(s) = p_i\, \pi_{0,i}(s)$ for $s \in S_i$, and transitions within each S_i are copied unchanged.

Empty trace. From (7) and the semantics of SPTs (Definition 4):

$$\mathcal{L}_{\mathcal{D}}(\epsilon) = 1 - \sum_{s \in S} \pi_0(s) = 1 - \sum_{i=1}^{n} \sum_{s \in S_i} p_i\, \pi_{0,i}(s) = 1 - \sum_{i=1}^{n} p_i \left(\sum_{s \in S_i} \pi_{0,i}(s) \right)$$
$$= \sum_{i=1}^{n} p_i \left(1 - \sum_{s \in S_i} \pi_{0,i}(s) \right) = \sum_{i=1}^{n} p_i\, \pi_{\epsilon,i} = \sum_{i=1}^{n} p_i\, \mathcal{L}_{Q_i}(\epsilon) = \mathcal{L}_{Q}(\epsilon).$$

Non-empty trace$\sigma = \langle a_1, \ldots, a_m \rangle$. Since transitions do not cross components (i.e., $P(s, s', l) = 0$ if $s \in S_i$ and $s' \in S_j$ for $i \neq j$), every state sequence ω that contributes to $G(\omega, \sigma)$ lies entirely within some S_i. A sequence $\omega = \langle s_0, \ldots, s_{m-1} \rangle \in S_i^m$ contributes

$$G(\omega, \sigma) = \pi_0(s_0) \prod_{k=1}^{m-1} P(s_{k-1}, s_k, a_k) \cdot T(s_{m-1}, a_m)$$
$$= p_i\, \pi_{0,i}(s_0) \prod_{k=1}^{m-1} P_i(s_{k-1}, s_k, a_k) \cdot T_i(s_{m-1}, a_m) = p_i\, G_i(\omega, \sigma).$$

Summing over all state sequences:

$$\mathcal{L}_{\mathcal{D}}(\sigma) = \sum_{i=1}^{n} p_i \sum_{\omega \in S_i^m} G_i(\omega, \sigma) = \sum_{i=1}^{n} p_i\, \mathcal{L}_{\mathcal{D}_i}(\sigma) \overset{\text{IH}}{=} \sum_{i=1}^{n} p_i\, \mathcal{L}_{Q_i}(\sigma) = \mathcal{L}_{Q}(\sigma).$$

- **Sequence:** $Q = \rightarrow (Q_1, Q_2)$.

Empty trace. Based on (10) and the fact that Q produces ϵ iff both Q_1 and Q_2 produce ϵ:

$$\mathcal{L}_{\mathcal{D}}(\epsilon) = 1 - \sum_{s \in S} \pi_0(s) = \pi_{\epsilon,1} - \pi_{\epsilon,1}(1 - \pi_{\epsilon,2}) = \pi_{\epsilon,1}\, \pi_{\epsilon,2} = \mathcal{L}_{Q}(\epsilon)\ .$$

Non-empty trace$\sigma = \langle a_1, \ldots, a_m \rangle$. Any state sequence $\omega = \langle s_0, \ldots, s_{m-1} \rangle$ in S^m that generates σ crosses from S_1 to S_2 at most once. Let $k \in \{0, \ldots, m\}$ be the number of steps performed from a state in S_1 (so the remaining $m - k$ steps are from a state in S_2). We split $\sigma = \sigma_1.\sigma_2$ where $\sigma_1 = \langle a_1, \ldots, a_k \rangle$ (possibly empty) and $\sigma_2 = \langle a_{k+1}, \ldots, a_m \rangle$ (possibly empty). For a pair of state sequences $\omega_1 = \langle s_0, \ldots, s_{k-1} \rangle \in S_1^k$ and $\omega_2 = \langle s_k, \ldots, s_{m-1} \rangle \in S_2^{m-k}$:

- If $0 < k < m$ (that is, $\omega_1.\omega_2$ contains states both from S_1 and S_2): the initial factor is $\pi_{0,1}(s_0)$; transitions within S_1 contribute $\prod_{t=1}^{k-1} P_1(s_{t-1}, s_t, a_t)$; the crossing step $s_{k-1} \rightarrow s_k$ contributes $T_1(s_{k-1}, a_k) \cdot \pi_{0,2}(s_k)$ (from the construction of P in (11)); transitions within S_2 contribute $\prod_{t=k+1}^{m-1} P_2(s_{t-1}, s_t, a_t)$; the final step contributes $T_2(s_{m-1}, a_m)$ (from (12)).
- If $k = 0$ (path entirely in S_2): the initial factor is $\pi_{\epsilon,1} \cdot \pi_{0,2}(s_0)$ (from (10)), and the remaining factors are entirely in $\mathcal{D}_2$.

- If $k = m$ (path entirely in S_1): the initial and transition probabilities among transient states are from $\mathcal{D}_1$; the transition to the absorbing state contributes with $T_1(s_{m-1}, a_m) \cdot \pi_{\epsilon,2}$ (as in (12)).

Factoring the contributions and summing over all splittings and state sequences:

$$\mathcal{L}_{\mathcal{D}}(\sigma) = \sum_{\sigma_1.\sigma_2=\sigma} \sum_{\omega_1 \in S_1^{|\sigma_1|}} \sum_{\omega_2 \in S_2^{|\sigma_2|}} G_1(\omega_1, \sigma_1) \cdot G_2(\omega_2, \sigma_2)$$

$$= \sum_{\sigma_1.\sigma_2=\sigma} \mathcal{L}_{\mathcal{D}_1}(\sigma_1) \cdot \mathcal{L}_{\mathcal{D}_2}(\sigma_2) \overset{\text{IH}}{=} \sum_{\sigma_1.\sigma_2=\sigma} \mathcal{L}_{Q_1}(\sigma_1) \cdot \mathcal{L}_{Q_2}(\sigma_2) = \mathcal{L}_Q(\sigma).$$

Consequently, $\mathcal{L}_{\mathcal{D}} = \mathcal{L}_Q$ for the binary case. The n-ary case follows by induction on n using the recursion $\rightarrow(Q_1, \ldots, Q_n) = \rightarrow(\rightarrow(Q_1, \ldots, Q_{n-1}), Q_n)$.
• **Loop:** $Q = \circlearrowleft(Q_1, Q_2, p)$. Recall the probability of a silent loop $\alpha = \pi_{\epsilon,1} \cdot p \cdot \pi_{\epsilon,2}$ and the derived factor $\alpha^\star = \sum_{i=0}^{\infty} \alpha^i = 1/(1-\alpha)$.
Step 1: Unrolling. From Definition 4 it follows that the SL of Q is equal to that of $Q' = \times(Q_1, \rightarrow(Q_1, Q_2, Q_1), \rightarrow(Q_1, Q_2, Q_1, Q_2, Q_1), \ldots; (1-p), p(1-p), p^2(1-p), \ldots)$. By the equivalences for choice and sequence, $\mathcal{L}_{Q'}$ equals the SL of the LADTMC $\mathcal{D}'$ obtained by applying the construction rules to Q', i.e., an infinite choice of chains of copies of $\mathcal{D}_1$ and $\mathcal{D}_2$. It suffices now to show $\mathcal{L}_{\mathcal{D}'} = \mathcal{L}_{\mathcal{D}}$.
Step 2: Folding. Thanks to the memoryless property of the geometric distribution (which implies that the number of future executions of Q_1 does not depend on the number of past executions), we can merge all copies of S_1 and S_2 across all branches of $\mathcal{D}'$ into a single shared copy, and derive the resulting probabilities by summing over branches. In each case, entries and re-entries into S_1 or S_2 may be preceded by any number $j \geq 0$ of silent loops, taken into account by $\alpha^\star$.
Initial probabilities. Entry into $s \in S_1$ is preceded by $j \geq 0$ silent loops, giving $\pi_0(s) = \alpha^\star \pi_{0,1}(s)$; entry into $s \in S_2$ additionally requires one silent completion of $\mathcal{D}_1$ and one loop-back, giving $\pi_0(s) = \pi_{\epsilon,1} \cdot p \cdot \alpha^\star \cdot \pi_{0,2}(s)$. These are exactly (17).
Transition and termination probabilities. Each entry of (18) and (19) follows by enumerating all paths between the merged states. We detail the two non-obvious cases of (18) (the remaining ones are analogous):

- $s, s' \in S_1$: transition happens directly inside $\mathcal{D}_1$ (probability $P_1(s, s', l)$), or via absorption in $\mathcal{D}_1$ ($T_1(s, l)$), loop-back (p), silent completion of $\mathcal{D}_2$ ($\pi_{\epsilon,2}$), any number of silent loops ($\alpha^\star$), and re-entry at s' ($\pi_{0,1}(s')$), yielding the first case of (18),
- $s \in S_1, s' \in S_2$: absorption in $\mathcal{D}_1$ ($T_1(s, l)$), any silent loops ($\alpha^\star$), loop-back (p), and entry at s' ($\pi_{0,2}(s')$), yielding the fourth case of (18).

The termination probabilities (19) follow similarly: from $s \in S_1$, absorption in $\mathcal{D}_1$ followed by any number of silent loops and exit $(1-p)$; from $s \in S_2$, absorption in $\mathcal{D}_2$ followed by any silent loops, silent completion of $\mathcal{D}_1$, and exit.
Consequently, $\mathcal{L}_{\mathcal{D}'} = \mathcal{L}_{\mathcal{D}}$, and hence $\mathcal{L}_Q = \mathcal{L}_{\mathcal{D}}$.
• **Parallel:** $Q = \wedge(Q_1, \ldots, Q_n, p_1, \ldots, p_n)$.
Product construction. By $S = \times_{i=1}^{n}(S_i \cup \{s_{i,end}\}) \setminus \{(s_{1,end}, \ldots, s_{n,end})\}$ we track

the current state of each component. At each step, component $\mathcal{D}_i$ is selected with probability $w(s,i) = p_i / \sum_{j \in \mathcal{A}(s)} p_j$, where $\mathcal{A}(s)$ is the set of still-active components, and executes one transition. The resulting transition probabilities (15) implement the shuffling distribution Sh of (5): the probability of any interleaving σ of $\sigma_1, \ldots, \sigma_n$ is $P(Sh(\sigma_1, \ldots, \sigma_n, p_1, \ldots, p_n) = \sigma)$.
Initial and termination probabilities. The initial probabilities (14) account for components that may produce the empty trace at the start: each such component i contributes $\pi_{\epsilon,i}$ (jumping immediately to $s_{i,end}$) while the others contribute $\pi_{0,i}(s_i)$. Termination (16) can only occur when exactly one component remains active ($\mathcal{A}(s) = \{i\}$), at which point it absorbs via $T_i(s_i, l)$ with no scheduling weight since no choice remains.

Consequently, $\mathcal{L}_{\mathcal{D}} = \mathcal{L}_Q$. □

4 Experiments

We assess the LADTMC semantics of SPTs against the SWN baseline on real-life event logs sourced from the Business Process Intelligence Challenge (BPIC) database (https://data.4tu.nl/), via two experiments: exact stochastic language computation on a single model (Table 1), and parameter optimisation, which repeatedly invokes language computation (Table 2). In both, exact computation of $\mathcal{L}_M(\sigma)$ for every $\sigma \in Supp(E)$ is performed via the unfolding procedure of [7]: a single breadth-first pass over the state-transition graph of M (the LADTMC for an SPT, the reachability graph for an SWN) produces a DAG whose leaves carry the probability values $\mathcal{L}_M(\sigma)$; perfect fitness of IM [10] guarantees coverage of $Supp(E)$, and values are normalised over $Supp(E)$ when $\mathcal{L}_M$ has infinite support. For optimisation (Table 2), this evaluation is wrapped in an outer loop minimising the restricted Earth Mover's Distance (rEMD, i.e. the EMD [14] restricted to the log-supported traces emitted by the model [7]); the same numerical optimiser (`scipy.optimize.minimize` with the Powell method) with identical hyperparameters is used on both sides, so the only difference is the underlying model. Beyond the SWN unfolding baseline of [7], the silenced-DFA plus linear-system approach of [12,13] is conceptually competing; as discussed in Sect. 1.1, a head-to-head experimental comparison is not feasible without aligning the discovery pipeline (their experiments use heavily noise-filtered IMf, ours perfectly fitting IM) and is left to follow-up work.

Exact Language Computation: SPT versus SWN. Table 1 reports the language-computation experiment. PT and WN models are discovered via the pm4py implementation of the IM and then enriched with stochastic parameters, whose counts are reported in the `#pars` columns for both SPT and SWN. Columns `log stats` report statistics of the logs: total number of traces, number of unique traces and activities. Columns `#nodes per type` under `SPT` give the number of SPT nodes per operator type, with L referring to leaf nodes. In columns `LADTMC` and `SWN`, resp., we report basic statistics (number of states and transitions) for the LADTMC and the reachability graph of the SWN, along with the `runtime` for exact computation of the probability of each trace in the log.

Table 1. Language omputation: SPT versus SWN.

Event log	log stats			SPT		LADTMC			SWN (reachability graph)			
	#ttr	#utr	#act	#nodes per type	#pars	#st	#tr	runtime	#pars	#st	#tr	runtime
BPIC13_c	1,487	183	4	$\rightarrow$:0, $\times$:4, $\wedge$:3, $\circlearrowleft$:3, L:11	10	16	63	0.040	19	46	176	0.092
BPIC13_i	7,554	1,511	4	$\rightarrow$:1, $\times$:6, $\wedge$:3, $\circlearrowleft$:3, L:12	11	12	45	1.06	23	90	352	4.64
BPIC13_o	819	108	3	$\rightarrow$:0, $\times$:4, $\wedge$:2, $\circlearrowleft$:3, L:10	9	8	27	0.0085	20	74	264	0.037
BPIC17_ol	42,995	16	8	$\rightarrow$:2, $\times$:5, $\wedge$:0, $\circlearrowleft$:0, L:11	7	9	24	0.00019	11	6	11	0.00021
BPIC20_dd	10,500	99	17	$\rightarrow$:5, $\times$:15, $\wedge$:3, $\circlearrowleft$:4, L:30	26	79	510	0.049	43	235	1,147	0.093
BPIC20_rfp	6,886	89	19	$\rightarrow$:6, $\times$:20, $\wedge$:5, $\circlearrowleft$:3, L:27	30	132	785	0.061	51	247	1,292	0.26
Roadfines	150,370	231	11	$\rightarrow$:3, $\times$:12, $\wedge$:6, $\circlearrowleft$:1, L:23	21	277	1,536	0.038	33	906	6,270	0.17
BPIC12	13,087	4,366	24	$\rightarrow$:9, $\times$:26, $\wedge$:7, $\circlearrowleft$:7, L:54	45	419	9,166	256.49	78	7,266	54,762	∞

The results indicate a substantial gain with the SPT approach, both in terms of the underlying model's dimensions (LADTMC models having substantially fewer states and transitions than the corresponding SWN reachability graphs) and of runtime (up to 4.47× faster on `Roadfines` and average gain of 3.25×). This gain is the unsurprising consequence of the LADTMC semantics, which, unlike SWN, generates silent-transition-free models; it also bypasses the non-termination issue of SWN unfolding in the presence of silent loops, as for `BPIC12` (corresponding ∞ runtime). The only outlier is `BPIC17_ol`, for which the LADTMC contains more states and transitions than the reachability graph of the SWN (though even in this case the runtime with the LADTMC is lower). This is because this log is atypical: its mined process tree contains neither parallel nor loop operators, and hence the resulting SWN suffers less from silent transitions and yields a more compact reachability graph.

Table 2. rEMD-based optimization of SWN models versus SPT models. Source code and results: https://github.com/DocPierro/optimised_spt.

Event log	SWN optimization			SPT optimization (LADTMC-unfolding)			Gain
	rEMD	time	iter	rEMD	time	iter	
BPIC13_c	0.050	602.83	11	0.087	**21.27**	4	28.3×
BPIC13_i	N/C[1]	N/C	N/C	0.181	**73,031.13**	7	≥ 2.36×
BPIC13_o	0.072	153.3	10	0.089	**52.24**	30	2.93×
BPIC17_ol	0.017	1.007	10	0.015	**0.86**	10	1.17×
BPIC20_dd	0.011	2,047.0	33	0.017	**1,804.63**	54	1.13×
BPIC20_rfp	0.0486	8,925.1	50	0.0423	**2,810.08**	52	3.17×
Roadfines	0.025	1,983.1	18	0.1344	**30.65**	2	64.7×

[1] The computation exceeded 48 h.

SPT versus SWN optimization. Table 2 reports the optimisation experiment. Results show that optimisation on SPT models is consistently faster than on SWN models, with runtime reductions ranging from moderate (1.13× for `BPIC20_dd`) to very significant (64.7× for `Roadfines`), while preserving a comparable level of accuracy (in terms of rEMD). The number of iterations varies across logs but does not exhibit a systematic disadvantage for the SPT-based approach.

5 Conclusion

Stochastic process discovery requires efficient procedures to compute the stochastic language emitted by a model. In this paper we introduced a Markovian semantics for SPTs in terms of LADTMCs, and proved that the language of a SPT Q equals that of its corresponding LADTMC $\mathcal{D}$. This enables exact language computation via unfolding of $\mathcal{D}$, entirely bypassing the silent-loop issue that affects SWN-based approaches, and allowing optimization engines to scale to larger real-world event logs.

Since a LADTMC of a SPT is a standard absorbing DTMC, beyond language computation, it is directly amenable to the full spectrum of analyses available for this class of models. These include reachability and expected-hitting-time analyses via the fundamental matrix, probabilistic model checking against PCTL or LTL specifications, and process-mining queries such as those of [12] and its follow-up [13]. Being silent-transition-free and with a bounded state space, such analyses simplify substantially compared to SWN-based approaches, making this a promising direction for future work. Further directions include extending to LADTMCs the memoization strategy of [7], which would further accelerate optimization, and lifting the framework to capture the time dimension of event logs as outlined in [8].

References

1. van der Aalst, W., Buijs, J., van Dongen, B.: Towards Improving the Representational Bias of Process Mining. In: Aberer, K., Damiani, E., Dillon, T. (eds.) SIMPDA 2011. LNBIP, vol. 116, pp. 39–54. Springer, Heidelberg (2012). https://doi.org/10.1007/978-3-642-34044-4_3
2. Alkhammash, H., Polyvyanyy, A., Moffat, A., García-Bañuelos, L.: Entropic relevance: a mechanism for measuring stochastic process models discovered from event data. Inf. Syst. **107** (2022)
3. Brockhoff, T., Uysal, M.S., Van Der Aalst, W.M.: Wasserstein weight estimation for stochastic petri nets. In: 2024 6th International Conference on Process Mining (ICPM), pp. 81–88 (2024). https://doi.org/10.1109/ICPM63005.2024.10680664
4. Burke, A., Leemans, S.J.J., Wynn, M.T.: Stochastic Process Discovery by Weight Estimation. In: Leemans, S., Leopold, H. (eds.) Stochastic process discovery by weight estimation. LNBIP, vol. 406, pp. 260–272. Springer, Cham (2021). https://doi.org/10.1007/978-3-030-72693-5_20
5. Burke, A., Leemans, S.J.J., Wynn, M.T.: Discovering Stochastic Process Models by Reduction and Abstraction. In: Buchs, D., Carmona, J. (eds.) PETRI NETS 2021. LNCS, vol. 12734, pp. 312–336. Springer, Cham (2021). https://doi.org/10.1007/978-3-030-76983-3_16

6. Cry, P., Horváth, A., Ballarini, P.: Stochastic process trees: A formal framework for stochastic process discovery. In: 7th International Conference on Process Mining, ICPM 2025, Montevideo, Uruguay, pp. 1–8. IEEE (2025)
7. Cry, P., Horváth, A., Ballarini, P., Le Gall, P.: An efficient stochastic process discovery framework based on optimization. Softw. Tools Technol. Transf. (2026). https://doi.org/10.1007/s10009-026-00850-4
8. Horváth, A., Ballarini, P., Cry, P.: Process trees with stochasticity and time. Process Sci. (2026)
9. Leemans, S.J.J., Fahland, D., van der Aalst, W.M.P.: Discovering Block-Structured Process Models from Event Logs - A Constructive Approach. In: Colom, J.-M., Desel, J. (eds.) PETRI NETS 2013. LNCS, vol. 7927, pp. 311–329. Springer, Heidelberg (2013). https://doi.org/10.1007/978-3-642-38697-8_17
10. Leemans, S.J.J., Fahland, D., van der Aalst, W.M.P.: Discovering Block-Structured Process Models from Event Logs Containing Infrequent Behaviour. In: Lohmann, N., Song, M., Wohed, P. (eds.) BPM 2013. LNBIP, vol. 171, pp. 66–78. Springer, Cham (2014). https://doi.org/10.1007/978-3-319-06257-0_6
11. Leemans, S.J.J., Li, T., Montali, M., Polyvyanyy, A.: Stochastic process discovery: Can it be done optimally? In: Guizzardi, G., Santoro, F., Mouratidis, H., Soffer, P. (eds.) Advanced Information Systems Engineering. CAiSE 2024. LNCS, vol. 14663, pp. 36–52. Springer, Cham (2024). https://doi.org/10.1007/978-3-031-61057-8_3
12. Leemans, S.J.J., Maggi, F.M., Montali, M.: Enjoy the silence: analysis of stochastic petri nets with silent transitions. Inf. Syst. **124**, 102383 (2024)
13. Leemans, S.J.J., Montali, M., Gersing, T., Engelhardt, F., Sidorova, N.: Enjoy the silence, part II: probability-based queries on stochastic labelled petri nets. In: Amparore, E., Mikulski, Ł. (eds.) Application and Theory of Petri Nets and Concurrency. PETRI NETS 2025. LNCS, vol. 15714, pp. 310–332. Springer, Cham (2025). https://doi.org/10.1007/978-3-031-94634-9_15
14. Leemans, S.J., van der Aalst, W.M., Brockhoff, T., Polyvyanyy, A.: Stochastic process mining: earth movers' stochastic conformance. Inf. Syst. **102**, 101724 (2021)

Temporal Object-Centric Event Data: A Property-Graph-Based Model with Consistency Constraints

Ava Swevels[1(✉)], Marco Montali[2], Francesca Zerbato[1], and Dirk Fahland[1]

[1] Eindhoven University of Technology, Eindhoven, The Netherlands
{a.j.e.swevels,f.zerbato,d.fahland}@tue.nl
[2] Free University of Bozen-Bolzano, Bolzano, Italy
montali@inf.unibz.it

Abstract. Data in process mining (PM) is inherently temporal. Yet, existing object-centric event data (OCED) models (e.g., OCEL and Event Knowledge Graphs) associate time only with events, leaving objects and relationships static. Existing extensions address isolated aspects (e.g., evolving attributes), but lack a unified approach to model time-varying objects, attributes, and relationships. This absence of explicit temporal semantics hinders the specification of constraints (e.g. forbidding a *Borrow Book* event when the book is already on loan), often leading to implicit or ad-hoc assumptions and inconsistent analyses. To address this limitation, this paper introduces *Temporal Object Centric Event Data* (T-OCED), a temporal data model based on temporal property graphs that captures time across events, objects, attributes, and relationships. We formalize T-OCED and define criteria aligning event occurrences with the temporal validity of referenced elements. We further show how T-OCED supports the formalization of structural and semantic constraints. Finally, we demonstrate feasibility through an implementation in Neo4j, tested on a simulated dataset with injected violations.

Keywords: Process mining · Temporal Data · Data Modeling · Consistency Constraints

1 Introduction

Object-centric process mining aims at providing models and methods for studying process behavior in terms of the co-evolution of multiple interacting objects (in possibly complex relations) instead of event sequences alone [7,36]. To support this aim, several representations for event data over objects have been proposed, including OCEL 1.0 event tables [20] and Event Knowledge Graphs (EKGs) [13].

These object-centric event data (OCED) models relate (atomic) events to objects in an object-relational structure [12] that describes valid event orderings from multiple vantage points of individual objects [13,36]. However, these models

F. M. Maggi et al. (Eds.): BPM 2026, LNCS 16936, pp. 184–201, 2027.
https://doi.org/10.1007/978-3-032-37777-7_13

assume objects and relations to be static, i.e., unchanging over time (cf. [33]). This assumption is easily violated in practice. For example, consider a library where a member may borrow a book *only if this book is not reserved (or borrowed) by another member.* Likewise, a member may cease membership *only if they have returned all their borrowed books and still hold a valid membership.* Because these relationships and membership status evolve over time, treating them as static makes it difficult to analyze how books and memberships synchronize. It may also distort statistics, such as the average number of simultaneously borrowed books per member, which are key for object-centric process discovery [6,33,38].

However, the issue goes beyond correct statistics. Process analysis assumes that events are embedded in the *timeline* of a meaningful process instance, providing a coherent temporal reference frame [14]. That is, the data model must provide *temporal grounding* for all involved data points. The classical single case notion and its sequential traces provide this grounding in terms of a case life-cycle, but only for processes that can be represented this way without error [36]. Proposals for object-centric case notions aim to provide such temporal reference frames [14] on OCED with static objects and relations. In general, the multiple life-cycles of objects and relations must be temporally grounded to ensure a consistent interpretation from any vantage point. In our library example, it is essential to determine unambiguously, for any time interval, which memberships existed (and in which state) and which books were borrowed by whom.

Object evolution cannot be represented unambiguously in OCED [15] without explicit external domain knowledge [8,21]. Common implicit assumptions – such as an object being created by the first event mentioning it and destroyed by its last [30] – do not hold in general. This motivates extensions of OCED that explicitly represent the temporal evolution of objects and relations. Current approaches, including EKGs and OCEL lack systematic support, while extensions with time-varying attributes (cf. OCEL2.0 [26], tEKGs [25]), and evolving object states [27] only partially address the problem and omit relation life-cycles.

To overcome these limitations, three complementary strategies can be pursued to unambiguously reconstruct the evolution of objects, attributes, and relationships in response to events. First, object-centric logs can be augmented with domain knowledge on event semantics [8] or with interpretation assumptions [21]. Second, the intensional schema underlying event data can be enriched with more expressive constructs for objects, events, attributes and relationships [24], thereby requiring a correspondingly richer logging framework. Third, and followed in this paper, the extensional level can be enhanced by explicitly encoding how events affect objects, attributes and object-to-object (O2O) relationships, enabling unambiguous reconstruction of changes. So far, such reconstruction has been studied mainly for object attributes [25].

To fully realize this extensional approach, we take inspiration from the long-standing literature in temporal data modeling [4]. We extend *temporal property graphs* (TPGs) [3], which model evolving objects, attributes, and O2O relationships, with events and event-to-object (E2O) relationships resulting in *Temporal Object Centric Event Data* (T-OCED).

Our contribution is threefold. First, we formalize T-OCED as a specialization of TPGs aligned with OCED [15] linking event occurrences to the temporal validity of objects, attributes, and relationships. Second, we formulate constraints over T-OCEDs to ensure interpretability and consistency, ranging from standard snapshot constraints (enforced at each time point) to richer semantic constraints. In particular, we capture event semantics by requiring that each occurrence of an event type induces consistent changes (e.g., any *Borrow book* event creates a *Borrowed By* relationship). Third, we demonstrate feasibility through an encoding of T-OCED in Neo4j, showing that semantic inconsistencies can be effectively detected using Cypher queries on a simulated dataset with controlled violations.

The paper is structured as follows. Section 2 presents a motivating example, followed by related work in Sect. 3. Section 4 introduces the T-OCED data model based on TPGs, provide formal definitions, and show how it can be coalesced into an interval-timestamped representation. We then formalize three types of concistency constraints. Section 5 demonstrates feasibility through an implementation in Neo4j and shows how semantic violations can be detected. Finally, Sect. 6 discusses the limitations of our approach and outlines future work.

2 Motivating Example

This section introduces a motivating example illustrating how the lack of explicit temporal semantics in current OCED models leads to inconsistencies and ambiguity, hindering reliable analysis.

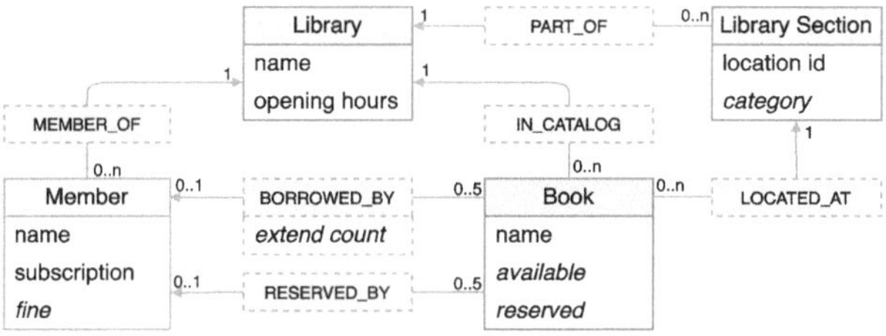

Fig. 1. Snapshot data model of the library with cardinality constraints

Consider the process of borrowing books in a library. Individuals may register as members, borrow books for a defined loan period, return or extend loans, incur fines for overdue returns, pay fines, and eventually deregister. A key rule is that a member may deregister only if they currently have no books on loan. The library wishes to analyze whether this rule is respected in practice.

Figure 1 depicts the snapshot data model underlying this process, i.e., the conceptual structure valid at any point in time. The model captures objects, their attributes, and relationships, with time-varying attributes shown in italics. For example, a member's fine is updated whenever new fines are issued or payments are made. Snapshot cardinality constraints are indicated on the relationships; for instance, a book can be borrowed by at most one member at a time, and a member can have at most five books on loan simultaneously.

To capture the evolution of this process, the library logs timestamped events for books (borrow, return, extend, reserve) and members (register, deregister, fine issued/paid). In OCED, books and members would be objects referenced by these events. However, the resulting object-centric representation would not be aligned with the snapshot data model, because all changes to objects and relationships are aggregated into a single time-agnostic view. Indeed, OCED does not explicitly represent *when* object states (e.g., membership validity) or relationships (e.g., member-book loan) start and end. The temporal aspects must therefore be reconstructed indirectly from event sequences.

This lack of explicit temporal representation becomes particularly problematic when verifying behavioral rules. Consider the following simplified fragment of the event log: a *Borrow Book* event for Book X by Member A at time t_1, followed by a *Deregister Member* event for Member A at time t_2, with no *Return Book* event recorded in between. To determine whether deregistration is valid, the analyst must *infer* the loan status of Member A at time t_2. Several interpretations are possible: (i) the analyst may assume that the log is complete, in which case the absence of a return implies that the book is still on loan and the deregistration violates the rule; (ii) alternatively, the analyst may assume that events are missing and infer that the book was returned but not logged, making the deregistration valid; or (iii) the analyst may assume that the *Deregister Member* event implicitly closes all open loans, even though this is never stated in the log. Each one of these three interpretations leads to different reconstructions of the member's loan status at time t_2. The log itself cannot resolve this ambiguity because the temporal validity of objects and relationships is not stored explicitly. As a result, analysts must reconstruct object states implicitly. This is an error-prone step that can fundamentally affect the conclusions drawn from the data, particularly in large-scale processes such as baggage-handling systems processing thousands of items per day [10].

This ambiguity remains unresolved in existing data sources or models. They attach time either to events or to objects and relationships, but none provide semantically consistent temporal information at both levels. OCED records time at the event level (cf. Sect. 1). In contrast, system logs like database redo logs [32] or SAP change logs timestamp only low-level object or attribute changes, without inherent process context. While the XOC data model [29] conceptually links events to object snapshots, real systems typically record discrete events, not the granular state changes needed to construct such snapshots reliably.

Consequently, we assume that the source of the data is, and remains, events: single-timestamped records with *semantic* meaning that implicitly or explicitly describe changes to objects and their relationships. The challenge is to complement these events with a representation of the *resulting* objects and relationships as they evolve over time. This requires (i) a representation of time periods during which specific objects and relationships hold, and (ii) a unified data model that semantically aligns events with these evolving structures, enabling reasoning over both event occurrences and object states at any moment in time.

3 Related Work

Object-centric process mining requires event data they go beyond case-centricity. A core requirement, identified in seminal works [36] is the inclusion of events, objects and event-to-object (E2O) relationships, enabling events to simultaneously refer to multiple objects of possibly different types. OCEL supports this feature through event tables since version 1 [20], while EKGs model events and objects as nodes linked via E2O relationship [13]. Extensions beyond E2O relationships vary widely. DOCEL [23] extends OCEL1.0 with richer data attributes, supporting multi-valued and time-varying attributes for both events and objects. Time-varying attributes are also supported in OCEL2.0 [26] and tEKGs [25]. OCEL2.0 further introduces explicit O2O relationships, which are also naturally supported in EKGs. However these relationships lack explicit temporal semantics, leaving their interpretation ambiguous (e.g., as only holding initially, or holding forever, i.e., being *rigid*). Additional extensions address dynamic objects. Specifically, [27] extends OCEL2.0 with object states and their changes over time, while [39] builds on DOCEL to support object states and alignment of object-centric event data across the process lifecycle.

The lack of explicit tracking of how objects and their relationships change over time has led to numerous assumptions or limitations in mining and analysis techniques. Discovery from OCELs has targeted object-centric Petri nets [38] that do not explicitly account for object identity and O2O relationships, leading to underspecified models [22,38] that admit a clear interpretation only when relationships are assumed to be rigid [33]. Similar limitations hold when discovering OC-DECLARE models, which extend the Declare language with object-centric features [28]. TOTeM [30] extracts from OCELs interval-based O2O temporal relationships, with two key assumptions: first, that objects are created and destroyed respectively by the first and last events mentioning them; second, that whenever two objects are connected to the same event, then an O2O relationship exists between them, and holds in the time interval where both objects exist.

Approaches for handling time-varying relationships either depend on manually specified reference models with explicit creation, deletion and update constructs [22], or require additional ad-hoc interpretations of the event data [11], since in general it is impossible to unambiguously reconstruct how relationships change from an OCEL or EKG [24].

4 Proposed Data Model

The data model of temporal property graphs (TPGs) [3] captures how objects and relationships *change over time*, but it does not define the concept of an event. We show how to adapt TPGs to object-centric event data by integrating events with the temporal evolution of objects, their attributes, and relationships.

Section 4.1 introduces the generic temporal property-graph model and its conversion to an interval-timestamp form. Section 4.2 builds on this model to define *Temporal Object Centric Event Data* (T-OCED), which can likewise be converted. Section 4.3 formalizes three types of consistency constraints for T-OCED.

4.1 Temporal Property Graphs

A temporal property graph (TPG) extends the notion of a property graph [1,2,18] to include explicit access to time, allowing us to capture how relationships and nodes evolve. A TPG is a theoretical model that explicitly states, for each point in time, whether an object/relation exists and which property values hold for those objects or relations. This point-based approach represents the evolution of a property graph by offering a snapshot at every moment in time. However, timestamping objects at individual time points can lead to considerable space overhead, motivating the use of interval-based representations instead [3].

We first present a formal definition of a temporal property graph (TPG), which can then be transformed into an interval-timestamped temporal property graph (ITPG). In our work, we adopt the TPG definition by Arenas et al. [3].

We consider a label set Λ partitioned into node labels Λ_N and relationship labels Λ_R such that $\Lambda = \Lambda_N \cup \Lambda_R$ and $\Lambda_N \cap \Lambda_R = \emptyset$, and a set of attribute names *Attr* whose values come from a domain Val. Temporal property graphs are defined over finite sets of time points, chosen according to the temporal granularity relevant to the application (e.g., seconds, weeks, or years). For convenience, we denote the set of all time points as $\mathbb{N}$, and define a temporal domain Ω which expresses a closed interval of discrete time points between a start time α and an end time β: $\Omega = \{ i \in \mathbb{N} \mid \alpha \leq i \leq \beta \}$, where $\alpha, \beta \in \mathbb{N}$ and $\alpha \leq \beta$.

Definition 1 (Temporal Property Graphs). *A temporal property graph (TPG) $G = (\Omega, N, R, \rho, \lambda, \xi, \#)$ is a graph where*

1. *Ω is a temporal domain; N is a finite set of nodes, R is a finite set of relationships (edges) and $N \cap R = \emptyset$.*
2. *Function $\rho : R \to (N \times N)$ relates a relationship r to its source and target nodes $\rho(r) = (n_{source}, n_{target})$. We also write $\overrightarrow{r} = (n_{source}, n_{target})$.*
3. *Function $\lambda : (N \cup R) \to 2^{\Lambda}$ assigns to each node $n \in N$ at least one* label *$\lambda(n) \subseteq \Lambda_N$ and $\lambda(n) \neq \emptyset$, and to each relationship $r \in R$ exactly one label $\lambda(r) \in \Lambda_R$, also called relationship type. We also write $N^L = \{ n \in N \mid \lambda(n) \in L \}$ and $R^L = \{ r \in R \mid \lambda(r) \in L \}$.*
4. *Predicate $\xi : (N \cup R) \times \Omega \to \{true, false\}$ states whether a node or relationship $x \in N \cup R$ exists at time t, i.e. $\xi(x, t) = true$. Moreover, if $\xi(r, t)$ and $\rho(r) = (n_1, n_2)$, then $\xi(n_1, t)$ and $\xi(n_2, t)$.*
5. *Partial function $\# : (N \cup R) \times Attr \times \Omega \nrightarrow Val$ states whether attribute a has a value v at time t for a node or relationship $x \in N \cup R$, i.e. $\#(x, a, t) = v$. Moreover, if $\#(x, a, t) = v$ then $\xi(x, t)$. We also write $x.a_t = v$ when $\#(x, a, t) = v$ and $x.a_t = \perp$ if $\#(x, a, t)$ is undefined.*

6. *An attribute a is* static *(i.e. rigid) for a node or relationship $x \in N \cup R$ iff there exists a value v s.t. $\forall t \in \Omega : \xi(x,t) \implies \#(x,a,t) = v$. We write $x.a = v$ when a is a static attribute with value v for x.*

TPGs are subject to two rules that ensure that they align conceptually with sequences of valid conventional property graphs. Specifically, an edge is only permissible at times when both nodes it connects exist (cf. Definition 1.4), and an attribute can only be assigned a value when its associated object exists (cf. Definition 1.5). Furthermore, by restricting $\#(x,a,t)$ to a *finite* collection of tuples (x,a,t), it is ensured that each TPG can be represented finitely.

Interval-Timestamped Temporal Property Graphs. Each TPG G can be converted into an Interval-timestamped Temporal Property Graph (ITPG) by *coalescing* consecutive time points for which the existence and attribute values remain unchanged. Specifically, an ITPG $I = (\Omega', N, R, \rho, \lambda, \xi', \#')$ that encodes G is characterized as follows: the time domain Ω is substituted by the interval $\Omega' = [\alpha, \beta]$. The components N, R, ρ, λ remain unchanged from G. ξ' is a function assigning each entity $x \in (N \cup R)$ to a set of maximal intervals during which x exists, as per the function ξ. For instance, for $\Omega = \{1,2,3,4,5\}$ and a node n where $\xi(n,1) = \xi(n,2) = \xi(n,3) = \xi(n,5) = true$ and $\xi(n,4) = false$, it results in $\Omega' = [1,5]$ and $\xi'(n) = \{[1,3],[5,5]\}$. We use the open interval notation (a,b) to indicate the discrete interval $[a+1, b-1]$ when convenient. The function $\#'$ is derived from $\#$ in a manner akin to ξ'. The detailed definition is available in [3].

4.2 Temporal Object Centric Event Data

To represent how objects and relationships evolve over time in relation to events, we define a specialized TPG: Temporal Object Centric Event Data (T-OCED), whose structure aligns with the core OCED meta-model [15].

Let OT be a set of object types and RT a set of O2O relationship types. Each relationship type $rt \in RT$ is *strictly typed* by the object types of the nodes it connects. Formally, we denote the typing of rt by $ot(rt) = (ot_1, ot_2) \in OT \times OT$.

In T-OCED, each node is labeled either as an *Event* or as an object type $ot \in OT$. Any event-to-object (E2O) relationship is labeled *observes*. We use $rt \in RT$ to label any O2O relationship.

Definition 2 (Temporal Object Centric Event Data). *A Temporal Object Centric Event Data over object types OT and O2O relationship types RT is a TPG $G = (\Omega, N, R, \rho, \lambda, \xi, \#)$ with node labels $\{Event\} \uplus OT \subseteq \Lambda_N$ and relationship labels $\{observes\} \uplus RT \subseteq \Lambda_R$ with the following properties.*

1. *Every node x is typed with exactly one label $\lambda(x) \in \Lambda_N$.*
2. *For every event node $e \in N^{Event}$, there exists exactly one time point $t \in \Omega$ s.t. $\xi(e,t) = true$. At this time point, the event node e records an activity name $e.act_t \neq \bot$ and a timestamp $e.timestamp_t = t$.*

3. *For every object node $n \in N^{ot}, ot \in OT$, all time points at which n exists are consecutive: for any $t, t' \in \Omega$ with $t < t'$ and $\xi(n,t) = \xi(n,t') = true$, there is no t'' with $t < t'' < t'$ such that $\xi(n,t'') = false$.*
4. *Every observes relationship $r \in R^{observes}$, $\overrightarrow{r} = (e,n)$ is defined from an event node $e \in N^{Event}$ to an object node $n \in N^{ot}, ot \in OT$ s.t. $\xi(e,t)$ iff $\xi(r,t)$ (i.e., r only holds at the time t of event e) and $\xi(r,t)$ implies $\xi(n,t)$ (i.e., n exists at time t).*
5. *Every other relationship $r \in R^{rt}$, $rt \in RT$ (with $ot(rt) = (ot_1, ot_2)$) is defined between two object nodes: $r = (n_1, n_2)$ with $n_1 \in N^{ot_1}, n_2 \in N^{ot_2}$. Additionally, all time points at which r exists are consecutive: for any $t, t' \in \Omega$ with $t < t'$ and $\xi(r,t) = \xi(r,t') = true$, there is no t'' with $t < t'' < t'$ such that $\xi(r,t'') = false$.*

We denote all object nodes by $N^{OT} = \bigcup_{ot \in OT} N^{ot}$ and all O2O relationships by $R^{RT} = \bigcup_{rt \in RT} R^{rt}$.

T-OCED is subject to several rules reflecting how events and objects behave in real processes. In T-OCED, events are instantaneous (occur at a single point in time), whereas objects and O2O relationships persist over a single continuous time interval, during which their type remains static – even though, in reality, objects or relationships may disappear, reappear, or change type (e.g., a library member becoming an employee). However, these limitations do not reduce modeling expressiveness: any type change can be represented by modeling the same real-world object as multiple object nodes, or the same O2O relationship as multiple edges, each with its own lifetime and type.

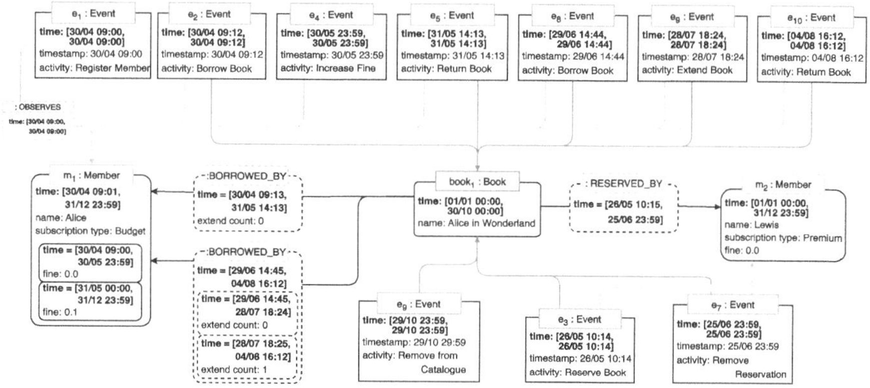

Fig. 2. An instance of IT-OCED showing a small fragment of the library process.

Interval-Timestamped Temporal Object Centric Event Data. Since every T-OCED instance is a specialization of a TPG, these graphs can be coalesced into interval-timestamped TPGs using the operation defined in Sect. 4.1.

As T-OCEDs constrain TPGs (Definition 2), the corresponding IT-OCED G' of an T-OCED has the following properties[1]:

1. The existence interval for event nodes $e \in N^{Event}$ is a single point: $\xi'(e) = \{[t,t]\}$ for exactly one time point $t \in \Omega'$.
2. The existence interval for object nodes $n \in N^{OT}$ is a consecutive interval: $\xi'(n) = \{[a,b]\}$ for some $a \leq b$ with $a, b \in \Omega'$.
3. The existence interval for observes relationships $r \in R^{observes}$ is a single point: $\xi'(r) = \{[t,t]\}$ where $t \in \Omega'$ is the timestamp of the corresponding event $e \in N^{Event}$ with $\xi'(e) = \{[t,t]\}$ and $r = (e,n)$.
4. The existence interval for every O2O relationship $r \in R^{RT}$ is a consecutive interval: $\xi'(r) = \{[a,b]\}$ for some $a \leq b$ with $a, b \in \Omega'$.

Figure 2 illustrates an instance of IT-OCED for the library process described in Sect. 2. It depicts several events related to members *Alice*, *Lewis*, and the book *Alice in Wonderland*, along with their validity intervals and the intervals of the related relationships and attributes.

4.3 Consistency Constraints

To maintain semantic consistency between events and evolving objects and relationships in T-OCED, we introduce *key constraints* ensuring unique identification of nodes, *cardinality constraints* governing the number of permitted relationships, and *semantic constraints* ensuring events correspond to valid changes in objects, attributes, and relationships. These constraints are not tied to a single purpose; instead, they characterize valid states or behaviors and can be used for data validation, conformance monitoring, and analysis. Under noise-free conditions, they may even be discovered from observed data. We do not consider other quantifications intended to express allowed behavior. For instance, one may require the eventual or universal existence of relations over time (or after a specific time point) [37]. Integrating both ideas, w.r.t. analytical reference frames such as object-centric cases [14], is left for future work.

Key Constraints. In a property graph, a node already has an inherent unique identity, i.e., the node itself. However, many applications designate certain node attributes as *external identifiers* that must be unique across all instances of a given type. Such attributes are called *keys* and they must be *static*, which we formalize as key constraints.

Node Key Constraints. Let G be an T-OCED and $K \subseteq Attr$ be a nonempty set of attribute names. We say that K is a key for type $l \in \Lambda_N$ if, for any two nodes $x_1, x_2 \in N^l$, $(\forall a \in K : x_1.a = x_2.a) \Rightarrow x_1 = x_2$.

[1] The full definition can be found in App. A (see: https://github.com/Ava-S/IT-OCED-Library).

For relationships, the edge between two nodes already provides an inherent unique identity, so defining keys for relationships is usually unnecessary. One-to-one and one-to-many relationships (cf. Sect. 4.3) are typically structural: they express how two objects are related, but do not have an independent identity. Examples include a book being borrowed by at a member, or a book located at a section in the library. In such cases, assigning keys to relationships offers no modelling benefits. In contrast, many-to-many relationships often represent a business object [19]. Here, the relationship carries meaning beyond simply connecting two objects, and is therefore commonly reified as its own object. For instance, a member holding memberships at multiple libraries correspond to a *Membership* object with its own identity and attributes (e.g., membership type). In these cases, the relationship should be reified as an object node, after which its attributes can be treated as keys like those of any other object.

Cardinality Constraints. We may wish to constrain the cardinality of relationship types between object nodes of a given type (e.g., one-to-one, one-to-many, or many-to-many). Adding time makes cardinality more nuanced: a relationship can be one-to-many at *every moment*, even if multiple such relationships occur over time on the 'one' side. For instance, a book may be borrowed by at most one member at any moment, yet over its lifetime be borrowed many times. Therefore, we differentiate between snapshot and lifespan cardinality.

Let G be an T-OCED and $rt \in RT$ be a relationship type with $ot(rt) = (ot_1, ot_2)$.

1. **Snapshot cardinality constraints** describe how many relationships of a given type may exist simultaneously and should hold at every time point (e.g. a book is borrowed by at most one member at any moment).
   Let $card_{min}, card_{max} \in \mathbb{N} \cup \{\infty\}$ be cardinality bounds. We say that relationship type rt has snapshot cardinality $[card_{min}, card_{max}]$ iff: for every time point $t \in \Omega$ and every object node $n_1 \in N^{ot_1}$ and $\xi(n_1, t)$, then $card_{min} \leq |\{ r = (n_1, n_2) \in R^{rt} \mid n_2 \in N^{ot_2}, \xi(r, t) \}| \leq card_{max}$, i.e., at any valid time point t, an object of type ot_1 must be connected to between $card_{min}$ and $card_{max}$ objects of type ot_2 through relationships of type rt that are valid at that time t.
2. **Lifespan cardinality constraints** describe how many relationships may occur in total across time and apply to the entire lifetime of an object (e.g., a book may be borrowed at most 10 times during their life).
   Let $life_{min}, life_{max} \in \mathbb{N} \cup \{\infty\}$ be lifetime bounds. We say that relationship type rt has lifespan cardinality $[life_{min}, life_{max}]$ iff: for every object node $n_1 \in N^{ot_1}$, $life_{min} \leq |\{ r = (n_1, n_2) \in R^{rt} \mid n_2 \in N^{ot_2} \}| \leq life_{max}$. Note that the lifespan cardinality constraint has no temporal constraint $\xi(r, t)$.
   Thus, over its entire lifetime, an object of type ot_1 must participate in between $life_{min}$ and $life_{max}$ relationships of type rt with objects of type ot_2 regardless of when those relationships occur.

Semantic Constraints. Semantic constraints capture the expected relationships between events and the objects and relationships they affect. For example, *Borrow Book* updates a `Book` object (updating *available*) and creates a `BORROWED BY` relationship to a `Member`. We base these constraints on CRUD operations (Create, Read, Update, and Delete) [31]. For this, we must know which operation each activity performs. Because activity semantics are often not explicitly represented in existing data models, we assume an oracle that provides this information, consistent with the OCED Core Meta-Model [15], where known activity semantics may be expressed as a qualifier on the E2O *observes* relationship.

For relationships, we introduce an additional *Replace* operator. Since relationships can be seen as non-rigid attributes (similar to ER relationships implemented via foreign keys), they may be *updated* from an object's perspective. From the relationship's perspective, however, such a change corresponds to deleting the old relationship and creating a new one. *Replace* therefore subsumes both *Create* and *Delete*, capturing an atomic replacement where an activity terminates an existing relationship and introduces a new one. For example, a *Relocate book* activity replaces the `LOCATED AT` relationship between a `Book` and its current `Library Section` with a new one pointing to the updated one. Importantly, *Replace* applies only to relationships, not objects, since object identity persists across updates, whereas relationship identity changes when an endpoint changes.

We consider an IT-OCED G', a finite set of activity names Act, a finite set of object types OT and a finite set of relationship types RT and an operator set $OP = \{Create, Read, Update, Delete, Replace\}$. For readability, the constraints are defined over an IT-OCED[2]. Some constraints are accompanied by an example to help anchor their interpretation.

Definition 3 (Oracle Function). *Partial function $f : Act \times (OT \cup RT) \nrightarrow OP$ assigns to an activity name $a \in Act$ and type $tp \in OT \cup RT$ an operation type $op \in OP$. We also write $f_{tp}(a) = op$ as shorthand. Additionaly, $f_{ot}(a) \neq Replace$ for any $ot \in OT$.*

Using the Oracle function, we can define the semantic constraints.

Semantic Constraints on Events and Objects. Let $e \in N^{\{Event\}}$ be an event occurring at time $t \in \Omega$, i.e. $\xi(e) = \{[t, t]\}$. Let $ot \in OT$. If the oracle assigns $f_{ot}(e.act) \neq \bot$, then the following must hold.

- **CREATE** If $f_{ot}(e.act) = Create$, then there exists an object $n \in N^{ot}$ such that $(e, n) \in R^{observes}$ and $\xi(n) = \{(t, b]\}$ for some $b \geq t$, i.e. e causes an object n of type ot to come into existence immediately after time t and to remain in existence until (and including) time b. For example, since $f_{Member}(Register\ Member) = Create$, a member is created when a *Register Member* event occurs and is related to that event via $R^{observes}$.
- **READ** If $f_{ot}(e.act) = Read$, then there exists an object $n \in N^{ot}$ such that $(e, n) \in R^{observes}$ and $\xi(n) = \{(a, b]\}$ for some $a \leq t \leq b$, i.e. event e does indeed read an object of type ot.

[2] An equivalent formulation over T-OCED is provided in App. B (see https://github.com/Ava-S/IT-OCED-Library).

- **UPDATE** If $f_{ot}(e.act) = Update$, then there exists an object $n \in N^{ot}$ such that $(e, n) \in R^{observes}$, and $\xi(n) = \{(a, b]\}$ for some $a \leq t \leq b$, and there exists an attribute $a \in Attr$ such that $n.a_t \neq n.a_{t+1}$, i.e. an update on an object n of type ot is observable as as attribute value change at time t. For example, since $f_{Book}(Borrow\ Book) = Update$, an observed book exists when an *Borrow Book* event occurs and one of its attributes (available) changes.
- **DELETE** If $f_{ot}(e.act) = Delete$, then there exists an object $n \in N^{ot}$ such that $(e, n) \in R^{observes}$ and $\xi(n) = \{(a, t]\}$ for some $a \leq t$, i.e. Hence, e marks the end of the validity interval of object n of type ot, which ceases to exist immediately after time t.

When formulating these constraints, it is crucial to interpret an event's timestamp and determine when its effects take place. In this paper, we assume that an event's effect becomes visible *immediately after* the event, i.e., an attribute updated at time t takes effect at $t + 1$, and the old value still holds at t. This addresses the well-known *divided instant problem* [16], which concerns the ambiguity of the exact moment of change and requires ensuring that *only one* value is valid at any given instant. This assumption also implies that an object will only come into existence just after a CREATE event, and that it still exists at the time of a DELETE event, ceasing to exist only one instant later. This interpretation aligns with common reasoning in event calculus [9].

Because the divided instant problem forces a design choice about when effects occur, one could alternatively decide that effects take place *at* the timestamp of the event, which would imply, for example, that an object no longer exists at the time of deletion. Either choice requires some flexibility in the TPG definition. In a TPG, a relationship may exist only when both endpoints exist, but under our assumption, during create the object does not yet exist. Therefore, for the *observes* relationship, we relax this constraint to allow such transitional cases.

Semantic Constraints on Events and Relationships. Since event nodes are related only to object nodes, and not directly to O2O relationships, semantic constraints over relationships must ensure that an event observes at least one object node n to which the relevant relationship is *incident*. An O2O relationship is incident to an object node n if $\overrightarrow{r} = (n, n') \in R^{RT}$ or $\overrightarrow{r} = (n', n) \in R^{RT}$.

Let $e \in N^{Event}$ be an event occurring at time $t \in \Omega$, i.e. $\xi(e) = \{[t, t]\}$. Let $rt \in RT$. If the oracle assigns $f_{rt}(e.act) \neq \bot$, then the following must hold.

- **CREATE** If $f_{rt}(e.act) = Create$, then there exist an object $n \in N^{OT}$ such that $(e, n) \in R^{observes}$, $r \in R^{rt}$ is incident to n and $\xi(r) = \{(t, b]\}$ for some $b \geq t$, i.e. e causes a r of type rt incident to the observed object n to come into existence immediately after time t and continues to exist until (and including) a time b. For example, since $f_{borrowed\ by}(Borrow\ Book) = Create$, a *borrowed by* relationship of an observed object (e.g., book) is created when a *Borrow Book* event occurs.
- **READ** If $f_{rt}(e.act) = Read$, then there exist an object $n \in N^{OT}$ such that $(e, n) \in R^{observes}$, $r \in R^{rt}$ is incident to n and $\xi(r) = \{(a, b]\}$ with $a \leq t \leq b$, i.e. event e does indeed read a relationship r of type rt that exists at time t.

- **UPDATE** If $f_{rt}(e.act) = Update$, then there exist an object $n \in N^{OT}$ such that $(e,n) \in R^{observes}$, $r \in R^{rt}$ is incident to n, $\xi(r) = \{(a,b]\}$ with $a \leq t \leq b$, and there exists an attribute $attr \in Attr$ such that $r.attr_t \neq r.attr_{t+1}$, i.e. an update on r of type rt incident to the observed node n is observable as as attribute value change at time t.
- **DELETE** If $f_{rt}(e.act) = Delete$, then there exist an object $n \in N^{OT}$ such that $(e,n) \in R^{observes}$, $r \in R^{rt}$ is incident to n and $\xi(r) = \{(a,t]\}$ for some $a \leq t$, i.e. r of type rt incident to the observed node n exists exactly from time a onwards and continues to exist until exactly time t.
- **REPLACE** If $f_{rt}(e.act) = Replace$, then there exist an object $n \in N^{OT}$ such that $(e,n) \in R^{observes}$, $r_{old} \in R^{rt}$ is incident to n, $r_{new} \in R^{rt}$ is incident to n (with $r_{old} \neq r_{new}$), $\xi(r_{old}) = \{(a,t]\}$, $\xi(r_{new}) = \{(t,b]\}$ with $a \leq t \leq b$, i.e., a relationship update of type rt for the observed object n is represented by terminating r_{old} at time t (so it no longer exists at $t+1$) and initiating r_{new} at time $t+1$. For example, since $f_{located\ at}(Relocate\ book) = Update$, a *located at* relationship of an observed object (e.g., book) is replaced by another *located at* relationship when a *Relocate book* event occurs.

From the semantic constraints, it follows that if an object n of type ot is observed by a *Create* event e, no earlier event e' with $e'.timestamp < e.timestamp$ may observe n. Otherwise, n would be observed before its creation, i.e. at a time when it does not yet exist. Correspondingly, if a *Delete* event e observes n, no later event may observe n. Moreover, any *Read* and *Update* event must be after the *Create* event and before the *Delete* event, if present.

We also observe that the Cardinality Constraints have implications for the Semantic Constraints. For example, if a snapshot cardinality constraint specifies that every *Book* must be in the catalogue of exactly one *Library* (i.e. a many-to-one relationship), then whenever an activity creates a *Book*, it must also create the corresponding *in catalogue of* relationship to ensure the cardinality constraint is satisfied at that moment in time.

5 Demonstration

We demonstrate feasibility using an off-the-shelf graph database on a simulated dataset[3]. Since no widely used graph database offers native temporal support, we implement IT-OCED in Neo4j using the following representation choices: object evolution is captured via global state nodes with validity intervals expressed using `validFrom` and `validTo` properties. These state nodes are attached to object nodes. Relationships remain incident to object nodes and relationship evolution is captured through multiple edges, with a counter tracking which edges correspond to the same relationship over time. Exploring the performance and conceptual implications of these choices (e.g., multiple object nodes, relations attached to states vs. object nodes) is beyond the scope of this paper.

[3] The implementation of the proposed data model and the simulation are available at https://github.com/Ava-S/IT-OCED-Library.

To demonstrate both the feasibility of generating an IT-OCED instance consistent with our modeling decisions and the verifiability of constraints over this data model, we generated a simulated dataset of the library process described in Sect. 2 using the SimPy library. The dataset takes the form of a raw event table recording each event together with its activity and timestamp, the identifier of the involved member, its outstanding fine, and its subscription type. It also includes the identifiers of up to five books along with their availability or reservation status. We injected controlled violations to test whether queries checking the semantic constraints could correctly identify these violations.

We then used OCED-PG [35] to transform the simulated data (event table) into an IT-OCED instance, thereby mapping event data into events, objects, and relationships. Events were assigned their validity intervals directly from their timestamps. For objects and relationships, we derived when they were created, updated, and deleted based on the known semantics of the activity. For example, a *Register Member* event occurring at time t observing member m creates member m at time t. For initial states, i.e., before the first observed event for an object, we inferred values where possible. For instance, if the first event for a book is *Reserve Book*, we infer that its initial value for *reserved* was *false*. Objects and relationships without an explicit create or DELETE event were assumed to hold from the start of the dataset until its end.

To evaluate where the constructed IT-OCED instance satisfies or violates the formal IT-OCED specification as well as the cardinality and semantic constraints, we translated all constraints to Cypher queries. The following example illustrates the translation principle; the implementation includes the full set of constraints together with their corresponding queries. The following Cypher query implements the instantiation of the UPDATE constraint for *Extend Book* of our running example, where *Extend Book* updates the `:BORROWED_BY` relationship by modifying the attribute `extendCount`. The query detects whether any feature of the relationship has changed at the timestamp of the event:

```
1   MATCH (e:Event {activity: 'Extend Book'})
2   RETURN EXISTS{
3    (e) - [:OBSERVES] -> (b:Book) - [r:BORROWED_BY] -> (:Member) <- [prev:BORROWED_BY]
        - (b)
4    WHERE prev.validTo = e.timestamp AND r.validFrom = e.timestamp
5    AND any(k IN keys(r) + keys(prev) WHERE
6       NOT k in ['validFrom', 'validTo'] //ignore temporal keys
7       AND properties(prev)[k] <> properties(r)[k]) // detect changed feature
8    }, count(e) as eventCount
```

When the specific feature to be updated is known, the generic comparison (lines 5–7) can be replaced with a direct value comparison. In our case, this simplifies to `AND r.extendCount <> prev.extendCount`.

To cover all types of constraints in a non-trivial setting, we generated datasets for a single library operating over several time periods, each preceded by a 50-day warm-up phase to simulate an already functional environment. For each period, we constructed an IT-OCED instance in Neo4j v5.26, using non-optimized queries, with indexing on unique identifiers for events and objects and on the validity intervals of State nodes. Performance measurements are reported in Table 1. For each simulation period, we recorded the time required to build the IT-OCED instance and to check all constraints (CC). In total, we defined 22 semantic constraints, 6 snapshot cardinality constraints, and 1 lifespan constraint.

Table 1. Simulation Size and Performance Metrics per Period Length

Simulated Period	#Events	#Members	#Books	#observes	#O2O relation	Build Time	CC Time
1 year	170,072	1216	7000	216,231	46,374	1.3 min	0.2 min
2 year	309,995	1242	7000	410,596	81,960	2.2 min	0.5 min
3 year	470,063	1258	7000	621,469	114,839	3.8 min	1.0 min

We injected one violation per constraint type into our simulated model; snapshot cardinality was violated by removing some *Return Book* events, causing consecutive borrows without intervening returns; lifespan cardinality by allowing some books to be borrowed more than ten times; and semantic constraints by omitting some fine updates when a member incurred a fine. Our implementation correctly identified all injected violations, except for some lifespan cardinality violations. This is due to *borrow* events executed during the warm-up phase not being logged, resulting in incomplete lifetime information.

Because of how the IT-OCED instance is constructed, some semantic constraints (e.g., each CREATE event producing an object) always hold. However, when CREATE or DELETE events are missing, other constraints (e.g., snapshot cardinality) are inevitably violated because the object's lifecycle becomes inconsistent. Certain semantic violations, such as missing updates, remain detectable since the expected value change is absent. Even when some semantic constraints cannot be violated, they remain useful for validating data transformation.

6 Conclusion

This work introduces T-OCED, a data model based on Temporal Property Graphs (TPGs), to explicitly represent time for events, objects, and their relationships. By defining three types of constraints, we ensure semantic alignment between events and evolving object states. We demonstrate feasibility through a Neo4j implementation validated on a simulated dataset with injected violations.

This work constitutes a first step toward jointly modeling events and evolving structures with semantic alignment, enabling process analysis based on temporal synchronization. Limitations remain: relying on events alone prevents full reconstruction of some object states. Furthermore, as current graph databases lack native temporal support, implementation choices must be made; evaluating alternatives is left for future work. Moreover, TPGs do not support event sequence navigation (e.g. Directly Follows in EKGs [13]), leaving this as future work. Another interesting direction is to complement T-OCED with intensional, temporal domain knowledge [5], thereby enabling T-OCED knowledge graphs equipped with systematic reasoning support to deal with ambiguity [17] and inference under incomplete information [34,37].

Acknowledgment. The research underlying this paper was partially supported by EuroFMX EU GA n. 101299128, and by the NextGenerationEU FAIR PE0000013 project MAIPM (CUP C63C22000770006). We thank Julia Stoyanovich for the insightful discussions and exchange of ideas that led to the conception of this paper.

References

1. Angles, R., et al.: G-Core: a core for future graph query languages. In: SIGMOD 2018, pp. 1421–1432. ACM (2018)
2. Angles, R., Arenas, M., Barceló, P., Hogan, A., Reutter, J., Vrgoč, D.: Foundations of modern query languages for graph databases. ACM Comput. Surv. (CSUR) **50**(5), 1–40 (2017)
3. Arenas, M., Bahamondes, P., Aghasadeghi, A., Stoyanovich, J.: Temporal regular path queries. In: ICDE 2022, pp. 2412–2425 (2022)
4. Artale, A., Franconi, E.: Foundations of temporal conceptual data models. In: Borgida, A.T., Chaudhri, V.K., Giorgini, P., Yu, E.S. (eds.) Conceptual Modeling: Foundations and Applications. LNCS, vol. 5600, pp. 10–35. Springer, Heidelberg (2009). https://doi.org/10.1007/978-3-642-02463-4_2
5. Artale, A., Kontchakov, R., Kovtunova, A., Ryzhikov, V., Wolter, F., Zakharyaschev, M.: Ontology-mediated query answering over temporal data: a survey (invited talk). In: TIME 2017, LIPIcs, vol. 90. Schloss Dagstuhl (2017)
6. Barenholz, D., Montali, M., Polyvyanyy, A., Reijers, H.A., Rivkin, A., van der Werf, J.M.E.: There and back again: on the reconstructability and rediscoverability of typed Jackson nets. In: Gomes, L., Lorenz, R. (eds) PETRI NETS 2023, pp. 37–58. Springer, Cham (2023). https://doi.org/10.1007/978-3-031-33620-1_3
7. Berti, A., Montali, M., van der Aalst, W.M.P.: Advancements and challenges in object-centric process mining: a systematic literature review. CoRR abs/2311.08795 (2023)
8. Chaki, R., Calvanese, D., Montali, M.: Generation of timelines from event knowledge graphs using domain knowledge. In: Saeed, K., Dvorský, J., Fukumoto, M., Nishiuchi, N. (eds) CISIM. LNCS, vol. 15927, pp. 275–289. Springer, Cham (2025). https://doi.org/10.1007/978-3-032-02406-0_20
9. Chesani, F., Mello, P., Montali, M., Torroni, P.: A logic-based, reactive calculus of events. Fund. Inform. **105**(1–2), 135–161 (2010)
10. Chu, V.: Using event knowledge graphs to model multi-dimensional dynamics in a baggage handling system. In: Master's thesis, Eindhoven University of Technology (2022)
11. van Detten, J.N., Schumacher, P., Leemans, S.J.J.: Modeling and discovering dynamic identity relations in object-centric process mining. In: ICPM, pp. 1–8 (2025)
12. Fahland, D., et al.: Towards a simple and extensible standard for object-centric event data (OCED) - core model, design space, and lessons learned. arXiv abs/2410.14495 (2024)

13. Fahland, D.: Process mining over multiple behavioral dimensions with event knowledge graphs. In: van der Aalst, W.M.P., Carmona, J. (eds) Process Mining Handbook, vol. 448, pp. 274–319. Springer, Cham (2022). https://doi.org/10.1007/978-3-031-08848-3_9
14. Fahland, D., Montali, M.: What is an object-centric case? an exploration. In: Mendling, J., Leemans, S., van Dongen, B.F., Reijers, H. (eds) Mining a Scientist's Process - Essays Dedicated to Wil van der Aalst on the Occasion of His 60th Birthday, vol. 16480, pp. 398–425. Springer, Cham (2026). https://doi.org/10.1007/978-3-032-17618-9_28
15. Fahland, D., et al.: Towards a simple and extensible standard for object-centric event data (OCED)–core model, design space, and lessons learned. arXiv preprint arXiv:2410.14495 (2024)
16. Fisher, M.D., Gabbay, D.M., Vila, L.: Handbook of Temporal Reasoning in Artificial Intelligence, vol. 1. Elsevier (2005)
17. Franceschetti, M., Buchegger, D.M., Seiger, R., Weber, B.: Toward IoT-based process analytics: extending event knowledge graphs with ambiguity. In: Guizzardi, R., Pufahl, L., Sturm, A., van der Aa, H. (eds) BPMDS/EMSSAD 2025. LNBIP 558. Springer, Cham (2025). https://doi.org/10.1007/978-3-031-95397-2_11
18. Francis, N., et al.: Cypher: an evolving query language for property graphs. In: Proceedings of the 2018 International Conference on Management of Data, pp. 1433–1445 (2018)
19. Frisendal, T.: Getting the Structure Right, pp. 45–60. Apress (2018)
20. Ghahfarokhi, A.F., Park, G., Berti, A., van der Aalst, W.M.P.: OCEL: a standard for object-centric event logs. In: Bellatreche, L., et al. New Trends in Database and Information Systems, pp. 169–175. Springer, Cham (2021). https://doi.org/10.1007/978-3-030-85082-1_16
21. Gianola, A., Hameed, Z., Montali, M., Seidel, A., Weske, M., Winkler, S.: Detecting dynamic relationships in object-centric event logs. In: CAiSE Forum 2026. LNBIP, vol. 587 (2026)
22. Gianola, A., Montali, M., Winkler, S.: Object-centric conformance alignments with synchronization. In: CAiSE 2024, vol. 14663, pp. 3–19 (2024)
23. Goossens, A., De Smedt, J., Vanthienen, J., van der Aalst, W.M.P.: Enhancing data-awareness of object-centric event logs. In: Montali, M., Senderovich, A., Weidlich, M. (eds) ICPM 2022 Workshops, vol. 468, pp. 18–30. Springer, Cham (2022). https://doi.org/10.1007/978-3-031-27815-0_2
24. Hooshyar, H., Fumagalli, M., Montali, M., Guizzardi, G.: Time and relations into focus: ontological foundations of object-centric event data. In: CAiSE 2026. LNCS, vol. 16559 (2026)
25. Khayatbashi, S., Hartig, O., Jalali, A.: Transforming object-centric event logs to temporal event knowledge graphs. In: Gdowska, K., Gómez-López, M.T., Rehse, JR. (eds) Business Process Management Workshops. BPM 2024. Lecture Notes in Business Information Processing, vol. 534. Springer, Cham (2024). https://doi.org/10.1007/978-3-031-78666-2_23
26. Koren, I., Adams, J.N., Berti, A., van der Aalst, W.M.P.: OCEL 2.0 resources. In: ICPM 2023 Demo Track (2023). www.ocel-standard.org
27. Kretzschmann, D., Berti, A., van der Aalst, W.M.P.: State-aware object-centric process mining: enhancing OCEL 2.0 with explicit state transitions. In: EDOC 2025, vol. 16213, pp. 103–118 (2025)
28. Küsters, A., van der Aalst, W.M.P.: OC-DECLARE: discovering object-centric declarative patterns with synchronization. In: Senderovich, A., Cabanillas, C., Van-

derfeesten, I., A. Reijers, H. (eds) BPM 2025, vol. 16044, pp. 162–179. Springer, Cham (2025). https://doi.org/10.1007/978-3-032-02867-9_11
29. Li, G., de Murillas, E.G.L., de Carvalho, R.M., van der Aalst, W.M.P.: Extracting object-centric event logs to support process mining on databases. In: CAiSE Forum 2018, pp. 182–199 (2018)
30. Liss, L., Adams, J.N., van der Aalst, W.M.P.: Totem: temporal object type model for object-centric process mining. In: BPM 2024 (2024)
31. Martin, J.: Managing the Data Base Environment, Prentice Hall PTR (1983)
32. González López de Murillas, E., Hoogendoorn, G., Reijers, H.A.: Redo log process mining in real life: data challenges & opportunities. In: Teniente, E., Weidlich, M. (eds) Business Process Management Workshops. BPM 2017. Lecture Notes in Business Information Processing, vol. 308, pp. 573–587. Springer, Cham (2017). https://doi.org/10.1007/978-3-319-74030-0_45
33. Seidel, A., Winkler, S., Gianola, A., Montali, M., Weske, M.: To bind or not to bind? Discovering stable relationships in object-centric processes. In: Cham ER 2025, vol. 16189, pp. 223–241. Springer, Cham (2025). https://doi.org/10.1007/978-3-032-08623-5_12
34. Swevels, A., Dijkman, R.M., Fahland, D.: Inferring missing entity identifiers from context using event knowledge graphs. In: Di Francescomarino, C., Burattin, A., Janiesch, C., Sadiq, S. (eds) BPM 2023. LNCS, vol. 14159, Springer, Cham (2023). https://doi.org/10.1007/978-3-031-41620-0_11
35. Swevels, A., Fahland, D., Montali, M.: Implementing object-centric event data models in event knowledge graphs. In: De Smedt, J., Soffer, P. (eds) ICPM 2023 Workshops, pp. 431–443. Springer, Cham (2023). https://doi.org/10.1007/978-3-031-56107-8_33
36. van der Aalst, W.M.P.: Object-centric process mining: Dealing with divergence and convergence in event data. In: Ölveczky, P., Salaün, G. (eds) SEFM 2019, vol. 11724, pp. 3–25. Springer, Cham (2019). https://doi.org/10.1007/978-3-030-30446-1_1
37. van der Aalst, W.M.P., Artale, A., Montali, M., Tritini, S.: Object-centric behavioral constraints: integrating data and declarative process modelling. In: Workshop on Description Logics, vol. 1879. CEUR-WS.org (2017)
38. van der Aalst, W.M.P., Berti, A.: Discovering object-centric petri nets. Fundamenta Informaticae **175**(1-4) (2020)
39. Verbruggen, C., Goossens, A., De Smedt, J., Vanthienen, J., Snoeck, M.: iDOCEM: defining a common terminology for object-centric event logging and data-centric process modelling. Software Syst. Model. **24**(1) (2025)

Footprint-Based Conformance Checking for Object-Centric Process Trees

Jan Niklas van Detten[1,2(✉)], Christian Rennert[1], Aaron Küsters[1], Wil M.P. van der Aalst[1,2], and Sander J. J. Leemans[1]

[1] RWTH University, Aachen, Germany
n.vandetten@bpm.rwth-aachen.de
[2] Celonis, Munich, Germany

Abstract. Object-centric process mining analyzes processes with interacting objects. This analysis is based on object-centric event logs, that can be extracted from enterprise systems. A common strategy is to encode the expected behavior of the process in a model. Subsequently, conformance checking techniques can be applied to identify deviations between that model and the behavior in the most recently extracted log. However, existing procedural object-centric approaches either lack automated methods for the model construction, allow the inclusion of behavioral deadlocks, or suffer from an exponential complexity. Respectively, this induces a dependency on domain knowledge, impedes further analysis, or consumes infeasible computational resources. In this paper, we close this gap by introducing a specialized conformance checking technique for object-centric process trees. These trees already have automated techniques for their construction available and provide guarantees on the absence of deadlocks. Additionally, they possess a recursive structure that allows us to abstract them into a set of relevant patterns in polynomial complexity, creating a behavioral footprint. Similar to constraints in declarative approaches, these footprint patterns can be efficiently validated on object-centric event logs for conformance checking. We evaluate our approach on ten public logs and observe it to provide useful conformance checking results significantly faster than existing approaches applicable to procedual object-centric process trees.

Keywords: process mining · object-centric · conformance checking

1 Introduction

Enterprise information systems produce digital track records of business processes, called event logs. Process mining encompasses techniques to analyze processes based on such logs. Traditionally, event logs referred to individual types of business objects, resulting in their isolated analysis. However, in real-life processes, objects are rarely isolated. Instead, they frequently interact with each other. Object-centric process mining accounts for this reality and provides analysis techniques for processes and logs with interacting business objects.

F. M. Maggi et al. (Eds.): BPM 2026, LNCS 16936, pp. 202–219, 2027.
https://doi.org/10.1007/978-3-032-37777-7_14

A common analysis strategy for such processes includes the initial construction of a model that encodes the behavior of the process [3,7,9,13]. Subsequently, object-centric conformance checking techniques can be applied to identify any potential deviations with regard to the most recently recorded object-centric event log [4,16,17,20]. Such information can then be used to either adjust the underlying business process or the corresponding process model.

In theory, the iterative application of conformance checking techniques in the aforementioned setting should keep the documented process model in synchronization with the underlying real-life process as far as the representational bias of the utilized modeling formalism allows it. However, in practice, this type of analysis is often impeded, as existing object-centric approaches either lack automated methods for the initial model construction [13,19], allow the inclusion of behaviorally problematic constructs without means to detect them [3], or suffer from an exponential complexity of applicable conformance checking techniques [16,20]. Respectively, these shortcomings induce a domain knowledge dependency, impede further algorithmic analysis, or require infeasible resources.

The only runtime efficient alternatives are declarative approaches [1,6,15,21], which model business processes by excluding unintended behaviour with a set of constraints. However, the translation of procedural models with soundness guarantees into such efficiently verifiable constraint sets has not been investigated.

In this paper we close this gap by introducing a specialized conformance checking technique for the modeling formalism of object-centric process trees. These trees can be algorithmically constructed by multiple approaches [7,9] and guarantee the absence of behaviorally problematic constructs such as deadlocks by restricting the representable processes to block-structured ones. At the same time, they possess a recursive structure that allows us to abstract them into a set of conformance-relevant patterns in polynomial complexity, creating a footprint of their behavior for conformance checking. The patterns in this footprint summarize all modeling aspects that can be expressed in object-centric process trees, i.e., control flow relations of individual objects [3,19], explicit restrictions on the multiplicities between objects during interactions [7], and connections between objects that arise due to their specific identities [9,13]. Similar to constraints in declarative approaches, those patterns can be efficiently verified.

We evaluate our approach on a range of publicly available object-centric event logs and observe it to exhibit runtime orders of magnitude lower than existing conformance checking techniques applicable to object-centric process trees [4,16,17]. At the same time, we find the utility of the generated insights to justify the conceptual limitations induced by object-centric process trees.

2 Preliminaries

In this section, we introduce background information. We write $\langle \dots \rangle$, $\{\dots\}$ and $(\dots)$ for sequences, sets, and tuples, we denote the universe of events, activities, objects, types, and timestamps with $\mathbb{U}_{eve}, \mathbb{U}_{act}, \mathbb{U}_{obj}, \mathbb{U}_{typ}, \mathbb{U}_{tim}$ each. The projections $\pi_{tim} : \mathbb{U}_{eve} \mapsto \mathbb{U}_{tim}$, $\pi_{act} : \mathbb{U}_{eve} \mapsto \mathbb{U}_{act}$, and $\pi_{typ} : \mathbb{U}_{obj} \mapsto \mathbb{U}_{typ}$ map each event to its execution time and activity and each object to its type.

2.1 Object-Centric Event Logs

In this paper, we use a simplified definition of object-centric event logs. Various properties, such as relation qualifiers and data attributes are omitted as they cannot be specified in any way by object-centric process trees.

As such, we define object-centric event logs as sets of events, objects, and relations. Events express the execution of an activity, while objects represent typed entities. Relations between them indicate the object's involvement in the event. We write Σ_L and Ω_L for all activities and object types in a given log L. We use the artificial order management log visualized in Fig. 1 as a running example. Customers can place orders for items that are then packed to be picked up by the customer. Items are refunded if they are not picked up in time.

Definition 1 (Object-Centric Event Log). *An object-centric event log* $L = (E, O, E2O)$ *is a tuple with events* $E \subset \mathbb{U}_{eve}$, *objects* $O \subset \mathbb{U}_{obj}$, *and relations* $E2O \subseteq E \times O$. *The universe of object-centric event logs is denoted with* $\mathbb{U}_{log}$.

2.2 Object-Centric Log Patterns

Next, we summarize properties of object-centric logs that can be restricted by object-centric process trees. These properties refer to the control flow, the multiplicities of objects per type, and the identity-based relations between them.

Each object type in an object-centric event log induces an individual control flow. To determine this control flow, the log can be projected onto each object of the corresponding type. For any given object $o \in O$, this *object projection* is defined as the sequence of activity labels $\sigma_o = \langle a_1 \dots a_n \rangle$ of related events, sorted by their timestamp. These traces are subsequently aggregated per object type into the so-called *directly-follows relation* that describes the order in which activities occur. Formally, the relation holds if two activities directly follow each other in any projection. Additionally, activities at the first and last position of an object projection are referred to as start and end activities, respectively.

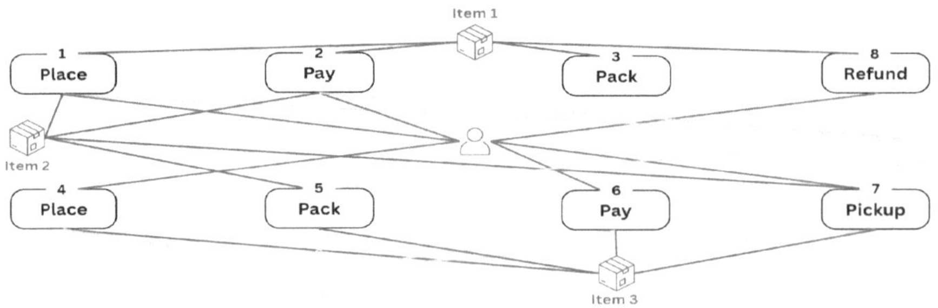

Fig. 1. Object-centric log visualization with customers (blue), orders (yellow), and items (green). Numbers at the event nodes (grey) indicate the order between them. (Color figure online)

Definition 2 (Projected Directly-Follows Relation). *The directly-follows relation* $(a, b) \in df_{ot}(L)$ *for the type* $ot \in \Omega_L$ *and the activities* $a, b \in \Sigma_L$ *holds if* $\exists_{o \in O} : \pi_{typ}(o) = ot \wedge \sigma_o = \langle \dots a, b \dots \rangle$. *An activity* $a \in \Sigma_L$ *is a start activity, denoted by* $a \in start_{ot}(L)$ *if* $\exists_{o \in O} : \pi_{typ}(o) = ot \wedge \sigma_o = \langle a \dots \rangle$ *holds and an end activity, denoted by* $a \in end_{ot}(L)$ *if* $\exists_{o \in O} : \pi_{typ}(o) = ot \wedge \sigma_o = \langle \dots a \rangle$ *holds.*

Figure 2 shows the relation for the example log from Fig. 1. Furthermore, interactions between objects in events can be categorized based on their multiplicity per object type and activity. In particular, we track all deviations from the trasditional 1:1 relation. That is, we check if objects of an object type can participate in none or multiple events with the activity and if an event with the activity can include none or multiple objects of the respective object type.

Definition 3 (Type-Activity Multiplicity). *The type-activity multiplicity in a log* $L = (E, O, E2O)$ *between* $ot \in \Omega_L$ *and* $a \in \Sigma_L$ *is a tuple* (m_{ot}, m_a) *with* $m_{ot}, m_a \subseteq \{0, 1, \mathtt{MANY}\}$. *The values in the multiplicity set* m_{ot} *arederived from the observed number of objects per event. That is, if* 0, 1 *or any* $n > 1$ *can be found in* $\{|\{o \mid (e, o) \in E2O \wedge \pi_{typ}(o) = ot\}| \mid e \in E \wedge \pi_{act}(e) = a\}$, *then* $0, 1, \mathtt{MANY}$ *are included in* m_{ot} *respectively. The set* m_a *is analogously derived from the observations* $\{|\{e \mid (e, o) \in E2O \wedge \pi_{act}(o) = a\}| \mid o \in O \wedge \pi_{typ}(o) = ot\}$.

Note that by using the `MANY` placeholder this multiplicity definition abstracts from the exact numbers in the case of multiple objects per event or multiple events per object, as those are not considered by object-centric process trees.

We consider an activity and a type as *related* if their multiplicity contains any non-zero value. For our example, all multiplicities are shown in Fig. 2.

Lastly, objects might occur in stable identity-based combinations across multiple events [8,9,13]. While various identity relation types have been discussed in literature [8,13], only a limited subset of them are supported in object-centric process trees [9] due to their structural restrictions. All of the relation types in this subset can be broken down into so called implications. Those express that the same identities of objects from one object type imply the same identities of another object type across all events with specified activities [9].

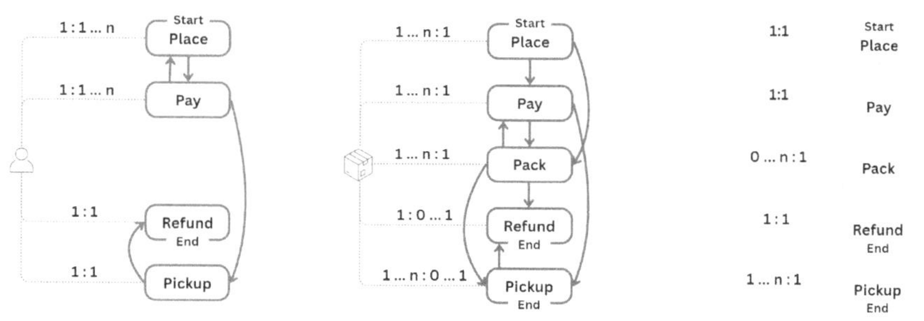

Fig. 2. Directly-follows relation (bold arcs) for customers (blue), items (green), and orders (yellow) from Fig. 1 and multiplicities ($m_{ot} : m_a$ on the dotted lines). (Color figure online)

Definition 4 (Implied Identity Relation). *An object type $ot \in \Omega_L$ implies another type $ot' \in \Omega_L$ across $a, b \in \Sigma_L$ if $\forall e, e' \in E : \pi_{act}(e), \pi_{act}(e') \in \{a, b\} \wedge (\{o \mid (e, o) \in E2O \wedge \pi_{typ}(o) = ot\} = \{o \mid (e', o) \in E2O \wedge \pi_{typ}(o) = ot\}) \implies (\{o \mid (e, o) \in E2O \wedge \pi_{typ}(o) = ot'\} = \{o \mid (e', o) \in E2O \wedge \pi_{typ}(o) = ot'\})$.*

In our example, there is a bidirectional implication between items and orders for placing and paying. That is, the order implies the items and vice versa.

2.3 Object-Centric Process Trees

Object-centric process trees define languages of object-centric event logs based on the control flow of individual object types, their multiplicities during interactions, and their identity relations [7,9]. The inner nodes in the tree structure represent operators that can be used to enforce control flow relations and identity relations in the subtrees below. The leaf nodes contain an activity label and specify the multiplicities between the corresponding activity and each type.

Definition 5 (Object-Centric Process Tree). *An object-centric process tree is recursively defined. A tree leaf node is a tuple $(a, rel_a, div_a, con_a, def_a)$ with an activity $a \in \mathbb{U}_{act}$ or $a = \tau$ to represent unobserved steps, a set of related object types $rel \subset \mathbb{U}_{typ}$ and the three subsets of object types $div_a, con_a, def_a \subseteq rel_a$ to indicate the respectively allowed activity-type multiplicities. An operator node is a tuple $\oplus(S_1 \dots S_n)$ with an operator $\oplus \in \{\rightarrow, \times, ||, \circlearrowleft\}$ for a sequence, choice, concurrency, or loop of the subtrees $S_1 \dots S_n$. An identity node is a tuple (ot, ot', S) denoting the implication from $ot \in \mathbb{U}_{typ}$ to $ot' \in \mathbb{U}_{typ}$ in the subtree.*

The language of an object-centric process tree $\mathcal{L}(S)$ is recursively defined to include logs that adhere to all restrictions of the nodes [7,9]. In case of a leaf node, object types can only interact in events according to their respective configuration. That is, all activity-type multiplicities (m_{ot}, m_a) must be zero $(\{0\}, \{0\})$ for all types that are not related to the leaf node with $ot \notin rel_a$. Related types $ot \in rel_a$ may only deviate from the multiplicity $(\{1\}, \{1\})$ if they are allowed to do so by inclusion in the respective set with $0 \in m_{ot} \implies ot \in con_a$, $0 \in m_{ot} \implies ot \in def_a$ and $0 \in m_a \vee \texttt{MANY} \in m_a \implies ot \in div_a$.

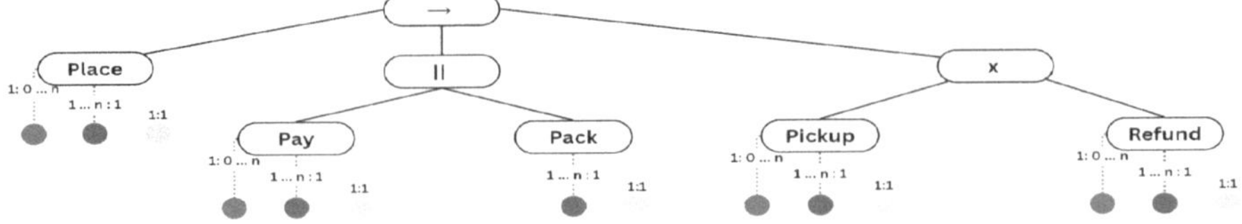

Fig. 3. Object-centric process tree with customers (blue), items (green), and orders (yellow). Object types with non-related multiplicities are omitted from the figure. (Color figure online)

Operator nodes and identity relation nodes recursively combine the logs in the language of their subtrees. That is, below an identity relation node, the specified implication must hold for all events in an accepted log. Additionally, each object must individually follow the control flow between activities related to its type defined by the operator nodes above. Objects of resource-like types, i.e., types that are allowed to have a multiplicity of m_{ot}, m_a with $m_a = \{0, 1, \texttt{MANY}\}$ by inclusion of $ot \in div_a$ in all leaf nodes of a tree, are excluded from the enforcement of a control flow [7]. That is, they can execute the activities in the leaf nodes below arbitrary often, in any arbitrary order. We write $res(S)$ for the set of corresponding object types for any object-centric process tree S.

Figure 3 shows an object-centric process tree for our example process. Note that the example log from Fig. 1 slightly deviates from the allowed behavior.

3 Footprint-Based Conformance Checking

In this section, we introduce footprint-based conformance checking for object-centric process trees and object-centric event logs. For this purpose, we define object-centric footprints as collections of patterns that describe present or absent log properties and then derive conformance values from them. Additionally, we show how all these footprint patterns can be extracted from object-centric process trees and object-centric event logs in polynomial runtime complexity.

3.1 Object-Centric Pattern Footprints

Technically, a footprint pattern is any function that maps a set of object-centric event logs, as defined by object-centric process trees, onto a boolean value of zero or one to indicate the presence or absence of certain behavior. A footprint is subsequently defined as the pattern values for a collection of such patterns.

Definition 6 (Object-Centric Pattern Footprint). *A footprint pattern is any function* $pat : \mathbb{P}(\mathbb{U}_{log}) \mapsto \{0, 1\}$ *that maps sets of logs to binary values. For a set of logs* $\mathbb{L} \subset \mathbb{U}_{log}$ *and patterns* $pat_1 \ldots pat_n$ *the object-centric footprint is defined as the values that result from evaluating all patterns* $pat_1(\mathbb{L}) \ldots pat_n(\mathbb{L})$.

Note that while only single object-centric event logs are typically used as input for conformance checking, object-centric process trees usually define languages that include multiple logs, requiring sets of logs in the footprint definition. Next, we define specific patterns that describe the presence or absence of all log properties that can be expressed by object-centric process trees. That is, we define patterns for control flow relations, multiplicities, and identity relations.

We utilize the directly-follows relation for the control flow patterns, following the line of research from [17]. We define our patterns for activity pairs and object types. The resulting value indicates whether the corresponding directly-follows relation, or start and end activity status, holds for the respective type.

Definition 7 (Control Flow Patterns). *We define patterns across all activity pairs* $(a, b) \in \mathbb{U}_{act}$ *and types* $ot \in \mathbb{U}_{typ}$*. The corresponding patterns* $df_{a,b,ot}(\mathbb{L})$ *evaluate to one if* $\exists L \in \mathbb{L} : (a, b) \in df_{ot}(L)$*. Similarly,* $start_{a,ot}(\mathbb{L})$ *and* $end_{a,ot}(\mathbb{L})$ *evaluate to one if* $\exists L \in \mathbb{L} : a \in start_{ot}(L)$ *and* $\exists L \in \mathbb{L} : a \in end_{ot}(L)$ *respectively.*

Similarly, we formalize patterns for interaction multiplicities. In particular, we introduce patterns that denote if object types are related to activities and whether the multiplicities between them deviate in any way from the 1:1 relation that is typically assumed in the traditional process mining setting. All such patterns evaluate to one if the corresponding deviation is present in a log.

Definition 8 (Interaction Multiplicity Patterns). *We define multiplicity patterns for activities* $a \in \mathbb{U}_{act}$ *and types* $ot \in \mathbb{U}_{typ}$*. Let* $(m_{ot,L}, m_{a,L})$ *be the multiplicity of* ot, a *in any log* $L \in \mathbb{U}_{log}$*. The multiplicity pattern* $rel_{a,ot}(\mathbb{L})$ *evaluates to one if there is a non-zero multiplicity,* $\exists L : \mathbb{L} : m_{ot,L} \neq \{0\} \wedge m_{a,L} \neq \{0\}$*. The remaining patterns* $rel_{a,ot>1}(\mathbb{L})$*,* $rel_{a,ot<1}(\mathbb{L})$*,* $rel_{a>1,ot}(\mathbb{L})$*, and* $rel_{a<1,ot}(\mathbb{L})$ *evaluate to one if there is a non-zero multiplicity that deviates correspondingly.*

Lastly, we introduce patterns for identity relations that specify if an implication between two object types and activities can be violated or not. That is, a pattern for identity relations between an activity pair and an object type pair evaluates to one if the corresponding implication is violated in any input log.

Definition 9 (Identity Relation Patterns). *We define identity relation patterns for all activity and type pairs* $a, b \in \mathbb{U}_{act} \wedge ot, ot' \in \mathbb{U}_{typ}$*.The corresponding pattern* $imply_{a,b,ot,ot'}(\mathbb{L})$ *for a set of logs* $\mathbb{L}$ *evaluates to one if there is a log* $L \in \mathbb{L}$ *for which a corresponding implication from* ot *to* ot' *is violated.*

Note that our specific selection and formalization of footprint patterns has several intentional consequences. First, all patterns are consistently formulated such that a value of one represents a behavioral relaxation, i.e. an increase in the number of logs that have the otherwise equivalent footprint, while zero represents a behavioral restriction, i.e. a reduction in the number of corresponding logs.

Additionally, the footprint properties are exclusively referring to semantic properties, in the sense that two trees with the same language are guaranteed to have the same footprint, even if they differ in syntax. This can be trivially proven for any two object-centric process trees S_1 and S_2, since $\mathcal{L}(S_1) = \mathcal{L}(S_2)$ directly implies $pat_1(\mathcal{L}(S_1)) \dots pat_n(\mathcal{L}(S_1)) = pat_1(\mathcal{L}(S_2)) \dots pat_n(\mathcal{L}(S_2))$ for any set of deterministic patterns $pat_1 \dots pat_n$. More interestingly, any difference in footprints between object-centric process trees, guarantees a difference in their language. This can be easily be proven by the following reasoning. Let us assume two footprints to be given across all previously introduced patterns $pat_1 \dots pat_n$ such that they evaluate differently on two trees S_1 and S_2 with regards to at least one pattern, i.e. $\exists_{1 \leq i \leq n} : pat_i(\mathcal{L}(S_1)) \neq pat_i(\mathcal{L}(S_2))$. As all patterns are binary, this directly implies that pat_i evaluates to one and zero on the language of the two trees respectively. In the following we assume $\exists_{1 \leq i \leq n} : pat_i(\mathcal{L}(S_1)) = 1 \wedge pat_i(\mathcal{L}(S_2)) = 0$ as the opposite case can be proven symmetrically by switching

S_1 and S_2. Due to $pat_i(\mathcal{L}(S_1)) = 1$, we can conclude that there must be at least one specific log in the language of S_1 that causes this pattern value, i.e. $\exists L \in \mathcal{L}(S_1) : pat_i(\{L\}) = 1$. However, this specific log cannot be in the language of S_2, since $pat_i(\mathcal{L}(S_2)) = 0$ implies $\forall L' \in S_2 : pat_i(\{L'\}) = 0$ and hence $\forall L' \in \mathcal{L}(S_2) : L' \neq L$. As such, we can deduct the log to be in the language of exactly one of the two trees, i.e. $L \in \mathcal{L}(S_1) \wedge L \notin \mathcal{L}(S_2)$. This proves $\mathcal{L}(S_1) \neq \mathcal{L}(S_2)$ under footprint differences. These properties ensure that footprints exclusively refer to semantic properties that are insensitive to syntactical differences.

Lastly, it should be noted that the three categories are mostly independent of each other with only very few cross-category pattern values implying each other. That is, it is not possible to deduct the truth values for one entire category of properties from the other ones. In case of identity relations, the fully orthogonal nature towards the control flow properties and interaction multiplicities is formally obvious. Between the control flow properties and the interaction multiplicities, few limited and only one-directional pattern implications exist. For example, a self loop in the directly-follows relation implies that the respective object type can be involved in multiple events with this activity. However, this limited overlap only meaningfully influences the conformance results in case of object-centric process trees with a lot of, otherwise isolated, such activities.

3.2 Footprint Conformance Values

Given two footprints, we derive conformance values based on the behavioral overlap between them. In line with existing literature, we differentiate these conformance values into fitness and precision [4,16,20]. These measures reflect the degree to which observed behavior in the log is captured by the model and the extent to which modeled behavior is actually observed in the log, respectively.

Conceptually, we group patterns into the three categories of control flow relations, interaction multiplicities, and identity relations. For each of these categories, we determine the subset of patterns that evaluate to one in the footprint of the tree, the model, or both. Then, we measure the relative overlap between these subsets to determine fitness and precision for each category. The overall fitness and precision values for the tree and log are then the category averages. Figure 4 illustrates the calculation of fitness and precision values per category.

Definition 10 (Object-Centric Footprint Conformance). *Let an object-centric process tree S be given and let L be an object-centric log. Additionally, let $pat_1 \dots pat_n$ all be patterns in one of the categories of control flow relations, interaction multiplicities, and identity relations. The category fitness and precision is defined as* $\frac{|\{i|pat_i(\{L\})=1\} \cap \{i|pat_i(\mathcal{L}(S))=1\}|}{|\{i|pat_i(\{L\})=1\}|}$ *and* $\frac{|\{i|pat_i(\{L\})\} \cap \{i|pat_i(\mathcal{L}(S))=1\}|}{|\{i|pat_i(\mathcal{L}(S))=1\}|}$. *The overall fitness and precision for S and L is the category average.*

We consider the initial aggregation per category necessary before the aggregation across categories, as the number of patterns per category differs from

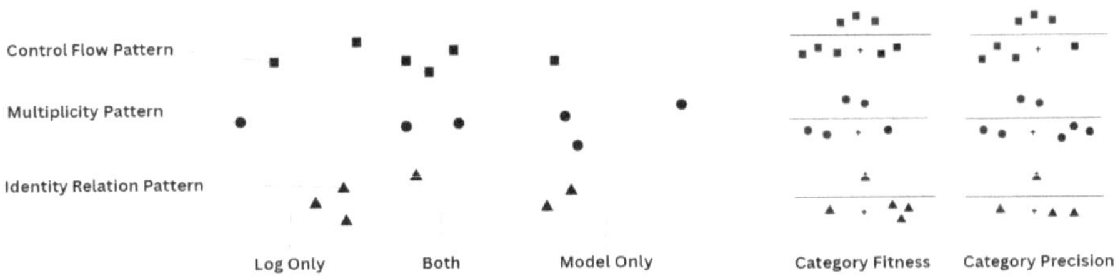

Fig. 4. Fitness and precision calculation across patterns per category that hold on log only, model only, or both. Overall fitness and precision values are the category averages.

$|\Sigma|^2 \cdot |\Omega| + 2 \cdot |\Sigma| \cdot |\Omega|$ for control flow relations to $5 \cdot |\Sigma| \cdot |\Omega|$ for interaction multiplicities and $|\Sigma|^2 \cdot |\Omega|^2$ for identity relations. Without the initial aggregation per category, identity relations would dominate the resulting values.

While customized weights between the categories could in principle make sense for the subsequent cross-category aggregation in certain settings, we generally assume them to be of equal importance. By doing so we follow recent object-centric literature, in which multiplicities and identity relations are considered an integral part of object-centric processes [5], next to the combination of object type control flows that was typically focused in earlier work [3].

However, we recommend to use the overall conformance average only when necessary, such as during optimizations or benchmarks with limited number of dimensions. Within the scope of the initially motivated analytical setting to verify and improve trees, we strongly suggest to check the standard deviation between categories and investigate each of them in depth should it be high.

3.3 Recursive Footprint Extraction

Next, we specify how to extract footprint patterns from object-centric process trees. First, we provide a recursive extraction strategy for the directly-follows relations from an object-centric process tree S per object type, denoted by the set of activity pairs $df_{ot}(\ldots)$ and the corresponding sets of start activities ($start_{ot}(\ldots)$) and end activities ($end_{ot}(\ldots)$). Besides these properties, we also determine several auxiliary constructs, such as the set of related activities ($\Sigma_{ot}(S)$) in a tree and a boolean indicator in case the tree can be skipped ($skip_{ot}(S)$).

For a single leaf node and every related object type, there is only a single activity that represents a local start and end. The corresponding directly-follows relation contains a self-loop if the type is resource-like as defined in Sect. 2.3. The leaf node can be skipped by types that can participate in zero events.

Formally, for an object-centric leaf node (a, rel, div, con, def) and object type $ot \in \mathbb{U}_{typ}$, we have $start_{ot}(\mathcal{L}(a, rel_a, div_a, con_a, def_a)) = end_{ot}(\mathcal{L}(a, rel_a, div_a, con_a, def_a)) = \Sigma_{ot}(\mathcal{L}(a, rel_a, div_a, con_a, def_a)) = \{a\}$ if $a \neq \tau \wedge ot \in rel_a$, and $df_{ot}(\mathcal{L}(a, m)) = \{(a, a)\}$ if $ot \in div_a$. Otherwise, all these sets are empty. Note that this self-loop on a is only induced by the resource-like types that can have arbitrarily many and arbitrarily ordered events with that activity. Hence, in the language of the tree with such a leaf, a corresponding self loop can always be observed in at least one accepted log. Similarly, we define the auxiliary function to indicate a potential skip with $skip_{ot}(a, rel_a, div_a, con_a, def_a) = ot \in div_a$.

Next, we show how each of the operators aggregates the directly-follows relation from its subtrees for any given type. These aggregations strictly follow the language definition of object-centric process trees from [7]. For the concurrent operator, the start (1) and end (2) activities are the start and end activities from all subtrees. The directly-follows relation combines all relations from the subtrees (3). Additionally, all cross-subtree activity pairs are added (4). A concurrent operator node can only be skipped if all subtrees can be skipped as well (5). The included activities of that operator node are those from all subtrees (6).

$$start_{ot}(\mathcal{L}(||, S_1 \ldots S_n)) = \cup_{1 \leq i \leq n}\, start_{ot}(\mathcal{L}(S_i)) \tag{1}$$
$$end_{ot}(\mathcal{L}(||, S_1 \ldots S_n)) = \cup_{1 \leq i \leq n}\, end_{ot}(\mathcal{L}(S_i)) \tag{2}$$
$$df_{ot}(\mathcal{L}(||, S_1 \ldots S_n)) = \cup_{1 \leq i \leq n}\, df_{ot}(\mathcal{L}(S_i)) \cup \tag{3}$$
$$\cup_{i \neq j \in \{1 \ldots n\}}\, \Sigma_{ot}(S_i) \times \Sigma_{ot}(S_j) \tag{4}$$
$$skip_{ot}(||, S_1 \ldots S_n) = \forall_{1 \leq i \leq n} : skip_{ot}(S_i) \tag{5}$$
$$\Sigma_{ot}(||, S_1 \ldots S_n) = \cup_{1 \leq i \leq n} : \Sigma_{ot}(S_i) \tag{6}$$

For the exclusive choice operator, activities, starts, and ends are aggregated as for the concurrent operator. The directly follows relation for the exclusive choice operator again combines all relations from each subtree (7). Additionally, for the subtrees for which the object type is acting like a resource, all cross-subtree activity pairs are included in the directly-follows relation (8). This is again caused by these types being allowed to execute arbitrarily many events in any arbitrary order, leading to all these directly-follows relations being present in at least one log in the language of the tree. For non-resource types, the exclusive choice is enforced between the subtrees, i.e., no new directly-follow relation is added between them. A choice can be skipped if one of its subtrees can be (9).

$$df_{ot}(\mathcal{L}(\times, S_1 \ldots S_n)) = \cup_{1 \leq i \leq n}\, df_{ot}(\mathcal{L}(S_i)) \tag{7}$$
$$\cup_{1 \leq i \neq j \leq n:} \{(a, b) \in \Sigma_{ot}(S_i) \times \Sigma_{ot}(S_j) \mid ot \in res(S_i) \cap res(S_j)\} \tag{8}$$
$$skip_{ot}(\times, S_1 \ldots S_n) = \exists_{1 \leq i \leq n} : skip_{ot}(S_i) \tag{9}$$

For the sequence operator, the start (10) and end (11) activities are those from all subtrees that are only preceded or followed by skippable subtrees. The auxiliary functions are handled as before for the concurrent operator.

$$start_{ot}(\rightarrow, S_1 \ldots S_n) = \cup_{1 \leq i \leq n: \forall 1 \leq j < i: skip_{ot}(S_j)}\, start_{ot}(S_i) \tag{10}$$
$$end_{ot}(\rightarrow, S_1 \ldots S_n) = \cup_{1 \leq i \leq n: \forall i < j \leq n: skip_{ot}(S_j)}\, end_{ot}(S_i) \tag{11}$$

The directly-follows relations of all subtrees are again merged (12). Additionally, ends and starts of following subtrees, and those only separated by optional subtrees, can follow each other in order (13). Lastly, resource-like types, as defined in Sect. 2.3, are handled by including cross-subtree activity pairs (14).

$$df_{ot}(\mathcal{L}(\rightarrow, S_1 \dots S_n)) = \cup_{1 \leq i \leq n} df_{ot}(\mathcal{L}(S_i)) \cup \tag{12}$$
$$\cup_{1 \leq i < j \leq n} \{(a, b) \in end_{ot}(S_i) \times start_{ot}(S_j) \mid \forall_{i<l<j} : skip_{ot}(S_l)\} \cup \tag{13}$$
$$\cup_{1 \leq i < j \leq n:} \{(a, b) \in \Sigma_{ot}(S_i) \times \Sigma_{ot}(S_j) \mid ot \in \cap_{i \leq l \leq j} res(S_l)\} \tag{14}$$

According to [7] the language of loop operator nodes $\circlearrowleft (S_1 \dots S_n)$ across the loop body S_1 and the redo part $S_2 \dots S_n$ expresses that sequentially, after each execution of the loop body, a choice is made between another iteration through one of the redo parts and the end of the loop. Conceptually, this results in the following aggregation of the directly-follows relation from the subtrees: The starts and ends always include those of the loop body. If the loop body can be skipped, the starts and ends of all other subtrees are added. For the directly-follows relations, we combine the relations in the subtrees and connect the end activities of the loop body with the start activities of all redo parts and vice versa. If any redo part can be skipped, the end activities of S_1 can lead to the start activities of S_1. Similarly, if S_1 is optional, end activities of $S_2 \dots S_n$ lead to start activities of $S_2 \dots S_n$. Lastly, we account for resource-like types as before.Figure 5 shows the overall result for the tree in Fig. 3.

Next, we specify the extraction of the multiplicities from object-centric process trees. The pattern for related object types and activities, i.e., $rel_{a,ot}(\mathcal{L}(S))$, can be directly evaluated on the leaf nodes of a given tree S, by checking if the object type is related to any leaf node with that label. The pattern for deviations from the 1:1 multiplicity requires some further processing. That is, the patterns $rel_{a,ot>1}(\mathbb{L})$, $rel_{a,ot<1}(\mathbb{L})$, $rel_{a>1,ot}(\mathbb{L})$, and $rel_{a<1,ot}(\mathbb{L})$ evaluate to true if any corresponding leaf allows it, as specified in Sect. 2.3. Additionally, $rel_{a<1,ot}(\mathbb{L})$ evaluates positively if there is a choice on the path from the root node to any leaf node with that activity. Similarly, $rel_{a>1,ot}(\mathbb{L})$ evaluates to true if there is an operator node with a loop on the path to any respective leaf node. The extracted multiplicity patterns for our example tree are shown in Fig. 5.

Lastly, the patterns for identity relations can be extracted from an object-centric process tree by iterating through all inner nodes and collecting all explicitly included identity relations. Then, for each activity pair, the transitive closure of implications between object types is used to determine if a corresponding pattern should evaluate positively during the conformance check.

Note that all extraction strategies can also be applied in case of duplicated activity labels. They will be treated as a single activity in the footprint that is allowed to perform any behaviour that one of the leaves with that label induces. Finally, we discuss the polynomial runtime complexity of our approach. The derivation of the footprints from object-centric event logs is trivially in polynomial complexity. For the object-centric trees, we sketch the runtime complexity for any given number of tree nodes N, associated to $|\Sigma|$ activities and $|\Omega|$ types. The complexity of the extraction of the directly-follows relation is limited by the number of object types. For each object type, it requires an iteration across all nodes in a tree ($|\Omega| \cdot N$). The complexity of the pattern extraction for interac-

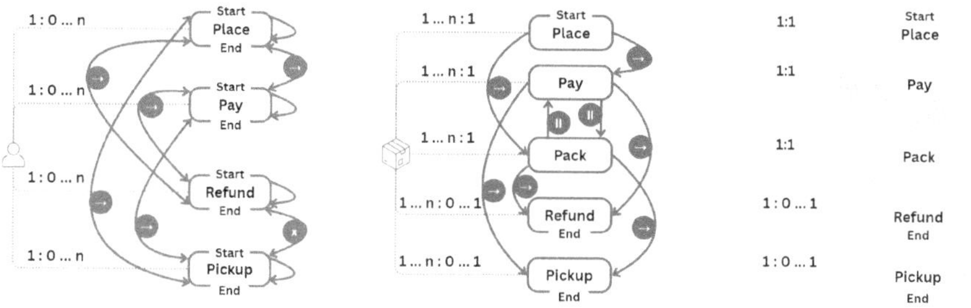

Fig. 5. Directly-follows relation and multiplicities, extracted from the tree in Fig. 3. The operator icons indicate which non-leaf node in the tree caused the relation.

tion multiplicities is limited by the number of activity-type combinations and the maximum number of nodes between each leaf and the root node ($N \cdot |\Sigma| \cdot |\Omega|$) . Lastly, the complexity of the identity pattern extraction is limited by the number of nodes in the tree and the squared number of object types and activities ($N \cdot |\Sigma|^2 \cdot |\Omega|^2$). The comparison between all footprint patterns is limited, as discussed in Sect. 3.2, by $|\Sigma|^2 \cdot |\Omega| + 2 \cdot |\Sigma| \cdot |\Omega|$ for control flow relations, $5 \cdot |\Sigma| \cdot |\Omega|$ for interaction multiplicities, and $|\Sigma|^2 \cdot |\Omega|^2$ for identity relations.

3.4 Conformance Analysis

We propose to use our approach to determine the overall conformance and then drill down into deviations when the standard deviation between averages of different groups is significant. We suggest investigating individual groups of patterns based on 1) the categories of control flow, multiplicitics, or identity relations, 2) the tree node that causes a negative evaluation or 3) the activity or type a pattern refers to. Note that all patterns included in our footprint provide a notion of locality, in the sense that one can easily trace back which category, node, event, type, or activity caused a decrease in conformance values to occur. Figure 5 indicates the tree nodes that introduce each of the relations.

For our example log and tree, the overall fitness and precision amount to 0.80 and 0.72. Splitting patterns up based on the pattern categories reveals that the control flow relations are reflected correctly (category fitness of 0.96) but imprecisely (category precision of 0.54). This imprecision is primarily caused by the incomplete observation of the resource-like behavior of the customer across activities in the exclusive-choice subtree. In practice, we would assume this to be intended behavior, which is simply not observed due to the small input log size. Furthermore, we find the identity relations to cause the most significant drop in fitness (category fitness of 0.54) while retaining high precision (category precision of 0.90). When drilling down on this issue, we find this to be primarily caused by the events that refer to the packing activity. We can observe both orders in our log to be packed jointly or partially. This deviation is also reflected in the corresponding multiplicity patterns, which do not show any other noticeable deviations. This might be intended behavior that is not reflected in the model. In practice, the model should be adjusted after a confirmation by stakeholders.

4 Evaluation

We evaluate the feasibility and expressiveness of our approach by checking the runtime and category standard deviation on public logs. Additionally, we compare conformance values against existing techniques using our implementation available at https://github.com/Nik314/OCPT-Conformance-Checking.

4.1 Runtime Demands and Category Standard Deviation

We compare the runtime demands of our approach against perspective-based [17], alignment-based [16] and context-based techniques [4]. Additionally, we measure the cross-category standard deviation of our approach as a proxy value for the expressiveness of the drill-down. All logs can be found in the log library of the OCEL format. First, we use the discovery techniques from [7,9] to mine object-centric process trees with identity relations. For the alignment-based and context-based techniques, we translate each tree to an object-centric Petri net. For the perspective aggregation, we project the tree on workflow nets for alignments on each type, as suggested in [17]. Then, we measure the runtime on an Intel I9-13980HX processor with 64GB of RAM and a timeout of one hour per log. For our approach we additionally report the category standard deviation.

Table 1 shows the results. The context-based and alignment-based techniques timed out on all logs with little progress ($\leq 1\%$). A follow-up in which we increased the timeout to a day on a server with 500 GB of RAM lead to memory errors. This is expected, as they both lie within exponential runtime complexity, as already discussed in [16]. Preliminary evaluations performed in [4,16] suggest that these and similar techniques [13,20] can only be used with heavy preprocessing to reduce the log connectivity. The perspective aggregation timed out on two logs and required 41 s on average on the remaining ones. Our approach did not time out and required 4.14 s on average per input log. The cross-category standard deviation on average was 0.05 and 0.17 for fitness and precision, suggesting that the drill-down analysis suggested in the previous section should provide meaningful insights upon the investigation per category.

4.2 Absolute Conformance Comparison

To confirm that our method remains in line with existing object-centric conformance checking approaches that address similar modeling aspects, we compare the resulting values. Given the previous experiment, we focus on the perspective aggregation, as the other methods are infeasible, and compare the resulting conformance values to our approach. For fitness, the difference between our approach and the perspective-based measure was 0.04 on average. For the precision

Table 1. Runtime in seconds for footprint-based and perspective-based [17] conformance checking. Additionally, category standard deviation, fitness and precision differences, and agreement percentages for the relative comparison are shown.

Log	Footprint (Runtime)	Perspective (Runtime)	Absolute Difference (Fit / Prec)	Relative Difference (Fit / Prec)	Category Deviation (Fit / Prec)
Container	4.77 s	30.77 s	(0.02 / 0.08)	(42 % / 76%)	(0.11 / 0.16)
Orders	6.13 s	206.86 s	(0.00 / 0.21)	(100% / 71%)	(0.00 / 0.13)
P2P	4.16 s	27.47 s	(0.17 / 0.12)	(100% / 88%)	(0.06 / 0.08)
LRMS O2C	5.78 s	5.57 s	(0.00 / 0.08)	(100% / 84%)	(0.00 / 0.20))
LRMS P2P	1.18 s	7.73 s	(0.04 / 0.02)	(88% / 78%)	(0.04 / 0.21))
LRMS HR	3.91 s	0.77 s	(0.05 / 0.12)	(93% / 76%)	(0.07 / 0.18)
Hospital	1.91 s	38.70 s	(0.01 / 0.05)	(88% / 82%)	(0.03 / 0.09)
Production	11.01 s	t/o	t/o	t/o	(0.04 / 0.21)
GitHub	0.76 s	t/o	t/o	t/o	(0.08 / 0.31)
Recruiting	1.78 s	9.44 s	(0.03 / 0.35)	(68% / 72%)	(0.05 / 0.14)

measure, the observed differences were larger, with an average of 0.16 per log. The values for each of the investigated input logs can be found in Table 1.

We further investigated the differences in conformance values and found those to be caused by conceptual differences in terms of what the two approaches are sensitive to. Most importantly, our approach considers object multiplicities and identity relations, which are ignored by the perspective aggregation. That is, each object trace is individually aligned by the perspective aggregation without checking if this causes any problem with regards to the allowed multiplicities during the necessary interaction with other objects in shared events or potential identity relations between them. This leads to an overestimation of fitness in comparison o our approach. Reversely it is also not checked if the tree allows additional multiplicities or identity-based interaction than those present in the log. Hence, a tree that precisely restricts these aspects is not rewarded in comparison to a tree that leaves them unspecified. For the logs investigated here, this caused some differences in the resulting precision values, since the algorithms from [7,9] are precise with regards to the allowed multiplicities and identity relations in the discovered trees. Our approaches rewarded that precision, while the perspective aggregation did not. Further, more minor, differences between the resulting values of the two approaches can be traced back to the differences in our control flow patterns and the alignments used per object type. In that regard, the perspective aggregation is more sensitive as the alignments per object capture additional aspects. Our approach does, for example, not account for multiple occurrence of the same pattern violation due to their binary values.

Examples for the above mentioned types of differences between the perspective aggregation and our approach could be consistently observed across all input logs. For the order management log, for example, the discovered trees precisely reflect that the placement of an order can contain multiple items, but only refer to a single customer. It also precisely specifies, that across confirming and paying, the same set of items implies the same order and customer. These aspects are rewarded by our approach, but not rewarded by the perspective aggregation. In our opinion, those differences correctly reflect intentional design choices.

4.3 Relative Conformance Comparison

Lastly, we investigate if our approach and the perspective aggregation agree on the relative effect of known changes to the input log. Specifically, we check if precision and fitness improve or worsen, when each object type and each activity is individually removed. In each case, we check if the two approaches agree on the direction of the conformance value change in comparison to the results on the full log. The resulting percentages of object types and activities per log can be found in Table 1. On average, we observed an agreement of 85% for fitness and 78% for precision. This confirms that while both approach result in different absolute values for the reasons discussed in the previous experiment, they mostly agree on the relative effect of known changes. More importantly, it verifies the usage of object types and activities with regards to the locality of deviations.

4.4 Discussion

Our evaluation showed that our approach can determine conformance results orders of magnitude faster than alignment-based and context-based techniques. In comparison to perspective-aggregating approaches, it still shows considerable runtime. While the absolute values between the approaches differ reasonably, especially with regards to precision, we found those differences to be mainly caused by intentional design choices. Specifically, the sensitivity of our approach to interaction multiplicities and identity relations. Further differences could be explained by the limited expressivity of control flow patterns in comparison to object trace alignments. Finally, we confirmed the reliability of our approach for relative deviations under known changes with regards to types and activities.

The obvious key limitation of our approach lies within the attachment to object-centric process trees and the abstraction into a limited set of properties. In principle, our footprints could also be extracted from any other object-centric modeling formalism, albeit without the guarantee for polynomial runtime complexity. However, the conformance results would cater heavily towards the representational bias of object-centric process trees, as included patterns were selected to exactly capture their semantic properties. Advantages of other formalisms that exceed the expressiveness of object-centric process trees will not be evaluated correctly with our approach. As such, it should be considered an in-formalism technique. Additionally, due to the abstraction into a limited set of properties there can be equivalence classes of object-centric process trees that map onto the same footprint despite having different languages. Within each of these classes, our conformance checking is insensitive to language differences.

5 Related Work

In this section we present related approaches on object-centric conformance checking. Their application typically relies on translating the trees to language-equivalent object-centric Petri nets, as outlined in [7]. Alternatively, pattern extraction strategies can be used to apply declarative approaches.

Token-based [14] conformance checking use the replay mechanism of object-centric Petri net formalisms, such as [3], to determine the fitness of an object-centric event log [14]. The measures result in non-deterministic values for object-centric process trees, depending on the object-centric Petri net translation. Alignment-based [10,13,18] approaches determine the closest entry in the language of the model in comparison to the log. While this provides theoretically optimal diagnostics, these methods have exponential complexity [10,13,18]. Context-based techniques are based on grouping events based on the previous behavior of all involved objects [4]. The approach relies on alignments in a pre-processing step for non-perfect fitness, inheriting the exponential complexity.

Besides the above mentioned approaches based on Petri nets, our approach also relates heavily to constraint-based and declarative approaches, which model processes by specifying explicitly disallowed behaviour [1,6,15,21]. In relation to these approaches, the core contribution of this paper is the efficient extraction strategy of the footprint patterns from object-centric process trees. With slight approach specific adjustments of this strategy, one could translate object-centric process trees into a set of constraints in each of the respective formalisms. The specific patterns proposed in this paper could, for example, also be expressed in OC-Declare [15] which supports control flows, multiplicities and implication-based identity relations. In this case, the expected runtime complexity of these approaches should be identical to the approach presented in this paper.

Conformance techniques for other formalisms include artifact-centric techniques [11], proclets [2,12] and other Petri net variants. Our footprint-based approach could in principle be applied to all these formalisms, although the footprint pattern extraction might require checking soundness properties. As such, we cannot guarantee the polynomial complexity for formalisms that do not provide the soundness guarantees of object-centric process trees.

Lastly, there are perspective-aggregating techniques, such as [17]. Those approaches are as runtime-efficient as the techniques per perspective. Most techniques discussed above are designed for model-to-log checking, while our technique could also be used for model-to-model and log-to-log comparisons. Footprints have been used similarly in traditional process mining. To the best of our knowledge, our approach is the first to extend them to the object-centric setting.

6 Conclusion

In this paper, we introduced a novel conformance checking approach for object-centric process trees based on pattern footprints. We specified patterns for all covered modeling aspects of control flow relations, interaction multiplicities, and identity relations and defined their extraction in polynomial complexity. As a result, the application of our approach to guide practitioners in identifying deviations is feasible even for large input logs. We evaluated our approach on a range of public logs and found significant runtime improvements in comparison to perspective-based, alignment-based, and context-based approaches.

Our approach remains limited by the reliance on object-centric process trees, inducing a representational bias. As such, our approach should not be applied in a cross-formalism setting without readjusting the selected patterns to account for additional expressiveness. In future work, we plan to pinpoint these differences precisely and offer extended footprints with corresponding extraction strategies.

Acknowledgment. The project on which this work is based upon was funded by the German Federal Ministry of Research, Technology and Astronautics (grant 01IS25010). The responsibility for the content lies with the authors.

References

1. van der Aalst, W.M.P., Artale, A., Montali, M., Tritini, S.: Object-centric behavioral constraints: integrating data and declarative process modelling. In: Workshop on Description Logics, 2017. CEUR Workshop, vol. 1879. CEUR-WS.org (2017)
2. van der Aalst, W.M.P., Barthelmess, P., Ellis, C.A., Wainer, J.: Proclets: a framework for lightweight interacting workflow processes. Int. J. Cooperative Inf. Syst. (2001)
3. van der Aalst, W.M.P., Berti, A.: Discovering object-centric petri nets. Fundam. Informaticae **175**(1–4), 1–40 (2020)
4. Adams, J.N., van der Aalst, W.M.P.: Precision and fitness in object-centric process mining. In: ICPM, pp. 128–135. IEEE (2021)
5. Anjo Seidel, E.A.: Object-centric process management: a research manifesto. Inf. Syst. **141**, 102728 (2026)
6. Artale, A., Kovtunova, A., Montali, M., van der Aalst, W.M.P.: Modeling and reasoning over declarative data-aware processes with object-centric behavioral constraints. In: Hildebrandt, T., van Dongen, B., Röglinger, M., Mendling, J. (eds) Business Process Management BPM, pp. 139–156. Springer, Cham (2019). https://doi.org/10.1007/978-3-030-26619-6_11
7. van Detten, J.N., Schumacher, P., Leemans, S.J.J.: Discovering compact, live and identifier-sound object-centric process models. In: ICPM, pp. 113–120. IEEE (2024)
8. van Detten, J.N., Schumacher, P., Leemans, S.J.J.: Object synchronizations and specializations with silent objects in object-centric petri nets. In: BPM Proceedings. LNCS, vol. 14940, pp. 57–74. Springer, Cham (2024). https://doi.org/10.1007/978-3-031-70396-6_4
9. van Detten, J.N., Schumacher, P., Leemans, S.J.J.: Modeling and discovering dynamic identity relations in object-centric process mining. In: ICPM. IEEE (2025)
10. van Dongen, B.F., Carmona, J., Chatain, T.: Alignment-based quality metrics in conformance checking. EMISA Forum **36**(2), 77–80 (2016)
11. Estañol, M., Munoz-Gama, J., Carmona, J., Teniente, E.: Conformance checking in UML artifact-centric business process models. Softw. Syst., Model (2019)

12. Fahland, D.: Describing behavior of processes with many-to-many interactions. In: Donatelli, S., Haar, S. (eds) Petri Nets. LNCS, vol. 11522, pp. 3–24. Springer, Cham (2019). https://doi.org/10.1007/978-3-030-21571-2_1
13. Gianola, A., Montali, M., Winkler, S.: Object-centric conformance alignments with synchronization. In: Guizzardi, G., Santoro, F., Mouratidis, H., Soffer, P. (eds) CAiSE. LNCS, vol. 14663, pp. 3–19. Springer, Cham (2024). https://doi.org/10.1007/978-3-031-61057-8_1
14. Carrasquel, J.C., K.M.: Object-centric replay-based conformance checking: unveiling desire lines and local deviations. J. Model. Anal. Inform. Sist., (2021)
15. Küsters, A., van der Aalst, W.M.P.: OC-DECLARE: discovering object-centric declarative patterns with synchronization. In: Senderovich, A., Cabanillas, C., Vanderfeesten, I., A. Reijers, H. (eds) BPM, pp. 162–179. Springer, Cham (2026). https://doi.org/10.1007/978-3-032-02867-9_11
16. Liss, L., Adams, J.N., van der Aalst, W.M.P.: Object-centric alignments. In: Conceptual Modeling. LNCS, vol. 14320 (2023)
17. Park, G., Adams, J.N., van der Aalst, W.M.P.: Conformance checking and performance analysis using object-centric directly-follows graphs. In: Marrella, A., Resinas, M., Jans, M., Rosemann, M. (eds) BPM (Forum). LNBIP, vol. 526, pp. 179–196. Springer, Cham (2024). https://doi.org/10.1007/978-3-031-70418-5_11
18. Sommers, D., Sidorova, N., van Dongen, B.F.: Conformance checking with model projections - rethinking log-model alignments for processes with interacting objects. In: Petri Nets. LNCS, vol. 14628, pp. 61–82. Springer, Cham (2024). https://doi.org/10.1007/978-3-031-70418-5_11
19. van der Werf, J.M.E.M., Rivkin, A., Polyvyanyy, A., Montali, M.: Data and process resonance - identifier soundness for models of information systems. In: Bernardinello, L., Petrucci, L. (eds) Proceedings, vol. 13288, pp. 369–392. Springer, Cham (2022). https://doi.org/10.1007/978-3-031-06653-5_19
20. Wrede, E., van Detten, J.N., Liss, L., Leemans, S.J.J.: Out of Babylon: object-centric conformance checking on graph-based abstractions. In: Cappiello, C., Hartig, O., Sellami, M., Ouni, A. (eds) CoopIS. LNCS, vol. 15535, pp. 183–201. Springer, Cham (2025). https://doi.org/10.1007/978-3-032-15538-2_11
21. Xiu, B., Li, G.: Diagnosing conformance between object-centric event logs and models. IEEE Access **11**, 110837–110849 (2023)

Modeling Analytic Provenance in Process Mining

Francesca Zerbato([✉]) and Dirk Fahland

Eindhoven University of Technology, Eindhoven, The Netherlands
{f.zerbato,d.fahland}@tue.nl

Abstract. Process mining analysis is a complex, human-in-the-loop activity where analysts iteratively choose what steps to perform based on the results. Due to its emergent and non-linear nature, the analysis is often difficult to replicate or revisit. As a result, analysts may lack transparency into their own workflows, while stakeholders may distrust the findings. We argue that these challenges stem from a lack of *analytic provenance*, i.e., the record of the steps, intermediate results, and reasoning that lead to findings. However, existing provenance models fail to capture both the analysis results and the underlying reasoning, which are essential aspects of process mining analysis. To address this gap, we propose a model of analytic provenance that integrates analysis steps, data artifacts, and human reasoning into a multiperspective provenance graph. We then frame the realization of provenance-aware support for meta-analysis as a new research problem in BPM. Using our model as a foundation, we identify challenges that must be overcome to develop systems for supporting and studying meta-analysis in process mining.

Keywords: Process Mining · Analytic Provenance · Provenance Graphs · Meta-analysis · Provenance Models

1 Introduction

Process mining (PM) provides analysts with tools and techniques to analyze process data [23]. However, in practice, PM analysis is a complex human-in-the-loop activity in which analysts iteratively perform steps, generate artifacts, and, based on the interpretation of these artifacts, choose what to do next until they reach a final insight [26]. This emergent and non-linear nature makes PM analysis difficult to replicate or revisit. As a consequence, analysts struggle to maintain an overview of their workflows and lack transparency into their own processes [27], while stakeholders may lack trust in the findings, particularly when the steps that lead to them are not reported [22].

In other domains, such as scientific workflows and visual analytics, similar challenges have been addressed through provenance, i.e., information that describes the process used to generate an end product [11]. In particular, *analytic provenance* studies the reasoning of analysts through their interactions with

F. M. Maggi et al. (Eds.): BPM 2026, LNCS 16936, pp. 220–237, 2027.
https://doi.org/10.1007/978-3-032-37777-7_15

a system [17], under the assumption that the analysis is a knowledge discovery process that is "as important as knowledge itself" [9]. We argue that a comprehensive record of analytic provenance would provide PM analysts with an overview of their steps, the artifacts they generated, and their analytic reasoning. Beyond transparency, such a record would support the replication of results, enable recall of analysis steps and their underlying rationale, and facilitate *meta-analysis*, i.e., the study of the analysis process itself to improve future practice [18].

However, capturing provenance in PM is a unique challenge, as it requires recording the operations performed in PM tools, the generated data artifacts, and the reasoning of the analyst, e.g., what hypotheses are made and why, and how insights are derived from testing them [20,26]. Existing provenance models fail to capture this combination, as they emphasize artifact reproducibility [15] or user interaction tracking [1,6], often disregarding human reasoning.

To address this gap, we first review the provenance literature in Sect. 2 and identify in Sect. 3 the conceptual requirements for modeling analytic provenance in PM, alongside a motivating example. Building on these requirements, we then make a twofold contribution. First, in Sect. 4, we propose a model that integrates three perspectives into a *provenance graph*:

(i) the analysis steps, derived from a history of operations performed in a PM tool;
(ii) the data artifacts generated from and used by these operations;
(iii) the analyst's reasoning, in the form of observations, hypotheses, and insights.

While steps and artifacts are automatically discovered from a history of operations, we introduce *parameters* that allow the analyst to enrich the graph with artifact features and reasoning. We show that graph discovery is monotonous and enrichment is commutative, together enabling iterative graph construction alongside an ongoing analysis.

Second, in Sect. 5, we frame the realization of provenance-aware support for meta-analysis in process mining as a new research problem in BPM that entails a circular dependency: studying analyst behavior requires systems that can represent and reason over provenance, yet designing such systems requires knowledge of how analysts would use them. By representing PM analysis as a multi-perspective process that integrates operations, artifacts, and reasoning, our model provides the core concepts (i) to discuss how prerequisites for meta-analysis, such as replication and recall, can be supported, and (ii) to articulate research challenges to overcome this circularity.

2 Background and Related Work

Provenance has been defined in various ways in different fields, including database management systems [11], scientific workflows [5], and visual analytics [18], each describing different types of provenance information and purposes.

In this work, we focus on research that conceptualizes the provenance of analytic processes to support analyst reasoning. We first review provenance categorizations in Sect. 2.1, then discuss existing provenance models to identify the concepts needed to represent provenance in PM analysis in Sect. 2.2.

2.1 Provenance: Categorizations and Purposes

The concept of provenance has its roots in the database management field, where *data provenance* (also called "lineage") was introduced to describe the "origin of a piece of data together with an explanation of how and why it got to the present place" [10]. In this setting, provenance records how individual data items, such as relational tuples, are derived through formally defined query operators. Because structured data models and declarative query languages have well-defined semantics, data provenance is typically fine-grained [11]. Furthermore, it is usually retrospective as it explains how an existing result was obtained. Beyond explaining results, its main purposes include tracking dependencies and maintaining materialized views in response to changes in database tables [4].

This perspective was later extended in the context of scientific workflows, where *workflow provenance* (viz. "process provenance") was defined. Workflow provenance focuses on the execution of computational tasks and captures information about the steps taken to derive a data product from a given input, the intermediate results, and the environment in which the workflow is executed [5]. Since users can specify computational tasks at different levels of abstraction, the granularity of workflow provenance ranges from coarse-grained, where tasks are treated as "black boxes", to highly detailed [11]. This depends both on the user specification and on the level of instrumentation of the workflow. Workflow provenance can be prospective, retrospective, or capture workflow evolution [7]. Specifically, prospective provenance captures the specification of a given workflow as a recipe; retrospective provenance captures a detailed log of the workflow's execution; and evolution provenance captures changes between different versions of the workflow. Workflow provenance is mainly used to support the reproducibility, collaboration, and reuse of computational experiments.

In contrast to these data- and workflow-centric views, our work focuses on a third notion introduced in the field of visual analytics, namely *analytic provenance*. Analytic provenance captures the interactive data exploration and reasoning processes of a human analyst during sensemaking tasks [17,24]. Analytic provenance encompasses both the history of user activities performed to obtain a specific result (interaction provenance), as well as aspects of reasoning such as the insights gained during an analysis and the intents behind it (insight and rationale provenance) [18]. Because it covers interaction events, activities, tasks, and reasoning episodes, analytic provenance can be defined at many levels of granularity [8]. Analytic provenance can be used to support the recall, replication, and meta-analysis of an analyst's reasoning process [17,18]. While tool instrumentation is needed to capture interaction events, e.g., UI clicks, the analyst's reasoning is often documented manually, e.g., through user annotations [6].

2.2 Modeling Analytic Provenance

As described in Sect. 2.1, analytic provenance captures two main perspectives: (i) the analyst's interactions with a tool and (ii) their reasoning process [8,17].

To represent tool interactions, retrospective workflow provenance models can be a useful starting point. In these models, provenance is represented as a directed acyclic graph, where nodes are execution events and edges capture the causal dependencies between them [11]. Two well-known models based on this formalism are the Open Provenance Model (OPM) [16] and the W3C PROV Data Model (PROV-DM) [15]. Both models have typed nodes used to represent: (i) artifacts (or entities), i.e., immutable data states, (ii) processes (or activities), i.e., the actions performed on these artifacts, and (iii) agents. Relationships are defined by directed, typed edges that capture causal dependencies. For example, a process *used* an artifact, which in turn *was generated by* that process. Because causality is topologically explicit, these models establish a partial order of events without relying on timestamps. This structure supports graph-based inference through queries that reconstruct the history of an artifact by traversing the edges backward to its original input. Graph elements can be annotated with textual information about workflow execution. However, OPM and PROV-DM focus on artifact derivation and, thus, they do not support the modeling of processes that do not generate (final) results nor consider the analyst's reasoning.

Analytic provenance models, instead, focus on tracking analyst interactions, observations, and insights. Groth and Streefkerk [9] propose a framework for analysts to annotate the nodes (i.e., states of the visualization system) and edges (i.e., state transitions) of an interaction graph. Such node-level annotations support insight provenance [18] by tracking the observations analysts may make as they explore. However, because the user interactions in the graph are fine-grained, observations remain equally fine-grained. To go beyond low-level interactions, Gotz and Zhou [8] propose a hierarchical model of analytic provenance rooted in activity theory. This model defines four key elements based on their semantic richness:

(i) *events* represent individual UI interactions with little or no semantic value (e.g., a click);
(ii) *actions* represent atomic, domain-independent analytical steps (e.g., filtering);
(iii) *sub-tasks* represent domain-specific goals ; and
(iv) *tasks* represent generic analytical goals.

Parameters define the functional scope and required system inputs. Actions, which are defined by a type, a user's intent, and parameters, are considered to be the main building blocks for insight provenance, as they carry "enough semantics" to be meaningful to an analyst. The model allows one to explicitly represent *trails*, that is, sequences of exploration actions that automatically capture how an insight was generated. However, despite its focus on insights, it lacks an explicit representation of artifacts and allows the analyst to link textual annotations only with trails; i.e., reasoning that does not ultimately lead to insights is not captured.

In summary, existing models lack the necessary components needed to capture the provenance of a PM analysis. Workflow provenance models can represent artifact derivation, but lack the means to represent user interactions and reasoning. Instead, analytic provenance models capture user interaction hierarchies but neglect artifacts. Most importantly, none of these models consider the analyst's reasoning as a first-class element, which is essential for enabling meta-analysis.

A Closing Note on Process Mining. Although event logs in PM are records of process execution, they are rarely treated as a form of provenance because they are intended for the automated analysis of operational processes. Similarly, user behavior mining [1] relies on low-level interaction logs that lack the semantic context needed for reasoning. Yet, some PM research has addressed the analysts's need to keep track of their analysis process and results, e.g., replication of data analysis steps can be achieved through workflow-based approaches, such as RapidProM [14] or discovery pipelines [12]. However, these approaches require analysts to predefine data transformation steps for automation purposes, leaving user interactions, intermediate results, and reasoning implicit. To our knowledge, the approach in [25] is the only one to address provenance in exploratory PM. The authors propose a framework that maintains a branching tree of user operations, called *provenance view*, alongside *data views* that capture states of the event log. While the tree supports the capture of analysis steps and user annotations, the framework has two main limitations relative to our goals: (i) it does not track the artifacts generated during the analysis as the analysis state is a single "working log," and (ii) it models reasoning as textual annotations attached to nodes.

3 Motivating Example: A Process Mining Analysis

In this section, we present a simple process mining analysis to illustrate how an analyst's steps and reasoning unfold. This example is based on a video recording that captured both the analyst's interactions with a PM tool and their reasoning in terms of **O**bservations, **H**ypotheses, and **I**nsights [19].

In our example, the analyst works with the Road Traffic Fine Management (RTFM) event log [13]. The goal of the analysis is to understand why a significant number of fines remain unpaid despite the application of penalties.

Figure 1 shows a history of the analysis including the steps the analyst performed and the artifacts used and generated. We assume that the steps in this history are analysis units that are "meaningful" to the analyst, similar to the *actions* in [8], or the operations typically described in reports [3] or think-aloud studies [26]. Choosing a level of abstraction that mirrors what analysts work with in a PM environment allows us to faithfully narrate their reasoning [17]. In the following text, we describe the process followed by the analyst, referring to the history shown in Fig. 1 and reporting their reasoning in quotations.

To investigate unpaid fines, the analyst starts by discovering a directly-follows graph (DFG) to understand the process flow (1). By adjusting the view to simplify the DFG and inspecting case frequency and median duration (2–4), they

Id	Analysis step	Input	Output	Time (s)
1	discoverDFG()	L0	DFG1	0
2	explore(90%, 10%, absCaseFreq)	DFG1		45
3	explore(90%, 40%, absCaseFreq)	DFG1		70
4	explore(90%, 40%, avgWaitingTime)	DFG1		95
5	activityFilter(trace, remove, "Payment")	L0	L1	395
6	attrStats(cases, #cases)	L1	#cases_L1	415
7	traceVariants()	L1	variantsL1	450
8	viewVariants("Variant1")	variantsL1	casesVariant1	480
9	caseDetails("A100", "A10010")	casesVariant1	cases_A100,A10010	540
10	explore("caseA100", "amount")	cases_A100,A10010		600
11	explore("caseA10010", "amount")	cases_A100,A10010		660
12	attrStats(cases, "vehicleClass")	L1	L1_vehicleClasses	810
13	attrStats(cases, "vehicleClass")	L0	L0_vehicleClasses	860
14	activityFilter(trace, remove, "Payment")	L0	L2	1000
15	traceVariants()	L2	variantsL2	1030
16	viewVariants("Variant2")	variantsL2	casesVariant2	1090
17	explore("activities")	casesVariant2		1150
18	dFilter({'CF','SF'}, traces, keep, 90days)	L0	L3	1350
19	discoverDFG()	L3	DFG2	1380
20	explore(100%, 30%, absCaseFreq)	DFG2		1420
21	explore(100%, 89%, absCaseFreq)	DFG2		1460

Fig. 1. Example history of a process mining analysis. The *analysis steps*, identified by their *id*, are assumed to represent meaningful analysis units, reported with their *input* and *output* artifacts. *Time* is the relative time in seconds from the start of the analysis.

observe that many cases end with *Send Fine (SF)* (**O1**). To explain this, the analyst forms an initial hypothesis: "these cases may be incomplete due to recording or to constraints imposed by the law" (**H1**). Indeed, they had learned from the police that if a fine is sent more than 90 days after *Create Fine (CF)*, the offender does not have to pay. In the DFG, they note that the median time between *CF* and *SF* is 87 days (4), which is near the 90-day deadline but "still on time" (**O2**).

To further focus on the question, the analyst filters the log to retrieve cases without activity *Payment* (5). An initial check of the case statistics (6) shows that "more than half of the cases are unpaid" (**O3**). To further characterize these cases, the analyst inspects trace variants one by one, checking a few example cases to see what data attributes are recorded (7–11). The most frequent variant shows that "the offender does not pay, although penalties are added to the fine amount to incentivize early payment" (**I1**). To explain this behavior, the analyst looks further into data attributes, hypothesizing that "vehicle classes may correlate with non-payment" (**H2**). However, when comparing the distribution of attribute *vehicleClass* in the filtered subset with the original log (12–13), they observe "no significant correlation" (**O4**). Since this contradicts **H2**, the analyst returns to inspecting variants of unpaid cases (14–17).

In the second most frequent variant, they observe that "the process ends after the fine is sent" (**O5**). This triggers a recall of **H1**: "perhaps the 90-day constraint is the reason for this early ending." To test it, the analyst applies a directly-follows filter to the original log to keep cases where the delay between

CF and *SF* exceeds 90 days (18). They discover a subset of about 50K cases violating the constraint, which they explore using a DFG (19–21). While concluding that "the offender does not pay because the fine has not been sent within 90 days and is thus expired" (**I2**), the analyst observes something unexpected for their future analysis: "some offenders in this group seem to pay nevertheless" (**O6**).

Requirements for Modeling Analytic Provenance. While brief, this example illustrates the iterative nature of PM analysis, where reasoning builds upon artifacts.

The history in Fig. 1 shows different types of *analysis steps*. Some steps use an existing artifact to generate a new one, either through transformations, such as filters that create event log subsets (5, 14, 18), or by computing abstractions, such as DFG discovery (1, 19), trace variant derivation (7, 15), or statistics (6, 12, 13). Other steps allow exploring interactive artifacts, typically visual ones, by adjusting their parameters (2–4, 10, 11, 17, 20, 21). These steps do not generate new artifacts, but establish the visual context in which observations are made. Analysis steps operate on *artifacts*, such as event logs, trace variants, or DFGs. Beyond presenting final results, artifacts serve as reusable evidence that underpins the analyst's *reasoning*. Drawing on visual analytics literature [19], we differentiate between observations, hypotheses, and insights. Observations and insights may support the formation of hypotheses or emerge from contradictions when testing them. For example, early observations of cases ending with *Send Fine* (**O1** and **O2**) informed hypothesis **H1**, which was only recalled later and tested with a filter (18) after a similar pattern was observed (**O5**). Similarly, insight **I2** emerges only after **H1** is tested. However, reasoning is not recorded in PM tools and is often only documented post-analysis [8], manually.

Our ultimate goal is to leverage analytic provenance to support analysts in reflecting on and improving their work through meta-analysis [18]. Realizing this goal requires first supporting two essential provenance purposes: replication, to reproduce specific findings, and recall, to reconstruct the train of thought for revisiting, validating, or extending the analysis. To support these purposes in PM analysis, we need a representation of provenance that integrates:

(i) Well-defined analysis steps that describe interactions at a meaningful level for the analyst;
(ii) The content and generation of artifacts;
(iii) A temporal history that combines a linear timeline with a branching history to track the logical derivation of artifacts and (successful) analysis paths;
(iv) The analyst's reasoning, such as observations, hypotheses, and insights, associated with the steps and artifacts.

4 A Model of Provenance for Process Mining Analysis

The provenance approaches discussed in Sect. 2 support some of the aspects identified in Sect. 3, but do not integrate them. Here, we introduce a provenance model for PM analysis that serves as a data structure for integrating all these aspects, and identify parameters for model construction that must be left open for analyst choices during model use as discussed in Sect. 5.

We define the model constructively through *operations* that transform input from a PM analysis (history of PM operations, artifacts created and used, analyst reasoning) into a *provenance graph* (PG) integrating all aspects. We first show how to encode the branching history of an analysis in the PG starting from an immutable history of operations as input. We then define enrichments of the PG with linear history, artifact features, and the analyst's reasoning.

Input is an (immutable) *history* log of parameterized invocations of operations; here, we restrict ourselves to *deterministic* operations taking a single input object and an arbitrary number of parameters; they may return a single output object; Sect. 5 discusses how to lift this assumption. Formally, we assume universes $Attr \supseteq \{op, in, out, time\}$ of attribute names, Val of values, a time domain $Time \subset Val$ (a totally ordered set of timestamps), operation names $Op \subset Val$, and function $params : Op \rightarrow 2^{Attr}$ specifying the parameters of an operation.

Definition 1 (History). *A* history $H(E, \#)$ *has a finite set* E *of* events *and an attribute-value function* $\# : E \times Attr \nrightarrow Val$ *(denoting* $e.a = v$ *for* $\#(e, a) = v$ *and* $e.a = \perp$ *if* a *is undefined for* e*) such that each event* $e \in E$ *logs a timestamp* $e.time \in Time$*, an operation* $e.op \in Op$ *with parameters* $e.p \neq \perp, \forall p \in params(op)$*, and an input object* $e.in \neq \perp$*.*

An event e logs the $call(e) = (e.op, [p \mapsto e.p]_{p \in params(e.op)})$ and the input (and output) artifacts at time $e.time$. For example, in Fig. 1 e_5 logs a call to "activityFilter" (parameter names omitted for brevity) at time 395 with input log $L0$ and filtered log $L1$ as output. The *object ids referenced* in H are $O_H = \{e.in, e.out \mid e \in E\} \setminus \{\perp\}$. Assuming no two operations are executed at the same time, all events $e_1, e_2 \in E$ are totally ordered $e_1 < e_2$ iff $e_1.time < e_2.time$; e_1 *directly-follows* e_2, $e_1 \lessdot e_2$, iff $\nexists e \in E : e_1 < e < e_2$.

Next, we derive from H the *branching history* of operations and artifacts. But we cannot rely on the object ids referenced in $e.in$ or $e.out$ to be semantically consistent over H: if the analyst repeats the same deterministic operation op on the same input object o_{in} at different points in time, the PM environment may return different output object ids $o_{out} \neq o'_{out}$ despite the "contents" of o_{out} and o'_{out} being identical, e.g., e_5 and e_{14} in Fig. 1. To represent branches consistent with semantics, we encode objects not in terms of logged identifiers, but in terms of the sequence of operations generating them expressing a sufficient condition for *semantic equivalence*; Sect. 5 discusses artifact semantics further. Likewise, we encode operation invocations in terms of parameters and respectively encoded objects, not in terms of time-stamped events, using a recursive definition:

Definition 2 (Branching Encoding). *Let* $H = (E, \#)$ *be a history.*
A root object $o \in O_H$ *that is not the output of an event, i.e.,* $\nexists e \in E : e.out = o$*, has branching* $id(o) = (Artifact, o)$*.*
An event $e \in E$ *has branching* $id(e) = (type(e), call(e), id(e.in))$*, where* $type(e) = Op$ *iff* $e.out \neq \perp$ *(operation generating an artifact),* $type(e) = Act$

iff $e.out = \perp$ *(action exploring input artifact* $e.in$*). If* $type(e) = Op$ *then the branching id of its output artifact* $e.out$ *is* $id(e.out) = (Artifact, call(e), id(e.in))$.

For example, the encoding unifies e_5 and e_{14} in Fig. 1 and their outputs: $id(e_5) = (Op, activityFilter(trace, remove, \text{“}Payment\text{”}, id(L_0)) = id(e_{14})$ and also $id(e_5.out) = id(e_{14}.out)$, and by recursion also unifies $id(e_7) = id(e_{15})$, etc., expressing that these operations and artifacts are semantically equivalent. History H is *semantically consistent* iff any two events $e, e' \in E$ with the same logged output object ids $e.out = e'.out \neq \perp$ have the same operation history: $id(e.out) = id(e'.out)$. H is *single-rooted* iff there exists exactly one root object $o \in O_H$ with $\nexists e \in E : e.out = o$.

We now use the branching encoding to translate the history into a provenance graph, which we base on *labeled property graphs* (LPGs) [2]. Given node labels Λ_N and relationship labels Λ_R, an LPG $G = (N, R, \lambda, \vec{\cdot}, \#)$ has *nodes* N and *relationships* R, $N \cap R = \emptyset$, where each $r \in R$ relates two nodes $\vec{r} = (n_1, n_2) \in N \times N$. Each node $n \in N$ has a label $\lambda(n) \in \Lambda_N$ and each relationship has a label $\lambda(r) in \Lambda_R$. Function $\# : (N \cup R) \times Attr \nrightarrow Val$ assigns each $x \in N \cup R$ properties as attribute-value pairs. We also write $n \in N^\ell$ iff $n \in N \wedge \lambda(n) = \ell$; $r \in R^\ell$ iff $r \in R \wedge \lambda(r) = \ell$; $(n_1, n_2) \in R^\ell$ iff $n_1, n_2 \in N \wedge r \in R^\ell, \vec{r} = (n_1, n_2)$. We write $G_1 \subseteq G_2$ if LPG G_1 is an element-wise subset of G_2.

The branching encoding (Def. 2) of a single-rooted, semantically consistent history H induces a graph of artifacts, their generating operations, and associated exploration actions.

Definition 3 (Provenance Graph). *Let* $H = (E, \#)$ *be a semantically consistent history. The* Provenance Graph (PG) *of* H *is an LPG* $G = (N, R, \lambda, \vec{\cdot}, \#)$ *with* $\Lambda_N = \{Artifact, Op, Act\}$ *and* $\Lambda_R = \{used, generated, explored\}$ *s.t.:*

- *for each* $o \in O_H$, G *has artifact node* $n = id(o) \in N^{Artifact}$ *with* $n.ids = \{e.out \mid e \in E : id(e.out) = id(o)\}$ *referring to all logged object ids;*
- *for each* $e \in E$, G *has an operation or action node* $n = id(e) \in N^{type(e)}$ *(depending on e's type) with* $n.op = e.op$ *and* $n.p = e.p, \forall p \in params(op)$ *with relationships to artifacts:*
  - *if* $type(e) = Op$, *then* $(id(e.in), id(e)) \in R^{used}, (id(e), id(e.out)) \in R^{generated}$, *and*
  - *if* $type(e) = Act$, *then* $(id(e), id(e.in)) \in R^{explored}$.

Figure 2 (left) shows the Provenance Graph of the history up to e_{17} in Fig. 1. It describes the branching history of the analysis recorded in H and unifies semantically identical operations (e.g., e_5, e_{14}) and artifacts (e.g., L_1, L_2).

We now extend the provenance graph G with other provenance perspectives (linear history, artifact features, reasoning) through enrichment operations. We integrate the branching history in G with the linear history in H by encoding H as a graph, linking events in H to the operations in G, and lifting temporal ordering from events to operations.

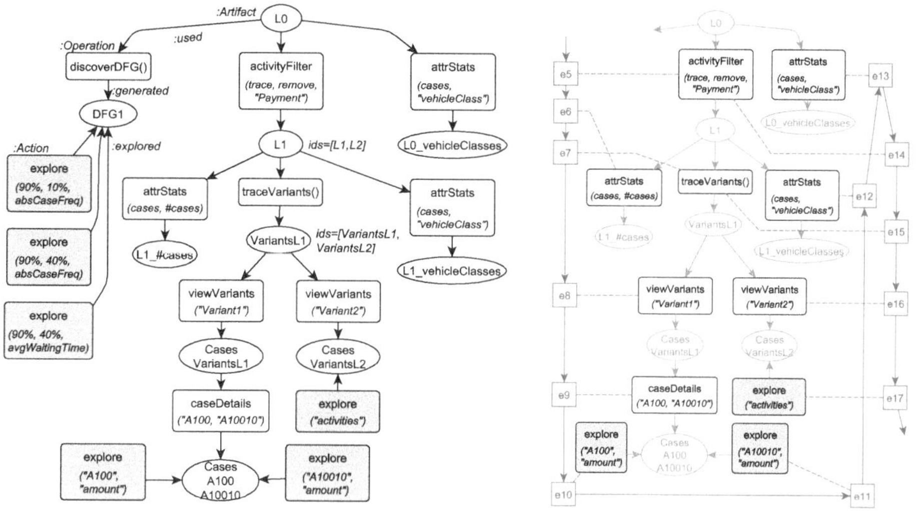

Fig. 2. Provenance Graph of the history up to e_{17} in Fig. 1 (left) and enrichment with linear history (right).

Definition 4 (Enrichment with Linear History). *Let $G = (N_P, R_P, \lambda_P, \vec{\cdot}_P, \#_P)$ be the PG of history $H = (E, \#_H)$. The* enrichment *of G with its history H is the LPG $G_H = (N, R, \lambda, \vec{\cdot}, \#)$ with $G \subseteq G_H$ s.t.*

- *for each $e \in E$: $e \in N^{Event}$ with $\#(e, a) = \#_H(e, a)$ for all $a \in Attr$ (each event is a node) and $(id(e), e) \in R^{occurred}$ (the event is referred to from the operation node in G);*
- *$(e_1, e_2) \in R^{df}$ iff $e_1 \lessdot e_2$ (any two directly following event nodes are connected via a relationship with label df); and*
- *for all $n_1, n_2 \in N^{Op} \cup N^{Act}$ with $occ(n_1, n_2) = \{(e_1, e_2) \in R^{df} \mid id(e_1) = n_1, id(e_2) = n_2)\} \neq \emptyset$ exists $r \in R^{df}$, $\vec{r} = (n_1, n_2), r.freq = |occ(n_1, n_2)|$ (df-relations are lifted from events to operations nodes).*

Figure 2 (right) shows the result of steps 1. and 2. from Def. 4 applied on the PG of Fig. 2 (left) for events e_5-e_{17}, visualizing the order and repetition of operations and actions in the branching history. The green, solid edges in Fig. 3 show the result of lifting the ordering from events to operation/action nodes.

The PG G of H describes artifacts only in terms of the operations that generated them. To aid the analyst in reflecting and reasoning over the "contents" of the artifacts, such as relevant log statistics or properties of a DFG, we enrich the artifact nodes with features of the artifacts themselves. As the specific features of interest evolve during the analyst's reflection and reasoning, specific feature enrichment of G must remain an *open parameter* for model construction. We introduce as a parameter the *feature extraction function* $p : O_H \rightarrow ((Attr \setminus \{ids\}) \rightarrow Val)$ that extracts for a given logged artifact id o a mapping $p(o)$, where $p(o).a = v$ states that o has feature a with value v. For example,

$p(cases_{Variant1}).trace = \langle CF, SF, IFN, AP, CC\rangle$, $p(cases_{Variant1}).freq = 56482$, etc., while for a DFG d, $p(d)$ could extract frequencies of (specific) directly-follows pairs. We discuss in Sect. 5 how to operationalize this parameter. Given p, each artifact node is enriched with the extracted features for one of the originating objects.

Definition 5 (Enrichment with Artifact Features). *The enrichment of a PG G of a history H by a feature extraction function p picks for each artifact $n \in N^{Artifact}$ one id $o \in n.ids$ and sets $n.a = p(o).a$ for all $a \in Attr$ with $p(o).a \neq \perp$. We write G^p for the enriched graph.*

As all objects in $n.ids$ are semantically identical (same operations on the same input objects), Def. 5 can pick any $o \in n.ids$, for instance depending on which o is still in memory or persisted. Figure 3 shows the PG of the entire history of Fig. 1 with some log and trace variant features providing some context to the analyst on the data seen at various steps. When p is chosen, G^p_H can be automatically discovered from H, resulting in a PG that reflects the linear history and branching history of the operations and actions performed by the analyst over the explicitly represented artifacts seen until that point.

Assuming that operations, actions, and artifacts are represented at a level of granularity that aligns with the analyst's reasoning (based on Sect. 3, further discussed in Sect. 5), G^p_H now provides the provenance data structure an analyst can (selectively) query [2], inspect, and enrich with their own reasoning. Aiming at having a consistent syntax for reasoning annotations [18], we define the following operations, leaving the choice and application by the analyst to enrich G^p_H with reasoning as a parameter in the graph construction.

Definition 6 (Enrichment with reasoning). *A PG G is* enriched with reasoning *by applying a sequence $\sigma \in \Sigma^*$ of enrichment operations Σ; we write G^σ for the enriched PG. We define the following enrichment operations Σ:*

- *addHypothesis(h, N') adds a new node $n_h \in N^{Hypothesis}$ with $n_h.text = h$ and relationships $(n_h, n) \in R^{formedOn}$ to existing nodes $n \in N' \subseteq N$, $\lambda(n) \in \{Artifact, Act\}$ indicating that hypothesis h was formed on N';*
- *addObservation(o, N') adds a new node $n_o \in N^{Observation}$ with $n_o.text = o$ and relationships $(n_o, n) \in R^{madeOn}$ to existing nodes $n \in N' \subseteq N$, $\lambda(n) \in \{Artifact, Act\}$ indicating that observation o was made on N';*
- *addInsight(i, N') adding $n_i \in N^{Insight}$ is defined respectively;*
- *add(n_{from}, ℓ, n_{to}) adds a new relationship $(n_{from}, n_{to}) \in R^\ell$; and*
- *supports(n_{io}, n_h) = add$(n_{io}, supports, n_h)$ and contradicts(n_{io}, n_h) = add$(n_{io}, contradicts, n_h)$ for insight or observation n_{io} and hypothesis n_h.*

Figure 3 shows the PG of the history of Fig. 1 enriched with history, artifact features, and reasoning based on Sect. 3. Note how it recalls the analysis narrated in Sect. 3, showing various branching and repetition to answer **H1** and documenting the genesis and conclusion of **H1**.

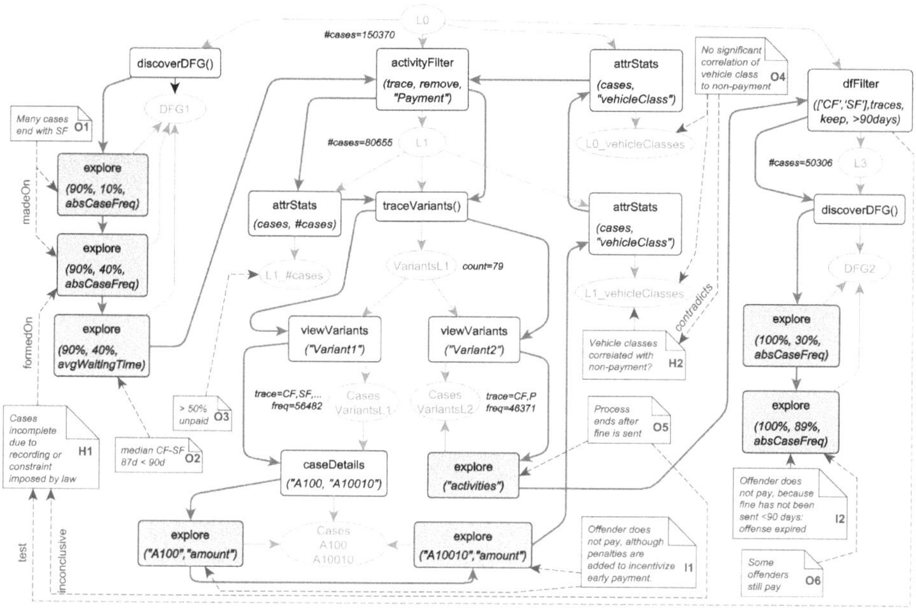

Fig. 3. Provenance Graph enriched with linear history and reasoning.

Properties. Our model enjoys properties in support of a human-in-the-loop construction and enrichment of G alongside a PM analysis. Formally, enriching G with history H (Def. 4), with artifact features p (Def. 5) and with reasoning σ (Def. 6) is commutative.

Proposition 1 (Commutativity). *Let G be the PG of history H. Let p and σ be chosen. Let G_H, G^p, and G^σ be the result of history, feature, and reasoning enrichment, respectively. Then $(G_H)^p = (G^p)_H = G^p_H$, $(G_H)^\sigma = (G^\sigma)_H = G^\sigma_H$, $(G^p)^\sigma = (G^\sigma)^p = G^{\sigma,p}$ and $(G^p_H)^\sigma = (G^\sigma)^p_H = G^{\sigma,p}_H$.*

Prop. 1 follows directly from the respective definitions as Def. 4-Def. 6 only add elements to G without depending on the elements added by other enrichments. While formally commutative, an analyst will base their reasoning enrichment G^σ on prior feature and history enrichment G^p_H (cf. Sect. 5). Enrichment and graph construction (Def. 3) are not commutative as the former depends on the latter.

The proposed model (Def. 2–6) is monotonous wrt. H in the sense that adding another event e only adds further nodes and edges to the graph G and its enrichments, allowing incremental construction. Formally, history H_1 is *contained in* history H_2, written $H_1 \subseteq H_2$ iff $E_1 \subseteq E_2$ and $\#_1 = \#_2|_{E_1}$, H_1 is a *prefix of* H_2, $H_1 \sqsubseteq H_2$ iff $H_1 \subseteq H_2$ and $e_1 < e_2$ for all $e_1 \in E_1, e_2 \in E_2 \setminus E_1$. Recall that we write $G_1 \subseteq G_2$ if LPG G_1 is a sub-graph (an element-wise subset) of G_2.

Proposition 2 (Monotonicity). *Let G_1 and G_2 be the PG of histories H_1 and H_2, respectively. Let $G_{H,i}, G_i^p, G_{H,i}^p$ be the (combined) enrichments of G_i with linear history and artifact features, respectively.*

- *If $H_1 \subseteq H_2$ then $G_1 \subseteq G_2$ and $G_1^p \subseteq G_2^p$.*
- *If $H_1 \sqsubseteq H_2$ then additionally $N_{H,1}^{Event} \subseteq N_{H,2}^{Event}$, $R_{H,1}^{occurred} \subseteq R_{H,2}^{occurred}$, and for each $r \in R_{H,1}^{df}$ holds $r \in R_{H,2}^{df}$ and if $\#_1(r, freq) \neq \perp$ then $\#_1(r, freq) \leq \#_2(r, freq)$ (correspondingly for $G_{H,1}^p$ and $G_{H,2}^p$).*

This proposition can be proven by inductively extending H_1 with a new event e. Then either $G_1 = G_2$ (if $id(e) = id(e')$ for some $e \in E_1$) or G_2 has either an additional action node compared to G_1 or additional operation and artifact nodes. Enrichment with artifact features (G_2^p) is idempotent for artifact nodes already in G_1 and thus only adds features for new artifact nodes. If additionally $e.time > e'.time$ for any $e' \in E_1$, then G_H^2 contains $e \in N_{H,2}^{Event}$ with $(e', e) \in R_{H,2}^{df}$ for the last event e' in H_1 (i.e., all other event orderings remain). Further, the df-relationship $(id(e'), id(e)) \in R_{H,2}^{df}$ between operation/action nodes $id(e')$ and $id(e)$ either gets created or increases in frequency by 1.

These properties render the PG a suitable basis for an interactive provenance artifact integrated with a PM tool: any step in the PM tool is logged as an event e appended to the history H, the current graph G_H^p is then monotonously extended according to Def. 2-5. The analyst may stop performing steps in the tool at any point in time, query [2] and inspect parts of G_H^p, and enrich it with reasoning σ (Def. 6), which also just adds nodes and relations. Informed by the inspection and enrichment, the analyst may return to the PM tool and continue their analysis, leading to further monotonous extensions of $G_H^{p,\sigma}$.

5 Provenance-Aware Support: Discussion and Challenges

The model introduced in Sect. 4 provides a foundation to represent a PM analysis as a multiperspective process. By integrating operations, artifacts, and reasoning into a single provenance graph (cf. Defs 3–6), it bridges the gap between artifact-centric models [15,16] and approaches focused solely on user interactions [6,17], providing the necessary vocabulary to discuss meta-analysis in PM. However, our model relies on four main assumptions: (i) operations have exactly one input and one output artifact, (ii) these operations are recorded at a granularity meaningful to the analyst and aligned with their reasoning, (iii) artifact nodes are enriched with meaningful information about the artifacts, and (iv) analysts actively enrich the provenance graph with their reasoning. Acknowledging these assumptions is critical for operationalizing our model into systems for provenance-aware support. While (i) is a simplifying limitation of our model, (ii)-(iv) explicate the ideal state in which analysts work in systems. Although this state still requires effort to be achieved (cf. [4,8,17]), it can be controlled by our model's input and parameters.

In the remainder, we first discuss how the integration of perspectives in our model supports replication and recall (Sect. 5.1). Then, in Sect. 5.2, we frame the development of provenance-aware support for meta-analysis in PM as a new problem in BPM, where our assumptions guide the definition of specific research challenges to operationalize our parameters.

5.1 Using Provenance Graphs to Support Replication and Recall

Provenance can be used for several purposes [18], including replication and recall.

Replication is the ability to reproduce a specific analysis result for validation or reuse as a starting point for further analysis [11,18]. Existing approaches, such as RapidProM [14] or discovery pipelines [12], support repeatable analyses but require prespecified workflows. Our model supports replication through the discovered semantically consistent branching history of artifacts and operations and exploration actions. Because an artifact's identity is defined by its derivation path (see Def. 2), an analyst can replicate any artifact by re-executing the sequence of operation edges ($type(e) = Op$) from the root. In our example of Sect. 3, an analyst can replicate the distribution of attribute *vehicleClass* for unpaid cases (*L1_vehicleClasses* in Fig. 2) by replaying the sequence of operations from *L0*. The branching history maintains *every* analytical path explored and thereby ensures that all the generated artifacts, including discarded ones, can be replicated. Artifact enrichment by a suitable feature extraction function (Def. 5) allows the analyst to compare how artifacts differ or whether artifacts in different branches share the same features, allowing one to understand the *semantic effects of operations*. At last, exploration actions ($type(e) = Act$) allow the analyst to replicate the artifact views that provided the analytic context for their reasoning.

Recall instead is about maintaining an overview of the rationale behind past steps and current results, i.e., the "why" that is often lost or remains implicit in the analyst's mind, particularly over extended or intermittent periods of analysis [18]. This forces analysts who want to reuse or validate a given artifact to retrace parts of the analysis to refresh their memory. Our model supports recall through operations that are meaningful to the analyst and enrichment with linear history and reasoning. By assuming that operations resemble meaningful steps done in the PM environment, the provenance graph can support analysts in recalling what they did at a functional level [8]. e.g., operation *discoverDFG()* in Fig. 2 recalls process discovery. Enrichment with the linear history (Def. 4) supports temporal recall by preserving the sequential order of the steps across the analysis branches. For example, during the analysis, after ending the unfruitful branch of finding a correlation between vehicle classes (e_{12}-e_{13} in Fig. 2 (right)), the analyst can recover the context from the preceding events e_7-e_{11} and return to inspecting variants (e_{14}-e_{17}). After the analysis, the linear history helps recall that **H2** and its path in Fig. 3 were discarded because *vehicleClass* distributions were compared between $L0$ and $L1$. Enriching the graph with reasoning nodes (Def. 5) allows analysts to offload observations, hypotheses, and insights from their working memory for later recall. For instance, materializing the observations about the process ending early (**O1** and **O5** in Fig. 3) and linking it to earlier hypothesis **H1** allows the analyst to recall the 90-day constraint as a possible cause. These nodes help preserve the rationale for decisions that would otherwise be lost after the analysis.

5.2 Supporting and Studying Meta-Analysis: Research Challenges

Supporting replication and recall is a prerequisite to enable meta-analysis. Although our model integrates the concepts needed for meta-analysis, achieving this goal requires developing provenance-aware systems that automate the construction of the provenance graph and provide interfaces for the analyst to enrich and review it. Such systems would bridge PM tools, which capture user interactions, with analysts, who need interfaces to conduct meta-analysis. This is a novel research problem in the field of BPM. The main novelty lies in treating the analyst's behavior as a process itself to be modeled and mined. As noted in Sect. 1, achieving this entails a circular dependency: we cannot build provenance-aware systems without an empirical understanding of analyst behavior, yet we cannot gain that understanding without systems that allow us to capture and study it. To unravel this circularity, we use our model to identify a series of concrete research challenges to resolve our assumptions by operationalizing the model's parameters with the aim to build systems that support meta-analysis in PM.

Research Challenges. A first challenge concerns **capturing and modeling human reasoning (C1)**. While PM analysts already externalize some of their reasoning by taking notes about their insights [21,26], manual documentation can disrupt the analytical flow [6] or result in omitted information [8,17] in practice. Yet, current PM tools do not provide systematic mechanisms to capture such reasoning. Our assumption (iv) that analysts actively enrich the graph with reasoning aims to close this gap. Earlier work suggests that analysts can externalize their reasoning through three distinct components: observations, hypotheses, and insights [19,20]. By formalizing these as different node types (cf. Def. 6), our model provides the syntax to represent reasoning in a provenance graph, going beyond simple textual annotations [9,15,16]. However, it remains unclear whether the proposed concepts are aligned with how PM analysts perceive their own reasoning, and in what form they are willing to document it. For this reason, our model only provides the reasoning syntax and no semantic constraints, rendering it a canvas for empirical studies, e.g., to explore different interfaces for capturing and linking specific reasoning statements, and to observe how analysts enrich (and revise) reasoning alongside or at the end of an analysis. Insights from these studies will inform refinements to the semantics of reasoning elements in our model and the development of inference over the analyst's reasoning.

PM tools typically record analyst behavior not as meaningful operations supporting recall but as lower-level interaction logs. For instance, a standard interaction log would capture the *activityFilter(traces, remove "Payment")* operation (cf. Step 5 in Fig. 1) as the sequence of UI events to select, configure, and apply the filter, and input/output logs (L_0 and L_1) would be recorded as parameters, if at all. Thus, the interaction logs derived from PM tools must be reconciled with operations meaningful to an analyst (cf. Sect. 5.1 and assumption (ii)). This reconciliation introduces the challenge of **operation mapping and maintaining a multilevel granularity (C2)**. First, we have to identify a "vocabulary" of meaningful PM operations, potentially by drawing on existing taxonomies [3], setting a base granularity for our model. However, a fixed granularity is insufficient to support meta-analysis. PM tools vary in operation granularity, e.g., some decouple model discovery from visualization, while others treat them as one step. We therefore need a mapping mechanism to translate fine-grained, tool-specific interactions into meaningful operations. Furthermore, while replication benefits from a fine granularity but the possibly included interaction noise may hinder recall [4], a coarser granularity can hide details relevant for replication [8]. Because meta-analysis relies on both replication *and* recall, provenance-aware systems must maintain a multilevel granularity that can be adjusted by aggregating and filtering the provenance graph. In this way, the graph's granularity could be aligned with the analyst's current reasoning or with a specific meta-analysis goal.

Analysts also rely on information about artifacts for replication and recall (cf. Sect. 5.1). Hence, operationalizing our model faces the challenge of **meaningful artifact information (C3)**. Similar to C2, by making the artifact feature enrichment function (see Def. 5) a tunable parameter, our model enables the analyst to align artifact feature extraction with their current reasoning or goal, i.e., our assumption (iii). An important aspect is thereby measuring artifact similarity. Since our model lacks direct knowledge of the underlying data, we need a way to determine when different sequences of operations lead to the same or similar artifacts. Understanding exact data similarity aids replications, e.g., whether the same artifact is revisited on different branches, whereas for recall and meta-analysis analysts may focus on "semantic similarity", i.e., if two artifacts support the same hypothesis. We need to research which artifact features are meaningful for analysts for which purposes. This requires developing a catalog of feature extraction functions and distance functions for artifact similarity that allows analysts to adjust artifact information (features and similarity) to their specific goals and the generation of data views [25]. Orthogonally, overcoming assumption (i) of a single input/output artifact per operation, e.g., joining/clustering logs, requires generalizing the history and our model to an object-centric setting.

Even if the abstractions in C2 and C3 are available, we still need **interaction and visualization mechanisms (C4)** that allow analysts to review their analysis. Indeed, the enriched provenance graph is not intended to be the final artifact shown to the analyst, as assuming that they would interact with such a complex and possibly large representation of their process is unrealistic. Therefore, we need to provide multiperspective views that synthesize branching paths, artifacts, and reasoning into visual artifacts that can be used by analysts to review their own analysis without cognitive overload.

Finally, we need to support **querying and inference (C5)** of the provenance graph to allow the analyst to improve their work. For example, by analyzing the links between the reasoning nodes, a system could infer and recommend subsequent analysis steps to validate a specific hypothesis based on previously tested ones. Similarly, inference could detect logical gaps or inconsistencies in the analysis, such as insights that lack a supporting artifact, prompting the analyst to revisit their reasoning.

Conclusion. In this paper, we have presented a model of analytic provenance that integrates analysis steps, data artifacts, and human reasoning into a multiperspective provenance graph. Using this model, we have framed the development of provenance-aware support for meta-analysis in PM as a novel research problem. By providing the necessary vocabulary to describe meta-analysis, our model unravels the circular dependency between system design and empirical research. Addressing the research challenges outlined above will operationalize our model into provenance-aware systems. In the long term, analysts and their collaborators will be able to engage in (self-)reflection and improve their analysis strategies and efficiency through these systems. Similarly, researchers will be able to capture and study analyst behavior at scale and use this knowledge to design methods and tools to improve PM practice.

References

1. Abb, L., Rehse, J.R.: A reference data model for process-related user interaction logs. In: International Conference on Business Process Management, pp. 57–74. Springer (2022)
2. Angles, R.: The property graph database model. In: Proceedings of the 12th Alberto Mendelzon International Workshop on Foundations of Data Management, pp. 1–10. CEUR-WS (2018)
3. Capitán-Agudo, C., Salas-Urbano, M., Cabanillas, C., Resinas, M.: Analyzing how process mining reports answer time performance questions. In: International Conference on Business Process Management (BPM), pp. 234–250. Springer, Cham (2022)
4. Carata, L., et al.: A primer on provenance. Commun. ACM **57**(5), 52–60 (2014)
5. Davidson, S.B., Freire, J.: Provenance and scientific workflows: challenges and opportunities. In: Proceedings of ACM SIGMOD International Conference Management Data, pp. 1345–1350 (2008)
6. Dou, W., Jeong, D.H., Stukes, F., Ribarsky, W., Lipford, H.R., Chang, R.: Recovering reasoning processes from user interactions. IEEE Comput. Graph. Appl. **29**(3), 52–61 (2009)
7. Freire, J., Koop, D., Santos, E., Silva, C.T.: Provenance for computational tasks: a survey. Computing Sci. Eng. **10**(3), 11–21 (2008)
8. Gotz, D., Zhou, M.X.: Characterizing users' visual analytic activity for insight provenance. Inf. Vis. **8**(1), 42–55 (2009)
9. Groth, D.P., Streefkerk, K.: Provenance and annotation for visual exploration systems. IEEE Trans. Vis. Comput. Graph. **12**(6), 1500–1510 (2006)
10. Gupta, A.: Data Provenance, pp. 608–608. Springer US, Boston, MA (2009)

11. Herschel, M., Diestelkämper, R., Ben Lahmar, H.: A survey on provenance: what for? what form? what from? VLDB J. **26**(6), 881–906 (2017)
12. Klinkmüller, C., Seeliger, A., Müller, R., Pufahl, L., Weber, I.: A method for debugging process discovery pipelines to analyze the consistency of model properties. In: International Conference on Business Process Management, pp. 65–84. Springer (2021)
13. de Leoni, M.M., Mannhardt, F.: Road traffic fine management process (2015)
14. Mans, R.S., van der Aalst, W.M., Verbeek, H.: Supporting process mining workflows with RapidProM. In: BPM Demo Sessions. pp. 56–60. CEUR-WS.org (2014)
15. Missier, P., Belhajjame, K., Cheney, J.: The W3C PROV family of specifications for modelling provenance metadata. In: EDBT 2013, p. 773–776. ACM (2013)
16. Moreau, L., Clifford, B., Freire, J., Futrelle, J., Gil, Y.: The Open Provenance Model core specification (v1.1). Futur. Gener. Comput. Syst. **27**(6), 743–756 (2011)
17. North, C., et al.: Analytic provenance: process+ interaction+ insight. In: CHI EA '11', pp. 33–36. ACM (2011)
18. Ragan, E.D., Endert, A., Sanyal, J., Chen, J.: Characterizing provenance in visualization and data analysis: an organizational framework of provenance types and purposes. IEEE Trans. Vis. Comput. Graph. **22**(1), 31–40 (2015)
19. Sacha, D., Stoffel, A., Kwon, B.C., Ellis, G., Keim, D.A.: Knowledge generation model for visual analytics. IEEE Trans. Vis. Comput. Graph. **20**(12) (2014)
20. Sorokina, E., Soffer, P., Hadar, I., Leron, U., Zerbato, F., Weber, B.: PEM4PPM: a cognitive perspective on the process of process mining. In: International Conference on Business Process Management, pp. 465–481. Springer (2023)
21. Suetendael, J.V., Depaire, B., Jans, M., Martin, N.: Understanding the behavior of process mining analysts: a catalogue of exploratory process mining behaviors. Process Sci. **2**(1), 16 (2025)
22. Tentina, I., Zerbato, F., Mannhardt, F., van Dongen, B.: Why do users struggle to get insights from process mining? In: International Conference on Process Mining 2025, pp. 1–8. IEEE (2025)
23. Van Der Aalst, W.: Data science in action, Springer (2016)
24. Xu, K., Attfield, S., Jankun-Kelly, T., Wheat, A., Nguyen, P.H., Selvaraj, N.: Analytic provenance for sensemaking: a research agenda. IEEE Comput. Graph. Appl. **35**(3), 56–64 (2015)
25. Zerbato, F., Burattin, A., Völzer, H., Becker, P.N., Boscaini, E., Weber, B.: Supporting provenance and data awareness in exploratory process mining. In: International Conference on Advanced Information Systems Engineering (CAiSE), pp. 454–470. Springer (2023)
26. Zerbato, F., Zimmermann, L., Vrotsou, K., Weber, B.: From analysis to findings: how do process mining analysts discover results? Inf. Syst. 102596 (2025)
27. Zimmermann, L., Zerbato, F., Weber, B.: What makes life for process mining analysts difficult? A reflection of challenges. Softw. Syst. Model. **23**(6), 1345–1373 (2024)

Track II: Engineering

Mind the Long Tail: Understanding the Difficulty of Delay Detection in Business Processes

Keyvan Amiri Elyasi[1(✉)], Lukas Kirchdorfer[1,2], and Heiner Stuckenschmidt[1]

[1] Data and Web Science Group, University of Mannheim, Mannheim, Germany
{keyvan.amiri,heiner.stuckenschmidt}@uni-mannheim.de
[2] SAP Signavio, Walldorf, Germany
lukas.kirchdorfer@sap.com

Abstract. The early detection of delayed cases in business processes is a critical capability for organizations. Predictive process monitoring (PPM) supports this task by using historical event logs to predict the remaining time of ongoing cases, enabling timely interventions to avoid missed deadlines and service level violations. Although remaining time prediction has advanced considerably through sophisticated deep learning architectures, little is known about the intrinsic difficulty of delay detection itself. Since performance is typically assessed using aggregate metrics, prior work provides limited insight into how models perform across the target distribution, especially on the operationally most critical cases with large delays. In this paper, we address this gap by analyzing the difficulty of delay detection. Across 14 event logs, we show that remaining times are typically strongly right-skewed, with only a small fraction of cases exhibiting large delays. Existing models capture the mode of this distribution well but perform poorly on high-delay cases. We further uncover pronounced heteroscedasticity, showing that predictive uncertainty increases with delay magnitude. Based on these findings, we evaluate approaches to mitigate the imbalance problem, but find only limited benefits, suggesting that the key underlying problem may not be imbalance but higher uncertainty associated with delayed cases. We show that this correlation can be exploited to substantially improve the identification of delayed cases. Overall, our work provides new insights into the sources of difficulty in delay detection and identifies uncertainty-aware modeling as a promising direction for future PPM research.

Keywords: Predictive Process Monitoring · Remaining Time Prediction · Imbalanced Regression · Uncertainty Quantification

1 Introduction

Organizations rely on business processes to deliver services to customers and ensure the timely execution of work. In this context, execution delays are a major

K. A. Elyasi and L. Kirchdorfer—Equal contribution.

F. M. Maggi et al. (Eds.): BPM 2026, LNCS 16936, pp. 241–258, 2027.
https://doi.org/10.1007/978-3-032-37777-7_16

concern, as they lead to missed deadlines, violated service level agreements, increased operational costs, and diminished customer satisfaction [12,23,29]. Consequently, the early identification of delay-prone cases is essential for effective process management. Such identification can leverage execution data recorded by enterprise information systems in the form of event logs, which document historical process executions and provide a basis for anticipating future behavior.

In business process management, this is addressed through predictive process monitoring (PPM), which aims to predict future characteristics of running cases based on historical event data [5]. Among PPM tasks, remaining time prediction is one of the most widely studied. Given the observed prefix of an ongoing case, it estimates the time until completion and provides an actionable signal for identifying potentially delayed cases and enabling timely interventions.

In recent years, substantial progress has been made in remaining time prediction, largely driven by advances in deep learning architectures such as LSTMs, Transformers, and graph neural networks [1,10,16]. While these approaches improve predictive performance, they are typically evaluated using aggregate metrics such as mean absolute error (MAE), which obscure where models perform well or fail. In particular, such metrics do not capture performance on rare delayed cases, despite their critical operational impact due to their potential to cause service level violations and increased costs.

As we show in this paper, distributions of remaining time in business processes are typically highly right-skewed: most cases complete within an expected time frame, whereas a small number exhibit extreme delays. Accurately identifying these rare delayed cases is essential for effective process management, yet it remains unclear whether existing models can reliably detect them. More fundamentally, there is still limited understanding of the intrinsic difficulty of delay detection in business processes, which hinders the development of more effective PPM approaches grounded in process-specific insights rather than relying primarily on advances in general-purpose machine learning models.

To address these challenges, this paper makes three contributions.

1. First, we study the intrinsic difficulty of delay detection in remaining time prediction. Our analysis shows that (a) remaining times typically follow a strongly right-skewed distribution, (b) existing approaches capture the mode of this distribution well but fail to predict cases with large delays accurately, and (c) predictive uncertainty is heteroscedastic and increases with remaining time, indicating that high-delay cases are substantially noisier than normal cases.
2. Second, building on findings (a) and (b), we hypothesize that a key reason why existing models fail on high-delay cases is the identified *target imbalance*. We therefore evaluate established imbalanced regression approaches and find, however, that their impact is limited.
3. Third, based on finding (c), we investigate the utility of heteroscedastic uncertainty estimates for delay detection. Specifically, we study whether the relationship between predictive uncertainty and remaining time magnitude can

be exploited to improve the identification of delayed cases. Our results show that uncertainty-aware models can substantially improve delay detection.

The remainder of the paper is structured as follows. Section 2 analyzes the difficulty of delay detection across a variety of processes. Section 3 describes the experimental setup, followed by the evaluation of imbalanced regression approaches in Sect. 4 and the assessment of heteroscedastic uncertainty in Sect. 5. Finally, we discuss related work in Sect. 6 and conclude the paper in Sect. 7.

2 Motivation and Analysis

We illustrate the difficulty of delay detection using the *Prepaid Travel Cost* process from the BPIC20 benchmark and then show that the observed patterns generalize across other event logs.

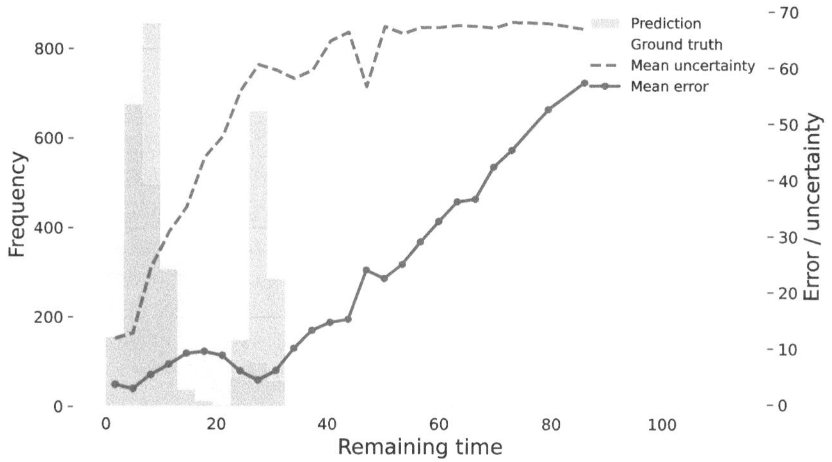

Fig. 1. BPIC20PTC distribution of ground-truth and predicted remaining times (in days), with average error and heteroscedastic uncertainty per target range.

Illustration. To illustrate the difficulty of delay detection, we analyze the BPIC20PTC process from Eindhoven University of Technology, which contains events related to two years of travel expense claims. Figure 1 shows the distribution of remaining times across all possible prefix positions in the process (orange histogram) and compares it with the distribution of predicted remaining times (blue histogram) produced by a standard LSTM-based approach [16].

The ground-truth distribution of remaining times is *strongly right-skewed*, with most observations concentrated around 10 days and only a small number extending to 40–100 days. In contrast, the predictions are compressed into a much narrower range and fail to capture high-delay cases. The model never predicts a remaining time above 35 days. As a result, a case that actually requires, for example, 80 days until completion would receive a prediction of only 30–35 days. Such errors render the prediction ineffective for timely delay identification and operational intervention. Consequently, prediction error increases substantially toward the tail of the distribution, as illustrated by the red line.

Another key insight is that predictive uncertainty increases with remaining time, exhibiting a clear *heteroscedastic* pattern. In other words, the figure suggests a correlation between the magnitude of uncertainty and the magnitude of remaining time. Heteroscedastic uncertainty is data-dependent and reflects the irreducible noise inherent in the data generation process, and arises from stochastic and unobserved factors that influence process execution [27]. In our example process, the uncertainty is low around the mode of the distribution, where the majority of normal cases are located, but increases sharply for larger remaining times. It reaches its highest values around 40 days and remains elevated thereafter. This indicates that high-delay cases are not only rare, but also substantially noisier and therefore more difficult to predict than normal cases.

Analysis Across Event Logs. To assess whether these observations are the rule rather than the exception, we analyze 14 publicly available event logs, focusing on the skewness of the remaining time distribution and the degree of heteroscedastic uncertainty.

Table 1 summarizes the characteristics of the event logs. It reports the Fisher moment coefficient of skewness (γ) of the remaining time distribution, which quantifies its asymmetry. Positive skewness indicates a long right tail, with larger values corresponding to stronger asymmetry (for example, $\gamma = 2$ for an exponential distribution). Most logs exhibit pronounced positive skewness, indicating heavy-tailed remaining time distributions where most cases complete within a normal time frame, while a small number experience extremely long durations.

Table 1. Characteristics of the 14 event logs used for evaluation. Case duration is in days. γ denotes the skewness of the remaining-time distribution.

Event log	Cases	Events	Case duration		γ	Event log	Cases	Events	Case duration		γ
			Avg.	Max					Avg.	Max	
P2P	608	9119	21.5	109.4	2.8	Helpdesk	4580	21348	40.9	60.0	−0.1
BPIC17W	30276	240854	12.7	288.9	4.2	Sepsis	1050	15214	28.5	422.3	3.0
BPIC15-1	1199	52217	95.9	1486.0	5.2	BPIC20ID	6449	72151	86.5	742.0	3.5
BPIC15-2	832	44354	160.3	1326.0	3.2	BPIC20DD	10500	56437	11.5	469.2	11.5
BPIC15-3	1409	59681	62.2	1512.0	10.0	BPIC20PTC	2099	18246	36.8	328.2	4.0
BPIC15-4	1053	47293	116.9	927.0	3.7	BPIC20TPD	7065	86581	87.4	1202.0	2.6
BPIC15-5	1156	59083	98.0	1344.0	5.2	BPIC20RFP	6886	36796	12.0	410.0	10.6

To examine how prediction error and predictive uncertainty relate to remaining time, we analyze their association with the ground-truth remaining time y

Table 2. Spearman correlation between ground-truth remaining time y and (i) absolute prediction error $|e|$ and (ii) prediction interval (PI) width produced by the survival model (Surv).

Dataset	$\rho(y, \|e\|)$	$\rho(y, \mathrm{PI}_{\mathrm{Surv}})$	Dataset	$\rho(y, \|e\|)$	$\rho(y, \mathrm{PI}_{\mathrm{Surv}})$
P2P	0.89	0.46	Helpdesk	−0.41	−0.53
BPIC17W	0.38	−0.43	Sepsis	0.51	0.39
BPIC15-1	0.30	0.55	BPIC20ID	0.64	0.80
BPIC15-2	−0.20	−0.22	BPIC20DD	0.47	0.21
BPIC15-3	0.68	0.64	BPIC20PTC	0.56	0.67
BPIC15-4	−0.69	0.09	BPIC20TPD	0.61	0.68
BPIC15-5	0.60	0.32	BPIC20RFP	0.55	0.22

on the test sets (final 20% of cases) of each event log. Specifically, we compute the Spearman correlation between y and two quantities: (i) the absolute prediction error $|\hat{y} - y|$ of a standard LSTM model trained with MAE [16], and (ii) the width of the prediction interval from an uncertainty-aware model based on a survival formulation [7]. Details on this model are provided in Sect. 5; here, it suffices to note that, unlike the standard LSTM, it produces both point estimates and associated uncertainty estimates in the form of prediction intervals.

Table 2 summarizes the correlations. There is a clear tendency toward heteroscedasticity: in 10 out of 14 event logs, both the absolute prediction error and the prediction interval width are positively correlated with the true remaining time. In several logs (e.g., BPIC15-3, BPIC20ID, BPIC20PTC, BPIC20TPD), these correlations are moderate to strong for both measures, indicating that long-running cases are substantially noisier than short-running ones. However, this pattern is not universal. Two datasets (BPIC15-2 and Helpdesk) show negative correlations for both measures, while BPIC17W and BPIC15-4 exhibit mixed behavior. This variation indicates that both the strength and direction of heteroscedasticity differ across processes. Nevertheless, the dominant trend is a positive association between remaining time and both prediction error and uncertainty. Consequently, remaining time prediction is inherently noisy, with certain regions–such as the long tail–being particularly difficult to predict. This noise can arise from a wide range of sources, including resource contention, synchronization effects, batching, task handovers, and system state fluctuations [2,13].

These results indicate clear heteroscedasticity in the right tail of the remaining time distribution. Long-delay cases are not only rare, due to target imbalance, but also substantially more variable and harder to predict. Prediction difficulty in the tail, therefore, arises from both data scarcity and increased noise.

3 Problem Setting and Experimental Setup

In this section, we define the remaining time prediction task and describe the experimental setup used to evaluate the proposed approaches.

Problem Definition. We formalize remaining time prediction as a supervised learning task based on event log data.
Event Log, Traces, Event Prefixes. Process execution data are recorded in an event log $\mathcal{E}$, which is a collection of traces. Each trace $\sigma = \langle e_1, e_2, ..., e_n \rangle$ represents the chronological sequence of events for a single process instance (case). An event is a tuple $e = (c, a, t, \Delta)$, where c is the case identifier, a the executed activity, t the timestamp, and Δ an attribute-value map with optional data payload. We define the projection $\pi_T(e) = t$ to extract an event's timestamp. A prefix of length $m \in [1, n-1]$ is denoted $\sigma^m = \langle e_1, \ldots, e_m \rangle$ and represents partial execution up to the m-th event.
Feature Extraction. Remaining time prediction is formulated as a regression task, starting with feature extraction. Event prefixes of varying lengths are encoded into feature vectors $\mathbf{x} = \Gamma(\sigma^m)$, and target values $y = \pi_T(e_n) - \pi_T(e_m)$. This yields the training dataset $\mathcal{D} = \{(\mathbf{x}_i, y_i)\}_{i=1}^N$, where $\mathbf{x}_i \in \mathbb{R}^d$ denotes the input and $y_i \in \mathbb{R}$ the target.
Problem Statement. Given the training dataset $\mathcal{D}$, the goal is to approximate the function $y = f(\mathbf{x})$ that maps inputs to remaining time. The predictive model typically follows an encoderhead structure: an encoder $\mathbf{z} = g(\mathbf{x}, \theta)$ produces a latent representation $\mathbf{z}$, which is mapped to the predicted remaining time $\hat{y} = h(\mathbf{z}, \psi)$. The parameters θ and ψ are optimized via supervised learning to minimize a regression loss:

$$\mathcal{L}(\theta, \psi) = \begin{cases} \frac{1}{N} \sum_{i=1}^{N} w_i \, |\hat{y}_i - y_i|, & \text{if absolute error is used,} \\ \frac{1}{N} \sum_{i=1}^{N} w_i \, (\hat{y}_i - y_i)^2, & \text{if squared error is used.} \end{cases} \quad (1)$$

Note that standard loss functions (e.g., MAE, MSE) assign equal weight ($w_i = 1$) to all training examples, biasing models toward frequent target values.

Experimental Setup. For the experiments conducted in the subsequent sections, we will use the following event logs, models, and metrics. The event logs, implementation details, and configurations are available in our repository.[1]
Event Logs and Data Split. We use 14 event logs for evaluation to ensure robust and generalizable results (see Table 1). We apply a temporal hold-out split, dividing each log into 64% training, 16% validation, and 20% test data.
Model. We use a data-aware LSTM model proposed by Navarin et al. [16], which predicts remaining time from event prefixes encoded as sequences of feature vectors (control-flow, temporal, and all data attributes), left-padded to a fixed length. The model consists of two LSTM layers (hidden size 150) with layer normalization and dropout (0.1), followed by a dense linear regression head that

[1] https://zenodo.org/records/20433645.

maps the final hidden representation to the predicted remaining time. Training uses AdamW, batch size 128, early stopping (patience 30), and up to 300 epochs.
Hyperparameter Tuning. Hyperparameters are tuned on the validation set using Bayesian optimization. We define approach-specific search spaces, while keeping the model architecture fixed (see details in the supplementary repository).
Evaluation Metrics. To analyze prediction difficulty across target regions, we partition test samples into many/medium/few groups using quantile-based splits. Thresholds are computed from the training and validation targets. By default, the few-shot region corresponds to the highest 10% of target values, the medium region to the next 30%, and the many-shot region to the remaining 60%. To enable comparisons across datasets with different cycle time scales, we use a normalized mean absolute error (nMAE) [21]. For a region r, it is defined as

$$\text{nMAE}_r = \frac{\frac{1}{|r|}\sum_{i\in r}|y_i - \hat{y}_i|}{\frac{1}{N}\sum_{i=1}^{N}|y_i - \text{median}(y_{\text{train}})|},$$

where y_i and $\hat{y}_i$ denote true and predicted remaining times, $|r|$ is the number of samples in region r, and N is the total number of test samples. The denominator corresponds to the mean absolute deviation from the training-set median, i.e., the error of a constant median predictor. This normalizes the MAE relative to a trivial baseline, enabling assessment of prediction difficulty and model performance across regions and datasets on a common scale. Consequently, an nMAE of 1 reflects performance comparable to this baseline, values below 1 indicate improved predictions, and values above 1 indicate worse performance. This interpretation enables direct comparison across many-, medium-, and few-shot regions.

4 Assessing the Effectiveness of Imbalanced Regression

The remaining time distribution is often highly skewed, with only a small fraction of cases exhibiting large delays. As a result, standard machine learning models tend to perform poorly on these rare cases, which naturally leads to the hypothesis that *target imbalance* is a key source of the problem. In this section, we therefore evaluate imbalanced regression approaches designed to address this issue, which have shown promising results in time series forecasting, computer vision, and natural language processing [4,19,22,28].

4.1 Data-Level Approaches: Resampling the Training Distribution

We describe data-level approaches for imbalanced regression and evaluate their effectiveness for remaining time prediction.

Approach. Data-level approaches mitigate imbalance via resampling the training data, either by *undersampling* frequent target ranges or *oversampling* rare ones. Oversampling methods generate synthetic samples in sparse regions of the target space, e.g., by interpolating between neighboring observations as in

SMOTER [26], or by perturbing rare samples with Gaussian noise [3]. Hybrid methods such as SMOGN combine both strategies by undersampling frequent regions and oversampling rare targets through interpolation and noise injection [4].

Applying data-level approaches to PPM is non-trivial. Remaining time prediction operates on event prefixes encoded as left-padded sequences, where the meaning of an input depends on both feature values and their position. Naive interpolation can produce unrealistic samples by mixing padding with valid events or combining prefixes of different lengths, distorting the sequential structure and violating admissible process behavior.

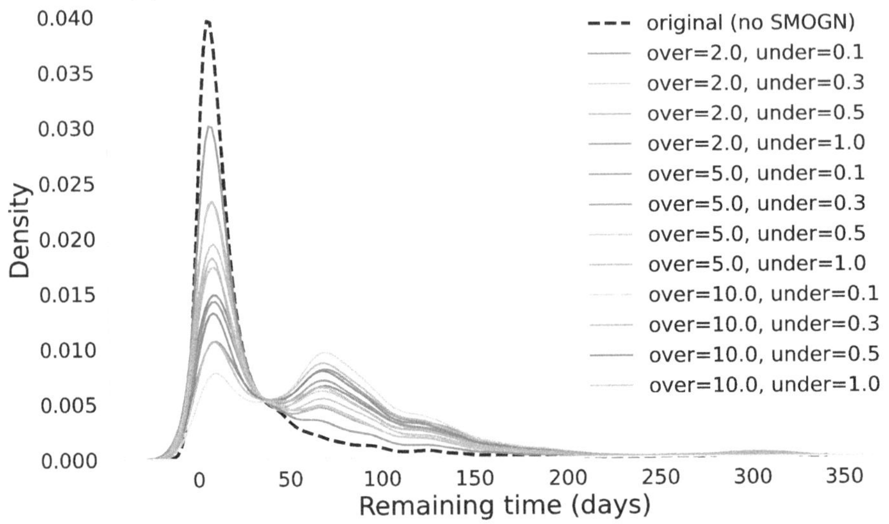

Fig. 2. BPIC20PTC target distribution for various configurations of SMOGN.

We adapt SMOGN to this setting as a feature-space augmentation procedure applied after prefix encoding. Rare samples are identified using a relevance function derived from the training target distribution, assigning higher importance to large remaining times. To avoid mixing prefixes of different temporal extent, we first group samples by their actual prefix length, computed from the non-padded part of the left-padded input sequence. Nearest neighbors are computed only within these prefix-length groups. For each prefix, the padded rows are discarded, the remaining encoded event-level feature vectors are flattened, and Euclidean distance is used to select neighbors. Thus, padding positions do not influence neighbor selection. Synthetic samples are generated in the encoded numerical input space, rather than from raw trace symbols or learned embedding vectors. For interpolation-based oversampling, we linearly interpolate the non-padded feature vectors of a rare prefix and one of its neighbors from the same prefix-length group; the padded part remains unchanged. The remaining-time target is interpolated analogously. Alternatively, Gaussian noise is added only to the non-padded part of rare prefixes, scaled by the empirical feature-wise standard deviation within the group. Frequent samples are reduced via random undersampling, while all rare samples are retained.

Experiments. Figure 2 shows how SMOGN reshapes the training distribution for BPIC20PTC under different configurations. We vary the oversampling factor for rare cases and the undersampling ratio for frequent ones. Increasing both shifts probability mass toward higher remaining times, increasing the density of delayed cases. However, this does not improve predictive performance: the MAE remains between 8.1 and 9.4 days across configurations.

We apply SMOGN to all 14 event logs and compare it to the same LSTM trained on the original (imbalanced) data. Averaged over five seeds per log, SMOGN reduces nMAE in the few-shot region by 8.0% but increases error in the medium and many regions by 20.5% and 31.4%, respectively, resulting in a 17.5% higher overall nMAE. Detailed results are provided in the supplementary GitHub repository.

Overall, even substantial changes to the training distribution do not improve predictive performance, suggesting that data scarcity in the tail is not the primary bottleneck. We therefore turn to algorithm-level approaches.

4.2 Algorithm-Level Approaches: Reweighting the Loss Function

Algorithm-level approaches mitigate imbalance by modifying the regression objective in Eq. 1 to assign greater weight to underrepresented target regions. We first introduce these approaches and then evaluate their effectiveness for remaining time prediction.

Approaches. We consider cost-sensitive weighting, distribution-balanced objectives, error-aware losses, and relevance-based error measures, and refer to the original publications for details.

Cost-Sensitive Weighting (CSW). CSW assigns static loss weights to different target ranges based on the empirical target distribution. The continuous target domain is discretized into bins, and their frequencies are estimated from the training data. Samples in less frequent bins receive higher weights, while those in dense regions are down-weighted. In practice, weights are set inversely proportional to the bin frequency or its square root [28].

Balanced Mean Squared Error (BMSE). BMSE incorporates the target distribution directly into the regression objective. We use the batch-based Monte Carlo (BMC) variant proposed in [19]. Instead of applying static weights, BMC contrasts the prediction error of a sample with the errors obtained for other target values within the same mini-batch. This is implemented through a softmax-style normalization over squared errors across batch targets. Consequently, prediction errors associated with dense target regions receive relatively less influence than those in sparse regions.

Error-Aware Loss (EAL). EAL adjusts sample contributions based on prediction error magnitude. We consider the Focal-R loss [28], which modulates the error by a nonlinear function that increases with its magnitude. This reduces the contribution of small errors while increasing the relative influence of larger errors. As a result, the optimization focuses more on difficult examples, which are often associated with rare or extreme target values.

Squared Error Relevance Area (SERA). SERA [22] is a relevance-based loss for imbalanced regression. It relies on a relevance function $\phi : Y \rightarrow [0, 1]$ that assigns each target value an importance score, allowing non-uniform preference over the continuous target domain [3,25]. In this work, ϕ is automatically derived from the training target distribution using box-plot statistics [20], assigning higher relevance to extreme remaining time values. Consequently, prediction errors on highly relevant targets receive greater weight, while errors on low-relevance targets contribute less to the overall loss.

Table 3. Comparison of algorithm-level approaches using average ranks (lower is better) and significant pairwise differences identified by the Wilcoxon signed-rank test. Results are reported for nMAE across target regions.

Mode	Average Ranks (lower is better)	Significant Performance Differences (Wilcoxon signed-rank test $p-values$)
Many	Vanilla (1.64), CSW (2.14), EAL (2.36), BMSE (3.86), SERA (5.00)	Vanilla > SERA (0.001); CSW > SERA (0.006); EAL > SERA (0.015)
Medium	CSW (2.21), BMSE (2.36), EAL (2.64), Vanilla (2.86), SERA (4.93)	CSW > SERA (0.012); BMSE > SERA (0.020)
Few	SERA (1.00), BMSE (2.14), CSW (3.57), EAL (3.79), Vanilla (4.50)	Vanilla vs BMSE (0.042); SERA > Vanilla (0.000); SERA > CSW (0.020); SERA > EAL (0.009)
All	EAL (1.93), CSW (2.00), Vanilla (2.36), BMSE (3.71), SERA (5.00)	Vanilla > SERA (0.015); CSW > SERA (0.004); EAL > SERA (0.003)

Experiments. We compare the above approaches to a baseline trained with standard MAE (Vanilla), using nMAE across different target regions averaged over five random seeds. A Friedman test reveals significant differences across approaches ($p < 0.001$ for all regions: all, many, medium, few), motivating a more detailed comparison. In Table 3, approaches are ranked per dataset by nMAE within each region and averaged across datasets, while significant pairwise differences are identified using the Wilcoxon signed-rank test.

In the many-shot region, Vanilla attains the best average rank (1.64), while SERA performs significantly worse than Vanilla, CSW, and EAL. In the medium-shot region, results are comparable, with SERA again underperforming relative to CSW and BMSE. A different pattern emerges in the few-shot region, where SERA achieves the best rank (1.00) and significantly outperforms Vanilla, CSW, and EAL; BMSE ranks second and also improves over Vanilla. However, these gains do not translate to overall performance: SERA performs significantly worse overall, while EAL and CSW achieve the best ranks.

The boxplots in Fig. 3 show nMAE distributions across datasets for each approach. SERA is omitted due to its substantially larger errors in the many- and medium-shot regions and overall. In the many-shot region, Vanilla, CSW, and EAL exhibit nearly identical medians and similar variability, while BMSE shows a slightly higher median and greater spread. A similar pattern appears in the medium-shot region, with increased variability and higher errors across all approaches. In contrast, the few-shot region exhibits substantially higher variability across datasets and consistently large errors, with nMAE often exceeding 1. Since nMAE is normalized by the error of a constant median predictor, values above 1 indicate performance worse than this trivial baseline. This highlights the difficulty and instability of prediction in the long-tail region.

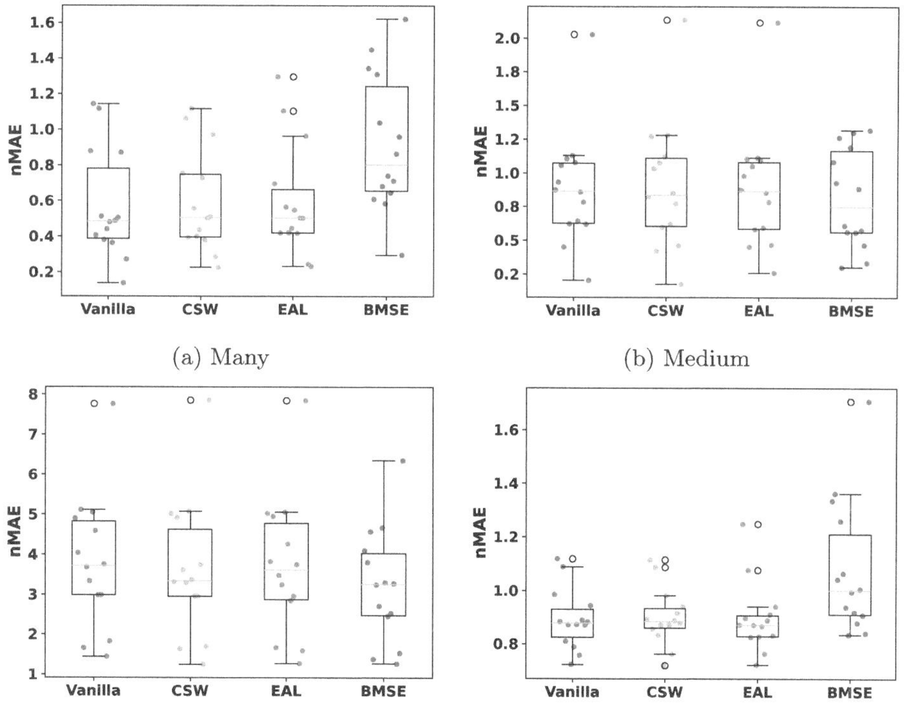

(a) Many (b) Medium

(c) Few (d) All

Fig. 3. Distribution of nMAE across datasets for algorithm-level imbalanced regression approaches, analyzed across different target-frequency regions.

We also evaluate feature and label distribution smoothing [28] in combination with these approaches, but observe only marginal and inconsistent effects (see supplementary repository).

Overall, algorithm-level approaches do not consistently improve prediction accuracy in the long tail. When improvements occur (e.g., SERA), they come at the cost of worse performance in the many-shot region and overall accuracy. Thus, if the goal is solely to identify high-delay cases, SERA may be effective. However, in most practical settings, models are expected to perform well across all cases.

5 Assessing the Effectiveness of Uncertainty Modeling

Building on the observation from Sect. 2 that heteroscedastic uncertainty increases with remaining time and is particularly pronounced in the long tail of the distribution, we investigate whether this relationship can be exploited to improve delay detection. To this end, we turn the previously used deterministic LSTM into a probabilistic model via survival analysis, enabling it to produce explicit uncertainty estimates. These estimates are then used as input to a downstream classifier that predicts whether a case will exceed a given delay threshold. This reformulation is not intended to replace remaining time regression, but rather to study whether uncertainty estimates can help distinguish highly delayed cases from normal executions. We first introduce the uncertainty modeling framework and then evaluate its utility for delay detection.

Modeling Uncertainty via Survival Analysis. To capture predictive uncertainty, we model remaining time using a discrete-time survival formulation, treating process completion as a time-to-event problem [7]. Given an input feature vector $\mathbf{x}$, instead of predicting a single remaining time value $\hat{y}$, the model estimates a distribution over future completion times. Concretely, the remaining time horizon is discretized into K ordered intervals using quantile-based binning of the training targets, where K is treated as a tunable hyperparameter. For each interval $k \in \{1, \ldots, K\}$, the model predicts a hazard value $h_k(\mathbf{x})$, corresponding to the conditional probability that the case completes in interval k given that it has not completed earlier.

The predictive model uses the same encoder as the Vanilla LSTM model to produce a latent representation $\mathbf{z} = g(\mathbf{x}, \theta)$, which is mapped to K hazard logits via a dedicated output layer. These logits are transformed into probabilities representing the discrete hazard function. Training minimizes the negative log-likelihood of the observed event time under this formulation.

From the predicted hazards, the full discrete remaining time distribution is reconstructed via the corresponding survival and event probabilities. This allows deriving both point estimates and predictive uncertainty measures. In our implementation, we extract summary statistics including the expected remaining time and prediction intervals directly from this distribution. Heteroscedastic uncertainty is quantified by the width of the central 80% prediction interval. Further implementation details are provided in the accompanying repository.

Delay Detection via Uncertainty Modeling. To evaluate the utility of uncertainty estimates, we formulate delay detection as a binary classification task. A case is labeled as delayed if its total duration exceeds a predefined quantile threshold (e.g., $q = 0.8$) of the cycle times in the training and validation sets. The goal is to predict this label for ongoing cases based on their observed prefixes.

As illustrated in Fig. 4, we develop an *uncertainty-aware model* that uses the outputs of the previously described survival model. Specifically, we extract distributional summaries (mean and median remaining time), uncertainty measures (prediction standard deviation and the widths of 80% and 90% prediction intervals), tail mass (i.e., the probability that the remaining time exceeds the modeled horizon), and temporal context features (elapsed time since the start of the case and time since the last event). These features are used as input to a gradient boosting classifier (CatBoost). This design enables the model to incorporate uncertainty signals about future completion times, rather than relying solely on prefix features.

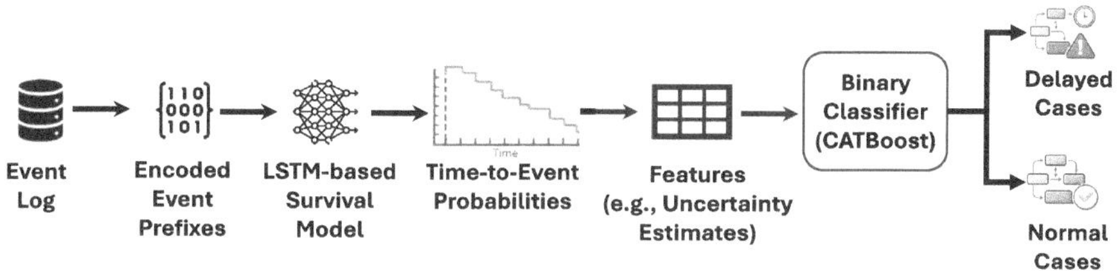

Fig. 4. Uncertainty-aware delay detection pipeline using survival-based predictions and a tabular classifier.

Results. We evaluate the performance of our uncertainty-aware model in detecting high-delay cases using recall, precision, F1-score, and PR-AUC across all 14 event logs in Table 4. We compare our model against a *baseline*, which shares the same LSTM encoder as the survival model, but—instead of modeling uncertainty—directly produces a point estimate, classifying the prefix either as normal or high-delay. The baseline achieves very low recall and therefore misses most delayed cases. In contrast, the uncertainty-aware model substantially improves recall across all datasets, increasing the average from 0.21 to 0.61. This means that a much larger proportion of delayed cases can be identified in advance, which is essential for timely intervention. For instance, on the BPIC20PTC log, the baseline correctly identifies only 1 out of 175 high-delay cases, whereas the uncertainty-aware model detects 132. A Wilcoxon signed-rank test ($p < 0.0001$) confirms that this improvement is statistically significant.

These gains are achieved while maintaining competitive precision: in most cases, precision remains comparable or improves (e.g., BPIC17W: 0.48 to 0.77), with no statistically significant difference overall ($p = 0.17$). Consequently, the F1-score consistently improves across datasets, reflecting a better balance

Table 4. Delay detection performance (threshold $q = 0.8$). Average Recall, Precision, F1-score, and PR-AUC over five random seeds for baseline and uncertainty-aware models.

Dataset	Baseline Classifier				Uncertainty-Aware Model			
	Recall	Precision	F1-Score	PRAUC	Recall	Precision	F1-Score	PRAUC
P2P	0.11	**0.99**	0.20	0.74	**0.83**	0.95	**0.89**	**0.91**
BPIC17W	0.34	0.48	0.40	0.38	**0.59**	**0.77**	**0.67**	**0.59**
BPIC15-1	0.49	0.34	0.39	0.23	**0.66**	**0.40**	**0.49**	**0.30**
BPIC15-2	0.38	0.11	0.17	0.07	**0.75**	**0.12**	**0.21**	**0.10**
BPIC15-3	0.31	0.39	0.34	0.23	**0.65**	**0.65**	**0.65**	**0.48**
BPIC15-4	0.40	0.04	0.07	0.04	**0.75**	**0.06**	**0.12**	**0.05**
BPIC15-5	0.33	**0.42**	0.36	0.24	**0.53**	0.41	**0.46**	**0.29**
HelpDesk	0.02	**0.32**	0.04	0.21	**0.66**	0.25	**0.36**	**0.24**
Sepsis	0.19	**0.59**	0.27	**0.26**	**0.51**	0.28	**0.36**	0.24
BPIC20ID	0.05	0.04	0.04	0.02	**0.39**	**0.12**	**0.18**	**0.06**
BPIC20DD	0.11	0.52	0.18	0.30	**0.47**	**0.65**	**0.55**	**0.45**
BPIC20PTC	0.01	0.18	0.01	0.06	**0.76**	**0.20**	**0.31**	**0.16**
BPIC20TPD	0.07	0.22	0.11	0.11	**0.46**	**0.28**	**0.35**	**0.18**
BPIC20RFP	0.18	0.73	0.29	0.42	**0.48**	**0.74**	**0.58**	**0.53**
Average	0.21	0.38	0.21	0.23	**0.61**	**0.42**	**0.44**	**0.33**

between precision and recall. Similarly, PR-AUC improves in nearly all cases, indicating improved ranking of delayed cases. These improvements are statistically significant (Wilcoxon test: $p < 0.0001$ for F1-score, $p < 0.0003$ for PR-AUC).

Overall, these results demonstrate that incorporating predictive uncertainty improves delay detection performance by substantially increasing the ability to identify delayed cases without sacrificing precision. This suggests that uncertainty provides actionable information for prioritizing high-risk cases in practice. We observe similar improvements under a stricter delay definition ($q = 0.9$), as reported in the supplementary repository. However, since delay detection abstracts the original regression problem into a threshold-based classification task, these findings should not be interpreted as improvements in remaining time estimation itself.

To further assess how early delayed cases can be identified during process execution, we perform an earliness analysis that evaluates model performance at different stages of a running case. Specifically, each prefix is interpreted as a partial observation of an ongoing process instance, and its length is normalized by the total number of events in that case (i.e., its full execution trace). This allows us to group prefixes according to how far the case has progressed (e.g., early, mid, and late stages), and to evaluate performance on subsets of prefixes corresponding to up to 20%, 40%, 60%, and 80% of the process execution. The uncertainty-aware model consistently outperforms the baseline across all prefix ratios, with the most pronounced gains at early stages. In particular, recall is substantially higher even when only a small portion of the case is observed, while precision remains comparable, resulting in consistently improved F1-scores and PR-AUC. Although the performance gap narrows as more of the case becomes available, the uncertainty-aware model remains superior throughout, indicating that incorporating predictive uncertainty enables earlier and more reliable detection of delayed cases.

6 Prior Work

In this section, we review prior work on imbalanced learning in predictive process monitoring (PPM) and discuss emerging uncertainty-aware approaches for remaining time prediction.

Data imbalance is a recurring challenge across a wide range of PPM tasks [17], including next activity prediction, outcome prediction, and anomaly detection. In these settings, imbalance arises from skewed class distributions, where rare but operationally critical events (e.g., exceptional activities, negative outcomes, or anomalies) are underrepresented. Existing work has addressed imbalance in PPM primarily using techniques from imbalanced classification.

In next activity prediction, cost-sensitive learning and focal-style losses have been proposed to emphasize hard and minority examples, improving performance on infrequent activities [9,11,18]. At the data level, oversampling techniques have been used to rebalance class distributions [15]. In outcome prediction, resampling approaches such as SMOTE and its variants are commonly applied to mitigate class imbalance [14,24]. Similar strategies have also been adopted in domain-specific applications, including loan default and payment delay prediction [23,29,30]. Finally, in process anomaly and deviation detection, where imbalance is inherent, prior work often combines representation learning with biased sampling or under-sampling to better detect rare behaviors [6,8].

Overall, these approaches can be broadly categorized into data-level methods (e.g., resampling [14]), algorithm-level methods (e.g., cost-sensitive learning [11]), and hybrid combinations [30]. However, all existing work focuses on *classification* settings. In contrast, imbalance in *regression* tasks–such as remaining time prediction–has not been systematically addressed, despite the strongly skewed and long-tailed nature of target distributions in this domain.

In parallel, a small body of work has begun to investigate uncertainty-aware modeling for remaining time prediction [2,27]. These approaches aim to quantify predictive uncertainty, which can provide additional information beyond point estimates. However, they do not explicitly address imbalance, nor do they study the interaction between skewed target distributions and predictive uncertainty.

In this work, we address these gaps by systematically studying imbalanced regression in PPM, considering both data-level and algorithm-level approaches, and by investigating uncertainty-aware modeling as a complementary perspective for delay detection.

7 Conclusion

In this paper, we investigated the intrinsic difficulty of delay detection in PPM. Our analysis showed that remaining times in business processes are typically strongly right-skewed, with existing models capturing normal cases well but struggling on the rare and operationally critical high-delay cases. We further found that predictive uncertainty increases with delay magnitude, revealing pronounced heteroscedasticity in the tail of the distribution. Thus, our findings identified *target imbalance* and *noise* as potential sources of difficulty. While established imbalanced regression approaches yielded only limited improvements, leveraging predictive uncertainty substantially enhanced the identification of delayed cases. These findings highlight the long tail of remaining time prediction as a key challenge for PPM and identify uncertainty-aware modeling as a promising avenue for future work.

This work has several limitations that open up avenues for future research. First, while we show that predictive uncertainty can substantially improve delay detection, the underlying sources of this uncertainty are not yet fully disentangled and may stem from multiple factors, such as process variability, missing context, or concept drift. Future work could therefore investigate richer uncertainty modeling approaches, integrate additional contextual information, and study how delay detection methods can be adapted online in changing process environments. Second, reformulating remaining time prediction as a binary delay detection task entails a loss of information and dependence on a threshold choice. A promising direction for future work is therefore to move beyond binary classification and instead model the full conditional distribution of remaining time, enabling delay detection and related problems (e.g., service level agreement violations) to be formulated as downstream decision rules rather than fixed thresholds. Third, our empirical evaluation is based on a set of publicly available event logs and LSTM-based predictive models. Although this reveals consistent patterns, it remains an open question how strongly these findings generalize to other process domains, industrial settings, and other architectures.

References

1. Amiri Elyasi, K., van der Aa, H., Stuckenschmidt, H.: PGTNet: a process graph transformer network for remaining time prediction of business process instances. In: Guizzardi, G., Santoro, F., Mouratidis, H., Soffer, P. (eds.) Advanced Information Systems Engineering, CAiSE 2024. LNCS, vol. 14663, pp. 124–140. Springer, Cham (2024). https://doi.org/10.1007/978-3-031-61057-8_8
2. Amiri Elyasi, K., van der Aa, H., Stuckenschmidt, H.: A simple and calibrated approach for uncertainty-aware remaining time prediction. In: Senderovich, A., Cabanillas, C., Vanderfeesten, I.A., Reijers, H. (eds.) Business Process Management, BPM 2025. LNCS, vol. 16044, pp. 217–234. Springer, Cham (2025). https://doi.org/10.1007/978-3-032-02867-9_14
3. Branco, P., Ribeiro, R.P., Torgo, L.: UBL: an R package for utility-based learning. arXiv preprint arXiv:1604.08079 (2016)

4. Branco, P., Torgo, L., Ribeiro, R.P.: SMOGN: a pre-processing approach for imbalanced regression. In: First International Workshop on Learning with Imbalanced Domains: Theory and Applications, pp. 36–50. PMLR (2017)
5. Francescomarino, D., Ghidini, C., C.: Predictive process monitoring. In: van der Aalst, W.M.P., Carmona, J. (eds) Process Mining Handbook. LNBIP, vol. 448, pp. 320–346. Springer, Cham (2022). https://doi.org/10.1007/978-3-031-08848-3_10
6. Elaziz, E.A.: Deep reinforcement learning for data-efficient weakly supervised business process anomaly detection. J. Big Data **10**(1) (2023)
7. George, B., Seals, S., Aban, I.: Survival analysis and regression models. J. Nucl. Cardiol. **21**(4), 686–694 (2014)
8. Grohs, M., Pfeiffer, P., Rehse, J.R.: Business process deviation prediction: Predicting non-conforming process behavior. In: ICPM, pp. 113–120. IEEE (2023)
9. He, X., Oyamada, R., De Smedt, J.: Tackling multi-class imbalance in next activity prediction with class-balanced focal loss. In: ICPM Workshops (2025)
10. Hennig, M.C., Schmidt, R.: Leveraging temporal graphs for enhancing transformer-based predictive process monitoring. In: Senderovich, A., Cabanillas, C., Vanderfeesten, I., A. Reijers, H. (eds.) Business Process Management, BPM 2025. LNCS, vol. 16044, pp. 291–307. Springer, Cham (2026). https://doi.org/10.1007/978-3-032-02867-9_18
11. Käppel, M., Jablonski, S., Schönig, S.: Cost-sensitive predictive business process monitoring. In: Bellatreche, L., et al. (eds.) New Trends in Database and Information Systems, ADBIS 2021. Communications in Computer and Information Science, vol. 1450, pp. 14–26. Springer, Cham (2022). https://doi.org/10.1007/978-3-030-85082-1_2
12. Kim, S., Kim, H., Park, Y.: Early detection of vessel delays using combined historical and real-time information. J. Oper. Res. Soc. **68**(2) (2017)
13. Kraus, A., Elyasi, K.A., van der Aa, H.: On the use of steady-state detection for process mining: achieving more accurate insights. In: Krogstie, J., Rinderle-Ma, S., Kappel, G., Proper, H.A. (eds.) Advanced Information Systems Engineering, CAiSE 2025. LNCS, vol. 15701, pp. 204–220. Springer, Cham (2025). https://doi.org/10.1007/978-3-031-94569-4_12
14. Márquez-Chamorro, A.E., Resinas, M., Ruiz-Cortés, A., Toro, M.: Run-time prediction of business process indicators using evolutionary decision rules. Exp. Syst. Appl. **87**, 1–14 (2017)
15. Mehdiyev, N., Fettke, P., Evermann, J.: A novel business process prediction model using a deeplearning method. BISE **62**(2), 143–157 (2020)
16. Navarin, N., Vincenzi, B., Polato, M., Sperduti, A.: LSTM networks for data-aware remaining time prediction of business process instances. In: 2017 IEEE Symposium Series on Computational Intelligence (SSCI), pp. 1–7. IEEE (2017)
17. Neu, D.A., Lahann, J., Fettke, P.: A systematic literature review on state-of-the-art deep learning methods for process prediction. Artif. Intell. Rev. **55**(2), 801–827 (2022)
18. Nguyen, A., Chatterjee, S., Weinzierl, S., Schwinn, L., Matzner, M., Eskofier, B.: Time matters: time-aware LSTMs for predictive business process monitoring. In: Leemans, S., Leopold, H. (eds.) ICPM 2020. LNBIP, vol. 406, pp. 112–123. Springer, Cham (2021). https://doi.org/10.1007/978-3-030-72693-5_9
19. Ren, J., Zhang, M., Yu, C., Liu, Z.: Balanced MSE for imbalanced visual regression. In: CVPR, pp. 7926–7935 (2022)
20. Ribeiro, R.P., Moniz, N.: Imbalanced regression and extreme value prediction. Mach. Learn. **109**(9), 1803–1835 (2020)

21. Roider, J., Nguyen, A., Zanca, D., Eskofier, B.M.: Assessing the performance of remaining time prediction methods for business processes. IEEE Access (2024)
22. Silva, A., Ribeiro, R.P., Moniz, N.: Model optimization in imbalanced regression. In: Pascal, P., Ienco, D. (eds.) International Conference on Discovery Science, pp. 3–21. Springer. Cham (2022). https://doi.org/10.1007/978-3-031-18840-4_1
23. Tater, T., Dechu, S., Mani, S., Maurya, C.: Prediction of invoice payment status in account payable business process. In: Pahl, C., Vukovic, M., Yin, J., Yu, Q. (eds.) ICSOC 2018. LNCS, vol. 11236, pp. 165–180. Springer, Cham (2018). https://doi.org/10.1007/978-3-030-03596-9_11
24. Teinemaa, I., Dumas, M., Maggi, F.M., Di Francescomarino, C.: Predictive business process monitoring with structured and unstructured data. In: La Rosa, M., Loos, P., Pastor, O. (eds.) BPM 2016. LNCS, vol. 9850, pp. 401–417. Springer, Cham (2016). https://doi.org/10.1007/978-3-319-45348-4_23
25. Torgo, L., Ribeiro, R.: Utility-based regression. In: Kok, J.N., Koronacki, J., Lopez de Mantaras, R., Matwin, S., Mladenič, D., Skowron, A. (eds.) PKDD 2007. LNCS (LNAI), vol. 4702, pp. 597–604. Springer, Heidelberg (2007). https://doi.org/10.1007/978-3-540-74976-9_63
26. Torgo, L., Ribeiro, R.P., Pfahringer, B., Branco, P.: SMOTE for regression. In: Correia, L., Reis, L.P., Cascalho, J. (eds.) EPIA 2013. LNCS (LNAI), vol. 8154, pp. 378–389. Springer, Heidelberg (2013). https://doi.org/10.1007/978-3-642-40669-0_33
27. Weytjens, H., De Weerdt, J.: Learning uncertainty with artificial neural networks for predictive process monitoring. Appl. Soft Comput. **125**, 109134 (2022)
28. Yang, Y., Zha, K., Chen, Y., Wang, H., Katabi, D.: Delving into deep imbalanced regression. In: ICML, pp. 11842–11851. PMLR (2021)
29. Zeng, S., Melville, P., Lang, C.A., Boier-Martin, I., Murphy, C.: Using predictive analysis to improve invoice-to-cash collection. In: KDD, pp. 1043–1050 (2008)
30. Zhang, X., et al.: Data-driven loan default prediction: a machine learning approach for enhancing business process management. Systems **13**(7), 581 (2025)

AI-Assisted Event Log Generation for Process Mining in Healthcare

Saimir Bala(✉), Zeinab Soleimani, and Mathias Weske

Hasso Plattner Institute, University of Potsdam, Potsdam, Germany
{saimir.bala,zeinab.soleimani,mathias.weske}@hpi.de

Abstract. Process mining in healthcare requires structured event logs, yet extracting such logs from complex clinical databases remains a manual, error-prone task requiring substantial SQL skills and domain expertise. This paper investigates the use of large language models to generate SQL queries for event log generation from healthcare data structured according to OMOP, a common data model in healthcare. We develop an artifact to evaluate four prompting strategies across three state-of-the-art LLMs using real-world healthcare data and six clinical treatment processes of varying schema complexity. Generated queries are compared against expert-written ground truth using five metrics: executability, schema conformance, event log completeness, process fidelity, and prompt sensitivity. Results show that prompting strategy has a stronger impact on performance than model choice, while schema complexity is the primary driver of task difficulty. This paper provides a framework as well as practical guidance for LLM-assisted event log generation for process mining from clinical relational data using SQL, further enabling process mining on healthcare data.

Keywords: event log generation · large language models · SQL generation · MIMIC-IV · process mining · healthcare · OMOP CDM

1 Introduction

Process mining enables organisations to discover, monitor, and improve real processes by analyzing event logs extracted from information systems [1]. Healthcare institutions increasingly apply process mining to analyze clinical workflows [2] such as patient pathways, medication administration, and emergency department flow. A prerequisite for any such analysis is an event log that records process instances, activities, and timestamps. In healthcare, this data often originates from electronic health record (EHR) systems and is stored in complex relational databases or clinical data warehouses, where extracting a usable event log requires formulating multi-table SQL queries that demand both clinical domain knowledge and deep familiarity with the underlying schema.

The adoption of standardised clinical data models, most notably the Observational Medical Outcomes Partnership Common Data Model (OMOP CDM) [17],

F. M. Maggi et al. (Eds.): BPM 2026, LNCS 16936, pp. 259–276, 2027.
https://doi.org/10.1007/978-3-032-37777-7_17

has partially eased this task by addressing the heterogeneity problem. Over 330 databases in more than 30 countries have been mapped to OMOP CDM [19]. Yet, the event log generation problem persists. In particular, generating reference queries in order to retrieve the data from the correct tables requires substantial manual effort and expertise [6]. To date, the dominant approach in event log generation (ELG) in healthcare remains *rule-based* and *manual*. Large language models (LLMs) have shown strong performance on general text-to-SQL tasks. Recently, LLMs have been applied to the generation of event logs with promising results on well-known databases used in process mining [21]. However, clinical databases introduce unique challenges that concern the resolution of concept vocabulary, indirect foreign-key paths through existing data models (such as the OMOP CDM tables), and domain-specific activity semantics. Whether and how LLMs can navigate these challenges remains an open question.

This paper addresses this gap by designing, implementing and evaluating a framework for LLM-driven SQL generation targeting OMOP CDM-structured clinical databases. The resulting artifact takes as input a natural-language clinical process description and an OMOP CDM schema context, and produces executable SQL that extracts an event log that can be used by process mining algorithms. It comprises a three-phase pipeline: an OMOP-CDM-aware Prompt Builder supporting four prompting strategies of increasing informativeness; an SQL execution layer; and a five-metric evaluation harness covering executability, schema conformance, event log completeness, process fidelity, and prompt sensitivity. We instantiate and evaluate the framework on six clinical treatment processes across three state-of-the-art LLMs and four prompt strategies, using real-world data and expert-authored ground-truth SQL as the reference. The framework is designed to be portable to any OMOP CDM-compliant database.

The remainder of the paper is structured as follows. Section 2 reviews related work. Section 3 formulates the problem statement, and derives research questions and design objectives. Section 4 describes the artifact design, including six clinical treatment processes that instantiate the problem on OMOP CDM. Section 5 presents the experimental evaluation. Section 6 concludes the paper.

2 Related Work

2.1 Event Log Generation in Healthcare

Event log generation (ELG) is a prerequisite for process mining, yet in healthcare it remains a manual, expert-intensive activity. While this task has also been referred to as event log extraction [21], we use the terminology defined by Diba et al. [7], who characterize this endeavor (i.e., ELG) as encompassing three phases: extraction, correlation, and abstraction. They identify healthcare as a domain where all three phases present significant challenges due to data volume, heterogeneity, and clinical complexity. Munoz-Gama et al. [15] confirm in a multi-author roadmap that event log construction is among the most persistent obstacles to adopting process mining in clinical settings.

In practice, this challenge is typically addressed by relying on domain experts to manually specify generation logic, often in the form of hand-written SQL queries. Rebuge and Ferreira [20] present an early methodology for process mining in hospital settings, where event logs are constructed through analyst-defined SQL queries over the hospital information system. Cremerius et al. [6] provide detailed reference queries and configurations for extracting event logs from MIMIC-IV. Their work illustrates that even with a structured clinical database, event log construction requires careful selection and combination of multiple tables and event definitions for each analysis scenario. Park et al. [19] propose a query-based method for constructing event logs from the OMOP CDM, covering inpatient, outpatient, and emergency room processes, and validate the approach using real-world data from a Korean tertiary hospital. More recently, Park et al. [18] extend this line of work to object-centric event logs by proposing a two-phase method for transforming OMOP CDM data into OCEL. Further healthcare ELG contributions include standards-based pipelines for FHIR [10] and the German Medical Informatics Initiative Core Data Set [9], where event logs are constructed through predefined transformation rules over standardized healthcare data representations.

2.2 LLMs as SQL Generators

The task of generating SQL from natural language, text-to-SQL, has emerged as a central benchmark in NLP. The Spider dataset [24], comprising 200 complex multi-table databases, introduced a standard benchmark for cross-domain SQL generation. The more recent BIRD benchmark [14] scales this setting to large, noisy databases with domain-specific values, more closely resembling real-world conditions. Experimental results show that even frontier LLMs still achieve limited execution accuracy, underscoring the difficulty of text-to-SQL on realistic database workloads.

Stein Dani et al. [21] provide the first systematic study of LLM-assisted *event log* extraction, evaluating the generation of SQL queries for process mining across several relational databases, including procurement-to-pay, ERP, and public administration systems. They use gold-standard event logs as references and evaluate the generated logs using precision, recall, and F1-score over event matches. Their setting represents the closest existing baseline to our work. However, the evaluated schemas are conventional relational databases which differ from healthcare data, with the latter adding additional complexity in terms of concept identifiers, use of an additional vocabulary layer and domain semantic.

In the broader process mining context, Vidgof et al. [23] discuss opportunities and challenges of applying LLMs across the BPM lifecycle, while Berti and Qafari [5] experiment with prompting strategies for analyzing process mining artifacts using LLMs. More recently, Berti et al. [4] introduce PM-LLM-Benchmark, evaluating the capabilities of LLMs on a range of process mining reasoning tasks. These works focus on process analysis over existing event logs and models, and do not address the problem of generating SQL queries for ELG from relational data.

2.3 Difficulties Associated With Clinical Data Models

The OMOP CDM [11,17] introduces structural properties absent from the generic schemas used in existing text-to-SQL benchmarks and ELG studies. One key problem is that clinical events are distributed across multiple domain tables, requiring multi-table `UNION ALL` patterns rather than simple joins. Furthermore, all clinical concepts are encoded as integer `concept_id` values that must be resolved against standardised vocabularies like SNOMED CT, LOINC, etc. It is a challenge to select the correct concept identifiers for the clinical process under study. Moreover, some tables lack a direct foreign key (FK) to the case identifier. For example, `drug_era` aggregates drug exposures at the person level and carries no `visit_occurrence_id`, requiring indirect joins via `person_id` and date-range overlap. These challenges compound: a single clinical treatment process may require all three simultaneously.

Recent work by Gómez-Noé et al. [8] applies LLMs to healthcare process mining, but does not address these challenges. They use LLMs to extract medication mentions from free-text EHR notes, an NLP task distinct from structured SQL generation against a relational schema. Illueca-Fernández et al. [12] present HealthProcessAI, an LLM-enhanced framework for healthcare process analysis that targets process model interpretation and conformance checking, not event log generation. Naeem et al. [16] propose a multi-level ML framework for correlating biomedical activities, but operate on pre-existing event logs rather than generating them from raw clinical data.

In summary, the related work informs us with three findings: i) healthcare event log generation remains manually intensive despite data standardization; ii)LLMs have demonstrated SQL generation capability on simpler schemas but have not been evaluated on clinical data models; and iii)clinical data models such as OMOP CDM introduce additional structural properties that are not captured yet by standard text-to-SQL approaches.

Only three papers [8,12,16] tackle all three dimensions of ELG, AI/ML, and healthcare, and none target LLM-driven SQL generation for clinical CDM data. This paper fills this gap.

3 Problem, Research Questions, Design Objectives

Given three challenges identified in the related work, we state the problem of generating an event log for process mining from healthcare data as follows. Given a clinical database $\mathcal{D}$ structured according to the OMOP CDM schema and a natural-language description of a clinical process P, generate an SQL query Q such that executing Q on $\mathcal{D}$ yields a valid event log L, described as a table with three columns: `case_id` (process instance identifier), `activity` (human-readable event name), and `timestamp` (event occurrence time). The generated event log L must be compatible with the XES [22] event log format and faithfully represent the clinical activities defined in P. This task is non-trivial due to the complexity of OMOP CDM, and technical knowledge and domain expertise required.

Given the discussed challenges, the following research questions are derived. ***RQ1***. Can LLMs generate executable and schema-conformant SQL for healthcare event log generation from OMOP CDM without manual schema mapping? ***RQ2***. How do prompt strategies of increasing informativeness (naive, zero-shot, schema-aware, few-shot) affect SQL executability and event log quality? ***RQ3***. How does OMOP CDM schema complexity affect LLM performance across different clinical treatment processes?

To build our artifact, we operationalize the research questions into the following four design objectives. **O1.** Generated SQL must be executable on a standard OMOP CDM instance without manual correction. *(→ RQ1)* **O2.** Generated SQL queries must produce output with the required three-column structure (`case_id`, `activity`, `timestamp`). *(→ RQ1)* **O3.** The resulting event log must cover the case population and activity types defined in expert ground-truth SQL, evaluated across clinical treatment processes of varying OMOP CDM schema complexity. *(→ RQ1, RQ3)* **O4.** The framework must quantify result sensitivity across prompt strategies of increasing informativeness, and be replicable with any instruction-following LLM and portable to any OMOP CDM-compliant database. *(→ RQ2)*

4 Artifact Design

Our artifact is a framework that includes a three-phase pipeline, as illustrated in Fig. 1. In the *Prompting* phase, a Prompt Builder is responsible for generating prompts. Its capabilities include the construction of a natural-language request from the OMOP CDM schema and a clinical process description, using four prompting strategies: naive, zero-shot, schema-aware, and few-shot. These prompts are then sent to an LLM via the API. In the *Execution* phase, the SQL query extracted from the LLM API response is run against an OMOP CDM database instance. The result is treated as a candidate event log. In the *Evaluation* phase, the candidate event log and its generating SQL are assessed against five metrics (M1–M5, Sect. 4.4). The assessment of M3–M5 includes a comparison against expert-written ground-truth SQL.

4.1 Clinical Treatment Processes

To instantiate the problem, we selected six clinical treatment processes that cover a range of OMOP-CDM event-log generation task complexities, clinical domains, and data volumes, summarized in Table 1.

T1 spans hospital-level and unit-level events, requiring a two-granularity union of `visit_occurrence`, `visit_detail`, and `procedure_occurrence` via direct FK paths (*Medium*). **T2** is the hardest structurally: `drug_era` carries no `visit_occurrence_id`, so the join to `drug_exposure` must be performed indirectly via `person_id` and date-range overlap (*High*). **T3** requires a four-table union with disease-specific concept filtering to temporally align diagnosis, lab, medication, and observation events under a single visit (*High*). **T4**

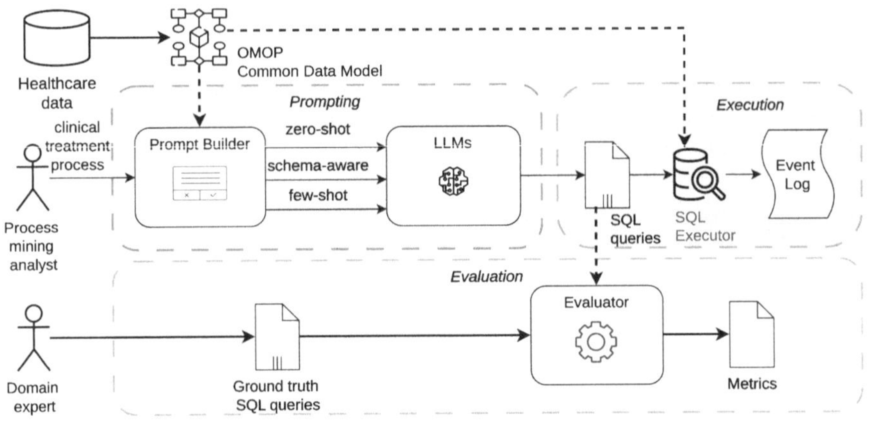

Fig. 1. Approach for LLM-driven SQL generation and evaluation.

presents a heuristic join challenge: `specimen` and `measurement` share no direct FK and must be linked via `person_id` and temporal proximity, risking Cartesian-product explosion (*Medium*). **T5** is the most self-contained: all tables join via `visit_occurrence_id` under a single ED concept filter (*Low*). **T6** was designed as a low-complexity reference: three tables with direct FK paths and a single `concept` vocabulary join to resolve condition names (*Low*). Together, T1–T6 exercise all three structural OMOP CDM challenges (i.e., multi-table composition, indirect joins, and vocabulary resolution).

4.2 Prompt Strategies and LLMs

We evaluate four prompt strategies of increasing informativeness. All prompts (except naive) include the instruction to return *only* the SQL query (no prose, no markdown fences). In all prompts, we ask to produce exactly the three required columns.

Naive. A minimal single-sentence natural-language request with no role framing, no schema hints, no table names, and no concept IDs. We only ask for the three-column (case_id, activity, timestamp) output requirement. Example: *"Generate a PostgreSQL query to extract a process mining event log for [process name], with columns case_id, activity, timestamp"*. This serves as a lower-bound baseline, testing performance in the absence of any clinical or structural context, consistent with standard NLP benchmarking practices [14,21].

Zero-Shot. It provides the clinical process description, the required output schema (`case_id`, `activity`, `timestamp`), and the OMOP CDM table names and relevant concept IDs to use. No column-level schema or foreign-key constraints are supplied. This tests whether process-level guidance alone is sufficient.

Table 1. Clinical treatment processes defined on MIMIC-IV OMOP CDM.

ID	Clinical Treatment Process	Primary OMOP Tables	Complexity
T1	ICU Patient Pathway	`visit_occurrence`, `visit_detail`, `procedure_occurrence`	Medium
T2	Medication Administration	`drug_exposure`, `drug_era`	High
T3	Sepsis Treatment Trajectory	`condition_occurrence`, `measurement`, `drug_exposure`, `observation`	High
T4	Lab-Order-to-Result Cycle	`measurement`, `specimen`	Medium
T5	Emergency Department Flow	`visit_occurrence`, `visit_detail`, `condition_occurrence`, `procedure_occurrence`	Low
T6	Inpatient Diagnosis Pathway	`visit_occurrence`, `condition_occurrence`, `concept`	Low

Schema-Aware. The zero-shot prompt is augmented with the Data Definition Language (DDL) schemas of all relevant OMOP CDM tables: column names, data types, primary keys, and foreign-key relationships. This tests whether explicit structural knowledge improves SQL correctness.

Few-Shot. The schema-aware prompt is further augmented with one worked example from a different clinical treatment process (T6 for T1–T5, and T1 for T6), including the full SQL query and its expected output structure. This tests whether in-context exemplars transfer across processes.

We evaluate three LLMs representing distinct capability tiers and access models. GPT-4o (OpenAI): proprietary frontier model, API access. Claude Sonnet 4 (Anthropic): proprietary frontier model, API access. Llama 3.3 70B (Meta): open-weight model, accessed via Groq API. All models are queried at temperature 0 for reproducibility, with a maximum of 4096 output tokens. Each (clinical treatment process, LLM, strategy) combination is repeated three times. This is sufficient to detect non-determinism at temperature 0 and consistent with prior LLM-based ELG evaluation practice [21]. In total, the combination yields 216 SQL queries (6 treatment processes $\times$ 3 LLMs $\times$ 4 strategies $\times$ 3 repetitions).

4.3 Execution Environment

Each LLM response passes through an SQL extractor that strips prose and markdown fences, saving the extracted query as a standalone `.sql` file. All queries are then executed against the MIMIC-IV OMOP CDM on a PostgreSQL instance (`search_path = cdm`) with a per-query timeout to prevent runaway scans. When a query fails, the SQL Executor classifies the error into six categories: *syntax errors*, *missing-table references*, *missing-column references*, *type mismatches*, *timeouts*, and *ambiguous column references*. This classification is used in the evaluation (Sect. 5.2) to distinguish structural from superficial LLM failures.

Our artifact, including all 216 generated SQL files, ground-truth queries, prompt templates, and annotated execution logs are available in the replication package [3]. The repository contains representative SQL outputs (including complete generated queries for each treatment process, LLM, and strategy, as well as illustrative success and schema-hallucination failure cases) and further steps to convert them to XES event logs, along with initial process mining results. The source code is maintained on GitHub https://github.com/s41m1r/elg-with-ai-experiment.

4.4 Evaluation Metrics

We define five metrics organised into two groups: i) *Execution metrics* (M1–M2, ground-truth independent) that assess the technical correctness of the generated SQL and its conformance to the required output schema, independent of any ground-truth reference; and ii) *Log quality metrics* (M3–M5, ground-truth dependent) that compare the generated event log against a reference log produced by expert-authored ground-truth SQL, assessing the completeness and fidelity of the extracted process data, as well as the sensitivity of results to prompt design.

Let Q denote a generated SQL query, $L = Q(\mathcal{D})$ the event log it produces when executed on the OMOP CDM database $\mathcal{D}$, and $L^* = Q^*(\mathcal{D})$ the reference log from the expert-authored ground-truth query Q^*. Each log row is a triple (*case_id*, *activity*, *timestamp*). We write $C(\cdot)$ for the set of distinct case identifiers in a log, $A(\cdot)$ for the set of distinct activity labels, and $\text{cols}(\cdot)$ for the set of column names in a result set. We define the metrics as follows.

M1 (Executability). $\text{M1}(Q) = 1$ if Q executes successfully on $\mathcal{D}$, and 0 otherwise. Failed queries are further classified by error category as described in Sect. 4.3.

M2 (Schema Conformance). $\text{M2}(Q) = 1$ if $\text{cols}(L) = \{\textit{case\_id}, \textit{activity}, \textit{timestamp}\}$; otherwise 0. This checks whether the query produces the required three-column structure. M2 is assessed only for queries with $\text{M1} = 1$.

M3 (Event Log Completeness) consists of three sub-metrics to capture different facets of coverage. **M3a** (case coverage): the fraction of ground-truth cases recovered, $\text{M3a} = |C(L) \cap C(L^*)| \,/\, |C(L^*)|$. **M3b** (activity type coverage): the fraction of ground-truth activity labels present, $\text{M3b} = |A(L) \cap A(L^*)| \,/\, |A(L^*)|$. For example, if the ground truth produces activities $\{\text{A}, \text{B}, \text{C}\}$ and the LLM produces $\{\text{A}, \text{D}, \text{E}\}$, then $\text{M3b} = 1/3$ even though both logs contain three activity types. **M3c** (volume ratio): the row count ratio, $\text{M3c} = |L| \,/\, |L^*|$, indicating whether the generated log over- or under-produces events.

M4 (Process Fidelity). Activity Jaccard similarity between the activity type sets: M4 $= |A(L) \cap A(L^*)| \,/\, |A(L) \cup A(L^*)|$. In the same example, M4 $= 1/5$ since $|\{A\}|/|\{A, B, C, D, E\}| = 1/5$; M4 is strictly $\leq$ M3b because the union in the denominator penalises spurious labels.

M5 (Prompt Sensitivity). For each (LLM, treatment process) pair, we first average each metric across the three repetitions per strategy to obtain $\bar{m}_{k,s}$ (the mean metric value for strategy s), then compute the standard deviation across the four strategies: $\mathrm{M5}_k = \sigma_{s \in \{S_0, S_1, S_2, S_3\}}(\bar{m}_{k,s}), \quad k \in \{\mathrm{M1},\ \mathrm{M3a},\ \mathrm{M3b},\ \mathrm{M4}\}$. M5 is computed over these four metrics because they are all bounded on a common $[0, 1]$ scale. For M1, the strategy-level mean corresponds to the executability rate, so its standard deviation captures how sensitive executability is to prompt strategy. For M3a, M3b, and M4, the standard deviations capture sensitivity in case coverage, activity-type coverage, and process fidelity, respectively. Their aggregate mean ($\overline{\mathrm{M5}}$) is therefore interpreted as a descriptive summary of prompt sensitivity across bounded evaluation dimensions. M2 is excluded because it is only defined for M1-passing queries, so its sample size varies across strategies and its σ would conflate prompt sensitivity with executability filtering. M3c is excluded because it is unbounded (values $\gg 1$ are common), so a single outlier would dominate the composite and mask meaningful variation in the bounded metrics. A composite $\overline{\mathrm{M5}}$ is the mean of the four individual sensitivities. Lower values indicate robustness to prompt design. For instance, if an LLM achieves M3a scores of 0.0, 0.83, 1.0, 0.89 across the four strategies for a given task, then $\mathrm{M5}_{\mathrm{M3a}} = \sigma(\{0.0, 0.83, 1.0, 0.89\}) \approx 0.46$, reflecting high sensitivity driven by the naive strategy's poor performance.

The introduced metrics are justified as follows: M1 and M2 follow standard text-to-SQL evaluation practice [14,24]; M3a–M3b adapt the case- and activity-level recall decomposition introduced by Stein Dani et al. [21]; M4 applies Jaccard set similarity, standard in process mining [19]; M5 is a prompt sensitivity measure based on cross-strategy standard deviation, introduced in this work.

5 Evaluation

Next, we present the results and a discussion of the research questions.

5.1 Experimental Setup

Our database setting is the following. We use MIMIC-IV [13] mapped to OMOP CDM, hosted on a PostgreSQL instance (schema `cdm`). The full database contains approximately 190,000 patients. Expert ground-truth SQL was validated on the full instance; ground-truth event-log sizes are 1.84M (T1), 30.9M (T2), 44.9M (T3), 215M (T4), 4.30M (T5), and 2.20M (T6) rows. For M3/M4/M5 evaluation, all LLM-generated and ground-truth queries are executed on a stratified 1,038-patient sample (12 OMOP CDM tables) drawn from the full database, covering all six treatment process cohorts with partial overlap: 400 ICU patients (T1),

350 sepsis patients (T3), 250 ED patients (T5), and 150 inpatient-only patients (T6), yielding 1,038 unique patients after deduplication; T2 and T4 are covered by the ICU/sepsis cohorts. This sample covers 1,038 of the ~190,000 patients in the full instance (≈0.55%), drawn by stratified cohort sampling to guarantee representation of all six treatment processes. M1 executability is assessed on the full database. The reference (ground-truth) queries for all six treatment processes were validaded by one author with combined SQL and clinical-data-modelling expertise with about two years of experience working with EHR data and knowledge of the MIMIC-IV OMOP CDM instance.

Our execution environment is set up as follows. M1 queries are executed from our Executor module, a Python script that uses `psycopg2` against PostgreSQL with a 60-second timeout per query; no manual correction is applied. M3–M5 are computed using DuckDB in-memory on the 1,038-patient CSV sample, using direct aggregate subqueries (`SELECT DISTINCT`) to avoid materialising Cartesian-product result sets. All queries are extracted from LLM responses by stripping prose and markdown fences.

Our experimental grid is the following. The full grid comprises 6 processes × 3 LLMs × 4 strategies × 3 repetitions = 216 query executions. The six treatment processes cover all three structural OMOP CDM challenges at two complexity levels; the three LLMs represent distinct capability tiers (open-weight vs. proprietary) and API access paradigms; the four strategies span the full spectrum from no context (naive) to full schema and in-context exemplar (few-shot). For reproducibility, the exact API model identifiers are `gpt-4o-2024-08-06` (GPT-4o), `claude-sonnet-4-20250514` (Claude Sonnet 4), and `llama-3.3-70b-versatile` (Llama 3.3 70B, served via Groq).

5.2 Results: Executability and Schema Conformance (M1, M2)

Overall, of the 216 generated queries, 179 execute without error on the full MIMIC-IV OMOP CDM PostgreSQL instance, yielding an overall M1 rate of 83%. Among all 179 executable queries, every result set contains the required `case_id`, `activity`, and `timestamp` columns (**M2 = 1.0 for all executed queries**). The collapse of M1 and M2 is a direct consequence of the prompt instruction mandating the three-column output schema. LLMs that generate executable SQL invariably also satisfy the structural constraint.

Next, we discuss the results by looking at the combination of LLM and prompt strategy. Table 2 reports M1 pass rates for all twelve (3 LLMs × 4 strategies) combinations. Claude Sonnet achieves the highest overall executability (96%), followed by GPT-4o (92%) and Llama 3.3 70B (61%). The few-shot strategy yields the highest overall rate (93%), with schema-aware at 91% and zero-shot at 85%. The naive strategy, as expected, performs worst at 63%: with only generic process-level guidance, LLMs generate queries that target incorrect tables or produce non-event-log structures. Notably, GPT-4o performs well under naive prompting (89%), while Llama 3.3 collapses to 17%, confirming that minimal prompting requires strong prior knowledge of OMOP CDM. Another notable observation is the Llama 3.3 schema-aware performance degradation (72% schema-aware vs. 78% zero-shot).

We also analyze executability by treatment process. Table 3 shows M1 executability pass rates per treatment process ELG task. T5 (ED Flow) is the easiest at 94%, with T1 (ICU) and T3 (Sepsis) close behind at 92%. T6 (Inpatient Diagnosis Pathway), the new low-complexity task, achieves 89% overall, validating its design intent. T2 (Medication Administration) remains the hardest at 64%, driven by consistent LLM hallucination of a non-existent `visit_occurrence_id` foreign key in the `drug_era` table. T4 (Lab-Order Cycle) at 67% reflects continued difficulty with the heuristic specimen-to-measurement join.

Table 2. M1 Executability (%): LLM and prompt strategy. Cell: pass/total (18 runs).

LLM	Naive	Zero-shot	Schema-aware	Few-shot	Overall
GPT-4o	89% (16/18)	78% (14/18)	100% (18/18)	100% (18/18)	92% (66/72)
Claude	83% (15/18)	100% (18/18)	100% (18/18)	100% (18/18)	96% (69/72)
Llama	17% (3/18)	78% (14/18)	72% (13/18)	78% (14/18)	61% (44/72)
Overall	**63%** (34/54)	**85%** (46/54)	**91%** (49/54)	**93%** (50/54)	**83%** (179/216)

Table 3. M1 Executability (%) by clinical treatment process (36 runs each).

Task	Process	M1 Pass	Rate
T1	ICU Pathway	33/36	92%
T2	Medication Admin.	23/36	64%
T3	Sepsis Trajectory	33/36	92%
T4	Lab-Order Cycle	24/36	67%
T5	ED Flow	34/36	94%
T6	Diagnosis Pathway	32/36	89%
Total		179/216	83%

The *error categories* of the 37 (= 216 − 179) failed queries fall into two main groups: (i) *schema hallucination* and (ii) *syntax errors and other failures*. Schema hallucination affected 20 queries. LLMs referenced columns absent from OMOP CDM, most commonly `drug_era.visit_occurrence_id` (T2, all models) and `specimen.visit_occurrence_id` (T4, Llama 3.3). Syntax errors and other failures affected 17 queries. They came largely from naive-strategy Llama 3.3 queries that generated malformed SQL without adequate structural guidance. Notably, no schema hallucinations involve native MIMIC-IV table names: all models correctly use OMOP CDM table names throughout.

5.3 Results: Log Completeness and Process Fidelity (M3, M4)

Case Coverage (M3a). Table 4 reports M3 and M4 results averaged across all six clinical treatment process tasks. Mean case coverage is 66.4% (see overall mean row in Table 4), indicating that LLM-generated queries recover the majority of patient populations defined in the ground-truth SQL. The naive strategy performs poorly on M3a (near 0% for Claude Sonnet and Llama 3.3), confirming that without clinical context the LLMs produce structurally valid but semantically incorrect queries that return wrong patient populations. The zero-shot and schema-aware strategies achieve 100% case coverage for Claude Sonnet, and few-shot achieves it for both GPT-4o and Claude Sonnet. Excluding the naive strategy, mean M3a rises to **87.5%** (versus 66.4% overall), demonstrating that the naive lower bound is the primary drag on the aggregate figure.

Table 4. M3 Completeness and M4 Fidelity results averaged across all six tasks.

LLM	Strategy	M3a (case)	M3b (act.)	M3c (vol.)	M4 (Jacc.)
GPT-4o	Naive	9.5%	0.0%	0.00	0.000
	Zero-shot	83.3%	0.1%	1.48	0.001
	Schema-aware	**100.0%**	0.1%	5.68	0.001
	Few-shot	88.9%	16.7%	5.23	0.089
Claude Sonnet	Naive	0.0%	0.0%	0.26	0.000
	Zero-shot	**100.0%**	16.8%	1.43	0.168
	Schema-aware	93.2%	16.8%	0.85	0.168
	Few-shot	**100.0%**	**17.2%**	1.18	**0.170**
Llama 3.3 70B	Naive	0.0%	0.0%	0.00	0.000
	Zero-shot	77.8%	0.1%	2.12	0.001
	Schema-aware	66.7%	0.1%	1.27	0.001
	Few-shot	77.8%	16.8%	1.43	0.090
Overall mean		**66.4%**	**7.1%**	**1.74**	**0.057**

Activity Label Coverage (M3b), Key Finding. Despite adequate case coverage for structured prompts, *exact* activity label coverage remains near-zero for T1–T5 across most of the combinations (mean M3b = 7.1% averaged over all six tasks, but effectively 0% for T1–T5). The exception is T6, where few-shot GPT-4o and all Claude Sonnet strategies achieve M3b = 1.0, and Llama 3.3 few-shot achieves M3b = 1.0 as well. The T6 result is explained by the fact that T6 explicitly requires a `concept` table join to resolve condition names, a natural-language pattern LLMs reproduce accurately when given structural guidance. For T1–T5, the mismatch arises from *activity label naming conventions*: ground-truth SQL uses OMOP `concept_name` lookups (e.g., `'Inpatient Visit'`, `'Blood pressure'`), while LLMs construct string literals (e.g., `'Hospital Admission'`,

'ICU Transfer'). These are semantically equivalent but lexically disjoint, yielding zero intersection under exact-match comparison. This confirms the label mismatch problem identified by [21], and extends it to the OMOP CDM setting where the underlying cause is controlled vocabulary dependency rather than schema naming conventions. The T6 result provides mechanistic evidence for this distinction.

Volume Ratio (M3c). LLM-generated queries tend to over-produce events relative to the ground truth. Claude Sonnet stays closest to the reference volume (M3c $\approx$ 0.85–1.43 for structured strategies), while GPT-4o schema-aware and few-shot over-produce by a factor of 5–6$\times$, largely driven by under-constrained joins on high-volume tasks (T2, T4) that produce Cartesian-product row inflation.

Process Fidelity (M4). M4 (Jaccard similarity) tracks M3b closely but is strictly lower, because the union denominator penalises spurious activity labels generated by LLMs that do not appear in the ground truth. The overall mean M4 is 0.057, compared to M3b = 7.1%. The gap between the two metrics is most visible for few-shot strategies, where LLMs achieve non-trivial M3b ($\approx$ 17%) but M4 drops to 0.09–0.17 because the LLM-generated queries introduce additional activity labels absent from the ground truth. As with M3b, T6 is the exception: M4 = 1.0 for the same configurations that achieve perfect M3b, confirming that when activity labels match exactly, no spurious labels are introduced.

Normalisation Experiment (M3b, M4). To quantify whether the near-zero M3b and M4 reflect a naming mismatch rather than a structural failure, we conduct a category-level normalisation on T5 (ED Flow). The ground-truth SQL for T5 produces 1,888 distinct activity labels across five categories by concatenating OMOP `concept_name` values (e.g., `'ED Diagnosis: Chest pain'`, `'ED Procedure: CT scan'`). LLM-generated queries produce exactly four category-level labels (`'ED Arrival'`, `'ED Discharge'`, `'ED Diagnosis Recorded'`, `'ED Procedure'`). We define a manual mapping of five rules: `'ED Arrival'` → `'ED Arrival'` (exact), `'ED Discharge'` → `'ED Departure'` (synonym), `'ED Diagnosis Recorded'` → `'ED Diagnosis: *'` (prefix), `'ED Procedure'` → `'ED Procedure: *'` (prefix), and no mapping for the GT-only category `'ED Sub-Stay: *'` (absent from LLM output). Under this category-level comparison, all 27 non-naive T5 queries yield identical exact M3b = 0.053% and normalised M3b = **80%**, a uniform 1,509$\times$ improvement (verifiable as 0.80 / 0.00053 $\approx$ 1,509). The remaining 20% gap is attributable to a single missing event category (`visit_detail` sub-stays), not to incorrect SQL. This demonstrates that the near-zero exact-match scores are an artefact of the evaluation granularity, not a fundamental limitation: category-level normalisation, achievable through a lightweight post-processing mapping, recovers the vast majority of semantic fidelity.

5.4 Results: Prompt Sensitivity (M5)

Table 5 reports M5 prompt sensitivity as the mean σ across M1, M3a, M3b, M4, broken down by task and LLM. With the inclusion of the naive strategy, sensitivity scores increase substantially compared to a three-strategy evaluation, as the naive lower bound widens the spread. Across LLMs, Llama 3.3 is the most sensitive overall ($\overline{\text{M5}} = 0.218$), driven by extreme variability between near-zero naive performance and reasonable structured-strategy results. GPT-4o is moderately sensitive ($\overline{\text{M5}} = 0.211$), while Claude Sonnet is the most consistent ($\overline{\text{M5}} = 0.188$), reflecting stronger robustness to prompt design. Across tasks, T6 (Inpatient Diagnosis Pathway) is the highest-sensitivity task for every model ($\overline{\text{M5}} = 0.382$): few-shot achieves M3b = 1.0 while naive achieves 0.0, creating a near-maximal spread. Conversely, T5 (ED Flow) is the most stable task, with identical sensitivity across all three LLMs ($\overline{\text{M5}} = 0.125$), indicating that the prompt strategy has a uniform and limited effect on performance for this task regardless of model choice.

Table 5. M5 Prompt Sensitivity per task and LLM: mean σ across M1, M3a, M3b, M4 (higher = more sensitive to prompt strategy choice). Bold = highest sensitivity per column.

Treatment Process	GPT-4o	Claude Sonnet	Llama 3.3 70B	Mean
T1 ICU Pathway	0.210	0.252	0.238	0.233
T2 Medication Admin.	0.268	0.125	0.096	0.163
T3 Sepsis Trajectory	0.167	0.118	0.250	0.178
T4 Lab-Order-to-Result	0.167	0.130	0.160	0.152
T5 ED Flow	0.125	0.125	0.125	0.125
T6 Inpatient Diagnosis	**0.331**	**0.375**	**0.441**	**0.382**
Mean	**0.211**	**0.188**	**0.218**	**0.206**

5.5 Discussion of Results Vs. Research Questions

Next, we discuss to what extent our results address each of the research questions.

RQ1: Can LLMs Generate Executable SQL for Healthcare ELG from OMOP CDM? Yes, with important caveats. Frontier proprietary models achieve up to 96% executability (Claude Sonnet) across all four prompt strategies; GPT-4o reaches 92% overall. The primary failure mode is schema hallucination (Sect. 5.2), consistent and model-agnostic in zero-shot settings. Open-weight models (Llama 3.3 70B, 61%) lag considerably and are less suitable for direct deployment without fine-tuning, particularly with naive prompting (17%).

Regarding event log quality: LLMs reliably identify the correct patient populations for structured prompts (M3a = 66.4% mean case coverage). Notably,

T6 demonstrates that low-complexity processes with explicit vocabulary joins are fully solvable, while the near-zero M3b/M4 for T1–T5 reflects activity label naming conventions rather than a fundamental limitation.

The near-zero M3b/M4 for T1–T5 is not addressed by the four prompt strategies evaluated. Schema-aware prompting conveys table structure but not vocabulary content: the OMOP `concept` table contains over 40 million concepts, and no static prompt can enumerate the task-relevant subset. A *vocabulary-aware* strategy, dynamically retrieving and injecting the relevant `concept_name` values, was deliberately excluded for two reasons. First, the four strategies evaluated represent the current standard in text-to-SQL benchmarking [14,21] and the primary goal was to establish a reproducible baseline. Second, healthcare vocabularies are high-volume databases requiring a dedicated experimental setting with higher memory and compute. Vocabulary-aware prompting is the most direct intervention and will be addressed in future work.

RQ2: How Do Prompt Strategies Affect SQL Quality? The naive strategy performs significantly worse than all structured strategies (63% M1 overall vs. 85–93%), confirming that clinical context is essential. Among structured strategies, few-shot achieves the highest rate (93%), with schema-aware close behind (91%). The few-shot strategy is the most reliable for frontier models; unexpectedly, Llama 3.3 degrades under schema injection (72% schema-aware vs. 78% zero-shot), suggesting that long DDL schemas exceed the effective instruction-following capacity of open-weight models. M5 sensitivity analysis (Table 5) confirms that the naive lower bound substantially increases variance; Claude Sonnet is the most stable model and Llama 3.3 the most sensitive. Schema-aware prompting is the recommended baseline; few-shot adds value for frontier models on complex tasks.

RQ3: How Does Schema Complexity Affect Performance? Complexity remains the primary driver of task difficulty. T5, T1, and T3 (low to high complexity but with direct FK paths) achieve the highest executability (92–94%), while T6 (low complexity) follows closely at 89%. T2 and T4 are the hardest: T2 due to the indirect `drug_era` join, T4 due to the heuristic specimen-to-measurement linkage. The T6 result provides a useful contrast: low-complexity processes with explicit concept joins are fully solvable by current frontier LLMs. These structural OMOP CDM properties (absent from Spider [24] and BIRD [14] benchmarks) are a distinct generalisation challenge for text-to-SQL models.

These findings have several implications. Our results show that LLMs perform well on SQL code generation. Non technical users, such as doctors, may now have access to event logs that can readily be used by process mining tools. Process mining practitioners may use our framework to explore datasets from different domains (such as healthcare). Research on event log generation, may use our framework as a systematic benchmark of LLM driven ELG from OMOP data.

Threats to Validity. *Internal*: M3/M4/M5 are computed on a stratified 1,038-patient sample; while larger than typical prototyping datasets, this may not capture the full distribution of clinical cases. The ground-truth SQL is authored by a single expert, so the reference logs encode one set of modelling choices on which experts could reasonably differ, and which propagate into M3–M5. For example, the T3 sepsis cohort uses a hand-curated 129-concept list; a different expert curating a narrower or broader concept set would change the reference case population $C(L^*)$ and thus the M3–M5 scores measured against it. Multi-expert ground truth, with agreement measured on the resulting cohort and activity definitions, would strengthen these claims. Activity label comparison uses exact string matching; fuzzy matching or embedding-based similarity would yield higher M3b/M4 and would be more informative. *External*: Three LLMs are evaluated at a single point in time; model updates may alter results. The six clinical treatment processes cover the most common OMOP domain tables but not `device_exposure`, `note`, or `episode`. *Construct*: M3b/M4 near-zero for T1–T5 may underestimate true semantic fidelity: LLM activity labels are clinically meaningful, just differently named. Normalisation could substantially improve these metrics.

Other Limitations. Beyond the threats discussed above, the proposed framework has a few practical limitations. The framework requires hospital data to be pre-mapped to OMOP CDM, which remains a significant ETL barrier for the majority of clinical institutions worldwide; results may further vary across OMOP instances where concept vocabulary mapping quality differs. General-purpose LLMs were evaluated throughout. Whether medically fine-tuned models yield better concept resolution on OMOP schemas remains an open question.

6 Conclusion

This paper presents a systematic benchmark of LLM-driven SQL generation for clinical ELG on the OMOP CDM. Across six clinical treatment processes, three LLMs, and four prompting strategies, frontier proprietary models achieve high SQL executability and medium-to-high case coverage. Low-complexity processes with direct vocabulary joins enable near-perfect event log generation, while complex tasks expose an activity labeling gap addressable through category-level normalization. OMOP CDM schema complexity is the primary determinant of task difficulty, largely independent of model choice; naive prompting confirms the importance of contextual information.

These findings provide practical guidance for practitioners. Schema information is essential at prompting time; zero-shot is reliable only for low-complexity processes with clear foreign key paths; frontier models perform best with few-shot prompting. Our framework is portable across OMOP CDM instances and the three-phase ELG approach generalizes beyond healthcare. Future work will address activity label normalization via OMOP vocabulary alignment, and object-centric log generation for processes such as concurrent ICU and ED stays.

Acknowledgements. We thank Lars Rückert, David Weese, and Pablo Guerrero from D4L data4life gGmbH for providing access to the MIMIC-IV OMOP CDM data.

References

1. van der Aalst, W.M.P.: Process Mining - Data Science in Action, 2nd edn. Springer, Heidelberg (2016). https://doi.org/10.1007/978-3-662-49851-4
2. Aversano, L., Iammarino, M., Madau, A., Pirlo, G., Semeraro, G.: Process mining applications in healthcare: a systematic literature review. PeerJ Comput. Sci. **11**, e2613 (2025)
3. Bala, S., Soleimani, Z., Weske, M.: AI-Assisted Event Log Extraction for Process Mining in Healthcare (Replication Package), May 2026. https://doi.org/10.5281/zenodo.20425398
4. Berti, A., Kourani, H., van der Aalst, W.M.P.: PM-LLM-Benchmark: evaluating large language models on process mining tasks. In: Delgado, A., Slaats, T. (eds.) Process Mining Workshops, ICPM 2024. LNBIP, vol. 533, pp. 610–623. Springer, Cham (2025). https://doi.org/10.1007/978-3-031-82225-4_45
5. Berti, A., Qafari, M.S.: Leveraging large language models (LLMs) for process mining (technical report). CoRR abs/2307.12701 (2023)
6. Cremerius, J., Pufahl, L., Klessascheck, F., Weske, M.: Event log generation in MIMIC-IV research paper. In: Montali, M., Senderovich, A., Weidlich, M. (eds.) Process Mining Workshops, ICPM 2022. LNBIP, vol. 468, pp. 302–314. Springer, Cham (2023). https://doi.org/10.1007/978-3-031-27815-0_22
7. Diba, K., Batoulis, K., Weidlich, M., Weske, M.: Extraction, correlation, and abstraction of event data for process mining. WIREs Data Min. Knowl. Discov. **10**(3) (2020)
8. Gomez-Noe, A.: Extraction and Identification of Medications from EHR Notes with LLMs: A Tool and a Case Study. In: ICPM Workshops (PODS4H). Springer, Heidelberg (2025)
9. Heidemeyer, H., et al.: A pipeline for the usage of the core data set of the medical informatics initiative for process mining - a technical case report. In: German Medical Data Sciences 2024. Studies in Health Technology and Informatics, vol. 317, pp. 30–37. IOS Press (2024)
10. Helm, E., Krauss, O., Lin, A., Pointner, A., Schuler, A., Küng, J.: Process mining on FHIR - an open standards-based process analytics approach for healthcare. In: Leemans, S., Leopold, H. (eds.) ICPM 2020. LNBIP, vol. 406, pp. 343–355. Springer, Cham (2021). https://doi.org/10.1007/978-3-030-72693-5_26
11. Hripcsak, G., Ryan, P.B. (eds.): The Book of OHDSI. OHDSI (2019)
12. Illueca-Fernandez, E., Chen, K., Seoane, F., Abtahi, F.: HealthProcessAI: a technical framework and proof-of-concept for LLM-enhanced healthcare process mining. Front. Artif. Intell. **9**, 1716819 (2026)
13. Johnson, A.E.W., et al.: MIMIC-IV, a freely accessible electronic health record dataset. Sci. Data **10**, 1 (2023)
14. Li, J., et al.: Can LLM already serve as a database interface? NeurIPS, a big bench for large-scale database grounded text-to-SQLs. In: NeurIPS (2024)
15. Munoz-Gama, J., et al.: Process mining for healthcare: characteristics and challenges. J. Biomed. Inform. **127**, 103994 (2022)

16. Naeem, M.R., Naeem, H., Aamir, M., Ali, W., Abro, W.A.: A multi-level process mining framework for correlating and clustering of biomedical activities using event logs. Int. J. Adv. Comput. Sci. Appl. **8**(3), 393–401 (2017)
17. Overhage, J.M., Ryan, P.B., Reich, C.G., Hartzema, A.G., Stang, P.E.: Validation of a common data model for active safety surveillance research. J. Am. Med. Inform. Assoc. **19**(1), 54–60 (2012)
18. Park, G., Lee, Y., Cho, M.: Enhancing healthcare process analysis through object-centric process mining: transforming OMOP CDM into OCEL. J. Biomed. Inform. **156**, 104682 (2024)
19. Park, K., et al.: Exploring the potential of OMOP common data model for process mining in healthcare. PLoS ONE **18**(1), e0279641 (2023)
20. Rebuge, Á., Ferreira, D.R.: Business process analysis in healthcare environments: a methodology based on process mining. Inf. Syst. **37**(2), 99–116 (2012)
21. Stein Dani, V., et al.: Event log extraction for process mining using large language models. In: Comuzzi, M., Grigori, D., Sellami, M., Zhou, Z. (eds.) Cooperative Information Systems. CoopIS 2024. Lecture Notes in Computer Science, vol. 15506, pp. 56–72. Springer, Cham (2025). https://doi.org/10.1007/978-3-031-81375-7_4
22. Verbeek, H.M.W., Buijs, J.C.A.M., van Dongen, B.F., van der Aalst, W.M.P.: XES, XESame, and ProM 6. In: Soffer, P., Proper, E. (eds.) CAiSE Forum 2010. LNBIP, vol. 72, pp. 60–75. Springer, Heidelberg (2011). https://doi.org/10.1007/978-3-642-17722-4_5
23. Vidgof, M., Bachhofner, S., Mendling, J.: Large language models for business process management: opportunities and challenges. In: Di Francescomarino, C., Burattin, A., Janiesch, C., Sadiq, S. (eds.) Business Process Management Forum, BPM 2023. LNBIP, vol. 490, pp. 107–123. Springer, Cham (2023). https://doi.org/10.1007/978-3-031-41623-1_7
24. Yu, T., et al.: Spider: a large-scale human-labeled dataset for complex and cross-domain semantic parsing and text-to-SQL task. In: EMNLP, pp. 3911–3921. ACL (2018)

HyProTwin: Hybrid Digital Process Twins for Predictive Simulation of Continuous Processes

Janik-Vasily Benzin(✉), Dominik Voigt, Jürgen Mangler, and Stefanie Rinderle-Ma

TUM School of Computation, Information and Technology, Technical University of Munich, Garching, Germany
{janik.benzin,dominik.voigt,juergen.mangler,stefanie.rinderle-ma}@tum.de

Abstract. Business process management has mostly studied discrete processes with clearly identifiable process outcomes, e.g., the package delivery by an online shop. In contrast, continuous processes are characterized by real-time, discrete control of complex, continuous physical dynamics to maintain stable operating conditions, e.g., a sewage water treatment process with constant desirable chemical composition. Digital twins (DTs) of continuous processes provide simulation and prediction capabilities to enable real-time decision making that is key for efficient, proactive operations. However, implementation of DTs is greatly hindered by interoperability, communication, orchestration, and modeling challenges. Moreover, real-time decision making could be enhanced by predictive simulation, i.e., the ability to simulate future what-if scenarios to consider future changes and their impact on predictions. To realize DTs of continuous processes and enable predictive simulation, we propose hybrid digital process twins (HyProTwin) with their architecture and predictive simulation techniques. We demonstrate the feasibility and performance of HyProTwin on three real-world continuous processes.

Keywords: Continuous processes · Predictive simulation · Digital process twins · Time-series forecasting

1 Introduction

The virtual replication of a real-world physical system with similar operational properties and real-time synchronization between the system and its replication results in a digital twin (DT) [5]. The DT paradigm can improve simulation [8], verification [3], operational control [9], and prediction [16]. While the integration of artificial intelligence (AI) and DTs is still at an early stage, it can enhance DTs solutions due to combining the statistical power of AI in anticipating the future [4] with the real-time operational properties of the DT [5,15]. For example, *explainable time series forecasting* (TSF) techniques [28] combined with a DT supports a human operator in real-time decision making [16].

F. M. Maggi et al. (Eds.): BPM 2026, LNCS 16936, pp. 277–294, 2027.
https://doi.org/10.1007/978-3-032-37777-7_18

As DTs are applied in heterogeneous domains such as manufacturing, healthcare, and energy, the physical systems and the technologies used to implement DTs are diverse [5] (cf. "technology" in Table 1). Therefore, the challenges of interoperability (C1) and standardized communications between the physical system and DTs (C2) arise as key challenges to DT implementation [5,15]. In addition, physical systems are typically modularized and orchestrated within larger physical systems, e.g., individual machines in a production facility, necessitating modularization and orchestration also for DTs. Yet, DTs largely remain focused on individual physical systems [17], leaving a gap for orchestrated networks of DTs (C3). As **process technologies** support standardized communications through service-oriented architectures [3], are modular and excel in interoperability as well as process orchestration [26], DTs can significantly benefit from them [13].

Table 1. Comparison of HyProTwin with related research areas.

Area	Process type	Approach	Technology	Scenarios	Time of Scenarios	State of Process
Digital Twin	All	All	Diverse	As-is/what-if	Current	Current
Stochastic Hybrid Systems	Continuous	Model-driven	Formal methods	What-if	Design time	Empty
Suffix Prediction	Discrete	Data-driven	AI	As-is	Current	Current
Short-term Business Process Simulation	Discrete	Model-driven	Process	What-if	Current	Current
Long-term Business Process Simulation	Discrete	Hybrid	Process, AI	What-if	Current	Empty
HyProTwin	Continuous	Hybrid	Process, AI	As-is/what-if	Future	Current

Taking a process perspective on physical systems, we distinguish between *discrete* and *continuous* processes. The discrete type is characterized by a clearly delineated process outcome for a case, e.g., a successfully-treated patient in a hospital, whereas the latter type of process constitutes the continuous, real-time control of physical quantities to maintain stable conditions [29]. Continuous processes often occur in the chemical [6], agriculture [5], and smart spaces domains [16]. Implementing DTs of continuous processes raises challenges C1-C3 [6]. In addition, a modeling challenge (C4) must be overcome: the DT must capture both complex, continuous dynamics and intertwined, discrete control signals, e.g., a thermostat is turned on to maintain a room's temperature. Model-driven approaches that explicitly represent continuous processes are *stochastic hybrid systems* (SHS) [14] (cf. "approach" in 1). However, SHS cannot be used at runtime, since they lack synchronization with the real-world system and suffer from computational intractability.

Recently, DTs for processes have been proposed, including DT composition in smart manufacturing [9], DTs for clinical trials [25], simulation [8], as well as monitoring [12], and a service-oriented digital process twin (ProTwin) architecture that builds on process technology and bridges the design-time modeling power of SHS with the runtime properties of DT [3]. [8,9,12,25] do not support continuous processes, reliability verification, and modeling of continuous dynamics (C4). ProTwin overcomes challenges (C1-C3) through Business Process Modeling and Notation (BPMN) and the *Cloud Process Execution Engine* (CPEE) [26] and offers verification methods for statistical reliability. Yet, a ProTwin suffers from high manual effort for building models of complex, continuous dynamics (C4), e.g., as the operator has to explicitly model non-linear room temperature dynamics as a differential equation system in buildings. In addition, a ProTwin lacks the ability to predict the short-term future behavior of the physical system and how it will react to future changes, i.e., future what-if scenarios that we term *predictive simulation*. Since predictive simulation anticipates the short-term future behavior of the physical system under different scenarios that happen in the future, it is crucial for optimal, real-time decision-making [4], e.g., understanding how future changes to the temperature boundaries affect a thermostat's ability to maintain room temperature. In particular, predictive simulation is distinct from *suffix prediction* [1,32], from short-term *business process simulation* (BPS) [2,22], and from long-term BPS with hybrid approaches [7,20] that enable real-time decision-making on the as-is scenario from the current state and on current what-if scenarios from the current and empty state [2] for discrete processes respectively (cf. Table 1). The contribution is a unique extension and combination of existing DTs with AI, suffix prediction, and BPS techniques to enable predictive simulation for continuous processes and to overcome C1-C4.

To enable predictive simulation for continuous processes, we propose a *hybrid digital process twin* (HyProTwin) that combines the anticipation of prediction with the what-if scenarios of simulation through DT and process technologies and AI. Our proposal transforms and extends ProTwin's architecture to allow predicting future behavior of a continuous process. By taking a *hybrid* approach [4,20] to overcome the modeling challenge C4 and to automate building models of continuous dynamics, HyProTwin integrates current data-driven TSF techniques [28] to capture complex, continuous dynamics with a model-driven approach for the discrete control of physical systems. As a result, our contribution is to (i) apply process technology in DTs of continuous processes that enable prediction ($\mapsto$ C1-C3), (ii) integrate AI and DT for automatically learning continuous dynamics in combination with discrete control ($\mapsto$ C4), and (iii) conceptualizing and enabling predictive simulation through a hybrid approach.

The paper is structured as follows. Section 2 presents the HyProTwin architecture and Sect. 3 conceptualizes predictive simulation in a HyProTwin. Section 4 evaluates HyProTwin with respect to the performance of predictive simulation. Section. 5 discusses related work and Sect. 6 concludes the paper.

2 Hybrid Digital Process Twin Architecture

We illustrate a ProTwin through a running example in Sect. 2.1 and we show how to transform and extend it into a HyProTwin in Sect. 2.2.

2.1 Digital Process Twins

A ProTwin [3] twins a continuous process using process technology (cf. Section 1). The ProTwin in Fig. 1 (b) twins the water tank continuous process (a). Both the ProTwin and water tank serve as the running example in the remaining sections, because formal approaches for continuous processes [14] use the water tank commonly. The continuous process in (a) shall maintain a water level between 10 and 30 litre (l). To that end, the continuous process consists of a microcontroller and a physical device. As the tank receives constant water inflow, a closed valve with a turned-off actuator (mode "Off", denoted in blue in Fig. 1) results in an increasing water level. A sensor measures the water level (denoted in green). If the level exceeds 30l, the microcontroller can turn the device's actuator on such that the valve is open and more water is released from the tank than is flowing in. If the level falls below 10l, the valve is closed again. Despite the running example's apparent simplicity, noise and uncertain delays in sensing and actuating can increase the complexity of the continuous dynamics as depicted in Fig. 1 (e) during the long and usually assumed infinite lifetime of the continuous process.

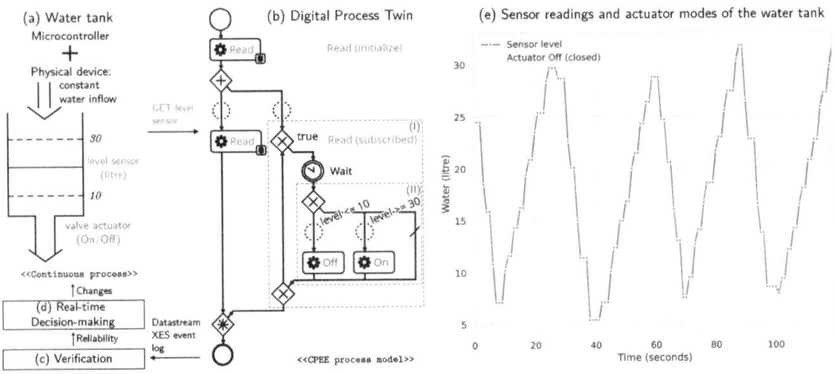

Fig. 1. A digital process twin of a water tank continuous process.

The ProTwin in Fig. 1 (b) is a CPEE process model, which is a serialized Refined Process Structure Tree of a BPMN process model with well-defined execution semantics. As continuous processes always run, the loop (I) that repeats the control logic (II) of the ProTwin is infinite. In every iteration, the current level sensor reading of the `Read` subscription service task is checked with respect to the 10l lower and 30l upper boundary levels. The ProTwin duplicates the `Read` to avoid race conditions. The ProTwin of the water tank continuous process is completed through the twinning of the valve actuator as `Off` and `On` service tasks that can actuate through HTTP put requests [3]. The goal of a ProTwin is to enable *seamless* reliability verification, i.e., reliability verification on the same CPEE process model (b) at design time and runtime. Since the ProTwin in Fig. 1 is connected to the real-world water tank, it enables runtime reliability verification. By replacing the real-world water tank with a manually-built simulation model consisting of differential equations specifying the continuous dynamics and similarly connected to the ProTwin via the same communication interface as in Fig. 1, the ProTwin would enable design time reliability verification.

Using BPMN addresses the interoperability and orchestration challenges (cf. C1 and C3) and can be expanded through extensions, e.g., to the Internet of Things (IoT) [10]. In addition, through loosely-coupled services, reliance on well-established, open standards such as HTTP and REST interfaces, and support for protocols such as *S7*, *OPC UA*, and *MQTT* protocols, the CPEE overcomes the communication challenge (cf. C2) and bridges BPMN's capabilities with respect to interoperability and orchestration into execution [3]. Lastly, ProTwin utilizes the Datastream eXtensible Event Stream (DXES) event log format [19] (cf. Fig. 1 b-c) that supports the systematic and continuous recording of sensor readings in event logs, enabling advanced analysis and mining algorithms [10]. Finally, ProTwins offer techniques for reliability verification and analysis.

However, modeling continuous dynamics in ProTwins is still left to domain experts. Hence, simulation is limited to the fidelity of manually-built simulation models similar to the situation within short-term business process simulation before automated approaches emerged more than 15 years ago [2]. Also, prediction is not supported at all. In the next section, we present HyProTwin as a transformation and extension of ProTwins to overcome limitations with respect to modeling and predictive simulation.

2.2 Requirements and Architecture

In this section, the requirements for a HyProTwin are specified and its resulting architecture is presented. The requirements are stated in the context of (i) a given physical system of type continuous process (cf. Section 1), (ii) a given, corresponding ProTwin (cf. Sect. 2.1), i.e., a CPEE process model twins the discrete control and does not have to be discovered from DXES event logs, and (iii) enabling short-term prediction and predictive simulation. Predictive simulation is introduced in the next section.

R1 Modeling: The continuous dynamics must be automatically learned and forecasted from historical continuous dynamics and discrete control. Continuous dynamics always happen and are defined by $\mathbf{X}(t) \in \mathbb{R}^N$ for real-valued time $t \in \mathbb{R}$ and N sensors. However, sensors measure their physical variables with a sampling rate $r(t)$ Hertz (hz). Thus, sensor readings capture continuous dynamics as a *time series* $\mathbf{X} \in \mathbb{R}^{H \times N}$ [28] of H time steps and N sensors. Following existing time series forecasting (TSF) techniques [23,28], we assume a constant sampling rate r for automatically learning continuous dynamics, i.e., the step width is constant. To capture the historical modes for M actuators, we define an augmented time series $\mathbf{X}^a = [\mathbf{X} \mid \mathbf{A}] \in \mathbb{R}^{H \times N+M}$[1] given a historical time series $\mathbf{X} \in \mathbb{R}^{H \times N}$ and historical actuator modes $\mathbf{A} \in \mathbb{R}^{H \times M}$ of H past time steps. Hence, the solution space to meet R1 is TSF f at time T: Forecast the next F future time steps of

[1] Any discrete control mode can be encoded in $\mathbb{R}$ with preservation of ordinal scales.

sensor readings, i.e., $\mathbf{Y} \in \mathbb{R}^{F \times N}$ by learning a forecasting model given H past time steps of sensor readings augmented with actuator modes $\mathbf{X}^a$. We refer to F and to H as the forecasting horizon and input size respectively. We differentiate between forecasting for anticipating future continuous dynamics and prediction for anticipating the future of the continuous process.

R2 Continuity: During the lifetime of a physical system, both predictive simulation and short-term prediction must be constantly enabled. Continuous processes are constantly running and have a long or theoretically-infinite lifetime (cf. Sect. 2.1). Thus, short-term prediction and predictive simulation results do not inform about the complete future of continuous process such that they must be constantly enabled.

R3 Preservation: The benefits of a ProTwin with respect to implementation challenges C1-C3 must be preserved. We refer to Sect. 2.1 for details.

R4 Scenarios: Future what-if scenarios must be specifiable without retraining. Decision-making in continuous processes must be done in real-time, i.e., strictly meeting early deadlines that may be due in minutes. In addition, an operator may want to specify and run multiple what-if scenarios in parallel. Hence, retraining a forecasting model is, in general, not possible in this setting. Next, we present the HyProTwin architecture.

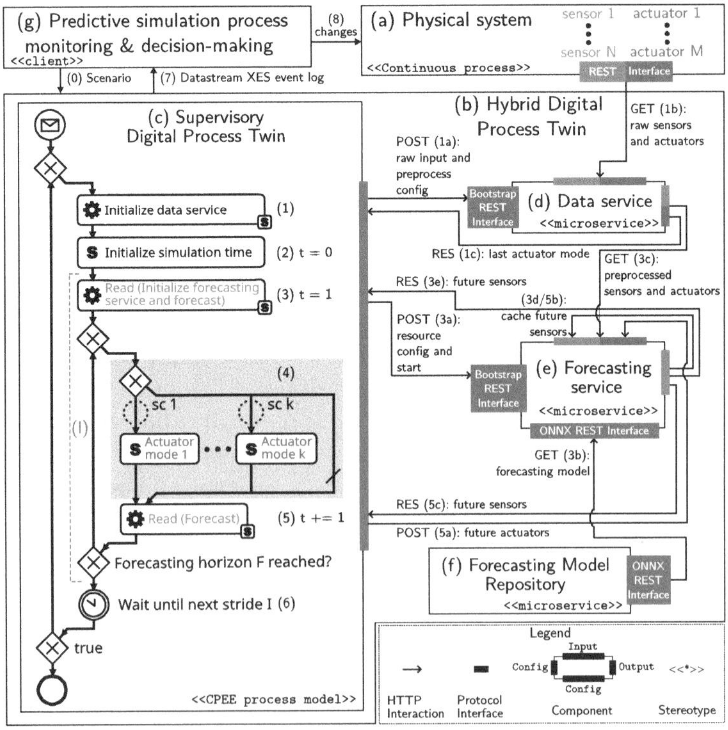

Fig. 2. The hybrid digital process twin architecture for predictive simulation.

Architecture. The HyProTwin architecture is depicted in Fig. 2. The physical system of type continuous process (a) with its N sensors (denoted in green) and M actuators (denoted in blue) is connected to the HyProTwin (b) through a REST interface. A HyProTwin consists of a supervisory ProTwin (c) and three microservices for data (d), forecasting (e), and a forecasting model repository (f). Monitoring through short-term prediction and predictive simulation for real-time decision-making (g) is detailed in the next section. At the core of HyProTwin is the *hard physical constraints* hybrid modeling pattern [27] that also underlies physics-informed neural networks: a TSF technique f (cf. Requirement R1) that forecasts the next sensors (e) is regulated by the discrete control logic of the supervisory ProTwin (c). The *recurrent composition* of the hard physical constraints pattern [27] through the infinite loop of the supervisory ProTwin shapes our proposed hybrid approach to meet the modeling requirement (req.) R1. To meet the continuity req. R2, we introduce an infinite cycle and a *stride* $I \in \mathbb{N}$ to forecast (cf. Fig. 2 I) and regulate with discrete control for F time steps every I time steps (*rolling forecasting* [23]). To meet the preservation req. R3, we transform a ProTwin into a supervisory ProTwin.

ProTwin Transformation. A supervisory ProTwin is resulting from a ProTwin (cf. Fig. 1 b) through replacing the two "Read" service tasks with their forecasting counterparts in Fig. 2 and linearizing the parallel "Read" and control loop of a ProTwin (cf. Fig. 1 I) to align with a constant sampling rate (cf. Fig. 2 I). Furthermore, the transformation embeds the iterative prediction every I steps (cf. the loop in Fig. 2 I with subsequent wait) into the infinite control loop, and adding initializations for the data service (1) and simulation time (2).

Summary. The structure of the architecture is established and is designed to meet the modeling, continuity, and preservation requirements R1 - R3. Because the runtime behavior of executing a HyProTwin results in short-term prediction and predictive simulation, we present the runtime behavior together with the predictive abilities in the next section.

3 Predictive Simulation for Continuous Processes

This section introduces predictive simulation and short-term prediction and shows how both are enabled by the runtime behavior of a HyProTwin. We write $\mathbf{X}_{T_0}$ and $\mathbf{X}_{T_1:T_2}$ for the data that is recorded at T_0 and ranges from T_1 to T_2 respectively. Predictive simulation solves the following problem:

Definition 1 (Predictive simulation problem statement). *Let T be the current time, $\mathbf{X}^a \in \mathbb{R}^{H\times N+M}$ be the augmented historical time series that is recorded from a continuous process, $g : \mathbb{R}^{H\times N+M} \rightarrow \mathbb{R}^{1\times M}$ be the function that is computed by the control logic of the microcontroller, and $h : \mathbb{R}^{H\times N+M} \rightarrow \mathbb{R}^{1\times N}$ be the function that approximates continuous dynamics through sensors. It holds that $\mathbf{X}^a_{T_P+1} = \left[h(\mathbf{X}^a_{T_P,H}) \mid g(\mathbf{X}^a_{T_P,H})\right]$, where $\mathbf{X}^a_{T_P,H} = \mathbf{X}^a_{T_P-(H-1):T_P}$ and $T_H \leq T_P < T$. A future what-if scenario is a change $c : (\mathbb{R}^{H\times N+M} \rightarrow$*

$\mathbb{R}^{1\times M}) \rightarrow (\mathbb{R}^{H\times N+M} \rightarrow \mathbb{R}^{1\times M})$ *on* g *that happens at time* $T_S > T$ *such that the new control logic* $g' = c(g)$ *has not been deployed to the microcontroller in the past. The predictive simulation problem is: Predict* F *future time steps of the augmented time series* $\mathbf{Y}^a = [\mathbf{Y} \mid \mathbf{A}_{T+1:T_F}] \in \mathbb{R}^{F\times N+M}$ *from* $\mathbf{X}^a$ *for* $T_S < T_F$.

Example 1. Let T be the last time step of the augmented historical time series $\mathbf{X}^a$ as depicted in Fig. 1 (e) for the water tank continuous process and $H = 1800$, i.e., 30 minutes for the sampling rate of 1Hz. g is defined by the process fragment in Fig. 1 (II) and h approximates the water level through the level sensor. The future what-if scenario c is to decrease the upper boundary of 30l (cf. Fig. 1 a) to 25l in 5 min, e.g., to test decreasing the likelihood of overflow. A decreased upper boundary has not been tested in practice yet. The problem is to predict at T the next 30 minutes ($F = 1800$) of level sensor values and actuator modes such that the new control logic $g' = c(g)$ regulates the water level from time $T_S = 300$ onwards. Because the ProTwin in Fig. 1 (b) twins the control logic of the microcontroller, the change c corresponds to a data flow change on the CPEE process model [24], i.e., c changes the "level >= 30" sensor condition in Fig. 1 II to "level >= 25" at time T_S.

Hybrid Solution. We propose a HyProTwin (cf. Fig. 2 b) to enable short-term predictive simulation and prediction for continuous processes. The predictive simulation problem without a future what-if scenario is short-term prediction: Predicting the future sensors and actuators without any changes in the future for $F \in \{20, \dots, 2000\}$ time steps depending on the sampling rate [28]. Our proposal focuses on data flow changes c [24], because changes to other perspectives often result in a fundamentally-different continuous process [29] that requires new data collection and training. As time T_S is relative to current time T and HyProTwin is always running, future what-if scenarios c can be provided multiple times when needed at any time. In the following, we elucidate how HyProTwin enables short-term predictive simulation and prediction through execution.

HyProTwin Runtime. Execution of a HyProTwin starting at T entails six steps as depicted in Fig. 2 (1-6). The first step is to initialize the data service (d), so current raw sensor readings and actuator modes from the continuous process are preprocessed and ready for forecasting. Initialization consists of three HTTP interactions: (1a) an HTTP POST configures the bootstrap interface such that the data service knows from which REST endpoint to get (1b) the raw sensors and actuators. From the preprocessed sensors and actuators, the data service returns an HTTP response (denoted by RES in 1c) that contains the last actuator mode, because the control logic (4) often executes the default branch. The second step initializes simulation time $t = 0$. From the constant sampling rate (cf. Req. R1), it follows that simulation time t is a step counter. The third step initializes the forecasting service (e) by sending an HTTP POST (3a) with a resource configuration and a start signal, so the forecasting service knows which forecasting model it must retrieve (3b) from the model repository (f) and which preprocessed sensors and actuators it must retrieve from the data service (3c). Due to Req. R4, a forecasting model for a continuous process is trained once and stored in the model repository (f) before a HyProTwin is executed. After retrieving the forecasting model and input, the future sensors for the next time step ($t = 1$) are forecasted, cached for subsequent, autoregressive forecasts (3d) and returned (3e). Given the forecasted sensor readings, the control logic (4) evaluates the k sensor conditions to determine which of the k actuator modes[2] the continuous process will be in at $t = 1$. The control logic is depicted as a single decision, but it can become any other, more complex control logic taken from a ProTwin of a continuous process. However, control logics must be executed in real-time and continuous processes and their ProTwins can be composed [3], so they are usually a single decision [29].

The fifth step invokes the next forecast by sending the future actuator modes (5a) at $t = 1$ for regulation to the forecasting service that augments the previously cached sensors (3d) with the future actuator modes, subsequently forecasts, caches (5b), and returns (5c) future sensor readings at time $t = 2$. The control logic (4) and regulated forecasting (5) are repeated until the forecasting horizon F is reached. After a prediction cycle, the supervisory ProTwin (c) waits until real execution time has reached stride I to trigger another prediction cycle. If a future what-if scenario c (cf. Fig. 2 0) is provided as input, the change c is applied to the k sensor conditions of the control logic (4) at time T_S (cf. Definition 1). Through our hybrid approach, the control logic (4) and regulated forecasting (5) predict the future behavior of the continuous process and result in the recording of a future DXES event log (7). The future DXES event log can be analyzed by the client for predictive simulation process monitoring and subsequent decision-making (g). The future DXES event log can be similarly analyzed as any other event log with sensor time series and actuator modes [3,10].

HyProTwin Formally. A single prediction cycle (1-6) predicts $\hat{\mathbf{Y}}^a = [\hat{\mathbf{Y}} \mid \hat{\mathbf{A}}] \in \mathbb{R}^{F \times N+M}$ at T iteratively through the hybrid approach. The sensor time series $\hat{\mathbf{Y}} \in \mathbb{R}^{F \times N}$ is forecasted and the actuator time series $\hat{\mathbf{A}} \in \mathbb{R}^{F \times M}$ is predicted by iteratively applying regulated forecasting (cf. Fig. 2 4 and e) on the respective iteration t's historical augmented time series $\mathbf{X}^a_{T+t-1,H}$ (cf. Def. 1) and executing the control logic (4) on the respective iteration's future sensors $\hat{\mathbf{y}}_{T+t}$. Let g_p be the function that represents the control logic. For presentational purposes, we omit future what-if scenarios c in Eqs. 1 and 2. For $t \in \{1, \ldots, F\}$:

$$\hat{\mathbf{Y}} = \begin{bmatrix} \hat{\mathbf{y}}_{T+1} \\ \vdots \\ \hat{\mathbf{y}}_{T+F} \end{bmatrix}, \quad \text{where } \hat{\mathbf{y}}_{T+t} = f(\mathbf{X}^a_{T+t-1,H}) = f(\mathbf{X}^a_{T+t-H:T+t-1}), \tag{1}$$

$$\hat{\mathbf{A}} = \begin{bmatrix} \hat{\mathbf{a}}_{T+1} \\ \vdots \\ \hat{\mathbf{a}}_{T+F} \end{bmatrix}, \quad \text{where } \hat{\mathbf{a}}_{T+t} = g_p(\hat{\mathbf{y}}_{T+t}). \tag{2}$$

[2] As an actuator mode in (4) represents setting multiple actuators to a desired value respectively, e.g., actuator 1, 2, and 10 are set to 0, 7, and 8, the value of k is at most the product of all M actuator's discrete domain sizes.

Remark 1. While actuator modes $\hat{\mathbf{A}}$ could also be forecasted by a TSF technique f_a, several limitations make this approach inappropriate. First, f_a would conflate continuous dynamics and discrete control, making it hard to tell apart learning of the physical process from control decisions, thus violating the modeling req. R1. Even if f_a could handle both, a model-driven approach for discrete control is important for the transparency underlying clear decision-making [16]. In addition, f_a would not inherit the benefits of a ProTwin in overcoming implementation challenges C1 - C3, therefore violating the preservation req. R3. Also, current TSF methods do not jointly predict dynamics and control [23,28,30]. Moreover, the scenario req. R4 may be violated: f_a can be conditioned as in [21] to enable what-if scenarios, but requires conditioning to happen before training. Finally, using suffix prediction with *a-priori knowledge*, like [11], is impractical with f_a because it needs (i) an efficient *beam* search on the much larger, predicted sensor time series $\hat{\mathbf{Y}}$, (ii) suffix prediction on traces of continuous processes whose events contain time series, and (iii) a timed scenario specification language defined on $\mathbb{R}^{H \times N+M}$ (cf. Def. 1).

Summary. Predictive simulation combines the ability to specify what-if scenarios as in short-term BPS for the future with the ability to anticipate the short-term future behavior as in suffix prediction for continuous processes. HyProTwin enables predictive simulation for continuous processes through its hybrid approach by allowing future changes to the model-driven control logic (future what-if scenarios) and, subsequently, regulating the data-driven forecasting of continuous dynamics given output from the changed control logic. If no future what-if scenario is specified, predictive simulation becomes short-term prediction, which is hence also supported by HyProTwin.

4 Evaluation

The aim of the following experiments is to evaluate HyProTwin's performance in forecasting continuous dynamics ($\mapsto$ R1) and constantly enabling ($\mapsto$ R2) predictive simulation on different continuous processes and corresponding future what-if scenarios ($\mapsto$ R4). To that end, the experiment configurations, continuous processes with corresponding scenarios, the TSF techniques f, and the quantitative performance metrics are reported in Sect. 4.1. The evaluation results and an overall assessment that combines the quantitative results with a qualitative assessment of the preservation req. R3 are reported in Sect. 4.2.

4.1 Evaluation Setup

Because forecasting sensor readings $\mathbf{Y}$ can be seen as a subproblem of predictive simulation (cf. Def. 1), we conduct a single set of rolling forecasting experiments that test HyProTwin's performance with respect to predictive simulation and project results to test forecasting sensor readings.

Rolling Forecasting Experiment. Figure 3 depicts a single rolling forecasting experiment. Predictive simulation starts at time T and rolls i times through

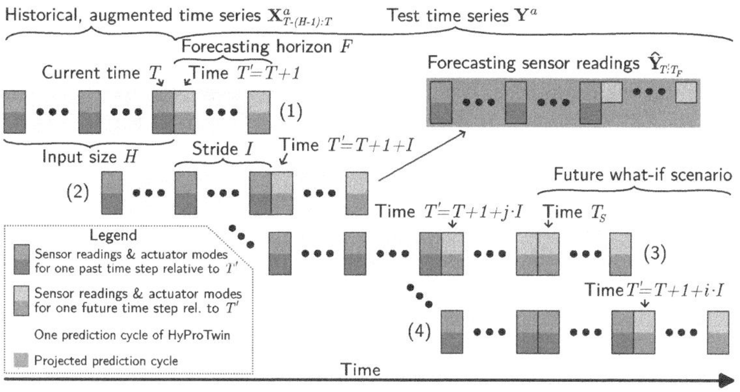

Fig. 3. A rolling forecasting experiment of predictive simulation.

Table 2. Statistics of continuous processes and future what-if scenarios.

Continuous Process	Domain	Sampling Rate r	Length	Dim.	Split	Test length	Stride I	Future what-if scenario: Change c	Time T_S
Water tank	Chemical	1	43,175	2	7:1:2	8,635	15	level >= 30 $\mapsto$ level >= 25	300
Heating	Smart city	1	43,197	2	7:1:2	8,639	15	temp >= 80 $\mapsto$ temp >= 90	1800
Irrigation	Agri-culture	1/3600	8,759	2	7:1:2	1,751	12	0.21 <= moisture <= 0.4 $\mapsto$ 0.25 <= moisture <= 0.35	168

HyProTwin's prediction cycle (cf. Sect. 3) with stride I, i.e., $T' = T + i \cdot I$ is the last relative prediction time such that $T' + F$ does not exceed the test time series $\mathbf{Y}^a$ as depicted in the fourth (4) prediction cycle (cf. Fig. 3). While the first (1) prediction cycle predicts future sensor readings and actuator modes $\hat{\mathbf{Y}}^a_{T':T_F}$ from the historical, augmented time series $\mathbf{X}^a_{T-(H-1):T}$, predictions $\hat{\mathbf{Y}}^a$ move closer to the future what-if scenario's time T_S with each additional prediction cycle. Time T_S is reached in roll j ($T_S < T' + F$) as depicted for the third (3) prediction cycle in Fig. 3. For every roll i, our evaluation projects the prediction $\hat{\mathbf{Y}}^a_{T':T_F}$ to the forecasting of continuous dynamics $\hat{\mathbf{Y}}_{T':T_F}$ as depicted for the second (2) prediction cycle. Rolling forecasting experiments never retrain (cf. Req. R4), i.e., T_S always happens during prediction and not during training. Next, we present the continuous processes and corresponding future what-if scenarios that are used in the rolling forecasting experiments.

Continuous Processes and Future What-if Scenarios. The evaluation includes three real-world continuous processes with three corresponding future what-if scenarios. First, the water tank exhibits linear continuous dynamics with fluid flow and storage as commonly occurring in chemical continuous processes (cf. Sect. 2.1). Next, a heating [14] and a crop irrigation continuous process [3] exhibit non-linear dynamics with random microcontroller failure and random environmental factors respectively. Statistics of the continuous processes such as the domain, the sampling rate r (Hz), the length of the whole time series recorded from the continuous process, the dimension (cf. $N + M$ in Def. 1), the temporal train-val-test split ratio [23], the resulting test time series length, and the stride I (cf. Sect. 2.2) are depicted in Table 2. Also, Table 2 reports the three future what-if scenarios c (cf. Def. 1) and the respective time T_S the scenario will happen. Hence, the upper boundary level is decreased in 5 min for the water tank (cf. Example 1), the lower boundary temperature is increased for the heating in 30 min, and the desired moisture interval for irrigating crops is tightened in 1 week. For every continuous process in Table 2, the rolling forecasting experiment is conducted with the following TSF techniques f.

Forecasting Techniques. The evaluation encompasses four TSF techniques f (cf. Req. R1) which are selected to balance high forecasting performance in recent benchmarkes [18,23,28] and representation of different architectures: LightGBM (GBM) [18] to represent non-deep-learning architectures, DLinear (DLin) [28] to represent linear architectures with multilayer perceptrons, LSTM [23] to represent recurrent architectures, and TCN [23] to represent convolutional architectures. All four TSF techniques f are trained with the hyperparameters of the respective benchmark. We do not include transformer-based TSF techniques f from recent benchmarks [18,23,28], because they do not support regulation with actuator modes (cf. Req. R1 and Remark 1). Adding transformer-based TSF techniques that support regulation with actuator modes [31] and that are not fairly implemented and evaluated in the benchmarks results in methodological inconsistencies. Next, we introduce the experiment configurations.

Baseline and Experiment Configurations. To forecast future sensors readings, the baseline [23,28] is a TSF technique f that *directly* forecasts all future time steps F in a single, non-recursive pass (*direct forecasting*). Consequently, we compare the performance of direct forecasting with HyProTwin's performance in the next section. As the DT with AI in [16] conceptualizes *forecasting-based prediction* such that it is equivalent to direct forecasting on sensor readings $\mathbf{X}$, taking direct forecasting as baseline is similar to including [16] as baseline. Further related DTs (cf. Sect. 5) cannot be applied as baselines, due to conceptual differences and the focus on discrete processes, e.g., Dolly [8] combines the IoT data (i.e., sensor readings $\mathbf{X}$) with their simulation model such that it is unclear how future IoT data should be generated. Analogous to common configurations [23,28], we set H to be a multiple of F and define that F is either 300 seconds (s) or 1800s with H either 300s, 900s, or 1800s respectively for the water tank and heating. In consideration of the hourly sampling rate of irrigation (cf. Table 2), we set F to 24 hours (h) or 78h with H either 12h, 24h, or 78h for irrigation.

During conducting the evaluation, we observed that HyProTwin's performance is sensitive to the timing of actuator mode switches, e.g., from "On" to "Off" (cf. Fig. 1). To further analyze the impact of mode switches and to counteract potential prediction error builtup in HyProTwin's recurrent architecture, we apply a hyperparameter gridsearch on the boundary values in the control logic (cf. Fig. 2 4). For example, the lower ($level_l = 10$) and upper level ($level_u = 30$)

of the water tank (cf. Fig. 1) are varied as follows: $level_l \in \{7, 8, \ldots, 13\}$ and upper level $level_h \in \{27, 28, \ldots, 33\}$. The respective boundary values are similarly varied in the heating and irrigation continuous processes. The best HyProTwin from the gridsearch is determined by evaluating on the validation set with the following performance metrics and referred to as *optimized* (opt) HyProTwin. Because direct forecasting does not iterate over future time steps as HyProTwin, we cannot apply the hyperparameter optimization to direct forecasting.

Performance Metrics. We follow [23], select *root mean squared error* (RMSE) for its sensitivity to outliers and *mean absolute error* (MAE) for its robustness, and refer to [23] for the definition of the metrics. Additionally, we compute the *accuracy* of predicting the actuator mode in the augmented time series $\hat{\mathbf{Y}}^a$ (cf. Fig. 3). We do not measure metrics for the control-flow or time perspective of the DXES event log recorded from the HyProTwin (cf. Fig. 2 7) as in BPS [2], because HyProTwin's control logic is inherited from a ProTwin (cf. Sect. 2.2), simulation time has a constant sampling rate, and future what-if scenarios target the data perspective (cf. Sect. 3).

The implementation of HyProTwin is available at https://github.com/janikbenzin/hyprotwin/ and contains details for replication.

4.2 Evaluation Results

The evaluation setup includes 3 continuous processes, 4 input size H/forecasting horizon F pairs per continuous process, and 4 TSF techniques f for the direct forecasting baseline, HyProTwin, and optimized HyProTwin. Thus the evaluation consists of $3 \cdot 4 \cdot 4 \cdot 3 = 144$ experiments in total. As there is no difference between the baseline and HyProTwin on the level of a roll in rolling forecasting (cf. Fig. 3), we always average the RMSE and MAE over all rolls [23].

Future Sensor Readings. Table 3 reports the RMSE and MAE performance results of the baseline direct forecasting, HyProTwin, HyProTwin (opt) with respect to forecasting future sensor readings. Considering RMSE, HyProTwin (opt) performs best in 7 of 12 experiments across all TSF techniques f, followed by HyProTwin with 4/12 and direct forecasting with 1/12 experiments respectively. Considering MAE, the results are almost similar except for a flip from HyProTwin to HyProTwin (opt), as TCN for heating with configuration 900/300 (H/F) performs very similar between HyProTwin and HyProTwin (opt). Without hyperparameter optimization and, thus, without HyProTwin (opt), the results change to 8/12 experiments led by HyProTwin for RMSE and 10/12 experiments led by HyProTwin for MAE. Turning to more detailed differences, the highest improvement from the best direct forecasting result to the best HyProTwin-based result is $0.2638 - 0.1895 = 0.0743$ for the RMSE on heating 1800/300, the least improvement is -0.0011 for the RMSE on irrigation 72/72, and the average improvement for RMSE is 0.0297 across all 12 experiments. Overall, HyProTwins outperform the baseline direct forecasting.

A more granular analysis of Table 3 yields further observations. First, DLin performs well on the water tank that exhibits linear dynamics. Second, both

Table 3. Performance of direct forecasting, HyProTwin, and optimized (opt) HyProTwin with respect to forecasting future sensor readings $\hat{\mathbf{Y}}$.

			Direct Forecasting				HyProTwin				HyProTwin (opt)			
Cont. Process	H/F	TSF f Metrics	GBM	DLin	LSTM	TCN	GBM	DLin	LSTM	TCN	GBM	DLin	LSTM	TCN
Water tank	300/300	RMSE	0.2055	0.2124	0.1933	–	0.2338	0.2146	*0.1980*	0.2093	0.2404	**0.1785**	0.2389	0.2293
		MAE	0.1734	0.1782	*0.1652*	–	0.1905	0.1724	0.1613	0.1690	0.1956	**0.1472**	0.1943	0.1868
	900/300	RMSE	0.2051	0.2179	0.1936	*0.1939*	0.2341	**0.1667**	0.1965	0.2084	0.2382	0.1954	0.2402	0.2236
		MAE	0.1731	0.1821	0.1657	0.1645	0.1910	**0.1392**	0.1588	0.1686	0.1939	*0.1606*	0.1956	0.1813
	1800/300	RMSE	0.2062	0.2266	0.2196	0.1964	0.2349	0.2118	0.2246	*0.2034*	0.2348	**0.1835**	0.2460	0.2146
		MAE	0.1740	0.1891	0.1830	*0.1664*	0.1914	0.1740	0.1808	0.1649	0.1915	**0.1507**	0.2006	0.1742
	1800/1800	RMSE	*0.1996*	0.2154	0.1940	–	0.2367	0.2653	0.2665	0.2135	0.2414	**0.1759**	0.2540	0.2135
		MAE	*0.1692*	0.1800	0.1656	–	0.1928	0.2152	0.2178	0.1730	0.1965	**0.1446**	0.2064	0.1730
Heating	300/300	RMSE	0.2763	0.3106	0.2762	–	**0.2401**	*0.2555*	0.3786	0.2486	0.3263	0.4022	0.2807	0.2932
		MAE	0.2353	0.2601	0.2335	–	**0.1942**	*0.2060*	0.3104	0.2037	0.2654	0.3331	0.2317	0.2391
	900/300	RMSE	0.2738	0.3113	0.2756	0.2673	*0.2180*	0.2301	0.3584	**0.2152**	0.3332	0.3240	0.3602	0.2155
		MAE	0.2334	0.2601	0.2333	0.2283	0.1752	0.1889	0.2926	*0.1752*	0.2699	0.2689	0.2951	**0.1750**
	1800/300	RMSE	0.2749	0.3249	0.2750	0.2638	*0.2369*	0.2386	0.3771	0.2145	0.3306	0.5764	0.3232	**0.1895**
		MAE	0.2344	0.2707	0.2320	0.2257	*0.1909*	0.1936	0.3089	0.1762	0.2676	0.4829	0.2655	**0.1562**
	1800/1800	RMSE	0.2841	0.3006	0.2752	–	*0.2163*	0.2139	0.3863	**0.2095**	0.3451	0.2139	0.3384	203.0715
		MAE	0.2408	0.2515	0.2363	–	*0.1766*	0.1742	0.3134	**0.1720**	0.2801	0.1742	0.2748	42.2000
Irrigation	24/24	RMSE	0.1091	0.0946	0.0932	–	0.0790	0.0962	0.0915	KS	*0.0824*	0.0956	**0.0741**	KS
		MAE	0.0893	0.0742	0.0813	–	0.0633	0.0768	0.0713	KS	*0.0660*	0.0761	**0.0545**	KS
	48/24	RMSE	0.1118	0.0947	0.0934	0.0931	0.0845	0.0964	0.0721	*0.0809*	0.0842	0.0953	**0.0711**	0.0995
		MAE	0.0918	0.0744	0.0788	0.0800	0.0675	0.0768	0.0551	0.0676	*0.0671*	0.0758	**0.0532**	0.0818
	72/24	RMSE	0.1120	0.0946	0.0921	0.1074	0.0856	0.0955	0.0930	0.0822	*0.0830*	0.0949	0.0916	**0.0762**
		MAE	0.0914	0.0743	0.0810	0.0890	0.0681	0.0760	0.0803	*0.0664*	0.0651	0.0755	0.0788	**0.0599**
	72/72	RMSE	0.1175	**0.1048**	0.1058	–	0.1114	0.1353	0.1162	0.1167	0.1111	0.1333	0.1139	*0.1059*
		MAE	0.0862	**0.0650**	0.0820	–	0.0774	0.0994	0.0967	0.0828	*0.0772*	0.0977	0.0936	0.0728

The best, second best, and third best performance result per row are highlighted in **bold**, gray and *italic* respectively. – reports that TCN does not support $H = F$. KS denotes that the kernel size of 25 [23] in TCN must be smaller than H.

GBM and TCN show good performance on the non-linear dynamics of heating and irrigation. Consequently, we can confirm the performance specificity of TSF forecasting techniques on different dynamics [23]. Also, regulation of the forecasting with actuator modes fails for HyProTwin (opt) on heating 1800/1800, as the model does not "react" to the changed actuator mode and starts to always hallucinate towards very large negative values. As as result, the RMSE is above 203. Nevertheless, hallucination has not occurred in any other of the 96 HyProTwin-based experiments.

Actuator Modes. Due to space constraints, we only report aggregate accuracies of the 11/12 experiments for which the HyProTwin-based results are best. The highest accuracy is 1.0 and is achieved by HyProTwin (opt) with LSTM for irrigation 48/24. The lowest accuracy is 0.0822 and is achieved by HyProTwin with TCN for heating 900/300. The average accuracy over 11 experiments is 0.6044. Low accuracies for heating are caused by the physical property of the water that is heated and cannot be heated above 100° Celsius (C) such that the upper boundary value of 100°C is often not reached exactly for the control logic to switch the actuator mode. Yet, the TSF technique embedded in the HyProTwin is robust enough to still exhibit good performance.

Real-Time Decision-Making. The stride I constitutes a hard real-time constraint on the prediction runtime that starts in relative current time T' and lasts until the respective prediction $\hat{\mathbf{Y}}^a_{T':T_F}$ for a given roll i is computed. As a single cycle of our HyProTwin (cf. Fig. 2 I) takes about 20-40 milliseconds, HyProTwin meets the constraint on prediction runtime in all experiments for water tank and heating with $F = 300$ and all experiments for irrigation, but violates the constraint for water tank and heating with $F = 1800$. As HyProTwin provides predictions in time for water tank with $F = 300$, an operator can take a real-time decision to apply change c (cf. Table 2) to the water tank continuous process based on satisfactory predicted behavior under change c.

Overall Assessment. We conclude that HyProTwin meets R1 modeling due to the best performance in forecasting sensor readings (cf. Table 3), R2 continuity due to the successful 96 rolling forecasting experiments (cf. Fig. 3), R3 due to the supervisory ProTwin as an integral component of HyProTwins (cf. Fig. 2 c), and R4 due to the well-performing predictive simulation of actuator modes and sensor readings on three future what-if scenarios (cf. Table 2) without retraining.

5 Related Work

HyProTwin is related to DTs, suffix prediction, and short-term BPS (cf. Table 1). We focus on recent DTs with process technologies, except for ProTwin as already discussed in Sect. 2.1; otherwise we refer to [5,6,15,17].

Digital Twin. To assign physical devices and actors to manufacturing tasks and orchestrate towards a common goal, [9] propose a DT composition that utilizes *markov decision processes* (MDP) for optimal assignment and orchestration of discrete processes. To develop, execute, and monitor discrete processes with physical devices and process technologies, FloBP [12] proposes a model-driven approach that results in a DT of the microcontroller's control logic. Both [9] and FloBP apply a model-driven approach and the focus is on optimized execution and monitoring. [25] explore responsible aspects of DT adoption for clinical trials and derive a framework with guidelines that lacks details to be assessed along Table 1. As decision-making benefits from explainability, [16] propose to combine explainable TSF techniques with DT in a data-driven approach for prediction of discrete and continuous processes. Thus, what-if scenarios cannot be defined, as the explainable TSF technique lack a suitable parametrization such as [21]. Dolly [8] is a conceptual framework to implement DTs of processes with a *MERODE domain model* for IoT sensor readings that supports decision-making through simulation. The domain model maintains the current state of IoT sensors and respective business objects that are relevant to the execution and simulation of discrete processes. Dolly does not distinguish between sensor readings and actuator modes and learns probabilities for decision branches from the current sensor readings that abstract our explicit control logic into a probabilistic representation. Overall, the five DT proposals [8,9,12,16,25] lack the ability for predictive simulation of continuous processes and do not apply a hybrid approach.

Suffix Prediction. [32] apply transformers to predict the sequence of activities and time attributes of an ongoing case. [1] apply bi-directional LSTM to predict the sequence of activities with start and end lifecycles for the suffixes of all ongoing cases in lockstep. [1,32] focus on discrete processes, as other SP techniques do, such that continuous processes are not supported. Also, what-if scenarios cannot be defined due to the data-driven approach.

Short-Term and Long-Term BPS. [22] propose a technique to estimate *steady states* of processes from which a short-term BPS can start. [2] propose a technique to estimate the current process state of ongoing cases from event logs that may not be steady. [20] propose long-term BPS technique RIMS to simulate discrete processes by integrating AI for predicting activity times and next activities with BPMN simulation models. [7] propose DSim to simulate discrete processes by integrating AI for predicting times similar to RIMS with BPMN simulation models, but instead of predicting times at simulation runtime, the times are predicted after the simulation has completed. Both techniques neither consider continuous processes, predicting future data such as sensor readings, nor predictive simulation. In addition, short-term BPS techniques do not apply a hybrid approach and long-term BPS techniques cannot estimate current process states.

6 Conclusion, Limitations, and Future Work

This work proposes a HyProTwin DT architecture that enables predictive simulation of continuous processes. Hence, HyProTwin can predict the future behavior of continuous processes under future what-if scenarios and outperforms baseline techniques. A HyProTwin implementation shows that through integrating DTs, process technologies, and AI, HyProTwin overcomes the interoperability (C1), communication (C2), orchestration (C3), and modeling (C4) challenges. Currently, HyProTwins are focused on continuous processes, assume a constant sampling rate, require the control logic from a ProTwin, and may not meet real-time constraints. In the future, we plan to generalize HyProTwins to discrete processes, to allow time-dependent sampling rates, to discover the control logic from DXES event logs, and to increase the runtime performance.

References

1. Ali, M.A., Dumas, M., Milani, F.: Predicting case suffixes with activity start and end times: a sweep-line based approach. In: CoopIS, pp, 88–106 (2026)
2. Avramenko, M., Chapela-Campa, D., Dumas, M., Milani, F.: What's coming next? Short-term simulation of business processes from current state. In: ICPM, pp. 1–8 (2025)

3. Benzin, J.V., Mangler, J., Rinderle-Ma, S.: A service-oriented digital twin architecture for seamless reliability verification of iot systems. In: ICSOC, pp, 175–190 (2026)
4. Benzin, J.V., Rinderle-Ma, S.: A survey on event prediction methods from a systems perspective: bringing together disparate research areas. ACM Comput. Surv. **57**(12), 301:1–301:37 (2025)
5. Bhatia, M., Kumar, V.: Artificial intelligence in digital twin technology: scientometric insights on architecture, applications, and tools. IEEE Internet Things J. **12**(16), 32653–32675 (2025)
6. Cabezas López, A.S., Calvo, F., Lozano Santamaria, F., Gómez, J.M.: Control of multi-unit chemical processes for production transitions: a review. Chem. Eng. Res. Des. **220**, 539–553 (2025)
7. Camargo, M., Dumas, M., González-Rojas, O.: Learning accurate business process simulation models from event logs via automated process discovery and deep learning. In: CAiSE, pp, 55–71 (2022)
8. Compagnucci, I., Re, B., Asensio, E.S., Snoeck, M.: A digital process twin conceptual architecture for what-if process analysis. In: EDOCW, pp. 373–388 (2025)
9. De Giacomo, G., Favorito, M., Leotta, F., Mecella, M., Silo, L.: Digital twin composition in smart manufacturing via markov decision processes. Comput. Ind. **149**, 103916 (2023)
10. De Luzi, F., Leotta, F., Marrella, A., Mecella, M.: On the interplay between business process management and internet-of-things. BISE **67**(2), 265–288 (2025)
11. Di Francescomarino, C., Ghidini, C., Maggi, F.M., Petrucci, G., Yeshchenko, A.: An eye into the future: leveraging a-priori knowledge in predictive business process monitoring. In: BPM, pp, 252–268 (2017)
12. Fedeli, A., Fornari, F., Polini, A., Re, B., Torres, V., Valderas, P.: FloBP: a model-driven approach for developing and executing IoT-enhanced business processes. SoSyM **23**(5), 1217–1246 (2024)
13. Fornari, F., et al.: Digital twins of business processes: a research manifesto. Internet Things **30**, 101477 (2025)
14. Hahn, E.M., Hartmanns, A., Hermanns, H., Katoen, J.P.: A compositional modelling and analysis framework for stochastic hybrid systems. Form. Methods Syst. Des. **43**(2), 191–232 (2013)
15. Kreuzer, T., Papapetrou, P., Zdravkovic, J.: Artificial intelligence in digital twins—a systematic literature review. Data Knowl. Eng. **151**, 102304 (2024)
16. Kreuzer, T., Papapetrou, P., Zdravkovic, J.: A meta-model for integrating explainable forecasting with digital twins. In: CAiSE Workshops, pp. 171–182 (2025)
17. Lehner, D., et al.: Model-driven engineering for digital twins: a systematic mapping study. SoSyM **24**(5), 1339–1377 (2025)
18. Makridakis, S., Spiliotis, E., Assimakopoulos, V.: M5 accuracy competition: results, findings, and conclusions. Int. J. Forecast. **38**(4), 1346–1364 (2022)
19. Mangler, J., et al.: Datastream XES extension: embedding iot sensor data into extensible event stream logs. Fut. Int. **15**(3), 109 (2023)
20. Meneghello, F., Di Francescomarino, C., Ghidini, C.: Runtime integration of machine learning and simulation for business processes. In: ICPM, pp. 9–16 (2023)
21. Oyamada, R.S., Marques Tavares, G., Barbon Junior, S., Ceravolo, P.: CoSMo: a framework to instantiate conditioned process simulation Models. In: BPM, pp. 328–344 (2024)
22. Pourbafrani, M., Lücking, N., Lucke, M., van der Aalst, W.M.P.: Steady state estimation for business process simulations. In: BPM Forum, pp. 178–195 (2023)

23. Qiu, X., et al.: TFB: towards comprehensive and fair benchmarking of time series forecasting methods. Proc. VLDB **17**(9), 2363–2377 (2024)
24. Reichert, M., Dadam, P.: Adeptflex—supporting dynamic changes of workflows without losing control. J. Intell. Inf. Syst. **10**(2), 93–129 (1998)
25. Ribeiro, L.R.B., de Pádua, S.I.D., Kremer, G., Stark, R.: Redpt-ct: designing responsible digital process twins for clinical trials. In: Responsible BPM Forum, pp. 102–116 (2025)
26. Rinderle-Ma, S., Mangler, J., Ritter, D.: Fundamentals of information systems interoperability, Springer (2024)
27. Rudolph, M., Kurz, S., Rakitsch, B.: Hybrid modeling design patterns. J. Math. Ind. **14**(1), 3 (2024)
28. Shao, Z., et al.: Exploring progress in multivariate time series forecasting: comprehensive benchmarking and heterogeneity analysis. IEEE Trans. Knowl. Data Eng. **37**(1), 291–305 (2025)
29. Strutzenberger, D., Mangler, J., Rinderle-Ma, S.: Evaluating BPMN extensions for continuous processes based on use cases and expert interviews. BISE **66**(6), 709–735 (2024)
30. Uremović, N., et al.: A new framework for multivariate time series forecasting in energy management system. IEEE Trans. Smart Grid **14**(4), 2934–2947 (2023)
31. Wang, Y., et al.: TimeXer: empowering transformers for time series forecasting with exogenous variables. NeurIPS **37**, 469–498 (2024)
32. Wuyts, B., Vanden Broucke, S., De Weerdt, J.: SuTraN: an encoder-decoder transformer for full-context-aware suffix prediction of business processes. In: ICPM, pp. 17–24 (2024)

Characterizing Process Variants with Natural Examples

Carlos Capitán-Agudo[1,2(✉)], Claudio di Ciccio[3], Cristina Cabanillas[1,2], and Manuel Resinas[1,2]

[1] SCORE Lab, Universidad de Sevilla, Seville, Spain
[2] I3US Institute, Universidad de Sevilla, Seville, Spain
{ccagudo,cristinacabanillas,resinas}@us.es
[3] Utrecht University, Utrecht, The Netherlands
c.diciccio@uu.nl

Abstract. To improve business processes, it is crucial to analyze why processes differ across process variants. To support this analysis, prior research has often relied on machine learning models trained to classify process variants and used model explanations to distinguish them. User studies in explainable artificial intelligence show that these explanations can be expanded through the inclusion of examples. To this end, we introduce an approach that uses declarative behavioural constraints to support variant classification tasks. Building on these constraints, we perform a post-hoc probability analysis and derive a set of multi-feature measures to characterize cases that are representative of each process variant (prototypes) alongside cases that are close to the classification boundaries between process variants (boundary cases). We assess the explanatory power of those natural examples with two real-world event logs. The results demonstrate that our approach shows the distinctive behavioural traits of variants effectively and intuitively.

Keywords: Process mining · Variant analysis · Explainability · Natural examples · Declarative specifications · Explainable Artificial Intelligence

1 Introduction and Motivation

The analysis of *process variants* has emerged as a relevant approach for gaining more fine-grained insights into process behavior. A process variant refers to a subset of cases that share a distinguishing characteristic such as a specific activity sequence, outcome, attribute value, or a combination of them, thereby differentiating them from other cases in the log [25]. Analyzing differences between such variants enables a deeper understanding of the process and potential improvement opportunities. For example, consider an event log related to the management of traffic fines [14]. Its activities include 'Send Fine', 'Payment', and 'Appeal to Judge'. Considering process variants that discriminate whether the

F. M. Maggi et al. (Eds.): BPM 2026, LNCS 16936, pp. 295–312, 2027.
https://doi.org/10.1007/978-3-032-37777-7_19

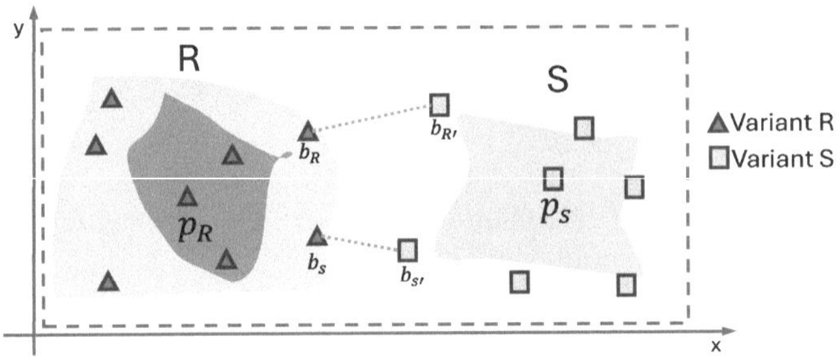

Fig. 1. Bidimensional representation of case examples for two process variants: R and S. Red triangles and yellow squares represent cases of their respective variants. The dashed purple rectangle represents the feature space of the entire log. Color intensity reflects the probability of belonging to a variant. p_R and p_S represent the prototypes and b_n highlight the case of a pair of boundary cases. X and Y are feature projections. (Color figure online)

payment is completed on time or with a delay, the differences between those variants should show what led to the differing outcomes (e.g., offenders appeal the fine rather than paying it directly).

A common approach to identify these differences between process variants is based on training machine learning (ML) models to classify cases into the different process variants considered [25]. From this model, a set of constraints can be extracted to understand what features characterize each process variant. These constraints are typically expressed in an "if X then Y" format, where X is a logical combination of features, and Y indicates a specific process variant. For instance, in the traffic fine scenario, a possible constraint might be: *"If 'Appeal to Judge' occurs in a case, then it belongs to the delayed process variant"*.

While such constraints provide concise and interpretable summaries, they remain abstractions of the underlying process behavior. In parallel, user studies in the field of explainable artificial intelligence (XAI) suggest that examples can complement rule-based explanations by illustrating how model behavior manifests in concrete instances [21]. This finding has motivated the development of example-based explainability methods, which aim to communicate model behavior through representative or contrastive samples, either drawn from the data or synthetically generated [20]. However, the use of examples as explanatory artifacts has received little attention in process variant analysis.

In this paper, we introduce an example-based approach for explaining differences between process variants, inspired by methods from XAI [10]. The approach is grounded in the use of natural examples, i.e., real cases extracted from event logs, as explanatory artifacts. Unlike synthetic examples, natural cases are inherently plausible, require no additional feasibility checks, and avoid dependencies on external generation or correction mechanisms. Our approach leverages an ML model that predicts the process variant of a case based on measurements of declarative behavioral constraints. Rather than treating the model as an end in itself, we use it to structure the behavioral space of cases and to retrieve informative examples. In particular, we extract two complementary types of cases: *prototypes*, which are highly representative of each process variant and thereby highlight the most salient differences between variants, and *boundary cases*, which are pairs of correctly classified cases that lie close to the model's decision frontier, thus exposing subtle behavioral divergences between variants that prototypes alone cannot capture. Figure 1 visually sketches the underlying idea. To support interpretation, we further relate these examples to the behavioral patterns that drive the model's predictions using SHAP [15].

The contribution of our work is threefold: *(i)* We introduce an example-based perspective on process variant explanation that treats real cases as first-class explanatory artifacts; *(ii)* We present a method for retrieving informative examples from event logs, including representative (prototype) and discriminative (boundary) cases derived from a behavioral feature space; *(iii)* We offer an open-source implementation of our approach, to test it with real-world event logs.

To evaluate our approach, we first quantitatively assess the quality of the retrieved examples by comparing them against baseline selection strategies with respect to properties such as intra- and inter-class distance. Secondly, we demonstrate through two use cases how the extracted examples complement rule-based explanations by providing concrete, case-level insights into variant differences.

The remainder of this paper is organized as follows. Section 2 presents the background and related work. Section 3 details our proposed approach. Section 4 presents the setting and results of the evaluation. Finally, Sect. 5 offers conclusions, discusses limitations, and outlines possible future directions.

2 Background and Related Work

This section reviews prior work related to our approach along three directions.

2.1 Process Variant Analysis

Process variant analysis has received increasing attention in the last years, leading to the development of a variety of approaches. A common way to characterize these approaches is based on the output they provide to describe how process variant differ [25]. A prominent class of approaches focuses on extracting sets of rules that provide interpretable differences that characterize each variant. These approches differ in the way used to identify the rules.

Some approaches rely on process models to identify them, which represent how a process is executed. For example, Swinnen et al. [24] compare process variants against a normative model to identify deviations and extract rules that explain how process variants differ. Statistical techniques have been applied to obtain such rules [2,5]. Among others, Cecconi et al. [2] use statistcal techniques to detect Declare constraints that differ significantly in the variants.

Other approaches encode cases as feature vectors labeled with the corresponding process variant. Then, a classifier is trained using these multidimensional representations of process executions. The resulting model is analyzed to extract explanatory differences. Bose and van der Aalst [1] extract features related to individual activities (e.g., repetitions of activity A) and their sequences (e.g., how often activity A is immediately followed by B). Each case is mapped to a feature vector, and association rules and decision tree algorithms are used to generate conditional statements that describe each variant. Cuzzocrea et al. [6] extend this idea by training an ensemble of classification models to distinguish case classes incorporating features derived from data attributes to the previous ones. Moreover, some approaches have enhanced case encodings by adding declarative features [9,22]. Di Francescomarino et al. [9] encode cases using declarative, activity and attribute features. They use DECLARE to describe their declarative behavior with the following mapping based on constraint activations: -1 if the constraint was violated, 0 if it was not activated, and n if it was fulfilled n times.

Closer to our work, Grigore et al. [12] define an agnostic framework to explain process variants globally through the use of prototypes and specific cases through its comparison with prototypes. The variants are automatically identified with unsupervised techniques. Particularly, they defined prototypes as the most central case of a process variant (i.e., the most frequent behaviour). They encoded cases at event and activity levels. We complement this work by including user-defined process variants, using a different case encoding, different prototype definition, and by using boundary cases to show subtle differences between variants.

2.2 Natural Example-Based Explainability Methods

Natural example-based explainability methods use instances from a dataset to explain model behavior or characterize specific classes [20]. They are categorized into different approaches depending on the semantic meaning of the examples.

A first line of work focuses on retrieving representative instances of a dataset, referred to as *prototypes*, which summarize the typical characteristics of a class or dataset. Prototypes are commonly used to provide *post-hoc interpretability* of ML models, as they reflect the most important features associated with a given class. In the context of process mining, prototypes have been explored to reduce the complexity of process discovery [11].

A second line of work considers *natural contrastive explanations*, which rely on examples to illustrate how differences in features lead to different model predictions. Among the various types of contrastive explanations, *natural counterfactual examples* identify, for a given instance, the closest instance in the dataset whose model prediction is different [20]. Such pairs highlight minimal changes that alter the model's decision, thereby revealing the boundaries between classes.

In this work, we build on these two perspectives by combining representative and contrastive examples to explain differences between process variants. In particular, we adopt the notion of prototypes to capture typical variant behavior and use pairs of nearby cases with different predictions, referred to as boundary cases, to expose subtle distinctions between variants.

2.3 Declarative Behavioral Constraints and Measures

Our approach relies on declarative behavioral constraints to represent process executions in a feature space used to train a ML model. These constraints describe permissible order relationships between activities without prescribing a fixed execution order [8]. The DECLARE language is a widely used formalism for specifying such constraints, offering templates to express temporal relations between activities [19]. For example, the DECLARE constraint RESPONSE(A, B) states that whenever activity A occurs, activity B must eventually follow.

Declarative constraints can be evaluated on individual cases to quantify their degree of satisfaction. A commonly used measure is *confidence* CONF(X), defined as the ratio of fulfilments over activations of a constraint X within a case, with values in $[0, 1]$ when at least one activation occurs [3]. When a constraint is not activated in a case, its confidence is undefined, corresponding to a vacuous satisfaction [16]. This distinction is relevant when encoding cases, as both activated and non-activated constraints contribute to characterizing behavioral patterns. Encodings based on DECLARE constraints have been used in prior work to support tasks such as classification and deviance detection [9,22]. In this work, we build on these representations by using declarative behavioral features to structure the space of cases, enabling the retrieval of representative and discriminative examples for explaining differences between process variants.

3 The Approach

Figure 2 illustrates our approach for retrieving *natural examples* that highlight differences between process variants. To illustrate it, consider the customer appeal process (cf. Section 2) is composed of activities such as *Send Appeal* (*SA*), *Evaluate Appeal* (*EA*), *Request information* (*RI*), *Take Decision* (*TD*), *Accept* (*A*) and *Reject* (*R*). Given the event log and its variants, our approach returns two types of samples that explain their differences: (i) prototypes and (ii) boundary cases. These samples are derived from the output of an ML classifier trained on declarative behavioral patterns. We also report the SHAP values of each sample, which identify the most influential features in the classifier's decisions, to improve their interpretability.

Step 1: Feature extraction. We extract features that represent each case of the input event log. In this paper, we focus on control-flow-based features. This choice ensures broad applicability, as control-flow information, i.e., sequences of activities, is the minimal requirement of event logs. Specifically, we use features derived from declarative behavioral contraints, due to their strong predictive performance in prior work [9]. Each case is described by the confidence values CONF(X) of the discovered DECLARE constraints X, obtained using standard declarative process discovery techniques [3,7,16]. When a constraint

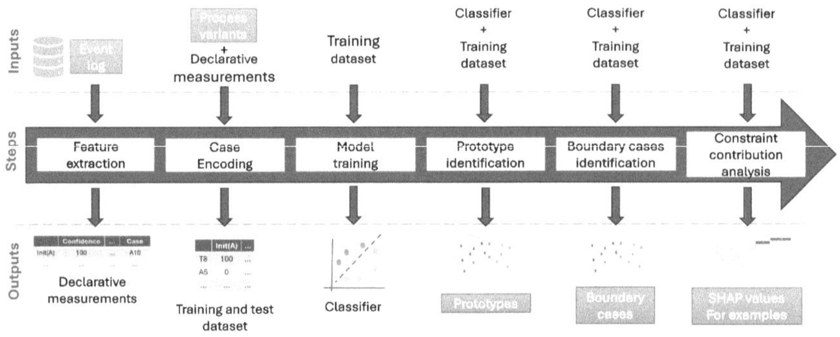

Fig. 2. Overview of our approach. Inputs highlighted in green are given as an external parameter, and outputs highlighted in purple constitute the final outcome. (Color figure online)

is not activated in a case, its value is set to *none*, reflecting vacuous satisfaction. For example, suppose that the travel declaration log contains cases $c_1 = \langle SA, NCR \rangle$, $c_2 = \langle NC, SA, RA, TDR \rangle$, and $c_3 = \langle RA, RI, A \rangle$. The constraint RESPONSE(SA, NC) has a confidence CONF(RESPONSE(SA, NC)) of 1 for c_1 (fulfilled), 0 for c_2 (violated), and *none* for c_3 (because SA does not occur).

Step 2: Case Encoding. Each case is encoded as a feature vector, where each dimension corresponds to a declarative feature obtained in the previous step. These vectors form a matrix X, whose rows represent cases and columns represent features, with each entry storing the corresponding confidence value (see Table 1). A label vector Y is also constructed to encode the process variant of each case. Finally, X and Y are randomly split into training and test sets (X_{train}, X_{test} and Y_{train}, Y_{test}), preserving the distribution of process variants.

Table 1. Excerpt of a matrix containing the feature vectors of cases

	CONF(INIT(SA))	CONF(RESPONSE(SA, NC))	CONF(RESPONSE(RA, RI))	...
c_1	1	1	*none*	...
c_2	1	0	*none*	...
c_3	*none*	*none*	1	...
...	...	...	...	...

Step 3: Model Training. We train a classifier M on X_{train} and Y_{train} to predict, for a given case c, the probability of belonging to each process variant based on its feature vector. Once trained, M can classify unseen cases using their DECLARE-constraint features. Importantly, M must be well-calibrated and output probability distributions over all variants rather than a single label.

Step 4: Prototype Identification. For each variant, we identify a prototype by selecting a case from X_{train}, following standard practice in example-based methods [20]. We define the prototype of a variant V as the training case c that maximizes the predicted probability of belonging to V. Such case exhibits the most distinctive features of its variant. In case of ties, we prefer shorter cases to favor more concise explanations, breaking remaining ties at random. For example, let the *time-compliant* variant's prototype be $c_{\text{compl}} = \langle SA, EA, TD, NR, A \rangle$, and the prototype of the *non-time-compliant* variant be $c_{\text{notcompl}} = \langle SA, TD, NR, R \rangle$. These cases exhibit distinct behavioral constraints. In the former, the appeal is evaluated (EA) before a decision is made (TD), reflecting that the appeal was performed on time, was reviewed and finally was accepted. Instead, in the latter, the decision (TD) follows immediately the submission because the appeal arrived too late, so it is rejected (R).

Step 5: Boundary Case Identification. *Boundary cases* are pairs of cases from different process variants that lie close to the decision frontier. They reveal subtle behavioral differences that may explain why a case belongs to one particular variant rather than another [20]. To identify such cases, we rely on the prediction of the trained classifier. Inspired by the *nearest unlike neighbors* concept [10], we search for pairs of cases in X_{train} that (i) are correctly classified in their actual process variant and (ii) are close to another variant according to the classifier output. We quantify this closeness as follows. Let c be a case whose true variant is V, and let $M(c)[V]$ and $M(c)[V']$ denote the probabilities assigned to c by classifier M to variants V and $V' \neq V$, respectively. We define the *Inter-Variant Boundary Proximity* (IVBP) as:

$$\text{IVBP}(M, c, V, V') = \frac{1}{1 + |M(c)[V] - M(c)[V']|}$$

A higher IVBP indicates that the two probabilities are closer, and thus that the case lies nearer to the decision boundary between V and V'. For each pair of variants, we select the top-k cases (e.g., $k = 10$) with the highest IVBP scores. Boundary pairs are then formed by choosing the two cases from different variants with the smallest Euclidean distance between their feature vectors. Since some confidence $\text{CONF}(x)$ may be *none* due to vacuous satisfaction, we map them to a large negative value (e.g., -100) when computing the Euclidean distance. This ensures that non-activated constraints are treated as behaviorally distinct from constraints with observed evidence.

Step 6: Constraint Contribution Analysis. To support the interpretation of prototypes and boundary cases, we analyze the contribution of each DECLARE-constraint feature to the classifier's predictions using SHAP (SHapley Additive exPlanations) values [15]. SHAP quantifies the impact of each feature by measuring the effect of including or excluding it from the model. Positive values indicate the constraint supports the assignment of a case to its predicted variant, while negative values indicate a shift toward alternative variants. For instance, in the *time-compliant* prototype c_{compl}, the constraint RESPONSE(SA, EA) with confidence 1.0 decisively contributes with a SHAP value of 3.16 toward the prediction of the *timecompliant* class. In contrast, in a boundary case of the same variant where this constraint is not fulfilled (confidence 0.0), its SHAP value of -0.41 weakens the evidence for the variant and moves the prediction closer to the decision boundary.

4 Evaluation

For the evaluation of our approach, we have implemented an open-source prototype, available at https://doi.org/10.5281/zenodo.20596012. We applied it to two real-world event logs. Our evaluation is threefold. First, we compare the retrieved samples against several baselines to assess their quality and interpretability. Second, we study how our approach behaves with different configurations to show the implementation-independence of our approach. Third, we illustrate how the extracted samples complement rule-based explanations. All experiments were run on a machine with a 12th-gen Intel Core i7-1265U CPU and 16 GB of RAM.

4.1 Datasets and Implementation

We use two event logs commonly used for explainability-oriented process mining [18]: *Road Traffic Fines* [14], and *Sepsis* [17]. The Road Traffic Fines (RTFM) log contains 150,370 cases describing the management of traffic fines in Italy. Each case starts with a fine creation, including an amount attribute. The offender is notified within 90 days, and may pay at any time. A penalty is added if payment is not made on time. Offenders may appeal to a judge or prefecture, and they will be notified of the outcome. If the fine is unpaid, an agency collects it.

Following Völzer et al. [26], we consider four outcome-based variants:

Fully paid: The fine is paid, including expenses and penalties (60,286 cases).
Dismissed: The fine is waived due to a successful appeal (2060 cases).
Collected: A collection agency recovers the unpaid amount (59,010 cases).
Unresolved: Fine is neither fully paid, dismissed, nor collected (29,010 cases).

Four cases related to multiple outcomes were removed, resulting in 150,366 cases.

The Sepsis log contains 1050 cases describing clinical activities in sepsis treatment. Most cases start with a patient being registered and diagnosed. Afterwards, various measurements are performed (e.g., leukocytes), and patients may be transferred to an unit (e.g., intensive care). Finally, they are discharged, but some may return. In this case, variants are defined by *Systemic Inflammatory Response Syndrome* (SIRS) criterion, indicated by a boolean attribute, which is one of the partition criteria suggested by Rubensson et al. [23]:

SIRS–True: SIRS conditions are met, and sepsis is diagnosed (853 cases).
SIRS–False: SIRS conditions are not met, and sepsis is not diagnosed (197).

These event logs were selected because they exhibit substantially different characteristics, as summarized in Table 2, including the number of variants and average case length. They also represent two distinct ML settings: binary classification in the Sepsis log, and multiclass classification in the RTFM log. Furthermore, the distribution of cases across process variants differ significantly; while the RTFM log contains three predominant classes (fully paid, collected, and dismissed), the Sepsis log exhibits a stronger imbalance between its two variants. These differences make the logs complementary for evaluating our approach.

For feature extraction, we used MINERful [13], due to its high performance on large event logs [7]. MINERful discovers DECLARE constraints and can compute, for each case, the confidence CONF of each constraint, with values in the range $[0, 100]$. This resulted in 2189 constraints for the RTFM log and 4704 for the Sepsis log, which we used as classification features. For classification, we employed XGBoost,[1] a gradient boosting framework based on ensembles of decision trees. We chose XGBoost because it outputs probability distributions over variants, supports missing values natively (which is beneficial when DECLARE constraints are vacuously satisfied), and demonstrated strong performance and scalability [4].

4.2 Configuration and Performance Evaluation of the Classifier

We trained the XGBoost classifier using default parameters with early stopping (patience of five rounds) to prevent overfitting. Each dataset was split into 80% training–validation and 20% test sets. Performance was evaluated on the test set using the per-class F1-score, which balances precision and recall and is suitable for imbalanced datasets like ours. With the RTFM log, the classifier achieved F1-scores of 1.00 (*collected*), 0.97 (*fully paid*), 0.91 (*unresolved*), and 0.77 (*dismissed*). With the Sepsis log, the F1-scores were 0.96 for *SIRS–true* and 0.82 for *SIRS–false*. These results indicate strong overall performance, with lower scores observed for less frequent or behaviorally overlapping variants. Figure 3 shows the confusion matrices. With the RTFM log, some *unresolved* cases are misclassified as *fully paid*. *Dismissed* and *unresolved* cases are rarely confused. In Sepsis log, almost all *SIRS–true* cases are classified correctly, whereas some *SIRS–false* cases are misclassified.

4.3 Comparison with Baselines

To evaluate the retrieved prototypes and boundary cases, we define a baseline approach for each task. For prototype identification, we use a length-based baseline: the prototype of a process variant is selected as the case whose trace length is closest to the mean length of that variant. In case of ties, one is chosen at random. The baseline is applied to training cases only, mirroring our approach. To assess the representativeness of the discovered prototypes, we compute two metrics based on DECLARE constraints confidence values: (i) *intra-class distance*,

[1] https://xgboost.readthedocs.io/en/stable/. Accessed: 31/05/2026.

i.e., the average Euclidean distance between the prototype and cases of the same variant, and (ii) *inter-class distance*, i.e., the distance to cases from other variants. In both cases, *none* values are mapped to -100 for consistency.

Table 2. Summary of log characteristics

Log	No. cases	No. activities	No. variants	Avg. case length
RTFM	150366	11	230	3.73
Sepsis	1050	16	846	14.48

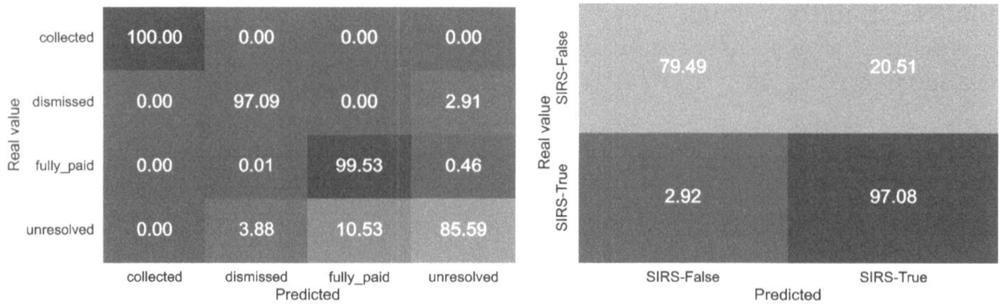

(a) Road Traffic Fines (RTFM) (b) Sepsis

Fig. 3. Prediction distribution for the process variants of each log on the test set. Rows correspond to actual classes, and columns to predicted classes.

For the identification of boundary cases, we also define a baseline based on length: Let V and V' be two process variants with $V \neq V'$. A boundary case is a case with label V whose trace length is the closest to the mean trace length of V'. Ties for closest trace length are broken randomly.

4.3.1 Prototype

Table 3 reports the prototypes identified by each approach, together with their intra- and inter-class distances. Lower intra-class distance indicates higher representativeness, while higher inter-class distance reflects better separation from other variants. For the *collected* variant, both approaches select the same prototype, likely because the variant is strongly characterized by the presence of the *Send for Credit Collection* (*SC*) activity[2]. For *fully paid*, *unresolved*, and *SIRS-false* variants, our prototypes exhibit lower intra-class distances and higher

[2] Full activity names and their abbreviations are listed in the caption of Table 3.

inter-class distances than the baseline prototypes, indicating improved representativeness and distinctiveness. The opposite trend emerges for the *SIRS–true* and *dismissed* variants, which may be influenced more by trace length than by distinctive behavioral patterns. It is important to note that Euclidean distance may not fully capture the quality of our prototypes. Our approach selects prototypes based on the classifier's probability space, which reflects learned decision boundaries rather than raw distances in an Euclidean space. Thus, a prototype based on probability might be representative for the classifier, but it could not be the best one in terms of Euclidean distance.

Beyond quantitative measures, our prototypes provide intuitive insights into variant behavior. For example, the *unresolved* prototype reflects the absence of *Payment* (*PYM*) and *SC*. Similarly, in the Sepsis log, the *SIRS–false* prototype shows no activities following *TGS*, indicating that no further clinical actions were required, consistently with the absence of a sepsis diagnosis.

By checking SHAP values, we can analyze the DECLARE-constraint features that contribute most positively to our prototypes. For instance, in the RTFM log, $\text{CONF}(\text{END}(AJ)) = 100$ strongly characterizes the *dismissed* prototype, indicating that the case ends in *AJ*. Conversely, $\text{CONF}(\text{ALTERNATERESPONSE}(PYM, AP)) = 0$ is highly characteristic of the *fully paid* prototype, indicating that *PYM* occurs but without a subsequent *AP*. These insights allow us to pinpoint and confirm the key features of our prototypes.

Table 3. Prototypes per variant, with intra- and inter-class distances. The distances highlighted in bold represent the best results for a process variant.

Approach	Log	Variant	Intraclass	Interclass	Prototype
Ours	Road Traffic Fines	Collected	107,98	4195,61	⟨ CF, SF, IF, AP, SC ⟩
		Dismissed	3860,16	3782,72	⟨ CF, SF, IF, AP, AJ ⟩
		Fully paid	**799,06**	**4303,50**	⟨ CF, PYM ⟩
		Unresolved	**1126,27**	**3449,54**	⟨ CF, SF ⟩
	Sepsis	SIRS-false	**3390,54**	**6947,33**	⟨ RG, TG, TGS ⟩
		SIRS-true	5412	4404,40	⟨ RG, TG, TGS, CRP, LC, AB ⟩
Case length	Road Traffic Fines	Collected	107,98	4195,61	⟨ CF, SF, IF, AP, SC ⟩
		Dismissed	**436,59**	**4362,17**	⟨ CF, SF, IF, ID, AP, SA ⟩
		Fully paid	2238,75	3736,05	⟨ CF, SF, PYM ⟩
		Unresolved	2510,67	3309,89	⟨ CF, PYM, SF ⟩
	Sepsis	SIRS-false	4956,33	6131,45	⟨ RG, TG, TGS, CRP, NC, NC, RA, RE ⟩
		SIRS-true	**2790,41**	**5728,74**	⟨ RG, TG, TGS, CRP, LA, LC, LQ, AB. NC, …, RA ⟩

Activity name abbreviations. (RTFM log) *CF*: Create Fine; *SF*: Send Fine; *IF*: Insert Fine Notification; *AP*: Add Penalty; *PYM*: Payment; *ID*: Insert Date Appeal to Prefecture; *SA*: Send Appeal to Prefecture; *RP*: Receive Result Appeal from Prefecture; *NR*: Notify Result Appeal to Offender; *AJ*: Appeal to Judge; *SC*: Send for Credit Collection. (Sepsis log) *RG*: ER registration; *TG*: ER Triage; *TGS*: ER Sepsis Triage; *LC*: Leucocytes; *AB*: IV Antibiotics; *CRP*: CRP; *LA*: LacticAcid; *LQ*: IV Liquid; *NC*: Admission NC; *RA*: Release A; *RE*: Return ER; *IC*: Admission IC.

4.3.2 Boundary Cases

Figure 3 shows that some variant pairs are naturally well separated. For instance, *collected* cases consistently receive predicted probabilities close to 99%, reflecting strong separability driven by the presence of *SC*, an activity absent from other variants. A similar pattern is exhibited by the *dismissed/fully paid* pair. As shown in Fig. 3(a), these variants are rarely confused by the classifier due to fundamentally different behavior captured by their DECLARE features: appeal handling versus payment completion. Figure 3(b) shows a similar observation for the Sepsis variants. This observation aligns with prior evidence indicating that identifying natural counterfactuals is inherently difficult due to the limited availability of cases that exhibit similar features yet belong to different classes [20].

Table 4 reports the boundary cases for the remaining variant pairs. The baseline boundary cases for *SIRS–false-SIRS–true* show some differences at the beginning (e.g., *CRP* instead of *LC*). However, the size dissimilarity makes it more difficult to figure out the frontiers. The same occurs for the baseline boundary cases of *dismissed-unresolved* process variants. On the other hand, the baseline boundary cases for *fully paid-unresolved* present similar sizes and subtle differences in *SF* and *PYM*. Nevertheless, their sequences match the baseline prototype sequences shown in Table 3. Thus, they are not adequate boundary cases, because they fail to reveal the subtle decision frontier between variants by considering the contrary prototype as a boundary case itself. In contrast, the boundary cases identified by our approach show more subtle divergences reflecting more accordingly the frontiers between the process variants. In the Sepsis log, the *SIRS–false-SIRS–true* pair initially follows similar treatment paths, but diverges after *NC*. The *SIRS–true* case includes additional diagnostic tests (such as *CRP* and *LC*) before release (*RA*). In the RTFM log, the *dismissed-unresolved* pair develops similarly but slightly differs in the final activities. In the *dismissed* case, *SA* is the final activity, whereas in the *unresolved* case finishes in *AJ*. Likewise, in the *fully paid-unresolved* pair, the cases share an identical prefix up to *IF*, but diverge afterward: *SA* occurs after *IF* in the *fully paid* case, but before *IF* in the *unresolved* one. Also, the *unresolved* case contains *PYM* at the end, suggesting inconsistent resolution patterns compared to *fully paid* cases.

Further insights emerge from SHAP explanations of the DECLARE-constraint features contributing to these predictions. For example, in the *fully paid-unresolved* pair, $\text{CONF}(\text{ALTERNATERESPONSE}(PYM, AP)) = 0$ (i.e., *AP* does not occur after every single occurrence of *PYM*) contributes negatively toward the *fully paid* class, which is consistent with the prototype, where no occurrences of *AP* precede *PYM*. Moreover, in the *unresolved-dismissed* pair, $\text{CONF}(\text{END}(AJ)) = 100$ in the *unresolved* case contributes negatively toward its variant, because it contradicts the *dismissed* prototype, which is characterized by ending in *SF* rather than *AJ*.

We notice that the differences showed by our boundary cases might not suffice to give a full account of the discriminants, because the differentiating features are control-flow centric. Other perspectives, such as the data attributes, could be required. We envisage this exploration as a crucial part of future work.

4.4 Configuration Study

To demonstrate implementation-independence, we have configured two steps differently and analyzed their results. Due to space constraints, we only report

Table 4. Boundary cases for the selected pairs of process variants. Activities are abbreviated as in Table 3.

Approach	Variant	Boundary cases
Ours	Dismissed	⟨ CF, SF, IF, AP, ID, SA ⟩
	Unresolved	⟨ CF, SF, IF, AP, ID, SA, RP, NR, AJ⟩
	Fully paid	⟨ CF, SF, IF, ID, SA, AP, PYM⟩
	Unresolved	⟨ CF, SF, ID, SA, IF, AP, PYM⟩
	SIRS-false	⟨ RG, TG, TGS, CRP, LC, NC, LC, CRP, LC, LC, CRP, CRP, LC, RA, CRP, LC⟩
	SIRS-true	⟨ RG, TG, TGS, CRP, LC, NC, LC, CRP, CRP, LC, CRP, LC, LC, CRP, CRP, LC, RA ⟩
Mean length	Dismissed	⟨ CF, SF, AJ ⟩
	Unresolved	⟨ CF, SF, IF, ID, AP, SA⟩
	Fully Paid	⟨ CF, SF, PYM⟩
	Unresolved	⟨ CF, PYM, SF⟩
	SIRS-false	⟨ RG, TG, CRP, LA, LC, TGS, IC, LA, LC, CRP, LA, LC, CRP, LA, NC, RA, RE⟩
	SIRS-true	⟨ RG, TG, TGS, LC, LA, CRP, LQ, AB⟩

a summary here. Further details are available in our repository.[3] Firstly, we changed Step 1 by encoding cases using a bag of activities (BOA), that is representing a case with sole activity frequencies. We identified a difference for the *Unresolved* variant, where BOA enconding has a worse F1-score (0.83) compared to the Declare encoding's F1-score (0.91). Analyzing SHAP values with BOA encoding, we have observed that it provides less descriptive explanations compared to the Declare-based ones. For example, the most contributing feature of the *Dismissed* prototype with the BOA encoding is $AJ = 1$ (i.e., an appeal to the judge occurs) while in the Declare encoding, it is $\textsc{conf}(\textsc{End}(AJ)) = 100$, clarifying that the prototype *concludes* with the appeal, hence providing a richer explanation.

Secondly, we changed Step 3 by replacing XGBoost with a Histogram Gradient Boosting Classifier (HIST).[4] No differences in F1-scores emerged. Prototype quality varied instead, as HIST achieved better intra- and inter-class distances with the *SIRS–False* variant, while XGBoost achieved better intra-class distances with the *collected*, *fully paid* and *dismissed* variants. HIST boundary cases for the Sepsis log are shorter, hence more intuitive. XGBoost boundary cases for the RTFM log represent frontiers better, since their differences are more subtle (see Table 5). Thus, our approach does not rely on a concrete classifier to find good

Table 5. Example of boundary cases by an alternative classifier for the Road Traffic Fines log. Activities are abbreviated as in Table 3.

Variant	Boundary cases
Fully paid	⟨ CF, SF, IF, AP, AJ, PYM, PYM, PYM, PYM⟩
Unresolved	⟨ CF, SF, IF, AJ, AP, PYM⟩

[3] Available at https://doi.org/10.5281/zenodo.20425194.
[4] scikit-learn/modules/HistGradientBoostingClassifier. Accessed: 31/05/2026.

examples. The classifier in use is a parameter to configure based on the desired explainability trade-off (e.g., intuitive or subtle boundary case differences).

4.5 Comparison with Rules

To illustrate how our approach complements rule-based explanations, we apply RIPPER, an algorihm of rule extraction, on X_{train}, replacing *none* values with -100. RIPPER has been previously used in process mining to distinguish variants [9]. Table 6 reports the extracted rules for each process variant, expressed as logic predicates over the confidence of DECLARE constraints. For readability, we express rules in disjunctive normal form. For example, three conjunctive clauses characterize *SIRS-false*. If any of those holds (e.g., CONF(ABSENCE(*IV Antibiotics*)) = 100), then the variant is *SIRS-false*.

Comparison with RTFM Rules.
To extract rules for the RTFM log, we applied RIPPER on a stratified 20% sample of X_{train} due to scalability constraints. We limited the maximum number of conditions of the rules to five. The resulting rules clearly characterize variants such as *collected* and *dismissed*. However, for *unresolved* and *fully paid*, the rule sets become harder to interpret: they involve multiple clauses, repeated predicates, and overlapping conditions (e.g., constraints involving *AddPenalty* and *SendForCreditCollection*). As a result, understanding how these variants unfold requires combining several predicates, which is cognitively demanding.

Our prototypes complement these rules by providing compact, concrete representations of variant behavior. For instance, while the rules for *fully paid* involve multiple constraints on penalties and ordering, the prototype directly captures the core behavior as a simple payment completion trace ($\langle CF, PYM \rangle$). SHAP values further align these two views by linking the classifier's decisions on concrete cases to the underlying declarative constraint features. In particular, for the *fully paid* prototype, the constraint ALTERNATERESPONSE(*Payment*, *AddPenalty*) with confidence 0 (i.e., no penalty following payment) is the strongest contributor to the prediction (1.21). This observation makes it explicit that the absence of penalties after payment is a key distinguishing factor. This insight is only implicitly distributed across multiple clauses. In this way, SHAP acts as a bridge between abstract rule-based explanations and concrete examples, enabling a coherent interpretation across representations.

The complementarity is also particularly evident when analyzing decision boundaries. For the *dismissedunresolved* pair, boundary cases reveal that real executions are more nuanced than what rule-based conditions suggest. Although the rules for the *dismissed* variant indicates that cases ending with *AppealToJudge* (CONF(END(*Appealtojudge*)) = 100) are classified as *dismissed*, the boundary case $\langle CF, SF, IF, AP, ID, SA, RP, NR, AJ \rangle$ belongs to the *unresolved* variant despite satisfying this condition. This shows that individual rule conditions are not always sufficient to determine the variant, as the broader execution context also plays a role. Boundary cases complement rules by grounding them in concrete executions and revealing where rule-based abstractions break down.

Table 6. Rules obtained for each process variant of each log. Rules span over different lines, each representing a conjunctive clause. When multiple clauses are listed for a variant, their logical disjunction entails that variant.

Log	Variant	Rules
RTFM	Collected	conf(Absence(*SendforCreditCollection*)) = 0
	Dismissed	conf(End(*Appealtojudge*)) = 100
		conf(CoExistence(*Payment*, *InsertDateAppealtoPrefecture*)) = 0 ∧ conf(Absence(*ReceiveResultAppealfromPrefecture*)) = 100
	Unresolved	conf(AlternateResponse(*SendFine*, *SendForCreditCollection*)) = 0 ∧ conf(End(*SendFine*)) = 100
		conf(AlternateResponse(*AddPenalty*, *SendForCreditCollection*)) = 0 ∧ conf(Alt.Prec.(*Payment*, *AddPenalty*)) = 100
		conf(Alt.Resp.(*AddPenalty*, *SendForCreditCollection*)) = 0 ∧ conf(AtLeast2(*Payment*)) = 0 ∧ conf(Alt.Prec.(*ReceiveResultAppealfromPrefecture*, *InsertDateAppealtoPrefecture*)) = 0 ∧
		conf(CoExistence(*ReceiveResultAppealfromPrefecture*, *SendAppealtoPrefecture*)) = 100 ∧ conf(End(*AppealToJudge*)) = 0
	Fully paid	conf(Alt.Resp.(*Payment*, *AddPenalty*)) = 0 ∧ conf(Absence(*SendFine*)) = 100 ∧ conf(Alt.Prec.(*CreateFine*, *Payment*)) = 50
		conf(Alt.Resp.(*Payment*, *AddPenalty*)) = 0 ∧ conf(Absence(*SendFine*)) = 100
		conf(Alt.Resp.(*Payment*, *AddPenalty*)) = 0 ∧ conf(End(*Payment*)) = 100 ∧ conf(Alt.Prec.(*CreateFine*, *Payment*)) = 50 ∧
		conf(Absence(*AppealtoJudge*)) = 100 ∧ conf(Absence(*AddPenalty*)) = 100
		conf(Alt.Resp.(*Payment*, *AddPenalty*) = 0) ∧ conf(End(*Payment*)) = 100 ∧ conf(Alt.Prec.(*AddPenalty*, *Payment*)) = 50
		conf(Alt.Resp.(*Payment*, *AddPenalty*)) = 0 ∧ conf(End(*Payment*)) = 100 ∧ conf(Absence(*AddPenalty*)) = 100
		conf(NotPrecedence(*AddPenalty*, *Payment*)) = 0 ∧ conf(Absence(*SendForCreditCollection*)) = 100 ∧ conf(ChainPrecedence(*InsertFineNotification*, *AddPenalty*)) = 100
		conf(Alt.Resp.(*Payment*, *SendForCreditCollection*)) = 0 ∧ conf(Alt.Prec.(*Payment*, *AddPenalty*)) = 0 ∧ conf(Alt.Resp.(*AddPenalty*, *ReceiveResultAppealFromPrefecture*)) = 0 ∧ conf(Alt.Resp.(*SendAppealtoPrefecture*, *ReceiveResultAppealfromPrecture*)) = 0
Sepsis	SIRS-false	conf(Absence(*IVAntibiotics*)) = 100 ∧ conf(NotPrecedence(*AdmissionNC*, *CRP*)) ≤ 14.29
		conf(Absence(*IVAntibiotics*)) = 100
		conf(ChainResponse(*Leucocytes*, *ReleaseC*)) = 33.33
	SIRS-true	conf(Absence(*IVAntibiotics*)) = 0

Comparison with Sepsis Rules. The extracted rules provide discriminative explanations suggesting that the distinction between *SIRS-false* and *SIRS-true* primarily depends on the presence of *IVAntibiotics*. While this offers a clear characterization, it reduces the difference between variants to a single condition and provides limited insight into how the process unfolds. Our prototypes further clarify how the classifier characterizes each variant. The shared prefix (⟨ *RG*, *TG*, *TGS* ⟩) appears in both prototypes, indicating that these initial steps are not discriminative for the model. Instead, the distinction emerges afterward: the *SIRS-false* prototype is characterized by the absence of subsequent activities, whereas the *SIRS-true* prototype includes additional diagnostics and treatment, most notably *IVAntibiotics*. Importantly, this distinction is not limited to a single activity. SHAP values show that the classification of *SIRS-false* is driven by the vacuous satisfaction of several declarative constraint features. Therefore, the model does not rely solely on the absence of *IVAntibiotics*, but on a broader lack of continued clinical actions, which remains implicit in the rule-based explanation.

The complementarity is further highlighted when analyzing decision boundaries. Boundary cases for the two variants exhibit highly similar sequences, including repeated diagnostic tests, yet differ in their assigned variant. Notably, both cases lack *IVAntibiotics*, though that is the key rule-based discriminator. This indicates that the rule condition alone is not sufficient to fully explain the classification, as additional contextual factors influence the outcome. Rules identify a dominant distinguishing feature, while prototypes and boundary cases contextualize that feature revealing where its explanatory power is insufficient.

5 Conclusions

In this work, we presented a new approach to analyze process variants using *natural examples*. Our results show that examples can complement rules to this end by grounding abstract conditions in concrete executions and by exposing how variant differences manifest in practice.

However, several limitations should be acknowledged. First, our approach inherits the typical challenges of classification algorithms. One of these is multicollinearity, which arises when two or more features are highly correlated. This can hinder interpretation, as the model may rely on one variable while the actual contribution comes from another. While not the focus of this work, our use of examples partially mitigates this issue by providing observable evidence of the model's behavior. Second, the evaluation does not involve domain experts, limiting our ability to assess the practical interpretability and usefulness of the explanations. Third, we observed that identifying boundary cases is not always feasible in real-world logs, particularly when variants are well separated. Finally, understanding Declare constraints and their associated confidence values may be challenging for users unfamiliar with declarative modeling.

Beyond these limitations, our findings also highlight opportunities. In particular, boundary cases not only explain differences but can reveal when variant distinctions cannot be fully captured by the chosen feature space or by rule-based abstractions. This observation suggests that example-based explanations can act as a diagnostic tool, indicating when additional perspectives (e.g., data attributes) are needed. Future work will explore several directions. We plan to investigate feature selection techniques to reduce multicollinearity and improve interpretability, as well as alternative modeling choices and parameter optimization strategies to enhance robustness. We also aim to incorporate data attributes alongside control-flow features to better capture variant differences that are not behaviorally explicit. Furthermore, we will study mechanisms to ensure the plausibility of boundary cases with respect to the original process behavior. Finally, we intend to conduct comprehensive user studies with process analysts to evaluate the practical usefulness and cognitive effectiveness of example-based explanations.

Acknowledgments. This paper is part of the projects PID2021-126227NB-C21, PID2022-140221NB-I00, and PID2024-156482NB-I00 funded by MICIU/AEI/10.13039/501100011033/FEDER, UE. This work was also funded by the Junta de Andalucía (Consejería de Universidad, Investigación e Innovación) and co-funded by the EU through the FEDER Andalucía 20212027 Programme under project DGP_PIDI_2024_01609. C. Capitán-Agudo is supported by the Spanish Ministry of Education under the FPU national plan (FPU21/03631).

References

1. Bose, R.J.C., van der Aalst, W.M.: Discovering signature patterns from event logs. In: CIDM, pp. 111–118 (2013)
2. Cecconi, A., Augusto, A., Di Ciccio, C.: Detection of statistically significant differences between process variants through declarative rules. In: BPM Forum, pp. 73–91. Springer International Publishing, Cham (2021)
3. Cecconi, A., De Giacomo, G., Di Ciccio, C., Maggi, F.M., Mendling, J.: Measuring the interestingness of temporal logic behavioral specifications in process mining. Inf. Syst. **107**, 101920 (2022)
4. Chen, T., Guestrin, C.: XGBoost: A scalable tree boosting system. In: SIGKDD 2022, pp. 785–794. Association for Computing Machinery, New York, NY, USA (2016)
5. Cremerius, J., Patzlaff, H., Rahn, V., Leopold, H.: Data-based process variant analysis. In: HICSS (2023)
6. Cuzzocrea, A., Folino, F., Guarascio, M., Pontieri, L.: A robust and versatile multi-view learning framework for the detection of deviant business process instances. Int. J. Coop. Inf. Syst. **25**(04), 1740003 (2016)
7. Di Ciccio, C., Mecella, M.: On the discovery of declarative control flows for artful processes. ACM Trans. Manage. Inf. Syst. **5**(4) (2015)
8. Di Ciccio, C., Montali, M.: Declarative process specifications: reasoning, discovery, monitoring. In: Process Mining Handbook, pp. 108–152. Springer (2022)
9. Di Francescomarino, C., Donadello, I., Ghidini, C., Maggi, F.M., Puura, J.: Business process deviance mining with sequential and declarative patterns. Bus. Inf. Syst. Eng. **67**, 877–894 (2025)
10. Doyle, D., Cunningham, P., Bridge, D., Rahman, Y.: Explanation oriented retrieval. In: Advances in Case-Based Reasoning, pp. 157–168 (2004)
11. Fani Sani, M., Boltenhagen, M., van der Aalst, W.: Prototype selection using clustering and conformance metrics for process discovery. In: BPM Workshops, pp. 281–294. Springer International Publishing, Cham (2020)
12. Grigore, I.M., Marques Tavares, G., Pasquadibisceglie, V., Seidl, T., Barbon Junior, S.: Towards trace variant explainability. In: ADBIS, pp. 95–110. Springer Nature Switzerland, Cham (2026)
13. Iacometta, C., Di Ciccio, C.: Declarative process mining with MINERful, reloaded. In: BPM Demos & Resources, pp. 328–335 (2025), open access

14. de Leoni, M.M., Mannhardt, F.: Road traffic fine management process (2015). https://doi.org/10.4121/uuid:270fd440-1057-4fb9-89a9-b699b47990f5
15. Lundberg, S.M., Lee, S.I.: A unified approach to interpreting model predictions. In: NeurIPS, p. 4768–4777. Curran Associates Inc., Red Hook, NY, USA (2017)
16. Maggi, F.M., Bose, R.P.J.C., van der Aalst, W.M.P.: Efficient discovery of understandable declarative process models from event logs. In: CAiSE, pp. 270–285. Springer Berlin Heidelberg, Berlin, Heidelberg (2012)
17. Mannhardt, F.: Sepsis cases - event log (2016). https://doi.org/10.4121/uuid:915d2bfb-7e84-49ad-a286-dc35f063a460
18. Nannini, L., Catala Bolos, A., Lama, M., Barro, S.: Explainable AI in process mining: a systematic literature review (2025)
19. Pesic, M., Schonenberg, H., van der Aalst, W.M.: DECLARE: full support for loosely-structured processes. In: EDOC, pp. 287–287 (2007)
20. Poché, A., Hervier, L., Bakkay, M.C.: Natural example-based explainability: a survey. In: XAI World Conference, pp. 24–47 (2023)
21. Rabold, J., Siebers, M., Schmid, U.: Generating contrastive explanations for inductive logic programming based on a near miss approach. Mach. Learn. **111**, 1799–1820 (2022)
22. Richetti, P.H.P., Jazbik, L.S., Baião, F.A., Campos, M.L.M.: Deviance mining with treatment learning and declare-based encoding of event logs. Expert Syst. Appl. **187**, 115962 (2022)
23. Rubensson, C., Mendling, J., Weidlich, M.: Variants of variants: context-based variant analysis for process mining. In: CAiSE, pp. 387–402. Springer Nature Switzerland, Cham (2024)
24. Swinnen, J., Depaire, B., Jans, M.J., Vanhoof, K.: A process deviation analysis - a case study. In: BPM Workshops, pp. 87–98 (2012)
25. Taymouri, F., La Rosa, M., Dumas, M., Maggi, F.M.: Business process variant analysis: survey and classification. Knowl.-Based Syst. **211**, 106557 (2021)
26. Völzer, H., Zerbato, F., Sulzer, T., Weber, B.: A fresh approach to analyze process outcomes. In: ICPM, pp. 97–104 (2023)

Agentic Business Process Modeling: Co-creating Business Processes with Human-Centric AI Agents

Seyyid A. Ciftci(✉) and Christian Janiesch

TU Dortmund University, Otto-Hahn-Str. 12, 44227 Dortmund, Germany
{seyyid.ciftci,christian.janiesch}@tu-dortmund.de

Abstract. Business process modeling is a highly complex, cognitively demanding task requiring deep domain understanding. While large language models make transforming natural language descriptions into formal models increasingly achievable, conventional single-prompt approaches often yield structural inconsistencies and computational inefficiency. To overcome these limitations, integrating agentic artificial intelligence (AI) systems offers promising capabilities. Deploying these autonomous systems can significantly enhance execution. Such architectures iteratively construct, control, and improve models by validating logic, resolving ambiguities, and inferring implicit requirements from unstructured tasks. Yet, integrating such proactive agents into organizational workflows introduces novel socio-technical risks. When deployed with a task-centric design, these applications often impose excessive cognitive load on human modelers, negatively affecting the psychological work experience and task performance, as autonomous execution inherently limits human agency. To address these challenges, we introduce an interactive AI system for agentic process modeling, developed and evaluated through an iterative design science approach. Shifting from monolithic generation to structured collaboration, we propose a self-reflective agent architecture. The system continuously validates internal logic and actively integrates human feedback to resolve structural discrepancies. By prioritizing human-centric design, this approach mitigates risks discovered in exploratory research, including agent-induced cognitive load and the complacent acceptance of AI-generated models. Our findings offer theoretical insights and practical guidelines for configuring human-AI collaborations that ensure efficiency without compromising human agency or cognitive well-being.

Keywords: Agentic AI · Business Process Modeling · Human-AI Collaboration · Large Language Model · LLM · Design Science Research

1 Introduction

Business process modeling is a highly complex, cognitively demanding task requiring deep domain understanding [26]. The formalization of organizational workflows into standardized representations, such as the Business Process Model and Notation (BPMN),

F. M. Maggi et al. (Eds.): BPM 2026, LNCS 16936, pp. 313–330, 2027.
https://doi.org/10.1007/978-3-032-37777-7_20

traditionally relies on the manual effort of human analysts [8]. Organizations increasingly turn to artificial intelligence (AI) to support such complex knowledge work. The use of generative AI (GenAI) and large language models (LLMs) has made the transformation of natural language descriptions into process models a viable approach to assist process discovery and automate process modeling [10, 14, 22].

However, conventional approaches to automated process modeling introduce systemic risks [19]. Relying on monolithic generation, where a single prompt is expected to synthesize a complete process model, frequently falters due to the probabilistic nature of LLMs. This approach culminates in structural inconsistencies, syntax errors, and severe computational inefficiency [17]. Consequently, by neglecting established guidelines for human-AI interaction [1], presenting human modelers with a complete, yet flawed, AI-generated model forces them into a cognitively taxing process of retroactive correction, hindering effective collaboration.

The paradigm of AI-generated content is shifting from reactive to agentic AI [9, 28]. Emerging as organizational partners, these autonomous systems – capable of multi-step reasoning, proactive planning, and independent execution – reshape traditional collaborative structures and introduce socio-technical challenges [29]. Guided by human-centric AI principles [31], these systems must be designed to augment human capabilities rather than replace human oversight. Deployed with a task-centric design, such systems impose excessive cognitive load on human modelers. This unstructured collaboration induces split-attention, degrading the psychological work experience, diminishing human agency, and compromising task performance [6, 33].

The question therefore no longer is whether agentic AI will be used in process modeling, but how such systems can be integrated into knowledge work in ways that increase performance while safeguarding human autonomy and well-being. Addressing these challenges leads us towards the following research question (RQ):

RQ: *How can an agentic AI system be designed to construct business process models autonomously in a way that maximizes task performance while mitigating cognitive load and preserving human agency?*

We applied design science research (DSR) to systematically answer this research question by developing and evaluating a functional AI application for agentic process modeling over three iterative design cycles [23]. Building on the first cycle – a prior explorative study of three initial AI configurations – we developed a more synergistic setting for human-AI collaboration. We propose an autonomous agent architecture that seamlessly integrates task planning, step-by-step execution, and validation. By engaging in continuous self-evaluation and transparently presenting potential conflicts to the human, the system ensures a collaborative modeling process. Grounded in cognitive load theory (CLT), our decoupled approach mitigates critical risks identified in prior research, specifically extraneous cognitive load and cognitive complacency towards AI-generated models [27]. It also yields better modeling results and higher user acceptance than earlier configurations. In solving engineering-related issues, we developed a lightweight data format to optimize communication efficiency with LLMs. This format reduces payloads compared to conventional JSON or XML, lowering the required context window and latency.

The remainder of this paper is structured as follows. Section 2 outlines the theoretical background, while Sect. 3 discusses related work in automated process modeling. Section 4 details the research methodology and artifact development. Section 5 presents the developed agentic application. Section 6 provides the summative evaluation results. Section 7 discusses the findings. Section 8 outlines the theoretical and practical implications as well as limitations, while Sect. 9 concludes our research.

2 Theoretical Background

2.1 Autonomous AI Agents

The integration of AI is fundamentally reshaping traditional collaborative structures. Organizations increasingly view AI not merely as a tool, but as a digital partner in organizational contexts [30]. This evolution necessitates a redesign of interaction dynamics, conceptualizing these collaborations as environments where rules, roles, and workflows are clearly defined. As AI systems assume complex cognitive tasks like process modeling, their application transitions towards a genuine delegation of responsibility.

A concept in this socio-technical evolution is AI agency [29]. Agency denotes the capability of an AI artifact to act intentionally and with authority within a defined social environment [3, 18]. Agentic systems are characterized by their capacity for multi-step reasoning, proactive planning, and independent execution [16]. Consequently, agency is perceived dynamically through interaction: humans delegate complex goals, and the agent orchestrates the necessary subtasks, uses tools, or queries input.

Applying this agentic paradigm to process modeling allows for the configuration of reflective architectures that simulate human iterative problem-solving [9]. Rather than attempting to process all constraints simultaneously in one monolithic pass, an autonomous system can transition through distinct, complementary cognitive phases within a workflow. The agent dynamically shifts between generating initial design proposals, independently evaluating its own structural compliance, and actively querying human feedback when ambiguities arise. This phased, self-correcting execution enables iterative refinement and error-resistant task fulfillment.

2.2 Cognitive Load Theory

The CLT provides a framework for understanding how human modelers process information and how interaction with autonomous AI agents affects their performance [33]. CLT assumes that human cognitive architecture consists of long-term memory with virtually unlimited capacity and working memory with very limited capacity. Every form of information processing generates *cognitive load*, which is the amount of mental effort required to hold and process information in one's working memory. This total cognitive load is divided into three categories:

First, intrinsic load is determined by the inherent complexity of the task itself. In the context of business process management, modeling workflows is a highly demanding cognitive activity [26]. Modelers must simultaneously process abstract business rules, maintain logical dependencies, and adhere to strict syntax constraints in working memory. The volume of interconnected elements in a business process requires continuous mental alignment, making the intrinsic cognitive baseline exceptionally high.

Second, extrinsic load is generated by the system's interaction design. In human-AI collaboration, inferior interfaces or unpredictable behaviors impose unnecessary mental effort. When systems present information in a disorganized manner or misalign with human cognitive processes, these setups induce a split-attention effect [6]. Users are forced to divide their limited working memory between executing the primary task and navigating or monitoring the AI, which finally leads to cognitive overload.

Third, germane load describes the mental resources expended on deep understanding, problem-solving, and building mental schemata in long-term memory. For process modelers, this involves synthesizing domain knowledge into coherent workflow structures [26]. The primary objective in designing effective human-AI interaction is to strictly minimize extrinsic load. By reducing the mental effort required to coordinate with the AI, valuable working memory capacity is freed up for intrinsic processing and germane schema construction, thereby enabling higher-quality modeling outcomes.

CLT has been used in relation to business process management and process modeling for over a decade [11, 25, 34, 38]. The NASA Task Load Index (NASA-TLX) [15] is a widely used subjective, multidimensional assessment to evaluate perceived workload.

3 Related Work

3.1 LLM-Based Process Modeling

Through the integration of natural language processing (NLP) and LLMs, the automated generation of BPMN models from textual descriptions has emerged as an established practice, building upon foundational extraction methodologies [12, 20, 22]. Leveraging these combined technologies enables the synthesis of process models directly from natural language and facilitates their immediate embedding into organizational workflows. A prevalent approach involves prompting the LLM to output standard serialization formats, such as XML, in a single generative pass to allow direct integration into existing process engines [39]. Advanced techniques, including refined prompting strategies, model fine-tuning, and the inclusion of few-shot examples, are frequently employed to enhance the comprehension and output accuracy of these models.

However, while this monolithic generation serves as a common baseline for process discovery, it consistently encounters structural limitations. The inherent verbosity of XML frequently exceeds the models' effective context window during generation [2]. This constraint leads to syntax errors, structural inconsistencies, and hallucinations, leading to invalid models. Consequently, users must either manually execute the necessary corrections or explicitly delegate structural adjustments through additional prompts. This continuous retroactive correction introduces significant extrinsic cognitive load, compounding the inherently high intrinsic complexity of process modeling.

3.2 Action-Based and Agentic Modeling Approaches

Addressing the structural failures of monolithic XML generation, the methodological focus has shifted toward more structured, intermediate representations. Certain frameworks conceptualize process modeling as a conversational interaction rather than a one-shot generation [17]. Instead of generating an entire XML tree at once, these approaches instruct the LLM to output process steps as a series of structured actions serialized in JSON format [21]. These JSON-based actions are subsequently parsed locally and compiled into the final valid XML structure. This action-based paradigm represents an improvement, as abstracting the graphical syntax into intermediate JSON objects reduces the immediate complexity handled by the LLM.

Furthermore, the evolution toward autonomous AI agents introduces distinct methodological advantages over standard conversational LLM applications. Agentic systems possess the capability to not merely generate process models, but to iteratively analyze task requirements, actively construct the workflow, and execute precise adjustments based on intermediate evaluations [5, 37]. Existing conversational frameworks predominantly rely on continuous, sequential outputs from a single model, lacking a decoupled architecture that separates task planning, execution, and validation. This leaves a critical gap for a framework that leverages agentic capabilities and optimized data formats to enhance both task performance and the construction of process models.

4 Research Methodology and Artifact Development

We followed an iterative DSR approach comprising three cycles with the aim of designing sustainable human-AI collaboration in process modeling [23], as illustrated in Fig. 1. The artifact was designed around three intervention points within active model creation, derived from recurring modeling needs identified in prior work on conversational process modeling and automated BPMN generation [17, 20]: (1) knowledge transfer and clarification through answering task-specific and BPMN-related questions to resolve uncertainties, (2) automated generation of BPMN models to reduce the user's manual effort, and (3) error detection and constructive feedback by reviewing models to highlight syntactic BPMN errors and semantic deviations from the task, providing concrete suggestions for improvement. In this context, syntactic errors refer to violations of the BPMN notation, metamodel, and structural modeling rules defined by the standard [35], whereas semantic deviations refer to conceptual mismatches between the BPMN model and the described task. These forms of support were operationalized as three interaction roles: *AI Assistant*, *AI Colleague*, and *AI Supervisor*. This role-based conceptualization allowed us to examine how different forms of AI involvement influence human cognition, agency, and behavior throughout the modeling process.

Cycle 1: Prior Exploratory Study and Problem Identification. The first cycle was conducted as a foundational prior study [7] and aimed at problem identification through an exploratory study involving 15 students from computer science, information systems, and data science. We summarize it in the following:

Prior completion of a BPMN course served as a prerequisite for participation. We implemented a software environment where participants solved randomly assigned modeling tasks in collaboration with three distinct AI archetypes. Using the AI Assistant, the

DSR Cycle Phases	1st Cycle	2nd Cycle	3rd Cycle
Problem Awareness	Human-centric AI for Process Modeling	Feedback Analysis	Benchmark Analysis
Suggestion	Explore AI Collaboration Archetypes	Avoiding Cognitive Overload and Architectural Redesign	Consolidation of Agents
Development	Prototyping AI Modeling Environment	Refinement of Worker and Supervisor Agent	Development of Self-Reflective Agentic AI
Evaluation	Evaluation through Exploratory Study	Evaluation through Benchmarks	Evaluation through Participants and Benchmarks
Conclusion	↗	↗	Final Artifact

Fig. 1. Overview of the design science research, illustrating the three iterative cycles

human modeled entirely alone and could complete the task independently, while the AI only answered questions. The AI Colleague modeled simultaneously with the human, requiring continuous coordination between both actors regarding task completion. The *AI Supervisor* focused on proactive error detection and feedback. Each participant completed three of five possible tasks with a time limit of 13 min per task. Data collection occurred via a think-aloud protocol, transcribed for analysis.

For the evaluation of the generated models, we developed a deductive scoring schema. Model solutions, which depicted the task requirements with high quality and a minimal number of elements, formed the assessment baseline. The schema penalized missing elements, incorrect assignments in pools or lanes, unresolved race conditions, and cosmetic errors. Since the five tasks featured varying element counts and maximum scores, normalizing the final values ensured comparability across all assignments. Measuring cognitive load occurred via NASA-TLX.

The formative results revealed critical limitations regarding user autonomy when interacting with proactive AI types. The AI Colleague delivered inconsistent performance and forced a high coordination overhead, which participants perceived as a severe cognitive burden. The AI Supervisor proactively checked progress and generated extensive amounts of textual feedback frequently, overwhelming the cognitive capacities of the participants. Because the AI Supervisor also held the authority to deny task completion, all participants in this configuration exceeded the time limit. Conversely, all participants preferred the reactive AI Assistant, as the system purposefully answered questions without undermining the decision-making authority of the users. This initial cycle concluded in November 2025 and grounds our current research [7].

Cycle 2: Human-in-the-Loop Refinement and Format Optimization. Driven by the critical limitations identified in our previous work, this current research commenced by directly addressing the participants' feedback. The second cycle therefore focused on restoring human agency and reducing cognitive load. Consequently, we revised the AI Colleague so that the system assumed the entire modeling task. This adjustment eliminated the split-attention effect, where users had to constantly switch focus between the task description, personal modeling efforts, and the chat with the AI. However, the initial modeling capabilities of the AI exhibited weaknesses. Integrating a preliminary planning phase resolved the problem of contradictory solution approaches during generation. The agent analyzes the task, formulates sub-goals, and continuously receives the current status of the BPMN model, creating a precise situational understanding.

The AI Supervisor also underwent an architectural adjustment. Participants perceived proactive, textual feedback as highly disruptive. Consequently, we transformed the AI Supervisor into a reactive system, checking models only upon user request. The system now highlights errors and annotations through visual markers directly within the process model. This localized presentation facilitates cognitive understanding without continuously interrupting the user's workflow. Both archetypes lost authority over task completion, transferring full control back to the human. The user decides independently when to complete, preventing frustration from endless AI Supervisor correction loops over minor details. Lastly, both roles integrate the AI Assistant's question-answering capability.

From an engineering perspective, the increased complexity of the prompts necessitated an optimization of token consumption. In response, we developed a lightweight data format that replaces the previous JSON format for action generation. The format eliminates repetitive keys through parameterized lists for similar objects and uses a simplified syntax. The resulting token reduction prevents exceeding the effective context length, a critical threshold where the attention mechanism of LLMs degrades and accuracy drops [2]. Concurrently, the lower token consumption decreases API costs and latency, as the language model generates fewer text components. This optimization was necessary, as continuous, multi-step interactions with the model would have otherwise led to unacceptable delays. We conducted benchmarks to evaluate the architectural adjustments by comparing XML, JSON, and our new lightweight data format. This assessment was completed in February 2026.

Cycle 3: Transition to an Agentic and Self-Reflective AI. Although the second cycle improved latency times, raw modeling performance still exhibited shortcomings. Letting solely the AI handle the modeling initially confronted the human with a massive verification effort, potentially increasing cognitive load. Furthermore, the first study revealed that the AI Colleague triggered systematic complacency, where participants accepted AI-generated solutions without critical examination.

The third cycle consolidated the AI Colleague and the AI Supervisor into a single, cohesive worker agent. Following the planning and modeling phase, this agent now controls the drafts through internal self-reflection. Upon detecting critical semantic deviations, inconsistencies with the task description, or syntactic errors, the agent iteratively revises the model. This autonomous preliminary correction measurably increases overall system performance and reduces the initial review effort for the human.

To mitigate previous issues, the revised workflow forces the human into an active reviewer role and counteracts automation bias [27]. The procedure secures human agency and authority while enforcing conscious quality control before task completion.

We evaluated the final artifact in two stages. First, an offline generation quantified the agent's raw performance without human involvement. Subsequently, the participants from the first study returned after five months to complete another run using the final system. This deliberate temporal gap mitigated potential habituation and learning effects. The evaluation captured modeling performance and perceived cognitive load, reusing our scoring schema once again. The last cycle completed in March 2026.

In summary, the final artifact developed in these three cycles provides an agentic system capable of automated model generation. This mitigates the cognitive burden and coordination overhead that hampered users during our prior study in the first cycle. Furthermore, the lightweight data format, initially engineered to ensure the artifact's operational feasibility, yielded substantial standalone benefits by reducing token consumption and increasing processing speed during continuous LLM interactions.

5 Artifact Design

Our approach introduces an autonomous AI agent capable of analyzing assigned tasks, modeling business processes within bpmn.io [4], critically reviewing generated output, and allowing human experts to focus on strategic oversight and final validation.[1] This decoupled architecture was chosen to overcome limitations of monolithic generation, specifically LLM context constraints and hallucination risks. It ensures efficiency and provides granular control for precise, targeted modeling adjustments.

The agent operates through four primary cognitive methods, which the application dynamically invokes based on the context: planning, executing, reviewing, and answering questions. To ensure robustness and minimize errors, we employ a modular, phase-specific prompting architecture that constrains the LLM to context-bound tasks, thereby improving the reliability of generated outputs. Rather than relying on rigid backend hardcoding, established best practices, specifically aligning with the "Seven Process Modeling Guidelines (7PMG)" [24] and the "Ten Principles of Good Business Process Management" [36], serve as explicit semantic guardrails within the context of the agent. Comprehensive BPMN rules, element definitions, layout constraints, and naming conventions are dynamically injected into the system prompts, guiding the semantic and syntactic decisions of the agent. A persistent memory module equips the agent to maintain context. The memory module continuously tracks the original task description, human instructions, dynamically generated plans, and a historical log of all previously executed modeling steps.

Upon receiving a task description, the agent initiates the planning phase to establish concrete sub-goals. The agent begins by generating the structural foundation, placing the appropriate pools and lanes derived from the task requirements, as Fig. 2a illustrates. The system does not attempt to generate a monolithic file representing the entire process model in a single pass. Instead, the application uses a one-shot prompting strategy to instruct the underlying LLM. Providing an exact structural blueprint of the expected output within the prompt forces the LLM to translate conceptual plans into a sequence of discrete, actionable operations, such as draw, connect, update, or delete. This command-based interaction paradigm is a deliberate design choice: it enables incremental validation, allowing the system to verify the BPMN syntax of every operation in real-time, which is a prerequisite for high-quality, error-free models.

The agent works autonomously until the application achieves all planned objectives, at which point the system transitions into the review phase. The agent critically evaluates the current draft against the task description for semantic alignment and checks for syntactic BPMN errors. If the agent detects discrepancies, the system initiates a new

[1] https://doi.org/10.5281/zenodo.20459385

internal iteration: the agent formulates a revised plan to address identified issues, executes the necessary adjustments via new actions, and reviews the model again. Capping the self-reflective cycle strictly at a maximum of three internal iterations prevents the system from getting caught in infinite loops or losing focus. This interaction mechanism functions as a critical safeguard against automation bias [27], ensuring that the system does not over-process and that human oversight remains the final authority. If the agent cannot resolve all detected issues within the specified limit, the system proactively halts the autonomous loop and hands the intermediate state over to the human. The safety mechanism ensures that the user can provide targeted instructions, clarify ambiguities, or offer course correction before the modeling process becomes inefficient.

Fig. 2. (a) Modeling phase depicting autonomous planning and process construction; (b) Review phase highlighting semantic and syntactic inconsistencies visually

When presenting the model to the user, the agent highlights any remaining observations using structured visual feedback, which Fig. 2b illustrates in detail. We implemented three categories of visual markers. Critical errors (red) indicate semantic deviations from the task or violations of BPMN syntax rules. Warnings (yellow) highlight suboptimal structures needing attention, such as missing labels on the outgoing paths of an exclusive gateway. Information markers (blue) represent non-mandatory suggestions, such as improving the naming convention of an element. If warnings still exist in the model, the system prevents the complacent completion of the task. The visual approach serves as a direct response to user feedback from the initial study. This visual feedback mechanism is grounded in CLT. By replacing extensive textual summaries, which often induce split-attention effects, with localized visual markers, we ensure users can assimilate information efficiently without cognitive overload.

In the final stage, the human modeler acts as the human-in-the-loop authoritative entity. The user must explicitly evaluate the markers and grant permission to ignore the highlighted issues or delegate the highlighted issues back to the agent for correction. Alternatively, the human can perform manual adjustments to the model at any time. Furthermore, the user can interact with the agent by asking clarifying questions or providing new textual instructions. Any new directive triggers a fresh cycle of planning, modeling, and reviewing by the agent before the application returns control to the human, ensuring a seamless, collaborative, and iterative design process.

Our optimized, lightweight data format serializes these operations. By stripping repetitive keys and nested structural overhead, the system achieves higher token efficiency, as shown in Table 1. The backend parses this syntax and dispatches commands to the frontend for sequential execution. The agent visualizes these actions step-by-step, using a cursor with intentional execution delays. Displaying the model's evolution incrementally, rather than instantaneously, is a transparency-enhancing interaction design. It helps users build a coherent mental model of the agent's logic and intent, significantly reducing cognitive burden.

Table 1. Syntax and token efficiency comparison between JSON and our format

JSON (316 Tokens)

```
{
  "actions": {
    "participate": [
      {
        "x": 160, "y": 80, "w": 1500, "h": 520, "label": "Retailer",
        "id": "PoolRet", "expanded": true, "lanes": ["Portal", "Whse"]
      },
      {
        "x": 160, "y": 660, "w": 1500, "h": 60, "label": "Customer",
        "id": "PoolCust", "expanded": false, "lanes": []
      }
    ],
    "draw": [
      {
        "type": "StartEvent", "x": 220, "y": 180, "label": "Registered",
        "id": "Start01", "parent": "LaneP", "connectTo": ["T01"], "event": "Msg"
      },
      {
        "type": "ServiceTask", "x": 360, "y": 158, "label": "Gen Label",
        "id": "T01", "parent": "LaneP", "connectTo": ["T02"], "event": null
      },
      {
        "type": "SendTask", "x": 520, "y": 158, "label": "Send Label",
        "id": "T02", "parent": "LaneP", "connectTo": ["Gw01"], "event": null
      }
    ]
  }
}
```

Our Format (180 Tokens)

```
actions: {
  participate(x, y, w, h, label, id, expanded, lanes): [
    {160, 80, 1500, 520, "Retailer", PoolRet, true, [Portal, Whse]},
    {160, 660, 1500, 60, "Customer", PoolCust, false, []}
  ],
  draw(type, x, y, label, id, parent, connectTo, event): [
    {StartEvent, 220, 180, "Registered", Start01, LaneP, [T01], Msg},
    {ServiceTask, 360, 158, "Gen Label", T01, LaneP, [T02], null},
    {SendTask, 520, 158, "Send Label", T02, LaneP, [Gw01], null}
  ]
}
```

6 Results and Evaluation

Our mixed-methods evaluation assesses the artifact's impact on human cognition and task performance alongside its technical capabilities. We present the results along two dimensions, ensuring an overview of our iterative findings. First, we analyzed the human-AI collaboration by directly contrasting the newly developed agentic archetype against the distinct AI configurations from our previous explorative study conducted in Cycle 1. Second, we isolate the underlying computational efficiency to demonstrate the impact of our new data serialization format compared to previous baseline approaches.

6.1 Human-AI Collaboration and Cognitive Load

The first phase of the evaluation focuses on the performance of the socio-technical system, examining the evolving dynamics of decision-making within human-AI teams. The execution of Cycle 3 took place using the same experimental settings and procedures as our prior study in Cycle 1 [7], with the distinction that participants exclusively interacted with the newly developed agentic AI.

To quantify task performance and ensure comparability across all design cycles, we applied the same scoring schema established in Cycle 1. We developed a sample solution for each of the five standardized tasks. Points were awarded for correctly matching elements, including their exact types, labels, and valid connections, as well as for successfully resolving underlying race conditions. Conversely, penalties were applied for aesthetic flaws, such as overlapping elements, or for utilizing suboptimal element types (e.g., repeatedly defaulting to Exclusive-ORs when Inclusive-OR or Event-based Gateways were structurally more appropriate). Depending on the required complexity, the maximum achievable score varied between 50 and 70 points per task. To ensure evaluation reliability, independent raters achieved an Intraclass Correlation Coefficient (ICC) [32] of 0.919 (Cycle 1) and 0.884 (Cycle 2 and Cycle 3).

Table 2 compares the empirical data from participants in the prior study (Cycle 1), who interacted with the isolated AI Assistant, AI Colleague, and AI Supervisor archetypes, against the results of the participants using the final agentic artifact (Cycle 3).

The comparison demonstrates that the developed agentic system successfully resolves the socio-technical deficits of the previous AI configurations. The AI Colleague yielded inadequate task performance alongside high coordination demands, as users struggled to align their mental models with the AI's continuous, unsolicited interventions. Similarly, the AI Supervisor stessed cognitive load through proactive, text-heavy feedback that forced users into extensive reading rather than active modeling.

Conversely, the final agentic system achieved superior results across all measured metrics. Combining the autonomous preparatory work of the AI with the targeted, on-demand visual highlighting of discrepancies substantially reduced perceived cognitive load, implying lower overall workload measured via NASA-TLX [15, 33]. Unlike the previous types, the agentic AI validates asynchronously in the background and restricts its output to actionable, visual conflict resolutions, preventing information overload.

Detailed analysis of the NASA-TLX measures reveals that a sharp decrease in each dimension primarily drove the overall reduction in cognitive load and workload. Participants no longer had to decode extensive textual blocks but merely executed targeted adjustments on an already structurally valid draft. Consequently, the total time required for task completion decreased. The final application thus preserves human autonomy and maximizes task performance while effectively safeguarding cognitive resources.

Table 2. Quantitative evaluation comparing the prior study results (manual baseline and three AI archetypes from Cycle 1) with the final agentic system (Cycle 3) across performance scores, completion times, and overall NASA-TLX workload

	Baseline	AI Assistant	AI Colleague	AI Supervisor	Agentic
Score	79.22%	80.90%	73.56%	82.59%	**99.02%**
Time	713.42 s	685.57 s	702.84 s	780 s	**348.53 s**
Overall Workload	3.53	3.74	3.94	4.28	**2.42**

6.2 Computational Efficiency and Modeling Accuracy

The second phase of the evaluation investigates the system's raw modeling capabilities without human intervention. For the underlying LLM, the architecture used OpenAI's

GPT-5.2, applying only default parameters without any configurations [13]. We deliberately retained default parameters, demonstrating the robustness of the custom format and the self-reflective loop under out-of-the-box LLM conditions and avoiding the necessity for complex, use-case-specific hyperparameter tuning.

During Cycle 2, we conducted automated benchmarks to compare the efficiency and accuracy of the different data formats systematically. Using our five standardized tasks, we evaluated the monolithic XML approach, the JSON format from Cycle 1, and our newly developed custom format. For each task and format, we executed 30 automated runs, yielding a total of 450 generated models. Additionally, to evaluate the fully autonomous agentic system in Cycle 3, we repeated the tasks 5 times each, adding 25 runs to the dataset. These models were evaluated entirely offline using the exact same scoring schema applied to the human participants in Cycle 1, ensuring comparability without human interference. Table 3 summarizes the average results across all test runs regarding accuracy, required time, token usage, and generation errors.

Table 3. Evaluating the three serialization formats against the autonomous agentic system, detailing average performance scores, completion times, token usage, and generation errors

Format	Score	Latency	Token Usage	Error
XML	87.03%	74.94 s	19,854.71	78
JSON	91.38%	25.13 s	8,765.54	18
Our Format	91.30%	**16.37 s**	**7,964.09**	5
Agentic	**98.42%**	145.78 s	53,132.16	**0**

The metrics reveal structural differences. The classical XML approach exhibited the highest token consumption and the longest generation times due to massive structural overhead, which directly resulted in the highest error rate. Such errors manifested as critical generation failures: the enormous token demand repeatedly exceeded the output limit of the LLM, causing the syntax structure to be cut off prematurely and resulting in incomplete, unparsable files. The JSON approach (Cycle 1) reduced token usage but still lacked the necessary reliability for complex process structures. Implementing the custom action-based format (Cycle 2) lowered token consumption by 10% compared to JSON and 60% compared to XML. The reduction in payload prevented the infrastructural truncation of outputs, immediately dropping the rate of corrupted files.

Activating the full agentic self-reflection (Cycle 3) achieved the highest overall modeling quality. Although the internal validation loop, comprising up to three iterations, increased the raw execution time compared to a single-pass generation, the accuracy of the generated BPMN models rose substantially while eliminating generation errors entirely. Fig. 3a-d visually contrasts these qualitative differences by illustrating a sample process generated via XML, JSON, our format, and the final self-reflective agent.

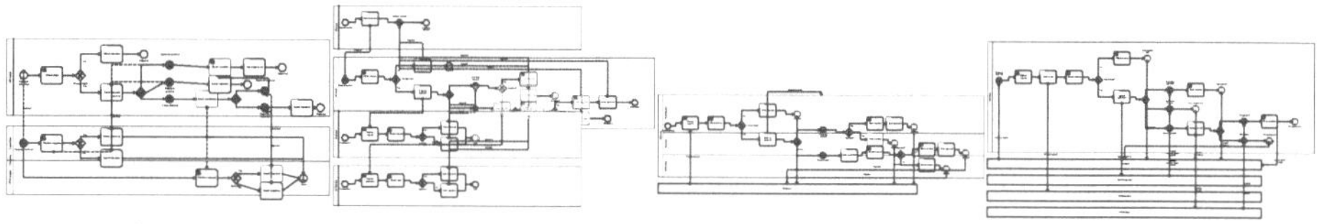

Fig. 3. Generated process patterns by approach: (a) Monolithic XML, (b) Action-based JSON, (c) our action-based format, (d) autonomous self-reflective agent

7 Discussion

Our research answers the question of how an agentic system should be designed to auto(no)mate the creation of process models without undermining human agency or inducing cognitive load [33]. The summative evaluation results illustrate that integrating generative AI into the modeling process does not inevitably lead to performance improvements. Instead, the exploratory study revealed that unstructured AI agency, particularly through proactive, text-heavy feedback mechanisms, drastically increases the extrinsic cognitive load of users. Statements from participants in the first evaluation phase impressively substantiated this dynamic. Participants described the constant, unsolicited interventions of the AI Supervisor as disruptive to concentration, tearing mental focus away from the actual modeling process. The interaction with the AI Colleague manifested even more critically. Parallel collaboration on a shared canvas led participants to stop questioning the outputs of the AI, uncritically adopting generated content instead. This behavior illustrates a classic automation bias [27]. The human becomes increasingly passive in this constellation, insidiously surrendering authoritative control as the examining and guiding entity to the AI.

Our final artifact resolves these socio-technical conflicts through an architectural realignment. Implementing agentic self-reflection shifts the resource-intensive verification process from the human to the internal processing loop of the machine. The agent no longer presents an intermediate result to the user, but a semantically and syntactically validated model. Replacing textual feedback with visual markers proved to be a decisive factor in reducing frustration and cognitive load. Visual anchor points enable an immediate cognitive mapping of the raised issues to the respective BPMN element. The human modeler no longer needs to mentally reconstruct the structure but instead acts efficiently as the final control authority. This systematically anchored human-in-the-loop approach specifically counters the previously observed automation bias. The system no longer delivers final, unassailable facts that provoke automation complacency. The asynchronous handover of the visually marked model forces the human into an active, critical review role. The user consciously assumes responsibility for error correction, which prevents passivity and secures human autonomy.

From an engineering point of view, the structural inefficiency of conventional serialization formats within agentic workflows was another issue. Our results unequivocally demonstrate that the monolithic generation of XML or JSON structures for complex process models represents an infrastructural source of errors. The strict limitation of the context window of LLMs inevitably leads to corrupted files and hallucinations when using verbose formats. Developing the lightweight, action-based format proves that reducing data overhead not only lowers latency but actually enables the fundamental reliability of the system in the first place. The application bypasses the limitations of LLMs by translating complex structures into discrete, lightweight modeling steps, laying the foundation for error-free, multi-step agentic interactions.

8 Implications

8.1 Theoretical Implications

The study provides substantial theoretical contributions for the research community in the fields of information systems and human-AI interaction. Our work expands the understanding of CLT within the context of human-AI collaborations [33]. The findings prove that cognitive load in agentic systems is not primarily determined by the complexity of the task (intrinsic load) but to a large extent depends on the interaction design (extrinsic load) of the AI. The investigation shows that proactive AI systems without adequate communication channeling led to a cognitive bottleneck. Furthermore, our research underscores the dangers of automation bias [27]. The empirical observations demonstrate that synchronous co-creation causes the human to fall into a passive observer role and to actively cease personal quality control. Future research must consequently consider the socio-technical dynamics of partnerships, as unstructured AI agency dilutes human accountability and fosters the loss of the critical review authority.

From an engineering perspective, developing the lightweight data format makes an important contribution to technical design research. The work challenges the established standard of forcing LLMs to output XML or JSON structures directly. The empirical data demonstrate that tailor-made, action-based serialization formats can increase the performance of creating graph structures. This approach provides a theoretical foundation for the architecture of future agent systems that rely on continuous data exchange and tool utilization.

8.2 Practical Implications

The developed system delivers actionable guidelines for corporate practice and the development of agentic modeling software. Software developers and tool providers receive empirical evidence that monolithic prompting approaches are insufficient for process modeling.

The results imply clear design principles for user interfaces in AI-supported modeling tools: Abandon traditional, text-based chat interfaces when users edit complex spatial or logical structures. Integrating localizable visual markers and establishing a human-in-the-loop concept, where the software makes suggestions but the human retains unrestricted decision-making authority, remain essential for user acceptance. A strictly sequential division of tasks, where the agent delivers the initial draft and the human subsequently validates the model asynchronously, decouples the user from the dynamics of synchronous generation. This temporal and spatial separation eliminates the tendency toward blind agreement and forces users to proactively assume objective quality control. Organizations can apply these design guidelines to implement AI systems that increase productivity, prevent uncritical automation bias, and anchor human autonomy as a non-negotiable constant within the work process [27]. From an engineering perspective, adapting token-optimized formats industry-wide will reduce API costs, minimize latency, and prevent system failures from corrupted syntax.

8.3 Limitations

Our research is not without limitations, which simultaneously open pathways for future work. The exploratory study and the subsequent evaluation relied on a qualitative approach involving 15 participants. While the selected sample size generated profound formative feedback for the iterative design of the artifact, a larger and more diverse participant pool would yield a broader distribution of quantitative data. Engaging a more extensive demographic could uncover additional usability preferences and further validate the socio-technical impacts across different levels of modeling expertise.

The implementation used the GPT-5.2 version from OpenAI. Newer or older versions thereof, systems from other vendors as well as smaller self-hosted systems may vary in their performance and the usability of the prototype and should therefore be reviewed comparatively before concluding towards a broader theory. Further, the new lightweight data format demonstrated superior performance regarding token consumption and latency compared to conventional XML and JSON structures within our specific modeling scenario. However, the current research did not benchmark the format against a wider array of data serialization methods, alternative structural schemas, or other LLM models. Consequently, the generalizability of the data format across other domains or highly divergent agentic tasks remains unverified. Future research must evaluate the custom format in broader contexts to ascertain universal applicability beyond business process modeling.

9 Conclusion

The integration of autonomous AI into complex knowledge work requires a paradigm shift from task-driven automation to human-centric collaboration. This research presents the design and evaluation of an agentic application for business process modeling. By systematically testing various AI configurations, ranging from reactive assistants to proactive supervisors, revealed critical thresholds where AI assistance degrades human agency and induces cognitive load. Addressing these socio-technical challenges, the artifact employs a self-reflective agent that iteratively plans, executes, and internally validates process models before presenting a refined draft to the human expert.

Evaluating the application through an exploratory study and benchmarks confirms the efficacy of the design approach. The developed application reduces the extrinsic cognitive burden on human users and enhances overall modeling performance without undermining psychological well-being or human agency. By emphasizing human-centric design [1] and preserving meaningful human control [31], these findings offer actionable guidelines for fostering sustainable human-AI partnerships in organizational contexts. Thus, our work ensures that agentic systems empower rather than overwhelm the human workforce. Nevertheless, in future research a systematic comparison with alternative interaction designs should be considered.

Lastly, from an engineering perspective our research introduces a highly optimized, lightweight data format. Developing the new serialization structure drastically reduced token consumption and LLM response times compared to established standards like JSON and XML. Optimizing data payloads represents a critical advancement for agentic AI, as autonomous systems heavily depend on continuous, semi-structured data exchanges for tool calling, internal reasoning, and state management.

Disclosure of Interests. The authors have no competing interests to declare that are relevant to the content of this article.

References

1. Amershi, S., et al.: Guidelines for human-AI interaction. In: Proceedings of the 2019 CHI Conference on Human Factors in Computing Systems, pp. 1–13, New York, NY (2019)
2. An, C., Zhang, J., Zhong, M., et al.: Why does the effective context length of LLMs fall short? In: Yue, Y., et al. (eds.) International Conference on Learning Representations, Singapore (2025)
3. Baird, A., Maruping, L.M.: The next generation of research on IS use: a theoretical framework of delegation to and from agentic IS artifacts. MIS Q. **45**(1), 315–341 (2021)
4. bpmn.io, https://bpmn.io/, last accessed 2026/06/12
5. Calvanese, D., et al.: Agentic business process management: a research manifesto. Inf. Syst. **140**, 102738 (2026)
6. Chandler, P., Sweller, J.: The split-attention effect as a factor in the design of instruction. Br. J. Educ. Psychol. **62**(2), 233–246 (1992)
7. Ciftci, S.A.: Designing synthetic work relationships: An exploratory study of agentic AI roles, cognitive load, and psychological need satisfaction. In: 5th International Conference on Hybrid Human-Artificial Intelligence (HHAI 2026), Brussels (2026)
8. Dumas, M., La Rosa, M., Mendling, J., Reijers, H.A.: process discovery. In: Dumas, M., et al. (eds.) Fundamentals of Business Process Management, pp. 159–212. Springer (2018)
9. Dumas, M., Milani, F., Chapela-Campa, D.: Agentic business process management systems. In: van de Weerd, I., Estrada Torres, B., van der Aa, H. (eds.) Business Process Management Workshops, pp. 3–14. Springer, Cham (2026)
10. Feuerriegel, S., Hartmann, J., Janiesch, C., Zschech, P.: Generative AI. Bus. Inf. Syst. Eng. **66**(1), 111–126 (2024)
11. Figl, K., Laue, R.: Cognitive complexity in business process Modeling. In: Mouratidis, H., Rolland, C. (eds.) Advanced Information Systems Engineering, pp. 452–466. Springer, Heidelberg (2011)
12. Friedrich, F., Mendling, J., Puhlmann, F.: Process model generation from natural language text. In: Mouratidis, H., Rolland, C. (eds.) Advanced Information Systems Engineering, pp. 482–496. Springer, Heidelberg (2011)
13. GPT-5.2 (2026), https://openai.com/index/introducing-gpt-5-2/, Accessed 16 Mar 2026
14. Hafner, A., Wittges, H., Rinderle-Ma, S.: GenAI in business process management: a systematic review of the current state. In: AMCIS 2025, Montreal (2025)
15. Hart, S.G., Staveland, L.E.: Development of NASA-TLX (task load index): results of empirical and theoretical research. In: Advances in Psychology, pp. 139–183 (1988)
16. Holldack, F., Banh, L., Strobel, G.: Agentic information systems. Electron. Mark. **36**(5) (2026)

17. Hörner, L.F., Möller, M., Reichert, M.: Automatically generating BPMN 2.0 process models from natural language process descriptions: challenges, framework, quality assessment. Business & Information. Syst. Eng. **68**(1) (2026)
18. Janiesch, C., Kowalkiewicz, M., Rosemann, M.: Process Autonomization: rethinking business process management. In: Senderovich, A., et al. (eds.) Business Process Management Forum, pp. 361–377. Springer, Cham (2026)
19. Kampik, T., Warmuth, C., Rebmann, A., et al.: Large process models: a vision for business process management in the age of generative AI. Künstl. Intell. **39**(2), 81–95 (2025)
20. Klievtsova, N., Benzin, J.-V., Kampik, T., Mangler, J., Rinderle-Ma, S.: Conversational process modelling: state of the art, applications, and implications in practice. In: Di Francescomarino, C., et al. (eds.) Business Process Management Forum, pp. 319–336. Springer, Cham (2023)
21. Köpke, J., Safan, A.: Introducing the BPMN-Chatbot for efficient LLM-based process Modeling. In: Proceedings of the BPM 2024 Demos & Resources Forum, pp. 86–90 (2024)
22. Kourani, H., Berti, A., Schuster, D., van der Aalst, W.M.: ProMoAI: process Modeling with generative AI. In: Proceedings of the 33rd International Joint Conference on Artificial Intelligence (IJCAI), pp. 8708–8712, Jeju (2024)
23. Kuechler, W., Vaishnavi, V.: A framework for theory development in design science research: multiple perspectives. Journal of the AIS. **13**(6) (2012)
24. Mendling, J., Reijers, H.A., van der Aalst, W.: Seven process modeling guidelines (7PMG). Inf. Softw. Technol. **52**(2), 127–136 (2010)
25. Nebel, M., Ciftci, S.A., Janiesch, C.: LACII: enhancing decision-making in business processes with the IIoT: a cognitive-load-aware design approach. In: Fortino, G., Mecella, M. (eds.) Internet of Things Meets Business Process Management: a Synergistic Integration, pp. 105–131. Springer, Cham (2026)
26. Neurauter, M., Pinggera, J., Martini, M., et al.: The influence of cognitive abilities and cognitive load on business process models and their creation. In: Davis, F.D., et al. (eds.) Information Systems and Neuroscience, pp. 107–115. Springer, Cham (2015)
27. Parasuraman, R., Manzey, D.H.: Complacency and bias in human use of automation: An attentional integration. Hum. Factors. **52**(3), 381–410 (2010)
28. Sapkota, R., Roumeliotis, K.I., Karkee, M.: Vibe coding vs. agentic coding: Fundamentals and practical implications of agentic AI. arXiv preprint https://arxiv.org/abs/2505.19443 (2025)
29. Schatzki, T.R.: Agency. Inf. Organ. **35**(1), 100553 (2025)
30. Seeber, I., Bittner, E., Briggs, R.O., et al.: Machines as teammates: a research agenda on AI in team collaboration. Inf. Manag. **57**(2), 103174 (2020)
31. Shneiderman, B.: Human-Centered artificial intelligence: reliable, Safe & Trustworthy. International Journal of Human–Computer Interaction 36(6), 495–504 (2020)
32. Shrout, P.E., Fleiss, J.L.: Intraclass correlations: uses in assessing rater reliability. Psychol. Bull. **86**(2), 420–428 (1979)
33. Sweller, J.: Cognitive load theory. In: Mestre, J.P., Ross, B.H. (eds.) Psychology of Learning and Motivation, vol. 55, pp. 37–76. Academic (2011)
34. Tallon, M., Winter, M., Pryss, R., et al.: Comprehension of business process models: insight into cognitive strategies via eye tracking. Expert Syst. Appl. **136**, 145–158 (2019)
35. The Object Management Group (OMG): BPMN™: Business Process Model and Notation (2014), https://www.omg.org/spec/BPMN/2.0.2/PDF, Accessed 30 May 2026
36. vom Brocke, J., et al.: Ten principles of good business process management. Bus. Process. Manag. J. **20**(4), 530–548 (2014)
37. Vu, H., Klievtsova, N., Leopold, H., Rinderle-Ma, S., Kampik, T.: Agentic business process management: practitioner perspectives on agent governance in business processes. In: ER, M., et al. (eds.) Business Process Management: Responsible BPM Forum, Process Technology Forum, Educators Forum, pp. 29–43. Springer, Cham (2026)

38. Weber, B., et al.: Measuring cognitive load during process model creation. In: Davis, F.D. et al. (eds.) Information Systems and Neuroscience, pp. 129–136. Springer, Cham (2015)
39. Ziche, C., Apruzzese, G.: LLM4PM: a case study on using large language models for process Modeling in Enterprise organizations. In: Di Ciccio, C., et al. (eds.) Business Process Management: Blockchain, Robotic Process Automation, Central and Eastern European, Educators and Industry Forum, pp. 472–483. Springer, Cham (2024)

Discovering Shared-Action Routines from Interleaved Executions for Robotic Process Automation

Massimiliano de Leoni(✉) and Faizan Ahmed Khan

Department of Mathematics, University of Padua, Padua, Italy
massimiliano.deleoni@unipd.it, faizanahmed.khan@phd.unipd.it

Abstract. Robotic Process Automation (RPA) aims to discover repetitive rule-based user procedures, a.k.a. routines, from UI interaction logs in order to automate them through software robots. A fundamental challenge in this context is the discovery of routine types from UI logs. Most existing approaches assume that routines are executed sequentially, i.e., a user completes one routine before starting another. However, in real-world settings, users frequently switch between partially-executed routines. As a result, UI logs often contain interleaved executions of multiple routines, making the identification of routine types considerably more challenging. At the same time, UI actions can also be shared across routine types, further increasing the complexity of their discovery. This paper presents a technique for discovering routine types from UI logs when executions can be interleaved and different routine types can share common UI actions. The technique combines label-splitting techniques to reduce shared-action ambiguity and community-detection algorithms to cluster actions into routine types. The technique was evaluated on synthetic and real-life UI logs, using state-of-the-art metrics, and also compared with state-of-the-art techniques. The results illustrate that the proposed technique outperforms the state of the art in identifying routine types with shared actions and interleaved executions.

Keywords: Robotic Process Automation · Routine Mining · Community Detection · Interleaving of Routine Executions

1 Introduction

Robotic Process Automation (RPA) aims to create software (SW) robots to partially or fully automate rule-based and repetitive *routines* performed by human users in their software user interfaces (UIs) [2,18]. The fundamental step in this pipeline is to identify the types of routines that are candidates for automation.

The starting point of a reliable, objective identification of routine types is the presence of a UI log, which captures the sequence of UI actions (e.g., clicks, keystrokes, copy-and-paste operations, menu selections) that users perform while working with software tools.

F. M. Maggi et al. (Eds.): BPM 2026, LNCS 16936, pp. 331–349, 2027.
https://doi.org/10.1007/978-3-032-37777-7_21

The availability of UI logs enables the automatic discovery of the routine types, where recent research abstracts a routine type as its set of actions (cf. Sect. 2). The problem of discovering the routine types in a UI log consists of assigning actions to routine types. Note that UI logs are fundamentally different from traditional event logs used in Process Mining: UI logs do not have instance identifiers and are conceptually represented as a single sequence of actions.

Note that actions can be shared among routine types (filling the same forms, copying the same data across systems, navigating the same windows/pages of software applications or websites, etc.). Each group comprises the routine executions of the same type (e.g., managing a ticket in a ticketing system). Also, the same resource can currently start multiple routine executions of the same or of different types. This can, e.g., be the case when a certain resource cannot complete an execution because (s)he needs to wait for further information: in the meantime, the resource can start a different routine.

Currently, the literature has overlooked the setting in which resources perform multiple routine-type executions concurrently and, at the same time, an action may belong to several routine types (cf. Sect. 2).

This paper proposes a novel technique for routine-type identification from UI logs containing interleaved routine executions and shared actions. The approach first refines candidate shared actions through context-based label splitting [24], so that identical UI actions occurring in different routine contexts can be distinguished. The refined UI log is then transformed into a *Normalized Directly-Follows Graph* (NDFG), where nodes represent UI actions and directed edges capture normalized frequencies of directly-following action pairs. Next, the NDFG is clustered to identify "communities" of actions, where each community contains the actions that correspond to different routine types. For our technique, the choice fell to *Infomap* [21], which aims to detect communities such that a typical walk through the graph (in our context, a routine execution) mostly stays within a single community (i.e., among actions of a specific routine type) and only infrequently moves out. These communities are then mapped back to the original action labels to reconstruct the final types of routine. By combining context-based label splitting with graph-based community detection, the proposed technique can effectively separate interleaved behaviors even when routine types share common UI actions.

We evaluated our approach on 400 interleaved UI logs derived from both real-life and synthetic UI logs. The real-life data originate from an industrial RPA provider, while the synthetic data are generated from nine benchmark datasets widely used in the literature. The generated UI logs contain controlled interleavings that involve up to three types of routine. The results, evaluated in terms of the Jaccard coefficient (JC), demonstrate that the proposed technique significantly outperforms existing approaches in accurately identifying routine types.

Section 2 discusses the gap in the related work. Section 3 illustrates the technique, while Sect. 4 reports on the experimental settings, namely the datasets employed, the evaluation methodology and the results. Finally, Sect. 5 concludes the paper, highlighting the limitations of our proposal and the consequent direction of future work.

2 Related Works

Robotic Process Automation consists of three main phases: UI log recording and preprocessing, discovery of routine executions and types, and RPA compilation [9,14]. In fact, when accurate routine types are discovered, existing approaches, such as Robidium [15] and SmartRPA [4], demonstrate the feasibility of compiling them into executable RPA scripts.

It follows that the key challenge is to discover the accurate routine types. Agostinelli [5] identifies three main cases of discovering routine types from UI log in RPA, each reflecting increasing complexity in how routines are recorded. In Case 1, the log contains multiple executions of a single routine type, making segmentation mainly about separating individual executions, even when they are interleaved. Case 2 introduces multiple distinct routine types, but assumes that each type uses its own exclusive set of user actions, so segmentation involves distinguishing between different routines and their executions, possibly under interleaving. Case 3 is the setting of this paper and represents the most challenging and realistic scenario where multiple routines are concurrently executed by the same resource Within Case 3, our focus is on Case 3.4, which involves interleaving between different routines with shared actions, making it difficult to assign actions to specific routine types. Agostinelli [5] does not list any technique that covers this case. He only lists two techniques that cover Case 3.3, which do not cover interleaving: these techniques are by Rebmann et al. [20] and by Agostinelli et al. [6]. The latter is also incorporated in SmartRPA [4]. After the publication of work [5], additional techniques have been devised that consider the presence of shared routine actions but assume sequential routine executions. de Leoni et al. [16] propose a clustering-based technique to identify routine traces from UI logs where routines may share actions. Hohenadl [11] proposes a similar method based on a clustering where UI logs are treated as time series, but explicitly mention the lack of robustness in the presence of interleaving.

Other techniques support interleaved executions, but do not allow for shared user actions, such as by Abb et al. [3], Urabe et al. [22] and Leno et al. [13]. In fact, the technique by Leno et al. [13] extends what is available in Robidium.

To illustrate the importance of developing techniques that explicitly aim at interleaving executions and shared actions, we experimentally compared our technique with those by Rebmann et al. [20], by Agostinelli et al. [6], by de Leoni et al. [16], which are representatives of Case 3.3.

With a similar goal in mind, we also aimed to compare with one candidate technique that supports shared routine actions but not interleaving: our choice fell onto Leno et al. [13], which is seminal and highly cited contribution in the domain.

3 Technique for Routine Type Identification

The starting point of our technique is an UI log:

Definition 1 (UI Log). *Let $\mathcal{A}$ be the finite set of UI actions of interest. An UI log is a finite sequence of actions Σ, such that $\Sigma \in \mathcal{A}^*$.*

The length of Σ is denoted by $|\Sigma|$, which represents the number of actions contained in the sequence. For each position i with $1 \leq i \leq |\Sigma|$, $\Sigma(i)$ denotes the i-th action in the sequence.

Note that UI logs can generally be richer than simply containing the names of actions. For instance, an UI log may also record timestamps, resources, and other relevant information. However, these aspects are abstracted in this work because they are not relevant for the method presented in the remainder of the paper. Our definition of an UI log does not consider resources as well. As discussed in Sect. 2, this is not limiting because, if the UI log contains actions of multiple resources, we can trivially extract the sublogs of actions of each resource, possibly then concatenating them.

Once we have the UI log, our goal is to identify routine types, i.e., to discover subsets of actions that correspond to distinct routine types. Given a set of actions $\mathcal{A}$, our aim is to group them into subsets $\mathcal{A}_1, \ldots, \mathcal{A}_n$, where n denotes the number of routine types, such that $\bigcup_{i=1}^{n} \mathcal{A}_i = \mathcal{A}$.

Each subset $\mathcal{A}_i$ represents the set of actions associated with the i-th routine type. Note that routine types are not necessarily disjoint, as the same action may occur in multiple routines.

Our technique consists of four main phases. In the first phase (Sect. 3.1), shared actions are identified in the UI log and their labels are refined according to their execution context using the context-based label-splitting technique [24], resulting in UI logs where shared actions are transformed into non-shared sub actions. In the second phase (Sect. 3.2), the UI log is transformed into a Normalized Directly-Follows Graph (NDFG). In the third phase (Sect. 3.3), the resulting NDFG is analyzed using the Infomap community detection algorithm to identify sets of actions (nodes) corresponding to distinct routine types. In the fourth phase (Sect. 3.4), the split action labels are mapped back to their original action labels in order to reconstruct the final routine types.

3.1 Context-Based Label Splitting for Shared Actions

In UI logs, the same UI action may occur in the execution of multiple routines. Such shared actions introduce ambiguity for routine-type identification because the routine to which an action belongs can only be determined from its execution context. To address this issue, we incorporate a preprocessing step that identifies candidate shared actions that are potentially shared among multiple routine types. To identify these candidate shared actions, we analyze the frequency of actions in the UI log.

Let $freq_\Sigma(a) = |\{i \in [1, |\Sigma|] \mid \Sigma(i) = a\}|$ denote the number of occurrences of action $a \in \mathcal{A}$ in Σ. The actions are ordered according to $freq_\Sigma$ and partitioned into quartiles Q_1, Q_2, Q_3, and Q_4. Actions whose frequency lies in $Q_2 \cup Q_3 \cup Q_4$ are considered candidate shared actions, as they are more likely to occur across multiple routines. Each candidate shared action a can potentially be split into $a_1, \ldots, a_n$ if it appears in different execution contexts, as proposed by van Zelst et al. [24]. An execution context consists of the multiset of k actions that precede a and that follow a in Σ, where k is a positive, integer number, that identifies the size of the window that is taken into consideration. The assumption here is that some different contexts may map to different routine types, although some others may refer to the same type. At the end of this phase, the UI log $\Sigma \in \mathcal{A}^*$ is transformed into a new UI log $\Sigma' \in \overline{\mathcal{A}}^*$ defined over a refined action set $\overline{\mathcal{A}}$. The relationship between the refined action set and the original is retained via a mapping $M_\Sigma : \overline{\mathcal{A}} \rightarrow \mathcal{A}$. This mapping will be necessary in the last phase, when the split action labels are mapped back to their original action labels to reconstruct the resulting routine types.

Example 1. *Consider the following UI log:*

$$\Sigma_{ex} = \langle A, X, B, Z, B, Z, X, A, C, A, B, C, B, C, A, X, B, Z, C, A \rangle.$$

For this running example, we use a context window of size $k = 1$, such that each occurrence of an action is characterized by its immediate predecessor and successor. Under this setting, the occurrences of B appear in the following contexts:

$$(\{A\}, B, \{C\}),\ (\{C\}, B, \{C\}),\ (\{X\}, B, \{Z\}),\ (\{Z\}, B, \{Z\}).$$

Contexts are compared using a similarity score $\phi \in [0, 1]$, computed based on normalized edit distance, with a predefined threshold (e.g., $t_s = 0.5$). Intuitively, $\phi = 1$ indicates identical contexts, while lower values indicate increasing dissimilarity.

For instance, comparing $(\{A\}, B, \{C\})$ and $(\{C\}, B, \{C\})$ yields $\phi = 0.5$, since the successor sequences coincide while the predecessors differ. Similarly, $(\{X\}, B, \{Z\})$ and $(\{Z\}, B, \{Z\})$ also yield $\phi = 0.5$. In contrast, contexts across these two groups yield $\phi = 0$, as both predecessor and successor differ.

Based on the threshold $t_s = 0.5$, the occurrences of B are thus partitioned into two groups:

$$\{(\{A\}, B, \{C\}), (\{C\}, B, \{C\})\} \quad \textit{and} \quad \{(\{X\}, B, \{Z\}), (\{Z\}, B, \{Z\})\}.$$

Each group is assigned a distinct split label, yielding B_1 and B_2. The resulting UI log after label splitting is:

$$\Sigma'_{ex} = \langle A, X, B_2, Z, B_2, Z, X, A, C, A, B_1, C, B_1, C, A, X, B_2, Z, C, A \rangle.$$

From the next phase onward, $\Sigma \in \mathcal{A}^*$ is meant to indicate a UI log where the relevant candidate shared actions have been split into non-shared sub-actions.

3.2 Construction of the Normalized Directly-Follows Graph

In this phase, we start by analyzing consecutive pairs of actions and measure how often each pair occurs within the refined UI log Σ obtained in the first phase of our technique. The frequency of these directly-follow pairs provides the basis for constructing the normalized directly-follows graph.

Definition 2 (Directly-Follow Frequency). *Let $\mathcal{A}$ be the finite set of UI actions, and let $\Sigma \in \mathcal{A}^*$ be a UI log. For every pair $(a, b) \in \mathcal{A} \times \mathcal{A}$, we define the frequency of the directly-follow pairs as: $freq_\Sigma(a, b) = |\{ i \in [1, |\Sigma|].\ \Sigma(i) = a,\ \Sigma(i+1) = b \}|$.*

Table 1. Directly-Follows frequencies extracted from the refined UI log Σ'_{ex} of Example 1.

$freq_\Sigma$	A	B_1	C	X	B_2	Z
A	0	1	1	2	0	0
B_1	0	0	2	0	0	0
C	3	1	0	0	0	0
X	1	0	0	0	2	0
B_2	0	0	0	0	0	3
Z	0	0	1	1	1	0

As an illustration, Table 1 shows the directly-follow frequencies for every pair of activities of the UI log Σ'_{ex} in Example 1.

Using the directly-follow frequency, we construct the Normalized Directly-Follows Graph (NDFG), which represents all pairs of actions observed in the UI log Σ and assigns a normalized weight to each pair.

Definition 3 (Normalized Directly-Follows Graph (NDFG)). *Let $\mathcal{A}$ be the finite set of UI actions, and let $\Sigma \in \mathcal{A}^*$ be a UI log. Let $V = \mathcal{A}$ be the set of vertices, where each vertex represents a distinct UI action observed in Σ. A Normalized Directly-Follows Graph is a triple $G_\Sigma = (V, E_\Sigma, w_\Sigma)$, consisting of*

- *a set of vertices $V = \mathcal{A}$,*
- *a set of arcs $E_\Sigma = \{(a, b) \mid a, b \in \mathcal{A},\ freq_\Sigma(a, b) > 0\}$,*
- *and an arc's weighting function $w_\Sigma : E \to \mathbb{R}_0^+$, such that, for each $(a, b) \in E_\Sigma$,*

$$w_\Sigma(a, b) = \frac{freq_\Sigma(a, b)}{\sum_{b' \in \mathcal{A}} freq_\Sigma(a, b')}$$

Note how the weight of any arc (a, b) is obtained through a normalization of the frequency of (a, b) with respect to the frequency of a, which is the summation of the frequency of every arc (a, b') with $b' \in \mathcal{A}$. This normalization builds on the concept of maximum-likelihood estimation in bigram language models, originally used in NLP and here adapted for routine identification [12]. It accounts for the probability of seeing a token (e.g., an action) $T2$ after $T1$, which is computed by counting how often $T1$ is followed by $T2$ and then normalizing by the total number of times $T1$ appears.

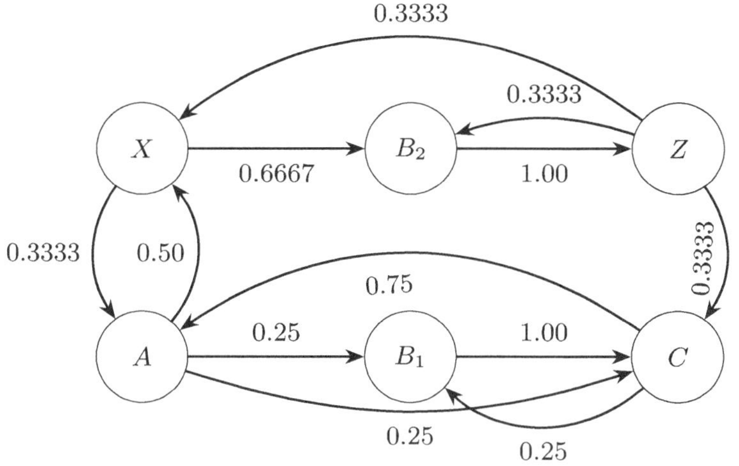

Fig. 1. Normalized Directly-Follows Graph G_Σ (NDFG) induced by the refined running example log Σ'_{ex}.

The directly-follow frequencies in the running Example 1 generate the NDFG in Fig. 1. As an example, let us consider the node A: as shown in the first row of Table 1, the sum of the frequency of all NDFG's outgoing arcs is $\sum_{b' \in \{A, B_1, C, X, B_2, Z\}} freq_{\Sigma'_{ex}}(A, b') = 4$. The frequency of the arc from A and B_1 is $freq_{\Sigma'_{ex}}(A, B_1) = 1$ and, thus, the NDFG's weight of the arc between A and B_1 is $w_{\Sigma'_{ex}}(a, b) = 1/4 = 0.25$. The weights of the other arcs are computed accordingly.

3.3 Graph Clustering via Community Detection

The third phase of our technique aims to cluster the NDFG's nodes. Each cluster corresponds to the actions that belong to a distinct routine type. To create clusters of NDFG's nodes, we leverage algorithms that detect communities of these nodes. A community is a group of nodes that are more strongly interconnected with each other than with the rest of the graph. Applied to our setting, the algorithm will discover the community of actions that are strongly connected via high normalized weights of these NDFG's nodes.

Several community detection algorithms are present in the literature, such as those based on modularity maximization (e.g. the Louvain algorithm [7]), spectral clustering [17], finding the minimal cut (or equivalently maximum flow) [8], or random walks (e.g., Walktrap [19] or Infomap [21]).

We ultimately opted for a community-detection algorithm based on random walks. Such algorithms identify communities by following a random walker on the graph and finding groups of nodes where the walker tends to remain for a long time before moving outside. The walker's movement is guided by the graph's links and weights.

Random-walk-based community-detection algorithms are particularly suitable for discovering routine types because they focus on the flow of movements, rather than only on the number of connections within or between communities. The goal is to identify communities, corresponding to routine types, among nodes, corresponding to actions. Ideally, when the flow moves from a node in one community c_1 to a node in another community c_2, it should then continue for a sufficiently long time within c_2. This reflects the intuition that a resource is not expected to continuously jump from one routine type to another, but should instead remain coherent for a period of time once a new routine type is entered.

After inspecting the literature on comparative analysis of random walk-based community detection algorithms, the choice fell on Infomap [21]. Specifically, Fortunato and Lancichinetti [10] indicate that Infomap consistently outperforms alternative algorithms, and Yang et al. [23] put Infomap among the most suitable algorithms in our settings. Indeed, Yang et al. [23] indicate that Infomap is a suitable choice if the number of nodes (i.e. actions in our setting) is below 6000 and the number of arcs going from nodes of one community to nodes of another community is lower than 50%. We believe that 6000 action labels are extremely high in an UI log setting, and the condition of 50% means that adjacent actions in UI logs belong to different routine types, which is also rather extreme.

Figure 2 shows the two communities built on the NDFG discussed in Example 1. The nodes of the two communities are visualized in red and blue colors.

3.4 Reconstructing Routine Types from Split Labels

Since the communities discovered in this previous phase contain split actions obtained during the label-splitting phase, these actions must be mapped back to their original labels in order to reconstruct the final routine types. Let $M : \overline{\mathcal{A}} \rightarrow \mathcal{A}$ denote the mapping function defined in the label-splitting phase. Let R be a routine type containing the split actions $\{a_1, a_2, \ldots, a_n\}$. The corresponding routine expressed in terms of the original actions is obtained as $R' = \{M(a_1), \ldots, M(a_n)\}$.

Applying this mapping to the communities in Fig. 2, the community $\{A, B_1, C\}$ yields the routine type $\{A, B, C\}$, while the community $\{X, B_2, Z\}$ yields the routine type $\{X, B, Z\}$. This step restores the original action labels while preserving the grouping of actions discovered by the community detection algorithm.

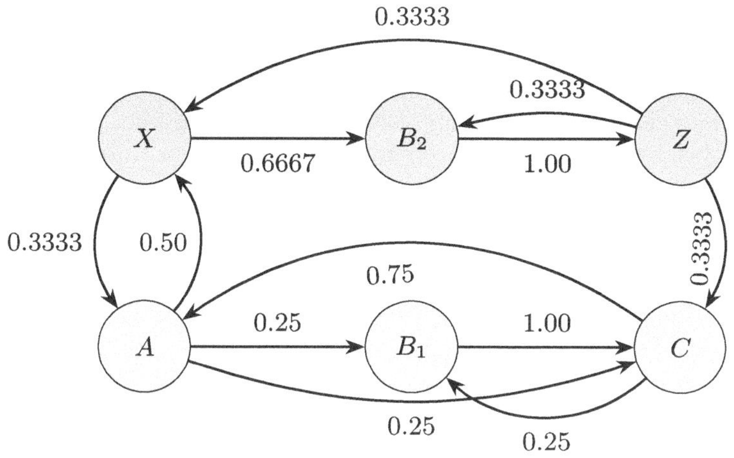

Fig. 2. Communities discovered by Infomap on the NDFG for Fig. 1, discussed in Example 1. The red and blue groups highlight the nodes belonging to two distinct routine types.

4 Evaluation

This section presents the datasets used in our experiments (cf. Sect. 4.1), the methodology and evaluation metrics applied (Sect. 4.2), as well as the results, which demonstrate the significant enhancement achieved by our approach (Sect. 4.3) with only modest computational requirements (Sect. 4.4). The implementation of the proposed technique in Python is available in the repository accompanying this submission [1].

4.1 Datasets

Our evaluation was based on one real-life UI log and nine synthetic UI logs:

Real-life UI Log. The real-life UI log was retrieved from the database of an Italian company that provides RPA solutions and is available in an anonymous repository (cf. Section 4). Each routine execution was recorded sequentially and started and ended with a special action, thus enabling us to accurately identify the routine executions. Each routine execution was also associated with a routine identifier and its real routine type.

Synthetic UI Logs. We also used the nine UI logs that Leno et al. [13] synthetically generated and used in the experiments for their technique. Leno et al. [13] also provide the ground truth models of the different routine types as Colour Petri nets in the CPN-Tools format.[1] Using this ground truth, we were able to determine, for each UI-log action, the routine execution to which it belongs and, consequently, its routine type.

[1] CPN Tools Website - https://cpntools.org/.

The choice fell on these logs because we could infer the ground-truth information that is necessary to compute the Jaccard coefficient (see Definition 4), which measures the quality of the identification of the routine types (see Sect. 4.2), and also to generate the UI logs with interleaved routine executions.

Action labels were shared between routine types and within single routine executions (i.e., where identical labels denoted different actions at different steps). Sharing within single routine executions occurred in five of the nine synthetic UI logs: four featured a single shared label, and an additional one featured four. Therefore, we were able to test our method against sharing within single executions, although we did not perform a formal ablation study on this specific condition.

4.2 Experimental Methodology

The experimental methodology can be divided into three steps, as shown in Fig. 3. Starting from the real-life and synthetic UI logs, we first filter out trivial routine types. The remaining non-trivial routine types are then used to generate interleaved UI logs involving either two or three routine types. Finally, we assess our technique on the resulting interleaved UI logs and compare it with existing state-of-the-art approaches.

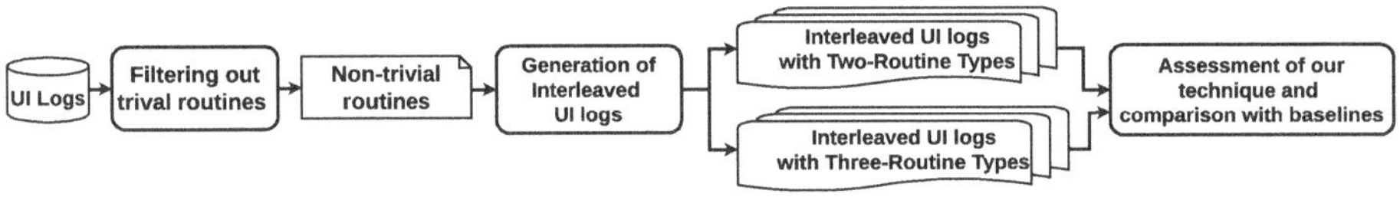

Fig. 3. Overview of the experimental methodology, which is composed by three phases, identified by rounded rectangles. The methodology applies to both the real-life and synthetic event logs.

1. Filtering Out Trivial Routines. For each routine type R identified from the UI logs discussed in Sect. 4.1, we compute the number of its executions, indicated by n_R, and the number of actions characterizing R.

For each of the datasets discussed in Sect. 4.1, we compute the median number of executions and the median number of actions across all routine types identified in that dataset. We then remove all executions of a routine type R if (i) n_R is lower than the median number of executions, or (ii) R contains fewer actions than the median number of actions. Note that these values can be computed because the ground-truth routine types and their constituent actions are known.

2. Generation of Interleaved UI Logs The real-life and synthetic UI logs used in our experiments do not originally contain interleaved routine executions. To evaluate routine identification techniques under interleaved executions with shared UI actions, we therefore introduce a strategy to synthetically inject interleaving among the non-trivial routine types retained after the filtering phase (cf. Section 4.2). The strategy combines executions of multiple routine types in a controlled manner, reproducing situations in which a user temporarily suspends the execution of a routine r_1 of a type R_1, performs the routine r_2 of a type R_2, and finally resumes and completes r_1. This models the situation in which the resource needs to divert attention from the task related to r_1 to a higher-priority task related to r_2.

Let us consider a UI log Σ. Knowing the ground truth, the routine types present in Σ are known with certainty. Furthermore, for each routine type, the corresponding routine executions in Σ are also known.. Furthermore, the set of routine executions in Σ is also known for each routine type. Our method aims to generate a UI log Σ' where routines of three types R_1, R_2 and R_3 within the set of routine types are interleaved. Routine type R_3 is of the highest priority, and R_2 is of higher priority than R_1. The UI log does not have interleaving; therefore, we can also extract the sequence L_1 and L_2 of routine executions for R_1 and R_2:

$$L_1 = \sigma_1^1, \ldots, \sigma_n^1 \qquad L_2 = \sigma_1^2, \ldots, \sigma_m^2 \qquad L_3 = \sigma_1^3, \ldots, \sigma_o^3$$

where $\sigma_j^i \in \mathcal{A}^*$ indicates the j-th execution of a routine of type i. The setting $n = m = o$ is the most challenging, because all executions in L_1, L_2 and L_3 are interleaved. We thus truncate the longer sequences:

$$L_1' = \sigma_1^1, \ldots, \sigma_p^1 \qquad L_2' = \sigma_1^2, \ldots, \sigma_p^2 \qquad L_3' = \sigma_1^3, \ldots, \sigma_p^3$$

where p is the minimum between n, m and o. Initially the resulting log Σ' is empty. For each i between 1 and p:

1. We take $\sigma_i^1 = \langle a_1, \ldots, a_q \rangle$ and select an index $j \in [1, q-1]$, uniformly at random.
2. Sequence $\langle a_1, \ldots, a_j \rangle$ is added to the end of Σ'.
3. We take $\sigma_i^2 = \langle b_1, \ldots, b_w \rangle$ and select an index $k \in [1, w-1]$, uniformly at random.
4. Sequence $\langle b_1, \ldots, b_k \rangle$ is added to the end of Σ'.
5. σ_i^3 is added to the end of Σ'.
6. Sequence $\langle b_{k+1}, \ldots, b_w \rangle$ is added to the end of Σ'.
7. Sequence $\langle a_{j+1}, \ldots, a_p \rangle$ to the end of Σ'

The case of two interleaving routine types corresponds to the same setting where, of course, L_3 and L_3' do not exist, and step 5 does not exist in the setting above.

3. Assessment of Our Technique and Comparison with the Literature. After applying the first two steps of the methodology, the experimental dataset consists of interleaved UI logs derived from both real-life and synthetic UI logs.

1. **Interleaving settings.** Following the interleaving strategy in Step 2 (cf. Section 4.2), we consider two scenarios: interleavings involving two routine types and interleavings involving three routine types.
2. **Baselines.** The superiority of our technique is assessed against the following baselines:
   - Our technique without label splitting (*No Split*) (namely the setup discussed in Sect. 3.1).
   - State-of-the-art techniques from the literature, namely the approaches proposed by Agostinelli et al. [6], de Leoni et al. [16], Leno et al. [13], and Rebmann et al. [20] (see Sect. 2). The approach by de Leoni et al. [16] is evaluated using three clustering variants based on DBSCAN, K-Means, and HDBSCAN, while the technique by Leno et al. [13] is evaluated using four variants based on length, frequency, cohesion, and coverage.
   - The aforementioned state-of-the-art techniques are also evaluated after applying label splitting (*Split*).

   This allows us to assess whether the superiority of our technique is due to label splitting alone, community detection alone, or the interplay of both.
3. **Number of logs.** For the real-life and synthetic datasets discussed in Sect. 4.1, we consider the non-trivial routine types retained after Step 1 in Sect. 4.2. For each interleaving setting involving either two or three routine types, we uniformly at random select two or three of these non-trivial routine types, depending on the interleaving setting, and apply Step 2 in Sect. 4.2 to generate one interleaved UI log. Repeating this procedure 100 times for each interleaving setting yields, for the real-life dataset, 100 interleaved UI logs for the two-routine-type setting and 100 interleaved UI logs for the three-routine-type setting; the same holds for the synthetic dataset. Overall, this results in 400 interleaved UI logs across both datasets.

At this stage, we applied our technique for routine-type identification, and we computed the quality of these identifications using the Jaccard Coefficient (JC), in line with what is proposed by Leno et al. [13]. The JC score measures the overlap between the sets of actions discovered by our technique and the corresponding ground-truth sets for each routine type:

Definition 4 (Jaccard Coefficient). *Let $\mathcal{A}$ be the set of actions of a UI log. Let $\{\mathcal{A}_1, \ldots, \mathcal{A}_n\}$ be the discovered sets of action sets where each $\mathcal{A}_i \subseteq \mathcal{A}$ represents the set of actions belonging to the i-th discovered routine type. Let $\mathcal{G} = \{G_1, \ldots, G_n\}$ denote the ground-truth sets of actions for each of the n routine types, where $G_i \subseteq \mathcal{A}$ for all $1 \leq i \leq n$. The Jaccard Coefficient is defined as:*

$$JC(\{\mathcal{A}_1, \ldots, \mathcal{A}_n\}, \mathcal{G}) = \frac{1}{n} \sum_{i=1}^{n} \max_{G_j \in \mathcal{G}} \frac{|\mathcal{A}_i \cap G_j|}{|\mathcal{A}_i \cup G_j|}.$$

The results obtained from the baselines are assessed by computing the Jaccard Coefficient (JC), which is then compared with the corresponding scores achieved by our technique.

During the first phase of label splitting, the technique by van Zelst et al. [24] requires one to provide the size k of the context window and the actions candidate for label splitting (cf. Sect. 3.1). In the experiments, we set $k = 1$, meaning that the context was determined solely by the immediately preceding and following actions. This is a reasonable choice. First, routine executions typically exhibit low interleaving because resources generally do not shift continuously between incomplete routine executions. Second, as far as we could observe in the UI logs that we inspected from the literature and in our real-life UI log, only a limited number of actions are shared across routine types. As a result, the adjacent actions are sufficient to provide a discriminative context for label splitting.

Table 2. Performance comparison of routine identification techniques under two-routine interleaving. The table reports the mean Jaccard coefficient (JC) together with the 95% confidence intervals (CI) in the form [CI_L, CI_U], where CI_L and CI_U denote the lower and upper bounds of the confidence interval, respectively, computed over the evaluated logs. Results are reported for both synthetic and real UI logs. The values in bold refer to our technique, which uses community detection (Infomap) together with label splitting. Our technique discovers routine types that are statistically more accurate than any baseline in any case study.

Technique	Synthetic UI logs				Real-life UI logs			
	No Split		Split		No Split		Split	
	Mean	CI	Mean	CI	Mean	CI	Mean	CI
Frequency in [13]	0.335	[0.296, 0.373]	0.533	[0.486, 0.580]	0.382	[0.356, 0.408]	0.411	[0.386, 0.436]
Cohesion in [13]	0.713	[0.702, 0.725]	0.721	[0.707, 0.736]	0.614	[0.590, 0.639]	0.623	[0.600, 0.646]
Length in [13]	0.715	[0.703, 0.727]	0.709	[0.696, 0.722]	0.606	[0.582, 0.631]	0.616	[0.592, 0.639]
Coverage in [13]	0.664	[0.645, 0.683]	0.679	[0.664, 0.694]	0.626	[0.598, 0.653]	0.627	[0.608, 0.646]
Rebmann et al. [20]	0.647	[0.630, 0.663]	0.647	[0.630, 0.663]	0.601	[0.573, 0.629]	0.601	[0.573, 0.629]
Agostinelli et al. [6]	0.664	[0.650, 0.678]	0.641	[0.629, 0.653]	0.564	[0.546, 0.582]	0.576	[0.558, 0.595]
DBSCAN in [16]	0.597	[0.589, 0.604]	0.596	[0.588, 0.603]	0.599	[0.584, 0.614]	0.603	[0.587, 0.619]
K-Means in [16]	0.679	[0.666, 0.691]	0.676	[0.666, 0.690]	0.634	[0.619, 0.648]	0.639	[0.625, 0.654]
HDBSCAN in [16]	0.637	[0.607, 0.667]	0.636	[0.606, 0.666]	0.617	[0.603, 0.632]	0.625	[0.612, 0.639]
Community Detection	0.708	[0.678, 0.738]	**0.830**	**[0.795, 0.865]**	0.562	[0.535, 0.588]	**0.755**	**[0.715, 0.794]**

4.3 Results

Tables 2 and 3 report the results for the two interleaving settings considered in our evaluation, namely those involving two and three routine types, as described in Step 2 (Sect. 4.2). Each table reports the mean Jaccard coefficient (JC) together with the corresponding 95% confidence interval (CI), expressed as [CI_L, CI_U], where CI_L and CI_U denote the lower and upper bounds of the interval, respectively, computed over the evaluated interleaved logs. The columns distinguish the two preprocessing settings applied to the UI logs, namely *No Split* and *Split*, for both synthetic and real-life datasets, while the rows report the baseline techniques considered in Step 3 (Sect. 4.2), together with our technique, i.e.,

Table 3. Performance comparison of routine identification techniques under three-routine interleaving. The table reports the mean Jaccard coefficient (JC) together with the 95% confidence intervals (CI) in the form [CI_L, CI_U], where CI_L and CI_U denote the lower and upper bounds of the confidence interval, respectively, computed over the evaluated logs. Results are reported for both synthetic and real UI logs. The values in bold refer to our technique, which uses community detection (Infomap) together with label splitting. Our technique discovers routine types that are statistically more accurate than any baseline in any case study.

Techniques	Synthetic UI logs				Real-life UI logs			
	No Split		Split		No Split		Split	
	Mean	CI	Mean	CI	Mean	CI	Mean	CI
Frequency in [13]	0.358	[0.312, 0.404]	0.482	[0.435, 0.529]	0.434	[0.386, 0.482]	0.543	[0.505, 0.581]
Cohesion in [13]	0.688	[0.664, 0.712]	0.705	[0.681, 0.728]	0.664	[0.640, 0.689]	0.647	[0.624, 0.670]
Length in [13]	0.699	[0.680, 0.717]	0.647	[0.628, 0.666]	0.663	[0.638, 0.688]	0.651	[0.628, 0.674]
Coverage in [13]	0.616	[0.593, 0.639]	0.640	[0.619, 0.662]	0.658	[0.631, 0.686]	0.654	[0.630, 0.677]
Rebmann et al. [20]	0.555	[0.537, 0.573]	0.555	[0.537, 0.573]	0.586	[0.562, 0.611]	0.583	[0.559, 0.607]
Agostinelli et al. [6]	0.588	[0.580, 0.596]	0.572	[0.564, 0.580]	0.600	[0.577, 0.623]	0.591	[0.572, 0.609]
DBSCAN in [16]	0.534	[0.526, 0.542]	0.533	[0.524, 0.541]	0.333	[0.293, 0.373]	0.336	[0.295, 0.376]
K-Means in [16]	0.556	[0.538, 0.574]	0.558	[0.542, 0.574]	0.377	[0.327, 0.426]	0.369	[0.320, 0.419]
HDBSCAN in [16]	0.659	[0.630, 0.688]	0.676	[0.649, 0.703]	0.366	[0.321, 0.410]	0.368	[0.322, 0.414]
Community Detection	0.648	[0.616, 0.679]	**0.801**	**[0.778, 0.823]**	0.698	[0.664, 0.702]	**0.735**	**[0.708, 0.762]**

community detection with label splitting. The mean JC captures the average performance of each technique in identifying routine types, whereas the confidence interval quantifies the uncertainty associated with the estimated mean.

Table 2 reports the results under the two-routine interleaving setting. The mean JC values show that our approach achieves higher accuracy in identifying routine types across both datasets, namely, synthetic and real-life UI logs. Moreover, the confidence intervals reveal a clear separation between our technique and the baselines: the lower bound CI_L of our method exceeds the upper bound CI_U of the baseline approaches. This separation indicates a consistent performance gap in favor of our technique under the two-routine interleaving setting in the presence of shared actions.

To better understand the observed differences, we analyze the behavior of the baselines. In particular, the approach by Agostinelli et al. [6] tends to extract an excessive number of routine types when applied to highly interleaved UI logs, resulting in fragmented routine structures. The technique by de Leoni et al. [16], evaluated using three clustering variants (DBSCAN, K-Means, and HDBSCAN), also shows limited performance in our setting. As discussed in Sect. 2, this approach can handle shared actions when routines are executed sequentially, but becomes less effective when multiple routine types are interleaved and actions from different routines overlap. Similarly, the four variants of the method by Leno et al. [13] (frequency, cohesion, length, and coverage) achieve comparatively lower JC values because the method assumes relatively sequential execution patterns and struggles when routine executions are interleaved. Furthermore, the technique by Rebmann et al. [20] consistently extracts a single routine type across the evaluated logs. Finally, our technique without label splitting also shows lower accuracy due to the limitations of Infomap in handling shared actions. The results of the baselines indicate that these approaches struggle to identify accurate routine types under the interleaving settings discussed in Step 3 (Sect. 4.2).

Although the pre-processing step with context-based label splitting is also applied to the state-of-the-art methods, their performance does not improve compared with the results without the step. This illustrates that the pre-processing step is only beneficial for our technique.

Table 3 shows a similar trend to that observed in Table 2, but under the more challenging three-routine interleaving setting. As the degree of interleaving and overlap among routine executions increases, the mean JC values decrease for all techniques, reflecting the increased difficulty of identifying routine types. Despite this degradation, our approach consistently achieves the highest mean JC across the evaluated baselines.

We further assessed the statistical significance of the observed performance improvements by performing two-sided paired t-tests between our technique and the considered baselines on the same UI logs. To account for multiple comparisons, the resulting p-values were adjusted using the Holm–Bonferroni correction method. The detailed results, including the t-statistics, raw p-values and Holm-corrected p-values, are available in the file `Statistical_Analysis_Results.xlsx` published in the repository accompanying this paper [1]. The smallest t-statistic was 4.09 with 99° of freedom, while the largest Holm-corrected p-value was ca. 0.00019. This demonstrated a consistently superior and statistically robust improvement of our technique over all baselines: the weakest comparison maintains a high degree of certainty ($p < 0.0002$) after adjusting for multiple tests.

4.4 Computational Performance

The technique introduced in this paper is highly efficient in terms of computation. Table 4 presents the 95% confidence intervals of the execution times (in seconds) for experimental scenarios involving two and three interleaving routine types, aggregated over the various experimental runs. Execution times are separately reported for the synthetic and real-life UI logs discussed in Sect. 4.2. The columns *Phase 1*, *Phase 2* and *Phase 3 & 4* report the execution times for the different phases (cf. Section 3), namely: (1) Context-Based Label Splitting for Shared Actions, (2) Construction of the Normalized Directly-Follows Graph, (3) Graph Clustering via Community Detection, and (4) Reconstructing Routine Types from Split Labels. Phases 3 and 4 are reported together because the last phase takes so little time that it is often difficult to actually measure the actual amount. The column *Total Time* reports the aggregated values over the phases. The experiments were executed on a machine equipped with an Intel Core i5-1135G7 CPU @ 2.40GHz, 32 GB RAM, running Windows 11 and Python 3.11.

The results indicate that execution times always remain within 7 s, thus remaining practically feasible across all evaluated settings. This supports the practical applicability of the technique to routine type identification from interleaved UI logs in RPA settings.

Table 4. Computational performance analysis of the proposed technique. The table reports the 95% confidence intervals of the execution times in the form [CI_L, CI_U], where CI_L and CI_U denote the lower and upper bounds of the confidence interval, respectively, expressed in seconds for the different phases of the technique under the evaluated settings.

Interleaving	Dataset	Phase 1	Phase 2	Phases 3 & 4	Total
Two	Synthetic	[0.5209, 0.7999]	[0.1519, 0.1989]	[2.5260, 3.6733]	[3.3320, 4.5390]
Two	Real-life	[0.0179, 0.0552]	[0.0333, 0.0488]	[5.3007, 5.9134]	[5.3899, 5.9794]
Three	Synthetic	[0.9074, 1.2423]	[0.1140, 0.1511]	[5.0408, 5.6343]	[6.1630, 6.9271]
Three	Real-life	[0.1743, 0.2512]	[0.0250, 0.3373]	[3.5753, 4.7097]	[3.8071, 4.9622]

5 Conclusion

This paper focuses on the identification of routine types in UI logs, a prerequisite step for employing RPA solutions to automate user routines. In fact, software robots require the provision of the routine types and their behavior to enable their automation.

Analyzing the literature on routine identification, it is evident that the vast majority of existing techniques assume that routines are executed sequentially. However, real operational environments are often characterized by users who are preempted while performing certain routines to attend to more urgent or important tasks. This behavior ultimately manifests itself as interleavings of routine executions in UI logs. In addition, actions belonging to different types of routine can also be shared between routines, which further increases the ambiguity in identifying the correct types of routines. These aspects pose a significant challenge that cannot be addressed by techniques assuming sequential routine executions: one cannot purely rely on the sequential adjacency of actions in the UI log, but this needs to be combined with the analysis of the underlying information flow and the contextual role of shared actions.

This paper proposes a four-phase technique for identifying routine types in interleaved executions where actions may be shared between routines. First, shared actions are refined through context-based label splitting, adopting the technique by van Zelst et al. [24]. Second, a normalized directly-follows graph is built using maximum-likelihood estimation from bigram language models. Third, Infomap community detection clusters actions of the normalized directly-follows graph into routine types by capturing information flow and coherent routine structures. Fourth, the action labels that were split are merged back to retain the same labels as the original UI logs.

The evaluation was conducted on real-life and synthetic UI logs, and the results were assessed using the Jaccard Coefficient (JC), which is a widely adopted metric for evaluating the accuracy of discovered routine types. The results demonstrate that the proposed technique consistently identifies routine types with higher accuracy than state-of-the-art methods, particularly in scenarios involving interleavings and shared actions.

The first phase of context-based label splitting requires setting the context-window length k, which was set to one in the experiments and deemed reasonable (cf. the end of Sect. 4.2). However, the accuracy of the discovered routine types may depend on this choice. Future work will assess the impact of varying k on accuracy. The first phase also requires specifying the actions that are candidates for splitting. Currently, the technique excludes actions in the first frequency quartile, a choice that may affect accuracy. Future work will assess how different q-quantile thresholds influence accuracy and compare them with treating every action as a potential candidate for label splitting. We also aim to investigate how the benefits of this technique scale with the frequency of shared actions across routine types and within single routine executions of a certain type.

The original UI logs did not feature interleaved executions of routines, and the interleavings were synthetically introduced in a manner that we consider both challenging and realistic. Nevertheless, we hope to gain access in the future to real-life UI logs that naturally exhibit interleaved routine executions and shared actions, which would allow further validation of the proposed technique.

References

1. Repository for this submission. https://anonymous.4open.science/r/Shared-Actions-RPA-45B0/README.md (2026)
2. van der Aalst, W.M.P., Bichler, M., Heinzl, A.: Robotic process automation. Bus. Inf. Syst. Eng. **60**(4), 269–272 (2018)
3. Abb, L., Bormann, C., van der Aa, H., Rehse, J.R.: Trace clustering for user behavior mining. In: ECIS (2022)
4. Agostinelli, S., Hohenadl, T., Marrella, A., Martínez-Rojas, A.: SmartRPA: Generating software robots from user interface logs. SoftwareX **29**, 101995 (2025)
5. Agostinelli, S.: Segmentation in RPA. In: Generating Executable Robotic Process Automation Scripts from Unsegmented User Interface Logs, pp. 33–44. Springer Nature Switzerland, Cham (2024)
6. Agostinelli, S., Leotta, F., Marrella, A.: Interactive segmentation of user interface logs. In: Hacid, H., Kao, O., Mecella, M., Moha, N., Paik, H. (eds.) ICSOC 2021. LNCS, vol. 13121, pp. 65–80. Springer, Cham (2021). https://doi.org/10.1007/978-3-030-91431-8_5
7. Blondel, V.D., Guillaume, J., Lambiotte, R.: Fast unfolding of communities in large networks: 15 years later. CoRR abs/2311.06047 (2023)
8. Cormen, T.H., Leiserson, C.E., Rivest, R.L., Stein, C.: Introduction to Algorithms. MIT Press (2009)

9. Dumas, M., Rosa, M.L., Leno, V., Polyvyanyy, A., Maggi, F.M.: Robotic process mining. In: van der Aalst, W.M.P., Carmona, J. (eds.) Process Mining Handbook, Lecture Notes in Business Information Processing, vol. 448, pp. 468–491. Springer, Cham (2022). https://doi.org/10.1007/978-3-031-08848-3_16
10. Fortunato, S., Lancichinetti, A.: Community detection algorithms: a comparative analysis: invited presentation, extended abstract. In: Proceedings of the Fourth International ICST Conference on Performance Evaluation Methodologies and Tools (VALUETOOLS '09 (2009)
11. Hohenadl, T.: Time series-based segmentation of noisy user interaction logs for robotic process automation. In: Business Process Management: Responsible BPM Forum. Process Technology Forum, Educators Forum, pp. 276–291. Springer Nature Switzerland, Cham (2026). https://doi.org/10.1007/978-3-032-02936-2_20
12. Jurafsky, D., Martin, J.H.: Speech and Language Processing, chap. N-gram Language Models and Maximum-Likelihood Estimation, 3 edn. Prentice Hall (2023)
13. Leno, V., Augusto, A., Dumas, M., Rosa, M.L., Maggi, F.M., Polyvyanyy, A.: Identifying candidate routines for robotic process automation from unsegmented UI logs. In: van Dongen, B.F., Montali, M., Wynn, M.T. (eds.) 2nd International Conference on Process Mining, ICPM 2020, Padua, Italy, October 4-9, 2020, pp. 153–160. IEEE (2020)
14. Leno, V., Polyvyanyy, A., Dumas, M., Rosa, M.L., Maggi, F.M.: Robotic process mining: vision and challenges. Bus. Inf. Syst. Eng. **63**(3), 301–314 (2021)
15. Leno, V., Polyvyanyy, A., Rosa, M.L., Dumas, M., Maggi, F.M.: Robidium: automated synthesis of robotic process automation scripts from UI logs. In: Proceedings of the BPM 2020 Demos and Resources Track. CEUR Workshop Proceedings, vol. 2673, pp. 102–106 (2020)
16. de Leoni, M., Khan, F.A., Agostinelli, S.: Accurate and noise-tolerant extraction of routine logs in robotic process automation. In: Cappiello, C., Hartig, O., Sellami, M., Ouni, A. (eds) Cooperative Information Systems, pp. 465–475. Springer Nature Switzerland, Cham (2026). https://doi.org/10.1007/978-3-032-15538-2_26
17. Ng, A.Y., Jordan, M.I., Weiss, Y.: On spectral clustering: analysis and an algorithm. In: Proceedings of the 15th International Conference on Neural Information Processing Systems: Natural and Synthetic, pp. 849–856. NIPS'01, MIT Press, Cambridge, MA, USA (2001)
18. Plattfaut, R., Borghoff, V.: Robotic process automation: a literature-based research agenda. J. Inf. Syst. **36**(2), 173–191 (2022)
19. Pons, P., Latapy, M.: Computing communities in large networks using random walks. In: Yolum, I., Güngör, T., Gürgen, F., Özturan, C. (eds.) ISCIS 2005. LNCS, vol. 3733, pp. 284–293. Springer, Heidelberg (2005). https://doi.org/10.1007/11569596_31
20. Rebmann, A., van der Aa, H.: Unsupervised task recognition from user interaction streams. In: Indulska, M., Reinhartz-Berger, I., Cetina, C., Pastor, O. (eds.) Advanced Information Systems Engineering - 35th International Conference, CAiSE 2023, Zaragoza, Spain, June 12-16, 2023, Proceedings. Lecture Notes in Computer Science, vol. 13901, pp. 141–157. Springer, Cham (2023). https://doi.org/10.1007/978-3-031-34560-9_9
21. Rosvall, M., Bergstrom, C.T.: Maps of random walks on complex networks reveal community structure. Proc. Natl. Acad. Sci. **105**(4), 1118–1123 (2008)
22. Urabe, Y., Yagi, S., Tsuchikawa, K., Oishi, H.: Task clustering method using user interaction logs to plan RPA introduction. In: Polyvyanyy, A., Wynn, M.T., Van Looy, A., Reichert, M. (eds.) BPM 2021. LNCS, vol. 12875, pp. 273–288. Springer, Cham (2021). https://doi.org/10.1007/978-3-030-85469-0_18

23. Yang, Z., Algesheimer, R., Tessone, C.J.: A comparative analysis of community detection algorithms on artificial networks. Sci. Rep. **6**, 30750 (2016)
24. van Zelst, S.J., Tai, J., Langenberg, M., Lu, X.: Context-based activity label-splitting. In: Proceedings of the 21st International Conference on Business Process Management (BPM 202). Lecture Notes in Computer Science, vol. 14159, pp. 232–248. Springer, Cham (2023). https://doi.org/10.1007/978-3-031-41620-0_14

SCOPE: Sequential Causal Optimization of Process Interventions

Jakob De Moor[1(✉)], Hans Weytjens[1,2], Johannes De Smedt[1], and Jochen De Weerdt[1]

[1] Research Centre for Information Systems Engineering (LIRIS), KU Leuven, Leuven, Belgium
{jakob.demoor,hans.weytjens,johannes.desmedt}@kuleuven.be
[2] School of Computation, Information and Technology, Technical University of Munich (TUM), Munich, Germany
hans.weytjens@tum.de

Abstract. Prescriptive Process Monitoring (PresPM) recommends interventions during running business processes to optimize key performance indicators (KPIs). In realistic settings, interventions are rarely isolated: organizations need to align sequences of interventions to jointly steer the outcome of a case. Existing PresPM approaches only partially address this challenge. Many focus on a single intervention decision, while others treat multiple interventions independently, ignoring how they interact over time. Methods that do address these dependencies depend either on simulation or data augmentation to approximate the process to train a Reinforcement Learning (RL) agent, which may create a reality gap and introduce bias. We introduce *SCOPE* (Sequential Causal Optimization of Process Interventions), a PresPM approach that learns aligned sequential intervention recommendations. *SCOPE* employs backward induction to estimate the effect of each candidate intervention action, propagating its impact from the final decision point back to the first. By leveraging causal learners, our method can utilize observational data directly, unlike methods that require constructing process approximations for RL. Experiments on both an existing synthetic dataset and a new semi-synthetic dataset show that *SCOPE* consistently outperforms state-of-the-art PresPM techniques in optimizing the KPI. The novel semi-synthetic setup, based on a real-life event log, is provided as a reusable benchmark for future work on sequential PresPM.

Keywords: Prescriptive Process Monitoring · Sequential Decision-Making · Causal Learning

1 Introduction

PresPM uses machine learning to provide case-specific recommendations at different decision points during the execution of business processes. These recom-

This work was supported by the Research Foundation - Flanders (11A6J25N and G039923N), and Internal Funds KU Leuven (C14/23/031).

F. M. Maggi et al. (Eds.): BPM 2026, LNCS 16936, pp. 350–368, 2027.
https://doi.org/10.1007/978-3-032-37777-7_22

mendations concern interventions, such as managerial escalations or customer communications, that aim to improve KPIs, for example, throughput time or cost efficiency. PresPM holds the potential to move organizations from merely describing and predicting process behavior towards actively steering process executions [8]. In this paper, we use the terms *intervention* for any controllable action taken on an ongoing case, and *intervention recommendation* for the action suggested by a PresPM method at a *decision point*.

In many processes, intervention decisions are not isolated. Typically, a process contains multiple, interdependent decision points that jointly determine the outcome of a case. The effect of an earlier (or later) intervention depends on which interventions are applied later (or earlier) in the same case. Optimizing intervention decisions one by one is therefore insufficient, because actions that look beneficial locally can undermine overall KPI performance. For example, in a marketing process aimed at maximizing revenue, the effectiveness of offering a client a discount may depend on an earlier choice to send a promotional email and the client's response to that email. This sequence of decisions together shapes the revenue. Optimizing each decision independently without considering the future might make it look like sending a promotional email is only moderately valuable for some clients, even though if you follow it with a discount, it could be highly effective. Methods for PresPM thus need to reason over sequences of decisions and their combined effect on the final KPI.

Most existing PresPM approaches do not yet offer such sequential support. First, many methods address only a single intervention scenario, even when multiple opportunities to intervene exist [3,4,8,19–22]. These approaches may improve performance for that single intervention scenario, but they do not coordinate multiple decisions over the full case. Second, some approaches handle sequential decisions but optimize each decision point in isolation. They focus on the immediate effect of the next intervention, for instance, on the remaining processing time, without explicitly aligning interventions across decision points. This can still lead to suboptimal end-to-end outcomes with respect to the final KPI [9]. Third, other methods for sequential intervention recommendations rely on process approximations, such as Markov decision processes (MDP) or data augmentation [1,5], to train an RL agent, and the resulting policies inherit any misspecification in these approximations, also known as the reality gap, which may lead to biased or underperforming recommendations in practice [16].

To address these limitations, this paper makes two key contributions:

1. We propose *SCOPE*, a PresPM approach that combines causal learners with backward induction to learn causally grounded sequential intervention policies that are aligned across multiple decision points. For each decision point, a causal model estimates the effect of alternative interventions on the target KPI, given the observed process execution history. During training, backward induction then propagates the impact of later intervention decisions back to earlier ones by starting from the final decision point and recursively deriving the optimal action and its expected outcome at each preceding decision point. Operating directly on observational event logs, *SCOPE* uses causal learning to

be able to identify actions rarely chosen historically but likely to enhance the target KPI, without requiring process-specific simulators or log augmentation.
2. We provide an empirical evaluation on existing synthetic data and a new semi-synthetic setup based on a real-life event log. Our code and the novel semi-synthetic benchmark are made publicly available as a reusable resource for sequential PresPM in our GitHub repository. The results show that *SCOPE* outperforms the considered baselines across a broad set of experimental settings, highlighting the importance of combining causal learning with backward induction for KPI optimization.

The paper is organized as follows: Sect. 2 reviews background and related work, Sect. 3 describes the methodology, Sect. 4 covers experiments, results, and discussion, and Sect. 5 concludes the paper.

2 Background

2.1 General

Current PresPM approaches are generally based on two streams of research. The first is Causal Inference (CI), which aims to estimate the effects of potential interventions from observational data and use these estimates to guide decisions. This is achieved through causal learners, which are model setups designed to use observational data to predict outcomes under an action and/or causal effects of choosing one action over the other. For example, an S-learner fits a single model that predicts the outcome using the intervention action as an input in addition to other features. PresPM approaches typically adapt causal learners to the process context by aggregating event data for use with standard CI setups [8], or by leveraging sequential models like LSTMs [26].[1] The second stream is RL-based, where an agent interacts with an environment that represents/approximates the real process, observes states, selects actions, and receives rewards, with the goal of learning a policy that maximizes cumulative rewards [19].

Similar to CI, PresPM typically relies on offline observational event logs, rather than data from controlled experiments. This approach avoids the cost and risk associated with randomized experiments, such as randomized controlled trials (RCTs) or an online RL agent exploring a real-life environment, which are often not feasible (e.g., when testing a loan assignment strategy). However, observational data introduces challenges. Because the data reflects an existing *historical decision policy* (e.g., a bank's current loan strategy), treatment/intervention assignment is not random as it would be in controlled trials. This complicates the estimation of optimal actions due to *confounding* factors affecting both intervention assignment and outcomes [12]. For example, if a bank offers optional financial literacy workshops (intervention) to improve repayment rates (KPI),

[1] CI typically follows causal discovery, which identifies causal relationships, while CI estimates their magnitude [26]. Most PresPM studies that optimize interventions (implicitly) adopt a known standard causal structure with time-varying confounders, interventions and an outcome.

individuals who choose to attend may already have stronger financial habits (confounder), making it difficult to isolate the workshop's true effect.

2.2 Related Work

This subsection reviews existing PresPM approaches, highlighting their key features and limitations. Table 1 summarizes how *SCOPE* compares to these methods, with the discussion below detailing the distinguishing dimensions.

Single-Interventional Approaches. Most existing PresPM methodologies focus on the impact of a single intervention scenario. For example, Bozorgi et al. [8] use causal effect estimation to identify which cases would benefit from one specific intervention decision, applying CI techniques after encoding event data into a suitable format. In their work, an example of such an intervention would be skipping a particular activity. Similarly, Shoush et al. [22] propose a white-box framework that integrates predictive, causal, and survival models to recommend single intervention decisions. In their case, an example intervention would be offering either multiple loan options to a client or just one. Yet, real-world business processes often require multiple (interdependent) interventions rather than a single one, as they consist of interconnected events. In Table 1, the first column shows whether a method is multi-interventional or not.

Multi-Interventional Approaches. Some PresPM approaches handle multiple interventions by recommending the next best activity to optimize a KPI. Typically, predictive models estimate KPIs for possible actions at each decision point; for instance, de Leoni et al. [9] use a transition-system to identify valid next activities and select the one maximizing the predicted KPI. However, optimizing only the immediate action without considering subsequent decision points can yield suboptimal results. Consider again the revenue-maximizing marketing process with two consecutive decisions: sending a promotional email (decision point 1), then offering a discount (decision point 2). Assessing actions at decision point 1 without accounting for decision point 2 yields predictions that are *averages over the historical distribution of actions taken at decision point 2.* Consequently, the model might undervalue an email that is only highly effective when followed by a discount, ignoring the cumulative effect of action sequences. Making optimal decisions requires aligning action sequences across all decision points. Column 2 in Table 1 categorizes whether methods align interventions across multiple decision points or optimize at a single point, as in [9].

A different line of work that addresses sequential interventions is presented by Abbasi et al. [1] and Branchi et al. [5]. Both aim to recommend the next best activity while explicitly accounting for all possible sequences of future actions. Branchi et al. achieve this by (1) applying KMeans clustering to group process prefixes in an event log; (2) defining an MDP where states are represented by the cluster label and previous activity, rewards are defined as the average reward observed for that state in the log, and transition probabilities are obtained by replaying the log; and (3) applying offline RL (Q-learning specifically) to derive a next-best-action policy [5]. Because Q-learning optimizes cumulative return over a full case, the resulting recommendations are aligned across decision points.[2] Abbasi et al. take a slightly different route by also applying offline RL, but combined with data augmentation [1]. They modify case timestamps and remove non-critical activities, arguing that these transformations help diversify the training data. However, both approaches rely on approximations of the real process through simulation (in the form of an MDP) or data augmentation. These approximations may lead to policies that underperform in practice, a phenomenon known as the reality gap [16]. For example, in Branchi et al., important prefix information may be lost when aggregating cases prior to clustering, and KMeans may struggle with high-dimensional or categorical data. In Abbasi et al., augmentation is constrained by manually crafted rules: timestamps are only varied within 10%, and three business rules (precedence, co-occurrence, conflict) guide modifications. Some processes may require different temporal bounds or a far richer set of constraints. In such situations, the learned policy may underperform. In Table 1, the last column indicates whether a method requires an approximation (e.g., a simulation or augmentation) of the process, or can directly use the data as-is.

We also note Weinzierl et al. [25], whose approach predicts the most likely future suffix and identifies similar historical cases to select actions with favorable KPIs. By only exploring the neighborhood of the (historically) most likely suffix, this method depends heavily on the historical decision policy and risks missing superior but rarely observed actions. We therefore do not pursue this method further.

In summary, Table 1 highlights a gap in current methods that *SCOPE* aims to address.

Sequential Decision-Making & CI. Sequential decision-making approaches are also used in CI, commonly studied under the framework of Dynamic Treatment Regimes (DTR). DTR methods aim to identify optimal treatment policies to maximize long-term outcomes, typically in medical settings. These methods rely on sequential optimization (similar to RL) and causal methods. For instance, in [18], the authors discuss Q-learning, essentially a form of backward induction, to determine the optimal treatment regime. Another approach models the DTR objective as a classification problem [27], using causal learners and a classifier to recommend the best treatment at each step. We argue that the sequential PresPM problem is related to DTRs and that DTR methods provide a promising foundation to address the gap identified in Table 1. However, mapping PresPM to the DTR framework is not straightforward. DTR typically assumes a fixed number of decision points (often medical stages), while PresPM operates on event-log prefixes from business processes with variable control flow. Case evolution depends on event-, case-, and activity-level variables, and may include

[2] Branchi et al. [5] extend their earlier work [6] by removing process-specific assumptions and thus improving generalizability, which is why we focus on it here.

Table 1. Comparison of PresPM methods for sequential interventions.

	Multi-interventional	Aligned interventions across decision points	No process approximations required
Why It Matters	Many business processes require a series of different intervention decisions, where each step shapes what happens next.	An intervention recommendation optimal at one decision point may be suboptimal across decision points.	Using an MDP or augmented data to approximate processes for training may misrepresent true behavior, leading to suboptimal policies.
Method			
[3,19]	×	×	×
[4,8,20–22]	×	×	✓
[9]	✓	×	✓
[1]	✓	✓	×
[5]	✓	✓	×
SCOPE (ours)	✓	✓	✓

loops, parallelism, and multiple branching structures. Consequently, the number of decision points varies across cases, and decision histories/prefixes can be highly heterogeneous at a decision point. We therefore build on the DTR framework, but adapt and configure it to the sequential PresPM setting to identify intervention recommendations that maximize process KPIs in business settings.

3 Methodology

This section introduces definitions, notation, and a detailed explanation of *SCOPE*.

3.1 Preliminaries

Definition 1 (Event, Event Log, Trace, Prefix). *An* EVENT *is a tuple* $e = (c, o, t, d, s)$*, comprising a case identifier* $c \in C$*, activity label* $o \in O$*, timestamp* $t \in \mathbb{R}^+$*, optional event-specific attributes* $d = (d_1, \ldots, d_{m_d})$*, and static case attributes* $s = (s_1, \ldots, s_{m_s})$*. An* EVENT LOG $L = \{e_i\}_{i=1}^N$ *is a collection of* N *observed events. A* TRACE $\sigma = \langle e_1, \ldots, e_{|\sigma|} \rangle$ *is the time-ordered sequence of all events for a single case. A* PREFIX *consists of the first* l *events of a trace.*

Definition 2 (Decision point, Intervention action, Policy, Outcome). *A* DECISION POINT $k \in \{1, \ldots, K\}$ *is a prespecified or discovered point in the process where an intervention may occur. For a case* c*, let* $\mathcal{K}^{(c)} \subseteq \{1, \ldots, K\}$ *denote the specific decision points it actually visits, noting that some may be skipped due to variable control flow. The history available at* k *is represented by the prefix* σ_k*, encompassing all events (and interventions) preceding* k*, and* *its length depends on the control flow it followed. An* INTERVENTION ACTION $a_k \in A_k$ *is chosen from a feasible space* A_k*, which maps to an existing attribute in the event log (i.e., activity label or any other event-specific attribute from s).*[3] *The historically observed action is denoted* a_k^{obs}*. A* POLICY $\pi = \{\pi_1, \ldots, \pi_K\}$ *is a set of decision rules mapping a prefix to an action:* $\pi_k : \sigma_k \rightarrow a_k$*. The* OUTCOME y *is the trace's final observed KPI. Consequently, an event log L can be encoded into the dataset:* $\mathcal{D} = \bigcup_{c \in C} \bigcup_{k \in \mathcal{K}^{(c)}} \left\{ \left(\sigma_k^{(c)}, a_k^{obs,(c)}, y^{(c)} \right) \right\}$.

Objective. The goal is to find the optimal policy $\pi^{opt} = \{\pi_1^{opt}, \ldots, \pi_K^{opt}\}$, given by $\pi^{opt} = \arg\max_\pi \mathbb{E}\big[y(\pi)\big]$, where $y(\pi)$ is the potential outcome under the policy π [17].

3.2 SCOPE

Below, we explain how *SCOPE* maximizes a KPI (use *min* instead of *max* for minimization). First, we outline backward induction theory. Next, we describe how causal learners are integrated. Finally, we present the full *SCOPE* algorithm.

Backward Induction. We now explain how backward induction (theoretically) identifies the optimal sequential intervention policy, adapted here to the business process context with DTR histories as prefixes with variable control-flow. Following DTR and PresPM literature, we begin with three standard assumptions: sequential ignorability, the stable unit treatment value assumption (SUTVA), and positivity.[4] Under these assumptions, the optimal policy π^{opt} is identifiable from observational data through backward induction, meaning it *can*, in principle, be obtained (see [18] for a detailed discussion on these assumptions and the identifiability).

Backward induction iterates through all decision points, starting from the last one K, where we define the Q-function and value function as follows:

$$Q_K(\sigma_K, a_K) = \mathbb{E}\big[y \mid \sigma_K, a_K\big]; \quad V_K(\sigma_K) = \max_{a_K \in A_K} Q_K(\sigma_K, a_K). \tag{1}$$

If we then continue recursively, we define the Q-function and value function at decision point $k \in \{K-1, \ldots, 1\}$ as

$$Q_k(\sigma_k, a_k) = \mathbb{E}\big[V_{k+1}(\sigma_{k+1}) \mid \sigma_k, a_k\big]; \quad V_k(\sigma_k) = \max_{a_k \in A_k} Q_k(\sigma_k, a_k). \tag{2}$$

Then, under a perfectly known Q-function, the optimal actions and policy are given by

$$\pi_k^{\text{opt}}(\sigma_k) = \arg\max_{a_k \in \mathcal{A}_k} Q_k(\sigma_k, a_k) = a_k^{opt}; \quad \pi^{opt} = \{\pi_1^{opt}(\sigma_1), \ldots, \pi_K^{opt}(\sigma_K)\} \tag{3}$$

[3] The action space can be predefined or discovered, e.g., as in [9].

[4] Sequential ignorability: no unmeasured confounders given a prefix; SUTVA: intervening on a case does not affect another case's outcome and interventions have well-defined versions; positivity: given a prefix, every feasible action has a nonzero chance of occurring historically.

which concludes the backward induction.

The value function can alternatively be expressed as:

$$V_k(\sigma_k) = \mathbb{E}\big[V_{k+1}(\sigma_{k+1}) + \big(Q_k(\sigma_k, a_k^{\text{opt}}) - Q_k(\sigma_k, a_k^{obs})\big) \mid \sigma_k\big] \tag{4}$$

with $V_{K+1} = y$. This formulation, known as the *regret-based* form, yields the same optimal policy.

Intuitively, backward induction works by starting at the last decision point K and asking: '*Given a current prefix σ_K, what is the expected outcome of each possible action?*'. This gives us the Q-function at K. The value function at K then tells us the best outcome achievable from any σ_K by selecting the highest Q-value (using the standard formulation at Eq. 1 here for simplicity). We then move one step backward to decision point $K-1$. Here, the Q-function asks: '*If I take action a_{K-1} now for a prefix σ_{K-1}, and then act optimally at K (with the new prefix σ_K resulting from a_{K-1}), what outcome can I expect?*'. This is answered by the value function at K for σ_K. The value function at $K-1$ again picks the action with the highest Q-value. We repeat this process all the way to the first decision point. In general, the Q-function answers: '*How good is it to take a specific action now, assuming we act optimally afterward?*', while the value function answers: '*How good is it to have this prefix, assuming optimal actions from here on?*'. By applying this procedure recursively, we can compute all Q-functions and thus determine the optimal policy.

Causal Learners. While backward induction provides the theoretically optimal solution, applying it in practice requires estimation. This is challenging because we rely on observational event logs, and, as discussed in Sect. 2, observed confounders make estimation difficult. To address this, we draw inspiration from the DTR approaches in [18,27] and use causal learners to estimate the Q-function at each decision point and, consequently, the optimal action. While *SCOPE* still uses function approximation to estimate Q-functions, it avoids explicit process approximations like MDP construction or data augmentation for generating training trajectories. Causal learners are designed to estimate outcomes and/or causal effects under different actions using observational data [2]. These techniques try to explicitly reduce dependence on the historical decision policy (in contrast to [25], see Sect. 2). Although *SCOPE* supports any suitable causal learner, we apply three widely used, model-agnostic options compatible with any base model (e.g., an LSTM): the S-, T-, and RA-learner [2]. We choose these for their stability, which is ideal for the PresPM setting. For example, unlike the IPW- and DR-learner, they avoid inverse propensity weights (IPW), which are used to handle confounding by reweighting inputs based on the inverse historical propensity of an action being taken for that input. However, in PresPM, the high input/prefix diversity (see Sect. 2) could produce near-zero probabilities, making IPWs highly unstable [2].

We adopt the regret-based formulation 4 of the value function, which is more robust to model misspecification (e.g., incorrect model assumptions) [14]. Incorporating the relative difference (regret) between optimal and observed actions can reduce error propagation compared to estimating absolute values (as in the standard form 2). This is particularly useful in business processes, again because inputs (prefixes) can be highly heterogeneous due to variable control flow (see Sect. 2). Under this formulation, we must estimate two key components at each decision point k: the Q-function Q_k and the corresponding optimal action a_k^{opt}. Once these are obtained, computing the value function becomes straightforward. Note that, while in the theoretical setting with a perfectly known Q-function, the optimal action is always given by the action that maximizes Q_k (see eq. 3), in practical approximation settings, the optimal action may be better estimated using alternative strategies to compensate for prediction error and model uncertainty, as we do below with the RA-learner.

*S-learner.*The S-learner estimates Q_k using a single model (e.g., an LSTM), taking as input the features in σ_k together with the historical action taken a_k^{obs} [2]. At inference time, the model is queried for each possible action a_k at decision point k to obtain an estimate of $Q_k(\sigma_k, a_k)$. The action with the highest estimate is selected.

T-learner. The T-learner builds a separate model for each possible action at decision point k [2], using only features σ_k. At inference, each model estimates $Q_k(\sigma_k, a_k)$ under its corresponding action, and the action with the highest estimate is selected.

RA-learner. The RA-learner works in two steps [2]: it first estimates Q_k like the S-learner, then constructs a pseudo-outcome for each possible action. For a prefix, observed action a_k^{obs}, decision point k, and action of interest a_k, the pseudo-outcome is:

$$\Phi_k^{(a_k)}(\sigma_k) = \mathbf{1}\{a_k^{obs} = a_k\}(y - \hat{Q}(\sigma_k, a_k)) + \sum_{f_k \neq a_k} \mathbf{1}\{a_k^{obs} = f_k\}(\hat{Q}(\sigma_k, f_k) - y)$$
$$+ \sum_{f_k \neq a_k} \mathbf{1}\{a_k^{obs} = f_k\}(\hat{Q}(\sigma_k, f_k) - \hat{Q}(\sigma_k, b_k)). \qquad (5)$$

where f_k runs over all actions other than a_k, and b_k is an arbitrary baseline action. These pseudo-outcomes (which still have the observed y in eq. 5, and thus cannot serve as estimation for new data) act as targets in a second predictive model, which outputs estimates of the causal effect of each action at decision point k. Intuitively, constructing pseudo-outcomes allows the second model to focus entirely on the causal effect (the difference in Q-values), which can be much simpler to learn than the Q-values (as in the other learners). The final action is chosen as the one with the largest predicted causal effect. The RA-learner is thus an explicit two-step approach for better causal effect estimation.

Algorithm. In summary, *SCOPE* combines regret-based backward induction with causal learning to derive effective policies from observational event logs, handling cases with varying numbers of decision points and heterogeneous inputs (prefixes) resulting from variable control flow. In Algorithm 1, we provide an overview of the full training and inference procedure for an S-learner variant, referred to as *SCOPE-S*. Lines highlighted in green mark the steps where the Q-function and optimal action at decision point k are estimated. When adopting a T- or RA-learner, these steps are the ones that change most substantially, being replaced by the corresponding estimation procedures described above. Full algorithms for these other learners are also provided in our repository (see footnote 5).

During training, the algorithm applies backward induction, starting from the last decision point. At each decision point, it uses a causal learner (here, an S-learner) to estimate the Q-function (line 4) and determine the optimal action for every case that reached that point (line 7). It then computes the estimated value for each relevant case using a regret-based formulation (lines 810). If a case did not visit the current decision point (due to variable control flow in business processes), the value from the subsequent decision point is carried backward (line 13). These values serve as targets for the preceding decision point. During inference, the process moves forward. For each case, the algorithm observes the current prefix and checks whether a decision point is reached (lines 1819). If so, it selects the model corresponding to that decision point to estimate and execute the optimal action (lines 2021). This continues until the case is completed.

4 Experiments and Discussion

This section outlines the (semi-)synthetic data, methods for comparison, the experimental procedure, and finally the results and implications[5]

4.1 Data

Synthetic and semi-synthetic data are widely used in causal machine learning because real-world datasets do not provide ground-truth counterfactuals [13]. In sequential PresPM, this challenge is even stronger, since for each case we only observe the KPI under the historically executed intervention sequence, not under the many alternative sequences that could have been recommended. Real-life event logs therefore do not allow direct ground-truth evaluation of learned sequential intervention policies. To our knowledge, SimBank is the only simulator designed specifically for PresPM, which we use in our evaluation [10]. We also introduce a new semi-synthetic setup based on the BPIC17_W event log [11] and release it publicly to support further research in sequential PresPM.

SimBank. SimBank models a bank's loan application process and has been thoroughly validated. The main KPI to optimize is the total profit generated

[5] The code, algorithms, simulators, and additional results are available on GitHub and at doi.org/zenodo.20627090.

Algorithm 1. SCOPE-S: Training and Inference

Training (Offline)

Input: Dataset $\mathcal{D} = \bigcup_{c \in C} \bigcup_{k \in \mathcal{K}^{(c)}} \left\{ (\sigma_k^{(c)}, a_k^{obs,(c)}, y^{(c)}) \right\}$, action spaces $\{A_k\}_{k=1}^{K}$

$\hat{V}_{K+1}^{(c)} = y^{(c)} \; \forall c \in \{1, \ldots, C\}$; $\mathbf{M} = [\,]$ ▷ Initialize
for $k = K, K-1, \ldots, 1$ **do** ▷ Iterate backwards
  $\mathcal{C}_k = \{c \mid k \in \mathcal{K}^{(c)}\}$ ▷ Identify cases that visited k
  $\mathcal{M}_k \leftarrow$ Regress $\hat{V}_{k+1}^{(c)}$ on $(\sigma_k^{(c)}, a_k^{obs,(c)})$ for all $c \in \mathcal{C}_k$ ▷ S-learner Q-function
  Append $\mathcal{M}_k$ to $\mathbf{M}$
  for each case $c \in \mathcal{C}_k$ **do**
    $\hat{a}_k^{opt,(c)} = \arg\max_{a_k \in A_k} \mathcal{M}_k(\sigma_k^{(c)}, a_k)$ ▷ Estimated optimal action
    $\hat{Q}_k^{obs,(c)} = \mathcal{M}_k(\sigma_k^{(c)}, a_k^{obs,(c)})$ ▷ Q-value of observed action
    $\hat{Q}_k^{opt,(c)} = \mathcal{M}_k(\sigma_k^{(c)}, \hat{a}_k^{opt,(c)})$ ▷ Q-value of optimal action
    $\hat{V}_k^{(c)} = \hat{V}_{k+1}^{(c)} + \hat{Q}_k^{opt,(c)} - \hat{Q}_k^{obs,(c)}$ ▷ Regret-based value update
  end for
  for each case $c \notin \mathcal{C}_k$ **do**
    $\hat{V}_k^{(c)} = \hat{V}_{k+1}^{(c)}$ ▷ Carry over value if point k is skipped
  end for
end for
return M

Inference (Online)

Input: Models $\{\mathcal{M}_k\}_{k=1}^{K}$, initial prefix $\sigma_{k_{first}}^{(c)}$

while case c is not completed **do**
  if case c reaches DECISION POINT $k \in \{1, \ldots, K\}$ **then**
    Observe current prefix $\sigma_k^{(c)}$
    $\hat{a}_k^{opt,(c)} = \arg\max_{a_k \in A_k} \mathcal{M}_k(\sigma_k^{(c)}, a_k)$
    Execute action $\hat{a}_k^{opt,(c)}$
  end if
end while

from loans in a test set. Among other options, we opt for the following two sequential interventions for this research: 1. *Choose procedure*: decide between a standard or priority procedure. 2. *Set interest rate*: choose one of 3 possible interest rate levels. The effects of these interventions vary across clients, making the decisions complex. The two decision points are highly interdependent. Selecting a standard procedure reduces costs, but increases the likelihood of a client refusal of the offer in the end. This refusal probability directly influences the optimal choice of interest rate, which must balance affordability for the client (to reduce refusals) against profitability (to cover costs, which are higher for priority procedures). The first decision thus affects both client refusal and costs, which are critical considerations in the second decision. Note that historically, the second decision point is sometimes (in minority of cases) skipped entirely due to the bank's policy. SimBank allows us to vary the level of confounding in the training data by adjusting the % of confounded data (generated fully under the bank's policy) versus full RCT data (generated by employing a random policy in the decision points). As shown in previous studies, higher confounding typically reduces the performance of PresPM methods [8,10].

SimBPIC17. We additionally introduce SimBPIC17, a semi-synthetic simulation based on the BPIC17_W event log, which records real-world loan application cases. In this simulation, we simplify the control-flow sligthly (by deleting 3 activities) and introduce an intervention on the existing activity 'call_incomplete_files'. At each decision point, the choice is whether to *call* the client to follow up on incomplete files or to *wait* for the client to provide the information themselves. The KPI to minimize is the total cost incurred while completing the files: $cost_{files} = cost_{tpt} * tpt_{files} + n\_calls * cost_{call}$, where tpt_{files} is the throughput time of completing the files, which is sampled from a uniform distribution, and $cost_{tpt}$ and $cost_{call}$ are fixed constants. There is a trade-off: waiting is cheaper but increases throughput time, whereas calling is more expensive but shortens the throughput time. The (causal) effect of a call on the throughput time depends on two factors: the loan type (with car' and loan takeover' having larger effects) and the average duration of all previous activities (longer durations increase the potential time saved by calling). This dependency on average duration creates strong inter-decision point interactions: a call in decision point 1 reduces throughput time but also lowers the average duration of activities after decision point 1, which in turn reduces the effect of a call in decision point 2. To generate datasets, variables except activity labels and activity durations are sampled from the original dataset. We then define the control-flow, the function determining the effect of a call, the function calculating the KPI based on the actions taken, and the bank policy, which governs whether the bank chooses to call or wait. The bank calls if the loan type is car' or loan takeover' and the average activity duration exceeds 4025 units. Unlike SimBank, SimBPIC17 makes it easy to vary the number of decision points by increasing the number of (obligatory) choices between call or wait. We mimic an observational event log by defining the bank policy, varying confounding levels in the same way as in SimBank.

4.2 Methods for Comparison

We compare *SCOPE* with the multi-interventional approaches of Branchi et al. [5] and de Leoni et al. [9], each missing a property from Table 1: [5] uses a process approximation, while [9] does not align interventions across decision points. Including both methods allows us to examine the role of these properties, which *SCOPE* explicitly includes.

We select Branchi et al. [5] as the representative method using a process approximation and exclude Abbasi et al.'s FORLAPS [1] (see Sect. 2). Although FORLAPS supports sequential interventions through process approximation, it is inflexible for adapting to our optimization problems. It defines MDP states solely by observed activity labels, making comparison unfair because our interventions require additional process variables. Incorporating these variables would require expanding the state space (producing an impractically large Q-table), redesigning the state representation, or introducing new augmentation methods–changes that would fundamentally alter the original method. In contrast, Branchi et al. approximate the process as an MDP and train a Q-learning algorithm using KMeans clustering, where each state is represented by the last executed activity and the cluster label assigned to the current prefix. While the original work defines states using only control-flow information, the method can be easily extended to include additional variables by incorporating them into the clustering inputs, which is how we apply it. We refer to this approach as *KMeans-Q*.

As our second baseline, the method of de Leoni et al. [9] represents an approach that does not align interventions across decision points. The original method trains a single model and selects actions by predicting the KPI for each possible action at every decision point, equivalent to using a single S-learner for all decision points. We adapt this approach by using a separate model per decision point, equivalent to applying an S-learner (separately) at each step. This modification is theoretically equivalent to the original method and we implement it for three reasons: (1) it simplifies the implementation of causal learners when decision variables vary across points; (2) despite this complexity of implementation, we initially implemented [9] as a single model, but it yielded worse results; and (3) it thus ensures a fairer comparison with *SCOPE*, which also uses a model per decision point. Because these models are applied independently, the action chosen at one decision point does not account for the impact of future actions. We refer to this approach as SEP-S, indicating the use of separate S-learners. More generally, *SEP* refers to any strategy in our experiments that trains separate models for each decision point.

We also include a simple *random policy* as a baseline, and an *upper bound* representing the best achievable KPI, computed by taking, for each test case, the maximum KPI obtainable across all possible action sequences (exhaustive search).

4.3 Preprocessing, Hyperparameter Tuning & Dataset Splits

Cases are truncated at decision points to form prefixes ending before the intervention. LSTMs use tensor encoding [24], with case-level features added at the final layer. Other models use last-event encoding for time features and aggregation encoding for other features, following [23]. Categorical features are one-hot encoded and continuous features standardized. Depending on the learner, an action feature may be included (S- or RA-learner) or not (T-learner). For KMeans-Q, we adopt the original preprocessing [5] for control-flow features and add noncontrol-flow features using last-event and aggregation encoding as before.

Hyperparameters are tuned via Bayesian optimization using 20% of the training set for validation. Models are then retrained on the full training set. For evaluation, we use 10,000 test cases for SimBank and 1,000 for SimBPIC17 (due to the exponential growth of sequences with decision points). Both *SCOPE* and SEP can use any base model, tuned identically for fair comparison. In KMeans-Q, we jointly tune the KMeans and RL components using a normalized metric that combines silhouette score and average reward, allowing twice the tuning budget compared to *SCOPE* and SEP due to the dual-model setup. The original paper tuned KMeans first using clustering metrics (e.g., silhouette score) and then tuned the RL model. In our experiments, this approach performed poorly, as optimizing KMeans for clustering quality alone does not support effective intervention decisions.

4.4 Experimental Setup

To evaluate *SCOPE*, we use the *Gain*, which measures the % improvement in total KPI achieved by a method's policy over the bank's historical decision policy on the test set. We vary key simulation parameters relevant to sequential PresPM, and do this especially in SimBank, as it is thoroughly validated in [10], contains more variables, and has a more complex causal structure than SimBPIC17 (where we focus on the number of decision points). In SimBank, we adjust:

- *Confounding level δ (0.9–0.99):* Reflects real-world observational data, which is usually heavily confounded (as there is generally already a decision policy in place in business processes). Varying this helps assess whether a method is robust to biases introduced by the historical decision policy (e.g., bank policy) and whether it can still select effective actions even when those actions are rarely observed in the data.
- *Training set size (1K, 10K, 50K):* Since each model at a decision point in *SCOPE* depends on the previous one, limited data may cause estimation errors that propagate through decision points. Varying training size helps measure this effect.
- *Learners and base models:* We test different learners (S-, T-, RA-learner) and base models (XGBoost, Random Forest, LSTM/MLP) for *SCOPE* and SEP, to examine whether *SCOPE*'s advantages (due to its aligned interventions) hold consistently across learner types and underlying base model. The default is S-learner + XGBoost (e.g., for varying training size).

In SimBPIC17, the focus is evaluating performance over more than 2 decision points. We vary the number of decision points from 2 to 6 and run experiments under 3 levels of confounding (0.9, 0.95, 0.99) to assess robustness across longer decision horizons. Each setting is run using 10 different random seeds in models.

In our repository (see footnote 5), we provide: (1) results across different learners and base models for SimBPIC17 as well, confirming consistency across datasets; and (2) a computational complexity analysis showing that, although *SCOPE* has the highest theoretical training cost, it matches SEP and outperforms the original implementation of [9] in inference complexity. While KMeans-Q is theoretically the most efficient, *SCOPE* trained significantly faster in practice, had only slightly slower inference than SEP simply due to obtaining more complex models from tuning, and can even outperform KMeans-Q at inference time.

4.5 Results

Figure 1 shows method performance across confounding levels and **training sizes** on SimBank. *SCOPE* and SEP are implemented as S-learners with XGBoost (SEP thus corresponds to de Leoni et al.'s SEP-S). *SCOPE* outperforms nearly all settings, except for the (1K, 0.99) case, demonstrating its strong ability to learn sequential intervention policies. Performance generally declines with higher confounding, except for KMeans-Q, and in the 50K condition, where having more data keeps performance high even for strong confounding. As expected, larger training sets improve performance (for KMeans-Q, this is expressed in less variability). *SCOPE*'s advantage grows with training size, as each model uses the outputs of other models in later decision points as targets through the value function. With more data, model errors decrease and propagate less across decision points. KMeans-Q performs the worst and most variable, suggesting that approximating the process for RL while still aligning all interventions can be less effective than treating interventions independently, as SEP(-S) does.

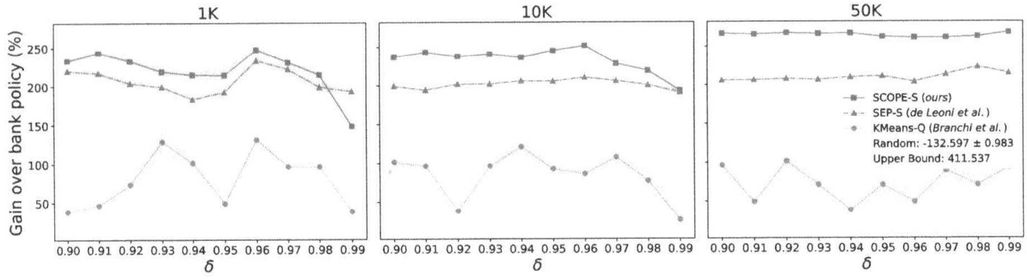

Fig. 1. Gain on SimBank across confounding levels (δ) for different training sizes. The shaded area shows one standard error over 10 iterations. *SCOPE* and Sep: S-learner with XGBoost.

Figure 2 compares *SCOPE* and SEP across confounding levels and **learner types** on SimBank (10K training cases, XGBoost). *SCOPE* consistently outperforms SEP, with only occasional exceptions at the 0.99 confounding level. Overall, the S-learner and RA-learner achieve the strongest performance, while the T-learner shows the largest degradation under high confounding. Similarly, Fig. 3 shows performance across confounding levels and **base models** (10K training cases, S-learner)[6]. *SCOPE-S* again outperforms SEP-S. XGBoost achieves the best overall performance, while Random Forest performs worst. Both these plots suggest aligning interventions across decision points is generally more effective than treating them independently, regardless of learner or base model type.

Figure 4 shows method performance across different **numbers of decision points** and confounding levels on SimBPIC17 (10K training cases, *SCOPE* and

[6] For the right-most plot, we use an MLP for the early decision point and and an LSTM in the later decision point, where sequential structure becomes important.

SEP are again S-learners with XGBoost). *SCOPE* consistently performs best, and its advantage over other methods increases as the number of decision points grows because it better captures interdependent decisions. Both *SCOPE* and SEP-S gain more over the bank policy with more decision points but drift further from the upper bound–*SCOPE* due to error accumulation across models, and SEP due to independent treatment of decision points. KMeans-Q degrades with more decision points, likely because the KMeans model increasingly aggregates

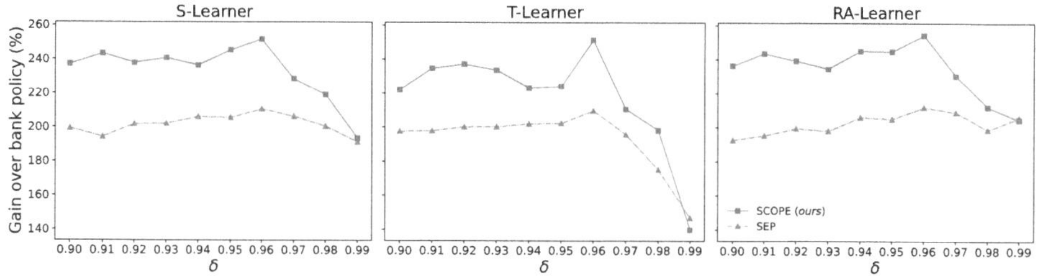

Fig. 2. Gain on SimBank across confounding levels (δ) for different learners. The shaded area shows one standard error over 10 iterations. *SCOPE* and Sep: XGBoost, trained on 10K cases.

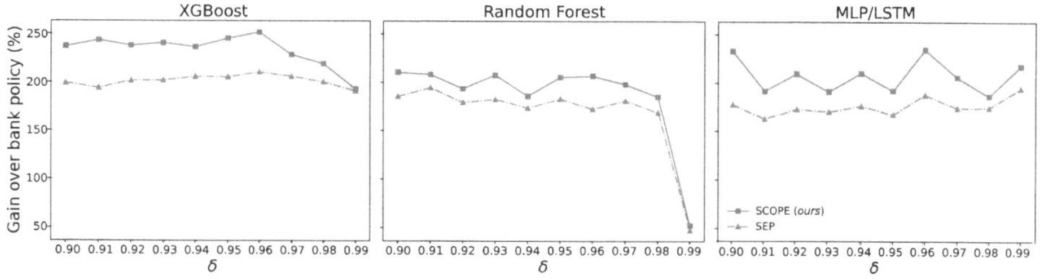

Fig. 3. Gain on SimBank across confounding levels (δ) for different base models. The shaded area shows one standard error over 10 iterations. *SCOPE* and Sep: S-learner trained on 10K cases.

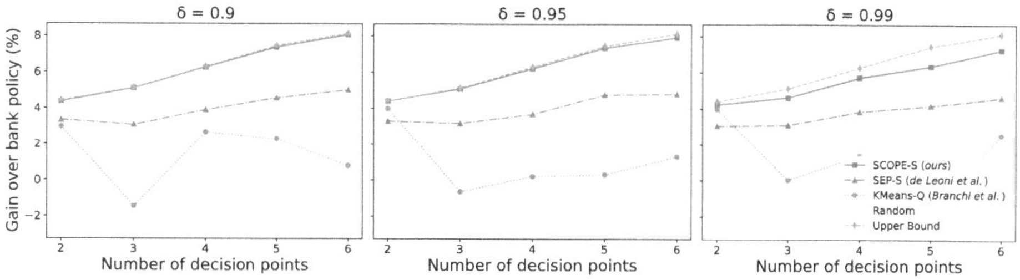

Fig. 4. Gain on SimBPIC17 across varying numbers of decision points for three confounding levels (δ), trained on 10K cases. The shaded area shows one standard error over 10 iterations. *SCOPE* and Sep: S-learner with XGBoost.

data relative to its original form, reducing the quality of the MDP approximation for Q-learning.

4.6 Discussion

In summary, the setup of *SCOPE* demonstrates clear advantages across a range of settings. Its performance improves with larger training sets by reducing error accumulation in backward induction and scales well with more decision points by aligning interdependent interventions, something separate optimization fails to exploit. *SCOPE* consistently outperforms separate optimization across learner types and base models. Practitioners should choose the learnermodel combination expected to perform best: S- and RA-learners suffice for our datasets, but T-learners struggle. While RL-based approaches that approximate the entire process as an MDP (like KMeans-Q) aim for joint optimization, they can still perform worse than treating decision points independently (as in SEP). *SCOPE* does rely on the predictive accuracy of underlying models: a highly unpredictable KPI can lead to strong error accumulation. However, this challenge affects all methods, and *SCOPE* still likely retains its advantage by leveraging observational event logs directly using causal learners and considering the interdependent decision point structure.

Limitations. *SCOPE* models sequential dependencies within individual process cases but assumes no interference between interventions across different cases, as required by the SUTVA assumption (Sect. 3). Similarly, sequential ignorability may not always hold in business processes. Despite these assumptions–common to most PresPM approaches-our experiments show that *SCOPE* can still identify effective sequential policies under realistic levels of confounding typical in business processes. Moreover, *SCOPE* theoretically increases training complexity by requiring a separate predictive model and backward induction at each decision point. However, as shown in our repository (see footnote 5), our practical results differ: despite being theoretically most demanding, *SCOPE* trained significantly faster than KMeans-Q. During inference, *SCOPE* matches SEP's theoretical complexity (and is more efficient than the original implementation of [9]), with only slightly longer runtimes in practice due to obtaining more complex tree models during hyperparameter tuning.

5 Conclusion

We introduce *SCOPE*, a method that combines backward induction with causal learning to optimize sequential interventions in business processes. Using a series of experiments, including a newly developed semi-synthetic simulator to support further research in sequential PresPM, we show that *SCOPE* outperforms existing sequential PresPM methods, which either handle each intervention independently or rely on approximating the process to train an RL agent.

Future work could extend SCOPE to account for interference (e.g., via causal methods for interference [7]) and relax sequential ignorability (e.g., using instrumental variables [15]). Computational cost could potentially be reduced via an end-to-end neural framework (e.g., [12]) adapted to backward induction.

References

1. Abbasi, M., Khadivi, M., Ahang, M., Lasserre, P., Lucet, Y., Najjaran, H.: An innovative data-driven and adaptive reinforcement learning approach for context-aware prescriptive process monitoring (2025)
2. Acharki, N., Lugo, R., Bertoncello, A., Garnier, J.: Comparison of meta-learners for estimating multi-valued treatment heterogeneous effects. ICML"23, JMLR.org (2023)
3. Bozorgi, Z.D., Dumas, M., Rosa, M.L., Polyvyanyy, A., Shoush, M., Teinemaa, I.: Learning when to treat business processes: Prescriptive process monitoring with causal inference and reinforcement learning (2023)
4. Bozorgi, Z.D., Teinemaa, I., Dumas, M., Rosa, M.L., Polyvyanyy, A.: Prescriptive process monitoring for cost-aware cycle time reduction. In: ICPM (2021)
5. Branchi, S., Buliga, A., Francescomarino, C.D., Ghidini, C., Meneghello, F., Ronzani, M.: Recommending the optimal policy by learning to act from temporal data (2023)
6. Branchi, S., Di Francescomarino, C., Ghidini, C., Massimo, D., Ricci, F., Ronzani, M.: Learning to act: a reinforcement learning approach to recommend the best next activities. In: BPM Forum, Springer (2022)
7. Caljon, D., Van Belle, J., Berrevoets, J., Verbeke, W.: Optimizing treatment allocation in the presence of interference. EJOR **328**(2), 620–632 (2026)
8. Bozorgi, D., Teinemaa, Z., Dumas, I., La Rosa, M., Polyvyanyy, A.: Prescriptive process monitoring based on causal effect estimation. Inf. Syst. **116** (2023)
9. de Leoni, M., Dees, M., Reulink, L.: Design and evaluation of a process-aware recommender system based on prescriptive analytics. In: ICPM (2020)
10. De Moor, J., Weytjens, H., De Smedt, J., De Weerdt, J.: Simbank: from simulation to solution in prescriptive process monitoring. In: BPM Forum, Springer (2026)
11. van Dongen, B.: Bpi challenge (2017). https://data.4tu.nl/articles/dataset/BPI_Challenge_2017/12696884
12. Hess, K., Frauen, D., Melnychuk, V., Feuerriegel, S.: IGC-net for conditional average potential outcome estimation over time. In: ICLR (2026)
13. Holland, P.W.: Statistics and causal inference. J. Am. Stat. Assoc. **81**(396), 945–960 (1986)
14. Huang, X., Choi, S., Wang, L., Thall, P.F.: Optimization of multi-stage dynamic treatment regimes utilizing accumulated data. Stat. Med. **34**(26) (2015)
15. Lousdal, M.L.: An introduction to instrumental variable assumptions, validation and estimation. Emerg. Themes Epidemiol. **15** (2018)
16. Ma, S., Flanigan, K.A., Bergés, M.: Bridging the reality gap in digital twins with context-aware, physics-guided deep learning. J. Comput. Civil Eng. (2026)
17. Rubin, D.B.: Estimating causal effects of treatments in randomized and nonrandomized studies. J. Educ. Psychol. **66**(5), 688–701 (1974)
18. Schulte, P.J., Tsiatis, A.A., Laber, E.B., Davidian, M.: Q- and a-learning methods for estimating optimal dynamic treatment regimes. Stat. Sci. (2014)

19. Shoush, M., Dumas, M.: Prescriptive process monitoring under resource constraints: a reinforcement learning approach. Künstl. Intell. **39**, 119–140 (2025)
20. Shoush, M., Dumas, M.: Intervening with confidence: conformal prescriptive monitoring of business processes (2022)
21. Shoush, M., Dumas, M.: Prescriptive process monitoring under resource constraints: a causal inference approach. In: Process Mining Workshops, Springer (2022)
22. Shoush, M., Dumas, M.: White box specification of intervention policies for prescriptive process monitoring. Data Knowl. Eng. (2025)
23. Teinemaa, I., Dumas, M., Rosa, M.L., Maggi, F.M.: Outcome-oriented predictive process monitoring: Review and benchmark. ACM Trans. Knowl. Discov. Data **13**(2) (2019)
24. Verenich, I., Dumas, M., Rosa, M.L., Maggi, F.M., Teinemaa, I.: Survey and cross-benchmark comparison of remaining time prediction methods in business process monitoring. ACM Trans. Intell. Syst. Technol **10**(4) (2019)
25. Weinzierl, S., Dunzer, S., Zilker, S., Matzner, M.: Prescriptive business process monitoring for recommending next best actions. In: BPM Forum, Springer (2020)
26. Weytjens, H., Verbeke, W., De Weerdt, J.: Timed process interventions: causal inference vs. reinforcement learning. In: BPM Workshops, Springer (2024)
27. Zhan, Z., Liu, Z., Lin, C., Yi, D., Liu, J., Yang, Y.: Censored c-learning for dynamic treatment regime in colorectal cancer study. Ann. Appl. Stat. (2025)

Fairness in Prescriptive Process Analytics: Integrating Fairness Constraints into the Learning Phase

Fabrizio De Santis[1(✉)], Alessandro Padella[2], Gyunam Park[3], Francesco Zanichelli[1], and Massimiliano de Leoni[2]

[1] University of Parma, Parma, Italy
{fabrizio.desantis,francesco.zanichelli}@unipr.it
[2] University of Padua, Padua, Italy
alessandro.padella@unipd.it, deleoni@math.unipd.it
[3] Eindhoven University of Technology, Eindhoven, The Netherlands
g.park@tue.nl

Abstract. Prescriptive Process Analytics is a core field in business process management that provides automated activity recommendations to improve Key Performance Indicators in business processes. Because such recommendations can directly affect individuals involved in the process, ensuring fairness is essential. Recent work has begun to address fairness in process monitoring by enforcing group-level parity across demographic groups at the predictive stage. However, these approaches leave the prescriptive phase entirely unexamined and rely solely on group fairness notions, failing to account for individual-level counterfactual fairness (i.e., the requirement that a recommendation should not change when sensitive attributes are altered). To address this gap, we propose to enforce counterfactual fairness directly in prescriptive process monitoring. We develop a novel neuro-symbolic framework based on Logic Tensor Networks that encodes counterfactual fairness as a logical axiom over predictions, seamlessly integrated into the learning process. We evaluate our approach on real-world event logs and find that it enforces counterfactual fairness at the prediction level and improves recommendation-level counterfactual consistency under the considered policy, while maintaining competitive predictive and prescriptive performance, establishing the first fairness-aware framework for Prescriptive Process Analytics.

Keywords: Fairness · Prescriptive Process Analytics · Predictive Process Monitoring · Logic Tensor Networks

1 Introduction

Process Mining is a discipline that bridges data science and business process management, extracting knowledge from event logs recorded by information systems to analyze, understand, and improve business processes. Within this field,

F. De Santis and A. Padella—Equal contribution.

F. M. Maggi et al. (Eds.): BPM 2026, LNCS 16936, pp. 369–387, 2027.
https://doi.org/10.1007/978-3-032-37777-7_23

Prescriptive Process Analytics (PrPA) aims to focus on recommending actions to mitigate "risky" executions that are predicted to lead to unsatisfactory outcomes, such as high costs or poor customer satisfaction.

Since such recommendations can directly affect the individuals involved in a process, ensuring that they are not driven by discriminatory criteria is essential. This paper addresses the problem of frameworks for PrPA that provide unfair recommendations, namely recommendations that are based on characteristics that discriminate in a form that is unacceptable from a legal and/or ethical point of view. For instance, in a system managing loan applications, an applicant might be requested for additional checks on the basis of their gender. Pohl et al. indicate monitoring, detecting and rectifying biased patterns to be the most significant challenge in Discrimination-Aware Process Mining [19]. Process characteristics are modeled as process attributes, where sensitive attributes are those that cannot be used to provide recommendations.

Despite growing interest in fairness in machine learning, existing work on fairness in Process Mining has focused exclusively on the predictive stage, aiming to ensure that sensitive attributes do not influence outcome predictions, adopting group fairness notions. This body of research work can certainly be adapted for PrPA, but they adopt the group fairness notions, which require statistical parity across demographic groups at the population level. Prescriptive settings also require asking whether the same running instance would receive a different recommendation under a counterfactual change of the sensitive attribute. This is captured by counterfactual fairness [13].

The importance of individual-level fairness can be expressed through an example of a process at a hypothetical hospital's emergency department. After assessing the symptoms and analyzing the lab results, the system recommends the next step in the medical process. Suppose that initially these recommendations may be unfair, and old patients are sent to cardiology, while young patients are sent elsewhere. In this scenario, when the daily "quota" for old-patient cardiology is reached, additional old patients at triage may be sent to other departments, regardless of the symptoms, and results: this would ensure group fairness. Although this may improve aggregate statistics, it could be dangerous for individual patients. Counterfactual fairness captures this individual-level concern: for a given patient, if we image a counterfactual version of the same individual where age is different, while symptoms and lab results remain unchanged, the recommendation should remain the same. This does not mean that counterfactual fairness is generally superior to group fairness. Group fairness remains essential for auditing and regulatory reporting, where aggregate disparities must be assessed. For example, a loan system may satisfy group-level parity by balancing individually unfair decisions across nationalities, still treating single applicants unfairly. Conversely, counterfactual fairness does not guarantee group parity.

This paper focuses on this individual-level layer and proposes the first fairness-aware framework for PrPA that enforces counterfactual fairness. In particular, it builds on Logic Tensor Networks (LTNs), a neuro-symbolic framework that maximizes the satisfaction of a set of logical formulas while fitting the observed data [2]. We define regression axioms for accurate KPI prediction and counterfactual-fairness axioms requiring prediction invariance under sensitive-attribute interventions, using counterfactuals identified through causal discovery and inference. Our fairness-aware framework has been evaluated on real-world logs and multiple sensitive attributes. Results show that it increases fairness with the same recommendation effectiveness as a framework not tailored for it.

The remainder of this paper is structured as follows: Sect. 2 discusses the previous literature pertaining to fairness in Process Mining and PrPA. Section 3 introduces the background concepts, and Sect. 4 proposes a framework that enables individually fair prescriptions. The case studies, the evaluation setup, and the results are described in Sect. 5, while Sect. 6 discusses the results and their limitations. Finally, Sect. 7 concludes the paper by summarizing our contribution and discussing future directions.

2 Related Work

Fairness has been identified as one of the four main challenges of Responsible Process Mining [16]. In parallel, PrPA has gained increasing attention in recent years [12]. Early approaches focused on recommending the next activity to execute by leveraging basic forms of counterfactual reasoning [8], with the primary goal of improving business process performance. Subsequently, in parallel with the rise of Predictive Process Monitoring (PPM) methods [7], and given that accurate predictions do not necessarily imply effective recommendations [6], the focus of PrPA expanded toward anticipating which ongoing process instances might lead to unfavorable KPIs [22]. The adoption of deep-learning techniques has enabled the enhancement of prescriptive capabilities, leading the research to explore the integration of causal inference [3] and proposing white-box approaches [20]. Recent advancements have focused on approaches that optimize a given KPI through recommending the optimal resource allocation [17], or jointly recommending both the activity to perform and the resource responsible for its execution [18], also integrating Large Language Models to enhance the interpretability of recommendations [11].

Despite the progress in PrPA, fairness within this field has received significantly less attention than in PPM where group fairness prediction approaches have been developed [14,15]. This is further confirmed in [21], a recent survey on fairness in PPM, which reports the evaluation of group fairness prediction methods and identifies fair prescriptive process analysis as an open direction for future work. Although counterfactual fairness has not yet been studied in PrPA, Heilmann et al. [9] proposed a neuro-symbolic approach for enforcing counterfactual fairness by encoding fairness constraints over model predictions using LTNs. Their work provides the fairness-aware learning mechanism that we build upon. However, their proposal is not directly applicable in our settings, where our framework provides recommendations by looking at the sequences of actions, namely the execution traces, that were executed for individual process instances.

Indeed, we generate a counterfactual trace for each trace, thus ultimately creating a counterfactual event log. This step is novel, if compared with Heilmann et al. [9], and it is crucial to enable fairness-aware prescriptive process analytics.

3 Preliminaries

Consider a loan application process. For a running process instance, the information system records the sequence of activities, such as *Application submitted*, *Documents requested*, *Documents verified*, *Risk assessment*, *Manual review*, *Additional document request*, *Fast-track assessment*, and *Application approved* or *Application accepted*. In addition to the control-flow, it also records the timestamps of the activities, the resources executing them, and they can also be associated with attributes such as *employment_type*, *requested_amount*, and *gender*. The goal of PrPA is to recommend the activity to perform as next to improve a pre-defined KPI. Suppose that, after a sequence of activities such as Application submitted, Documents requested, and Documents verified, a PrPA framework recommends the next activity to minimize the remaining processing time between the eligible next activities that are Manual review, Fast-track assessment, and Additional document request. For a male applicant, the system predicts remaining times of 6, 4, and 9 days, respectively, and recommends Fast-track assessment. Now consider the counterfactual version of the same case, where the applicant is female and gender-dependent attributes are updated accordingly. If the predicted remaining times become 5, 7, and 8 days, the recommendation changes to Manual review. Counterfactual fairness captures this individual-level concern: changing gender alone should not change the recommendation for the same case. The rest of this section outlines the core concepts behind our PrPA framework.

3.1 Prescriptive Process Analytics

An event log $\mathcal{L}$ is a collection traces, each describing one process instance as an ordered sequence of events $\sigma = \langle e_1, e_2, ..., e_n \rangle$ sharing the same case identifier c, where every event $e = (a, c, t, attr_1, ..., attr_m)$ captures the executed activity a, its timestamp t, and optional payload attributes. A running trace is represented by a prefix $p = (\sigma, k)$ that collects the first k events of σ. We denote the set of all prefixes extracted from an event log $\mathcal{L}$ as $\mathcal{P}$, and as $\mathcal{K} : \mathcal{P} \rightarrow \mathbb{R}$ a function that maps each prefix p to its own KPI.

Given a running trace p, the goal of a PrPA framework is to recommend the optimal next activity to execute. However, this activity has to be practically executable from the domain viewpoint. To ensure it, we leverage a *transition systems* [1]. Given an event log $\mathcal{L}$, each prefix in $\mathcal{P}$ is mapped to an abstract state based on the last few activities, leveraging a *trace-to-instance* encoding function. Given a prefix $p = (\sigma, k)$ in this paper it is formulated as $\rho_a(p) = \langle a_1, \ldots, a_k \rangle$. Given this state, the transition system, built from $\mathcal{L}$, identifies which activities are eligible as next activities to recommend.

In parallel with the transition systems, a PPM model $f : \mathcal{P} \to \mathbb{R}$ learns to estimate the KPI from a running trace. For a prefix p, the model f tries to forecast the value of $\mathcal{K}(p)$. Then, once the set of eligible next activities for p has been extracted from the transition system, the same PPM model is used to evaluate the expected KPI of the valid next activities, recommending the activity that is predicted to lead to the most favourable KPI.

3.2 Counterfactual Fairness

The PPM model f may learn to associate sensitive attributes with process outcomes, producing different predictions for individuals who differ only in these attributes. This form of algorithmic discrimination can propagate from predictions to recommendations, leading to unfair treatment of specific individuals.

Existing approaches to fairness in PPM adopt group fairness notions (cf. Sect. 2). However, it provides no guarantee about the recommendations provided to specific individuals and does not address the question: *would this running process execution p have received a different prediction/prescription if its sensitive attribute had been different?* Counterfactual fairness [13] addresses this limitation by stipulating that a decision or treatment must remain invariant when the sensitive attribute is intervened upon, provided that all other variables are propagated in accordance with the underlying causal mechanisms of the model.

Counterfactual fairness in Process Mining requires that for a running prefix p and its counterfactual version p^{cf}, the PPM predictor f predicts the same KPI, while the PrPA recommender prescribes the same activity for both cases. In the loan example, suppose the prefix is $\langle$ *Application submitted, Documents requested, Documents verified* $\rangle$ for a male applicant, and the transition system returns as eligible next activities *Manual review, Additional document request,* and *Fast-track assessment,* with predicted remaining times of 6, 9, and 4 days respectively. The system recommends *Fast-track assessment.* In the counterfactual version where the applicant is female and causally affected attributes are updated accordingly, the predicted remaining times become 5, 8, and 7 days, and the recommendation changes to *Manual review.* This illustrates the fairness problem addressed in this work: changing the applicant's gender from male to female should not lead to a different recommendation.

The main challenge is that meaningful counterfactuals cannot be obtained by simply flipping the sensitive attribute, as doing so may also affect causally downstream attributes. Counterfactual fairness therefore requires a causal model.

A Structural Causal Model is a tuple $\mathcal{M} = (U, V, F)$ where U is a set of exogenous variables representing background factors that are not caused by any variable in the model, such as personal or institutional circumstances not recorded in the event log; V is a set of endogenous variables, whose values are ultimately determined by the exogenous variables, such as gender, employment type, and requested loan amount; F is a set of structural equations that determine the value of each endogenous variable [13]. For instance, one equation may specify how employment type depends on gender and other background factors, while another may specify how the requested amount depends on employment type.

An SCM provides a formal description of how variables of a system are generated and how they influence each other. In a counterfactual query, we ask what the rest of the case would look like if a variable were changed, while the causal mechanisms of the system remain the same. In the loan example, changing gender may affect downstream attributes such as employment type, but not the requested loan amount. Intervening on gender would also update the employment type, while leaving the requested amount unchanged. An SCM specifies how these downstream variables should be recomputed after this intervention.

4 Counterfactual Fairness for Prescriptive Process Analytics

Figure 1 provides an overview of the proposed approach. Before training, the event log is used to construct a transition system encoding valid next activities for any process state, and to generate a counterfactual event log in which each trace is paired with counterfactual counterparts obtained by intervening on the sensitive attribute. Original and counterfactual prefix pairs are then used to ground two logical axioms: a regression axiom for accurate KPI prediction, and a counterfactual fairness axiom requiring invariant predictions across counterfactual pairs. Joint optimization of the two axioms yields a fair predictor. At inference time, the transition system determines the candidate next activities for a running process prefix, the model predicts the KPI for each, and the activity with the best predicted KPI is recommended.

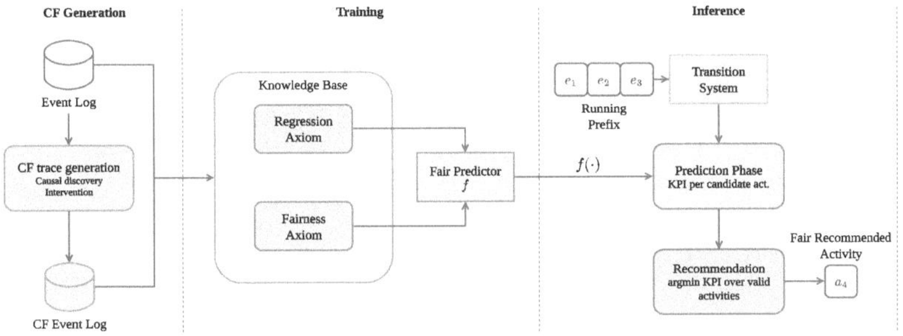

Fig. 1. Overview of the proposed framework. Counterfactual traces are generated from the event log via causal discovery and intervention, and used to ground a counterfactual fairness axiom alongside a regression axiom in the knowledge base; joint optimization of the two yields a fair predictor. At inference time, the model predicts the KPI for each candidate activity from the transition system, and the activity with the best predicted KPI is recommended.

4.1 Counterfactual Trace Generation

Enforcing counterfactual fairness in Process Mining requires, for each trace $\sigma \in \mathcal{L}$, one or more corresponding counterfactual traces in which the sensitive attribute has been altered while everything else is updated consistently with the underlying causal structure. Generate such counterfactual traces from an event log $\mathcal{L}$, requires knowledge of the causal structure under the data-generation process, encoded as an SCM. Since it is rarely known in advance, we employ a causal discovery algorithm to infer it from the data. Specifically, we use the Fast Causal Inference (FCI) [4], a constraint-based causal discovery algorithm that takes as input a dataset of variables and returns a Partial Ancestral Graph (PAG): a graphical representation encoding the set of all causal structures consistent with the data, including the possible presence of latent confounders.

We adopt two assumptions that constrain which parts of a trace need to be modified when intervening on the sensitive attribute.

Control-Flow Independence. The transition system constructed from the event log (cf. Sect. 3.1) is assumed to be independent of the sensitive attribute, meaning that the set of possible next activities for a given prefix is the same regardless of the demographic characteristics of the process instance. Therefore, we assume that the sequence of activities executed is determined by organizational rules and process logic rather than individual demographic characteristics. The sensitive attribute may, however, influence which activity is recommended by the predictor, since the predicted KPIs can differ across counterfactual pairs.

Case-Level Causal Dependence. We assume that only case-level attributes can causally depend on the sensitive attribute. This assumption is grounded in a principled distinction between *stable* and *unstable* attributes. Case-level attributes are stable, meaning that they are defined once at the case level and do not change across events. This makes them suitable anchors for a causal model whose structure is assumed fixed across a trace. Event-level attributes, on the other hand, can take different values and are therefore considered unstable. Including such unstable attributes in the causal model would destabilize the causal structure itself, since the causal relationships between the variables would have to be recalculated for each event.

FCI is applied to the tabular case-level representation extracted from the data, where each row represents a case with its corresponding case-level attributes. From the resulting PAG, we extract causal dependencies between the sensitive attribute and the case-level attributes, identifying which attributes are causally downstream of the sensitive attribute and must be updated when generating counterfactual traces. To reduce the risk of propagating uncertain causal relations, we only recompute case-level attributes for which the discovered PAG supports a directed or possibly directed path from the sensitive attribute. All other attributes are kept fixed. In the presence of latent confounders, indicated by bi-directed edges in the PAG returned by FCI, the affected attributes are treated as causally independent of the sensitive attribute and therefore left unchanged. In the loan example, the sensitive attribute is gender and the case-level attributes include employment type and requested amount. If FCI returns a directed edge from gender to employment type, reflecting historical bias in employment patterns, but no path to the requested amount, then employment type is the sole downstream attribute and the requested amount is treated as causally independent of gender.

Once the causal structure is identified, counterfactual traces are generated fitting the SCM to the data in order to learn the structural equations F under the relationship between variables. In our SCM, the sensitive attribute is always modeled as a root node in V and is therefore treated as an exogenous variable with respect to the other observed variables. Each structural equation is estimated independently from the training data by fitting a causal mechanism for the corresponding attribute, using its causal parents in the discovered as input. In practice, continuous attributes are modeled with regressors, while categorical attributes are modeled with classifiers.

Therefore, for each trace σ, we build the counterfactual part σ^{cf}. In the loan example, for σ related to a self-employed *female* applicant requesting $10,000, we generate its counterfactual version σ^{cf} by setting the gender to *male*. Since employment type is causally downstream of gender, the fitted structural equations predict that a male applicant with identical characteristics would be more likely be employed than self-employed. The requested amount remains $10,000 as it is causally independent of gender. If FCI reveals no causal dependencies between the sensitive attribute and any other case-level attributes, the counterfactual trace σ^{cf} is generated by simply flipping the value of the sensitive attribute.

From each original trace and its counterfactual part, we then extract prefix pairs. If a running prefix p contains the first k events of an σ, its counterfactual version p^{cf} is defined as the prefix containing the first k events of the σ^{cf}. In this way, each prefix p is paired with one or more counterfactual prefixes.

4.2 Encoding Fairness as a Learning Objective

LTNs ground logical formulas over neural-network outputs and optimize their satisfaction through gradient-based learning. This neuro-symbolic framework use a fuzzy interpretation of first-order logic, where each formula has a truth value in $[0, 1]$. The learning objective is defined as the degree to which the model satisfies the logical axioms, i.e., declarative statements that the model must satisfy. Given a set of axioms $\{\phi_1, ..., \phi_n\}$, each axiom receives a satisfaction level $\phi_i \in [0, 1]$, measuring how well the current model satisfies it. The satisfaction levels are aggregated into a single scalar value $\Phi \in [0, 1]$, and the loss is defined as:

$$L = 1 - \Phi \tag{1}$$

Minimizing L is equivalent to maximizing the joint satisfaction of all axioms. Since satisfaction levels are computed through differentiable operations, gradients flow back to the model parameters, enabling gradient-based optimization.

We define two logical axioms for the predictor f: a regression axiom, enforcing accurate KPI prediction for each prefix $p \in \mathcal{P}$, and a counterfactual fairness axiom requiring predictions to remain invariant under interventions on the sensitive attribute s.

To ground both axioms over the space of process prefixes $\mathcal{P}$, we define a grounded predicate Eq that measures the degree of similarity between two scalar KPI values z and z' using a Gaussian kernel:

$$\mathrm{Eq}(z, z') = e^{-\alpha\sqrt{(z-z')^2+\epsilon}}$$

The parameter $\alpha = 0.05$ controls the sensitivity of the similarity measure, and $\epsilon = 10^{-10}$ is a small constant for numerical stability. This predicate returns a truth value in $[0, 1]$, where a value of 1 indicates that the two KPI values are identical and values close to 0 indicate large disagreement. We use a Gaussian kernel as a similarity measure as it maps similarity in $[0, 1]$, making it directly interpretable as a fuzzy truth value within the LTN framework.

Regression Axiom. The regression axiom states that the predictive model f should accurately predict the target KPI for every process prefix $p \in \mathcal{P}$:

$$\forall p \in \mathcal{P} : \mathrm{Eq}(f(p), \mathcal{K}(p)) \tag{2}$$

The satisfaction level $\phi_{\mathrm{regr}} \in [0, 1]$ of the regression axiom is computed by aggregating the truth values of the Eq predicate evaluated over all prefix-KPI pairs. A value of $\phi_{\mathrm{regr}} = 1$ indicates that the model predicts the remaining time perfectly for every prefix, while values close to 0 indicate large disagreement between predictions and ground truth. For example, for a running prefix of a loan application, if the model predicts 5.1 days whose true remaining time is 5 days, then $\mathrm{Eq}(5.1, 5) \approx 0.99$, which is close to 1. In contrast, if the model predicts 12 days, then $\mathrm{Eq}(12, 5) \approx 0.49$, which penalizes the model.

Counterfactual Fairness Axiom. The counterfactual fairness axiom requires predictions to remain invariant under interventions on sensitive attributes s. In process context, cases differing only in protected characteristics, such as gender or age, should receive the same predicted remaining time. Formally, for each prefix $p \in \mathcal{P}$, the set $cf(p, s)$ contains the counterfactual prefixes generated by intervening on s while keeping all causally unrelated attributes unchanged. The axiom is defined as:

$$\forall p \in \mathcal{P}, p^{cf} \in cf(p, s) : \mathrm{Eq}(f(p), f(p^{cf})) \tag{3}$$

The satisfaction level $\phi_{\mathrm{fair}} \in [0, 1]$ is computed over all original and counterfactual prefix pairs in the counterfactual event log. Higher ϕ_{fair} values indicate more consistent predictions across counterfactual pairs, i.e., lower sensitivity to the sensitive attribute. Maximizing the satisfaction of this axiom penalizes prediction differences between an instance and its counterfactual, discouraging reliance on its sensitive attribute. For example, suppose the model predicts 4.8 days for a male applicant and 9.1 days for the counterfactual prefix corresponding to the same case with a female applicant. Then $\text{Eq}(4.8, 7.1) \approx 0.65$, indicating poor satisfaction of the fairness axiom.

Learning Objective. Training combines the two requirements introduced above. For both universal quantification and aggregation of the two axioms, we use the pMeanError aggregator. Given truth values $v_1, ..., v_n \in [0, 1]$, pMeanError is defined as $1 - (\frac{1}{n}\sum_i (1 - v_i)^p)^{\frac{1}{p}}$, where $p = 2$ controls the penalization of low-satisfaction grounding. The overall objective, therefore, aims to find the set of optimal parameters of model f that maximizes the aggregated satisfaction level Φ, obtained by aggregating the values of ϕ_{regr} and ϕ_{fair}, assigning equal importance to predictive accuracy and counterfactual fairness. Unlike [9], who adopt a sequential training strategy, we jointly optimize the two axioms, preventing the regression objective from dominating the fairness signal during early training.

Table 1. Dataset-specific causal setup and sensitive attribute distributions.

Dataset	Sensitive attr.	Distribution	Case-level attrs.	Causal structure
bpi12	gender	male: 55.83%, female: 44.17%	requested amount	gender→requested amount
hb-gen	gender	male: 45.37%, female: 54.63%	age, case type	gender→case type
hb-age	age	age<65: 89.71%, age≥65: 10.29%	gender, case type	age→case type
hb-age-gen	age, gender	age<65: 94.82%, age≥65: 5.18% male: 45.99%, female: 54.01%	age, gender, case type	age→case type gender→case type
bpi13i	organization country	se: 38.34%, pl: 23.35%, us: 10.30% in: 6.50%, be: 6.38%, other: 15.13%	organization involved, impact, product	organization country →product impact →product

5 Evaluation

5.1 Datasets

We evaluate our approach on three different event logs:

- *BPI Challenge 2012 (bpi12)*: The event log pertains to a Dutch bank loan application process. We focus on the *A_* subprocess, enriched with the categorical sensitive attribute *gender*, which is correlated with trace duration.
- *Hospital Billing (hb)*: This event log is sourced from the financial modules of the ERP system of a regional hospital, and contains events related to the billing of medical services provided by the hospital. The traces are enriched with the case-level attributes *age* and *gender*. Three scenarios are considered:
  - *hb-gen*: unfair bias in gender, no unfair bias in age
  - *hb-age*: unfair bias in age, no unfair bias in gender
  - *hb-age-gen*: unfair bias in both age and gender
- *BPI Challenge 2013 Incidents (bpi13i)*: An event log of the Volvo IT incident management system. We choose the case-level attribute *organization country* as the sensitive attribute as highlighted in [15].[1]

[1] Available at https://data.4tu.nl/articles/dataset/BPI_Challenge_2013_incidents/12693914.

The *bpi12* and the *hb* event logs have been introduced in [14].[2] The authors preprocessed and adapted the event logs to include sensitive attributes for evaluating group fairness in PPM. The code to reproduce our experiments is available.[3]

Table 1 summarizes the dataset-specific causal setup used for counterfactual trace generation. For each scenario, it reports the sensitive-attribute distribution, the case-level attributes considered by FCI, and the causal structure, including the attributes identified as causally downstream of the sensitive attribute.

5.2 Experimental Setup

In our experiments, we compare three architectural variants to evaluate the impact of incorporating fairness constraints into a recommendation system:

- **LSTM**: Purely data-driven baseline trained with MSE loss.
- **LTN-Data**: LTN [2] with LSTM backbone and the regression axiom (Eq. 2).
- **Fair-LTN**: LTN [2] where the regression axiom (Eq. 2) and counterfactual fairness axiom (Eq. 3) are jointly optimized with equal weight. Unlike the training strategy adopted in [9], we opt for joint optimization of the two axioms to prevent the regression axiom from dominating the fairness axiom.

The LSTM model is leveraged as a baseline and consists of two stacked layers with a hidden dimension of 128, trained using the Adam optimizer with a learning rate of 0.001 and a batch size of 32. All event logs undergo a preprocessing pipeline implemented in the released code: categorical variables are encoded as integers; numerical attributes are normalized to the unit interval; and missing values are removed. Each prefix is represented as the ordered sequence of its events up to the current point, padded with zeros to the maximum sequence length. Categorical attributes are embedded into 64-dimensional vectors, while numerical attributes are normalized and linearly projected before being passed to the backbone. We train the model for a maximum of 100 epochs with early stopping. While more complex architectures could be considered, we adopt a simplified design to isolate and evaluate the impact of the fairness constraint independently of architectural complexity. The LTN framework is architecture-agnostic, and alternative architectures, such as Transformers or graph neural networks over process traces, could be seamlessly substituted for the LSTM.

5.3 Train Test Split and Metrics

The event log $\mathcal{L}$ is temporally split into a training set $\mathcal{L}^{\text{train}}$ and a test set $\mathcal{L}^{\text{test}}$, ordering cases by the timestamp of their first event. The split point t_{split} is defined as the earliest time at which 80% of traces in $\mathcal{L}$ are completed. Traces still running at t_{split} form $\mathcal{L}^{\text{test}}$, while traces starting after t_{split} are excluded to preserve a realistic evaluation setting, where only information available up

[2] Available at https://zenodo.org/records/15387576.

[3] Available at https://github.com/FabrizioDeSantis/Fair-PrescriptiveProcessMonitoring.

to t_{split} is used. From $\mathcal{L}^{\text{test}}$, a running trace log $\mathcal{L}^{\text{run}}$ is derived by truncating each trace at t_{split}, simulating incomplete, ongoing executions against which the quality of recommendations can be assessed. We further extract a validation set composed by the 20% of $\mathcal{L}^{\text{train}}$, and use it to tune the hyperparameters, while $\mathcal{L}^{\text{run}}$ and $\mathcal{L}^{\text{test}}$ are dedicated to assessing recommendations.

We evaluate predictive performance using the Mean Absolute Error (MAE):

$$\text{MAE} = \frac{1}{|\mathcal{L}^{run}|} \sum_{p \in \mathcal{L}^{run}} |f(p) - \mathcal{K}(p)|$$

where $f(p)$ is the predicted KPI for prefix p, $\mathcal{K}(p)$ is the actual KPI value for p, and $|\mathcal{L}^{run}|$ is the number of prefix in the test log $\mathcal{L}^{run}$. Lower MAE values indicate better predictive performance.

We measure fairness at prediction and recommendation levels. For predictive fairness, we introduce the Counterfactual Mean Squared Error (CF-MSE):

$$\text{CF-MSE} = \frac{1}{|\mathcal{L}^{run}|} \sum_{p \in \mathcal{L}^{run}, p^{cf} \in cf(p,s)} |f(p) - f(p^{cf})|^2$$

A lower CF-MSE indicates a fairer model, with a value of zero corresponding to perfect predictive counterfactual fairness.

For prescriptive fairness, we introduce the Counterfactual Recommendation Score (CF-RECC), defined as the fraction of prefixes for which the recommended activity is equal to the one recommended for its counterfactual version:

$$\text{CF-RECC} = \frac{1}{|\mathcal{L}^{run}|} \sum_{p \in \mathcal{L}^{run}, p^{cf} \in cf(p,s)} 1[\pi(p) = \pi(p^{cf})]$$

where $\pi(p)$ denotes the recommended activity for the prefix p according to the defined policy. A higher CF-RECC indicates a fairer recommender system, with a value of one corresponding to perfect prescriptive counterfactual fairness.

Training uses all prefixes of each trace, whereas testing uses one running prefix per test trace, obtained by randomly choosing t_{split} within the trace. Therefore, metrics are averaged over evaluated running prefixes.

5.4 Evaluation Through Simulations

Since A/B testing on ongoing processes is infeasible for evaluation purposes, we follow common practice in PrPA and rely on *business process simulation* to assess the effect of following or not following the recommendations under realistic conditions [5]. Each trace σ_i' is a prefix of a corresponding trace σ_i in $\mathcal{L}^{\text{test}}$, where σ_i can be seen as $\sigma_i' \oplus \langle (a_i', c_i', t_i', attr_{1,i}', ..., attr_{m,i}') \rangle \oplus \sigma_i''$. If the recommendation is the activity a, we move the simulation to the state reached after performing $\sigma_i' \oplus \langle (a, c_i', t_{\text{split}}, attr_1, ..., attr_m) \rangle$, namely the recommendation is assumed to occur at the simulation time t_{split}. We then simulate the continuation of the trace to determine how the trace would complete as a result of following the recommendation, comparing then the *trace duration* if a given recommendation has been followed. Since our goal is not only to asses the performance in terms of trace duration prediction, but also the fairness level, we repeat this procedure leveraging the counterfactual version of $\mathcal{L}^{test}$, i.e., $\mathcal{L}^{test}_{cf} = \cup_{\sigma \in \mathcal{L}} \sigma^{cf}$. This allows us to understand if the recommendations are still effective in a test set in which the value of the sensitive attributes (and what it is correlated with it) changes. As a *baseline* scenario, we assume no recommendation is provided, letting the simulation simulate the timestamps, the attributes and carrying out the process instance. In each setting, each simulation is repeated ten times to mitigate the potential consequences of simulation stochasticity, as the results are reported in terms of mean and 95% confidence interval. Simulations were conducted using the ProSiT simulator [23], using the metrics defined in Sect. 5.3.

5.5 Experimental Results

Figure 2 reports the predictive and fairness performance of all models expressed as means and 95% confidence intervals over 10 independent runs.

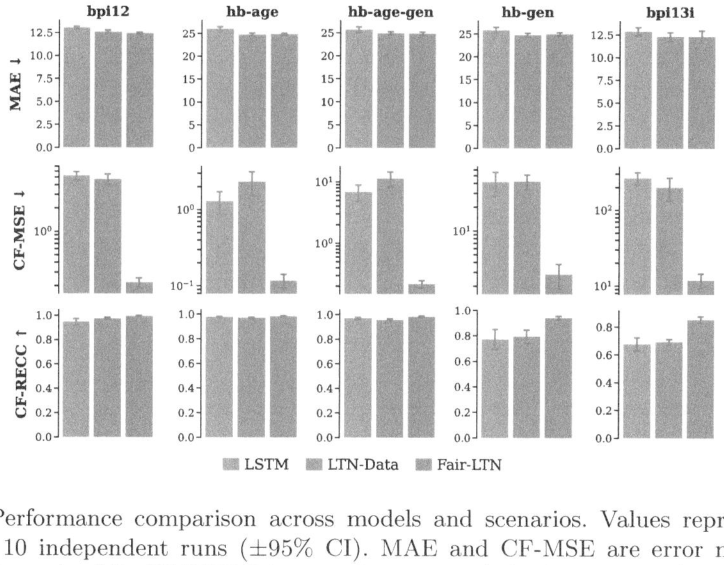

Fig. 2. Performance comparison across models and scenarios. Values represent the mean of 10 independent runs (±95% CI). MAE and CF-MSE are error metrics (↓ lower is better) while CF-RECC is a quality metric (↑ higher is better). CF-MSE is reported on a logarithmic scale and axes vary across datasets due to differences in absolute magnitude.

Predictive Performance. The baselines LSTM and LTN-Data achieve comparable MAE (reported in days) across most configurations, with LTN-Data consistently showing marginal improvements over the vanilla LSTM, suggesting that the LTN regression axiom provides a slightly more effective signal than standard

MSE minimization. Fair-LTN achieves similar or improved predictive performance compared to both baselines across all configurations. On *bpi12*, Fair-LTN achieves the lowest MAE (12.46, CI [12.37, 12.55]) compared to 13.08 for LSTM and 12.64 for LTN-Data. On the *hb* datasets, all three models achieve similar MAE values. This indicates that joint optimization preserve accuracy and may act as a regularizer by limiting reliance on spurious correlations between sensitive attributes and the target KPI.

Predictive Fairness. The results show a clear and consistent ranking across all configurations. The vanilla LSTM and LTN-Data models exhibit the highest CF-MSE values, indicating that the models trained without explicit fairness constraints learn to exploit sensitive attributes in their predictions. Fair-LTN substantially reduces CF-MSE compared to both baselines across all configurations. The most challenging dataset is *hb-gen*, where LSTM and LTN-Data exhibit high CF-MSE values of 41.53 and 42.20, respectively, indicating a strong influence of the gender attribute on predictions. Fair-LTN substantially reduces this to 2.85 (CI [2.13, 3.70]). The largest absolute improvement is observed on *bpi13i*, where CF-MSE drops from 265.76 (LSTM) and 199.71 (LTN-Data) to 11.83 (CI [9.78, 14.19]) for Fair-LTN. This demonstrates that the joint optimization of the two axioms is effective at enforcing predictive counterfactual fairness. Moreover, the confidence intervals of Fair-LTN are consistently narrower than those of the baselines, indicating that fairness improvements are stable.

Prescriptive Fairness . The relationship between predictive and prescriptive fairness reveals an important nuance. On *bpi12*, *hb-age*, and *hb-age-gen*, the LSTM and LTN-Data models already achieve relatively high CF-RECC scores (above 0.95) despite their substantially higher CF-MSE values. This follows from two complementary factors. First, a model can be consistently biased yet still yield consistent recommendations if the bias affects all candidate actions equally without altering their relative ranking. Second, when few candidate activities are proposed for a given prefix, recommendations are forced to be consistent regardless of fairness, trivially inflating CF-RECC scores. In contrast, on *hb-gen* and *bpi13i*, the sensitive attribute has a stronger effect. On *hb-gen*, LSTM and LTN-Data achieve CF-RECC scores of 0.77 and 0.79, respectively, indicating that predictive unfairness translates into prescriptive unfairness when the sensitive attribute affects the relative ranking of candidate actions sufficiently to change the recommended activity. On this dataset, Fair-LTN recovers CF-RECC to 0.94 (CI [0.93, 0.95]). Similarly, on *bpi13i*, LSTM and LTN-Data achieve CF-RECC scores of 0.68 and 0.69, respectively, while Fair-LTN improves this to 0.85 (CI [0.83, 0.87]). Fair-LTN achieves the highest CF-RECC scores, confirming that predictive counterfactual fairness effectively propagates to the prescriptive level.

Prescriptive Performance. Table 2 reports the average simulated trace duration (in days) for each method, computed as described in Sect. 5.4. For each model, the *Original* rows refer to simulations on $\mathcal{L}^{test}$, while the *Counterfactual* rows refer to simulations on $\mathcal{L}_{cf}^{test}$, the *Baseline* corresponds to simulations in which

Table 2. Average simulated trace duration (days) of across 10 simulations. The rows labeled "Original" and "Counterfactual" report simulations on the original and counterfactual running prefixes, respectively.

Method	Test Log	bpi13i	hb-age	hb-age-gen	hb-gen	bpi12
LSTM	Original	7.23 ±0.11	107.45 ±0.12	107.02 ±0.14	105.39 ±0.10	5.97 ±0.03
	Counterfactual	7.38 ±0.14	106.72 ±0.14	107.26 ±0.12	105.63 ±0.13	6.33 ±0.03
LTN-Data	Original	7.33 ±0.13	105.91 ±0.10	105.97 ±0.11	105.26 ±0.12	4.38 ±0.02
	Counterfactual	7.40 ±0.12	106.74 ±0.13	106.08 ±0.13	107.02 ±0.14	4.32 ±0.08
Fair-LTN	Original	7.44 ±0.10	106.81 ±0.13	106.51 ±0.12	105.44 ±0.11	4.54 ±0.06
	Counterfactual	7.34 ±0.13	105.19 ±0.11	106.78 ±0.14	106.58 ±0.13	4.87 ±0.01
Baseline	Original	11.04 ±0.02	110.72 ±2.12	111.28 ±1.3	112.32 ±2.4	9.86 ±0.02

no recommendation is followed. All proposed methods reduce the average simulated duration relative to the baseline, although the magnitude of the reduction depends on the dataset. On *bpi12* and *bpi13i*, the reductions reach 40–56% on *bpi12* and 33–35% on *bpi13i*, while on the three hospital-billing variants the reductions are smaller in relative terms (3% to 7%). Crucially, Fair-LTN obtains nearly the same values with respect to the unfair LTN-Data and LSTM, with absolute differences within ±0.3 days on *bpi12* and *bpi13i* and within ±1.7 days on the *hb* variants. In fact, on the *Counterfactual* rows of *hb-age* and *bpi13i*, Fair-LTN even outperforms both baselines. Taken together with the fairness improvements shown in Fig. 2, these findings demonstrate that the proposed framework delivers fair predictions and prescriptions while preserving predictive quality. The following examples illustrate the difference in how individual recommendations may differ between the Fair-LTN and the LSTM baseline.

Example 1. For a running process instance of *hb-gen* with prefix [`NEW`, `FIN`] and gender set to male, the LSTM baseline recommends `RELEASE`, but after the counterfactual intervention changing gender to female, it recommends `REOPEN`. This means that the same billing case would proceed toward release in the original instance, but would be sent back for further handling in the counterfactual one, potentially delaying closure and requiring additional administrative work. Fair-LTN instead recommends `RELEASE` in both cases, preserving the billing decision. In one seed, this patter occurs for 19 running instances in the test set.

Example 2. In a *bpi12* case with prefix [`A_SUBMITTED`, `A_PARTLYSUBMITTED`, `A_PREACCEPTED`, `A_ACCEPTED`], changing gender from female to male makes LSTM switch from `A_DECLINED` to `A_FINALIZED`, leading the same loan application to a different terminal outcome, whereas Fair-LTN keeps `A_DECLINED` in both cases. In one seed, this patter occurs for 18 running instances in the test set.

6 Discussion

6.1 Predictive and Prescriptive Fairness

A central finding is that enforcing counterfactual fairness at the predictive level is empirically sufficient to achieve it at the prescriptive level. Since recommendations are determined by the relative ranking of predicted KPI values, a counterfactually fair predictor produces identical rankings for a process instance and its counterfactual, guaranteeing the same recommended activity. This implication is, however, specific to the recommendation policy adopted here and does not necessarily generalize to others. The recommendation policy itself may introduce additional sources of unfairness that are independent of the predictive model. For instance, if the policy incorporates additional decision criteria beyond the predicted KPI, such as resource allocation, sensitive attributes could influence the recommendation through these additional criteria, even when the predictor is perfectly fair. In such settings, predictive fairness alone would not guarantee prescriptive fairness. This can be addressed by adding an axiom that directly encodes counterfactual fairness at the recommendation level, providing a direct signal on the recommendation policy, and enforcing fairness across the pipeline.

6.2 Fairness as Regularization

Fair-LTN achieves comparable or lower MAE than both baselines across all evaluated event logs. This suggests that sensitive attributes can act as spurious shortcuts in KPI prediction, with the model exploiting correlations between sensitive attributes and the target KPI present in the training data. By preventing reliance on these shortcuts, the fairness axiom forces the model toward more robust and generalizable predictive patterns, a phenomenon sometimes referred to as fairness for free [10], where fairness constraints simultaneously improve both fairness and predictive performance when sensitive attributes are correlated with process outcomes due to historical bias.

6.3 Limitations

While the proposed approach demonstrates promising results, several limitations should be acknowledged. First, the implication from predictive to prescriptive fairness relies on the assumption that the recommendation policy relies exclusively on the predicted KPI. It also relies on the assumption that the transition system is fair, meaning that the set of possible next activities, as well as the control-flow and event-level attributes, are independent from the sensitive attribute. Violations of these assumptions could break this implication and require additional constraints on the recommendation policy. Future work will relax these assumptions by considering the effects of sensitive attributes on control-flow and event-level attributes. Second, the counterfactual trace generation procedure relies on the discovered causal structure, which may be unreliable when the event log is small or when latent confounders are present. The pipeline is also sensitive to the quality of the learned causal structure: omitted dependencies yield incomplete counterfactuals, while spurious ones cause unnecessary attribute recomputation, in both cases optimizing the fairness constraint against imperfect counterfactuals. Furthermore, the learned graph captures statistically supported dependencies but does not determine which causal pathways are normatively admissible; fairness improvements should therefore be interpreted relative to the learned structure, with robustness analysis and domain-constrained causal discovery left for future work. Third, counterfactual fairness can conflict with group fairness. In process contexts where regulatory requirements need statistical parity across demographic groups, the framework would need to be extended with additional group-level constraints. Finally, we focus on remaining time for prediction and trace duration for prescription as KPIs, and counterfactual generation considers one sensitive attribute at a time, leaving interaction effects between multiple sensitive attributes for future work.

7 Conclusion

Prescriptive Process Analytics focuses on recommending activities to improve process performance, but these recommendations can be biased by sensitive attributes such as gender, nationality, or age. Prior work in PPM has largely addressed this through group fairness, which ensures statistical equality across the population but offers no guarantee at the individual level: a person may receive a different recommendation solely due to a change in a sensitive attribute, even when all other case characteristics remain unchanged. Counterfactual fairness addresses this gap directly. This paper proposed a framework for enforcing counterfactual fairness at the prediction level, ensuring that predictions remain invariant to changes in sensitive attributes. The framework generates causally consistent counterfactual traces via FCI-based causal discovery and intervention. Using LTNs, predictive accuracy and counterfactual fairness are encoded as jointly optimized logical axioms, yielding a predictor that is both accurate and fair. Under the recommendation policy adopted in this work, counterfactual fairness propagates to the prescriptive level, yielding consistent recommendations across counterfactual instances.

As future work, we aim to extend the repertoire of KPIs on which we perform experiments and assess results. Although our framework enforces counterfactual fairness through the predictive model, recommendation policies may still introduce bias by prioritizing certain activities, groups or resources. Therefore, an important future direction is to consider constraints directly over policies and also relax the current assumptions that the transition system, control-flow, and event-level attributes are independent of sensitive attributes. We also plan to strengthen the causal component by assessing robustness to causal misspecification, latent confounding, and alternative causal structures. This includes combining causal discovery with domain constraints and conducting ablation studies on how attribute selection and learned causal graphs affect predictive performance, counterfactual fairness, and recommendation quality.

Acknowledgments. F. De Santis is supported by the Italian Ministry of University and Research (MUR) under the National Recovery and Resilience Plan (NRRP), Mission 4, Component 1, Investment 4.1, CUP D91I23000080006, funded by the European Union - NextGenerationEU.

References

1. Aalst, W.: Data science in action. In: Process Mining, pp. 3–23. Springer, Heidelberg (2016). https://doi.org/10.1007/978-3-662-49851-4_1
2. Badreddine, S., d'Avila Garcez, A.S., Serafini, L., Spranger, M.: Logic tensor networks. Artif. Intell. **303**, 103649 (2022)
3. Bozorgi, Z.D., Teinemaa, I., Dumas, M., Rosa, M.L., Polyvyanyy, A.: Prescriptive process monitoring based on causal effect estimation. Inf. Syst. **116**, 102198 (2023)
4. Burr, T.: Causation, prediction, and search. Technometrics **45**(3), 272–273 (2003)
5. De Moor, J., Weytjens, H., De Smedt, J., De Weerdt, J.: SimBank: from simulation to solution in prescriptive process monitoring. In: Proceedings of the 23rd International Conference on Business Process Management (BPM 2025), pp. 165–182 (2025)
6. Dees, M., de Leoni, M., van der Aalst, W.M.P., Reijers, H.A.: What if process predictions are not followed by good recommendations? In: Proceedings of the Industry Forum at BPM 2019 Co-located with 17th International Conference on Business Process Management, vol. 2428, pp. 61–72 (2019)
7. Di Francescomarino, C., Ghidini, C.: Predictive process monitoring. In: van der Aalst, W.M.P., Carmona, J. (eds.) Process Mining Handbook, pp. 320–346, LNBIP. Springer International Publishing, Berlin-Heidelberg (2022). https://doi.org/10.1007/978-3-031-08848-3_10
8. Gröger, C., Schwarz, H., Mitschang, B.: Prescriptive analytics for recommendation-based business process optimization. In: Abramowicz, W., Kokkinaki, A. (eds.) BIS 2014. LNBIP, vol. 176, pp. 25–37. Springer, Cham (2014). https://doi.org/10.1007/978-3-319-06695-0_3
9. Heilmann, X., Manganini, C., Cerrato, M., Belle, V.: A neurosymbolic approach to counterfactual fairness. In: Proceedings of The 19th International Conference on Neurosymbolic Learning and Reasoning, vol. 284, pp. 1004–1025. PMLR (2025)
10. Islam, R., Pan, S., Foulds, J.R.: Can we obtain fairness for free? In: AIES '21: AAAI/ACM Conference on AI, Ethics, and Society, Virtual Event, USA, May 19–21, 2021, pp. 586–596. ACM (2021)
11. Kubrak, K., Botchorishvili, L., Milani, F., Nolte, A., Dumas, M.: Explanatory capabilities of large language models in prescriptive process monitoring. In: Marrella, A., Resinas, M., Jans, M., Rosemann, M. (eds.) Proceedings of the 22nd International Conference on Business Process Management (BPM 2024), pp. 403–420. Springer Nature Switzerland, Cham (2024). https://doi.org/10.1007/978-3-031-70396-6_23
12. Kubrak, K., Milani, F., Nolte, A., Dumas, M.: Prescriptive process monitoring: quo vadis? PeerJ Comput. Sci. **8**, e1097 (2022)
13. Kusner, M.J., Loftus, J.R., Russell, C., Silva, R.: Counterfactual fairness. In: Advances in Neural Information Processing Systems 30: Annual Conference on Neural Information Processing Systems 2017, December 4–9, 2017, Long Beach, CA, USA, pp. 4066–4076 (2017)
14. Käppel, M., Neuberger, J., Möhrlein, F., Weinzierl, S., Matzner, M., Jablonski, S.: A human-in-the-loop approach for improving fairness in predictive business process monitoring. In: Senderovich, A., Cabanillas, C., Vanderfeesten, I., A. Reijers, H. (eds.) Business Process Management - 23rd International Conference, BPM, vol. 16044, pp. 343–360. Springer, Cham (2025). https://doi.org/10.1007/978-3-032-02867-9_21

15. de Leoni, M., Padella, A.: Achieving fairness in predictive process analytics via adversarial learning. In: Comuzzi, M., Grigori, D., Sellami, M., Zhou, Z. (eds.) 30th International Conference, CoopIS 2024, Porto, Portugal, Proceedings, vol. 15506, pp. 346–354. Springer, Cham (2024). https://doi.org/10.1007/978-3-031-81375-7_21
16. Mannhardt, F.: Responsible process mining. In: van der Aalst, W.M.P., Carmona, J. (eds.) Process Mining Handbook, pp. 373–401. Springer, Cham (2022). https://doi.org/10.1007/978-3-031-08848-3_12
17. Meneghello, F., et al.: Optimizing resource allocation policies in real-world business processes using hybrid process simulation and deep reinforcement learning. In: Marrella, A., Resinas, M., Jans, M., Rosemann, M. (eds.) Proceedings of the 22nd International Conference on Business Process Management (BPM 2024), vol. 14940, pp. 167–184. Springer, Cham (2024). https://doi.org/10.1007/978-3-031-70396-6_10
18. Padella, A., Mannhardt, F., Vinci, F., de Leoni, M., Vanderfeesten, I.: Experience-based resource allocation for remaining time optimization. In: Marrella, A., Resinas, M., Jans, M., Rosemann, M. (eds.) Proceedings of the 22nd International Conference on Business Process Management (BPM 2024), pp. 345–362. Springer Nature Switzerland, Cham (2024). https://doi.org/10.1007/978-3-031-70396-6_20
19. Pohl, T., Qafari, M.S., van der Aalst, W.M.P.: Discrimination-aware process mining: a discussion. In: Montali, M., Senderovich, A., Weidlich, M. (eds.) Process Mining Workshops. Springer, Cham (2023). https://doi.org/10.1007/978-3-031-27815-0_8
20. Shoush, M., Dumas, M.: White box specification of intervention policies for prescriptive process monitoring. Data Knowl. Eng. **155**, 102379 (2025)
21. da Silva, M.C., Fantinato, M., Peres, S.M.: Towards fairness-aware predictive process monitoring: Evaluating bias mitigation techniques. In: Comuzzi, M., Grigori, D., Sellami, M., Zhou, Z. (eds.) Cooperative Information Systems - 30th International Conference, CoopIS 2024, Porto, Portugal, November 19–21, 2024, Proceedings, pp. 150–166. Springer, Cham (2024). https://doi.org/10.1007/978-3-031-81375-7_9
22. Teinemaa, I., Tax, N., de Leoni, M., Dumas, M., Maggi, F.M.: Alarm-based prescriptive process monitoring. In: Weske, M., Montali, M., Weber, I., vom Brocke, J. (eds.) BPM 2018. LNBIP, vol. 329, pp. 91–107. Springer, Cham (2018). https://doi.org/10.1007/978-3-319-98651-7_6
23. Vinci, F., Park, G., van der Aalst, W., de Leoni, M.: Prosit: a tool for interactive and transparent process simulations. In: Proceedings of the Workshops and Satellite Events of the 23rd International Conference on Service-Oriented Computing (ICSOC 2025). Springer (2025)

Respecting Object Relationships: An Object-Centric Sampling Algorithm

Frederik Fonger[1,2(✉)] and Agnes Koschmider[1]

[1] Chair of Business Informatics and Process Analytics, University of Bayreuth, Bayreuth, Germany
agnes.koschmider@uni-bayreuth.de
[2] Department of Computer Science, Kiel University, Kiel, Germany
ffo@informatik.uni-kiel.de

Abstract. Recently, object-centric process mining has emerged as a new process mining direction that places the relationships among multiple data objects at the center, with the Object-Centric Event Log (OCEL) serving as the standard exchange format. OCEL can contain a high volume of events and complex object relationships. Particularly, the size of real-life OCELs presents a challenge for conformance checking, hindering the scalability. This makes it necessary to extract representative subsets of the log for analysis purposes, thereby calling for appropriate sampling algorithms. Currently, only random sampling exists to allow for scalability. However, it disregards dependencies between multiple objects and therefore does not produce reliable conformance checking results. Thus, it cannot be considered as an appropriate sampling strategy. To overcome this, we propose the sampling algorithm RP-OCEL that incorporates object relationships, while reducing the size of the OCEL. RP-OCEL applies weighting according to the connectivity of object types and selects only object types that exhibit below a connectivity score, while the final sample still contains all object types. We applied our approach on a well-known OCEL dataset and compared it to random sampling. The results show that, in contrast to random sampling, RP-OCEL is suited for conformance checking, can perform sampling without altering the objects associated with an event and can significantly reduce processing time.

Keywords: object-centric event log · sampling · algorithm · process mining · conformance checking

1 Introduction

Object-centric process mining has become an essential part of process analytics particularly in use cases involving multiple interconnected objects or multidimensional relationships [20]. For instance, consider the order handling process of an online shop. There, multiple orders placed by the same customer within a short period of time are combined into a single package. Representing this requires a more complex structure than traditional case-oriented event logs. For

F. M. Maggi et al. (Eds.): BPM 2026, LNCS 16936, pp. 388–404, 2027.
https://doi.org/10.1007/978-3-032-37777-7_24

this, object-centric process mining provides a structure to appropriately record such processes. The Object-Centric Event Log standard (OCEL standard) serves as the exchange format for object-centric process mining. Processing OCELs requires significant computational time and power, particularly for conformance checking [12]. For example, calculating fitness and precision for the conformance of the well-known OCELs from OCEL-Standard.org [19,20] has often took up to multiple days in our evaluation on current workstation computers. This hinders adaptability of these methods, as they are not scalable for real-live sized dataset. To improve this, sampling seems an appropriate solution. For this, it needs to be ensured that the conformance checking results from the sample closely match the results of the original log, while speeding up the computation. One OCEL sampling algorithm has already been proposed [12], relying on a random sampling strategy. However the generated samples are not suited for conformance checking, as random sampling splits related objects apart and thus the samples do not produce conformance checking results that closely match the results of the unsampled OCEL. This leads to unknown discrepancies between the calculated and the actual conformance.

In this paper, we propose RP-OCEL, which is a deterministic OCEL sampling approach that preserves the structure of the original log by selecting a subset without breaking apart related events and object types. Figure 1 shows the difference between random sampling and our approach. A random sampling algorithm does not consider the relationship or connectivity between object types (e.g., a package and order) when splitting the log into subsets but rather splits the event log randomly. This would be similar to a sampling algorithm for traditional event logs that considers only a subset of events from a case. In contrast, RP-OCEL uses the relationships between object types as splitting criteria, thereby improving the representativeness of the resulting subsets.

Table 1. Example of a process execution for the object types order (o), package (p), invoice (i) and worker (w) as all events are directly or indirectly connected.

event ID	Activity	Related Objects
e1	Create Order	o1
e2	Process order	o1, w1
e3	Pack Order	o1, p1
e3	Send Invoice	o1, i1
e5	Create order	o2
e6	Process order	o2, w1
e7	Pack order	o2, p2
e8	Send invoice	o2, i2

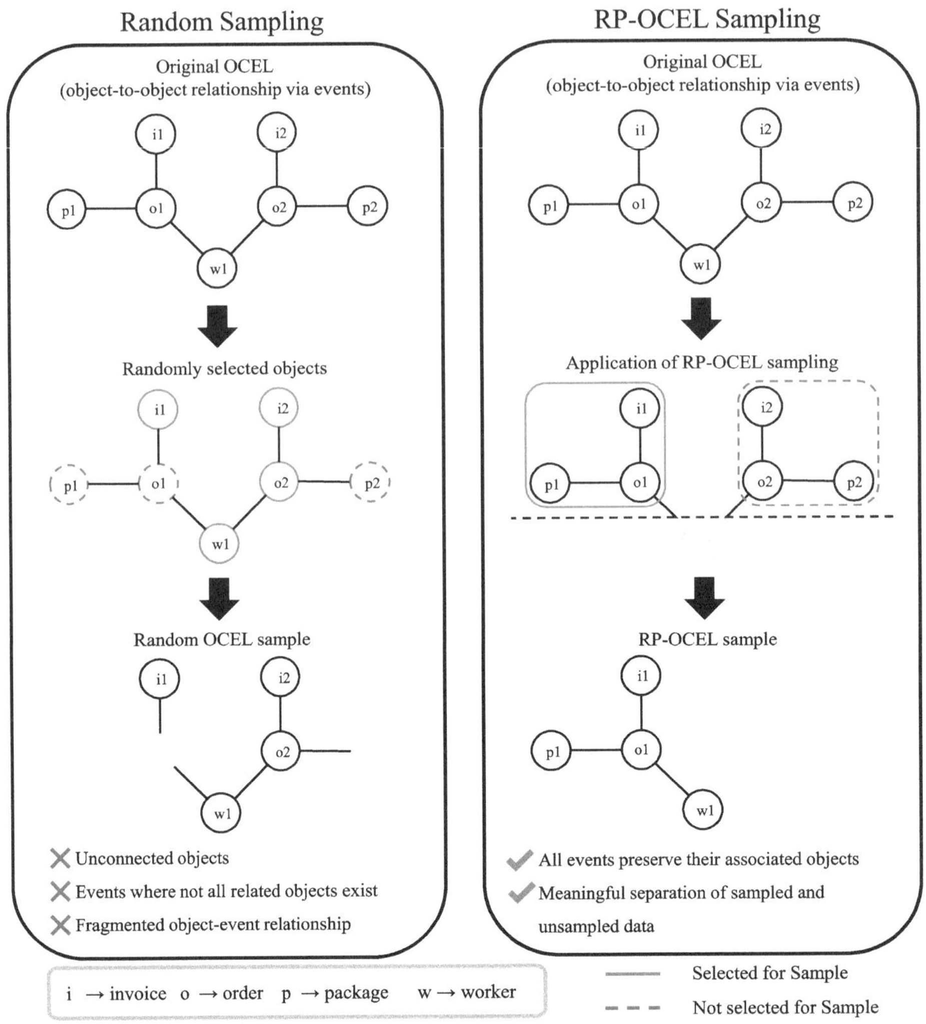

Fig. 1. Comparison of random and RP-OCEL sampling. Here, random sampling splits the event log randomly, while RP-OCEL preserves structure by taking the relationships between object types into account.

RP-OCEL relies upon four steps that work as follows:

1. The algorithm analyses the event log in terms of process executions and determines the degree of connectivity. It counts the number of object-to-object relationships of each object related to an object type. A process execution is a selection of events (i.e., a subset) from a complete object-centric event log. In this subset all events are connected either directly (e.g., through the same object) or indirectly (e.g., through a chain of objects or events) like connected components in a graph. Let us consider an example with a package, an order,

an invoice and a worker as object types, see Table 1. An order is linked to a package and an invoice. The worker is linked to all orders. The related events *Create Order, Process Order, Pack Order* and *Send Invoice* are connected in the OCEL graph. The events e1 to e8 represent a process execution, as all events are directly or indirectly connected (e1 is indirectly connected by e8).
2. To balance the inclusion of objects in the sampling procedure with the need for a usable set of process executions, object types above a connectivity threshold are temporarily disregarded for the sampling process. This is done to ensure that a sufficient amount of process executions remains for the subsequent sampling step. In this example, the worker object type is temporarily disregarded. Next, the connectivity degree is calculated again for process executions.
3. The sampling algorithm RemainderPlus [18], initially suggested for traditional event logs and modified to satisfy OCEL characteristics, is then applied. RemainderPlus aims at generating a sample with a similar distribution of process execution and Directly-Follows pairs as in the original OCEL.
4. The object types that have been temporarily disregarded are added back to the events in the sample.

With our OCEL sampling we are able to achieve better results than with current OCEL sampling approaches. The evaluation results indicate that RP-OCEL effectively addresses the issues related to incomplete object relationships. Additionally, the evaluation results show that RP-OCEL sampling produces results that are more representative for conformance checking than samples from random sampling. Furthermore, we show that the RP-OCEL sampling allows for significant runtime improvements.

The remainder of the paper is structured as follows. Section 2 provides definitions to which we refer throughout the paper. In Sect. 3 the sampling algorithm is presented, followed by the evaluation in Sect. 4. The related work is summarized in Sect. 5 and the paper concludes in Sect. 6.

2 Preliminaries

2.1 Definiton of an OCEL

An OCEL has, compared to a flat event log, no traditional case notion [1]. Instead it consists of objects in addition to events and activities. We define an OCEL L and its sample S_L as follows.

Definition 1 (Object-centric event log, sample). *Let $\mathcal{E}$ be the universe of events, $\mathcal{T}$ the universe of object types, $\mathcal{O}$ the universe of objects and $\mathcal{A}$ the universe of activities. Then let OCEL L be a tuple $(E, T, O, \varphi_{ot}, \varphi_{eo}, \varphi_{ea}, \leq_e)$ with $E \subseteq \mathcal{E}, T \subseteq \mathcal{T}, O \subseteq \mathcal{O}, \varphi_{ot} : O \to T$ a function mapping objects to object types, $\varphi_{eo} : E \to \mathcal{P}(O)$ a function mapping events to sets of objects, $\varphi_{ea} : E \to \mathcal{A}$ a function mapping events to activities and a total ordering $\leq_e$ for events [7]. The cardinality of L is defined as $|L| = |E|$. The set of objects of a specific object type $t \in T$ is denoted as $O_t = \{o \in O \mid \varphi_{ot}(o) = t\}$.*

To simplify the notation, for $L = (E, T, O, \varphi_{ot}, \varphi_{eo}, \varphi_{ea}, \leq_e)$ *we denote the set of events as* E_L*, the set of object types as* T_L *and the set of objects as* O_L*. For the merging of two OCEL* $\cup_{ocel}$ *is used as operator.*

For any L*, we call* S_L *a sample of* L *if* $S_L = (E_S, T_S, O_S, \varphi_{ot}, \varphi_{eo}, \varphi_{ea}, \leq_e)$ *with* $E_S \subseteq E, T_S \subseteq T$ *and* $O_S \subseteq O$*. The sample ratio* c *of a sample is defined as* $c = \frac{|E_S|}{|E_L|}$*, with* $0 < c \leq 1$*.*

A process execution is a subset of an OCEL where all events are connected either directly or indirectly. A process execution can be considered as a connected part of a graph, where nodes represent events and edges represent connections through shared objects. We define a process execution as follows:

Definition 2 (Process execution). *Let* L *be an OCEL* $(E, T, O, \varphi_{ot}, \varphi_{eo}, \varphi_{ea}, \leq_e)$ *and* $R \subseteq E \times O$ *be the event-object reference relation. Then a process execution* p *is a tuple* $(E', T, O', \varphi_{ot}, \varphi_{eo}, \varphi_{ea}, \leq_e)$ *where* $E' \subseteq E$ *and* $O' \subseteq O$ *such that the undirected bipartite graph* $G = (E' \cup O', R')$ *with* $R' = R \cap (E' \times O')$ *is connected. The cardinality of* p *is defined as* $|p| = |E'|$*.* P_L *is the set of all possible process executions of OCEL* L *and* $f_L(p_1) = |\{p_2 \in P_L \mid p_1 = p_2\}|$ *is the frequency of identical process executions to* p*.*
Let L' *be an OCEL with* $L' \subset L$ *and* e *an event with* $e_1 \in L \setminus L'$*. Then* $rel_{P_{L'}}^{P_L}(e_1) = \{p_1 \in P_{L'} \mid \exists e_2 \in E_{p_1} \wedge \exists o \in \varphi_{ot}(e_2) \wedge \exists p_2 \in P_{L \setminus L'} : o \in O_{p_2} \wedge e_1 \in E_{p_2}\}$ *is the set of all process execution of* $P_{L'}$ *where* e *is related but not part of.*

2.2 Characteristics of an OCEL

To find a representative sample of the OCEL, we need some information of the content of the OCEL. Therefore, we define the following projections and measures that are used by the RP-OCEL sampling algorithm.

Definition 3 (Behavior of the OCEL). *Let* L *be an OCEL and* $A = \{a \in \mathcal{A} \mid \exists e \in E_L : \varphi_{ea}(e) = a\}$ *be the activities of* L*. Then we define* $\mathcal{B}(L)$ *as the set of all Directly-Follows pairs in* L *which will also be called the behavior of* L*.* $\mathcal{B}(L) = \{(a_1, a_2) \in A \times A | \exists e_1, e_2 \in E_L : (\varphi_{ea}(e_1) = a_1) \wedge (\varphi_{ea}(e_2) = a_2) \wedge (\varphi_{eo}(e_1) \cap \varphi_{eo}(e_2) \neq \emptyset \wedge e_1 \leq_e e_2) \wedge (\neg \exists e_3 \in E_L : \varphi_{eo}(e_1) \cap \varphi_{eo}(e_2) \cap \varphi_{eo}(e_3) \neq \emptyset \wedge e_1 \leq_e e_3 \leq_e e_2) \wedge e_1 \neq e_2 \wedge e_1 \neq e_3 \wedge e_2 \neq e_3\}$*. Furthermore, for a Directly-Follows pair* b *let* $f_L(b_1) = |\{b_2 \in \mathcal{B}(L) | b_1 = b_2\}|$ *be the frequency of* b *in* L*.*

Definition 4 (Connectivity of an OCEL). *Let* $o_1 \in O$*. Then we define* $\theta(o_1) = |\{o_2 \in O | \exists e \in E : o_1, o_2 \in \varphi_{eo}(e) \wedge o_1 \neq o_2\}|$ *as the function that calculates the number of object to object relationships of object* o_1*.*

Definition 5 (Oversampled and undersampled behavior). *Let* L *be an OCEL and* S_L *a sample from* L *and* c *be the sample ratio of* S_L*, and let* $\tau_L^{S_L}(b) = \frac{f_{S_L}(b)}{f_L(b)}$ *be the sample ratio of a Directly-Follows pair* b*. Then, we call* $\mathcal{O}_L^{S_L} = \{b \in \mathcal{B}(L) \mid \tau_L^{S_L}(b) > c\}$ *the oversampled behavior and* $\mathcal{U}_L^{S_L} = \{b \in \mathcal{B}(L) \mid \tau_L^{S_L}(b) < c\}$ *the undersampled behavior from* L*.*

Definition 6 (Frequency of behavior). *Let L be an OCEL and (a_1, a_2) be a Directly-Follows pair from L. Then we define the frequency of (a_1, a_2) in L as* $f_L(a_1, a_2) = |\{(e_1, e_2) \in E \times E : (\varphi_{ea}(e_1) = a_1 \wedge \varphi_{ea}(e_2) = a_2 \wedge (a_1, a_2) \in \mathcal{B}(L)) \wedge (e_1 \leq_e e_2) \wedge (\neg \exists e_3 \in E : e_1 \leq_e e_3 \leq_e e_2) \wedge (e_1 \neq e_2 \wedge e_1 \neq e_3 \wedge e_2 \neq e_3)\}|$. *Let p be a process execution, then we define the frequency of (a_1, a_2) in p as* $f_p(a_1, a_2) = |\{(e_1, e_2) \in E \times E : (\varphi_{ea}(e_1) = a_1 \wedge \varphi_{ea}(e_2) = a_2 \wedge (a_1, a_2) \in \mathcal{B}(P)) \wedge (e_1 \leq_e e_2) \wedge (\neg \exists e_3 \in E : e_1 \leq_e e_3 \leq_e e_2) \wedge (e_1 \neq e_2 \wedge e_1 \neq e_3 \wedge e_2 \neq e_3)\}|$.

2.3 Filtering an OCEL

RP-OCEL requires techniques for filtering an OCEL since a list of selected object types and events is calculated by RP-OCEL, and the OCEL needs to be filtered based on these. For this, we define the following filters.

Definition 7 (Filter techniques). *Let L be an OCEL and $T' \subseteq T_L$. Then we define a function that filters an OCEL L by object types T' as* $filter_{ot}(L, T') = (E', T', O', \varphi_{ot}, \varphi_{eo}, \varphi_{ea}, \leq_e)$ *with* $O' = \{o \in O \mid \varphi_{ot}(o) \in T'\}$ *and* $E' = \{e \in E \mid \exists o \in O' : o \in \varphi_{eo}(e)\}$. *We additionally define a function that filters an OCEL L by the events E' as* $filter_e(L, E') = (E', T', O', \varphi_{ot}, \varphi_{eo}, \varphi_{ea}, \leq_e)$ *with* $O' = \{o \in O \mid \exists e \in E' : o \in \varphi_{eo}(e)\}$ *and* $T' = \{t \in T \mid \exists o \in O' : \varphi_{ot}(o) = t\}$.

2.4 Quality Metrics

We introduce the two well-known metrics coverage and mean absolute error (MAE) [25] for the evaluation of a sample to which we will also refer in the evaluation section. The coverage metric indicates how many of the Directly-Follows pairs from the original OCEL are preserved in the sample. Thus, the sample is checked to see whether such a pair occurs at least once. It is sufficient if a DFP occurs once to count it as "covered" regardless of how often the pair occurs.

Definition 8 (Coverage). *Let L be an OCEL and S_L a sample of L. Then, the coverage is defined as* $C_L(S_L) = \frac{|\mathcal{B}(S_L)|}{|\mathcal{B}(L)|}$.

The MAE indicates how much the frequency of a DFP in the sample differs on average across all DFPs from the original frequency of the original log.

Definition 9 (Mean absolute error (MAE)). *Let $A \subseteq \mathcal{A}$ and L be an OCEL, let S_L be a sample of L with a sample ratio of $c \in (0, 1]$. Then $\mathcal{B}(L)$ is the Behavior of L with $\mathcal{B}(L) = \{b_1, ..., b_n \in A \times A\}$ and $n = |\mathcal{B}(L)|$. Then the MAE is defined as* $M_L(S_L) = \frac{1}{n} \sum_{i=1}^{n} | f_{S_L}(b_i) - (c * f_L(b_i)) |$.

3 Method

For a flat event log, sampling algorithms use cases as elements for sampling. However, in an OCEL, as already introduced, interconnected and multidimensional object relationships exist, which make sampling difficult, see step 1 in Fig. 2. Due to the interconnectivity, there are no clearly distinct elements to sample from. In many OCELs, all events are indirectly connected to each other. RP-OCEL seeks to preserve the integrity of interconnected objectssuch as an order and its corresponding package, as illustrated in the introductory example. This requires the algorithm to create elements to sample from.

The RP-OCEL algorithm is summarized as pseudocode in Algorithm 1, implemented in Python and available on GitHub[1]. The algorithm works as follows.

1. Firstly, the OCEL needs to be transformed in such a way that process executions are divided into smaller ones, for example, splitting one or two process executions into 200 smaller executions to allow sampling. To achieve this, object types with many object-to-object connections are filtered out. Consider the order example from the introduction. The introduced object types (*order*, *package* and *invoice*) have a limited number of related objects. In contrast, *worker* w1 is related to every *order*, thus indirectly related to every object. As a result, a single, monolithic process execution is created, making it impossible to extract a sample without cutting object relationships. To avoid this, for each object type, a metric (connectivity) is calculated that counts how many objects-to-object relationships each object of that type has. These values are averaged, which indicate the average degree of connectivity for an object type. Object types above a connectivity threshold are temporarily disregarded from the OCEL, as depicted in step 2 of Fig. 2. Let us assume we have three objects o3, o4 and o5 of the object type *order* with 4, 5, and 3 directly related objects. Then, the connectivity for the object type *order* is 4 = (4+5+3)/3. We then normalize the value based on the highest connectivity, resulting in a value between 0 and 1. The default value of the connectivity threshold is 0.5, but can be adjusted accordingly by the user. Object types that were initially disregarded for sampling are considered later with regard to adding them back to the events. Referring to our example in Fig. 2, the *worker* object type would be excluded from the sampling process, as it connects all *order-package-invoice* combinations and thus prevents the creation of a large number of disconnected process executions. Despite possible many-to-many relationships between the remaining object types, the OCEL then contains a large number of disconnected process executions, enabling the sampling algorithm to select meaningful subsets without breaking its logical coherence. Then, considering Fig. 1, p1 is connected with o1 and i1 and p2 is connected with o2 and i2. The combinations o1, p1 and i1 and o2, p2 and i2 do not depend on each other, they are independent process executions. RP-OCEL

[1] https://github.com/Frederik-Fonger/RP_OCEL.

decides which combination to choose, however, the choice does not break the logical coherence.

2. Secondly, RP-OCEL selects existing process executions, as shown in step 3 in Fig. 2. For this purpose, we rely on a modified version of the RemainderPlus sampling that was introduced for traditional event logs [18]. It selects process executions with the aim of generating the same distribution of Directly-Follows pairs (DFPs) in the sample as in the original log. RemainderPlus work as follows:
   I. *(Preparation and initial sample)* First, the expected frequency for each process execution is calculated and rounded down to the nearest whole number. Process executions with an expected frequency ≥ 1 are added to the initial sample. However, as process executions can only be added as a whole to the sample, it has to be decided which process execution is added (based on an expected frequency $0 < f < 1$) and which is not further considered. For this, it is determined which process execution is added next. This requires the following two steps.
   II. *(Ranking process executions by expected frequency)* In the first step, the remaining expected frequency of all process executions is updated. These values lie between 0 and 1, as all process executions with an expected frequency > 1 have been added in the previous step. Next, the process executions need to be ranked by their importance to the sample. For this, they are sorted based on the first decimal place of their remaining expected frequencies. It is possible that multiple process executions are ranked highest, for this, a second ranking is performed in the next step.
   III. *(Ranking process executions by undersampled and oversampled DFPs)* All process executions that have the highest ranking in the previous step are ranked according to the normalized difference between the undersampled and oversampled DFPs. The goal is to select the process execution with the highest proportion of undersamples DFPs. The process execution with the highest normalized difference is added to the sample.
   IV. *(Catching discarded object types)* To avoid disregarded events that are only related to temporarily filtered objects, a list is created to track events that are connected solely to the chosen process executions but are not part of them. These are added to the sample in the last step.
   V. *(Loop until desired sample size)* The algorithm is repeated again starting at the II. step until the desired sample size is reached. The sample then consists of all events that belong to the chosen process executions.

3. As a final step, the disregarded object types, from the first step of RP-OCEL, are added back to the events in the sample, see step 4 of Fig. 2. Referring again to our example process: a sample consists only of the events related to p1, o1 and i1. The object type *worker* is added back to the "process order" event in this last step.

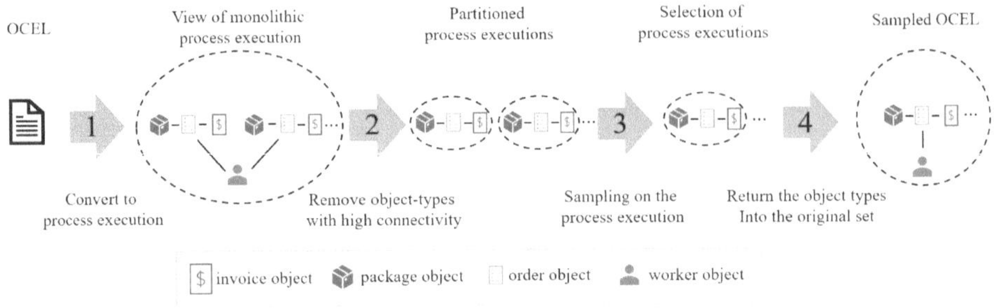

Fig. 2. Pipeline of RP-OCEL sampling

4 Evaluation

We applied our sampling algorithm to a well-known OCEL dataset, namely the "Procure-to-Pay" dataset from the LRMS collection [13]. This dataset covers the procurement lifecycle, recorded from requisition creation to the supplier payment. Next, we compared our approach with the random object sampling algorithm presented in [14]. The random sampling implementation is available in PM4PY [15]. However, only a comparison with random sampling on objects was possible, as the available implementations of random sampling on events and object types do not allow to specify the sample size. For random sampling, the evaluation was repeated five times and the results were averaged. Due to the complexity of an OCEL, the conformance checking algorithms have a very high computational demand [12]. In the evaluation setup, mining a process model took seconds, whereas calculating the conformance took hours or even days. Therefore, sampling is most beneficial for conformance checking. Consequently, we evaluated the usability of the samples for conformance checking and whether they are able to speed up the computation. Additionally, we calculated the standard sampling metrics of coverage and MAE. In this regard, coverage validated the existence of DFPs in the sample, where MAE evaluated the distribution. To demonstrate a wide applicability of the approach, a variety of sample ratios were tested ($c = \{0.05, 0.1, 0.2, 0.3, 0.4, 0.5\}$). To calculate a baseline, we need to perform conformance checking on an OCEL before sampling. However, most existing OCELs cannot handle conformance checking on the full dataset within a feasible runtime (i.e. for certain datasets, the computation time would extend to nearly a month, even when executed on a computing cluster). For each sample, the conformance checking was performed on an Intel Xeon with 32 cores and 240 GB of memory. We tried all available datasets from https://www.ocel-standard.org/ and only the "Procure-to-Pay" log had a computation time for the conformance checking of less than 10 d. Further highlighting the need for sampling. Therefore, only this log has been selected for evaluation. This OCEL file contains 8300 events from 12 event types (activity classes) and 4493 objects for 6 object types.

Algorithm 1: OCEL sampling

Data: $L = (E_L, T_L, O_L, \varphi_{ot}, \varphi_{eo}, \varphi_{ea}, \leq_e)$, // OCEL
$c \in (0, 1]$, // sample ratio
$k \in (0, 1]$ // connectivity threshold
Result: S_L
// First phase: calculate connectivity
$\gamma_{norm}(t) \leftarrow \underset{o \in O_t}{avg}\, \theta(o) \mid t \in T_L$
$T' \leftarrow \{t \in T_L | \gamma_{norm}(t) \leq k\}$
$L' \leftarrow filter_{ot}(L, T')$
// Second phase: initialize preliminary sample
$S \leftarrow [p^{\lfloor f_L(p) \cdot c \rfloor} \mid p \in P_L \wedge \lfloor f_L(p) \cdot c \rfloor \geq 1]$
$E' \leftarrow \emptyset$
// Third phase: Sampling on PEs
while $|E_s| + |E'| < \lfloor |E_L| \cdot c \rfloor$ **do**
  $\hat{F} \leftarrow \emptyset$
  for $p \in P_L$ **do**
    $f_e \leftarrow f_L(p) \cdot c - f_S(p)$ // Remaining expected variant frequency
    $sgn \leftarrow 1 \text{ } if \text{ } f_e \geq 0 \text{ } else \text{ } -1$
    $\hat{f} \leftarrow sgn \cdot \lfloor 10 \cdot (|f_e| - \lfloor |f_e| \rfloor) \rfloor$ // Signed first decimal
    $\hat{F} \leftarrow \hat{F} \cup (p, \hat{f})$
  end
  $\Lambda \leftarrow \arg\max_{p \in P_{L'}} rem(p) \quad with \quad rem(p) = \hat{f} \;\; \forall (p, \hat{f}) \in \hat{F}$

  $\Delta_{norm} \leftarrow \emptyset$
  for $\lambda \in \Lambda$ **do**
    $n_{\mathcal{U}} \leftarrow \sum_{b \in \mathcal{B}(\lambda) \cap \mathcal{U}^S_{L'}} f_\lambda(b)$ // Frequency of undersampled Directly Follows Relations
    $n_{\mathcal{O}} \leftarrow \sum_{b \in \mathcal{B}(\lambda) \cap \mathcal{O}^S_{L'}} f_\lambda(b)$ // Frequency of oversampled Directly Follows Relations
    $\delta_{norm} \leftarrow \frac{n_{\mathcal{U}} - n_{\mathcal{O}}}{\sum_{b \in \mathcal{B}(\lambda)} f_\lambda(b)}$ // Normalized difference
    $\Delta_{norm} \leftarrow \Delta_{norm} \cup (\lambda, \delta_{norm})$
  end

  $S \leftarrow S \cup_{ocel} \arg\max_{\lambda \in \Lambda} normdiff(\lambda) \quad with$
    $normdiff(\lambda) = \delta_{norm} \;\; \forall (\lambda, \delta_{norm}) \in \Delta_{norm}$
  $E' \leftarrow [e \mid e \in E_L \wedge \exists p \in rel_{P_{L'}}(e) : p \in S \wedge \nexists p \in rel_{P_{L'}}(e) : p \in P_S \setminus P_{L'}$
end
$S_L \leftarrow filter_e(L, E_S \cup E')$
return S_L

4.1 Event Alteration

One limitation of random sampling is that it leads to alters the event data. Because the selection of objects is random, not all objects related to an event are included in the sample. For instance, we inspected that an event such as

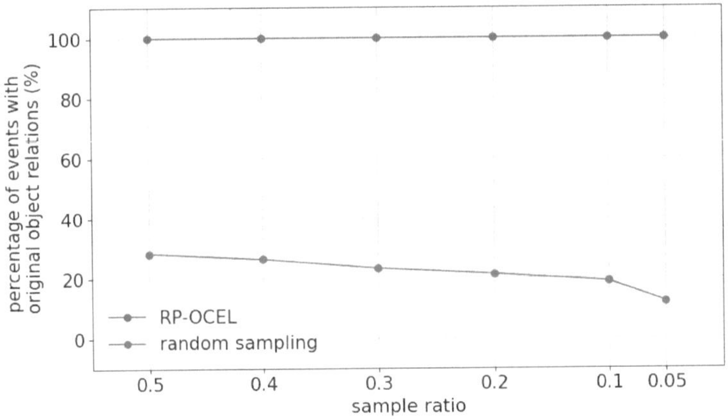

Fig. 3. Percentage of events in the sample that have their original object relationships.

"send invoice" is separated from an invoice and a customer, and instead it is only related to an invoice in the sample. Therefore, there is no record indicating which customer the invoice was sent to. Consequently, essential information or a defined relationship is missing. Compared to a sample generated by RP-OCEL, approximately 70–80% of the events in the random sample missed at least one event. This leads to around 20–30% of the sampled events maintaining their original information, which hints to missing information, such as incomplete item lists for a package. In contrast, RP-OCEL does not alter events at any of the tested sample sizes (see Fig. 3).

4.2 OCEL-Sampling for Conformance Checking

We compared RP-OCEL to random sampling for the purpose of conformance checking. For this, we evaluated how closely the calculated conformance from the sample matched that of the original OCEL. To do this, we mined an object-centric Petri net using the object-centric Inductive Miner algorithms with the original OCEL as input. Then, we used OCPA [8] to calculate the conformance between the model and the original OCEL. The conformance value is the reference conformance, which is to be reached by sampling. Figure 4 shows the absolute deviation of the calculated fitness from the sample to the reference fitness. This figure illustrates that, for all tested sample ratios, the deviation of the fitness from the RP-OCEL samples is notably smaller than the deviation from the random samples. For five of the six tested sample ratios, the deviation with the RP-OCEL sample is less than half that for the random samples. RP-OCEL is especially likely to increase the gap for smaller sample ratios. For a sample ratio of $c = 0.05$, the fitness calculated from RP-OCEL differs from the reference by only 0.04, where the fitness of the random sample deviates by 0.22 from the reference. We assume that the increase in performance for $r = 0.1$ and $r = 0.05$ comes from the fact, that for small samples our algorithm focuses on the main process behavior and includes less deviations. Thus leading to easier process discovery for the mining algorithm.

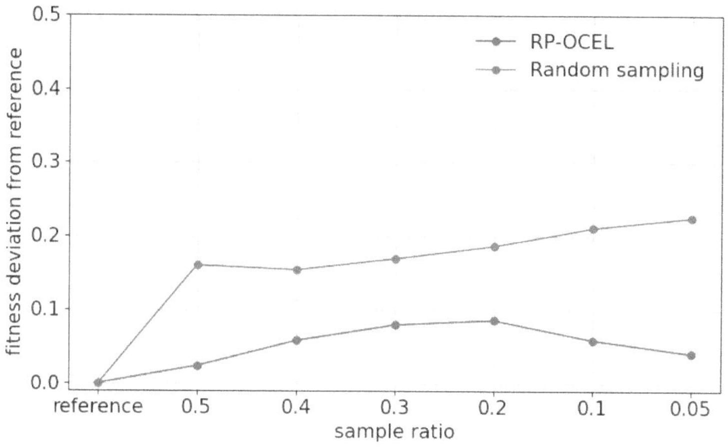

Fig. 4. Deviation of the fitness calculated from the original OCEL and the sample.

Figure 5 shows the absolute deviation of the calculated precision from the sample against the reference precision. For higher sample ratios ($r = 0.3, 0.4, 0.5$) RP-OCEL performs better, while for smaller samples ratios ($r = 0.05, 0.1, 0.2$) random samples have a small advantage. Overall RP-OCEL achieves a slightly lower average deviation of 0.18, while random sampling has an average deviation of 0.22.

4.3 Quality Metrics

To further support the evaluation, we applied the sample metrics MAE and coverage. Typically, strong results on these two metrics correspond to strong performance in both process discovery and conformance checking. However, the considerable performance improvements for conformance checking is not apparent in the results of the MAE and coverage. This gap between real-world applications and metrics may indicate that MAE and coverage are not well suited for evaluating OCEL samples. This probably stems from the fact that there are no suited elements (like cases in traditional event log) for random sampling to sample on. This results in strong performance according to simple quality metrics, however the complex nature of OCELs is not incorporated. Nevertheless, we applied them since they are widely accepted evaluation metrics. MAE results for RP-OCEL sampling are slightly better than random sampling across all sampling ratios. For coverage, both algorithms can achieve an optimal coverage of 1 for the sample ratios of 0.4 and 0.5. Both algorithms struggle to achieve optimal coverage at low sample ratios; however, random sampling performs better at larger sample ratios (Figs. 6 and 7).

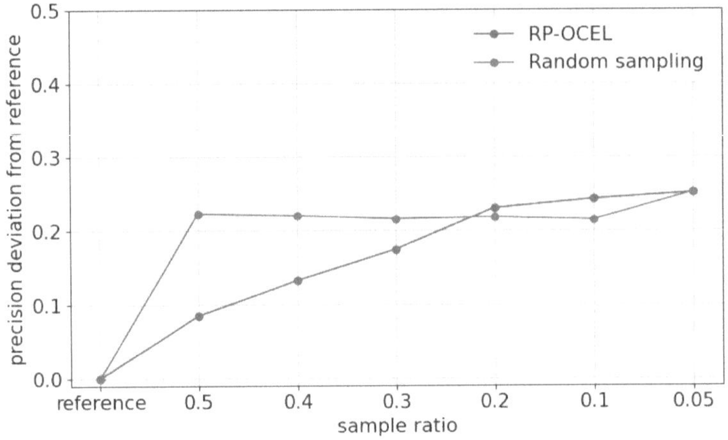

Fig. 5. Deviation of the precision calculated from the original OCEL and the sample.

4.4 Runtime

The generation of the samples was performed on an end-user PC with an Intel i5 processor and 16 GB of memory. The sample generation with RP-OCEL took for all sample sizes between 28 and 51 s. As expected, the runtime for random sampling was much faster, with less than one second for all sampling sizes. This characteristic stems from the fact, that it bypasses the need to interpret the content of the event log. Nonetheless, RP-OCEL reduces both time and computational effort; for instance, conformance checking on the previously mentioned compute cluster is considerably faster. For calculating the fitness on a sample with a sampling ratio of $c = 0.1$, the runtime is only 63 min, whereas the computation with the original OCEL took more than six days. Therefore, the

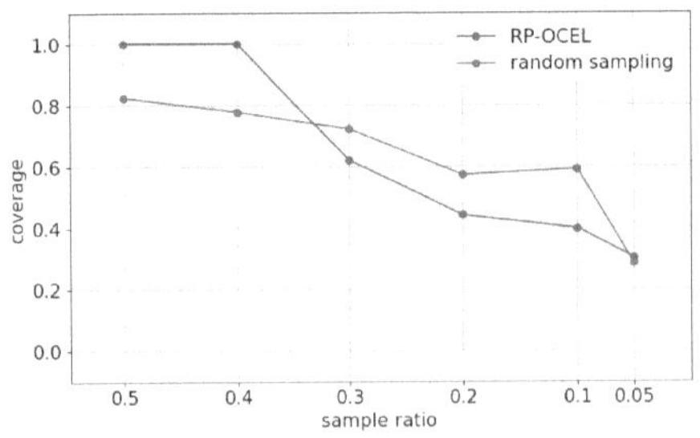

Fig. 6. The coverage of the sample compared between RP-OCEL and random sampling.

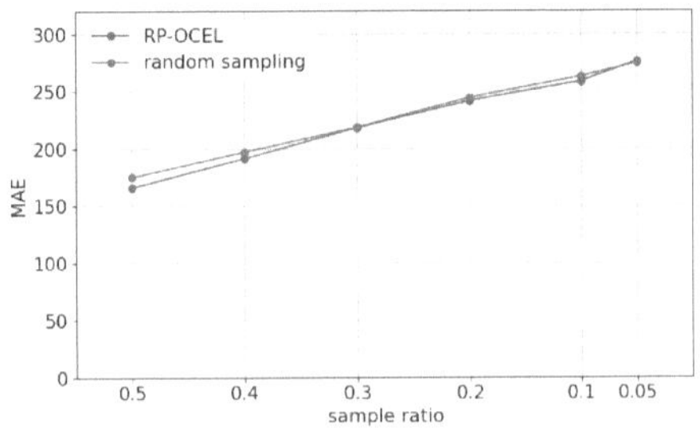

Fig. 7. The MAE of the sample compared between RP-OCEL and random sampling.

calculation time of less then a minute for RP-OCEL sampling can be neglected for the improvements in the conformance checking quality.

5 Related Work

To date object-centric process mining is an essential part of process analytics [4]. Recent efforts have explored the integration of diverse data sources such as unstructured text, streaming data, and video for process mining [21,23]. Therefore, techniques are required to efficiently leverage such heterogeneous data types for process mining by means of sampling strategies. It departs from traditional case-IDs and instead uses objects to related events to each other [20]. In the context of object-centric process mining, techniques for discovery, conformance checking, and enhancement have been explored [3]. For instance, several approaches have been suggested for process discovery of OCEL [2,5,14]. They include approaches to discover an object-centric Petri net, such as the object-centric Inductive Miner [5], which uses the flat event log of each object type and applies the traditional inductive miner to discover Petri nets. Subsequently, the Petri nets are merged to one object-centric Petri net. Techniques for discovering an object-centric Directly-Follows-graph are presented in [2,15].

Similar to the approach presented in our paper, there are other approaches in the literature that rely upon process executions. Adams et al. state that process executions (which they call "connected components") are the most useful way to process object-centric event data, as it is the only way to keep free of convergence, deficiency and divergence [6]. They also describe the challenge of OCELs with one monolithic process execution and present a graph based method to reduce the size of the process executions [9].

A different approach is the MANTA framework, it is designed for exploratory purposes [10]. It generates process views to provide a better understanding of the OCEL. For this, it also relies upon process executions [10]. Additionally, Küsters et al. presented the approach OCPQ for querying OCELs [22]. The approach generates a subset of an OCEL based on set of constrains [22]. Furthermore, Berti proposes several preprocessing techniques, including diverse filtering and early sampling methods, there impact on the quality of the resulting OCEL has not been yet evaluated [12].

For traditional event logs sampling has been applied for process discovery and conformance checking [11,17]. For process discovery, using whole event logs is no longer feasible [17]. That is why different sampling approaches for process discovery were evaluated. The results have shown that comparable models could be mined in a shorter time and that variant-based sampling is better suited than trace-based sampling for this application [17]. Bauer et al. evaluated the usage of sampling for conformance checking on traditional event logs [11]. They found that a small fraction of a event log can be sufficient to compute the conformance [11].

Although quality metrics for samples of traditional event logs, such as MAE and coverage, are well established in the literature [25], there meaningfulness for

OCELs has still to be evaluated. To verify the conformance of a mined process model, standard conformance checking measures such as fitness and precision have been developed for OCEL and are accessible in the Python package OCPA [7,8]. Also for calculating the alignments a related approach was proposed [24]. However, its current implementations in not suited for processing whole OCEL files due to the processing requirements [24]. Furthermore, there are new metrics proposed for conformance checking of OCELs. A quality measure, utilizing a case notion similar to process executions, was presented by Van Detten et al. [16]. It is a theoretical concept for evaluating simplicity and correctness of OCELs [16]. Additionally, Wrede et al. proposed a new metric for a graph bases conformance measure for OCELs [26]. A review of the literature indicates that the sampling of OCELs remains largely unexplored. Instead, existing research primarily focuses on improving process mining applications such as process discovery algorithms or conformance checking.

6 Conclusion

This paper suggested a novel approach for sampling object-centric event logs to face the challenge of high computational effort when processing OCELs. The algorithm focuses on object relationships and aims to preserve structure by taking the relationship between objects types into account. The algorithm evaluates the number of related objects for each object type and prioritizes those with fewer connections when selecting events. Nevertheless, all object types remain represented in the final sample. The evaluation results show that our RP-OCEL algorithm is superior to random sampling in terms that relationships between objects are preserved in sampling. Moreover, when the conformance checking results of the RP-OCEL samples are compared with random sampling, the results obtained from RP-OCEL are considerably closer to the results from the original log. Additionally, we showed that substantial runtime improvements can be made with OCEL sampling, as conformance checking can be speeded up from days to minutes. Such a reduction in computational effort can lead to improved adaptability, as well as reducing cost and gaining sustainability benefits. One potential limitation of our study is the limited number of OCEL logs used in the evaluation, which may affect the generalizability of the results. However, with the current selection of available OCEL datasets, the used dataset was the only which allowed to compute a base line on the original OCEL. Nevertheless, the proposed algorithm allows OCEL sampling that respects object relationships and produces representative samples.

Acknowledgments. This work was supported by the German Research Foundation (DFG, Deutsche Forschungsgemeinschaft) in the SPP 2422 "Data-Driven Process Modelling in Forming Technology" [grant number 500936349].

References

1. Aalst, W.M.P.: Object-centric process mining: dealing with divergence and convergence in event data. In: Ölveczky, P.C., Salaün, G. (eds.) SEFM 2019. LNCS, vol. 11724, pp. 3–25. Springer, Cham (2019). https://doi.org/10.1007/978-3-030-30446-1_1
2. van der Aalst, W.M.P.: Object-centric process mining: An Introduction. In: Formal Methods for an Informal World (ICTAC 2021), LNCS, vol. 13490, pp. 73–105. Springer, Cham (2023)
3. van der Aalst, W.M.P.: Object-centric process mining: unraveling the fabric of real processes. Mathematics **11**(12), 2691 (2023)
4. van der Aalst, W.M.P., Carmona, J. (eds.): Process Mining Handbook, LNBIP, vol. 448. Springer, Cham (2022)
5. van der Aalst, W.M., Berti, A.: Discovering object-centric petri nets. Fundamenta informaticae **175**(1–4), 1–40 (2020)
6. Adams, J.N., van der Aalst, W.M.P.: Addressing convergence, divergence, and deficiency issues. In: Business Process Management Workshops, LNBIP, vol. 492, pp. 496–507. Springer, Cham (2024)
7. Adams, J.N., van der Aalst, W.M.: Precision and fitness in object-centric process mining. In: 3rd International Conference on Process Mining (ICPM 2021). pp. 128–135. IEEE, Eindhoven, Netherlands (2021)
8. Adams, J.N., Park, G., van der Aalst, W.M.: ocpa: a Python library for object-centric process analysis. Softw. Impacts **14**, 100438 (2022)
9. Adams, J.N., Schuster, D., Schmitz, S., Schuh, G., van der Aalst, W.M.: Defining cases and variants for object-centric event data. In: 4th International Conference on Process Mining (ICPM 2022), pp. 128–135. IEEE, Bolzano, Italy (2022)
10. Basmer, M., Ueck, H., Fahland, D., Weidlich, M.: MANTA: materializing views on event data for context exploration in process analysis. In: Business Process Management (BPM 2025), LNCS, vol. 16044, pp. 51–68. Springer, Cham (2026)
11. Bauer, M.: Sampling and approximation techniques for efficient process conformance checking. Inf. Syst. **104**, 101666 (2022)
12. Berti, A.: Filtering and sampling object-centric event log. arXiv:2205.01428 [cs] (2022)
13. Berti, A.: Simulated object-centric event logs (OCEL 2.0) for order-to-cash, procure-to-pay, hiring, and hospital patient lifecycle processes (2024). https://doi.org/10.5281/zenodo.13879980
14. Berti, A., van der Aalst, W.M.P.: OC-PM: analyzing object-centric event logs and process models. Int. J. Softw. Tools Technol. Transfer **25**(1), 1–17 (2023)
15. Berti, A., van Zelst, S., Schuster, D.: PM4Py: a process mining library for Python. Softw. Impacts **17**, 100556 (2023)
16. van Detten, J.N., Schumacher, P., Leemans, S.J.J.: A framework for advanced case notions in object-centric process mining. In: ICPM 2024 Workshops, LNBIP, vol. 533, pp. 402–414. Springer, Cham (2025)
17. Fani Sani, M., van Zelst, S.J., van der Aalst, W.M.P.: The impact of biased sampling of event logs on the performance of process discovery. Computing **103**(6), 1085–1104 (2021). https://doi.org/10.1007/s00607-021-00910-4
18. Fonger, F., et al.: Representative sampling in process mining: two novel sampling algorithms for event logs. In: ICPM 2024 Workshops, LNBIP, vol. 533, pp. 44–56. Springer, Cham (2025)

19. Ghahfarokhi, A.F., Park, G., Berti, A., van der Aalst, W.M.P.: OCEL: a Standard for Object-Centric Event Logs. In: Bellatreche, L., Dumas, M., Karras, P., Matulevičius, R., Awad, A., Weidlich, M., Ivanović, M., Hartig, O. (eds.) ADBIS 2021, CCIS, vol. 1450, pp. 169–175. Springer, Cham (2021). https://doi.org/10.1007/978-3-030-85082-1_16
20. Koren, I., Adams, N., Berti, A.: OCEL 2.0 Resources. arXiv:2403.01982 [cs] (2024).www.ocel-standard.org
21. Koschmider, A.: Process mining for unstructured data: challenges and research directions. In: Modellierung 2024, LNI, P-348. GI (2024)
22. Küsters, A., van der Aalst, W.M.P.: OCPQ: object-centric process querying and constraints. In: Research Challenges in Information Science (RCIS 2025), LNBIP, vol. 547, pp. 383–400. Springer, Cham (2025)
23. Lepsien, A., Koschmider, A., Kratsch, W.: Analytics pipeline for process mining on video data. In: Business Process Management Forum. LNBIP, vol. 490, pp. 196–213. Springer, Cham (2023)
24. Liss, L., Adams, J.N., van der Aalst, W.M.P.: Object-centric alignments. In: Conceptual Modeling (ER 2023), vol. 14320, pp. 201–219. Springer, Cham (2023)
25. van der Werf, J.M.E., Polyvyanyy, A., van Wensveen, B.R., Brinkhuis, M., Reijers, H.A.: All that glitters is not gold: Four maturity stages of process discovery algorithms. Inf. Syst. **114**, 102155 (2023)
26. Wrede, E., van Detten, J.N., Liss, L., Leemans, S.J.: Out of babylon: Object-centric conformance checking on graph-based abstractions. In: International Conference on Cooperative Information Systems (CoopIS 2025), LNCS, vol. 15535, pp. 183–201. Springer (2025)

BRIDGE: A Bayes-Risk-Informed Process Decoder for Suffix Prediction

Xiaomeng He[1(✉)], Johannes De Smedt[1], Seppe vanden Broucke[1,2], and Jochen De Weerdt[1]

[1] Research Center for Information Systems Engineering (LIRIS), KU Leuven, Naamsestraat 69, 3000 Leuven, Belgium
{xiaomeng.he,johannes.desmedt,jochen.deweerdt}@kuleuven.be, seppe.vandenbroucke@ugent.be

[2] Department of Business Informatics and Operations Management, Ghent University, Tweekerkenstraat 2, 9000 Ghent, Belgium

Abstract. In Predictive Process Monitoring, activity suffix prediction predicts the remaining sequence of activities given a prefix. As a multi-step sequence prediction task with many plausible suffixes, decoding is critical to predictive accuracy. However, the performance of existing decoding strategies is highly dataset-dependent, and no method is consistently reliable. A key limitation is a recurring mismatch between how suffixes are selected during decoding and how prediction quality is evaluated: current approaches typically rely on likelihood to guide decoding, without explicitly accounting for sequence-level prediction error. In this work, we characterize this mismatch and propose BRIDGE, a Bayes-risk-informed decoding framework that samples candidate suffixes and ranks them by expected quality under the evaluation metric, thereby aligning decoding more closely with the task objective. As a post-training inference-time method, it can be applied across predictive models without retraining. We evaluate BRIDGE on eight real-life event logs against multiple decoding strategies and established suffix prediction models. BRIDGE is consistently competitive and often best-performing, with particularly strong gains on logs with long traces and for simple statistical models.

Keywords: Predictive Process Monitoring · Activity Suffix Prediction · Decoding Strategy · Bayes Risk

1 Introduction

Anticipating how an ongoing business process instance will evolve based on observed events and case characteristics is a crucial capability for organizations, as it enables early identification of needs and proactive intervention [5]. Many Predictive Process Monitoring (PPM) works focus on activity suffix prediction [1,15–17], i.e., predicting the remaining sequence of activities, to characterize the future process trajectory. This is particularly useful when decisions

F. M. Maggi et al. (Eds.): BPM 2026, LNCS 16936, pp. 405–424, 2027.
https://doi.org/10.1007/978-3-032-37777-7_25

depend on expected subsequent steps, such as allocating capacity across upcoming tasks or managing potential rework loops.

Most activity suffix prediction approaches require a decoding strategy to make predictions for ongoing cases at inference time. Given an observed prefix, the predictive model typically generates the suffix iteratively, assigning probabilities to alternative next activities at each step and thereby inducing many plausible future continuations. The decoding strategy determines which continuation is returned as the final prediction.

Existing work largely relies on likelihood-guided decoding strategies conventionally used in other sequence generation domains, which can be problematic for process prediction. A common approach is to output the most likely suffix via argmax decoding [20,23] or beam search [7]. In control-flow prediction, however, this approach tends to favor frequent variants and fails to predict suffixes that deviate from the "happy path", which may contain more critical and actionable information. By considering only the most likely continuations, it may also struggle to capture the often weak end-of-case signal in long traces, generating excessively long suffixes with large prediction errors. As an alternative, some studies adopt random sampling to allow exploration of less frequent activities [4]. However, without stable quality control, random sampling may sometimes produce good predictions but can also generate predictions far from the true suffix, making it unreliable for process prediction.

These theoretical limitations are also reflected in empirical findings: the effectiveness of existing decoding strategies is highly dataset-dependent and no single strategy is consistently reliable [1]. This motivates the need for a decoding strategy that is better aligned with the requirements of process suffix prediction, which are typically captured by a specified evaluation metric such as Damerau-Levenshtein distance. In this work, we propose BRIDGE, a Bayes-Risk-Informed Decoding framework for GEnerating suffix predictions. BRIDGE samples a set of candidate suffixes from a predictive model and ranks them by expected quality under a given evaluation metric, thereby aligning decoding more closely with the task objective. By aggregating information across multiple plausible suffixes, BRIDGE reduces the chance of selecting a likely but error-prone "outlier" that is poorly supported by the broader posterior mass.

The main contributions of this work are as follows: (i) we identify and empirically illustrate a recurring likelihood–quality mismatch in suffix decoding, showing that likelihood, usually used to guide decoding, is only weakly correlated with predictive quality that is measured by a sequence-level accuracy metric, (ii) we introduce BRIDGE as a generally applicable decoding framework that can be applied post-training across predictive models without additional retraining, and (iii) we provide an empirical study that evaluates BRIDGE against multiple decoding strategies and competitive suffix prediction models on eight real-world datasets, showing that it is a robust inference-time solution with best or near-best performance across datasets and model families.

The remainder of the paper is structured as follows. Section 2 introduces suffix prediction and decoding strategies, and discusses related work. Section 3 empirically illustrates the mismatch between likelihood-guided decoding and evaluation. Section 4 presents the methodology of BRIDGE. Section 5 describes the experimental setup. Section 6 reports the results. Section 7 concludes the paper, discusses limitations and points to elements for future work.

2 Background and Related Work

2.1 Background

This section introduces activity suffix prediction and decoding strategies.

Activity Suffix Prediction. In event logs, a case is a business process instance, which consists of a series of events. An event can be represented as a tuple $e = (a, t, c, d_1, \ldots, d_m)$, where $a \in A$ is the activity label drawn from the set of all distinct activity labels A, t is the timestamp, c is the case ID, and $d_1, \ldots, d_m$ are optional case or event attributes. A trace is a non-empty sequence consisting of temporally ordered events with the same case ID: $\sigma = \langle e_1, \ldots, e_{|\sigma|} \rangle$. Based on these concepts, we define activity suffix prediction as follows:

Definition 1 (Activity Suffix Prediction). Let e_i be the event at position i in trace σ and a_i be the activity label of e_i. Given a trace σ and a positive integer $k < |\sigma|$, we define $\text{Prefix}(\sigma, k) = \langle e_1, \ldots, e_k \rangle$ and $\text{ActSuffix}(\sigma, k) = \langle a_{k+1}, \ldots, a_{|\sigma|} \rangle$. Activity suffix prediction involves predicting $\text{ActSuffix}(\sigma, k)$ given $\text{Prefix}(\sigma, k)$. For brevity, activity suffix prediction is abbreviated as *suffix prediction* in the remainder of the paper.

To assess the prediction quality, the normalized Damerau-Levenshtein (DL) distance is the most widely used evaluation metric in the literature (e.g., [4,14,18,20,23]). It quantifies the sequence-level prediction error as the minimum number of insertions, deletions, substitutions, and transpositions required to transform the predicted suffix into the ground-truth suffix, normalized by the length of the longer sequence. DL distance ranges from 0 to 1, and smaller distances indicate more accurate predictions.

Decoding Strategies. *Decoding strategy* is the method used at inference time to select the final prediction from a probability distribution over possible suffixes. Suffix prediction usually involves two phases: training and inference. During training, a predictive model is learned from historical traces. At inference time, the trained model generates suffix predictions for ongoing cases, typically in an autoregressive manner: at each generation step t, the model conditions on the observed prefix and the activities generated so far to output a probability distribution over the activity label set A (e.g., via a softmax function in neural networks), denoted by $p_\theta(\cdot \mid x,\ a_{k+1}, \ldots, a_{t-1})$, where $x = \langle e_1, \ldots, e_k \rangle$ is the observed prefix and $\langle a_{k+1}, \ldots, a_{t-1} \rangle$ denotes the previously generated activities[1]. By chaining these step-wise probabilities, the model ultimately outputs

[1] When $t = k + 1$, the sequence of previously generated activities is empty, hence the distribution is $p_\theta(\cdot \mid x)$.

a probability distribution over the set of possible suffixes, where the model-assigned likelihood of a candidate suffix $s = \langle a_{k+1}, \dots, a_{|\sigma|} \rangle$ can be written as the product of step-wise probabilities:

$$p_\theta(s \mid \mathrm{x}) = \prod_{t=k+1}^{|\sigma|} p_\theta(a_t \mid x,\, a_{k+1}, \dots, a_{t-1})\,.$$

During decoding, the model distribution $p_\theta(\cdot \mid x)$ is converted into a concrete final prediction to allow practical use. Decoding can therefore be viewed as a decision-making process that determines which candidate suffix should be returned as the final prediction, with the decoding strategy defining the corresponding decision rule.

A common approach is to return the mode, i.e., the most likely suffix under the model. Since exhaustive search over all possible suffixes is often infeasible, this objective is commonly approximated using argmax decoding or beam search. *Argmax decoding* selects the most probable activity at each generation step, while *beam search* keeps multiple promising partial sequences and returns the completed suffix with the highest $p_\theta(s \mid x)$. A less frequently used alternative in PPM is *random sampling* from $p_\theta(\cdot \mid x)$, implemented by randomly drawing an activity from the output distribution at each step. *Top-k* and *top-p sampling* further restricts the choice to a subset of activities (the top k activities with the highest probability, or the smallest set of activities whose cumulative probability exceeds a threshold p).

2.2 Related Work

Most suffix prediction studies aim to improve predictive accuracy through training-time optimization, broadly along two directions. One direction focuses on architecture design for parameter learning: architectures that predict suffixes autoregressively have evolved from LSTM-based next-activity predictors [9,20] to encoder-decoder models built on Generative Adversarial Networks (GANs) [21], Transformers [23] and recurrent Graph Neural Networks (GNNs) [17]. Some works predict the full suffix directly rather than autoregressively [10,15]. [16] applies reinforcement learning with DL distance as the reward. Non-deep-learning alternatives such as subtrace-based methods [18] have also been proposed. The second line of work expands the information used for prediction beyond control-flow and temporal features. [4] incorporates resource roles as both features and targets, [10,23] study fully data-aware models using all available case and event attributes, and inter-case features have also been explored [11].

These studies usually treat the decoding strategy as an implementation detail and select a standard strategy with limited justification. Argmax decoding is the most common choice (e.g., [16,20,23]), while [17,21] adopt beam search. [4] compares argmax decoding with random sampling and finds that the latter generally performs better.

In contrast to extensive training-time advances, inference-time optimization remains relatively underexplored, with only a few works targeting it. [7] incorporates a-priori knowledge by applying mined LTL rules at inference time. [1] proposes a Daemon Action approach for decoding, which aims to discourage unrealistic activity repetitions using historical activity frequencies. Our method differs from this work in that it optimizes decoding with respect to the evaluation metric rather than using a heuristic that penalizes repeated activities.

Beyond predictive accuracy, [14] studies uncertainty by training an uncertainty-aware LSTM and sampling multiple suffixes as final predictions. Unlike [14], which outputs an unranked set of suffixes without prioritization by prediction quality, our approach combines sampling with ranking suffixes by expected prediction error, enabling an explicit recommendation of the final prediction.

Some works extend suffix prediction beyond activities, for example to timestamps [23] and resources [4]. As a scope clarification, this paper focuses on activity suffix prediction because it remains the most widely studied suffix prediction task in PPM. However, our method can be directly applied to other discrete sequence prediction tasks such as resource suffix prediction, as well as to the activity prediction head of multi-task models that also predict timestamps or remaining time.

3 The Likelihood-Quality Mismatch in Decoding

Based on the literature review, the dominant suffix decoding approach is to approximate the mode-seeking objective of selecting the most likely suffix. But users ultimately want predictions with desirable quality (e.g., close to the ground truth), rather than high likelihood per se. This approach relies on the assumption that likelihood correlates with quality: among the suffixes a model can possibly generate, those with higher likelihood should be of higher quality than those with lower likelihood. However, this assumption may not hold in real-world processes.

To illustrate this issue, we conduct an exemplary experiment on three datasets including BPIC12[2], NASA[3] and Sepsis[4]. We first train LSTM models as described in Sect. 5.2. Conditioned on each test prefix, we randomly sample 50 candidate suffixes from the model, and compute two scores per candidate: its likelihood $p_\theta(s \mid x)$, and its quality, measured by its DL distance to the ground-truth suffix. We then assess Spearman's rank correlation between the two scores over the 50 candidates. Spearman's coefficient ranges from -1 (perfect negative correlation) to 1 (perfect positive correlation), with 0 indicating no association [19]. As shown in Fig. 1, the coefficients vary across instances and are generally weak; the strong negative correlation assumed by selecting the most likely suffix occurs in fewer than 30% of instances.

---

[2] https://doi.org/10.4121/uuid:3926db30-f712-4394-aebc-75976070e91f.
[3] https://doi.org/10.4121/uuid:60383406-ffcd-441f-aa5e-4ec763426b76.
[4] https://doi.org/10.4121/uuid:915d2bfb-7e84-49ad-a286-dc35f063a460.

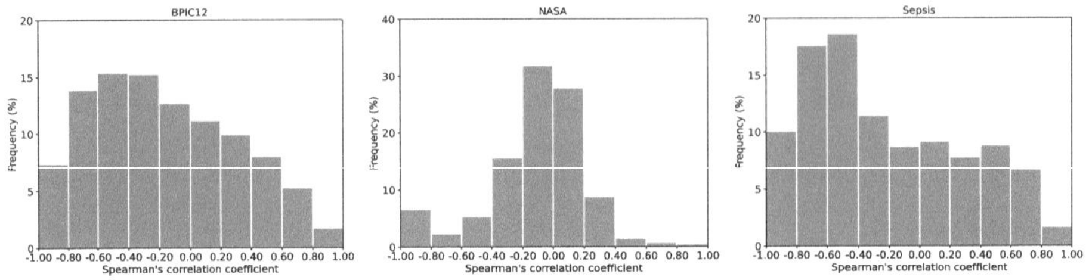

Fig. 1. Distribution of Spearman's correlation coefficients between candidate likelihood and DL distance on three datasets.

To further illustrate the likelihood-quality mismatch, Fig. 2 presents two concrete examples from BPIC12. Each subfigure corresponds to one input prefix, and each point represents a candidate suffix generated from that prefix. The x-axis shows the candidate's likelihood $p_\theta(s \mid x)$, and the y-axis shows its DL distance to the ground-truth suffix. The ideal candidate lies at the lowest position on the y-axis (smallest DL distance), but existing decoding strategies select candidates along the x-axis without considering their corresponding positions on the y-axis. Specifically, beam search (BS) selects the rightmost candidate with the highest likelihood, which may still have high prediction error. Argmax decoding targets the same objective, but its local decisions may produce a low-quality sequence with lower overall likelihood. The limitation of sampling-based methods such as top-p sampling is also evident in Fig. 2: they choose candidates stochastically according to the likelihoods, so the final output may lie anywhere along the x-axis with no guarantee of quality. These observations suggest that likelihood alone cannot reliably guide decoding toward high-quality predictions and motivate explicitly assessing the expected quality of candidate suffixes during decoding.

4 Methodology

In this section, we begin by introducing Bayesian decision theory that serves as the theoretical foundation for our method. We then describe the method in detail.

4.1 Bayesian Decision Theory

Decision theory studies how to rationally choose among alternative *actions* when the resulting consequence (or *cost*) depends on both the chosen action and the *state of nature* whose true status is unknown. Among the various decision rules applicable to different settings, Bayesian decision theory [3] provides a framework for problems in which the goal is to minimize a cost measure and a probability estimate over possible states of nature is available. It determines the optimal decision by considering, for each action, how costly it would be under different

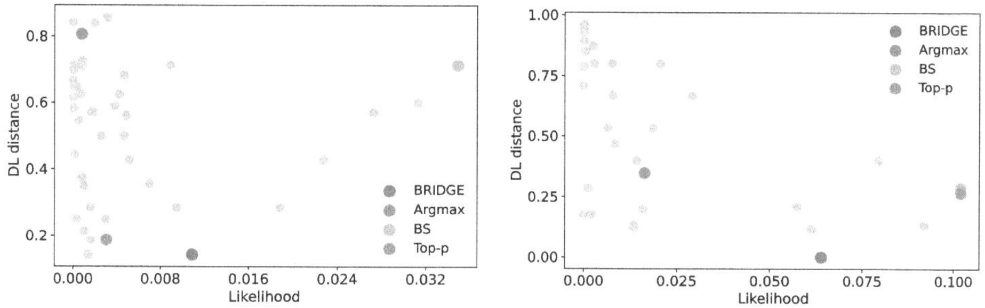

Fig. 2. Illustrative examples of candidate suffixes plotted by model likelihood and DL distance to the ground truth. Each decoding strategy's selected suffix is annotated. Spearman's coefficient is -0.27 for the left and -0.57 for the right subfigure.

possible states of nature and how likely each state will materialize given the available evidence. For example, in medical diagnosis, the clinician needs to choose a treatment with the least adverse consequences, while the patient's disease status is unknown. A Bayes-optimal decision prescribes treatment by considering the disease probability estimated from symptoms and patient history, and the costs of missed treatment and unnecessary medication.

Formally, let $u \in \mathcal{U}$ denote a candidate action, and $z \in \mathcal{Z}$ the unknown state of nature. Given evidence x, uncertainty about z is quantified by the posterior probability $p(z \mid x)$. To compare actions, a cost function $\ell(u, z)$ is defined, which measures the cost of taking action u when the true state of nature is z. The *Bayes risk* of an action u is the expected cost under $p(\cdot \mid x)$, obtained by averaging the cost $\ell(u, z)$ over all $z \in \mathcal{Z}$ weighted by how likely each z is given x (i.e., the posterior probability). The Bayes risk can be written as:

$$\mathcal{R}(u \mid \mathrm{x}) = \mathbb{E}_{z \sim p(\cdot \mid x)}\big[\ell(u, z)\big] = \sum_{z \in \mathcal{Z}} \ell(u, z)\, p(z \mid x).$$

The Bayes-optimal decision rule selects the action that minimizes the Bayes risk:

$$u^{\star}(x) = \arg\min_{u \in \mathcal{U}} \mathcal{R}(u \mid x).$$

Bayesian decision theory has been applied post-training in various contexts to align final outputs with task-specific objectives, as an alternative to simply selecting the highest-probability output. Examples include minimizing monetary losses in fraud detection [2] and generating text with desirable quality [8,12]. Although there is theoretical potential to apply this principle to optimize suffix decoding by formulating decoding as a decision-making problem (Sect. 2.1), notable challenges remain in establishing an effective PPM technique based on it, given the unique characteristics of process sequences such as variable length, repeated activities and loops. Our work addresses this gap by formulating a Bayes-risk-informed decoding framework for suffix prediction and demonstrating its effectiveness through extensive experiments.

4.2 BRIDGE: A Bayes-Risk-Informed Decoding Framework

We propose BRIDGE, a decoding framework that aligns inference with the task goal by using a user-specified cost function as the decoding criterion and selecting the suffix prediction with the lowest expected cost under the model's posterior distribution. BRIDGE is flexible in the choice of cost function and can therefore be adapted to different user-defined objectives. In this work, we instantiate the framework with the objective of minimizing prediction error, which is usually the central goal of suffix prediction studies.

The overall idea is to extract, from the model's output distribution, the suffix that minimizes the expected prediction error (Bayes risk). Specifically, we treat the ground-truth suffix as the unknown state of nature $z \in \mathcal{Z}$, and each candidate prediction produced by the model as an action $u \in \mathcal{U}$. We define the cost function $\ell(u, z)$ using the DL distance, so that the cost corresponds to the prediction error resulting from choosing u when the true suffix is z. We then compute the Bayes risk $\mathcal{R}(u \mid x)$ for each candidate by averaging its DL distances to all possible suffixes $z \in \mathcal{Z}$, weighted by the posterior probability $p(z \mid x)$, and select the one with the lowest risk as the final prediction.

In principle, determining the Bayes-optimal decision $u^\star(x)$ requires evaluating the Bayes risk $\mathcal{R}(u \mid x)$ of every candidate suffix $u \in \mathcal{U}$, where $\mathcal{R}(u \mid x)$ is defined as an expectation over all possible true suffixes $z \in \mathcal{Z}$. In practice, this is intractable for most suffix prediction models because both $\mathcal{U}$ and $\mathcal{Z}$ are very large. Therefore, BRIDGE approximates the Bayes risk using a subset of $\mathcal{Z}$, referred to as the *reference set*, and selects the final prediction from a subset of $\mathcal{U}$, referred to as the *candidate set*. In the current implementation, BRIDGE uses $\mathcal{H}_{\text{uniq}}$ as both the reference set and the candidate set, which contains unique randomly sampled suffixes. Algorithm 1 presents the overall procedure of BRIDGE, and the methodological choices are discussed below.

Constructing the Reference Set. We construct the reference set by drawing Monte Carlo (MC) samples from the model's output distribution conditioned on the prefix x. The motivation is that MC sampling provides a practical approximation of the Bayes risk over the full space, and this approximation improves as N increases [8]. Specifically, we draw N suffix samples from $p_\theta(\cdot \mid x)$ and collect them in $\mathcal{H}$, allowing duplicates. The set of unique samples, denoted by $\mathcal{H}_{\text{uniq}}$, is then used as the reference set for approximating the Bayes risk. Meanwhile, we compute an estimate of $p(h \mid x)$ for each $h \in \mathcal{H}_{\text{uniq}}$, denoted by $\hat{p}(h \mid x)$. Under MC sampling, a straightforward estimator is the empirical frequency, where $\hat{p}(h \mid x)$ is defined as the relative frequency of h among the N samples, i.e., the number of times h occurs in $\mathcal{H}$ divided by N [8]. Alternatively, a model-based estimator can be used, which defines $\hat{p}(h \mid x)$ as the model-assigned likelihood $p_\theta(h \mid x)$ (renormalized over $\mathcal{H}_{\text{uniq}}$) [12]. While the model-based estimator can differentiate suffixes even when they occur only once in $\mathcal{H}$, it can be affected by length bias, as shorter sequences tend to have higher likelihood

Algorithm 1. Decoding Procedure of BRIDGE

Require: Prefix x; model $p_\theta(\cdot \mid x)$; DL distance function $\ell(\cdot, \cdot)$; sample size N; estimator type MC or MODEL
Ensure: Predicted suffix $h^\star(x)$

```
Draw N i.i.d. suffix samples {h^(i)}_{i=1}^N ~ p_θ(· | x)
Let H ← {h^(i)}_{i=1}^N                                   ▷ multiset; duplicates allowed
Let H_uniq ← uniq(H)                                       ▷ set of unique samples
if estimator = MC then
    for all h ∈ H_uniq do
        c(h) ← Σ_{i=1}^N 𝕀[h^(i) = h]
        p̂(h | x) ← c(h)/N                                  ▷ empirical proportion of the suffix
else                                                       ▷ estimator = MODEL
    for all h ∈ H_uniq do
        w(h) ← p_θ(h | x)
    for all h ∈ H_uniq do
        p̂(h | x) ← w(h) / Σ_{h'∈H_uniq} w(h')               ▷ renormalize over H_uniq
for all h ∈ H_uniq do
    R(h | x) ← Σ_{h'∈H_uniq} ℓ(h, h') p̂(h' | x)
return h*(x) ← arg min_{h∈H_uniq} R(h | x)
```

scores. Since the impact of this bias varies across datasets, we treat the estimator as a hyperparameter to tune for each dataset.

Constructing the Candidate Set. Compared with the reference set, the construction of the candidate set is more flexible, since the purpose is to include high-quality candidates instead of approximating the full space. In the initial experiments, we considered several sampling strategies, including random, top-k and top-p sampling. Since top-k and top-p sampling does not result in consistent performance gains, we adopt random sampling and reuse $\mathcal{H}_{\text{uniq}}$ as the candidate set. We make this choice because constructing a separate candidate set would require an additional forward pass and increase computational cost, while not necessarily producing better candidates, as the candidates would be generated by the same random sampling procedure.

Under this design, the sample size N is a key hyperparameter that should be sufficiently large to ensure both adequate candidate coverage and precise ranking. In addition, the capacity of the predictive model itself, a factor external to BRIDGE, may also influence BRIDGE's effectiveness, since stronger models may generate better candidates and produce more reliable likelihoods for ranking. We empirically assess the influence of these factors in Sect. 6.

Ranking Candidates by Bayes Risk. For a candidate h, its Bayes risk is computed as the likelihood-weighted average DL distance to all $h' \in \mathcal{H}_{\text{uniq}}$:

$$\mathcal{R}(h \mid \mathrm{x}) = \sum_{h' \in \mathcal{H}_{\text{uniq}}} \ell(h, h')\, \hat{p}(h' \mid x).$$

The final prediction is the candidate with the minimal Bayes risk:

$$h^{\star}(x) = \arg\min_{h \in \mathcal{H}_{\text{uniq}}} \mathcal{R}(h \mid \mathrm{x}) = \arg\min_{h \in \mathcal{H}_{\text{uniq}}} \sum_{h' \in \mathcal{H}_{\text{uniq}}} \ell(h, h')\, \hat{p}(h' \mid x).$$

Intuitively, BRIDGE penalizes candidates that deviate from other probable suffixes, with larger penalties when those dissimilar suffixes receive higher probability mass. When the posterior mass is concentrated on a single suffix, BRIDGE will often agree with beam search and select that dominant candidate. However, when probability is distributed across many similarly likely suffixes, as is common in a large output space, BRIDGE avoids simply selecting a marginally most likely but isolated sequence and instead prefers the candidate that is, on average, closest to the broader set of likely suffixes. This helps avoid high-likelihood suffixes that are poorly supported by the surrounding posterior mass and are therefore more likely to be error-prone.

5 Experimental Setup

This section describes the datasets, preprocessing, and benchmark methods used for empirical evaluation. To ensure reproducibility, the code for BRIDGE and our reimplementations of all baseline methods are publicly available at https://github.com/Xiaomeng-He/BRIDGE, with an archived version available on Zenodo at https://doi.org/10.5281/zenodo.20449533.

5.1 Datasets and Preprocessing

We conduct experiments on eight public datasets commonly used for suffix prediction. To broadly assess BRIDGE's effectiveness, we include event logs with challenging long suffixes such as BPIC17[5], NASA, BPIC12, and Sepsis, and logs with shorter traces such as BPIC13 Incidents (BPIC13)[6], BPIC19[7] and BPIC20 Permit Log (BPIC20)[8]. Following [1,16,18], we use both BPIC12 and BPIC12-W (a sub-process tracking work item states): BPIC12 covers the full process, while BPIC12-W mainly contains manually executed events [20]. Table 1 summarizes the key characteristics of the event logs[9].

[5] https://doi.org/10.4121/uuid:5f3067df-f10b-45da-b98b-86ae4c7a310b.

[6] https://doi.org/10.4121/uuid:500573e6-accc-4b0c-9576-aa5468b10cee.

[7] https://doi.org/10.4121/uuid:d06aff4b-79f0-45e6-8ec8-e19730c248f1.

[8] https://doi.org/10.4121/uuid:ea03d361-a7cd-4f5e-83d8-5fbdf0362550.

[9] For BPIC12, BPIC12-W, BPIC17, and BPIC13, we define the activity label as the concatenation of `concept:name` and `lifecycle:transition`, as in [18,22].

Table 1. Summary statistics of event logs.

Event log	Num.cases	Num.events	Num.activities	Max./Median/Min. case length	Avg.case length	Avg.case duration
NASA	2,566	73,638	47	50/28/12	28.70	0.0014 s
Sepsis	1,050	15,214	16	185/13/3	14.49	28.47 days
BPIC12	13,087	262,200	24	175/11/3	20.04	8.62 days
BPIC12-W	9,658	170,107	19	156/14/2	17.61	11.68 days
BPIC13	7,554	65,533	13	123/6/1	8.68	12.08 days
BPIC17	31,509	1,202,267	66	180/35/10	38.16	21.90 days
BPIC19	251,734	1,595,923	42	990/5/1	6.34	71.52 days
BPIC20	7,065	86,581	51	90/11/3	12.25	87.39 days

For preprocessing, we remove potentially incomplete cases that start near the end of the log, i.e., within a trailing window whose length equals the 75th-percentile case duration, and retain all remaining cases to preserve true process behavior. For BPIC19 only, we additionally remove extreme outliers above the 99.5th percentile of case length. For each trace σ, we consider all prefixes with lengths $0 < k < |\sigma|$, and treat each resulting prefix-suffix pair as an instance. We add an EOC token to every suffix; the model stops generating activities when EOC is produced or when it reaches the maximum length, set as the maximum trace length in the preprocessed event log.

We order cases by start time and split them temporally into training/test sets (80%/20%). By default, we use strict temporal splitting [22], excluding from training any cases that are still running when the first test case begins. For BPIC13, BPIC19, and BPIC20, this yields only short training traces; therefore, we retain these overlapping cases in the training set. We further split the training set into train/validation (80%/20%) for model learning and hyperparameter tuning.

5.2 Base Model Architectures

BRIDGE is a general decoding framework that is compatible with a broad range of predictive models which output activity label probabilities. To test whether BRIDGE benefits different model families, we implement it with three base architectures: a simple statistical Markov model and two complex, high-capacity deep learning models (LSTM and Transformer). Implementation details are described below.

Markov Model. Markov models are a class of probabilistic models that assume the next state depends only on a limited history of previous states. In a jth-order Markov model, the next activity is predicted from the most recent j activities in the prefix. Let $context = \langle a_{k-j+1}, \ldots, a_k \rangle$ and $\#(context, a)$ be the number of training instances in which $context$ is immediately followed by activity a. From the training set we estimate the transition probability:

$$P(a \mid context) = \frac{\#(context, a)}{\sum_{a'} \#(context, a')}$$

At prediction time, given a *context*, $P(\cdot \mid context)$ defines the predicted distribution for the next activity. Suffixes are generated recursively by predicting the next activity, appending it, updating the context to the most recent j activities, and repeating until termination. We fit first- and second-order models and use a back-off procedure [13]: we apply the second-order model when the two-activity context is observed in training data; otherwise we drop the oldest activity and use the first-order model; if this context is still unseen, we fall back to a uniform distribution.

LSTM. We implement a data-aware encoder-decoder LSTM based on the ED-LSTM architecture from [23] while retaining only the activity suffix prediction head. The model uses all available features including: (i) activity labels, (ii) two temporal features (time since the previous event and time since case start), and (iii) additional case/event attributes. Numerical features are standardized using training set statistics, while categorical features are passed through an embedding layer; the resulting representations are concatenated and fed to the encoder LSTM. The decoder LSTM generates the suffix autoregressively, with each output conditioned on the prefix and prior outputs.

We train the model with AdamW (batch size 64) using an initial learning rate of 0.002 for up to 200 epochs. The learning rate is reduced if validation performance does not improve for 15 epochs, and training stops after 20 unimproved epochs. We train the model using cross-entropy loss and adopt teacher forcing to accelerate training. For each dataset, we grid search the hidden size ($\{32, 64, 128\}$) and number of layers ($\{1, 2, 4\}$) on the validation set.

Transformer. We implement a data-aware encoder-decoder Transformer based on SuTraN [23] while retaining only the activity suffix prediction head. We use the same input features, training setup, and autoregressive decoding as for the LSTM model. For each dataset, we fix the model dimension to 32 and grid search the feed-forward layer size ($\{64, 128\}$), number of attention heads ($\{4, 8\}$), and number of Transformer layers ($\{2, 4\}$) on the validation set.

5.3 BRIDGE Configuration and Decoding Strategy Benchmarking

As discussed in Sect. 4, there are two important configuration choices for BRIDGE: the estimator and the sample size. We treat the estimator (MC vs. model-based) as a hyperparameter and tune it on the validation set. In our experiments, the MC estimator is selected for most datasets because it yields better performance, while the model-based estimator is selected only for BPIC13. As for the sample size, we use a fixed sample size of 50 when comparing BRIDGE with other decoding strategies or suffix prediction models. We further study the performance and runtime resulting from different sample sizes ($\{10, 30, 50, 100\}$) in an ablation study; for runtime reporting, we evaluate all test instances in batches (size 64) on an NVIDIA Tesla V100 GPU and report average per-instance time.

We benchmark BRIDGE against four representative decoding strategies, tuning each method's key hyperparameters to ensure a fair comparison: (i) argmax decoding, (ii) beam search (BS) with beam width $\in \{5, 10\}$ and length normalization [17], (iii) top-p sampling with $p \in \{0.7, 0.8, 0.9\}$ [6], and (iv) the Daemon Action (D-Act) approach [1]. In our experiments, we first fit a predictive model; the same model is then used for inference on the test set with each decoding strategy. For each base model, we evaluate each decoding strategy across 15 runs to account for stochasticity[10].

5.4 Reimplemented Suffix Prediction Models

To better position BRIDGE within the literature, we also benchmark BRIDGE (LSTM-based) against suffix prediction models from prior works [4,10,16,18, 20,23]. The reimplemented models cover a diverse and competitive set of approaches, including deep-learning and non-deep-learning models, autoregressive and non-autoregressive approaches, and several state-of-the-art methods with strong reported performance (e.g., [10,16,18,23]). We especially include the prior work of [16] that uses reinforcement learning (RL) to learn an optimal decoding strategy, which is related to our study. Table 2 provides an overview of each model's key characteristics.

For each model, we keep the original architecture (including any multi-task settings) and use the hyperparameter configuration that achieves the best overall results as reported in the original paper. For comparability, we follow BRIDGE's data preprocessing and split procedure when reimplementing other methods; accordingly, all models are trained, tuned, and tested on the same training, validation, and test sets. For each model, we perform 5 training runs with different random seeds.

Table 2. Overview of the reimplemented suffix prediction models. In the Input Features column, abbreviations denote: A–activity label, T–temporal features, R–resource features, C–all categorical features and All–all available features.

	Input Features	Deep Learning	Architecture	Autoregressive	Decoding Strategy
Rauch et al. [18]	A	×	Subtrace Tree	\	\
Tax et al. [20]	A,T	✓	LSTM	✓	Argmax
Camargo et al. [4]	A,T,R	✓	LSTM	✓	Random*
Gunnarsson et al. [10]	All	✓	LSTM	×	Argmax
Wuyts et al. [23]	All	✓	Transformer	✓	Argmax
Rama-Maneiro et al. [16]	A,T,C	✓	RL	✓	Argmax

* [4] experiments with both argmax decoding and random sampling. We reimplement the random sampling version since it's reported by [4] to have better performance.

[10] For the LSTM and Transformer, we train five models with different random seeds, and for each trained model, we run inference three times with different seeds (15 runs total). For the Markov model, training is deterministic, so we run inference with 15 random seeds.

6 Results and Analysis

In this section, we first report benchmark results comparing BRIDGE with baseline decoding strategies and suffix prediction models. We then analyze BRIDGE's sensitivity to sample size in terms of predictive performance and computational efficiency. Finally, we present a qualitative study on BPIC12 analyzing how BRIDGE improves performance by addressing typical error patterns.

6.1 Benchmark Results

We use DL distance to measure predictive performance and report average distances across multiple runs (with standard deviations provided in the repository). To assess whether the performance difference between BRIDGE and the best-performing competitor is statistically significant, we apply a Wilcoxon signed-rank test, with a p-value below 0.05 indicating significance; italics indicate that the difference is not significant. The best results are in bold and underlined, while the second-best results are underlined.

Comparison of Decoding Strategies. Table 3 reports that BRIDGE is overall the best-performing decoding strategy among the evaluated approaches. In particular, two advantages are noteworthy.

Table 3. DL distances of decoding strategies with different base models.

NASA	Argmax	BS	Top-p	D-Act	BRIDGE
Markov	0.5249	0.5144	0.5765	0.5233	**0.4910**
LSTM	0.5152	0.5251	0.5139	0.5108	**0.4746**
Transf.	0.5385	0.5253	0.5385	0.5339	**0.4910**

Sepsis	Argmax	BS	Top-p	D-Act	BRIDGE
Markov	0.8866	0.5840	0.5889	0.5541	**0.5200**
LSTM	0.4529	0.4367	0.4516	0.4398	**0.4265**
Transf.	0.4989	0.4578	0.4668	0.4477	**0.4307**

BPIC12	Argmax	BS	Top-p	D-Act	BRIDGE
Markov	0.8093	0.7702	0.6208	0.6022	**0.5354**
LSTM	0.5938	0.5780	0.5663	0.5671	**0.5259**
Transf.	0.7045	0.5807	0.5736	0.5960	**0.5149**

BPIC12-W	Argmax	BS	Top-p	D-Act	BRIDGE
Markov	0.8284	0.6148	0.6395	0.5544	**0.5327**
LSTM	0.6014	0.6053	0.5951	0.5891	**0.5676**
Transf.	0.6698	0.5887	0.5987	0.5957	**0.5449**

BPIC13	Argmax	BS	Top-p	D-Act	BRIDGE
Markov	0.6296	***0.2834***	0.3643	0.3758	***0.2834***
LSTM	0.2698	0.2358	0.2854	0.2502	**0.2307**
Transf.	0.2870	0.2620	0.2895	0.2567	**0.2526**

BPIC17	Argmax	BS	Top-p	D-Act	BRIDGE
Markov	0.8898	0.6359	0.6036	0.7009	**0.5379**
LSTM	0.5145	0.4496	0.4698	0.4560	**0.4409**
Transf.	0.6284	0.4421	0.4838	0.4568	**0.4286**

BPIC19	Argmax	BS	Top-p	D-Act	BRIDGE
Markov	0.3505	0.3421	0.3575	0.3414	**0.3387**
LSTM	0.2988	0.2978	0.3057	0.2985	**0.2964**
Transf.	0.3058	0.3001	0.3082	0.2983	**0.2963**

BPIC20	Argmax	BS	Top-p	D-Act	BRIDGE
Markov	**0.3052**	0.3053	0.3552	**0.3052**	0.3065
LSTM	0.2064	0.2083	0.2233	**0.2060**	0.2101
Transf.	0.1967	0.2034	0.2140	**0.1959**	0.2020

First, BRIDGE is consistently among the top-performing methods across datasets and base models; such consistency and robustness is not observed for any other decoding strategies. BRIDGE achieves the best results on seven of the eight event logs, while on the remaining log its performance is comparable to the strongest baseline (DL distance gap < 0.01). The strong performance also holds across base models, suggesting its broad applicability. In contrast, all other decoding strategies can perform much worse than the best-performing method on certain datasets. For example, argmax decoding on BPIC17 with Transformer ($\uparrow$ 0.20); beam search on BPIC12 with Markov model ($\uparrow$ 0.23) and Transformer ($\uparrow$ 0.06); top-p sampling on BPIC12-W with Markov model ($\uparrow$ 0.10); and D-Act on BPIC17 with Markov model ($\uparrow$ 0.16) and on BPIC12 with Transformer ($\uparrow$ 0.08).

Second, the performance gains from BRIDGE are substantial rather than marginal, especially for certain datasets and base models. The improvements are most evident for event logs with long, challenging-to-predict suffixes, such as BPIC12, BPIC12-W, and NASA. For logs with shorter suffixes including BPIC13, BPIC19 and BPIC20, most decoding strategies perform similarly well. Moreover, BRIDGE appears to benefit simpler models even more: the largest absolute reductions in DL distance are observed for the Markov models on BPIC12 ($\downarrow$ 0.07) and BPIC17 ($\downarrow$ 0.07). This highlights BRIDGE's practical value as an inference-time enhancement that can yield sizable improvements even for simple models when retraining is not feasible.

Further investigation shows that the performance gains are not solely due to the use of more random samples; Bayes-risk-informed ranking plays a critical role. Specifically, when we sample 50 suffixes as in BRIDGE but rank them by likelihood instead, the resulting performance is worse than that of BRIDGE. Due to space limitations, detailed results of this supporting analysis are provided in our repository.

Another notable finding is that while the Markov model with limited input information (only the previous two activities) significantly underperforms data-aware deep learning models under naïve argmax decoding, BRIDGE largely narrows this gap. On certain datasets such as NASA, BPIC12, and BPIC13, simple models can still achieve competitive performance when paired with an optimal decoding strategy, whereas on other datasets, richer inputs and more complex architectures provide clearer benefits.

Benchmark of Suffix Prediction Models. As shown in Table 4, BRIDGE applied to a data-aware LSTM achieves state-of-the-art results and outperforms all competing suffix prediction models. Notably, while [16] proposes learning an optimal decoding strategy during training, we find that separately optimizing the decoding procedure still provides substantial gains and outperforms their approach.

6.2 Sensitivity Analysis

We analyze BRIDGE's sensitivity to sample size in terms of performance (Fig. 3) and runtime (Table 5) using LSTM as the base model. Larger sample sizes gen-

Table 4. DL distances of different suffix prediction models.

	Rauch et al. [18]	Tax et al. [20]	Camargo et al. [4]	Gunnarsson et al. [10]	Wuyts et al. [23]	Rama-Maneiro et al. [16]	BRIDGE (LSTM)
NASA	0.5964	0.5455	—*	0.5146	0.5647	0.8157	**0.4746**
Sepsis	0.5810	0.6339	0.5791	0.5200	0.6134	0.5357	**0.4265**
BPIC12	0.6209	0.7412	0.5723	0.6466	0.7276	0.6753	**0.5259**
BPIC12-W	0.6413	0.6711	0.6066	0.6857	0.7946	0.6589	**0.5676**
BPIC13	0.4971	0.6392	0.3903	0.3863	0.3697	0.3186	**0.2307**
BPIC17	0.6023	0.7545	0.6498	0.6096	0.6960	0.7299	**0.4409**
BPIC19	0.5089	0.3546	0.4214	0.3244	0.3057	0.3810	**0.2964**
BPIC20	0.4310	0.2607	0.4026	0.2553	0.2726	0.3713	**0.2101**

*This dataset does not include an informative resource attribute required to run this model.

erally reduce prediction error but increase runtime. In most datasets, the performance differences are small: NASA is the most sensitive dataset, with DL distance dropping from 0.493 ($N = 10$) to 0.469 ($N = 100$); elsewhere, differences are typically within 0.01. This indicates that BRIDGE can improve performance with feasible computational cost, since a sample size of 10 (runtime shown in orange) already provides near-best accuracy with runtime comparable to the fastest baselines. Depending on the accuracy-efficiency priority, the sample size can be flexibly adjusted, though our experiments show that beyond $N = 50$ (runtime shown in blue), gains diminish while runtime may become too costly.

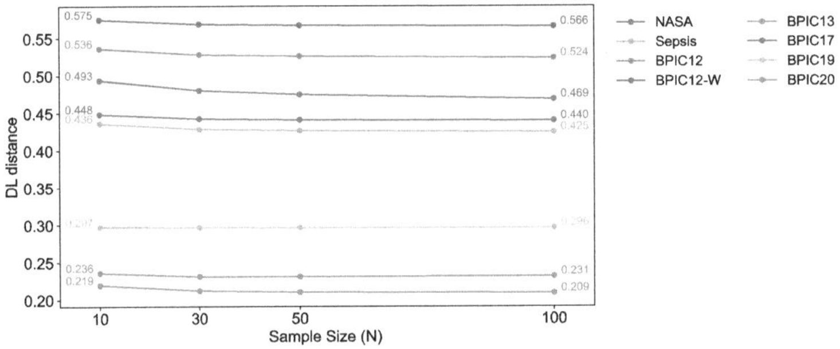

Fig. 3. Effect of BRIDGE sample size on DL distance.

6.3 Qualitative Analysis

We conduct a case study on BPIC12 using LSTM as the base model to analyze how BRIDGE improves performance. As Fig. 4 shows, BRIDGE increases the average accuracy by reducing extreme errors (DL distance > 0.8), partly due to more accurate suffix length control with fewer overly long or short suffixes. Limiting these extreme deviations may also strengthen perceived model reliability and user trust. In addition, by analyzing actual predictions from BPIC12,

Table 5. Runtime per test instance (in milliseconds).

	BRIDGE				Argmax	BS	Top-p	D-Act
	$N = 10$	$N = 30$	$N = 50$	$N = 100$				
NASA	1.49	7.19	17.92	64.59	0.51	0.88	0.76	7.58
Sepsis	1.66	6.07	14.40	51.43	1.74	3.14	2.77	28.38
BPIC12	3.50	12.59	29.40	101.79	2.21	3.36	3.12	25.77
BPIC12-W	3.29	7.41	13.66	37.57	3.05	4.12	3.88	25.24
BPIC13	1.56	3.94	8.21	25.91	1.15	2.07	1.88	18.68
BPIC17	1.17	14.71	34.47	123.92	3.53	4.80	4.75	29.96
BPIC19	0.93	1.49	2.27	5.13	0.67	1.02	0.95	7.83
BPIC20	1.13	2.85	5.35	15.47	0.86	1.48	1.35	14.00

we identify two patterns in Fig. 5 that may explain BRIDGE's improvements over choosing the most likely suffix. In Example 1, when the most likely candidate deviates from the structure shared by other high-likelihood candidates (it lacks $<$ J,H,K,C $>$), BRIDGE selects a candidate that is more consistent with the others and follows this shared structure, which is often less error-prone. In Example 2, in processes with frequent repetition, the model may misestimate the number of repeats; by selecting the candidate that is on average closest to the others, BRIDGE tends to produce a more appropriate repetition count. The two concrete examples also suggest BRIDGE's improvements are qualitatively meaningful rather than artifacts of metric over-optimization.

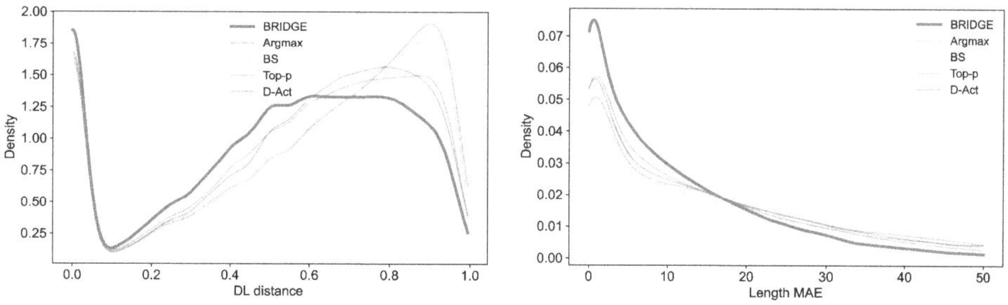

Fig. 4. Distribution of DL distances and Mean Absolute Error (MAE) between lengths of predicted and ground-truth suffixes.

Example 1

Candidates	Likelihood	Bayes Risk	True Error
Z Y H Z C	1 / 0.042	4 / 0.703	4 / 0.500
I Z J H K C	2 / 0.035	1 / 0.666	1 / 0
K J H K C	3 / 0.031	3 / 0.671	3 / 0.333
F G O C	4 / 0.011	5 / 0.807	5 / 0.833
Y Z J H K C	5 / 0.004	2 / 0.668	2 / 0.167
Ground Truth	I Z J H K C		

Example 2

Candidates	Likelihood	Bayes Risk	True Error
L H M C	1 / 0.076	4 / 0.712	4 / 0.330
L M L H M C	2 / 0.070	1 / 0.667	1 / 0
L M L M L H M C	3 / 0.040	2 / 0.671	3 / 0.250
L M L W M C	4 / 0.037	3 / 0.680	2 / 0.170
L W M C	5 / 0.034	5 / 0.753	5 / 0.500
Ground Truth	L M L H M C		

Fig. 5. Examples comparing ranking by likelihood and Bayes risk. For each of the five candidate predictions, we report its rank and score under three criteria: likelihood, Bayes risk score, and true prediction error. The highest-ranked candidate is in red. (Color figure online)

7 Conclusion

When suffix prediction models are deployed online for ongoing cases, the decoding strategy plays a critical role in determining output quality. Yet most existing strategies select suffixes using likelihood alone, without explicitly assessing quality under the evaluation metric. To address this limitation, we introduce BRIDGE, a Bayes-risk-informed decoding framework that aligns inference more closely with a specified evaluation metric. BRIDGE is inherently metric-agnostic and architecture-agnostic. In this work, we instantiate it with DL distance as the cost function and apply it to statistical model and deep learning architectures.

Extensive experiments demonstrate that BRIDGE is broadly applicable across model architectures and consistently ranks among the top-performing methods (best or close to best) on all eight datasets, making it a strong default choice for decoding. The performance gains are particularly notable on event logs with long traces and for relatively simple base models. A qualitative case study indicates that BRIDGE improves average accuracy by reducing extreme errors via more accurate suffix length control and by avoiding likely but atypical candidates that deviate from the posterior mass (e.g., missing shared patterns or under/over-repeating activities).

A main limitation of BRIDGE concerns its computational time at large sample sizes. In our implementation, we observe that the primary bottleneck is the computation of pairwise DL distances, which scales nearly quadratically with the sample size and becomes slower for longer sequences. A promising direction is to prune the candidate set (e.g., remove overly long or low-quality suffixes) or to speed up DL distance computation via vectorized or batched implementations.

In future work, we will further explore BRIDGE in a multi-task suffix prediction setting, for example by jointly predicting suffixes of activities, timestamps and resources.

Acknowledgments. This work was supported by the Research Foundation Flanders (FWO) under grant number G039923N, and Internal Funds KU Leuven under grant number C14/23/031. The authors gratefully acknowledge Euroclear's financial support through the Euroclear Research Chair at KU Leuven.

References

1. Ali, M.A., Dumas, M., Milani, F.: Enhancing the accuracy of predictors of activity sequences of business processes. In: RCIS, vol. 513, pp. 149–165. Springer, Cham (2024)
2. Bahnsen, A.C., Stojanovic, A., Aouada, D., Ottersten, B.: Cost sensitive credit card fraud detection using bayes minimum risk. In: ICMLA, vol. 1, pp. 333–338 (2013)
3. Berger, J.O.: Statistical decision theory and Bayesian analysis, Springer (1985)
4. Camargo, M., Dumas, M., González-Rojas, O.: Learning accurate LSTM models of business processes. In: BPM, vol. 11675, pp. 286–302. Springer, Cham (2019)
5. Ceravolo, P., Comuzzi, M., De Weerdt, J., Di Francescomarino, C., Maggi, F.M.: Predictive process monitoring: concepts, challenges, and future research directions. Process Sci. **1**(1), 2 (2024)
6. DeLucia, A., Mueller, A., Li, X.L., Sedoc, J.: Decoding methods for neural narrative generation. In: GEM, pp. 166–185 (2021)
7. Di Francescomarino, C., Ghidini, C., Maggi, F.M., Petrucci, G., Yeshchenko, A.: An eye into the future: leveraging a-priori knowledge in predictive business process monitoring. In: BPM, vol. 10445, pp. 252–268. Springer, Cham (2017)
8. Eikema, B., Aziz, W.: Is MAP decoding all you need? the inadequacy of the mode in neural machine translation. In: COLING, pp. 4506–4520 (2020)
9. Evermann, J., Rehse, J.R., Fettke, P.: Predicting process behaviour using deep learning. Decis. Support Syst. **100**, 129–140 (2017)
10. Gunnarsson, B.R., Broucke, S.V., De Weerdt, J.: A direct data aware LSTM neural network architecture for complete remaining trace and runtime prediction. IEEE Trans. Serv. Comput. **16**(4), 2330–2342 (2023)
11. Gunnarsson, B.R., vanden Broucke, S., De Weerdt, J.: LS-ICE: a load state inter-case encoding framework for improved predictive monitoring of business processes. Inf. Syst. **125**, 102432 (2024)
12. Jinnai, Y., Morimura, T., Honda, U., Ariu, K., Abe, K.: Model-based minimum Bayes risk decoding for text generation. In: ICML, pp. 22326–22347. PMLR (2024)
13. Katz, S.: Estimation of probabilities from sparse data for the language model component of a speech recognizer. IEEE Trans. Acoust. Speech Signal Process. **35**(3), 400–401 (1985)
14. Kunkler, M., Mustroph, H., Rinderle-Ma, S.: Probabilistic suffix prediction of business processes. In: ICPM, pp. 1–8. IEEE (2025)
15. Pasquadibisceglie, V., Appice, A., Malerba, D.: LUPIN: A LLM approach for activity suffix prediction in business process event logs. In: ICPM, pp. 1–8. IEEE (2024)
16. Rama-Maneiro, E., Patrizi, F., Vidal, J., Lama, M.: Towards learning the optimal sampling strategy for suffix prediction in predictive monitoring. In: CAiSE, vol. 14663, pp. 215–230. Springer, Cham (2024)
17. Rama-Maneiro, E., Vidal, J.C., Lama, M., Monteagudo-Lago, P.: Exploiting recurrent graph neural networks for suffix prediction in predictive monitoring. Computing **106**, 3085–3111 (2024)

18. Rauch, S., Frey, C.M.M., Maldonado, A., Seidl, T.: BEST: Bilaterally expanding subtrace tree for event sequence prediction. In: BPM 2025, vol. 16044, pp. 415–432. Springer, Cham (2026)
19. Schober, P., Boer, C., Schwarte, L.A.: Correlation coefficients: appropriate use and interpretation. Anesthesia Analgesia **126**(5) (2018)
20. Tax, N., Verenich, I., La Rosa, M., Dumas, M.: Predictive business process monitoring with LSTM neural networks. In: CAiSE, vol. 10253, pp. 477–492. Springer, Cham (2017)
21. Taymouri, F., Rosa, M.L., Erfani, S.M.: A deep adversarial model for suffix and remaining time prediction of event sequences. In: SDM, pp. 522–530 (2021)
22. Weytjens, H., De Weerdt, J.: Creating unbiased public benchmark datasets with data leakage prevention for predictive process monitoring. In: Business Process Management Workshops. BPM 2021, vol. 436, pp. 18–29. Springer, Cham (2021)
23. Wuyts, B., Vanden Broucke, S., De Weerdt, J.: SuTraN: an encoder-decoder transformer for full-context-aware suffix prediction of business processes. In: ICPM, pp. 17–24. IEEE (2024)

From Global Policies to Local Strategies: Multi-objective Optimization of Resource-Specific Handover Policies

Lukas Kirchdorfer[1,2(✉)], Artemis Doumeni[3], Han van der Aa[4], and Hugo A. López[3]

[1] SAP Signavio, Walldorf, Germany
lukas.kirchdorfer@sap.com
[2] Data and Web Science Group, University of Mannheim, Mannheim, Germany
[3] DTU Compute, Technical University of Denmark, Kongens Lyngby, Denmark
hulo@dtu.dk
[4] Faculty of Computer Science, University of Vienna, Vienna, Austria
han.van.der.aa@univie.ac.at

Abstract. Efficient resource allocation is a key challenge in business process management, with direct implications for cost, throughput time, and utilization. While recent Reinforcement Learning (RL) approaches have shown promise in deriving adaptive allocation policies, they typically neglect inter-resource collaboration patterns that can strongly influence real-world task handovers. Recognizing this, this paper introduces the first approach for multi-objective optimization of resource-level decision-making, enabling the recommendation of person-specific handover policies. To achieve this, our work combines an existing Multi-Agent System-based process simulator with a multi-objective evolutionary algorithm. The resulting approach produces Pareto-optimal, resource-specific policies that optimize the process across multiple objectives. Experimental results on synthetic and real-world datasets show that our approach reduces costs by an average of 37% and waiting time by 58%, consistently outperforming heuristic baselines and demonstrating the potential of leveraging collaboration-aware optimization to improve process performance.

Keywords: Process mining · Multi-objective optimization · Resource allocation · Simulation · Multi-agent systems

1 Introduction

Efficient resource allocation—the task of assigning resources, such as employees, machines, and systems, to process activities—is a central challenge in business process management, as it directly affects key performance indicators such as cost, throughput time, and resource utilization [24]. However, finding effective allocations is difficult due to the complexity and interdependence of concurrent process instances, the need for decisions under dynamic conditions, and the

F. M. Maggi et al. (Eds.): BPM 2026, LNCS 16936, pp. 425–443, 2027.
https://doi.org/10.1007/978-3-032-37777-7_26

long-term consequences of allocation choices [14]. Accordingly, a wide range of approaches has been proposed, ranging from rule- and scheduling-based methods [24] to more recent Reinforcement Learning (RL)-based approaches [2,14,23], which showed promising results by balancing short- and long-term objectives.

Despite this progress, existing RL-based approaches exhibit a key limitation: they learn *global* resource-allocation policies based only on (i) which tasks are awaiting execution and (ii) which resources are available to perform them. This allows them to exploit differences in skill, execution time, and cost, but it treats resources as passive recipients of assignments and ignores an important aspect of real organizational behavior: collaboration patterns among resources. In practice, however, resources—especially human resources—are active decision-makers during process execution, influencing not only which work they perform, but also with whom they collaborate. These inter-resource dependencies give rise to characteristic handover and collaboration patterns that shape how work flows through an organization. Concretely, we are interested in *handover policies* that specify, for a given resource and completed activity, to which resource the case should be passed next for the subsequent activity. While accounting for such *local*, resource-specific handovers has been shown to improve the fidelity of as-is process simulations [9], existing optimization approaches do not account for these local handovers, and therefore overlook collaboration structure as a leverage for process improvement.

This paper introduces the first approach for grounded multi-objective optimization of resource-specific handover policies in business processes. Our approach takes as input a Multi-Agent System (MAS)-based simulation model of the process that explicitly captures resource interdependencies, building on prior work [9]. On top of this model, we develop an analysis-and-design optimization framework that integrates the widely used multi-objective evolutionary algorithm NSGA-II [4] to search for Pareto-optimal handover policies. This approach offers two key advantages. First, it can be grounded in data, since the MAS simulation model taken as input can be discovered from event logs (cf. [9]). Second, it is generic with respect to any set of objectives, yielding a Pareto front of handover policies that offer practitioners actionable alternatives that reflect different operational trade-offs. To increase solution diversity, we incorporate four mutation variants into the genetic algorithm. Experiments on synthetic and real-world datasets, using cost and time as example objectives, demonstrate substantial improvements over the as-is process, with average reductions of 37% in cost and 58% in waiting time, also outperforming established heuristic baselines.

The remainder of this paper is structured as follows: Sect. 2 reviews related work, Sect. 3 defines the problem, Sect. 4 presents our approach, and Sect. 5 reports the evaluation. Finally, Sect. 6 concludes the paper.

2 Related Work

Our approach to optimizing resource-specific handover policies relates to three main research directions.

Resource Allocation. A large body of work has addressed resource allocation in business processes, mostly focusing on optimizing task–resource assignments during process execution, i.e., at runtime. Early approaches rely on rule-based assignment and scheduling heuristics [24], which depend on predefined rules or short-term objectives and therefore adapt poorly to dynamic process conditions. Prescriptive Process Monitoring (PrPM) approaches [18,22] extend these methods by recommending resources for ongoing cases, yet they still optimize decisions locally and often ignore their broader process-level impact.

To address longer-term and process-level effects, more advanced work integrates simulation and applies metaheuristic or learning-based optimization methods, such as genetic algorithms [6,10,17], heuristic search [13], or RL [8,14,15,23]. Yet, only a subset considers multi-objective optimization as addressed in this paper. For example, Huang et al. [8] use RL to optimize cost and cycle time.

While our approach can be employed at runtime to determine handover decisions (which we will do in our evaluation), in principle, it optimizes a static model of resource collaboration, independently of runtime case information. Thus, we consider it as design-time. Within this design-time line of work, several approaches optimize the number of required resources per resource pool [7,13,16], solving a staffing problem rather than optimizing handovers. Closest to our work, Bejarano et al. [1] also address a design-time, simulation-based, multi-objective optimization problem, but their decision variables remain activity-to-resource assignments, whereas ours are handover decisions between resources.

In summary, a crucial distinction sets our approach apart from all existing work: the underlying representation of resources. Existing methods treat resources as passive entities that are assigned to activities; optimization thus centers on determining which resource should execute which task. In contrast, we adopt a MAS perspective in which resources are first-class citizens, modeled as autonomous decision-making agents capable of choosing their collaboration partners. Consequently, instead of learning a single global allocation rule, our approach discovers *resource-specific handover policies* that capture and optimize the collaborative structure of the process. This enables a form of collaboration-aware optimization that has not been addressed in previous research.

Organizational and Agent System Mining. Complementary to optimization approaches, organizational and resource-oriented process mining methods analyze how resources collaborate [19]. More recently, agent system mining has been proposed to discover process control-flow models from an agent perspective [21]. However, these approaches remain descriptive: they reveal collaboration structures but do not treat them as decision variables for optimization.

Agent-Based Business Process Simulation. Traditional business process simulation typically follows a control-flow-first perspective, extending a process model (e.g., a Petri net) with temporal and resource-related parameters [11]. In contrast, agent-based approaches adopt a resource-first perspective in which resources are modeled as autonomous entities at the core of the simulation [9]. While such agent-based simulations can be used in a *what-if* manner to explore alternative collaboration patterns, this requires manually defining and evaluating different configurations. Our approach automates this process by systematically exploring handover policies and presenting the user with a set of Pareto-optimal collaboration patterns and their corresponding trade-offs.

3 Problem Definition

Event Logs. Event logs represent a key artifact of our approach, generated during the search for optimal resource policies. We define an event log L as a collection of traces. A trace $\sigma \in L$ is a finite sequence of events, $\langle e_1, ..., e_n \rangle$, recording the execution of activities performed for a single case in a process. Each event e_i is a tuple $(case, act, ts_{start}, ts_{end}, res)$, where $case$ is the case's identifier, *act* is the activity to which the event corresponds, ts_{start} and ts_{end}, respectively, are the start and end timestamps of the activity's execution, and res is the resource that executed the activity.

Pareto Optimization. Our approach aims to improve resource handover policies by optimizing a set of $k > 1$ objective functions. In our setting, an event log L captures the realization of the resource handover policy applied during the execution of its cases. Thus, we define the cost vector of an event log L as

$$\mathbf{c}(L) = [f_1(L), \ldots, f_k(L)], \qquad \text{where } f_i(L) = \sum_{\sigma \in L} f_i(\sigma),\ f_i : L \to \mathbb{R}. \tag{1}$$

Since our goal is to minimize all components of the cost vector $\mathbf{c}(L)$, we seek to minimize the corresponding set of objective functions $\mathcal{K} = \{f_1(L), \ldots, f_k(L)\}$. As these objectives are frequently in conflict, we focus on identifying solutions that represent optimal trade-offs, formalized through the notion of *Pareto optimality*. Let $\mathcal{L}_C$ denote the set of all event logs defined over a set of cases C. For two logs $L, L' \in \mathcal{L}_C$, we follow [5] and say that L *dominates* L' if

$$\forall i \in 1, \ldots, k : f_i(L) \leq f_i(L') \quad \wedge \quad \exists j \in 1, \ldots, k : f_j(L) < f_j(L'). \tag{2}$$

In other words, L performs at least as well as L' on all objectives and strictly better on at least one. A log $L \in \mathcal{L}_C$ is *Pareto optimal* if no log $L' \in \mathcal{L}_C$ exists that dominates it. The collection of all non-dominated logs constitutes the *Pareto front*, representing the set of optimal trade-offs across the k objectives.

The Resource Allocation Problem. Let $\mathcal{A}$ denote a set of agents, where each agent $a \in \mathcal{A}$ corresponds to a resource $res \in \mathcal{RES}_L$ involved in the process. Likewise, let $\mathcal{ACT}_L$ denote the set of all activities in the process, where $act, act' \in \mathcal{ACT}_L$. We denote each tuple (a, act, act') a *decision point* for agent a after completing activity act, in the sense that it needs to hand over to some agent a' to work on act'. To model this decision, we define a *handover policy* π as a conditional probability distribution:

$$\pi(a' \mid a, act, act') \qquad \text{where } a, a' \in \mathcal{A}, \text{ and } act, act' \in \mathcal{ACT}_L. \tag{3}$$

Optimizing π thus corresponds to searching for a handover pattern that yields improved process performance with respect to the multi-objective cost vector defined in Eq. 1.

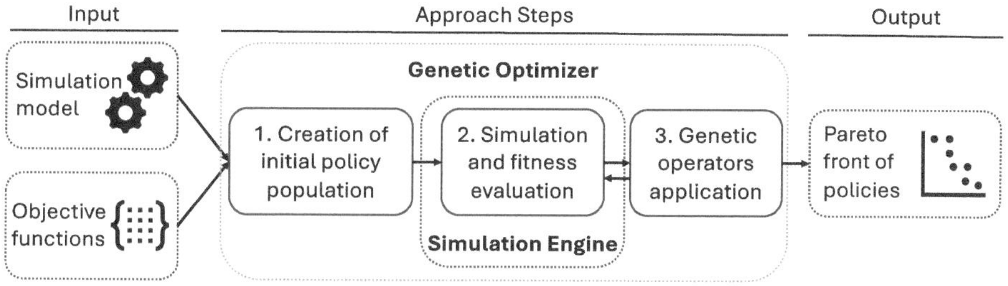

Fig. 1. Overview of the proposed approach.

4 Approach

This section introduces our genetic algorithm-based approach for discovering Pareto-optimal resource handover policies, as illustrated in Fig. 1. Its input, main steps, and output are as follows:

Input. Our approach takes as input i) an agent-based simulation model $\mathcal{M}$ and ii) a set of objective functions $\mathcal{K} = \{f_1(L), \ldots, f_k(L)\}$, which instantiate the cost vector (see Eq. 1) that guides the multi-objective optimization. We define the simulation model as $\mathcal{M} = (\mathcal{A}, \mathcal{S}, \pi_0, \beta)$, where $\mathcal{A}$ is the set of agents, $\mathcal{S}$ captures global simulation parameters, π_0 is the as-is handover policy, and β is the control-flow policy that determines the next activity in a running case. Each agent $a \in \mathcal{A}$ is characterized by its weekly availability calendar and its capabilities, comprising the set of activities it can perform and the corresponding execution-time distributions. The tuple $\mathcal{S}$ captures global simulation parameters such as the case interarrival distribution and the arrival calendar. The simulation model used in our approach can either be specified manually with domain experts or automatically discovered from an event log L. In this work, we employ

AgentSimulator [9], which automatically derives a multi-agent simulation model from an event log and supports the simulation of the corresponding process, thereby also serving as the simulation engine. For a detailed description of the quality of the discovered simulation models, we refer to the original publication. Note that other approaches may also be used to instantiate the engine.

Approach Steps. The genetic optimizer is the central component of our approach. It takes as input the as-is model $\mathcal{M}$ with its handover policy π_0 and optimizes the policy with respect to the specified set of objective functions $\mathcal{K}$. The optimization workflow consists of three steps:

1. *Creation of initial policy population.* Based on the as-is policy π_0, we generate an initial population of handover policies, denoted by $\mathcal{P} = \{\pi_1, \pi_2, \ldots, \pi_N\}$, consisting of N candidate policies.
2. *Simulation and fitness evaluation.* For each candidate policy $\pi_n \in \mathcal{P}$ in the population, the optimizer calls the simulation engine to simulate multiple runs. These runs yield event logs that are used to compute a multi-objective fitness score for the underlying policy, as defined in Eq. 1.
3. *Genetic operators application.* The optimizer uses the feedback from the fitness evaluation to apply evolutionary operators to its population, generating a new, potentially superior set of candidate policies.

We employ NSGA-II [4], a genetic algorithm specialized for multi-objective optimization, to guide the evolutionary search. We instantiate NSGA-II to operate on individuals in the population that encode full probabilistic handover policies, and adapt the algorithm through custom genetic operators designed to preserve and meaningfully modify these probability distributions. It is important to note that while the selection of NSGA-II has been driven by speed and usability concerns against alternatives (e.g., NSGA-I [20], SPEA [25]), our approach is independent of the chosen algorithm.

Output. Our approach repeats steps 2 and 3 a predefined number of iterations, controlled by a hyperparameter $G \in \mathbb{N}$. After G generations, the algorithm yields a Pareto front of handover policies $\mathcal{P}^{\star} = \{\pi_1, \pi_2, \ldots, \pi_M\}$, that indicate how resources should collaborate to achieve an optimal trade-off.
In the remainder, we describe the details of its core steps: the creation of an initial population, the simulation and fitness evaluation, and the genetic operators.

4.1 Creation of the Initial Policy Population

The first step of the genetic optimizer involves constructing an initial population of handover policies, derived from the as-is policy π_0. In this setting, each policy constitutes an *individual* (also known as the *chromosome*) in the population of the genetic algorithm.

Algorithm 1. Simulation and Fitness Evaluation

procedure EVALUATEFITNESS(var. policy π_n, fix. policy π_{fix}, $\mathcal{M}$, T, $\mathcal{K}$)
  $\pi_n \leftarrow$ MERGEPOLICYPARTS(π_n, π_{fix}) ▷ Reconstruct full handover policy
  $\mathcal{C} \leftarrow \emptyset$ ▷ Initialize set of cost vectors for all runs
  for $t \leftarrow 1$ **to** T **do**
    $\mathcal{M}_t \leftarrow$ COPY($\mathcal{M}$) ▷ Instantiate simulation model for run t
    $\mathcal{M}_t.\pi \leftarrow \pi_n$ ▷ Overwrite initial handover policy
    $L_n^t \leftarrow$ RUNSIMULATION($\mathcal{M}_t$) ▷ Simulated event log
    $\mathbf{c}_t \leftarrow$ CALCULATEMETRICS($L_n^t, \mathcal{K}$) ▷ Cost vector for run t
    $\mathcal{C} \leftarrow \mathcal{C} \cup \{\mathbf{c}_t\}$ ▷ Append to set of cost vectors
  $\bar{\mathbf{c}} \leftarrow$ VECTORMEAN($\mathcal{C}$) ▷ Compute fitness as average cost vector
  return $\bar{\mathbf{c}}$

Search Space Pruning. To reduce computational complexity, we prune the search space through a preprocessing step. We analyze the discovered handover policy and classify each decision point—i.e., each situation where a resource finishes an activity and must select a successor (see definition in Sect. 3)—based on the determinism of that decision point:

1. **Fixed:** Decision points with only one successor. These reflect non-negotiable business rules and are excluded from optimization.
2. **Variable:** Decision points with multiple successors. These represent flexible parts of the process and form the search space for optimization.

Therefore, we reduce the search space while preserving all meaningful degrees of freedom. Note that the fixed part of the policy π_{fix} will be used again during the simulation and fitness-evaluation steps.

Initial Population. To balance *exploration* and *exploitation*, we construct an initial population $\mathcal{P}$ of size N based on the variable part of the as-is handover policy π_0. As a heuristic, half of the individuals are replaced with fully random policies to ensure broad coverage of the search space. The other half is generated by mutating the as-is policy, producing candidates that remain close to the observed process behavior. Finally, one individual is set to an exact copy of the as-is policy to ensure it is included. This hybrid strategy provides both diverse and domain-informed starting points for the evolutionary search.

4.2 Simulation and Fitness Evaluation

After generating the initial population of handover policies $\mathcal{P}$, the next step evaluates the fitness of each individual by means of simulation. The fitness evaluation treats the simulator as a black-box function whose outputs are then assessed. As summarized in Algorithm 1, the **fitness of a single chromosome** is computed through three stages: policy reconstruction, simulation, and evaluation.

Policy Reconstruction. Each $\pi_n \in \mathcal{P}$ represents only the variable part of the handover policy. To obtain a complete policy compatible with the simulator, we merge this variable portion with the fixed part π_{fix} (Line 2) identified during the search space pruning. The result is a full, comprehensive handover policy that specifies all resource handover decisions in the process.

Simulation. Using the reconstructed policy $\pi_n \in \mathcal{P}$, we instantiate a new simulation model $\mathcal{M}_t$ (Line 5). Specifically, we begin with the parameters of the initial simulation model $\mathcal{M}$—preserving all settings such as arrival rates and resource schedules—and overwrite only the handover policy parameter (Line 6). This yields a complete, valid model that can be executed to generate a simulated event log (Line 7). To mitigate stochastic variance and obtain stable performance estimates, the simulation is run multiple times for each policy, independently generating several event logs. Thus, per policy π_n, we generate T event logs, denoted by $L_n^1, \ldots, L_n^T$.

Evaluation. For each simulated event log L_n^t, we compute the cost vector $\mathbf{c}_t$ (Line 8) defined in Eq. 1. The final fitness score of the policy π_n is obtained by averaging its cost vectors across all simulated logs $L_n^1, \ldots, L_n^T$, resulting in a robust multi-objective fitness estimate (Line 10).

4.3 Genetic Operators

Having computed a fitness score for each policy in the population, the next step in our approach is to apply genetic operators to the current population $\mathcal{P}$. In evolutionary computation, genetic operators are mechanisms that modify or recombine individuals to drive the search process. Inspired by biological evolution, they introduce variation, explore the search space, and steer the population toward higher-quality solutions [4]. In our approach, we employ NSGA-II's three standard operators—selection, crossover, and mutation—applied sequentially, and customize the latter two to operate on probabilistic handover policies.

Selection. First, the algorithm performs a selection step to retain only the most promising individuals for survival and reproduction through the subsequent crossover and mutation operators. Individuals are ranked into non-dominated fronts based on their fitness scores. Afterwards, pairs of randomly drawn policies are compared, and the one with the superior rank is selected. If both belong to the same front, ties are broken using the *crowding distance*, which measures how isolated a policy is from its neighbors in the objective space, thereby favoring solutions in less-explored regions. This combination of non-domination ranking and crowding-based tie-breaking ensures that NSGA-II selects individuals that promote both convergence toward the Pareto front and diversity along it.

Crossover. Given the selected subset of policies, the crossover operator combines two candidates to produce new policies (*offspring*) that inherit characteristics from both *parents*. In our context, a handover policy consists of a set of decision points, each represented by a probability distribution over possible successor resources. Because mixing individual probabilities can yield invalid distributions, crossover must preserve each decision point as a coherent unit. Our custom operator therefore works at the level of entire decision points: for each decision situation, it randomly swaps the full probability distribution between the two parents. This ensures that the offspring inherit valid decision behaviors while still enabling the exploration of new combinations of strategies within the probabilistic constraints of the handover model. The crossover operation is applied to two parents with probability p_c. If there is no crossover, the offspring are exact clones of their parents. This hyperparameter p_c controls the degree of exploitation by determining how frequently the algorithm recombines strategies from high-performing solutions.

Mutation. Mutation serves as the main driver of novelty in our evolutionary search, applied to the previously generated offspring. Whereas crossover recombines existing decision strategies, mutation introduces new behavioral patterns by perturbing selected decision points within a handover policy. Its purpose is to prevent premature convergence and to support continuous exploration of alternative resource allocation behaviors. In the context of resource handover policies, different business environments may benefit from different mutation dynamics. Some scenarios require strong exploratory jumps to escape local optima, while others benefit more from careful refinements once the population has converged toward a promising region of the search space. To promote robust performance across diverse process structures, we implement several complementary mutation strategies. Mutation is applied to each offspring with probability p_m. If selected for mutation, one of its decision points is chosen at random, and one of four mutation variants is applied to the probability distribution at that point.

Random. The probability distribution at the selected decision point is reinitialized, providing a strong exploratory step by generating a completely new handover strategy for that situation. Concretely, let the decision point specify a categorical distribution over its admissible successor resources in the variable part of the policy. The operator assigns a new positive weight to each successor, drawn independently from a uniform distribution, and normalizes these weights so that they again sum to one. Because this operator does not consult $\mathcal{M}$ or the objective set $\mathcal{K}$, it can realize handover behaviors that are not reachable through small, heuristic-guided shifts. In our implementation, it plays the same role in initializing the population, where fully random variable policies are mixed with mutated variants of π_0.

Greedy. Using a local heuristic derived from $\mathcal{M}$, the operator identifies the best and worst successor resources at the selected decision point and moves all probability mass away from the worst option—an aggressive exploitation step. Among the objectives in $\mathcal{K}$, the operator first draws one objective uniformly at random. It then scores each successor resource that carries a non-negligible probability at this decision point. The successors with the minimum and maximum scores are retained as best and worst, respectively. The entire probability allocated to the worst successor is set to zero and transferred to the best successor. If the best successor is associated with several successor activities, the incoming mass is distributed across them in proportion to their current probabilities (or uniformly if it currently receives no mass).

Guided. Also relying on the local heuristic above, this variant transfers only a small, random fraction of probability from the worst to the best successor. It is designed for fine-grained adjustments in an already promising region of the search space. After selecting the same guidable objective from $\mathcal{K}$ and ranking successors as in greedy mutation, guided mutation draws a transfer share uniformly between 10% and 50% of the probability mass currently assigned to the worst successor. This amount is subtracted from the worst successor's activities in proportion to their shares and added to the best successor's activities using the same proportional rule as in the greedy operator. The worst successor, therefore, retains most of its former probability, yielding a conservative, heuristic-biased step along one dimension of $\mathcal{K}$ while preserving a valid distribution at the decision point.

Hybrid. To balance exploration and exploitation, the operator selects one of the above strategies with predefined probabilities: 60% guided, 20% greedy, and 20% random. This weighting reflects the intuition that, as the search progresses, small refinements are often most effective, while still allowing for larger changes.

Together, these mutation strategies enable the genetic algorithm to flexibly adapt its search behavior, supporting both incremental improvements and substantial shifts in handover logic when needed. While we employ these four mutations, other heuristics can also be incorporated.

Thus, the genetic operators generate a new population of handover policies, which is then again used for simulation and fitness evaluation. We repeat this evolutionary cycle for a predefined number of G iterations, ultimately yielding a Pareto front of resource handover policies $\mathcal{P}^{\star}$.

5 Evaluation

This section presents the experiments used to evaluate the performance of our multi-objective optimization approach. We instantiate the set of objective functions with *waiting time* and *labor cost*, thus assessing whether we can learn resource-specific policies that improve both cost and time compared to the as-is scenario and other baselines. For internal validity, we assess the effectiveness of our approach in a controlled setting designed to test its performance across diverse scenarios. For external validity, we evaluate the effectiveness of our approach on seven widely used event logs from business process simulation research, providing an ideal competitive benchmark that lets us assess whether our approach can learn more optimal policies for these processes. In the remainder, Sect. 5.1 describes the experimental setup, followed by the results in Sect. 5.2. Implementations and additional results can be found in our repository[1].

5.1 Experimental Setup

The experimental setup is identical across both experiments, differing only in the datasets used.

Internal Validity. To investigate the performance of our approach, we study 5 variants of an existing synthetic *Loan Application* process [3], allowing us to assess the capability of our approach to learn effective policies across varying process configurations. The process consists of 12 activities, including parallelism, choices, and a loop. By default, the process is executed by 19 distinct resources, divided into 6 roles. To create the variants, we employ *AgentSimulator* [9] to discover a simulation model of the as-is process from the original event log, which we then modify. For each variation of the simulation model, we generate an event log containing 1000 cases, consistent with the size of the original log. Specifically, we create the following variants:

- Loan_{AD}: Cost and duration of activities are equal for all resources.
- Loan_{JS}: For each activity, resources are split into junior and senior employees, where juniors are cheaper but require more time to execute the task.
- Loan_{RC}: Creating resource contention by halving the available resources.
- Loan_{SCDT}: All resources cost the same, but differ in execution times.
- Loan_{ARR}: Increasing workload through a higher rate of case arrivals.

External Validity. To study how our approach could optimize logs in the process mining community, we employ seven widely used event logs (see Table 1) from business process simulation research [9,12] as a basis for optimization scenarios. They are well-suited for our analysis as they provide both start and end timestamps for each event, which is required for simulation. The datasets span multiple domains, including financial services, education, and procurement, and differ in size and complexity (datasets available in our repository).

A limiting factor for external validity is that existing logs do not include information on resource cost allocation. This information is essential for our multi-objective analysis, for which we decided to provide an approximation. We define hourly cost values for each resource. While assigning uniform costs to all resources provides a simple baseline, it prevents the exploration of cost-time trade-offs. Therefore, we derived a performance-correlated cost model for each event log. This reflects the real-world principle that more efficient or specialized resources are typically more expensive. To achieve this, we created a performance

[1] https://github.com/lukaskirchdorfer/BPS-MAS-Handover-Optimizer.

Table 1. Description of event log properties.

Log	Type	Traces	Events	Activities	Agents	Avg. cycle time
Loan App	syn	1000	7492	12	19	10
P2P	syn	608	9119	21	27	515
C1000	syn	1000	38160	42	14	22
C2000	syn	2000	77418	42	14	20
ACR	real	954	6870	18	432	357
BPI12W	real	8616	59302	6	52	214
BPI17W	real	30276	240854	8	136	304

profile for each resource based on skills and activity durations. We then clustered the profiles into five distinct tiers using k-means, assigned an hourly cost range to each tier with higher-performing tiers associated with higher costs (e.g., Tier 1: 10–25\$/hr, ..., Tier 5: 76–90\$/hr). We randomly sampled a value for each resource from its corresponding cost range.

Benchmark Approaches. We compare against several baselines[2]:

- **As-Is**: The as-is resource handover patterns discovered by AgentSimulator.
- **Avail**: Always hand over to the earliest available resource.
- **Random**: Select a random resource.
- **LC**: Always choose the cheapest resource.
- **SPT**: Select the resource with the shortest expected processing time.

Hyperparameters. We use the following hyperparameter configuration for all experiments. We use a population size $N = 100$ and run $G = 100$ iterations. To account for simulation stochasticity, each policy is evaluated using $T = 3$ simulated logs. The probability that crossover is applied is set to $p_c = 70\%$. The probability that a mutation is applied to an offspring is set to $p_m = 30\%$. As detailed in Sect. 4, we employ four mutation variants for our genetic operator (*Random, Greedy, Guided, Hybrid*), all of which we evaluate in our experiments.

Metrics. We use two types of metrics. The first assesses overall process performance under different handover policies by measuring total process cost in \$ (sum of event durations in hours multiplied by the resource's hourly rates) and waiting time (WT), defined as the sum of idle times. The second type assesses the

[2] While we initially wanted to include [8] in our evaluation, their method employs tabular Q-learning, which requires storing all stateaction values and thus does not scale to realistic BPM state spaces. In addition, the approach lacks a reproducible implementation and was evaluated only in a custom simulator. We therefore exclude it from the empirical comparison.

quality of the discovered Pareto fronts, evaluating how effectively each mutation variant explores a diverse and high-quality set of optimal solutions, following the methodology of López-Pintado et al. [13]. These metrics include the Hyperarea Ratio, Purity, Averaged Hausdorff Distance, and Delta Spread. As this analysis is not our primary focus, we summarize the key findings here, while detailed metric definitions and results are available in our repository.

Table 2. Results for the *Loan Application* variants. The best results per variant are highlighted in bold. Cost measured in thousand $, WT in hours.

	As-Is		Avail		Random		LC		SPT		Ours	
	Cost	WT	Cost	WT	Cost	WT	Cost	WT	Cost	WT	Cost	WT
Loan_{AD}	213	16.81	161	**0.43**	238	1.18	145	6.22	164	23.93	**120**	0.52
Loan_{JS}	246	1.46	231	0.65	333	1.39	194	26.79	212	1.00	**166**	**0.59**
Loan_{RC}	67	2.30	74	2.20	76	2.10	59	4.14	85	2.57	**49**	**2.03**
Loan_{SCDT}	254	2.78	209	0.71	287	1.27	204	0.65	237	26.85	**179**	**0.60**
Loan_{ARR}	88	0.74	83	**0.59**	81	0.73	57	3.84	97	1.38	**53**	0.69

5.2 Results

Internal Validity. Table 2 shows the performance of the different approaches in terms of WT and cost across the variants of the *Loan Application* process. Unlike the baseline methods, our approach produces an entire set of trade-off solutions. For comparability, we report the performance of the *most balanced solution* from the discovered Pareto front, defined as the solution with the smallest Euclidean distance to the ideal point in the objective space (i.e., zero cost and zero WT).

Overall, our approach consistently outperforms the as-is baseline in both cost and time across all five scenarios. For example, in the Loan_{SCDT} scenario—where resources share identical hourly costs but differ in execution times—our approach reduces the total cost from $254k to $179k and the WT from 2.78 to 0.60 h. This demonstrates that our approach effectively enhances individual handover policies by prioritizing faster resources while still engaging slower ones to prevent queues and excessive waiting times.

Compared to the other baselines, our approach achieves the lowest cost in all scenarios and the lowest waiting time in three scenarios, while ranking second-best in the remaining two. These results demonstrate that we learn more effective resource handover policies than those currently implemented. Notably, it also surpasses greedy heuristics such as *LC*, which always selects the cheapest resource, by accounting for the long-term consequences of assignment decisions and jointly optimizing cost and temporal performance.

To better understand these effects, we analyze the resulting handover patterns for the Loan_{JS} log in Fig. 2, which captures how often resources hand over work to others in the simulated event logs based on the *As-Is*, *LC*, and

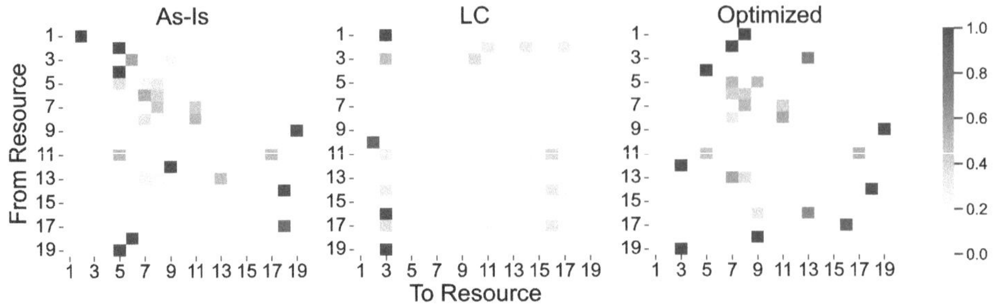

Fig. 2. Resource handover frequencies in the Loan$_{JS}$ log based on the *As-Is*, *LC*, and our optimized policy. Darker cells indicate more frequent handovers.

our optimized policy. The differences between the three are striking. In the *As-Is* baseline, for example, resource 1 always hands over to resource 2. In the *LC* baseline, which optimizes purely for cost, resource 1 instead always hands over to resource 3—who performs the same role but at a lower hourly rate. However, as all resources follow this greedy strategy, costly but faster resources are excluded entirely, leading to severe queue formation, as evident from the many blank cells in Fig. 2. Our approach, in contrast, learns a more balanced policy. Again, looking at resource 1, it now primarily hands over to resource 8—a senior costing \$150/hr compared to resource 2's \$50/hr, but completing tasks six times faster. This makes resource 8 both quicker and more cost-efficient overall. Moreover, unlike the *LC* baseline, where work accumulates at cheap but slow resources, our approach learns individual policies for each resource that, through multi-agent interactions, lead to more complementary collaboration patterns and a more balanced task distribution, resulting in a globally superior outcome.

External Validity. The results across the seven existing event logs are presented in Table 3. Overall, consistent with the findings from the first experiment, our approach achieves the best overall performance across most event logs. Importantly, it improves cost and waiting time compared to the *As-Is* baseline in all datasets, underscoring the robustness of our approach in balancing efficiency and cost-effectiveness across diverse real-world settings. While our approach yields particularly robust results in terms of WT, it does not always attain the lowest cost, ranking above the *LC* baseline in four of the seven datasets. Moreover, these gains come at the cost of higher computation time: whereas the baselines require about three minutes on average, our approach is roughly 35 times slower (cf. Table 5). Like the baselines, runtime increases with the number of events and agents in the log. At the same time, the optimization component scales comparably: for example, BPI17W takes 39 times longer than LoanApp for the baselines and 41 times longer for our approach.

We also used these event logs to investigate the performance of the four different mutation variants of our genetic operator in Table 4, reporting the most balanced solution per variant. This comparison reveals that the *Guided* variant

Table 3. Results over the existing logs. The best results per log are highlighted in bold. Cost measured in thousand $, WT in hours.

	As-Is		Avail		Random		LC		SPT		Ours	
	Cost	WT	Cost	WT	Cost	WT	Cost	WT	Cost	WT	Cost	WT
LoanApp	91	0.71	91	0.64	90	0.70	**62**	2.70	88	0.85	70	**0.61**
P2P	98	17.10	116	**0.79**	144	2.91	61	15.26	103	2.45	**44**	3.62
C1000	160	1.68	133	1.58	170	1.81	120	1.53	121	1.49	**98**	**0.75**
C2000	316	1.05	251	1.03	297	1.05	247	1.22	237	1.03	**211**	**0.71**
ACR	103	8.97	140	0.89	232	1.01	**23**	3.50	96	1.64	45	**0.51**
BPI12W	244	0.03	177	**0.02**	265	0.20	**24**	0.04	113	0.07	176	**0.02**
BPI17W	654	0.38	714	0.02	675	0.16	**209**	2.61	317	0.43	480	**0.01**

achieves the greatest average improvement across all datasets for both cost and WT relative to the *As-Is* baseline. However, the performance gap between variants is small, indicating that the approach is stable with respect to the mutation strategy. An analysis of the quality of the resulting Pareto fronts reveals that, while no single variant consistently dominates, the *Guided* and *Hybrid* variants stand out, whereas *Random* and *Greedy* tend to produce less dominant solutions.

Table 4. Relative improvement in cost and WT per mutation variant compared to the As-Is baseline, averaged across the seven existing logs.

Random		Greedy		Guided		Hybrid	
Cost	WT	Cost	WT	Cost	WT	Cost	WT
33.46%	51.64%	32.84%	54.56%	**36.88%**	**57.71%**	33.07%	55.07%

Table 5. Runtimes for all 7 logs (in minutes).

	LoanApp	P2P	C1000	C2000	ACR	BPI12W	BPI17W	Average
Baselines	0.28	0.35	1.78	3.58	1.72	2.17	10.83	2.96
Ours	9.33	16.03	70.67	142.60	41.83	67.31	379.17	103.85

We further illustrate the solution space and Pareto front for the real-life ACR event log in Fig. 3, comparing our solutions with the baselines. The *Guided* and *Hybrid* variants clearly dominate, offering Pareto-optimal trade-offs with process costs between $22k and $85k and waiting times from 0.2 to nearly 2 h. While the most balanced solution reported in Table 3 corresponds to $45k and 0.51 h, our approach offers a diverse set of high-quality alternatives, enabling practitioners and process owners to select policies aligned with their specific operational goals. For instance, if minimizing waiting time is prioritized over cost, one could choose a policy achieving around 0.2 h, representing an improvement of 0.3 h over the most balanced solution, 0.7 h over the heuristic policy that always selects the next available resource, and a remarkable 8.7 h improvement compared to the as-is status. Likewise, the Pareto front also includes solutions that outperform the *LC* heuristic in terms of cost, showing that our approach can offer both broad coverage and best-in-class solutions.

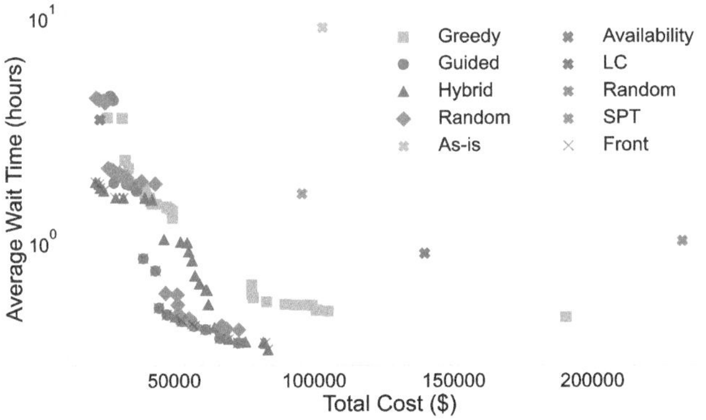

Fig. 3. Solution space and Pareto front for ACR event log.

6 Discussion and Conclusion

In this work, we presented a novel approach for optimizing resource-specific handover policies in business processes, enabling the discovery of more effective collaboration strategies. Building on a multi-agent business process simulation model where resources and their interactions are represented as autonomous agents, we integrated a genetic algorithm to optimize process performance across multiple dimensions. The approach is designed to be fully generic: it can incorporate any set of optimization objectives. Its output is a Pareto front of optimized handover policies, providing decision-makers with a transparent set of trade-offs. We validated the approach on synthetic and real logs with cost and time as objectives, achieving substantial improvements compared to the as-is process.

Threats to Validity. Our approach relies on a simulation model to evaluate candidate handover policies, which introduces several validity threats. First, optimization outcomes depend on simulation fidelity: if the simulator does not accurately reproduce real process dynamics, estimated improvements may not fully transfer to reality. To mitigate this risk, we deliberately use a data-driven simulation approach that has been shown in prior work to achieve state-of-the-art accuracy [9]. Moreover, the simulation model itself is not part of our optimization method but serves as an interchangeable input artifact; therefore, improvements in simulation fidelity can directly benefit optimization outcomes without changes to the proposed approach. Nevertheless, future work should explicitly quantify simulator accuracy and analyze the sensitivity of optimization results to modeling errors. Second, policies are optimized and evaluated within the same model, which creates a risk of overfitting to simulator-specific dynamics. In practice, this could lead to policies that perform well in simulation but less so in reality. We partially mitigate this threat by grounding simulations in discovered models derived from real event logs and by validating the approach on multiple datasets. A promising direction for future research is robustness analysis across alternative simulation models or perturbed model parameters to assess the stability of discovered policies. Third, our experiments augment event logs with hourly resource cost information that is typically not recorded in standard logs. While such augmentation is common in process optimization studies, it introduces assumptions that may affect absolute performance estimates. We partially mitigate this threat by using realistic cost assumptions that correlate with performance and by relying on the same assumptions for as-is and optimized policies. Future work could incorporate empirically observed cost data or uncertainty-aware optimization to reduce this dependency further. Furthermore, our approach assumes that resources can choose whom to collaborate with, or at least that handovers between resources are subject to change. While this assumption holds for many real-world processes—particularly knowledge-intensive processes in which resources have some degree of decision-making autonomy—it may be less applicable to highly orchestrated workflow processes. We refer to [9] for details on the types of processes benefiting from resource-specific modeling.

Future Work. Beyond validity considerations, the approach entails higher computational effort due to evolutionary search. Although NSGA-II provides a strong baseline for multi-objective optimization, other algorithms may be preferable as the number of objectives grows. Reducing computational overhead—e.g., by narrowing the search space, identifying the most influential decision points, or optimizing only selected handovers—remains an important direction. Furthermore, extending the policy space towards richer collaboration strategies (e.g., dynamic parallelization or adaptive routing) is a promising avenue for future work. Finally, our evaluation relies on heuristic baselines for comparison. This is partly due to the lack of suitable existing benchmarks addressing the same problem of resource handover optimization. Future work may add stronger optimization-oriented resource allocation baselines that optimize a single objective but add further insights into the value of our approach.

Acknowledgments. This work received support from VILLUM FONDEN (grant VIL57420) through the Center for Digital Compliance (DICE), the Innovation Fund Denmark projects "Explainable Hybrid-AI for Computational Law and Accurate Legal Chatbots" (XHAILe, grant 4355-00018B) and Predictive and Prescriptive Process Analytics for Industry 4.0 (P3AI4, grant 4105-00045B).

References

1. Bejarano, J., Barón, D., González-Rojas, O., Camargo, M.: Discovering optimal resource allocations for what-if scenarios using data-driven simulation. Front. Comput. Sci. **5** (2023)
2. Branchi, S., Buliga, A., Di Francescomarino, C., Ghidini, C., Meneghello, F., Ronzani, M.: Recommending the optimal policy by learning to act from temporal data. arXiv preprint arXiv:2303.09209 (2023)
3. Chapela-Campa, D., Benchekroun, I., Baron, O., Dumas, M., Krass, D., Senderovich, A.: A framework for measuring the quality of business process simulation models. Inf. Syst. **127**, 102447 (2025)
4. Deb, K., Pratap, A., Agarwal, S., Meyarivan, T.: A fast and elitist multiobjective genetic algorithm: NSGA-II. IEEE TEVC **6**(2), 182–197 (2002)
5. Diaz, J.F., López, H.A., Quesada, L., Rosero, J.C.: Pareto-optimal trace generation from declarative process models. In: BPM Workshops, pp. 314–325 (2024)
6. Djedović, A., Žunić, E., Avdagić, Z., Karabegović, A.: Optimization of business processes by automatic reallocation of resources using the genetic algorithm. In: BIHTEL, pp, 1–7 (2016)
7. Filatov, V.O., Yerokhin, M.A.: Improved multi-objective optimization in business process management using R-NSGA-II. Radio Electron. Comput. (3) (2023)
8. Huang, Z., van der Aalst, W.M., Lu, X., Duan, H.: Reinforcement learning based resource allocation in business process management. DKE **70**(1), 127–145 (2011)
9. Kirchdorfer, L., Blümel, R., Kampik, T., van der Aa, H., Stuckenschmidt, H.: Discovering multi-agent systems for resource-centric business process simulation. Process Sci. **2**, 4 (2025)
10. Lee, H., Kim, S.S.: Integration of process planning and scheduling using simulation based genetic algorithms. IJAMT **18**, 586–590 (2001)
11. Ling, S., Schmidt, H.: Time petri nets for workflow modelling and analysis. In: SMC, vol. 4, pp. 3039–3044. IEEE (2000)
12. López-Pintado, O., Dumas, M.: Business process simulation with differentiated resources: does it make a difference? In: Di Ciccio, C., Dijkman, R., del Río Ortega, A., Rinderle-Ma, S. (eds.) Business Process Management. BPM 2022. Lecture Notes in Computer Science, vol. 13420. Springer, Cham (2022). https://doi.org/10.1007/978-3-031-16103-2_24
13. López-Pintado, O., Dumas, M., Berx, J.: Discovery, simulation, and optimization of business processes with differentiated resources. Inf. Syst. **120**, 102289 (2024)
14. Meneghello, F., et al.: Optimizing resource allocation policies in real-world business processes using hybrid process simulation and deep reinforcement learning. In: Marrella, A., Resinas, M., Jans, M., Rosemann, M. (eds.) BPM, pp. 167–184. Springer, Cham (2024). https://doi.org/10.1007/978-3-031-70396-6_10
15. Middelhuis, J., Bianco, R.L., Sherzer, E., Bukhsh, Z., Adan, I., Dijkman, R.M.: Learning policies for resource allocation in business processes. Inf. Syst. **128**, 102492 (2025)
16. Peters, S.P.F., Dijkman, R.M., Grefen, P.W.P.J.: Resource optimization in business processes. In: EDOC, pp. 104–113 (2021)
17. Si, Y.W., Chan, V.I., Dumas, M., Zhang, D.: A petri nets based generic genetic algorithm framework for resource optimization in business processes. Simul. Model. Pract. Theory **86**, 72–101 (2018)
18. Sindhgatta, R., Ghose, A., Dam, H.K.: Context-aware analysis of past process executions to aid resource allocation decisions. In: CAISE, pp. 575–589 (2016)

19. Song, M., van der Aalst, W.M.P.: Towards comprehensive support for organizational mining. Decis. Support Syst. **46**(1), 300–317 (2008)
20. Srinivas, N., Deb, K.: Muiltiobjective optimization using nondominated sorting in genetic algorithms. Evol. Comput. **2**(3), 221–248 (1994)
21. Tour, A., Polyvyanyy, A., Kalenkova, A.: Agent system mining: vision, benefits, and challenges. IEEE Access **9**, 98336–98350 (2021)
22. Wibisono, A., Nisafani, A.S., Bae, H., Park, Y.-J.: On-the-fly performance-aware human resource allocation in the business process management systems environment using naïve bayes. In: Bae, J., Suriadi, S., Wen, L. (eds.) AP-BPM 2015. LNBIP, vol. 219, pp. 70–80. Springer, Cham (2015). https://doi.org/10.1007/978-3-319-19509-4_6
23. Zbikowski, K., Ostapowicz, M., Gawrysiak, P.: Deep reinforcement learning for resource allocation in business processes. In: ICPM, pp. 177–189 (2022)
24. Zhao, W., Yang, L., Liu, H., Wu, R.: The optimization of resource allocation based on process mining. In: ICIC, pp. 341–353 (2015)
25. Zitzler, E., Thiele, L.: An evolutionary algorithm for multiobjective optimization: the strength pareto approach. TIK Rep. **43** (1998)

Toward Automated Ticket Escalation in Customer Support Processes: From Emails to Explainable Predictions

Furkan Kocak[1] and Gyunam Park[2](✉)

[1] University of Padova, Padova, Italy
[2] Eindhoven University of Technology, Eindhoven, The Netherlands
g.park@tue.nl

Abstract. Business processes in customer support often remain undocumented, with critical decisions such as ticket escalation relying on tacit knowledge rather than explicit process models. Automating such process decisions is challenging because the underlying data is unstructured email text, not structured event logs assumed by existing process mining and predictive monitoring approaches. We present a novel pipeline artifact that addresses this gap: a semi-automated four-stage pipeline that extracts structured features from customer support emails using large language models (LLMs) together with lightweight domain-expert validation, and employs explainable machine learning to predict the optimal timing of ticket creation, a key decision point in the support process. Our pipeline generates an initial feature schema from sample threads, consolidates semantically similar features through embedding-based clustering under domain-expert review, extracts cumulative feature values at each email point, and constructs a structured dataset ready for machine learning. We introduce a SHAP-based active feature learning strategy that identifies which missing features, when acquired, would most improve prediction confidence, enabling principled decisions about when to request additional information versus making immediate predictions. We evaluate our approach in two stages: first, on a reproducible synthetic dataset where comparison against six baselines demonstrates that LLM-extracted features provide complementary semantic information, with combined features achieving the best overall performance (ROC-AUC 0.90) while end-to-end LLM classification performs worst (ROC-AUC 0.46), validating our design of using LLMs as feature extractors rather than decision makers; second, on 1,500 email instances from 800 real-world email threads from Nosco, a commercial CRM platform, where we demonstrate effective feature extraction, predictive performance, and active feature learning for practical deployment.

Keywords: Business Process Automation · Feature Extraction · Explainable AI · Active Learning · Large Language Models

F. M. Maggi et al. (Eds.): BPM 2026, LNCS 16936, pp. 444–463, 2027.
https://doi.org/10.1007/978-3-032-37777-7_27

1 Introduction

Business processes in customer support are often undocumented and rely on tacit knowledge rather than explicit process models. Despite the significant body of work on process automation in the Business Process Management (BPM) community, many real-world processes, particularly in small- to medium-sized businesses, generate only unstructured communication data (e.g., emails) rather than the structured event logs assumed by most process mining and monitoring techniques. A critical decision point in the customer support process is determining at which point in an email conversation a support ticket should be created for formal tracking and resolution. This timing decision requires assessing the cumulative technical complexity, urgency and business impact, multi-step resolution requirements, and historical patterns as they evolve throughout the conversation, knowledge that typically resides in experienced agents' expertise rather than explicit documentation. Automating such implicit process decisions from unstructured data presents a key challenge for BPM.

Existing approaches to ticket escalation typically assume access to pre-existing structured data or require extensive manual feature engineering with substantial domain expert involvement. Montgomery et al. [24] achieved strong recall on IBM support tickets but required analyst interviews to convert implicit business knowledge into explicit features, an approach that small- to medium-sized businesses cannot afford. More broadly, existing BPM approaches such as prescriptive process monitoring [25] and predictive process monitoring [9,26] assume structured event logs or explicit process models. Our domain has neither. Meanwhile, recent advances in large language models (LLMs) have shown strong capabilities in natural language understanding, but their direct deployment as end-to-end decision makers in business-critical applications faces several challenges: hallucination problems that undermine trust [11], opaque decision-making that resists auditing [17], and privacy concerns when processing sensitive communications through cloud services [27].

These challenges motivate an approach that leverages LLMs' language understanding without relying on them for final decisions. Specifically, there is a need for methods that can: (1) transform raw, unstructured email data into structured representations suitable for traditional machine learning, (2) build explainable models that support transparency and enable human oversight, (3) handle missing or uncertain information gracefully by identifying when additional data collection is warranted, (4) operate effectively with moderate-sized datasets typical of real-world business contexts, and (5) serve as a reusable artifact that can be adapted to other business process domains where decisions rely on unstructured communication data.

This paper addresses the challenge of automating process decisions from unstructured data by proposing a novel pipeline artifact, a reusable system that combines LLM-based feature extraction with explainable machine learning and active feature learning to automate ticket escalation in customer support processes. The novelty lies not in the individual components but in the integrated pipeline design: specifically, how LLMs, semantic clustering, and SHAP-based active learning are orchestrated to address the unique challenge of process automation from unstructured data. Our approach uses LLMs as feature extractors rather than end-to-end decision makers, enabling transparency while leveraging their natural language understanding capabilities to capture the evolving context within email conversations. Our key contributions are as follows:

- A semi-automated pipeline artifact that identifies features from sample threads, consolidates them through embedding-based clustering with bounded domain-expert validation, and extracts values across all threads, reducing 785 initial features to 35 consolidated representations (95.5% reduction) on real-world data.
- A SHAP-based two-stage confidence assessment that withholds process decisions when specific missing features would change the outcome, improving precision from 0.67 to 0.74 while preserving interpretability and auditability.
- Experimental validation against six baselines (end-to-end LLM, metadata, TF-IDF, sentiment/linguistic, rule-based) on a reproducible synthetic dataset (combined ROC-AUC: 0.90) and on 1,500 email instances from 800 real threads from Nosco, a commercial CRM platform.

2 Related Work

Our work bridges four research areas: process automation and decision support in BPM, LLM-based structured data extraction, machine learning for ticket escalation, and active feature learning.

Process Automation and Decision Support in BPM. BPM research has developed predictive [9] and prescriptive [25] process monitoring for automated decision support, achieving strong results in domains with structured event logs and explicit process models [5]. However, these approaches assume well-defined case identifiers, activities, and timestamps. These assumptions fail in customer support for small- to medium-sized businesses, where processes are implicit and generate only unstructured communication. Our pipeline addresses this gap by transforming unstructured email data into structured representations amenable to process decision support.

LLM-Based Feature Extraction from Unstructured Text. LLMs can automate transformation of unstructured text into structured representations. EVAPORATE-DIRECT [2] separates schema identification from value extraction but assumes a fixed schema and does not consolidate redundant features that emerge across diverse documents. GraphRAG [6] and KGGen [23] build knowledge graphs from extracted triples and prune for redundancy; healthcare applications [4] reveal accuracy limitations in 38% of reviewed studies. These approaches lack mechanisms for systematically consolidating semantically similar features and handling missing or uncertain extractions, which our pipeline addresses through embedding-based clustering and SHAP-based active feature learning.

Process and Activity Mining from Email Data. A dedicated line of BPM research mines process knowledge directly from email corpora. Elleuch et al. [8] discover frequent activities from email logs using pattern-discovery techniques and construct event logs amenable to process mining, and further propose a meta-model formalizing the relationship between emailing systems and business process elements [7]. Khandaker et al. [15] apply transformer models to classify sender intents in emails and map them to process activities in knowledge-intensive processes. Earlier, Mitchell et al. [22] extracted knowledge about users' activities from raw workstation contents, including email. These approaches focus on reconstructing activities, intents, or process models from email, i.e., recovering the implicit *control flow* of a process. Instead, our work targets *decision point*, i.e., when an ongoing thread should be escalated into a ticket, and produce an ML-ready structured dataset with explainable, confidence-aware predictions.

Machine Learning for Ticket Escalation. Montgomery et al. [24] trained XGBoost on 2.5M IBM support tickets (87.36% recall, 2.79% precision, 88.23% workload reduction), but converted implicit business knowledge into explicit features through analyst interviews, an effort small- to medium-sized businesses cannot afford. Gupta [10] combined exploratory analysis with SHAP for interpretability but still required manual feature engineering. These studies validate ML feasibility for ticket prediction yet assume pre-existing structured data; our work automates feature extraction from raw emails using LLMs while maintaining explainability.

Active Learning and Feature Acquisition. Active learning traditionally selects instances to label [28], while *active feature acquisition* (deciding which missing values to query) has received less practical attention [16]. EDDI [20] uses Bayesian experimental design with variational autoencoders but its complexity challenges interpretability, and recent LLM-based approaches [3,13] are black-box. Our SHAP-based approach provides transparent, auditable criteria by assessing both missing-feature importance and prediction stability.

Research Gap. No prior work combines LLM-based feature extraction from unstructured process data, semantic consolidation through embedding-based clustering, explainable tree-based prediction, and SHAP-based active feature acquisition. In particular, the email-mining stream above reconstructs activities and process models but does not address operational decision points such as escalation timing, nor does it produce explainable, confidence-aware predictions. Our pipeline fills this gap.

3 Automated Feature Extraction Pipeline

We present a systematic pipeline for transforming raw customer support emails into structured datasets suitable for machine learning. This pipeline serves as a reusable artifact for transforming unstructured process data into structured representations, applicable to any domain where business processes are implicit and documented only through communication records. As illustrated in Fig. 1, the pipeline consists of four stages: (1) feature identification produces an *initial feature schema* from sample threads, (2) semantic clustering and consolidation refines this into a *consolidated feature schema*, (3) feature extraction generates *feature mappings* for all threads, and (4) dataset construction produces the final *structured dataset*. Each stage processes the output of the previous stage, progressively refining raw email data into a structured representation.

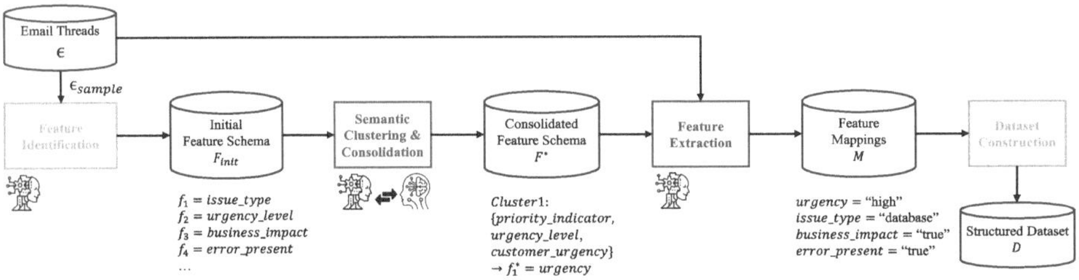

Fig. 1. Automated feature extraction pipeline with four stages. The pipeline transforms raw email threads into a structured dataset.

The input to the pipeline is a collection of email threads $\mathcal{E} = \{E_1, E_2, \ldots, E_N\}$, where each thread $E_i = \{e_{i,1}, e_{i,2}, \ldots, e_{i,n_i}\}$ is a sequence of emails exchanged between customer and support agent. Each email $e_{i,j}$ contains fields $(sender, subject, body, timestamp)$. The output is a structured dataset $D = \{(\mathbf{x}_{i,j}, y_{i,j})\}$ where each instance corresponds to the cumulative knowledge available up to email $e_{i,j}$ in thread E_i. Here, $\mathbf{x}_{i,j} \in \mathbb{R}^q$ is a feature vector representing the cumulative information from emails $\{e_{i,1}, e_{i,2}, \ldots, e_{i,j}\}$, and $y_{i,j} \in \{0, 1\}$ indicates whether a ticket should be created at this point in the conversation. We use forward-filling within each thread, where $y_{i,k} = 0$ for all emails $e_{i,k}$ with $k < j^*$ if ticket creation occurs at email e_{i,j^*}.

To illustrate the pipeline, we use a running example throughout this section. Consider an email thread where a customer contacts support about a software issue:

> ***Email 1 (Customer):*** *"Hi, since yesterday's system update we have been unable to access the inventory module in your system. Our business depends on this module for daily operations, and every hour it is unavailable we lose orders. Please treat this as urgent."*
> ***Email 2 (Support):*** *"Thank you for contacting us. Could you tell us which browser you are using, and whether you see a specific error message when accessing the inventory module?"*

Email 3 (Customer): *"We have tried both Chrome and Firefox, and the inventory module fails to load in both. Our IT team has looked into it and suspects the module cannot connect to the database."*

In this thread, ticket creation was triggered after Email 3 ($y_{i,3} = 1$) due to the persistent technical nature and business impact, while the earlier emails had $y_{i,1} = y_{i,2} = 0$. We will trace how this example flows through each stage of the pipeline, showing how features are extracted cumulatively at each email point.

3.1 Stage 1: Feature Identification

The first stage takes a sample of email threads $\mathcal{E}_{sample} \subseteq \mathcal{E}$ and uses an LLM to identify features characterizing support interactions relevant to ticket creation decisions, producing an *initial feature schema* $\mathcal{F}_{init}$.

We define a feature schema as a set $\mathcal{F}_{init} = \{f_1, f_2, \ldots, f_m\}$, where each feature f_k is a tuple:

$$f_k = (name_k, type_k, desc_k, values_k)$$

with $name_k$ as the feature identifier, $type_k \in \{\text{categorical, numerical, boolean}\}$ as the data type, $desc_k$ as semantic description, and $values_k$ as an optional set of possible values for categorical features (e.g., $values_k = \{\text{"high"}, \text{"medium"}, \text{"low"}\}$ for urgency levels), or $\emptyset$ for numerical and boolean features. The LLM receives a prompt that includes both the email thread and the corresponding ticket information (name and description) for threads that resulted in ticket creation, requesting: *"Analyze the following email threads along with their associated ticket details and identify features that would be relevant for deciding whether to create a support ticket of this type. For each feature, provide: name, type, description, and for categorical features, also provide a set of possible values."* Because a single thread typically exhibits only a subset of the possible categories, the value set $values_k$ produced by the LLM for that thread is necessarily partial. The value set is therefore not finalized here: it is formed in Stage 2 by taking the union of values observed across all sample threads in $\mathcal{E}_{sample}$ and is then refined by domain experts, who ensure it is comprehensive yet bounded.

For our running example, the LLM might identify features including:

$$f_1 = (\texttt{issue\_type}, \text{cat.}, \text{"Type of technical problem"}, \{\text{"DB\_conn."}, \text{"auth."}, \text{"perf."}\})$$
$$f_2 = (\texttt{urgency\_level}, \text{cat.}, \text{"Level of issue's urgency"}, \{\text{"high"}, \text{"medium"}, \text{"low"}\})$$
$$f_3 = (\texttt{business\_impact}, \text{bool.}, \text{"Affects business operations"}, \emptyset)$$
$$f_4 = (\texttt{error\_present}, \text{bool.}, \text{"Contains error message"}, \emptyset)$$

These identified features form the initial schema $\mathcal{F}_{init}$. Since the LLM analyzes multiple diverse threads, $\mathcal{F}_{init}$ may contain hundreds of features, many of which are semantically similar or redundant. The next stage addresses this redundancy through consolidation.

3.2 Stage 2: Semantic Clustering and Consolidation

The second stage consolidates the potentially large and redundant initial feature schema $\mathcal{F}_{init}$ into a smaller, more coherent *consolidated feature schema* $\mathcal{F}^*$. We employ a sentence embedding model to map each feature to a dense vector representation based on its name and description:

$$\psi : f_k \to \mathbf{e}_k \in \mathbb{R}^d$$

where $\mathbf{e}_k$ is the embedding of the concatenation of $name_k$ and $desc_k$. We then apply HDBSCAN (Hierarchical Density-Based Spatial Clustering of Applications with Noise) [21] to identify clusters:

$$\mathcal{C} = \text{HDBSCAN}(\{\mathbf{e}_1, \ldots, \mathbf{e}_m\}, \epsilon, \text{min\_samples})$$

producing a set of feature clusters $\mathcal{C} = \{C_1, C_2, \ldots, C_k\}$ where features within each cluster C_j are semantically similar.

For each cluster C_j, a domain expert reviews the features $\{f_k : f_k \in C_j\}$ and either selects the most representative feature $f_j^* \in C_j$, or creates a new consolidated feature f_j^* that better captures the cluster's semantic content. For categorical features, domain experts also review and refine the set of possible values, ensuring they are comprehensive yet constrained to prevent value explosion during extraction: they merge near-duplicate categories and add values known to occur in practice but absent from the sample. In our evaluation (Sect. 5), support managers from the Nosco platform independently reviewed the clusters and resolved disagreements through discussion, requiring substantially less time than manual feature engineering

Following clustering, domain experts perform post-processing based on feature correlation analysis and missingness patterns. Specifically, *missingness patterns* refer to the proportion of instances for which a feature cannot be extracted: features with missingness exceeding 80% across all email instances are flagged for review, as persistent absence suggests the feature is either too specific or ambiguously defined for reliable LLM extraction. Features with ambiguous definitions that could lead to inconsistent annotations are removed (e.g., subjective features like `issue_described_clearly`), and highly correlated features (Cramér's $V > 0.85$) are merged into unified representations. The consolidated schema is:

$$\mathcal{F}^* = \{f_1^*, f_2^*, \ldots, f_p^*\}$$

where $p \ll m$, significantly reducing the feature space dimensionality.

Returning to our running example, suppose the LLM identified three similar features across different threads in $\mathcal{E}_{sample}$:

$$f_a = (\texttt{customer\_urgency}, \text{cat.}, \text{“Level of urgency expressed”}, \{\text{“high”}, \text{“medium”}, \text{“low”}\})$$

$$f_b = (\texttt{priority\_indicator}, \text{cat.}, \text{“Importance of request”}, \{\text{“high”}, \text{“medium”}, \text{“low”}\})$$

$$f_c = (\texttt{urgency\_level}, \text{cat.}, \text{“How urgent the issue is”}, \{\text{“high”}, \text{“medium”}, \text{“low”}\})$$

After embedding and clustering, these features form cluster C_1 due to their semantic similarity. The domain expert consolidates them into:

$$f_1^* = (\texttt{urgency}, \text{cat.}, \text{“Customer-expressed urgency level”}, \{\text{“high”}, \text{“medium”}, \text{“low”}\})$$

This consolidation process is repeated for all clusters, producing the consolidated schema $\mathcal{F}^*$ that serves as the template for systematic feature extraction in the next stage.

3.3 Stage 3: Feature Extraction

With the consolidated feature schema $\mathcal{F}^*$ established, the third stage extracts actual values from *all* individual emails in the dataset. For each email $e_{i,j}$ in thread E_i and each feature $f_k^* \in \mathcal{F}^*$, we use the LLM as an extraction function to obtain the email-specific value.

Formally, we define $\phi_k : e_{i,j} \times f_k^* \to \mathcal{V}_k \cup \{\perp\}$ as the extraction function that maps an individual email and feature to a value from the feature's domain $\mathcal{V}_k$ or to $\perp$ (missing) if the information is not present in that email:

$$v_{i,j,k} = \phi_k(e_{i,j}, f_k^*) = \text{LLM}(\text{prompt}(e_{i,j}, f_k^*))$$

where $\text{prompt}(e_{i,j}, f_k^*)$ constructs a query asking the LLM to extract the value of feature f_k^* from email $e_{i,j}$.

The complete set of *feature mappings* for all emails is:

$$\mathcal{M} = \{(e_{i,j}, f_k^*, v_{i,j,k}) : i = 1, \ldots, N, j = 1, \ldots, n_i, k = 1, \ldots, p\}$$

Applying this extraction to our running example with the consolidated schema $\mathcal{F}^*$ yields email-specific values. For instance, at email $e_{i,1}$ (the customer's initial message):

$$v_{i,1,1} = \phi_1(e_{i,1}, f_1^*) = \texttt{``high''} \quad \text{(urgency)}$$
$$v_{i,1,2} = \phi_2(e_{i,1}, f_2^*) = \perp \quad \text{(issue\_type)}$$
$$v_{i,1,3} = \phi_3(e_{i,1}, f_3^*) = \texttt{true} \quad \text{(business\_impact)}$$
$$v_{i,1,4} = \phi_4(e_{i,1}, f_4^*) = \perp \quad \text{(error\_present)}$$

Email $e_{i,1}$ conveys urgency and business impact but does not yet identify the technical issue or report an error message, so ϕ_2 and ϕ_4 return $\perp$. The issue type is determined only at $e_{i,3}$, where the customer traces the problem to the database: $\phi_2(e_{i,3}, f_2^*) = \texttt{``database\_connection''}$. The `error_present` feature is never expressed in the thread and stays $\perp$ at every email. These email-level extracted values are stored in $\mathcal{M}$ and serve as the raw input for dataset construction in the final stage.

3.4 Stage 4: Dataset Construction

The final stage transforms the email-level feature mappings $\mathcal{M}$ into a structured dataset D ready for machine learning. For each email $e_{i,j}$ in thread E_i, we construct a raw feature vector by applying forward-filling to the email-specific extracted values. The forward-filling mechanism differs by feature type:

For categorical and numerical features, we use the most recently extracted non-missing value from emails $\{e_{i,1}, e_{i,2}, \ldots, e_{i,j}\}$:

$$v_{i,j,k}^{cumulative} = \begin{cases} v_{i,j,k} & \text{if } v_{i,j,k} \neq \perp \\ v_{i,j-1,k}^{cumulative} & \text{if } v_{i,j,k} = \perp \text{ and } j > 1 \\ \perp & \text{otherwise} \end{cases}$$

For boolean features, once a value becomes `true`, it persists in all subsequent emails within the thread. This ensures that if a boolean feature is observed as `true` at any point in the conversation, later emails are influenced by this knowledge even if the feature is not explicitly mentioned or extracted as `false`.

The cumulative feature vector is then:

$$\mathbf{x}_{i,j}^{raw} = (v_{i,j,1}^{cumulative}, v_{i,j,2}^{cumulative}, \ldots, v_{i,j,p}^{cumulative})$$

where $p = |\mathcal{F}^*|$ is the number of consolidated features.

We then apply standard preprocessing transformations T: categorical features undergo one-hot or ordinal encoding, numerical features are standardized to zero mean and unit variance, boolean features are converted to binary values, and missing values are handled appropriately for the target machine learning algorithm. The preprocessing function produces:

$$\mathbf{x}_{i,j} = T(\mathbf{x}_{i,j}^{raw}) \in \mathbb{R}^q$$

where $q \geq p$ due to encoding expansion (e.g., one-hot encoding of categorical features).

The final *structured dataset* is:

$$D = \{(\mathbf{x}_{i,j}, y_{i,j})\}$$

where each instance pairs a preprocessed feature vector $\mathbf{x}_{i,j}$ (cumulative knowledge up to email $e_{i,j}$) with label $y_{i,j} \in \{0, 1\}$ indicating ticket creation timing.

Returning to our running example, from email $e_{i,1}$ we extract `urgency` = `high` and `business_impact` = `true`, while `issue_type` and `error_present` are not yet determinable and return $\perp$. The support reply $e_{i,2}$ asks a clarifying question and adds no new feature values (all extractions return $\perp$), so forward-filling carries the $e_{i,1}$ values unchanged into $\mathbf{x}_{i,2}^{raw}$. At $e_{i,3}$ the customer traces the problem to the database, so `issue_type` = `database_connection` is extracted for the first time; `urgency` retains `high` from its most recent non-missing extraction at $e_{i,1}$, and the boolean `business_impact`, once `true`, persists regardless of re-extraction. The feature `error_present` is never expressed in the thread and remains $\perp$ throughout. This yields three cumulative instances $\mathbf{x}_{i,1}, \mathbf{x}_{i,2}, \mathbf{x}_{i,3}$ with labels $y_{i,1} = y_{i,2} = 0$ and $y_{i,3} = 1$, which enter the structured dataset D.

4 Explainable Prediction with Active Feature Learning

Building on the structured dataset D from the extraction pipeline, we develop an explainable prediction system that determines at each email point whether a ticket should be created based on the cumulative conversation history, while also identifying which missing features, when acquired, would most improve prediction confidence.

4.1 Binary Classification Model

Given a feature vector $\mathbf{x}_{i,j} \in \mathbb{R}^q$ from the structured dataset D, we learn a classification function $h : \mathbb{R}^q \rightarrow \{0, 1\}$ that predicts whether a ticket should be created at this point in the conversation based on the cumulative information available up to email $e_{i,j}$.

We employ gradient boosting decision trees (GBDT) [14], which build an ensemble sequentially:

$$h(\mathbf{x}) = \text{sign}\left(\sum_{t=1}^{T} \alpha_t h_t(\mathbf{x})\right)$$

where h_t is the t-th decision tree, α_t is its weight, and T is the number of trees. The model is trained on the structured dataset $D = \{(\mathbf{x}_{i,j}, y_{i,j})\}$ using standard gradient boosting optimization.

4.2 SHAP-Based Local Interpretability

To explain each prediction, we compute SHAP (SHapley Additive exPlanations) [19] values that decompose the model output into contributions from individual features. We choose SHAP because it is the only feature-attribution method with axiomatic guarantees, such as local accuracy, consistency, and missingness. This makes the per-feature contributions $\phi_j(\mathbf{x})$ directly comparable and auditable, enabling the active feature learning strategy that ranks missing features by $|\phi_j(\mathbf{x})|$. Crucially, for tree models that handle missing values natively (via default split directions), TreeSHAP attributes the prediction impact of a missing feature to that feature, yielding a non-zero $|\phi_j(\mathbf{x})|$ without any prior imputation. This lets Algorithm 1 rank missing features by importance directly, rather than having to impute surrogate values first.

For each prediction $h(\mathbf{x})$, SHAP assigns a contribution $\phi_j(\mathbf{x})$ to feature j based on cooperative game theory, satisfying:

$$h(\mathbf{x}) = \phi_0 + \sum_{j=1}^{q} \phi_j(\mathbf{x})$$

where ϕ_0 is the expected model output (baseline) and $\phi_j(\mathbf{x})$ represents the impact of feature j on the prediction. Features with larger $|\phi_j(\mathbf{x})|$ have greater influence on the model's decision. For tree-based models, TreeSHAP [18] computes these values efficiently in polynomial time.

For our running example, at the decision point $e_{i,3}$ TreeSHAP might produce the following contributions: $\phi_1(\mathbf{x}_{i,3}) = +0.35$ (urgency = high), $\phi_2(\mathbf{x}_{i,3}) = +0.28$ (issue_type = database), $\phi_3(\mathbf{x}_{i,3}) = +0.15$ (business_impact = true), $\phi_4(\mathbf{x}_{i,3}) = +0.22$ (error_present = missing).

Even for missing features, SHAP computes importance values based on how the tree ensemble handles missing data (e.g., through default split directions). These SHAP values flow into the active learning module, which uses them to assess prediction confidence and identify informative features to query.

4.3 Active Feature Learning Strategy

In practice, extracted features often contain missing values (either because the LLM extraction was uncertain or because the information is genuinely absent from the email thread). Requesting additional information from customers incurs time and effort costs. Our active feature learning strategy identifies which features, when queried, would most improve prediction confidence, enabling the system to make informed decisions about when to request more information versus making an immediate prediction.

Algorithm 1 presents the complete prediction refinement procedure. The algorithm takes as input an instance $\mathbf{x}$, the trained model h, and confidence thresholds, then decides whether to make an immediate prediction or request additional information. Intuitively, before committing to a prediction the algorithm asks two questions: *(i) are important features missing?* and *(ii) would plausible values for those missing features change the decision?* It requests additional information only when the answer to either question is yes; otherwise it predicts immediately. The algorithm operates in three phases, each assessing a different aspect of prediction confidence.

Phase 1: Missing Feature Impact. Phase 1 (Algorithm 1, lines 4–11) flags instances whose missing features are jointly important: it computes $\text{MissingImpact}(\mathbf{x}) = \sum_{j \in M(\mathbf{x})} |\phi_j(\mathbf{x})|$ over $M(\mathbf{x}) = \{j : x_j = \perp\}$ and, if it exceeds θ_{miss}, withholds the prediction and returns $j^* = \arg\max_{j \in M(\mathbf{x})} |\phi_j(\mathbf{x})|$ as the feature to request. For our running example, only error_present is missing, so $\text{MissingImpact}(\mathbf{x}_{i,3}) = |\phi_4| = 0.22$. If we set $\theta_{miss} = 0.3$, the impact is below threshold and the algorithm proceeds to the next phase.

Phase 2: Prediction Stability. Phase 2 (lines 13–20) catches instances whose prediction is sensitive to the unknown values: it imputes each missing feature with values sampled from $P(x_j|D)$, computes the variance $\sigma^2 = \text{Var}(\{h(\mathbf{x}^{(j,v)})\})$ of the resulting predictions, and withholds with the same j^* rule if $\sigma^2 > \theta_{stab}$. For our running example, we might simulate error_present $\in$ {true, false} and observe that both imputations yield $h(\mathbf{x}_{i,3}) = 1$, resulting in low variance. The prediction is stable despite the missing feature.

Algorithm 1. Active Feature Learning for Prediction Refinement

Input: Instance $\mathbf{x}$, model h, thresholds θ_{miss}, θ_{stab}
Output: Prediction decision and optional feature query

Compute SHAP values $\{\phi_j(\mathbf{x})\}_{j=1}^{q}$ using TreeSHAP
Identify missing features $M(\mathbf{x}) = \{j : x_j = \perp\}$
Compute MissingImpact$(\mathbf{x}) = \sum_{j \in M(\mathbf{x})} |\phi_j(\mathbf{x})|$

if MissingImpact$(\mathbf{x}) > \theta_{miss}$ **then**
    Select $j^* = \arg\max_{j \in M(\mathbf{x})} |\phi_j(\mathbf{x})|$
    return (WITHHOLD, j^*)
end if

Simulate imputation: For each $j \in M(\mathbf{x})$, create $\mathbf{x}^{(j,v)}$ with $x_j = v$
for v sampled from training distribution
Compute predictions $\{h(\mathbf{x}^{(j,v)})\}$ and calculate variance σ^2

if $\sigma^2 > \theta_{stab}$ **then**
    Select $j^* = \arg\max_{j \in M(\mathbf{x})} |\phi_j(\mathbf{x})|$
    return (WITHHOLD, j^*)
end if

return (PREDICT, $h(\mathbf{x})$)

Phase 3: Confident Prediction. If both checks pass, the algorithm returns $h(\mathbf{x})$. For our running example, both criteria pass and the algorithm returns (PREDICT, $h(\mathbf{x}_{i,3}) = 1$), recommending ticket creation based on the high urgency, database-related issue type, and business impact, even without explicit confirmation of an error message.

Design Rationale and Thresholds. The two checks address complementary failure modes: missing-impact flags absent influential features, while stability flags decision boundaries passing through the uncertainty region of those features. Both prioritize queries by SHAP importance, minimizing customer interaction per query. Lower thresholds increase caution; higher thresholds increase automation. As a concrete starting point, we recommend calibrating θ_{miss} from the empirical distribution of MissingImpact on a held-out set (e.g., the 75th percentile, flagging the most uncertain quartile) and θ_{stab} from the observed prediction variance on the same set; both can then be shifted along the automation–quality trade-off.

5 Application and Experimental Evaluation

We validate our pipeline artifact through application to Nosco[1], a commercial CRM platform that manages customer support interactions for small- to medium-sized businesses. This section presents the experimental setup, evaluates our approach on a reproducible synthetic dataset, and reports results from applying the full pipeline to real-world customer support data[2].

5.1 Experimental Setup

Nosco processes email communications between businesses and customers. Support agents manually decide whether incoming emails should be converted into formal support tickets, requiring understanding of technical issues, urgency, business impact, and historical context. This business process is not explicitly documented and relies on agent experience; the email data constitutes a form of unstructured process execution data from which no structured event log can be directly derived.

We collected 800 email threads from Nosco's database (only threads that resulted in ticket creation). The dataset was structured with one instance per email, representing cumulative knowledge up to that point. Using forward-filling within each thread, emails prior to ticket creation receive label $y_{i,j} = 0$, while the email triggering ticket creation receives $y_{i,j} = 1$, resulting in 1,500 total instances with natural class imbalance. We used stratified 5-fold cross-validation.

For feature extraction, we employed `Llama 3.1 8B Instruct` (primary) and `Qwen 2.5 7B Instruct` (validation) via Openrouter API. We deliberately selected models in the 7–8B range: they can be deployed locally so that sensitive customer communications need not be sent to third-party services, and they keep the pipeline reproducible. `Qwen 2.5 7B` serves as an independent second model to cross-check extraction robustness. Feature embedding used `mxbai-embed-large-v1` (1024 dimensions), chosen for its strong performance on semantic-similarity benchmarks and its suitability for the short *name+description* strings being embedded. For classification, we benchmarked six models: CATBoost, XGBoost, LightGBM, Random Forest, Explainable Boosting Machine (EBM), and Naive Bayes (baseline), using default hyperparameters for each model. Gradient-boosted decision trees are well suited to this setting: they handle mixed categorical and numerical features with frequent missing values natively, are consistently strong on tabular data of this scale, and admit exact TreeSHAP explanations. Benchmarking six models rather than fixing one keeps the choice empirical; EBM and Naive Bayes additionally serve as interpretable references.

For the baseline comparison, we additionally use a publicly available synthetic dataset of 350 email threads covering the same domain of industrial equipment

[1] https://leanbit.eu/en/.

[2] The source code for the experiments is available at https://doi.org/10.5281/zenodo.20392524.

and CRM support. The dataset includes 238 threads that resulted in ticket creation and 112 threads (32%) where issues were resolved without escalation, providing both positive and true-negative instances. Thread lengths vary from 2 to 9 emails (mean $\approx$ 5, std $>$ 1.5), with ticket creation occurring at variable points in the conversation (not always the last email), yielding 1,692 total instances with a positive rate of ~14%. This design avoids the trivial correlation between email position and label that would bias metadata-only baselines, enabling reproducible comparison against alternative approaches.

5.2 Evaluation on Synthetic Dataset

To assess the value of our LLM-based feature extraction approach, we compare against six baselines on the publicly available synthetic dataset (350 threads, 1,692 instances, ~14% positive rate). We apply the same pipeline to this dataset, producing LLM-extracted features, and compare against: (1) *TF-IDF*: term frequency–inverse document frequency [1], bag-of-words features that weight each term by how characteristic it is of a message (500 features, 1–2 g) with classifiers, (2) *Sentiment/Linguistic*: VADER (a lexicon- and rule-based sentiment analyzer) [12] sentiment scores, urgency/frustration keyword counts, and linguistic indicators (question/exclamation counts, all-caps ratio), (3) *Rule-Based*: handcrafted heuristic rules based on thread length, urgency keywords, and consecutive customer messages, (4) *Metadata-only*: conversational features (email count, sender patterns, word counts), and (5–6) *End-to-End LLM*: directly prompting Llama 3.1 8B in both zero-shot and few-shot (with chain-of-thought and escalation criteria) variants. All ML-based models use 5-fold stratified cross-validation with default hyperparameters.

Table 1 presents results across all baselines. The **Model** column reports, for each feature set, the best-performing classifier among the gradient-boosting models evaluated (CATBoost, XGBoost, LightGBM); rows therefore differ (e.g., CATBoost for LLM-based features, LightGBM for metadata). The *Rule-Based* and *End-to-End LLM* rows are not learned classifiers and are labeled by their method (heuristic, zero-shot, few-shot). Three key findings emerge. *First, LLM-extracted features provide strong complementary value.* Combining all feature sources achieves the best performance (AUC: 0.90 $\pm$ 0.05, F1: 0.85 $\pm$ 0.02), with LLM features improving metadata-only performance by 0.13 AUC and TF-IDF-only by 0.27 AUC, demonstrating that semantic, structural, and lexical signals capture distinct predictive information. *Second, standalone baselines show limited discriminative power.* TF-IDF (AUC: 0.61), metadata (AUC: 0.70), and sentiment features (AUC: 0.69) each carry moderate but insufficient predictive value. The rule-based heuristic achieves high precision (0.80) but very low recall (0.24). Critically, LLM-extracted features uniquely enable SHAP-based explanations that identify *which specific factors* drive predictions, enabling the active feature learning strategy described in Sect. 4. *Third, end-to-end LLM classification performs worst.* Both zero-shot (AUC: 0.46) and few-shot (AUC: 0.48) variants fail to make nuanced timing decisions, validating our design choice of using LLMs as feature extractors rather than end-to-end decision makers.

Table 1. Baseline comparison on publicly available synthetic dataset (350 threads, 1,692 instances, ~14% positive rate; 5-fold CV, mean ± 95% CI). Best result per column in bold.

Features	Model	AUC	Precision	Recall	F1
LLM-Extracted Features (Ours)					
LLM + Metadata	CATBoost	0.83 ± 0.05	0.72 ± 0.04	0.84 ± 0.05	0.77 ± 0.05
LLM + TF-IDF	CATBoost	0.88 ± 0.07	**0.80 ± 0.04**	0.87 ± 0.04	0.83 ± 0.02
All Combined	CATBoost	**0.90 ± 0.05**	0.79 ± 0.04	**0.91 ± 0.02**	**0.85 ± 0.02**
Text-Based Baselines					
TF-IDF	CATBoost	0.61 ± 0.09	0.56 ± 0.09	0.59 ± 0.10	0.57 ± 0.09
Sentiment/Ling.	XGBoost	0.69 ± 0.08	0.65 ± 0.05	0.65 ± 0.09	0.65 ± 0.06
Structural Baselines					
Metadata only	LightGBM	0.70 ± 0.02	0.62 ± 0.02	0.72 ± 0.05	0.66 ± 0.02
Rule-Based	Heuristic	–	0.80 ± 0.13	0.24 ± 0.05	0.37 ± 0.06
Direct LLM Classification					
End-to-End LLM	zero-shot	0.46 ± 0.00	0.46 ± 0.00	0.45 ± 0.00	0.46 ± 0.00
End-to-End LLM	few-shot	0.48 ± 0.00	0.49 ± 0.00	0.79 ± 0.00	0.60 ± 0.00

5.3 Application to Real-World Customer Support

We now apply the complete pipeline to real-world customer support interactions from Nosco.

Feature Extraction Results. We evaluated the feature extraction pipeline by applying it to 100 randomly sampled email threads and measuring the reduction in features through semantic clustering.

The LLM generated 785 initial features across diverse categories: damage and repair indicators, equipment specifications (including product models), return and shipping logistics, follow-up requirements, documentation requests, and contextual metadata (e.g., `urgency`, `warranty_status`). Only `company` and `product` were pre-existing CRM metadata; all other features were extracted from email content.

We embedded each feature using `mxbai-embed-large-v1` and applied HDBSCAN with $\epsilon = 0.5$ and min_samples $= 2$, which identified 24 semantically coherent clusters from 257 features. The remaining 528 features (67.26% of all features) were classified as noise by HDBSCAN. Domain experts reviewed both clusters and noise, identifying 11 additional informative features from noise that had insufficient density for automatic clustering.

Following clustering, domain experts performed post-processing based on feature correlation analysis and missingness patterns. Features with ambiguous definitions were removed (e.g., `issue_described_clearly`, `follow_up_required`). Highly correlated features (Cramér's $V > 0.85$) were merged (e.g., `repair_action_by_servien dce_provider`, `repair_action_taken` with correlation up to 0.93 were merged into `customer_service_action`). The two domain experts (support managers from the Nosco platform with 3+ years of customer support experience) completed this review in approximately 4 h of total effort, substantially less than manual feature engineering from scratch.

The consolidation process reduced the feature set from 785 to 35 features (95.5% reduction in feature space). The 35 consolidated features comprised: 12 categorical features, 22 boolean features, and 1 numerical feature. We used `Llama 3.1 8B Instruct` to extract values for all 35 features across 1,500 email instances (cumulative knowledge at each email point), producing the final dataset. Missing values were present in all of the features, primarily in features requiring specific technical details not always mentioned in emails (e.g., `damage_consequences`, `cause_of_damage`, `error_code`).

To assess LLM extraction robustness, we validated a subset of extractions using `Qwen 2.5 7B Instruct` as a secondary model. The two models showed consistent extraction patterns, with discrepancies primarily occurring in ambiguous cases where the email text did not clearly indicate a feature value, exactly the cases that result in missing values in the final dataset.

Predictive Performance. We benchmarked six classification models on the Nosco dataset (800 threads, 1,500 instances) for predicting ticket creation timing using LLM-extracted features. We report precision, recall, F1, and ROC-AUC with 95% confidence intervals (5-fold CV).

Table 2 presents results from 5-fold cross-validation. CATBoost achieved the best overall performance with ROC-AUC of 0.73 ± 0.04, F1 score of 0.67 ± 0.04, and balanced precision (0.67 ± 0.03) and recall (0.67 ± 0.04). Its native handling of categorical features through ordered target encoding contributed to superior performance. CATBoost obtained narrow confidence intervals across all metrics, demonstrating stability across data subsets. XGBoost (AUC: 0.65 ± 0.05) and LightGBM (AUC: 0.66 ± 0.05) showed performance comparable to Naive Bayes baseline (AUC: 0.65 ± 0.05), while EBM (AUC: 0.66 ± 0.04) showed competitive performance. Random Forest provided lower performance (AUC: 0.63 ± 0.05). These results confirm CATBoost's superior performance for predicting ticket creation timing.

Active Feature Learning. We evaluated whether the SHAP-based active feature learning strategy (Algorithm 1) improves prediction quality by selectively withholding predictions when confidence is low. We analyzed 150 test instances from the Nosco dataset, applying the algorithm to each to determine whether to predict immediately or withhold and request additional features.

Table 3 presents the results. The baseline configuration uses CATBoost making predictions on all 150 test instances without filtering, achieving performance consistent with the 5-fold cross-validation results (AUC: 0.73, Precision: 0.67, Recall: 0.67). With confidence filtering via Algorithm 1, predictions were made on only 76 instances (50.7% coverage) that satisfied both criteria. Precision

Table 2. Performance comparison of classification models on Nosco dataset (5-fold CV, mean ± 95% CI)

Model	AUC	F1	Precision	Recall
CATBoost	**0.73 ± 0.04**	**0.67 ± 0.04**	**0.67 ± 0.03**	**0.67 ± 0.04**
XGBoost	0.65 ± 0.05	0.59 ± 0.03	0.59 ± 0.03	0.59 ± 0.03
LightGBM	0.66 ± 0.05	0.61 ± 0.05	0.61 ± 0.05	0.61 ± 0.05
Random Forest	0.63 ± 0.05	0.57 ± 0.04	0.57 ± 0.04	0.57 ± 0.04
EBM	0.66 ± 0.04	0.62 ± 0.05	0.62 ± 0.04	0.62 ± 0.04
Naive Bayes	0.65 ± 0.05	0.61 ± 0.04	0.61 ± 0.04	0.61 ± 0.04

increased from 0.67 to 0.74 (10.4% relative), reducing false positives. ROC-AUC improved from 0.73 to 0.75. Recall decreased from 0.67 to 0.61 (9.0% relative), reflecting reduced coverage from withholding uncertain predictions.

Table 3. Impact of SHAP-based active feature learning on 150-instance test set (Nosco)

Configuration	AUC	Precision	Recall	F1	Coverage
Baseline (all predictions)	0.73	0.67	0.67	0.67	100%
With Algorithm 1 filtering	**0.75**	**0.74**	0.61	0.67	50.7%

To illustrate Algorithm 1, we present a case where CATBoost predicted ticket creation (probability = 0.521) with weak confidence. The algorithm computes missing feature impact (Phase 1). With θ_{miss} calibrated from the SHAP value distribution, this impact remains below the threshold, so the algorithm proceeds to Phase 2.

Table 4. Selected simulation results showing how different values for missing features `errors` and `user_action` affect prediction (baseline probability: 0.521). Label flip occurs when probability drops below 0.5.

errors	User_Action	Probability	Label Flip
Alarm_Indicated	no	0.47	Yes
Alarm_Indicated	yes	0.48	Yes
Software_Mentioned	no	0.38	Yes
Software_Mentioned	yes	0.39	Yes
Welding_Performance	yes	0.50	No

The algorithm simulates plausible imputations. Table 4 shows selected results. Predicted probabilities range from 0.38 (Software_Mentioned) to 0.50

(Welding_Performance), with a variance exceeding θ_{stab}. The stability criterion being violated, the algorithm returns (WITHHOLD, $j^* =$ errors), where errors has the highest $|\phi_j| = 0.14$ among missing features. This demonstrates that missing critical technical details prevented the model from making an accurate decision. Notably, this type of targeted information request is only possible with interpretable LLM-extracted features; metadata baselines cannot identify which specific information to request from customers.

6 Discussion

Interpreting the Results. Combining LLM-extracted, structural, and lexical features gives the best performance (ROC-AUC 0.90 on the synthetic dataset), and each source adds over the others, so the three capture different signals: semantic context, conversational structure, and surface lexical cues. End-to-end LLM classification performs worst (ROC-AUC 0.46–0.48), because deciding *when* to escalate requires weighing cumulative, partly quantitative evidence across a thread, which a single generative pass does not handle reliably. This supports our positioning of the LLM as a feature extractor rather than a decision maker: the predictive model itself uses established gradient-boosting methods, so the baseline comparison isolates the value added by LLM-extracted features rather than advancing classification. The drop from synthetic (0.90) to real-world (0.73) performance reflects the noise, class imbalance, and missingness of real email data; we treat 0.73 as a realistic lower bound for deployment.

Treating Withheld Predictions as Abstentions. When Algorithm 1 withholds, it issues a *request for information*, not a "do not escalate" decision. Scoring a withheld instance as a negative prediction would conflate uncertainty with a substantive decision and penalize the system for recognizing that it lacks information. We therefore evaluate withheld instances as abstentions and report coverage alongside precision and recall, making the precision–coverage trade-off explicit (Table 3).

Scalability of the Human-in-the-Loop Step. Expert involvement is limited to schema construction, takes roughly four hours in our study (well below the manual feature engineering required by prior work), and is incurred once rather than per thread. It can be reduced further by letting the LLM pre-rank clusters by coherence so that experts confirm only low-confidence ones.

7 Conclusion and Future Work

This paper presented a novel pipeline artifact for automating process decisions from unstructured data, demonstrated through ticket escalation in customer support processes. The pipeline combines LLM-based feature extraction with semantic clustering and SHAP-based active feature learning, addressing a gap in BPM tooling where existing approaches require structured event logs. Validated on 1,500 email instances from 800 real-world threads from Nosco, our approach consolidates 785 LLM-extracted features into 35 representations (95.5% reduction), achieves ROC-AUC of 0.90 when combining all feature sources on a synthetic dataset and 0.73 with CATBoost on real-world data, and improves precision from 0.67 to 0.74 through confidence-based filtering.

Our work has several limitations. First, LLM-based feature extraction is stochastic across models and prompts; we mitigated this by validating with Qwen 2.5 7B and involving experts in consolidation. Second, the pipeline was validated on one domain (industrial equipment and CRM support); transferability requires further study. Third, threads are treated independently, ignoring cross-thread process state (e.g., queue load, SLA deadlines). Fourth, the active feature learning strategy is evaluated indirectly via downstream precision and coverage; a direct comparison against an oracle is left for future work. Fifth, the stability check imputes missing values from the marginal training distribution rather than conditioning on thread context, which may mis-state instability for atypical threads. Sixth, the real-world Nosco dataset contains only threads that resulted in ticket creation (within-thread negatives arise through forward-filling), which may bias the real-world evaluation. Finally, the real-world data cannot be publicly released due to privacy constraints; we mitigate this by releasing the code and a public synthetic dataset.

References

1. Aizawa, A.N.: An information-theoretic perspective of TF-IDF measures. Inf. Process. Manag. **39**(1), 45–65 (2003)
2. Arora, S., et al.: Language models enable simple systems for generating structured views of heterogeneous data lakes. Proc. VLDB Endow. **17**(2), 92–105 (2023)
3. Bayer, M., Lutz, J., Reuter, C.: ActiveLLM: large language model-based active learning for textual few-shot scenarios. arXiv e-prints arXiv:2405.10808 (2024)
4. Chen, D., Alnassar, S.A., Avison, K.E., Huang, R.S., Raman, S.: Large language model applications for health information extraction in oncology: scoping review. JMIR Cancer **11**, e65984 (2025)
5. Dumas, M., La Rosa, M., Mendling, J., Reijers, H.A.: Fundamentals of Business Process Management, 2nd edn. Springer (2018). https://doi.org/10.1007/978-3-662-56509-4
6. Edge, D., et al.: From local to global: a graph RAG approach to query-focused summarization. CoRR abs/2404.16130 (2024)
7. Elleuch, M., Assy, N., Laga, N., Gaaloul, W., Ismaili, O.A., Benatallah, B.: A meta model for mining processes from email data. In: SCC, pp. 152–161. IEEE (2020)
8. Elleuch, M., Ismaili, O.A., Laga, N., Gaaloul, W., Benatallah, B.: Discovering activities from emails based on pattern discovery approach. In: Fahland, D., Ghidini, C., Becker, J., Dumas, M. (eds.) BPM Forum, pp. 88–104. Springer, LNBIP (2020). https://doi.org/10.1007/978-3-030-58638-6_6
9. Francescomarino, C.D., Ghidini, C.: Predictive process monitoring. In: van der Aalst, W.M.P., Carmona, J. (eds.) Process Mining Handbook, LNBIP, vol. 448, pp. 320–346. Springer (2022). https://doi.org/10.1007/978-3-031-08848-3_10

10. Gupta, S.: A hybrid machine learning framework of gradient boosting decision tree and sequence model for predicting escalation in customer support. In: IEEE International Conference on Big Data, pp. 5511–5518. IEEE (2020)
11. Huang, L.: A survey on hallucination in large language models: Principles, taxonomy, challenges, and open questions. ACM Trans. Manag. Inf. Syst. **43**(2) (2025)
12. Hutto, C.J., Gilbert, E., Adar, E., Resnick, P., Choudhury, M.D., Hogan, B.: VADER: a parsimonious rule-based model for sentiment analysis of social media text. Presented at the (2014)
13. Jeong, D.P., Lipton, Z.C., Ravikumar, P.K.: LLM-select: feature selection with large language models. Trans. Mach. Learn. Res. (2025)
14. Ke, G., et al.: LightGBM: a highly efficient gradient boosting decision tree. In: Guyon, I., et al. (eds.) NeurIPS, pp. 3146–3154 (2017)
15. Khandaker, F., et al.: Transformer models for mining intents and predicting activities from emails in knowledge-intensive processes. Eng. Appl. Artif. Intell. **128**, 107450 (2024)
16. Li, Y., Oliva, J.: Active feature acquisition with generative surrogate models. In: ICML Proceedings of Machine Learning Research, vol. 139, pp. 6450–6459 (2021)
17. Li, Y., Tan, Z., Liu, Y.: Privacy-preserving prompt tuning for large language model services. arXiv e-prints arXiv:2305.06212 (2023)
18. Lundberg, S.M., et al.: From local explanations to global understanding with explainable AI for trees. Nat. Mach. Intell. **2**(1), 56–67 (2020)
19. Lundberg, S.M., Lee, S.: A unified approach to interpreting model predictions. In: Guyon, I., et al. (eds.) NeurIPS, pp. 4765–4774 (2017)
20. Ma, C., et al.: EDDI: efficient dynamic discovery of high-value information with partial VAE. In: ICML, vol. 97, pp. 4234–4243. PMLR (2019)
21. McInnes, L., Healy, J., Astels, S.: HDBSCAN: hierarchical density based clustering. J. Open Source Softw. **2**(11), 205 (2017)
22. Mitchell, T.M., Wang, S.H., Huang, Y., Cheyer, A.: Extracting knowledge about users' activities from raw workstation contents. In: AAAI, pp. 181–186. AAAI Press (2006)
23. Mo, B., et al.: KGGEN: extracting knowledge graphs from plain text with language models. CoRR abs/2502.09956 (2025)
24. Montgomery, L., Damian, D., Bulmer, T., Quader, S.: Customer support ticket escalation prediction using feature engineering. Requir. Eng. **23**(3), 333–355 (2018)
25. Padella, A., de Leoni, M., Dogan, O., Galanti, R.: Explainable process prescriptive analytics. In: Burattin, A., Polyvyanyy, A., Weber, B. (eds.) ICPM, pp. 16–23. IEEE (2022)
26. Park, G., van der Aalst, W.M.: Operational process monitoring: an object-centric approach. Comput. Ind. **164**, 104170 (2025)
27. Sun, T., Shao, Y., Qian, H., Huang, X., Qiu, X.: Black-box tuning for language-model-as-a-service. In: ICML Proceedings of Machine Learning Research, vol. 162, pp. 20841–20855. PMLR (2022)
28. Xu, Y., Sun, F., Zhang, X.: Literature survey of active learning in multimedia annotation and retrieval. In: Lu, K., Mei, T., Wu, X. (eds.) ICIMCS, pp. 237–242. ACM (2013)

What Do Process Embeddings Learn? A Probing Investigation

Franziska Köllschen(✉), Michael Grohs, and Jana-Rebecca Rehse

University of Mannheim, Mannheim, Germany
{franziska.koellschen,michael.grohs,rehse}@uni-mannheim.de

Abstract. Process representation learning transforms event data into numeric vectors, called embeddings, that machine learning models can use for downstream tasks like prediction, clustering, or anomaly detection. To support these tasks, the goal is to create general-purpose embeddings that encode all relevant data properties. However, embeddings are typically evaluated by downstream task performance, assessing only task-specific quality. Thus, it remains unclear which properties of the event data are encoded in the embeddings and how accessible they are, making it difficult to design, improve or compare task-agnostic general-purpose embeddings. To address this issue, we present a probing evaluation framework that investigates embeddings of process instances by training small classifiers, called probes, to recover predefined process properties. We define seven probing tasks, each dedicated to one property, ranging from simple ones, like duration, to more complex ones, like order relations. We apply the framework to embeddings from three existing representation learning approaches on synthetic and real-life event data. The results show that the embeddings reliably encode simple properties, but have difficulties with more complex ones.

Keywords: Process Mining · Representation Learning · Probing

1 Introduction

Representation learning aims to transform data into low-dimensional vectors, known as embeddings, that can be used to solve downstream tasks [5]. Embeddings are typically learned using machine learning (ML) models such as neural network autoencoders [12], trained to capture relevant data properties for tasks like classification or prediction [5]. Following its success in other domains, representation learning was applied in process mining [23], where the primary data source are event logs that record process executions as traces. Learned embeddings can represent different elements of a log, including activities, traces, or the log itself [9]. Prior research suggests that suitable embeddings improve ML performance in downstream process mining tasks, including trace clustering, anomaly detection, and prediction [23]. Accordingly, the goal in process representation learning is to learn general-purpose embeddings that encode properties

F. M. Maggi et al. (Eds.): BPM 2026, LNCS 16936, pp. 464–484, 2027.
https://doi.org/10.1007/978-3-032-37777-7_28

of the process that are task-agnostically relevant in a compact and accessible form [5].

Since general-purpose process embeddings are intended to encode relevant event data properties independent of downstream tasks, they are ideally evaluated directly on these properties [21]. In practice, however, they are often evaluated only indirectly through task-specific performance [9,12,19,26], which can be influenced by many factors beyond the embedding. Such evaluations only assess how well embeddings support the downstream task, but not which properties they encode how accessibly [7]. As this provides little insight into their representational quality, i.e., the ability to represent features in a condensed form [8], learned embeddings largely remain a black box. The lack of transparency makes it difficult to determine which information is captured, complicating explanations of why downstream models produce certain outcomes [23]. Also, it hinders the systematic design, improvement, and comparison of embeddings as it is unclear if relevant properties are captured, particularly in unexplored settings.

The lack of clarity about what properties are encoded in embeddings is a general problem in representation learning, not only in process mining [5]. One option to address this is via *probing*. Probing refers to the idea of creating small classifiers, called probes, that aim to recover properties of the original data from embeddings [4]. If probe performance indicates that a probe can accurately recover the property, it is concluded that the embedding encodes this property in an accessible way [8]. Thus, probing is a first step towards opening the black box of learned embeddings [1]. Besides its application in natural language processing (NLP) [1,8], it was used in computer vision [3].

In this paper, we transfer the idea of probing to process mining and introduce a probing evaluation framework that analyzes which properties learned trace embeddings encode. The framework defines seven probing tasks, each aiming to recover one property. They range from simple properties like trace duration to more complex ones like activity order relations. Recoverability is measured by the probe performance. Since that can also be influenced by other factors such as probe architectures, training procedures, or data characteristics [4], the framework introduces control mechanisms to ensure that probe performance only indicates recoverability. We experimentally apply the framework by generating trace embeddings with existing approaches and training a separate probe for each probing task. Our results indicate that simple properties can be recovered better than more complex ones.

The paper is structured as follows: After background and related work in Sect. 2, Sect. 3 introduces the probing evaluation framework. The experimental application is illustrated in Sect. 4 and discussed in Sect. 5. Finally, Sect. 6 concludes the paper.

2 Background and Related Work

We discuss background and other work on process representation learning and probing.

2.1 Process Representation Learning

Representation learning describes the task of "learning representations of the data that make it easier to extract useful information when building classifiers or other predictors" [5, p.1798]. In many modern approaches, these representations are learned in form of embeddings, i.e., dense numerical vectors that encode relevant properties of the data. A common use case is NLP, where words or sentences should be represented in a semantically meaningful manner [1]. The idea was translated to process data, aiming to encode process-related properties relevant for a multitude of downstream tasks.

Existing Process Representation Learning Approaches. Early work introduced architectures for learning embeddings of activities (*act2vec*), traces (*trace2vec*), event logs (*log2vec*), and process models (*model2vec*) [9]. These approaches train neural networks to predict activities from their context, thereby learning embeddings of process entities as model parameters. Building on this idea, the *case2vec* approach incorporates additional event and trace attributes such as executing resources [18]. Other approaches learn trace embeddings from activities and additional event attributes by predicting trace attributes [26] or propose context-based event embeddings [12] and graph-based event log embeddings that incorporate domain knowledge [19]. Another approach, called Multi-Perspective Process Network (MPPN), encodes all possible types of trace and event attributes by transforming them into image-like representations and combining them into a single trace embedding with convolutional neural networks (CNNs) [23].

Investigation of Learned Embeddings. Although the presented approaches aim to produce general-purpose embeddings, they are typically evaluated only through downstream tasks. To address this, several studies have investigated learned representations more directly. One study assesses whether neural networks can learn process structure [20] by training a recurrent neural network on next-event prediction and using it to simulate a new log to analyze whether the model reproduces known behavior and generates unseen traces. However, the focus is on the behavior of the trained model rather than on embeddings. Further, some studies compare multiple trace encoding techniques, including manually designed ones as well as learned encodings like embeddings, with respect to classification performance and feature space characteristics [2], as well as expressivity, scalability, and correlation power [27]. Recent work considers multi-perspective encodings, including learned trace embeddings, that incorporate event attributes beyond activities [25]. Although these studies guide the selection of encoding techniques, they do not assess which information learned embeddings encode.

Use of Learned Embeddings in Downstream Tasks. Beyond explicitly studying learned embeddings, many studies investigate them implicitly by proposing encoding schemes and learning architectures for tasks such as anomaly detection, clustering, and process prediction [21]. Some also study whether existing representation learning approaches such as MPPN can be used for predictive purposes [15]. Others propose performing tasks such as process discovery and conformance checking directly in embedding spaces derived from ML models [13]. However, these approaches focus on improving downstream task performance rather than analyzing embeddings.

2.2 Probing

Probing aims to assess whether learned embeddings encode properties of the represented data by testing if these properties can be recovered from the embedding [8]. A small classifier, called probe, receives an embedding as input and is trained to recover a specific property [4]. Probes are intentionally simple, e.g., logistic regression or shallow neural networks, so that good performance indicates accessibility of the property in the embedding rather than the power of the model [4,24].

Probing has mainly been studied in NLP, where it is used to analyze sentence embeddings [8]. As an example, Fig. 1 shows the pipeline for probing a sentence embedding, generated by any representation learning model [4]. Using this embedding as input, probes are trained for three different probing tasks that refer to linguistic properties: content ("Is the sentence about sports?"), tense ("Is the sentence in present tense?'), and length ("Is the sentence long?"). Probing has also been used to investigate linguistic properties such as word order [1] or the number of subjects and objects [8]. As it is agnostic to the used embedding, it was also applied in domains like computer vision [3].

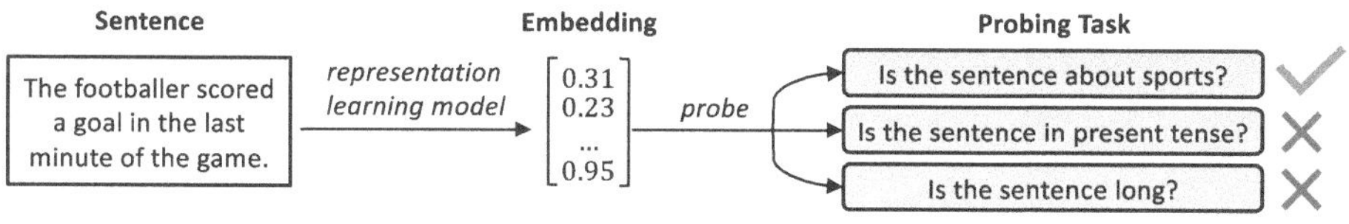

Fig. 1. Illustration of probing in natural language processing (inspired by [3])

3 Probing Evaluation Framework

Transferring the idea of probing to the process mining domain, this section presents a probing evaluation framework for learned trace embeddings that assesses whether the embeddings encode specific process properties. As shown in Fig. 2, the framework can be applied to trace embeddings generated by any representation learning model. Probes are trained on probing tasks, each implemented as a supervised classifier with embeddings as input and discrete property labels as output [1,4,8]. Each probe aims to recover a specific property. Representation learning model and probes are trained on traces from an event log (or their embeddings, respectively) and tested on a disjoint set of traces from the same log. Probe performance on the test traces indicates how well the respective property can be recovered, but can also be influenced by other factors. To ensure results reflect only the recoverability of probed properties, the framework includes control mechanisms. We now introduce probing tasks and control mechanisms.

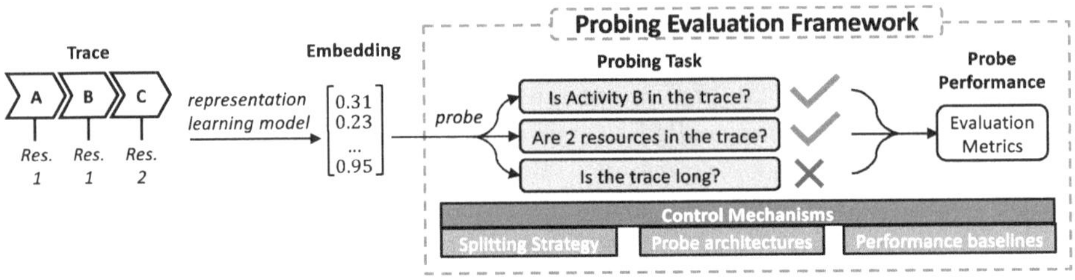

Fig. 2. Illustration of probing process embeddings

3.1 Probing Tasks

To assess which process-related properties are encoded in learned trace embeddings, we define seven probing tasks. Each task investigates a specific property by training one probe to predict the corresponding label. Following the common probing paradigm, all probing tasks are formulated as supervised classification problems [4] . Alternatives like regression are possible for numerical properties as event count, but our goal is to assess if information about a property is recoverable and not to reconstruct exact values. Thus, classification is suited better to evaluate if numerical properties are accessible.

Table 1 shows the tasks and their label types. We select those tasks because they refer to different process perspectives (control-flow, time, resources) and range from simple (activity presence) to more complex (variant identification). Together, they cover a diverse set of properties. However, this selection is not exhaustive: it represents an initial suggestion, but the framework readily accepts additional tasks.

Activity Presence (AP). This task evaluates whether the embedding encodes the presence of specific activities. We exclude activities that occur in over 85% or under 15% of traces to avoid highly imbalanced labels. Imbalance allows probes to perform well through majority predictions, but provide little evidence about the encoding. This restriction to intermediate frequencies thus yields a better assessment of recoverability. The task is formulated as a multi-label binary classification, with one label per activity.

Table 1. Overview of probing tasks, their questions, example labels, and types.

Probing Task	Probing Question	Example Label	Type
Activity Presence	Are these activities in the trace?	[0,1,0,0,1,0,1]	multilabel
Event Count	How many events are in the trace?	$[0 :< 5, 1 : 5 - 8, 2 :> 8]$	multiclass
Order Relation	Is Act. A before Act. B in the trace?	[0:false, 1:true]	binary
Variant ID	What is the control-flow variant of the trace?	[0--17]	multiclass
Loop Presence	Is there a loop in the trace?	[0:false, 1:true]	binary
Trace Duration	What is the total trace duration?	$[0 : q_{0.2}, \ldots, 4 : q_1]$	multiclass
Resource Count	How many resources participated?	[0:3--5, 1:6, 2:7-9]	multiclass

Event Count (EC). This tasks evaluates whether the embedding encodes the number of events in a trace. We discretize the count into bins to reduce class imbalance and mitigate skewed value distributions, which frequently occur in event logs [7]. The number of bins and their boundaries are determined by selecting the most balanced class distribution among all possible configurations with three to nine bins. For example, we may use three bins covering counts of <5, 5-8, and >8. This allows us to evaluate if different event counts are separable in the embedding, rather than reconstructing exact values. The task is formulated as a multiclass classification, with one class per bin.

Order Relation (OR). This task evaluates whether the embedding encodes the pairwise ordering between two activities. We select one activity pair, denoted as A and B, that has the highest co-occurrence frequency and is present in both ordering configurations. This ensures adequate label support and a non-trivial prediction task as well as avoids highly skewed label distributions, especially since few pairs typically are suitable candidates. Only traces containing both activities are considered to strictly evaluate ordering information and not require the probe to also detect activity presence. The task is formulated as a binary classification, predicting if activity A precedes activity B or vice versa.

Variant ID (VID). This task evaluates whether the embedding encodes the control-flow variant of a trace. Traces have the same variant if their sequence of executed activities is identical. This task serves as a diagnostic indicator of potential shortcut behavior in other probes: If an embedding accessibly encodes variant identity, another probe (e.g., for event count) could implicitly identify the variant and then rely on a fixed association between variant and label learned from the training data. Thus, it would not recover the target property directly from the embedding, but instead use variant identity as a lookup signal. The task is formulated as a multiclass classification, with one class per variant.

Loop Presence (LP). This task evaluates whether the embedding encodes the presence of loops, defined as the repetition of at least one activity within a trace.

The task is formulated as a binary classification, where the label 1 indicates loop presence.

Trace Duration (TD). This task evaluates whether the embedding encodes trace durations. As EC, TD is discretized to mitigate skewed distributions and not recover exact values. Durations are binned into five equal-frequency quantile bins because they are continuous. The task is formulated as a multiclass classification, with one class per bin.

Resource Count (RC). This task evaluates whether the embedding encodes the number of distinct resources involved in a trace. Again, this numeric property is discretized into bins to mitigate skewed class distributions, using the same procedure as EC. The task is formulated as a multiclass classification, with one class per bin.

3.2 Control Mechanisms

Probing is conducted in an ML setting where multiple components interact, including representation learning models, probes, datasets, and training procedures. Measuring probe performance may thus reflect factors other than the recoverability of a property from an embedding. For example, a highly expressive probe may recover a property from indirect signals in the embedding, potentially overstating its direct accessibility. This threatens the internal validity of probing, which aims at linking probe performance and encoded properties. To mitigate external influences and preserve that link between performance and property, our framework introduces control mechanisms. Concretely, it specifies the splitting strategy, probe architectures, and performance baselines.

Splitting Strategy. In ML experiments like probing, models are trained on one portion of the data (train set) and evaluated on a separate portion (test set) to assess whether they generalize to unseen instances. If this split into train and test set is not carefully designed, probe performance may be inflated because overlap between train and test data allows the probe to memorize recurring label patterns observed during training. In addition, insufficient label support or severe class imbalance can make probe training unstable or bias evaluation metrics toward dominant classes. Shifts in label distributions between the training and test set additionally challenge the probe. To account for that, the framework defines a splitting strategy, organized into general splitting principles, two complementary splitting conditions, and a defined split selection procedure.

General Splitting Principles. Splitting is performed once for each used event log. Each trace is assigned to either train or test set to ensure that none appears in both. This split is fixed consistently for all experiments, both for representation learning and probing. This ensures that probes are evaluated only on traces not used to learn the embeddings.

Splitting Conditions. Splitting should strive for comparable data distributions and sufficient label support in train and test set, but avoid direct overlap (called leakage). Here, event logs introduce an additional challenge as many traces typically follow the same control-flow variant [22]. If identical variants occur in both train and test set, probes may exploit this leakage to learn associations between variants and labels and use these for the task instead of recovering the probed property from the embedding. Prior work discusses similar risks of leakage and shortcut learning on process data, including the specific case of variant leakage [22, 28]. The split design thus directly affects whether performance reflects accessibility or trivial solutions based on variant memorization.

This issue creates a tension between two objectives: On the one hand, we aim to avoid variant leakage, which requires separating control-flow variants between train and test set so that probes cannot rely on variantlabel associations. On the other hand, we aim to maintain similar distributions between train and test set to provide a controlled setting in which the probe is not additionally challenged by shifts in label distributions. Such distribution shifts between train and test data may alter correlations in the training set, causing the probe to associate the label with other trace characteristics rather than the probed property. Because enforcing variant separation may introduce such shifts between splits, both objectives cannot be fully satisfied with a single splitting strategy. To account for that, the framework employs two complementary splitting conditions:

- The *distribution-controlled condition* prioritizes balanced label distributions and stable probe training. Variants may appear in both train and test sets, which preserves similar data distributions across the splits. Probe performance under this condition therefore evaluates recoverability of the probed property in a controlled laboratory-style setting, explicitly allowing the probe to memorize variant-label associations.
- The *variant-disjoint condition* assigns each variant to either training or test split. This prevents variant leakage and ensures that probes cannot rely on variantlabel associations. Probe performance therefore evaluates whether the probed property can be recovered when the probe must generalize beyond variants seen during training.[1]

The two conditions offer complementary insights: The distribution-controlled condition assesses recoverability under controlled distributions, whereas the variant-disjoint condition assesses recoverability when probes must generalize beyond observed variants.

Split Search. To find splits with sufficient label support and comparable distributions, the framework defines a split search procedure. As traces can be assigned to either split, the number of possible splits grows with the number of traces, making exhaustive enumeration infeasible. At the same time, random splits frequently violate label support requirements or introduce distribution shifts. With

[1] The *VID* task is not applicable here, since a classifier cannot predict never-observed variants.

a principled selection mechanism, the framework generates multiple candidate splits and evaluates each in two stages.

(1) *Hard constraints:* Candidate splits must satisfy minimum label support requirements for all probing tasks in both train and test sets. Otherwise, they are discarded.
(2) *Scoring:* A scoring function compares distributional quality of feasible candidates. Per probing task, we evaluate the distance between train and test label distributions, the degree of class imbalance within each split, and the margin by which label counts exceed minimum support. These factors are combined into a weighted objective function, selecting the lowest-scoring split for the respective condition.

The search is identical for both splitting conditions, but the admissible candidate space differs. For distribution-controlled, candidate splits are generated randomly at trace level. For variant-disjoint, candidate splits are restricted to assignments in which control-flow variants are exclusively allocated to one split, reducing the search space.

Probe Architectures. To ensure that probe performance reflects task-specific accessibility, the framework separately trains one probe per task. Probes are implemented as neural network classifiers whose input dimensionality corresponds to the embedding size and whose output dimensionality is determined by the label structure of the respective task. Their capacity is deliberately constrained by limiting their size, in particular their depth and width, which preserves the link between property recoverability and probe performance for two reasons. First, large classifiers may perform well even when the probed property is not directly encoded in the embedding, instead reconstructing it through complex transformations or correlations [4]. In such cases, performance reflects the probes' power rather than representational accessibility. Second, from the perspective of embedding quality, a useful embedding should encode relevant properties in a form that can be accessed without requiring a highly complex decoder [17,24].

To investigate both linear and limited non-linear accessibility without introducing excessive model capacity, the framework employs two probe architectures:

- *Linear classifier (Lin.)* consists of a single affine transformation from the input embedding space to the label space. Such linear probes are commonly used to test whether a property is linearly accessible from embeddings [8]. Good performance indicates that the property is directly separable by a linear decision boundary [14]. Thus, the linear architecture serves as a minimal-capacity diagnostic.
- *Multi-layer perceptron (MLP)* allows limited non-linear interactions between embedding dimensions with restricted depth and width. We employ an MLP with a single hidden layer of 128 neurons and ReLU activation.

Performance Baselines. To establish reference points for for interpreting probe performance [4], the framework includes both trivial and handcrafted baselines.

- The *random baseline* predicts labels uniformly at random according to the label space, providing a chance-level reference independent of distributional skew and establishing minimal expected performance in the absence of any meaningful signal.
- The *majority baseline* always predicts the most frequent class in the training split. It shows what can be achieved by relying on dominant labels and class imbalance.
- *Bag of Activities (BoA)* constitute handcrafted representations of traces as count vectors over the activity vocabulary, inspired by Bag of Words from NLP.

Probe performance that substantially exceeds the prediction baselines indicates that the embeddings provide information beyond trivial distributional regularities, while improvements over handcrafted representations indicate benefits of learned embeddings.

4 Experimental Application of the Probing Framework

In this section, we present an experimental application of the probing framework with the seven proposed probing tasks. All experiments and results can be found online.[2]

4.1 Experimental Setup

The following specifies the event logs, split search, representation learning models, probe training, and evaluation metrics used in the experimental application.

Event Logs. To assess the framework under different levels of control, we conduct experiments with both synthetic and real-life event logs. Synthetic logs are generated from constructed process models. This allows us to examine well-defined conditions with full control over control-flow and distributional properties and no disruptive factors such as temporal drift, dependencies between cases, or logging errors. Complementing this, we evaluate the framework with real-life logs. This enables us to examine if insights obtained under controlled conditions hold in realistic and potentially noisier environments.

[2] https://doi.org/10.6084/m9.figshare.32456568.

Table 2. Characteristics of synthetic logs. *Mid* denotes activities present in 15% to 85% of traces used in AP; A and B denote activity pair used in OR.

Log	Size		AP	EC		OR		VID	LP	TD (h)		RC	
	#Traces	#Events	#Mid	Avg.	Sd.	$A \rightarrow B$	$B \rightarrow A$	#Var	%	Avg.	Sd.	#	Avg.
Simple	10,000	102,783	14 9	10.3	2.1	5,052	4,948	18	0	7.8	6.5	9	6.3
Complex	20,000	262,979	14 3	13.1	3.5	9,076	8,964	548	48.1	30.2	16.0	7	4.5

Table 3. Characteristics of real-life logs. *Mid* denotes activities present in 15% to 85% of traces used in AP; A and B denote activity pair used in OR.

Log	Size		AP	EC		OR		VID	LP	TD (h)		RC	
	#Traces	#Events	#Mid	Avg.	Sd.	$A \rightarrow B$	$B \rightarrow A$	#Var	%	Avg.	Sd.	#	Avg.
BPI17	31,509	475,306	24 10	15.1	4.6	6,376	1,619	5,623	48.5	524.3	310.5	144	4.3
BPI20	10,500	56,437	17 2	5.4	1.5	641	239	99	9.7	276.6	408.5	2	2.0

Synthetic Logs. To evaluate varying complexity, we generate two synthetic scenarios: a *simple* one and a more *complex* one. Per scenario, we define a process model that specifies the control-flow structure. The simple scenario is modeled without loops, resulting in strictly acyclic traces. The complex scenario extends the simple model with optional loops, allowing repeated activity executions and increasing structural variability.

To generate the synthetic logs, we perform a playout of the predefined process models using the tool *CauseCheck* [16]. Time, resource, and trace-level attributes are generated according to predefined distributions to introduce controlled variability across traces. In addition, a trace-level attribute deterministically influences one decision gateway in the process model by requiring managerial approval for cases exceeding a specified threshold. Since CauseCheck does not natively support loops, they are subsequently added using *PM4PY* [6]. Table 2 shows descriptive statistics of the logs.

Real-Life Logs. To complement the synthetic scenarios, we evaluate probing under realistic process conditions using two real-world event logs, summarized in Table 3.

- *BPI17* [10] captures a loan application process with multiple decision points, repeated interactions between roles, and substantial structural variability. To ensure event-activity correspondence, we only consider events with state "complete".
- *BPI20 Domestic Declarations* [11] captures a travel expense reimbursement process with multi-step approvals and possible rejectionresubmission cycles.

The used label definitions and bin boundaries per log and task are shown in our repository. In the complex synthetic log, the highest-frequency pair was excluded for OR due to a loop-induced dependency and replaced by the next most frequent eligible pair.

Split Search. We instantiate the split search with specific parameters. Thereby, we target a traintest ratio of 80:20. For the variant-disjoint split, we allow a deviation of up to five percentage points as complete variants must be assigned exclusively to either training or test. The search procedure is conducted separately for both conditions, starting with the generation of candidate splits. For the distribution-controlled split, we generate 200,000 candidates using different random seeds. For the variant-disjoint split, candidates must have zero variant overlap. As this makes random sampling is inefficient, we generate 2,000 variant-level initializations and iteratively refine them by swapping variants.

For ensuring the hard constraints, each probe label must occur at least 100 times in the train split and 50 times in the test split. We discard candidate splits that violate these thresholds. For the VID probe, whose label distribution is highly skewed, we restrict the set of variants considered for the minimum support to those with a frequency of at least 150% of the minimum required support across both splits (225 occurrences). Other variants remain in the log but are not required to satisfy the minimum support. This prevents rare variants from dominating feasibility and allows the optimization to balance distributions across all probes rather than for VID alone. Among candidate splits that satisfy the hard constraints, we select the one with the lowest objective score.

Representation Learning Models. To generate embeddings for probing, we select three representation learning approaches to assess techniques that differ in the process perspectives they incorporate, their training objectives, and their architectural design.

trace2vec (T2V) [9] adapts Doc2Vec by treating traces as documents and activities as words. It employs the unsupervised PV-DBOW training objective that learns a trace embedding by predicting sampled activities from the current embedding, which is updated after each pass. The model is trained for 100 epochs using a decreasing learning rate starting at 0.025 and negative sampling with five not present activities. The resulting trace embedding has a dimensionality of 16. T2V acts as a lightweight embedding that relies solely on activity existence. Thus, we expect it to capture activity occurrence and process variant identity but not temporal or resource information.

case2vec (C2V) [18] extends T2V by incorporating additional process attributes into the event representation. Concretely, activity labels are concatenated with the corresponding resources before training. Model training including PV-DBOW objective, learning rate, epochs, negative sampling, and resulting embedding size are applied as in T2V. We selected C2V because it extends T2V with resource information while retaining the same architecture, dimensionality, and training objective. Consequently, we expect it to capture both activity and resource patterns.

MPPN [23] learns trace embeddings from multiple process perspectives. Activity, resource, and timestamp attributes are transformed into image based representations and processed by a CNN. The model is then trained to predict all attributes of the next event using a pre-trained CNN which is then fine-tuned on the image representations. It is trained for up to 20 epochs using a learning rate finder and early stopping, resulting in a trace embedding of dimensionality 128. We selected MPPN because it integrates control-flow, resource, and temporal features through a predictive training objective. Thus, we expect it to encode activity, resource, and temporal patterns.
Although all three approaches aim at general-purpose embeddings, they differ in input modality (control-flow vs. multi-perspective), training objective, and architectural design. This provides a suitable basis for our experiments, allowing us to analyze how the different modeling choices influence the information encoded in the embeddings. All embeddings are generated using the original implementations and hyperparameter settings of the corresponding publications, including embedding dimensionalities and training objectives. This allows us to evaluate the methods in their intended out of the box configuration without introducing additional task-specific optimization effects.

Probe Training. All probes are trained with the Adam optimizer, a learning rate of 0.001, a batch size of 64, and 20 epochs. Each probe is trained with five different random seeds, averaging evaluation metrics and reporting the standard deviation. The same configuration is used for all tasks and models. To ensure comparability across tasks, no task-specific hyperparameter tuning is performed. Detailed results for all runs like training loss and validation performance across epochs are available in our repository.

Evaluation Metrics. To assess probe performance, we use accuracy, precision, and recall. For multiclass and multilabel classification tasks, precision and recall are reported as macro averages, assigning equal weight to each class or label irrespective of its frequency to prevent majority classes from dominating the metrics. For multilabel tasks, accuracy is computed for each label and aggregated as an unweighted macro average.

4.2 Results

In this section, we present the experimental results for synthetic and real-life event logs.

Synthetic Logs. Table 4 shows accuracy (Ac.), precision (Pr.), and recall (Re.) for the random (Rd.) and majority (Maj.) baseline (Basel.) as well as T2V, C2V, and MPPN. LP is inapplicable in the *simple* log as it contains no loops. VID is inapplicable for *variant-disjoint* due to full variant separation between training and test set.

Table 4. Average evaluation metrics for synthetic logs. Standard deviations are reported using four bins: < 0.001 (no symbol), 0.001 to 0.01 (†), 0.01 to 0.1 ($*$), and > 0.1 (§).

				AP			EC			OR			VID			LP			TD			RC		
				Ac.	Pr.	Re.	Ac.	Pr.	Re.	Ac.	Pr.	Re.	Ac.	Pr.	Re.	Ac.	Pr.	Re.	Ac.	Pr.	Re.	Ac.	Pr.	Re.
Distribution-Controlled	simple	**Basel.**	Rd.	.50†	.58†	.50†	.33*	.33*	.33*	.51†	.51†	.51*	.05†	.05†	.06†				.20†	.20†	.20†	.25†	.25†	.25†
			Maj.	.74	.44	.56	.52	.17	.33	.50	.00	.00	.11	.01	.06				.20	.04	.20	.31	.08	.25
		BoA	Lin.	**1.0**	**1.0**	**1.0**	**1.0**	**1.0**	**1.0**	.51†	.51†	.30§	.30*	.08†	.28				.33†	.21*	.33†	.52†	.40*	.46*
			MLP	**1.0**	**1.0**	**1.0**	**1.0**	**1.0**	**1.0**	.50†	.30§	.29§	.30*	.08†	.28				.33†	.23*	.33†	.51†	.41*	.46*
		T2V	Lin.	.78	.65†	.65	.59†	.85†	.43†	.84†	.93†	.73*	.56†	.30†	.37†				.32†	.29†	.31†	.51†	.41§	.43
			MLP	.98†	.99	.96†	.97†	.98†	.97†	**.98**†	**.98**†	**.99**†	**.98**†	**.98**†	**.97**†		*n.a.*		.33†	.32*	.33†	.51†	.52†	.47†
		C2V	Lin.	.93†	.93	.85†	.83†	.91†	.79†	.53†	.53†	.46*	.21†	.14*	.14†				.33†	.31†	.32†	.57†	.66†	.52†
			MLP	**1.0**†	**1.0**†	.99†	**1.0**†	**1.0**	**1.0**†	.52†	.51†	.49*	.30†	.26*	.28†				.33†	.31†	.32†	**.81**†	**.81**†	**.80**†
		MPPN	Lin.	.98†	.97†	.97†	**1.0**†	**1.0**†	**1.0**†	.50†	.49*	.35§	.47*	.41*	.46†				.75*	.76*	.75†	.52†	.56*	.48*
			MLP	**1.0**	**1.0**	**1.0**†	**1.0**	**1.0**	**1.0**	.50†	.50†	.48§	.49†	.44*	.49†				**.82**†	**.83**†	**.82**†	.53†	.57†	.49*
	complex	**Basel.**	Rd.	.50†	.40†	.50*	.20†	.20†	.20†	.50†	.50†	.50†	.00	.00	.00†	.50*	.50*	.50*	.20†	.20†	.20†	.25†	.25†	.25†
			Maj.	.60	.00	.00	.30	.06	.20	.50	.00	.00	.04	.00	.00	.50	.00	.00	.20	.04	.20	.40	.10	.25
		BoA	Lin.	**1.0**	**1.0**	**1.0**	.91†	.92†	.91†	.51†	.51*	.53§	.15†	.00†	.02†	.98*	**1.0**	.97*	.35†	.36*	.35†	.52†	.47*	.45†
			MLP	**1.0**	**1.0**	**1.0**	**1.0**	**1.0**	**1.0**	.51†	.51*	.51§	.20†	.02†	.10†	**1.0**	**1.0**	**1.0**	.36†	.40*	.36†	.52†	.56†	.49*
		T2V	Lin.	.83†	.96†	.54†	.52†	.56*	.52†	.62†	.63†	.59*	.33†	.01	.03†	.72†	.80†	.60†	.30†	.30†	.30†	.48†	.41†	.38†
			MLP	.93	.92	.89†	.75†	.75†	.76†	.75*	.75*	.74*	**.74***	**.18***	**.16***	.81†	.81†	.80*	.32†	.33*	.32†	.49†	.52†	.44*
		C2V	Lin.	.98†	**1.0**	.95†	.71†	.71†	.71†	.50†	.50†	.48*	.13†	.01	.01	.80†	.85†	.73†	.33†	.31†	.33†	.78†	.79†	.68†
			MLP	**1.0**	**1.0**	**1.0**†	.85†	.86†	.86†	.50†	.50†	.50*	.18†	.04†	.04†	.91†	.91†	.91†	.34†	.33†	.34†	**.97**†	**.97**†	**.97**†
		MPPN	Lin.	.95†	.95†	.95†	.74†	.77†	.75*	.74†	.74*	.75*	.35†	.07†	.09†	.77†	.80*	.72*	.65†	.67*	.65†	.53*	.58†	.50*
			MLP	**1.0**†	**1.0**†	**1.0**†	.86*	.87†	.87*	**.89**†	**.90***	**.87***	.46†	.13†	.15†	.90*	.93*	.88*	**.67**†	**.70**†	**.67**†	.54†	.60*	.52*
Variant-Disjoint	simple	**Basel.**	Rd.	.50†	.45†	.50†	.33†	.33†	.33†	.51†	**.56**†	.50*							.20†	.20†	.20†	.25†	.25†	.25†
			Maj.	.53	.32	.67	.26	.09	.33	.45	.00	.00							.16	.03	.20	.23	.06	.25
		BoA	Lin.	.99*	**1.0**	.96*	**1.0**	**1.0**	**1.0**	.00	.00	.00							.44†	.21*	.35*	.65†	.38*	.49*
			MLP	**1.0**	**1.0**	**1.0**	**1.0**	**1.0**	**1.0**	.00	.00	.00							.45†	.21*	.36*	.64†	.35*	.49*
		T2V	Lin.	.55†	.37*	.67	.26	.09	.33	.07*	.13*	.12*							.43†	.28*	.35†	.61†	.39§	.46†
			MLP	.89†	.87†	.90†	.94†	.93†	.95†	**.55**	.55	**1.0**		*n.a.*			*n.a.*		.44†	.32*	.35†	.64†	.55*	.51†
		C2V	Lin.	.90†	.89†	.83†	.80†	.85†	.77†	.10†	.18†	.18†							.44†	.32†	.35†	.67†	.63†	.55†
			MLP	**1.0**	**1.0**	.99†	**1.0**	.99	.99	.03*	.05*	.05*							.45†	.34†	.35†	**.83**†	**.79**†	**.78**†
		MPPN	Lin.	.93†	.91†	.85†	**1.0**	**1.0**	**1.0**	.02†	.04*	.03*							.84†	.81†	.81†	.64†	.54†	.52*
			MLP	**1.0**†	.99†	.99†	**1.0**	**1.0**	**1.0**	.00†	.01†	.01†							**.88**†	**.86**†	**.85**†	.65*	.55*	.53*
	complex	**Basel.**	Rd.	.50†	.42†	.50†	.20†	.20†	.20†	.50*	.50*	.49*				.50†	.51†	.50†	.20†	.20†	.20†	.24†	.24†	.24†
			Maj.	.58	.00	.00	.19	.04	.20	.50	.00	.00				.49	.00	.00	.18	.04	.20	.38	.10	.25
		BoA	Lin.	**1.0**	**1.0**	**1.0**	.90†	.90†	.89*	.27†	.25*	.23*				.97*	**1.0**	.94*	.40*	.39*	.37*	.53†	.43†	.50†
			MLP	**1.0**	**1.0**	**1.0**	**1.0**	**1.0**	**1.0**	.21†	.10*	.09*				**1.0**	**1.0**	**1.0**	.40†	.34*	.36†	.54†	.53*	.55*
		T2V	Lin.	.73†	.86†	.50†	.39†	.51†	.43†	.28†	.27†	.25*				.67†	.88†	.42†	.36†	.37†	.34†	.46†	.38†	.38†
			MLP	.90†	.91†	.86†	.57†	.58†	.58*	.29†	.23*	.17*		*n.a.*		.76†	.85*	.65*	.38†	.35*	.36†	.50†	.51†	.49*
		C2V	Lin.	.95†	.97†	.91†	.67†	.67†	.67†	.32†	.32†	.33†				.67†	.66†	.72†	.37†	.34†	.35†	.67†	.71†	.62†
			MLP	**1.0**	**1.0**	**1.0**	.85†	.84†	.84†	.29†	.26*	.23*				.89†	.90†	.88*	.39†	.35†	.36†	**.96**†	**.96**†	**.95**†
		MPPN	Lin.	.99	.99†	.98†	.74†	.75†	.76*	.59*	.61*	.51*				.77†	.79†	.75*	.66†	.66†	.64†	.53†	.55†	.54*
			MLP	**1.0**	**1.0**	**1.0**	.80*	.82†	.80*	**.75**†	**.84***	**.64***				.88*	.91*	.84*	**.68**†	**.68***	**.66**†	.55*	.59*	.57*

Baseline and Approach Comparison. T2V, C2V, and MPPN achieve higher metrics than the random and majority baselines across most probing tasks. The only exception occurs for OR in the simple log under the variant-disjoint condition. Here, the random baseline achieves best precision and T2V has best recall and accuracy, whereas C2V and MPPN perform significantly worse. This is because the two parallel activities in the log lead to pairs of variants which differ only in the order of these activities, whereby one variant of each pair is assigned to the training and the other to the test set. Then, the probe learns a correlation between the label and a set of activities, but fails to capture their order. As the order of the activity pair is reversed at test time, probe performance is very low. Similarly, MPPN underperforms for OR in the simple log and distribution-controlled split, whereas T2V and C2V underperform in the complex log in the variant-disjoint split. Mostly low standard deviations across runs indicate reproducible behavior not caused by probe initialization noise. However, some OR results additionally exhibit high variance across runs, indicating unstable probe behavior when little signal is present in the embedding. This indicates that OR is hard to recover from the embeddings. For OR and LP, Maj. achieves non-zero accuracy despite zero precision and recall since the majority class is the negative class. BoA achieves (near-)perfect performance for AP, EC, and LP in most settings. This is expected since activity presence, event count, and loop occurrence are directly encoded in the activity count vector. In contrast, BoA achieves moderate performance for TD, RC, and VID and near-baseline performance for OR.

Across representation learning approaches, MPPN mostly outperforms C2V and T2V. Concretely, MPPN achieves the highest accuracy, precision and recall in 8, 7, and 7 cases, T2V in 1, 0, and 3 cases, and C2V in 0, 2 and 2 cases, respectively. MPPN particularly performs strong for TD. C2V performs best for RC and also performs well for LP. T2V performs best for OR in the simple log and for VID in both logs.

Tasks and Splitting Conditions. Tasks only related to the control-flow like AP or EC show the best metrics. TD and RC have the clearest separation between approaches. OR is less clear, with T2V performing best for the simple and MPPN for the complex log. Both splitting conditions yield similar results, besides the OR exception discussed above.

Probing Architectures. We see that the MLP probe often improves performance, particularly for T2V and C2V. The largest gap for C2V occurs for RC, where the MLP substantially outperforms the linear probe. That indicates that the relevant information is not linearly separable in the embedding space. BoA already achieves perfect performance with the linear probe for AP and EC, whereas the MLP provides almost no additional benefit. For MPPN, the differences between the probes are generally smaller, as the linear probe already achieves very high performance for several tasks.

Real-Life Logs. As visible in Table 5, the patterns observed in the synthetic logs largely persist for the real-life logs. Performance is lower and more variable across tasks, whereas near-perfect performance is less frequent. Likely, real-life logs contain additional variability and noise, making property recovery more challenging.

Baseline and Approach Comparison. BoA achieves near-perfect performance for AP and LP across most settings and outperforms all learned approaches for EC in BPI17. In contrast to the synthetic logs, BoA also performs strongly for OR and RC, likely by exploiting correlations between activity occurrences and labels. For BPI20, performance decreases in the variant-disjoint split, likely since one activity is absent from the training set, indicating BoA's inability to generalize. MPPN outperforms the random and majority baselines across most tasks. However, two notable exceptions emerge. For LP in BPI17, the majority baseline achieves perfect recall as loops represent the majority class, meaning they are always predicted and thus recalled. A similar effect is visible for RC in BPI20, where only two resources combined with strong label imbalance lead to recall of 1.0. T2V does not consistently outperform the baselines. In several cases, the linear probe effectively makes majority predictions, e.g., for AP, LP, and RC in BPI20 as well as for OR in both logs. This suggests that linear probes revert to trivial decision strategies when T2V does not encode discriminative information. Further, for EC in the BPI20 variant-disjoint condition, T2V and C2V perform similar or worse than Rd. and Maj., likely due to a substantial label distribution shift between training and test data, which could lead to the learning of spurious correlations that do not generalize. Also, in LP in the BPI20 variant-disjoint setting, the MLP probes of both MPPN and C2V perform below Maj., with MPPN even falling below Rd., likely as the minority class is overpredicted under strong class imbalance. The consistently low standard deviations indicate that observed behavior is systematic rather than influenced by instability.

Among the embeddings, MPPN performs best. It achieves the highest accuracy, precision, and recall in 8, 7, and 7 cases, whereas T2V does in 1, 0, and 3 cases, and C2V in 0, 1, and 2 cases. MPPN still shows clear advantages for TD across settings. C2V is competitive in LP and OR, especially in BPI20, but the strong RC results from the synthetic logs largely disappear. T2V is competitive in some OR and VID settings.

Tasks. Control-flow properties such as AP and LP again achieve the highest metrics. VID is even more challenging in real-life logs due to the large number of rare variants and their long-tailed distributions. This leads to a relatively high accuracy combined with relatively low precision and recall, arising as accuracy is dominated by frequent variants, whereas macro averaged precision and recall penalize poor performance on rare variants. EC also shows worse metrics, especially in BPI17 because of varying trace lengths that lead to more bins, increasing classification complexity. In contrast, BoA achieves near-perfect performance, as the number of events corresponds to the sum of the vector entries. TD again shows the clearest separation between approaches as MPPN consistently performs best. For RC, we observe near-perfect results for BPI20 as traces contain only one or two resources. In contrast, BPI17 has way more resources, making RC more difficult and reducing C2V's advantage in synthetic logs.

Splitting Conditions. The effect of splitting conditions is stronger than in the synthetic logs. In particular, the variant-disjoint split shows lower performance for some tasks, e.g., in BPI20 for AP and EC. This is since it introduces a distribution shift, where one class is underrepresented in the training set but dominates the test set. Thus, probes either overfit to dominant training patterns or learn spurious correlations. This suggests that structural differences between training and test data appear more prominently.

Probing Architectures. Again, MLP probes perform better, especially for T2V.

Table 5. Average evaluation metrics for real-life logs. Standard deviations are reported using four bins: < 0.001 (no symbol), 0.001 to 0.01 (†), 0.01 to 0.1 (∗), and > 0.1 (§).

				AP			EC			OR			VID			LP			TD			RC		
				Ac.	Pr.	Re.	Ac.	Pr.	Re.	Ac.	Pr.	Re.	Ac.	Pr.	Re.	Ac.	Pr.	Re.	Ac.	Pr.	Re.	Ac.	Pr.	Re.
Distribution-Controlled	BPI17	**Basel.**	Rd.	.50†	.54†	.50†	.11†	.11†	.11†	.51†	.24*	.51*	.00	.00	.00	.49†	.50†	.49*	.20†	.20†	.20†	.20†	.20†	.20†
			Maj.	.61	.37	.60	.13	.01	.11	.77	.00	.00	.07	.00	.00	.50	.50	**1.0**	.20	.04	.20	.24	.05	.20
		BoA	Lin.	**1.0**	**1.0**	**1.0**	.72*	.73*	.71*	.93*	**1.0**	.70§	.61†	.04†	.08†	**1.0**	**1.0**	**1.0**	.43†	.44†	.43†	.70†	.69†	.68†
			MLP	**1.0**	**1.0**	**1.0**	**1.0**†	**1.0**†	**1.0**†	**.94**†	.99*	.75†	**.73**†	**.16**†	**.20**†	**1.0**	**1.0**	**1.0**	.44†	.46*	.44†	**.76**†	**.75**†	**.75**†
		T2V	Lin.	.86	.86†	.90†	.38†	.37†	.36†	.77	.00	.00	.17†	.00	.00	.75†	.74†	.77†	.37†	.37†	.37†	.54†	.55†	.51†
			MLP	.91†	.92†	.92†	.58†	.57†	.58†	.79†	.63†	.18*	.62†	.06†	.08†	.84†	.85*	.84*	.39†	.41†	.39	.60†	.59†	.58†
		C2V	Lin.	.80	.79†	.85†	.25†	.21*	.23†	.77†	.57*	.03*	.12†	.00	.00	.69†	.68†	.72†	.33†	.31†	.33†	.48†	.41†	.45†
			MLP	.85†	.85†	.86†	.31†	.28†	.30†	.82†	.71*	.35*	.17†	.00†	.00	.74†	.75*	.73*	.40†	.42†	.40†	.58†	.57†	.56†
		MPPN	Lin.	.88†	.88†	.92†	.45†	.45†	.45†	.87†	.76†	.66*	.29†	.01	.02†	.82†	.80†	.85*	.78†	**.78**†	.78†	.56†	.54†	.54†
			MLP	.90†	.90†	.93†	.52†	.52†	.52†	.91†	.81*	**.80***	.32†	.03†	.03†	.93†	.93†	.92†	**.79**†	**.78**†	**.79**†	.60†	.59†	.58†
	BPI20	**Basel.**	Rd.	.51†	.51†	.50†	.34*	.34*	.34*	.51*	.27*	.50*	.01†	.01†	.00†	.50†	.11†	.49*	.20†	.20†	.20†	.50*	.95*	.50*
			Maj.	.74	.37	.50	.47	.16	.33	.74	.00	.00	.41	.01	.02	.89	.00	.00	.20	.04	.20	.94	.94	**1.0**
		BoA	Lin.	.94*	.94*	.99	.98	.97	.97	.76*	.40§	.10§	.94	.15†	.16	.93†	.69*	.71*	.32†	.31*	.32†	.99†	.99†	**1.0**
			MLP	**1.0**	**1.0**	**1.0**	**1.0**	**1.0**	**1.0**	**.97**†	**.97***	**.90**†	**.99**†	**.55***	**.58***	**1.0**	**1.0**	**1.0**	.33*	.33*	.32*	**1.0**	**1.0**	**1.0**
		T2V	Lin.	.74	.37	.50	.73†	.53†	.59†	.74	.00	.00	.64*	.04†	.04†	.89	.00	.00	.30†	.30†	.30†	.94	.94	**1.0**
			MLP	.98†	.98†	.98†	.98†	.98†	.97†	.81*	.94*	.30*	.94†	.20†	.21*	.97†	.84†	.86*	.31†	.30†	.31†	.96†	.96†	**1.0**†
		C2V	Lin.	.90†	.91†	.94†	.81†	.77†	.69†	.75*	.39§	.06§	.71†	.09*	.07†	.91†	.67†	.36*	.28†	.29†	.28†	.94	.94	**1.0**
			MLP	.99†	**1.0**†	.99†	.98†	.98†	.98†	.96†	.96*	.89	.97†	.32*	.37*	.98†	.93*	.93†	.31†	.32*	.31†	.99†	**1.0**†	**1.0**
		MPPN	Lin.	.83†	.79*	.59*	.74*	.78*	.76*	.83*	.85*	.43*	.78*	.16*	.17*	.92*	.73*	.35§	.89†	.90†	.89†	**1.0**	**1.0**	**1.0**
			MLP	.93†	.92*	.92*	.97†	.97†	.97†	.84†	.85*	.50*	.88†	.23*	.25*	.99†	.97*	.95*	**.90**†	**.91**†	**.90**†	**1.0**	**1.0**	**1.0**
Variant-Disjoint	BPI17	**Basel.**	Rd.	.50†	.52†	.50†	.11†	.11†	.11†	.51†	.34†	.50*	*n.a.*			.50†	.51†	.50†	.20†	.20†	.20†	.20†	.20†	.20†
			Maj.	.56	.34	.60	.12	.01	.11	.66	.00	.00				.51	.51	**1.0**	.18	.04	.20	.24	.05	.20
		BoA	Lin.	**1.0**	**1.0**	**1.0**	.28†	.32*	.30†	.74§	.80§	.23§				**1.0**	**1.0**	**1.0**	.52†	.51†	.50†	.52†	.58†	.54†
			MLP	**1.0**	**1.0**	**1.0**	**.96***	**.93***	**.93***	**.90***	.87§	**.84**†				**1.0**	**1.0**	**1.0**	.52†	.53†	.50†	**.70**†	**.70**†	**.68**†
		T2V	Lin.	.85†	.86†	.89†	.21†	.23†	.24†	.66	.00	.00				.77†	.78†	.77†	.32*	.40*	.35†	.57†	.52†	.52†
			MLP	.89†	.89†	.92†	.27†	.30†	.30†	.67†	.57*	.05*				.78†	.77†	.83†	.37*	.39†	.37*	.58†	.56†	.55†
		C2V	Lin.	.81	.81†	.86†	.22*	.18*	.20†	.66†	.42§	.00†				.73†	.74†	.73†	.30*	.33*	.31†	.46†	.39†	.41†
			MLP	.84†	.86†	.86†	.19†	.19†	.19†	.72†	.81*	.21*				.74†	.78*	.69*	.40*	.45†	.41†	.57†	.54†	.53†
		MPPN	Lin.	.86†	.88†	.90†	.18†	.18†	.20†	.89†	**.93**†	.73*				.84†	.82†	.88†	**.79***	**.80***	**.80***	.54†	.48†	.50†
			MLP	.88†	.89†	.92†	.23*	.23*	.23†	**.90**†	**.93***	.76*				.89†	.85†	.96†	.77*	.79*	**.80***	.56*	.51*	.52*
	BPI20	**Basel.**	Rd.	.50†	.06†	.49*	.34*	.34*	.33*	.51*	.56*	.50*	*n.a.*			.50*	.08†	.48*	.20*	.20*	.20*	.50†	.96†	.49†
			Maj.	.52	.04	.50	.26	.09	.33	.44	.00	.00				.92	.00	.00	.18	.04	.20	.96	.96	**1.0**
		BoA	Lin.	.53†	.53†	.65*	.23§	.29§	.48§	.49*	.60§	.09§				.96†	.72*	.74†	.27*	.19*	.26†	.97†	.97†	**1.0**
			MLP	.68	**.55**	**.99**†	**.36**†	**.68**†	**.64***	**.93***	**.92***	**.96***				**.98**†	**.99**†	.71	.24*	.21*	.25*	**1.0**	**1.0**	**1.0**
		T2V	Lin.	.52	.04	.50	.06†	.05†	.23†	.44	.00	.00				.92	.00	.00	.20†	.23*	.22†	.96	.96	**1.0**
			MLP	.60*	.10*	.55*	.08†	.25*	.33†	.44	.00	.00				**.98**†	.78†	**.97**†	.26†	.16*	.23†	.94†	.96	.98†
		C2V	Lin.	.53	.54	.80*	.26*	.27*	.53*	.48*	.53§	.06§				.94†	.82*	.28*	.25†	.18†	.24†	.96	.96	**1.0**
			MLP	.67	.54†	.89†	.08†	.21*	.28†	.88†	.84†	.95*				.76*	.21*	.70*	.24*	.22*	.24†	.98	.99†	**1.0**†
		MPPN	Lin.	.31§	.09*	.78	.34†	.56†	.58†	.47*	.53§	.09*				.62*	.11*	.50†	.81†	.85†	.82†	**1.0**	**1.0**	**1.0**
			MLP	**.79***	.52*	.80*	.35†	.59†	.63†	.68*	.73*	.66*				.43†	.11†	.86†	**.91**†	**.93**†	**.92**†	**1.0**	**1.0**	**1.0**

5 Discussion

The experimental application shows that probing can be applied to process mining and provides insights into the ability of three representation learning approaches to encode process-related properties. In this section, we discuss these insights.

Baseline and Approach Comparison. We see that several properties are highly recoverable from simple handcrafted activity counts: BoA achieves competitive or superior performance for multiple tasks, especially AP, EC, LP, and, for real-life logs, also OR and RC. This indicates that these properties correlate strongly with activity occurrence. However, this only holds for simpler properties or cases where log characteristics allow for shortcuts, and will likely not translate to superior downstream task performance. Learned embeddings retain advantages for properties that are less reflected in activity counts. Most notably, MPPN consistently performs best for TD, suggesting that explicitly using time features benefits the embedding here. T2V captures information about time and resources and C2V about time despite not explicitly targeting these properties in their architecture, indicating that they are partially recoverable via the control-flow. The strong RC performance of C2V largely disappears in the real-life logs, likely because the joint activity-resource vocabulary of C2V is much larger, causing many combinations to occur rarely and therefore to be encoded less reliably.

Differences Between Tasks. Simple properties such as AP, EC, or TD are easier to recover, but tasks that require separating similar variants, such as OR or VID, remain challenging. BoA fails for OR in the synthetic logs, where variants differ mainly in activity ordering (rather than occurrence), but performs well in real-life logs due to shortcuts from activity occurrence to execution order. Although T2V does not explicitly model order, weak differences in its embeddings appear sufficient to separate ordering in the simple log. However, this does not hold in more complex settings, where embeddings tend to overlap. MPPN is more robust for OR in complex logs due to its explicit modeling of orders, although convolutional processing may reduce sensitivity to positional differences. VID results indicate that some real-life variants are identifiable through activity frequencies alone, whereas the synthetic logs intentionally contain many variants that differ only in local ordering. This suggests that embeddings encode coarse grained patterns, whereas fine-grained distinctions are less prominently captured.

Splitting Conditions. Overall, the splitting conditions did not change performance significantly, indicating that variant-label memorization is unlikely. Only a small performance drop is visible for real-life logs, meaning that probes might rely on memorizing a label for some variants. However, bad performance for VID opposes this view. Further, the performance drop may also be due to distribution shifts between training and test data. Regardless, influence of variant-level shortcuts cannot be entirely excluded.

Probe Architectures. The superiority of MLP probes, especially for T2V and C2V, indicates that some properties are encoded non-linearly and require expressive probes. For MPPN, this is less visible, suggesting that properties are linearly accessible. However, MLP probes may be less robust under strong distribution shifts, e.g., for LP in the BPI20 variant-disjoint split. Also, repeated runs provide insights into stability, where low variance mostly indicates strong recoverable signals but high variance sometimes suggests near-chance performance, e.g., for the linear probe of BoA.

Impact of Log Characteristics. We see that class imbalance affects both the interpretation of metrics and task difficulty. For example, strongly skewed label distributions in multiclass tasks often lead to low precision and recall as minority classes are difficult to recover. This is particularly visible for VID in real-life logs, where the variant distribution is highly long-tailed. Further, we see that results strongly depend on how tightly probed properties are coupled to activity occurrences, as visible for BoA. It performs bad in synthetic logs, where properties are decoupled from activity frequencies, but well in real-life logs, where activity frequencies correlate with probed properties.

These observations show that probing results depend not only on the embeddings but also on the experimental setup. A central concern is if probe performance reveals more about the probe than about the embedding, e.g., if probes memorize variant-label associations instead of recovering the intended property. To address this, prior work from NLP proposed controls that estimate how much of performance is due to memorization [17,24]. Here, we introduce such controls, designed according to characteristics of event logs. Thus, it is important to interpret results with control mechanisms in mind.

6 Conclusion

In this paper, we propose a probing evaluation framework to assess which properties are encoded in learned trace embeddings. Based on seven initial probing tasks, we apply the framework to embeddings created by three different approaches. Three control mechanisms maximize confidence that performance can be attributed to recoverability of the probed property. We see that simple properties are easier to recover than more complex ones. Overall, our results highlight both potential and limitations of learned embeddings, as not all properties are encoded accessibly. Here, probing can help to explain downstream-task behavior, support the analysis and comparison of process representations, and guide the development of improved general-purpose embeddings. Such general-purpose embeddings would ideally capture universal knowledge about the process, making them very valuable for process management. Eventually, this could lead to process-independent embeddings, similar to pre-trained embeddings from NLP.

Our work is subject to a few limitations. First, finding a split that produces stable distributions across all tasks proved infeasible in some cases, so performance differences between random and variant-disjoint splits may partly reflect distribution shifts. We mitigated this by using multiple logs to produce suitable label distributions for all tasks at least once. Second, label imbalance may affect the interpretation of results, especially for tasks with highly skewed or long-tailed distributions. We addressed this with the (typically outperformed) majority baseline as lower bound. Third, we cannot rule out that probes exploit shortcuts beyond those covered in the control mechanisms, e.g., by relying on embedding magnitude or event-count-related cues for trace duration.

In the future, we aim to extend the framework to embeddings of other elements, such as process models or event logs. Likely, this will require adaptions to the control mechanisms. In this context, future work could add control mechanisms that restrict memorization effects beyond variant-level associations. Additional baselines could further help disentangle information encoded in the embeddings from simple characteristics. Another idea is to investigate the influence of representation learning design choices, such as embedding dimensionality. Further, the set of probing tasks could be extended to include rare activities or a semantic component that assesses the meaning of activities and traces. For example, probing tasks could aim to recover whether a process is related to purchasing or whether the trace outcome was positive. This also shows future work for representation learning as current ideas do not consider the meaning of activities.

References

1. Adi, Y., Kermany, E., Belinkov, Y., Lavi, O., Goldberg, Y.: Fine-grained analysis of sentence embeddings using auxiliary prediction tasks. In: ICLR (2017)
2. Barbon Junior, S., Ceravolo, P., Damiani, E., Marques Tavares, G.: Evaluating trace encoding methods in process mining. In: DataMod, pp. 174–189 (2021)
3. Basaj, D., et al.: Explaining self-supervised image representations with visual probing. In: IJCAI, pp. 592–598 (2021)
4. Belinkov, Y.: Probing classifiers: promises, shortcomings, and advances. J. Comput. Linguist. **48**, 207–219 (2022)
5. Bengio, Y., Courville, A., Vincent, P.: Representation learning: a review and new perspectives. PAMI **35**(8), 1798–1828 (2013)
6. Berti, A., van Zelst, S.J., van der Aalst, W.M.P.: Process mining for python (PM4Py): bridging the gap between process- and data science. ICPM Demos (2019)
7. Ceravolo, P., Barbon Junior, S., Damiani, E., van der Aalst, W.: Tuning machine learning to address process mining requirements. IEEE Access **12**, 24583–24595 (2024)
8. Conneau, A., Kruszewski, G., Lample, G., Barrault, L., Baroni, M.: What you can cram into a single $&!#* vector: probing sentence embeddings for linguistic properties. In: ACL, pp. 2126–2136 (2018)
9. De Koninck, P., vanden Broucke, S., De Weerdt, J.: act2vec, trace2vec, log2vec, and model2vec: representation learning for business processes. In: BPM, pp. 305–321 (2018)
10. van Dongen, B.: BPI challenge 2017 (2017). https://data.4tu.nl/articles/_/12696884/1
11. van Dongen, B.: BPI challenge 2020 (2020). https://data.4tu.nl/collections/_/5065541/1
12. Gamallo-Fernandez, P., Vidal, J.C., Lama, M.: Learning context-based representations of events in complex processes. ICWS, pp. 214–223 (2023)
13. Gavric, A., Bork, D., Proper, H.A.: Beyond logs: AI's internal representations as the new process evidence. In: BPM Forum, pp. 232–246 (2025)
14. Goodfellow, I., Bengio, Y., Courville, A.: Deep Learning. MIT Press (2016)
15. Grohs, M., Pfeiffer, P., Rehse, J.R.: Proactive conformance checking: an approach for predicting deviations in business processes. Inf. Syst. **127**, 102461 (2025)
16. Hake, F., et al.: CauseCheck: a tool for simulating deviations in event logs with known root causes. In: ICPM Demos (2024)
17. Hewitt, J., Liang, P.: Designing and interpreting probes with control tasks. In: EMNLP-IJCNLP, pp. 2733–2743 (2019)
18. Luettgen, S., Seeliger, A., Nolle, T., Mühlhäuser, M.: Case2Vec: advances in representation learning for business processes. In: ICPM Workshops, pp. 162–174 (2021)

19. Neubauer, T.R., Peeperkorn, J., Peres, S.M., De Weerdt, J., Fantinato, M.: Vector representation for business process: graph embedding for domain knowledge integration. ICMLA, pp. 588–594 (2023)
20. Peeperkorn, J., vanden Broucke, S., De Weerdt, J.: Can deep neural networks learn process model structure? An assessment framework and analysis. In: ICPM Workshops, pp. 127–139 (2021)
21. Pfeiffer, P.: Business Process Representation Learning, Saarland Univ. (2025). Ph.D. thesis
22. Pfeiffer, P., Abb, L., Fettke, P., Rehse, J.R.: Learning from the data to predict the process. BISE **67**(3), 357–383 (2025)
23. Pfeiffer, P., Lahann, J., Fettke, P.: Multivariate business process representation learning utilizing gramian angular fields and convolutional neural networks. In: BPM, pp. 327–344 (2021)
24. Pimentel, T., Valvoda, J., Hall Maudslay, R., Zmigrod, R., Williams, A., Cotterell, R.: Information-theoretic probing for linguistic structure. In: ACL, pp. 4609–4622 (2020)
25. Rullo, A., Alam, F., Serra, E.: Trace encoding techniques for multi-perspective process mining: a comparative study. WIREs Data Min. Knowl. Discov. **15**(1), e1573 (2025)
26. Seeliger, A., Luettgen, S., Nolle, T., Mühlhäuser, M.: Learning of process representations using recurrent neural networks. CAiSE, pp, 109–124 (2021)
27. Tavares, G.M., Oyamada, R.S., Barbon Junior, S., Ceravolo, P.: Trace encoding in process mining: a survey and benchmarking. Eng. Appl. Artif. Intell. **126**, 107028 (2023)
28. Weytjens, H., De Weerdt, J.: Creating unbiased public benchmark datasets with data leakage prevention for predictive process monitoring. In: BPM Workshops, pp. 18–29 (2022)

Process Routing and Decision Driver Discovery from Relational Databases Without Prior Event Logs

Zhi Quan Joel Lim(✉), Kar Way Tan, and Hoong Chuin Lau

School of Computing and Information Systems, Singapore Management University, Singapore, Singapore
joellim.2022@engd.smu.edu.sg, {kwtan,hclau}@smu.edu.sg

Abstract. Classical process mining derives process models from event logs, but in many operational systems such logs are unavailable or incomplete. Instead, process execution is recorded implicitly in relational databases, making it difficult to determine how process instances progress between tables or why alternative process paths occur. This paper introduces **Process ROuting Discovery from relational data (PROD)**, a framework that infers process routing behaviour directly from relational data-bases without requiring event logs. PROD identifies candidate successor relations between tables and assigns row-level routing across linked rows. It then learns routing decisions and their associated driving attributes at branching points, and constructs a routing graph whose branching behaviour is explained using SHAP-based attributions and expressed via a Large Language Model (LLM). We evaluate PROD on six datasets spanning synthetic, benchmark, and industrial relational databases. Results show that process behaviour and its driving attributes can be recovered, achieving routing accuracy ranging from 0.79–0.96 and Structural F1-scores of 0.83–0.95, while exhibiting gradual performance degradation as row-level data corruption increases. These findings suggest that PROD enables practical process analysis directly from relational data, broadening the scope of process analysis beyond traditional event-log-based settings.

Keywords: Process Analysis · Routing Analysis · Explainable Process

1 Introduction

Classical process mining techniques derive process models from event logs, where each step is recorded as an event with a case identifier, activity label, and ordering information [1]. These models support key analytical tasks such as identifying bottlenecks, analyzing decision points where process paths diverge, and checking conformance. Extensions such as data-aware process mining [19] and object-centric process mining [2] incorporate attributes and multi-object interactions to explain routing decisions [7], but still assume the availability of an

F. M. Maggi et al. (Eds.): BPM 2026, LNCS 16936, pp. 485–502, 2027.
https://doi.org/10.1007/978-3-032-37777-7_29

event log. In practice, however, event logs are often unavailable or incomplete [10,13]. Constructing them from relational databases requires schema knowledge, key mappings [20], and often involves substantial domain expertise and design choices in defining event abstractions [3], making it a non-trivial effort.

In many operational systems, process execution is reflected implicitly through linked rows across tables. For example, while an *Orders* table may reference both *Delivery* and *Cancellation* tables, schema-level relationships do not reveal how orders actually progress through these tables or what factors drive these outcomes with different business impact. Understanding such routing behaviour as observed in actual process execution is essential: if cancellations are consistently associated with insufficient wallet balance or remote delivery locations, organisations can redesign payment flows or adjust delivery policies to reduce failures. This motivates the practical need to infer routing structure and understand its driving attributes directly from relational data, enabling process analysis without preparing an event log. While full process model discovery remains valuable, routing behaviour and its drivers are often sufficient for decision-focused analyses such as process path diagnostics, policy evaluation, and bottleneck identification.

This paper proposes **Process ROuting Discovery from relational Data (PROD)**, a framework for inferring process routing behaviour directly from relational database tables without requiring event logs. In this setting, routing is inferred from linked rows across tables rather than explicit event logs. For each row in a source table, PROD identifies its successor set, namely the downstream tables containing related rows that occur later in the same reconstructed process instance. Aggregating these row-level successor sets reveals table-level routing structure, including branching behaviour where rows from the same table are observed to route to different downstream tables.

PROD consists of four stages. First, it identifies candidate successor relations between tables using linkage evidence derived from statistical signals (e.g., Jaccard similarity, Kolmogorov–Smirnov (KS) statistics), and neural-network-based temporal ordering analysis. Second, for each row, a successor set is inferred from linked rows belonging to the same reconstructed process instance, while rows lacking successors are marked as process-ended. These successor sets serve as input for routing graph construction in the next stage. Third, for tables with multiple successor tables, it learns routing drivers by training a Random Forest classifier from row-level feature vectors, including attributes of the current row and preceding tables. This step models the **routing decisions** and identifies the **driving attributes** associated with these decisions using TreeSHAP based on global importance ranking. The **routing graph** is then constructed by aggregating observed successor assignments, routing decisions, and driving attributes. Finally, the routing graph is semantically enriched using a Large-Language Model (LLM), which generates **natural-language explanations** based on the identified driving attributes and transition probabilities.

We evaluate PROD on six datasets, including two synthetic, two benchmark, and two industrial relational databases. The results show that routing behaviour, measured via Structural F1-score of inferred routing edges, achieves scores ranging from 0.83 to 0.95, and routing accuracy ranges from 0.79 to 0.96, while the driving attributes (measured via Spearman rank correlation) achieve scores of 0.80 to 0.94. PROD degrades gradually under row-level data corruption, with routing accuracy declining predictably and remaining at or above 0.65 even under 30% corruption. These findings suggest that PROD is a viable framework for supporting practical process analysis by enabling the understanding of process routing behaviour and its drivers in real-world settings where data are imperfect, schemas are unknown, and process execution is directly captured in operational data. While PROD does not remove the need for domain understanding when interpreting results, it lowers the entry barrier by surfacing routing structure and its drivers directly, without first committing to a case notion, activity abstraction, and full event-log construction.

Our contributions are twofold. First, we introduce Process ROuting Discovery from relational Data (PROD), a framework for automatically discovering the **routing structure** of business processes directly from relational databases, represented as a graph. By enabling routing inference from relational data, PROD supports process analysis in a data-driven manner without requiring event-log construction. We further make the framework and implementation available to the community to support reproducibility and adoption. Second, PROD identifies **routing decisions** at branching points, discovers the associated **driving attributes**, and generates **explanations** for why cases diverge across alternative paths. This enables practitioners to understand and act on routing behaviour through interpretable, data-driven insights. Together, these contributions broaden the scope of process analysis by enabling practical analysis of process flows, supporting organisations in understanding, diagnosing, and improving them through routing behaviour without event logs.

The rest of the paper proceeds as follows: Sect. 2 presents the formal definition of PROD; Sect. 3 reviews related work; Sect. 4 describes the PROD framework in detail; Sect. 5 presents the experiments and results; Sect. 6 provides practical implications, and Sect. 7 concludes the paper.

2 Formal Problem Definition

Let $\mathcal{R} = \{T_1, \dots, T_n\}$ denote the set of relational tables in a database snapshot. Assume that from $\mathcal{R}$ it is possible to reconstruct (or approximate) a multiset of process instances $\mathcal{C}$. Each instance (or case) $c = (i_1, i_2, \dots, i_k) \in \mathcal{C}$ is an ordered sequence of record instances drawn from tables in $\mathcal{R}$. Given only the relational tables $\mathcal{R}$ and the process instance set $\mathcal{C}$, the objective of PROD is to construct:

1. A directed routing graph G, capturing empirically observed successor relations between tables;
2. A set of routing functions $\{f_{T_i}\}$ defined for branching tables, where each f_{T_i} maps row-level feature vectors to observed tables in the successor set;
3. A set of driving attribute sets $\{A_{T_i}\}$, where each A_{T_i} identifies the top-ranked attributes associated with successor selection according to global SHAP importance ranking.

A solution therefore consists of the tuple $(G, \{f_{T_i}\}, \{A_{T_i}\})$. This enables the reconstruction of process flow and allows business stakeholders to understand routing behavior and its driving attributes directly from relational data, where G **captures the routing structure,** $\{f_{T_i}\}$ **capture routing decisions at branching tables, and** $\{A_{T_i}\}$ **identify the key attributes associated with these routing decisions.** These components further serve as inputs for semantic enrichment, where an LLM is used to **generate natural-language explanations** of routing behavior based on the identified driving attributes and transition structure.

3 Related Work

Decision points in business processes, where cases diverge across alternative paths, have been extensively studied in data-aware process mining. Decision mining approaches learn predictive models at branching points in a discovered process model to explain successor selection using event and case attributes [14,18,22]. Extensions such as object-centric and artifact-centric process mining further capture interactions among multiple business objects, enabling richer analysis of process execution and its branching behavior [2,8]. These approaches highlight the importance of understanding routing behaviour and its associated drivers for process analysis and improvement. However, they rely on event logs as input, where process behavior is represented as events that are explicitly defined, linked, temporally ordered and constructed, which can be a costly process.

A substantial body of work addresses the construction of event logs from relational databases to enable subsequent process mining. These approaches transform operational data into classical or object-centric event logs using information such as database schema, keys, timestamps, mappings, and user guidance. Techniques include schema-driven extraction [20], query-based case construction [23], semi-automated approaches [4,11], and object-centric log generation [3,15]. While these methods make process mining applicable to relational data, event-log construction remains a non-trivial and often manual step. It typically requires well-defined schema information, primaryforeign key mappings, or domain expertise to specify case notions, activity definitions, and event ordering. In many real-world systems, particularly operational Online Transaction Processing (OLTP) systems, where schema information may be incomplete, poorly documented, or difficult to access, this limits the applicability of these approaches.

Recent work has explored the use of Large Language Models (LLMs) for interpreting relational data, including tasks such as schema understanding, semantic labeling, and data profiling [5,24]. These approaches typically provide the LLM with database schema information, and in some cases sample values, with the expectation that it can infer data relationships and generate human-readable descriptions of the structure [12]. However, recent evaluations suggest that such capabilities remain limited when applied to complex operational databases, as LLMs tend to rely heavily on surface-level signals such as column names and small samples of attribute values [27].

In contrast to prior work discussed above, this work directly infers process routing behaviour at the relational level using data-driven methods, without requiring the availability of event logs or their reconstruction. LLM is used as a semantic layer to generate interpretable explanations, guided by structured inputs from the routing graph and the associated decisions and key attributes. This provides a complementary perspective that connects relational representations with interpretable process behaviour.

4 Methodology

In this section, we describe the PROD framework, which consists of four stages:

1. Identification of candidate successor relations
2. Successor assignment
3. Routing driver learning (using Random Forest and TreeSHAP) and routing graph construction
4. Semantic interpretation using an LLM

Figure 1 provides an overview of the PROD framework. The four stages are described in detail in the following subsections. To illustrate the methodology, a worked example based on a synthetic dataset, Orders, is provided in Section 5.7.

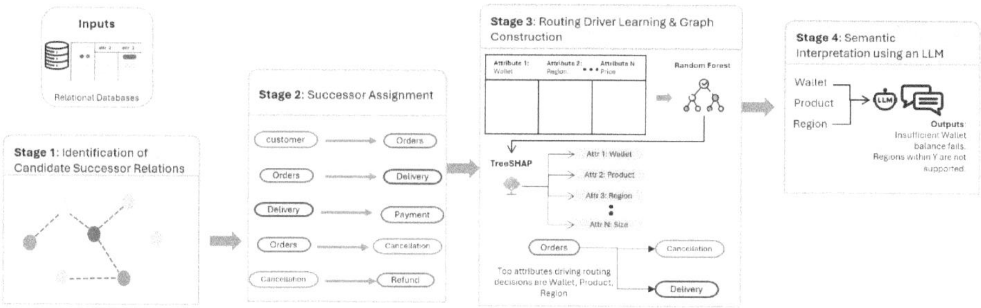

Fig. 1. Overview of the PROD Framework

4.1 Stage 1: Identification of Candidate Successor Relations

We aim to identify whether two tables are part of the same underlying process by inferring inter-table successor relations from statistical patterns in the data. Given two tables $T_i, T_j \in \mathcal{R}$, we evaluate their relatedness using three linkage signals.

- **Jaccard similarity (shared values)**: Measures the overlap of values (e.g., identifiers). A high overlap suggests both tables refer to the same entities.

- **Kolmogorov–Smirnov (KS) statistic (distribution similarity):** For each numerical column in one table, we compare its distribution with comparable numerical columns in another table using the two-sample KS statistic. As the raw KS statistic measures distributional difference, we convert it into a similarity score by taking one minus the statistic, so that higher values indicate greater similarity. The highest KS similarity across comparable numerical column pairs is used as the table-level distributional-similarity signal.
- **Overlap-based inclusion metrics (subset relationship)**: Measures whether values in one table are largely contained in another. This suggests one table may be a filtered or derived subset of the other.

These three signals capture different types of evidence that two tables may be related: shared values, distributional similarity, and containment. Since each signal may independently produce false positives, relying on any single signal may result in incorrect matches. At the same time, we do not require all three signals to pass, because a genuine table relationship may not show every type of evidence. Therefore, we retain a candidate successor relation ($T_i \rightarrow T_j$) when at least two of the three signals exceed their predefined thresholds.

For each retained pair, we compare datetime attributes to check temporal consistency. If records in T_i occur more frequently before those in T_j, this supports a successor relationship. When ordering is unclear, we apply simple precedence rules by ordering tables by the timestamp of their linked records, and further refine the inferred ordering using a Temporal Convolutional Network (TCN). This stage produces a candidate successor relation set:

$$E_c \subseteq \mathcal{R} \times \mathcal{R}.$$

4.2 Stage 2: Successor Assignment

Given candidate relations E_c, routing is assigned at the **row level**. Each table T_i consists of a collection of rows or records. For each row $r \in T_i$, we define its **successor set** $\text{succ}(r)$ as the set of tables that row r flows into. In other words, for the process instance that r belongs to, the successor set contains all tables that have a linked row occurring after r. In a sequential flow where only one path is observed, the successor set contains a single table. When multiple paths are possible, the successor set may contain multiple tables. Mathematically,

$$\text{succ}(r) = \{\, T_j \mid \text{a linked row in } T_j \text{ exists for the same } c \in \mathcal{C} \,\}.$$

If no linked row exists, then $\text{succ}(r) = \emptyset$, and the row is treated as terminal.

At the end of this stage, each table T_i is associated with a successor set $S(T_i)$, which is the union of successors sets across all rows in that table, capturing all the destinations that table ever routes to across all observed process instances. That is,

$$S(T_i) = \bigcup_{r \in T_i} \text{succ}(r).$$

A table T_i is considered **branching** (and hence a candidate for further routing logic in the next stage) if the size of $S(T_i)$ is greater than 1.

Routing is therefore inferred at the row level and summarized at the table level through successor sets. This defines structural connectivity between tables based on the observed routing of reconstructed process instances, and serves as input for routing graph construction in the next stage.

4.3 Stage 3: Routing Driver Learning and Routing Graph Construction

Routing Driver Learning. For each branching table T_i, we construct the routing dataset

$$D_{T_i} = \{(x_r, y_r) \mid r \in T_i,\ \mathrm{succ}(r) \neq \emptyset\},$$

where x_r denotes the feature vector of row r, constructed from the attributes of the current row together with attributes derived from preceding tables within the same reconstructed process instance. $y_r \in S(T_i)$ is the successor table that r routes to, derived from its successor set $\mathrm{succ}(r)$ assigned in Stage 2. We then train a Random Forest [6] classifier

$$f_{T_i} : \mathcal{X}_{T_i} \to S(T_i),$$

which learns the relationship between row-level attributes and observed routing outcomes, with $\mathcal{X}_{T_i}$ denoting the space of attribute vectors for rows in T_i, and $S(T_i)$ the set of successor tables serving as the target classes. The classifier is not used for routing prediction; rather, it serves to quantify, via TreeSHAP, the attributes that drive the selection among successor tables in $S(T_i)$. The feature vector also includes pairwise interaction terms between encoded categorical attributes such as *location* × *risk*, allowing the Random Forest to model combined effects and TreeSHAP to attribute their contribution to successor selection. We adopt a Random Forest model due to its ability to capture nonlinear relationships in heterogeneous tabular data and its compatibility with TreeSHAP for feature attribution.

Next, we introduce TreeSHAP [17] values $\phi_{k,m}$ for each feature m and each sample k. Global feature importance is obtained using the mean absolute SHAP value over all $N = |D_{T_i}|$ samples:

$$H_m = \frac{1}{N} \sum_{k=1}^{N} |\phi_{k,m}|,$$

Features are ranked in descending order of H_m. We then select the smallest set of top-ranked features whose cumulative importance exceeds a predefined threshold τ (e.g., 75%, 80%, 85%), determined by the analyst for the given dataset. In our experiments, we set $\tau = 80\%$. This threshold controls the trade-off between keeping explanations concise and retaining the most influential attributes. At least three features are retained to avoid trivial or overly narrow explanations. The resulting set A_{T_i} represents the set of **top-ranked driving attributes** which are associated with successor selection at table T_i.

Routing Graph Construction. Using the successor sets defined in Stage 2 and the empirical row-level assignments, we now construct the routing graph.

The directed routing graph is defined as $G = (\mathcal{R}, E)$ where the node set corresponds to the set of tables $\mathcal{R}$, and $E = \{(T_i, T_j) : T_j \in S(T_i)\}$.

For each edge (T_i, T_j), we define the transition probability

$$p_{T_i \to T_j} = \frac{|\{r \in T_i : T_j \in \text{succ}(r)\}|}{|\{r \in T_i : \text{succ}(r) \neq \emptyset\}|}.$$

These probabilities correspond to the maximum likelihood estimates of the transition probabilities from source table T_i to each successor table T_j. Each edge (T_i, T_j) is therefore associated with its transition probability $p_{T_i \to T_j}$ and its driving attribute set A_{T_i}.

The routing graph G thus captures both the structural routing behavior inferred from relational data and the attributes that drive the successor selection.

4.4 Stage 4: Semantic Interpretation Using an LLM

To support business interpretability, we generate natural-language explanations for the routing behavior identified in Stage 3. For each branching table T_i and successor $T_j \in S(T_i)$, we provide a large language model (LLM) with:

- the driving attribute set A_{T_i},
- their global importance scores H_m, and
- the transition probability $p_{T_i \to T_j}$.

The LLM generates explanations in natural language describing the routing behavior. For example: "Wallet balance is a primary driver of routing from table *orders* to *cancellation*."

To ensure that explanations remain grounded in empirical results, the prompt restricts the model to attributes in A_{T_i} and their corresponding global SHAP importance scores H_m. The model is configured to produce deterministic outputs. Notably, this stage does not modify the routing graph structure. Instead, it augments each edge (T_i, T_j) with a semantic annotation. The final routing graph therefore enriches each edge with:

- its transition probability $p_{T_i \to T_j}$,
- its driving attribute set A_{T_i}, and
- its corresponding natural-language explanation.

The resulting representation provides a data-derived routing graph, along with routing decisions, their driving attributes, and is annotated with interpretable semantic descriptions.

5 Experiment and Results

5.1 Datasets

We evaluate PROD on six datasets (2 synthetic, 2 benchmark and 2 real-world industrial datasets) designed to assess correctness, scalability, benchmarking robustness, and translational potential. The first is a small synthetic 'Orders' process, constructed to verify that PROD correctly recovers routing structure and decision drivers in a fully controlled setting. Its simplicity allows clear visualisation and step-by-step validation of the inferred routing graph. The second is a medium-scale synthetic dataset, developed to evaluate scalability and stability under increased structural complexity while retaining known routing logic and ground-truth attributes.

The third and fourth datasets are TPC-H [25] and TPC-E [26], standard decision-support and transaction-processing benchmarks defined by the Transaction Processing Performance Council (TPC). These benchmarks are widely used for evaluating database systems and transactional workloads. We derive reference process flows from their documented schema relationships and canonical order-fulfilment (TPC-H) and trade-execution (TPC-E) lifecycles, which serve as structural ground truth for routing validation.

The final two datasets, Industrial A and Industrial B, originate from real operational systems. Domain experts provided validated reference transitions, routing points, and key risk-driving attributes. These datasets enable assessment of PROD's applicability in practice and its potential for real-world decision auditing and operational diagnostics. Table 1 summarises the datasets, their sizes (in millions of rows) and attributes.

Table 1. Summary of datasets used in PROD evaluation

Dataset	#Tables	#Rows	#Attributes
Synthetic (Small)	23	1.3M	110
Synthetic (Medium)	50	3M	350
TPC-H	8	1.2M	61
TPC-E	33	8M	188
Industrial A	25	1.2M	257
Industrial B	28	1.8M	601

5.2 Evaluation Metrics

We evaluate PROD from two main perspectives: (1) performance across the four stages using five metrics covering structural recovery, routing behaviour, and explanation quality; and (2) robustness to row-level data corruption that may affect relational linkages and routing inference. Results for these two perspectives are reported in Sects. 5.4 and 5.5, respectively. The metrics are defined as follows:

1. **Discovery Rate:** Fraction of ground-truth routing edges that are successfully recovered in the inferred routing graph G.
2. **Structural F1:** Harmonic mean of precision and recall of the inferred routing edges in G, penalizing both missing and incorrect edges.

3. **Routing Accuracy (RA):** Accuracy of successor table assignment per record, i.e., the proportion of records for which the inferred successor table matches the ground-truth successor table.
4. **Attribution Coherence (AC):** Spearman rank correlation between ground-truth routing drivers and SHAP-based feature rankings, indicating agreement between them.
5. **Explanation Faithfulness (EF):** Proportion of explanations that correctly reflect SHAP-identified routing drivers, assessed by two raters with agreement measured using Cohen's Kappa κ.

5.3 Baselines

To the best of our knowledge, no existing approach performs routing inference directly from relational databases. We therefore compare PROD against the closest available methods that address related components of the problem.

For structural reconstruction (the routing graph G), we compare against **StarStar** [4], which reconstructs event logs from relational databases to enable downstream process mining. StarStar operates on activity sequences extracted from event data, where eventobject relationships are already available, and captures structural relationships between database tables within event logs. While it does not directly infer routing decisions and their driving attributes at the relational level, it provides a baseline for recovering connectivity between tables. We evaluate structural reconstruction using **Discovery Rate** and **Structural F1**, comparing PROD against StarStar in recovering routing relationships from underlying schema and relational linkages.

For routing decision modelling, we compare against a **Decision Tree** classifier trained on the same routing datasets [9,21]. Decision trees are widely used in decision mining and data-aware process mining to model branching behaviour in process models. They provide an interpretable baseline for evaluating routing decision modelling. Since PROD learns routing decisions using a Random Forest, this comparison evaluates whether the Random Forest captures routing behaviour more accurately than a Decision Tree baseline. We evaluate predictive performance using **RA**.

No direct baseline exists for **AC** (Stage 3) and **EF** (Stage 4), as existing relational process discovery approaches do not learn routing decision models or produce attribute-level explanations for routing behaviour. Consequently, they do not produce feature attribution outputs comparable to SHAP-derived drivers. We therefore report AC only for PROD, as all 6 evaluation datasets provide ground-truth or validated routing drivers (see Sect. 5.1), and evaluate EF through human evaluation.

5.4 Main Results

Table 2 summarises the performance of PROD across all datasets using the five evaluation metrics. For structural routing recovery, PROD achieves Discovery Rate of 0.79–0.98 and Structural F1 of 0.83–0.95. Performance is highest on the synthetic datasets (F1 up to 0.95), where routing relationships are well-defined. On TPC-H and TPC-E, Structural F1 remains high (0.83–0.87), with slight degradation likely due to increased schema complexity. On the industrial datasets, performance remains stable (F1 of 0.83–0.87) despite real-world data imperfections. Compared with the baseline (StarStar), which achieves an average Discovery Rate of 0.75 and Structural F1 of 0.64, PROD consistently attains higher scores on both metrics, indicating that it not only recovers a larger proportion of ground-truth routing relationships but also maintains a better balance between precision and recall in the recovered structure. For routing decision modelling, PROD achieves RA ranging from 0.79 to 0.96 across datasets, outperforming the decision tree baseline of 0.67. This shows that the Random Forest models effectively capture routing patterns derived from relational attributes.

Table 2. Summary of performances across all baselines and datasets. StarStar and Decision Tree scores are averaged across all datasets; the remaining rows report per-dataset results

Baseline/Dataset	Discovery Rate	Structural F1	Routing Accuracy	Attribution Coherence	Explanation Faithfulness
StarStar	0.75	0.64	–	–	–
Decision Tree	–	–	0.67	–	–
Synthetic (Small)	0.98	0.95	0.96	0.94	1.00
Synthetic (Medium)	0.92	0.90	0.91	0.88	0.96
TPC-H	0.89	0.87	0.89	0.85	0.94
TPC-E	0.86	0.83	0.86	0.83	0.92
Industrial A	0.84	0.87	0.84	0.82	0.92
Industrial B	0.79	0.83	0.79	0.80	0.91

For attribute driver identification, PROD achieves AC scores of 0.80–0.94, indicating strong agreement between SHAP-derived feature attributions and known routing drivers in datasets where ground truth or validated routing drivers are available. In many cases, SHAP analysis indicates that multiple attributes jointly contribute to predicted routing decisions rather than a single dominant feature. This shows that routing decisions in actual process execution depend on combinations of attributes, highlighting the value of data-driven modelling in uncovering multi-factor patterns beyond simple rule-based analysis.

For semantic interpretation, EF remains above 0.90 across all datasets. Inter-rater agreement measured using Cohen's κ was 0.78, indicating substantial agreement between the two evaluators. This shows that the LLM-generated explanations faithfully reflect SHAP-derived routing drivers without introducing unsupported or extraneous factors.

Taken together, the results show that process routing can be inferred directly from relational data, with stable performance across synthetic, benchmark, and real-world datasets.

5.5 Robustness to Row-Level Data Corruption

In this section, we evaluate the robustness of PROD under row-level data corruption. We inject noise into row values at varying levels from 10–30% of entries to simulate data quality issues such as incorrect or missing values. Such corruption may affect linkage detection between tables, successor assignment, and routing decisions inference. We then measure RA and compare it to the original setting (0%), as shown in Table 3. As corruption increases, RA degrades gradually across all datasets. Even under the highest corruption level of 30%, RA remains above 0.65, with benchmark and synthetic data maintaining RA above 0.7. These results indicate that PROD degrades gradually rather than abruptly. Routing accuracy declines predictably with corruption, staying close to clean-data levels under mild noise at 10% and falling to 0.65–0.78 at 30%. PROD is therefore well suited to databases with moderate quality issues. Under heavier corruption, the reduced accuracy means results should be treated as indicative and validated against domain knowledge.

Table 3. Routing accuracy under increasing attribute corruption (0–30%)

Dataset	0%	10%	20%	30%
Synthetic	0.96	0.92	0.85	0.78
Medium-scale	0.91	0.87	0.80	0.74
TPC-H	0.89	0.84	0.77	0.71
TPC-E	0.86	0.82	0.75	0.70
Industrial A	0.84	0.80	0.74	0.68
Industrial B	0.79	0.76	0.70	0.65

5.6 Ablation Study

To evaluate key components in the framework, we conduct an ablation study by removing individual elements of the PROD pipeline and measuring performance. The evaluated variants include removing the routing classifier, excluding upstream attributes, disabling SHAP attribution, and omitting the semantic interpretation stage.

Table 4. Ablation study across all datasets

Variant		Syn-Small	Syn-Med	TPC-H	TPC-E	Real-A	Real-B
	RA	0.96	0.91	0.89	0.86	0.84	0.79
Full PROD	AC	0.94	0.88	0.85	0.83	0.82	0.80
	EF	1.00	0.96	0.94	0.92	0.92	0.91
	RA	0.41	0.38	0.32	0.28	0.34	0.32
A1: No Classifier	AC	–	–	–	–	–	–
	EF	–	–	–	–	–	–
	RA	0.68	0.61	0.55	0.53	0.57	0.55
A2: No Upstream Attrs	AC	0.92	0.88	0.82	0.80	0.86	0.84
	EF	0.94	0.88	0.85	0.82	0.80	0.79
	RA	0.96	0.88	0.89	0.86	0.82	0.78
A3: No SHAP	AC	0.76	0.79	0.81	0.77	0.64	0.60
	EF	0.60	0.55	0.50	0.50	0.45	0.43
	RA	0.96	0.91	0.89	0.86	0.84	0.79
A4: No Semantic	AC	0.94	0.88	0.85	0.83	0.82	0.80
	EF	–	–	–	–	–	–

Table 4 summarises the ablation results across all datasets. Removing the routing classifier leads to a substantial reduction in RA, with RA dropping from 0.79–0.96 to 0.28–0.41 across datasets. This highlights the importance of supervised learning for modelling routing decisions. Removing upstream attributes reduces RA by approximately 30–40% across datasets, while having minimal effect on AC, indicating that contextual information from upstream tables is critical for modelling routing decisions. Disabling SHAP attribution mainly affects AC and EF, while RA remains relatively unchanged. This suggests that SHAP-based attributions are important for identifying drivers and generating faithful explanations by focusing explanations on higher-importance attributes. Finally, the LLM-based semantic interpretation operates after routing decision modelling and SHAP-based attribution. Removing it therefore leaves RA and AC unchanged and only removes explanation generation (EF unavailable), confirming its role is to translate attributions into human-readable form rather than to influence routing or driver identification. Overall, the ablation results show that each component contributes to routing decision modelling, driver identification, and explanation.

5.7 Case Study Order Process

To illustrate how PROD operates end-to-end, we examine the synthetic Orders dataset, which contains relational tables such as *customers*, *orders*, *payment*, *delivery*, *cancellation*, and *refund*. Each table stores records describing a specific type of business event or entity. For example, an orders row represents an order, a payment row represents a payment attempt, and a delivery row represents a delivery outcome. The process paths presented below are used only to explain the underlying process behaviour and are not provided to PROD as input. Instead, PROD receives only the separate relational tables and must infer the routing behaviour from the linked records.

For exposition, suppose the following three order process paths exist in the data:

- c_1: $customers(C_1) \rightarrow orders(O_1) \rightarrow payment(P_1) \rightarrow delivery(D_1)$
- c_2: $customers(C_2) \rightarrow orders(O_2) \rightarrow cancellation(K_1) \rightarrow refund(R_1)$
- c_3: $customers(C_3) \rightarrow orders(O_3) \rightarrow payment(P_2) \rightarrow delivery(D_2)$

In Stage 1, PROD identifies candidate successor relations between tables. In this example, it discovers that *customers* is followed by *orders*, and that *orders* can be followed by either *payment* or *cancellation*. *payment* can be followed by *delivery*, and *cancellation* can be followed by *refund*. In Stage 2, PROD assigns successor sets at the row level for all tables. For illustration, following the notation in Sect. 4.2, $\text{succ}(O_1) = \{payment\}$, $\text{succ}(O_2) = \{cancellation\}$, and $\text{succ}(O_3) = \{payment\}$. Thus, O_1 and O_3 route to *payment*, while O_2 routes to *cancellation*. Aggregating these row-level assignments gives $\text{S}(orders)=\{payment,\ cancellation\}$, showing that *orders* is a branching table.

In Stage 3, PROD learns the relationship between row-level attributes and routing decisions using attributes from the order and its linked customer record. For each branching table, PROD produces transition probabilities and SHAP-based feature attributions that quantify the contribution of each attribute to the routing decision. The SHAP summary plot in Fig. 2a shows that customer location (e.g., `cust_addr_north`), customer risk indicators (e.g., `cust_risk_high`), product category (e.g., `ord_cat_furn`), and interaction features such as customer location × customer risk contribute to the routing decision. This suggests that cancellation is not explained by one attribute alone, but by combinations of customer and order characteristics.

Finally, in Stage 4, PROD uses the SHAP-derived feature attributions and transition probabilities obtained in Stage 3 to generate semantic explanations and integrates them into the routing graph shown in Fig. 2b. In this graph, nodes represent tables and edges represent routing paths between them. The resulting graph shows the inferred routing behaviour and the attributes associated with alternative routes, helping users understand why some orders proceed to *payment* and *delivery* while others are routed to *cancellation* and *refund*.

5.8 Reproducibility

All reproduction materials are archived on Zenodo [16].

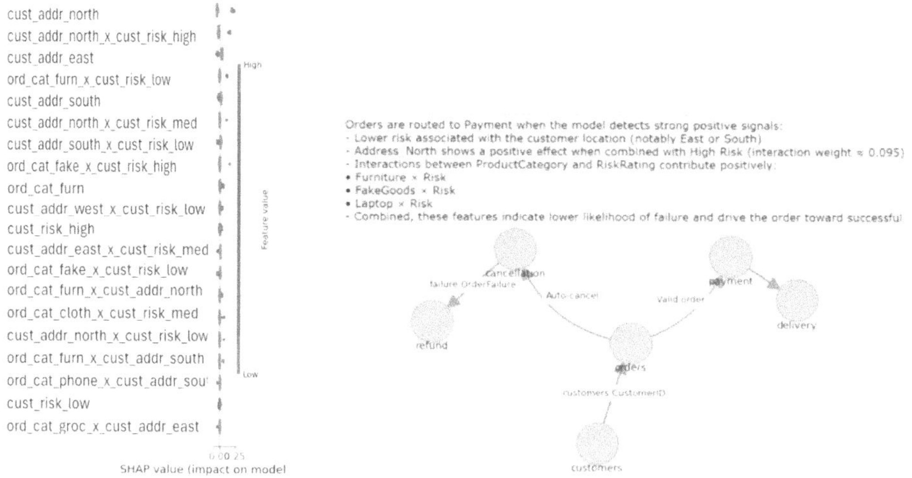

(a) SHAP summary plot highlighting key drivers of the cancellation route.

(b) Process routing graph inferred by PROD with semantic explanations.

Fig. 2. Outputs for the order-to-delivery dataset: routing drivers and the inferred process routing graph

6 Discussion

6.1 Implications for Process Analysis

The results highlight a setting in process analysis where event logs are not readily available and reconstructing them from relational databases is non-trivial. In such settings, PROD enables process routing to be inferred directly from relational data, allowing analysis without explicit event log reconstruction. This complements existing process mining approaches by supporting analysis in operational systems where process execution is implicitly recorded across tables. The focus on routing behaviour and its drivers provides a practical perspective for process analysis. Identifying how cases are routed and the attributes influencing these decisions can support operational insights, such as understanding factors associated with cancellations or successful transactions, thereby helping organisations make decisions such as adjusting risk thresholds or intervening earlier in the fulfilment process. In addition, integrating semantic explanations with an LLM enables routing decisions to be interpreted in natural language. This supports broader accessibility of process analysis, allowing business users to understand and act on routing patterns without requiring expertise in complex database schemas or detailed process mining techniques.

6.2 Threats to Validity

Threats to validity are acknowledged. First, the framework assumes that relational databases (RDBs) contain sufficient structural signals, such as meaningful table linkages or attribute relationships, to support the inference of routing relationships between tables. In systems where such signals are weak or ambiguous, routing inference may be less reliable, although the robustness experiments indicate some tolerance to imperfect linkages. A systematic sensitivity analysis of the linkage thresholds is left to future work. Second, the semantic explanation layer relies on prompt-based LLM explanations, which may be sensitive to prompt design and future model updates. Third, the evaluation includes only two industrial datasets, which may limit the generalisability of the findings to other operational domains. Fourth, direct baselines for routing discovery from relational databases without pre-existing logs are limited, as most existing process mining approaches assume event logs as input. To compensate, PROD was evaluated on multiple synthetic and benchmark datasets with known routing structures. Finally, the approach analyses routing behaviour observable in database snapshots; process paths that have not occurred in the observed data cannot be discovered.

7 Conclusion

This paper presented Process ROuting Discovery from relational Data (PROD), a framework for discovering routing behaviour from relational databases without requiring event logs. PROD reconstructs a routing graph across relational tables, learns routing decisions at branching points, and identifies the attributes that drive these decisions, supported by semantic explanations. Evaluation across synthetic, benchmark, and industrial datasets shows that process routing and its drivers can be recovered from relational data with consistent performance, while degrading gradually under moderate data corruption. These results suggest that relational databases can serve as a valuable source for process analysis beyond traditional event-log-based settings. They further enable practitioners to understand and act on routing behaviour through its underlying paths, decisions, and drivers, supporting more informed operational decision-making.

Future work will extend PROD to handle more complex process patterns, including loops, concurrency, and synchronisation, and to analyse routing behaviour across successive database snapshots. Evaluating the framework on a broader range of operational systems would further assess its generalisability.

References

1. van der Aalst, W., Weijters, T., Maruster, L.: Workflow mining: discovering process models from event logs. IEEE Trans. Knowl. Data Eng. **16**(9), 1128–1142 (2004)
2. van der Aalst, W.M.P.: Object-centric process mining: Dealing with divergence and convergence in event data. In: Ölveczky, P.C., Salaün, G. (eds.) Software Engineering and Formal Methods, pp. 3–25. Springer International Publishing (2019)
3. Berti, A., van der Aalst, W.: Extracting multiple viewpoint models from relational databases. In: Ceravolo, P., van Keulen, M., Gómez-López, M.T. (eds.) Data-Driven Process Discovery and Analysis, pp. 24–51. Springer International Publishing, Cham (2020)

4. Berti, A., van der Aalst, W.M.: Starstar models: Using events at database level for process analysis. In: International Symposium on Data-Driven Process Discovery and Analysis (2018)
5. Bodensohn, J.M., Brackmann, U., Vogel, L., Sanghi, A., Binnig, C.: Unveiling challenges for LLMs in enterprise data engineering. Proc. VLDB Endowment **19**(2), 196–209 (2025)
6. Breiman, L.: Random forests. Mach. Learn. **45**(1), 5–32 (Oct2001)
7. de Leoni, M., van der Aalst, W.M., Dees, M.: A general process mining framework for correlating, predicting and clustering dynamic behavior based on event logs. Inf. Syst. **56**, 235–257 (2016)
8. Fahland, D.: Artifact-Centric Process Mining, pp. 108–117. Springer International Publishing (2019)
9. Goossens, A., De Smedt, J., Vanthienen, J.: Extracting process-aware decision models from object-centric process data. Inf. Sci. **682**, 121263 (2024)
10. Günther, C.W., van der Aalst, W.M.P.: Fuzzy mining – adaptive process simplification based on multi-perspective metrics. In: Business Process Management. pp. 328–343. Springer Berlin Heidelberg (2007)
11. Hernandez-Resendiz, J.D., Tello-Leal, E., Ramirez-Alcocer, U.M., Macías-Hernández, B.A.: Semi-automated approach for building event logs for process mining from relational database (2022). https://doi.org/10.3390/app122110832
12. Kayali, M., Lykov, A., Fountalis, I., Vasiloglou, N., Olteanu, D., Suciu, D.: Chorus: Foundation models for unified data discovery and exploration (2023). https://doi.org/10.48550/arXiv.2306.09610
13. Leemans, S.J.J., Fahland, D., van der Aalst, W.M.P.: Discovering block-structured process models from event logs - a constructive approach. In: Colom, J.M., Desel, J. (eds.) Application and Theory of Petri Nets and Concurrency, pp. 311–329. Springer, Berlin Heidelberg, Berlin, Heidelberg (2013)
14. de Leoni, M., van der Aalst, W.M.P.: Data-aware process mining: discovering decisions in processes using alignments. In: Proceedings of the 28th Annual ACM Symposium on Applied Computing, pp. 1454–1461. SAC '13, Association for Computing Machinery (2013)
15. Li, G., de Murillas, E.G.L., de Carvalho, R.M., van der Aalst, W.M.P.: Extracting object-centric event logs to support process mining on databases. In: Mendling, J., Mouratidis, H. (eds.) Information Systems in the Big Data Era, pp. 182–199. Springer International Publishing, Cham (2018)
16. Lim, J.: Building-Process-Knowledge-Graph: Artifact for BPM 2026 paper (2026). https://doi.org/10.5281/zenodo.20428775
17. Lundberg, S., et al.: From local explanations to global understanding with explainable ai for trees. Nature Mach. Intell. **2** (01 2020)
18. Mannhardt, F., de Leoni, M., Reijers, H.A., van der Aalst, W.M.P.: Decision mining revisited - discovering overlapping rules. In: Nurcan, S., Soffer, P., Bajec, M., Eder, J. (eds.) Advanced Information Systems Engineering, pp. 377–392. Springer International Publishing, Cham (2016)
19. Mannhardt, F., de Leoni, M., Reijers, H.A., van der Aalst, W.M.P.: Data-driven process discovery - revealing conditional infrequent behavior from event logs. In: Dubois, E., Pohl, K. (eds.) Advanced Information Systems Engineering, pp. 545–560. Springer International Publishing, Cham (2017)
20. Murillas, E.G.L.d., Reijers, H.A., Aalst, W.M.P.v.d.: Connecting databases with process mining: a meta model and toolset. Softw. Syst. Model. **18**, 1209–1247 (2018). https://doi.org/10.1007/s10270-018-0664-7

21. Qafari, M.S., van der Aalst, W.: Fairness-aware process mining. In: Panetto, H., Debruyne, C., Hepp, M., Lewis, D., Ardagna, C.A., Meersman, R. (eds.) OTM 2019. LNCS, vol. 11877, pp. 182–192. Springer, Cham (2019). https://doi.org/10.1007/978-3-030-33246-4_11
22. Rozinat, A., Aalst, van der, W.: Decision mining in business processes. BETA publicatie : working papers, Technische Universiteit Eindhoven (2006)
23. Schönig, S., Solti, A., Cabanillas, C., Jablonski, S., Mendling, J.: Efficient and customisable declarative process mining with SQL (06 2016). https://doi.org/10.1007/978-3-319-39696-5_18
24. Sui, Y., et al.: TAP4LLM: Table provider on sampling, augmenting, and packing semi-structured data for large language model reasoning. In: Findings of the Association for Computational Linguistics: EMNLP 2024, pp. 10306–10323. Association for Computational Linguistics (2024)
25. Transaction Processing Performance Council: Tpc-h benchmark specification. https://www.tpc.org/tpch/ (2014)
26. Transaction Processing Performance Council: TPC-e benchmark specification. https://www.tpc.org/tpce/ (2015)
27. Wu, Z., Chen, J., Paton, N.W.: Schema inference for tabular data repositories using large language models (2025). https://doi.org/10.48550/arXiv.2509.04632

Neuro-Symbolic Prescriptive Process Analytics: Learning Compliant and Effective Process Recommendations Through Constraint Distillation

Gyunam Park(✉)

Eindhoven University of Technology, Groene Loper 5, 5612 AZ
Eindhoven, The Netherlands
g.park@tue.nl

Abstract. Given a running process case, prescriptive process analytics (PrPA) recommends the next activity that best improves a desired outcome, such as reducing cycle time. Recommendations must also be operationally feasible: an activity that improves the predicted outcome but violates process constraints cannot be enacted. The core challenge is therefore not compliance alone, which inference-time filtering can enforce, but achieving high outcome quality *within* the compliant space. Existing approaches fall short: post-hoc filtering never learns which compliant activities are outcome-improving, while logic-as-loss penalties balance unstable competing gradients from data fit and compliance. Both also treat all constraints equally, so combining mandatory hard constraints with relaxable soft constraints can leave many cases with no feasible recommendation. We propose **NeSyPA**, a neurosymbolic PrPA method with two contributions: *Hierarchical Constraint Relaxation*, which drops soft constraints only when the full constraint set is unsatisfiable, and *Constraint-Compiled Outcome Distillation*, which compiles per-prefix feasible sets into feasibility-masked, outcome-weighted soft targets that teach the neural prescriptor to rank feasible activities by outcome merit. On four real-world event logs, NeSyPA achieves the highest Effective Compliance Rate on all test sets, with gains of +6.7% to +9.9%, driven by higher outcome quality within the compliant space.

Keywords: Prescriptive process analytics · Neurosymbolic AI · Constraint compliance · Knowledge distillation

1 Introduction

Given a running process case, prescriptive process analytics (PrPA) recommends the next activity that best improves a desired outcome, for example reducing cycle time or avoiding an SLA breach [22]. Unlike predictive monitoring, which forecasts what *will* happen, PrPA must recommend what *should* happen under

F. M. Maggi et al. (Eds.): BPM 2026, LNCS 16936, pp. 503–520, 2027.
https://doi.org/10.1007/978-3-032-37777-7_30

process constraints that restrict the set of permissible activities. A recommendation to "skip the compliance check" may optimize predicted cycle time but is operationally infeasible. The core challenge is therefore not compliance alone (inference-time filtering can enforce that) but achieving high outcome improvement *within* the compliant space, so that recommendations are both feasible and effective.

Current PrPA approaches handle constraints in ways that sacrifice outcome quality. Post-hoc filtering trains an unconstrained model and discards infeasible recommendations at inference time. While this can maintain compliance, the model never learns *which compliant activities are outcome-improving*: the best recommendation is often discarded, and the fallback may be weak, since no learning signal flows back from the constraint structure. Penalty-based methods, sometimes called logic-as-loss approaches, add a constraint violation penalty to the training loss [13]. This forces the model to balance competing gradients from data fit and logic fit: the gradient tells the model *that* it is wrong but not *how* to fix it, and tuning the penalty weight λ is inherently unstable, often degrading both compliance and outcome quality.

Beyond these mechanism-level limitations, a fundamental practical challenge remains unaddressed. Real-world process constraint knowledge bases combine rules that operate at different levels of authority. Some constraints are *hard*: they derive from external regulations or domain expertise and must never be violated. Others are *soft*: they capture frequent behavioral patterns, e.g., discovered from historical data and reflect statistical regularities that may not hold for every case. When hard and soft constraints are enforced equally, they can conflict, causing the feasible set, i.e., the set of activities satisfying all constraints simultaneously, to become empty. In our experiments, the full constraint knowledge base renders 5.9% to 35.1% of prefixes irreparable across datasets. No existing method distinguishes constraint priority or handles these over-constrained states gracefully.

We propose NeSyPA, a neurosymbolic method for prescriptive process analytics that addresses both challenges through two contributions. First, *Hierarchical Constraint Relaxation* (HCR) partitions the constraint knowledge base into hard constraints that are never relaxed and soft constraints that may be dropped when the full set is unsatisfiable, reducing irreparable rates dramatically while preserving hard constraint compliance. Second, *Constraint-Compiled Outcome Distillation* (CCOD) pre-computes the hierarchical feasible set for each training prefix, constructs feasibility-masked soft targets weighted by predicted outcome improvement, and fine-tunes the neural prescriptor on these targets in a single phase. Unlike logic-as-loss methods, CCOD has no λ tradeoff parameter: the targets are zero outside the feasible set and concentrate probability mass on the compliant activities with the highest outcome lift. This teaches the model to optimize outcomes directly within the compliant space, rather than learning outcome improvement and compliance as competing objectives.

The method can be viewed as a form of knowledge distillation [11] where the teacher is a symbolic constraint-outcome oracle composed of deterministic finite automata and an outcome evaluator, and the student is the neural prescriptor that learns to approximate the oracle's feasibility-masked, outcome-weighted distributions. The key difference from standard distillation is that the teacher is not a larger neural model but a symbolic reasoner that computes exact constraint satisfaction.

We evaluate NeSyPA on four real-world event logs from two domains, comparing against four baselines spanning unconstrained, post-hoc filtering, reinforcement learning, and logic-as-loss approaches.[1] NeSyPA achieves the highest Effective Compliance Rate (ECR), the fraction of recommendations that are both outcome-improving and constraint-compliant, on all four test sets, driven by substantially higher outcome improvement while maintaining comparable constraint satisfaction. An ablation study confirms that CCOD is the dominant contributor, while a sensitivity analysis reveals the complementary roles of hard and soft constraints.

The remainder of this paper is organized as follows. Section 2 reviews related work. Section 3 introduces preliminaries. Section 4 presents the NeSyPA method. Section 5 describes the experimental design. Section 6 presents results and discussion. Section 7 discusses threats to validity. Section 8 concludes.

2 Related Work

Prescriptive process analytics has been approached through several distinct paradigms, each with characteristic strengths and limitations regarding constraint-aware recommendation.

Alarm-Based PrPA. Alarm-based approaches frame PrPA as a binary decision: whether or not to intervene in a running case [8,20]. These methods train classifiers to predict undesirable outcomes and trigger alarms when the predicted probability exceeds a threshold, potentially subject to resource constraints. While effective for intervention timing, alarm-based methods decide *whether* to act, not *which* activity to recommend. They are fundamentally complementary to, rather than competitive with, activity-level prescriptive methods, and they incorporate no notion of process constraint compliance.

Causal Inference PrPA. Causal approaches estimate treatment effects of predefined interventions and decide on whom and when to apply them [2,3,18,19]. The treatment set is fixed by domain experts, and the methods focus on identifying cases where intervention would be most beneficial. Since treatments are predefined rather than discovered, process constraints are implicitly satisfied by construction of the treatment set. However, this secures feasibility only by restricting recommendations to a fixed, pre-enumerated set of expert-defined interventions, which limits the method to domains where the intervention space

[1] Code and data are available at https://github.com/gyunamister/nesy_prpa.

can be fully enumerated in advance and provides no mechanism for adapting to new or complex constraint structures.

Reinforcement Learning PrPA. Reinforcement learning approaches model PrPA as sequential decision-making, where an agent learns a policy that maps process states to activity recommendations [1,4]. Constraint handling in these methods is either post-hoc, filtering infeasible actions from the policy output, or penalty-based, incorporating constraint violations as negative rewards. Neither mechanism provides a constructive approach to producing the nearest feasible alternative when the top recommendation is infeasible. Furthermore, all constraints are treated equally, with no distinction between hard constraints (mandatory regulatory requirements) and soft constraints (relaxable organizational patterns), leaving no principled way to handle over-constrained states.

Neurosymbolic Predictive Monitoring. Recent work has explored neurosymbolic integration for process monitoring, though primarily in the predictive rather than prescriptive setting. This emerging line spans compliance-aware predictive monitoring [15], next-activity prediction via two-stage logic tensor networks with rule pruning [17], process anomaly detection [9], and domain-specific applications such as real-time sepsis mortality prediction [16]. These efforts confirm the value of injecting symbolic knowledge into neural process models, but they address predictive or diagnostic tasks rather than constraint-aware activity recommendation. Mezini et al. [13] compile LTL_f constraints into a differentiable loss function for autoregressive next-activity prediction. This is the closest neurosymbolic approach to ours, but it differs in several important ways: it uses a logic-as-loss formulation that creates a λ tradeoff between data fit and constraint compliance; it is purely predictive rather than prescriptive; it treats all constraints equally without hierarchical prioritization; and it cannot constructively generate compliant alternatives when the top prediction violates constraints.

In contrast to all of the above, our approach pre-computes the feasible activity set exactly via deterministic finite automata [10] and distills it into soft training targets, so compliance is enforced by construction and outcome quality is optimized within the compliant space, with no λ tradeoff. To our knowledge, no existing method distinguishes hard and soft constraint priority to handle irreparable states from over-constrained knowledge bases, or compiles constraint feasibility into training targets as soft outcome-weighted distributions.

3 Preliminaries

Event Logs, Traces, and Prefixes. An event log L consists of a set of traces, where each trace $\sigma = \langle e_1, e_2, \ldots, e_n \rangle$ records the sequence of events for a single process case. Each event $e_i = (a_i, t_i, r_i, \mathbf{d}_i)$ comprises an activity label $a_i \in \mathcal{A}$, a timestamp t_i, a resource r_i, and a data payload $\mathbf{d}_i$. A prefix $\sigma_k = \langle e_1, \ldots, e_k \rangle$ is the subsequence of the first k events of a trace. The activity alphabet $\mathcal{A}$ denotes the set of all possible activity labels in the log.

Declare Constraints and Automata. Declare [14] is a declarative process modeling language that specifies constraints as Linear Temporal Logic over finite traces (LTL_f) formulas. Key templates used in this work include `existence` (A must occur at least once), `response` (if A occurs, B must eventually follow), `precedence` (B can only occur if A has occurred before), `not-co-existence` (A and B cannot both occur in the same trace), `alternate-response` (each occurrence of A must be followed by B before A can occur again), and `chain-precedence` (B can occur only if A occurred immediately before). Each Declare constraint can be compiled into a deterministic finite automaton (DFA) $\mathcal{M}_j$ [10] with states classified as accepting (F), pending, or permanently violated (V). Given a prefix σ_k, the automaton tracks constraint status incrementally: appending activity a to σ_k transitions the automaton from state q_k^j to successor state $q_{k+1}^j = \delta_j(q_k^j, a)$, where δ_j is the transition function of $\mathcal{M}_j$.

Data-Aware Constraints. Data-aware constraints take the form $\phi(\mathbf{d}) \Rightarrow \psi$, where ϕ is a predicate over case or event attributes and ψ is a Declare constraint. For example, "if `loan_amount` > 10K then `existence(manual_review)`." When the condition $\phi(\mathbf{d}_k)$ evaluates to false for the current case, the constraint is vacuously satisfied.

Constraint Hierarchy. The constraint knowledge base Φ is partitioned into two tiers based on their enforcement level: *hard constraints* Φ_{hard}, which must never be violated and are never relaxed; and *soft constraints* Φ_{soft}, which may be relaxed when they create infeasible states. In practice, hard constraints typically derive from external regulations and domain expertise (e.g., the EU Consumer Credit Directive mandating specific review steps), while soft constraints capture statistically frequent behavioral patterns that may not hold universally, e.g., discovered from training traces via Declare mining [5]. Because these two tiers originate from different authorities and operate at different levels of generality, conflicts between them are common: a soft constraint may prohibit an activity that a hard constraint requires, leaving no activity that satisfies both simultaneously. The full knowledge base is $\Phi = \Phi_{\text{hard}} \cup \Phi_{\text{soft}}$. The *hierarchical feasible set* for a prefix σ_k resolves such conflicts by prioritizing hard constraints:

$$\mathcal{F}_{\text{hier}}(\sigma_k) = \begin{cases} \mathcal{F}_{\text{full}}(\sigma_k) & \text{if } \mathcal{F}_{\text{full}}(\sigma_k) \neq \emptyset \\ \mathcal{F}_{\text{hard}}(\sigma_k) & \text{otherwise} \end{cases} \tag{1}$$

where $\text{Compliant}_\Psi(\sigma_k, a)$ holds iff appending activity a to prefix σ_k does not violate any constraint in Ψ, $\mathcal{F}_{\text{full}}(\sigma_k) = \{a \in \mathcal{A} \mid \text{Compliant}_\Phi(\sigma_k, a)\}$, and $\mathcal{F}_{\text{hard}}(\sigma_k) = \{a \in \mathcal{A} \mid \text{Compliant}_{\Phi_{\text{hard}}}(\sigma_k, a)\}$.

Knowledge Distillation. Knowledge distillation [11] transfers knowledge from a teacher to a student via soft probability distributions. Given a teacher distribution q over classes, the student is trained to minimize the cross-entropy $H(q, p)$

where p is the student's output distribution. Soft targets encode richer information than hard labels: they convey relative preferences among classes, not just the top-1 choice.

4 The NeSyPA Method

4.1 Problem Statement

Given a prefix σ_k of a running case, the goal is to recommend the next activity $a^* \in \mathcal{A}$ that maximizes estimated outcome improvement subject to hierarchical constraint compliance:

$$a^* = \arg \max_{a \in \mathcal{F}_{\text{hier}}(\sigma_k)} \text{Score}(\sigma_k, a) \tag{2}$$

The method operates in three phases: (1) warm-up training of the neural prescriptor and compilation of constraints into DFA automata, (2) constraint-compiled outcome distillation that constructs feasibility-masked targets and fine-tunes the prescriptor, and (3) hierarchical inference that generates candidates and applies constraint filtering. Figure 1 provides an overview.

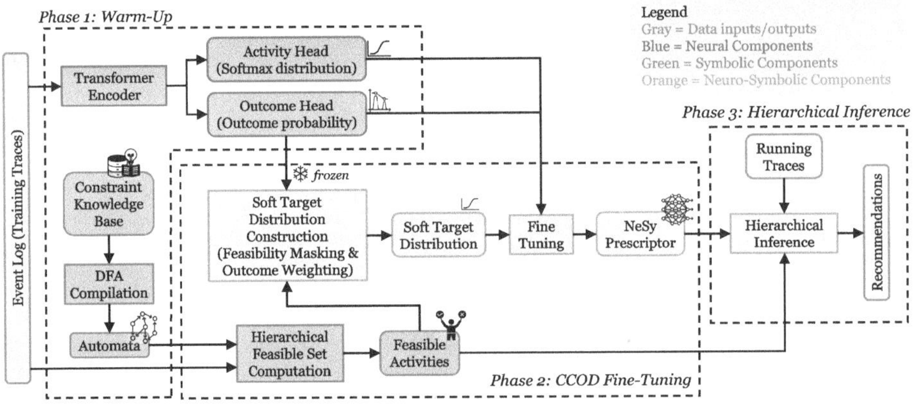

Fig. 1. Overview of the NeSyPA method. Phase 1 (Warm-Up) trains the neural prescriptor and compiles constraints into DFA automata. Phase 2 (CCOD Fine-Tuning) constructs feasibility-masked, outcome-weighted soft targets and fine-tunes the prescriptor. Phase 3 (Hierarchical Inference) generates candidates and applies hierarchical constraint filtering

4.2 Phase 1: Warm-Up Training

Prescriptor Architecture. The neural prescriptor takes as input a prefix σ_k and produces both an activity distribution and an outcome prediction. Each event

e_i is encoded as the concatenation of an activity embedding, a resource embedding, an MLP-encoded data payload, and a temporal encoding based on inter-event duration. The sequence of event encodings is processed by a Transformer encoder [21] to produce a context vector $\mathbf{c}_k$.

Two output heads operate on $\mathbf{c}_k$. The activity head produces a softmax distribution $\mathbf{p}_k \in \mathbb{R}^{|\mathcal{A}|}$ over all activities. The outcome head produces a predicted positive-outcome probability $\hat{y}_k \in [0,1]$. The prescriptive score for a candidate activity a combines both signals:

$$\text{Score}(a \mid \sigma_k) = \alpha \cdot p_k(a) + (1-\alpha) \cdot \Delta\hat{y}_k(a) \tag{3}$$

where $\Delta\hat{y}_k(a) = \hat{y}(\sigma_k \cdot a) - \hat{y}(\sigma_k)$ is the estimated outcome lift from appending a, and $\alpha \in [0,1]$ weights process likelihood against outcome improvement.

Warm-Up Loss. The model is trained on a supervised loss combining next-activity prediction (cross-entropy) and outcome prediction (binary cross-entropy):

$$\mathcal{L}_{\text{data}} = \mathcal{L}_{\text{CE}}(a_{k+1}, \mathbf{p}_k) + \beta \cdot \mathcal{L}_{\text{BCE}}(y, \hat{y}_k) \tag{4}$$

where a_{k+1} is the ground-truth next activity, y is the binary outcome label, and β balances the two loss terms.

Constraint Compilation. Each constraint $\phi_j \in \Phi$ is compiled to a DFA $\mathcal{M}_j$. Given a prefix σ_k, the automaton is in state q_k^j. Appending a candidate activity a yields the successor state $q_{k+1}^j = \delta_j(q_k^j, a)$, from which the compliance status $\text{Status}_j(\sigma_k, a)$ is determined; specifically, $\text{Status}_j(\sigma_k, a) = \text{violated}$ if q_{k+1}^j is a permanently violated state of $\mathcal{M}_j$. An activity is compliant with the full knowledge base if and only if no constraint enters a permanently violated state:

$$\text{Compliant}_{\Phi}(\sigma_k, a) = \bigwedge_j \text{Status}_j(\sigma_k, a) \neq \text{violated} \tag{5}$$

For data-aware constraints, the condition predicate $\phi(\mathbf{d}_k)$ is evaluated first; if the condition is false, the constraint is vacuously satisfied and does not restrict the feasible set.

4.3 Phase 2: Constraint-Compiled Outcome Distillation

CCOD constructs feasibility-masked, outcome-weighted soft training targets and fine-tunes the prescriptor. For each training prefix σ_k:

Step 1: Compute hierarchical feasible set. The constraint reasoner computes feasibility in three stages using the compiled DFA automata. First, it computes $\mathcal{F}_{\text{full}}(\sigma_k)$ by evaluating all activities against all automata. If $\mathcal{F}_{\text{full}}(\sigma_k) \neq \emptyset$, this set is used and all constraints are satisfied. If $\mathcal{F}_{\text{full}}(\sigma_k) = \emptyset$, the reasoner computes $\mathcal{F}_{\text{hard}}(\sigma_k)$ by evaluating activities against only hard constraint automata, relaxing all soft constraints as a group. If $\mathcal{F}_{\text{hard}}(\sigma_k) = \emptyset$, the prefix is flagged as irreparable and excluded from CCOD training. Hard constraints are never relaxed; soft constraints are relaxed as a group: either all are enforced or all are dropped. We adopt this binary scheme for simplicity; selective per-constraint relaxation is feasible but introduces a combinatorial constraint selection problem that is orthogonal to our core contribution. We discuss this trade-off and a path toward selective relaxation in Sect. 6.4.

Step 2: Compute outcome lift. For each activity $a \in \mathcal{F}_{\text{hier}}(\sigma_k)$, the estimated outcome lift is computed as $\Delta\hat{y}(a) = \hat{y}_{\text{frozen}}(\sigma_k \cdot a) - \hat{y}_{\text{frozen}}(\sigma_k)$ using a frozen copy of the warm-up outcome head. The frozen copy ensures stable targets throughout fine-tuning, preventing a feedback loop where changing outcome predictions shift the targets that drive further training.

Step 3: Construct soft target distribution. The target distribution concentrates probability mass on feasible activities proportionally to their outcome lift:

$$q(a) = \begin{cases} \dfrac{\exp(\Delta\hat{y}(a)/\tau)}{\sum_{a' \in \mathcal{F}_{\text{hier}}(\sigma_k)} \exp(\Delta\hat{y}(a')/\tau)} & \text{if } a \in \mathcal{F}_{\text{hier}}(\sigma_k) \\ 0 & \text{otherwise} \end{cases} \tag{6}$$

where $\tau > 0$ is a temperature parameter. Equation 6 is a standard softmax over the feasible set, so q is a valid probability distribution supported only on $\mathcal{F}_{\text{hier}}(\sigma_k)$ for any τ; the temperature is therefore a design choice that controls target sharpness, not a correctness condition. Low τ concentrates mass on the highest-outcome-lift feasible activity; high τ spreads mass more uniformly across feasible activities.

Step 4: Fine-tune. The prescriptor is fine-tuned by minimizing the cross-entropy between the soft target and the model's full (unmasked) softmax output, with the encoder and outcome head frozen:

$$\mathcal{L}_{\text{CCOD}} = -\sum_{a \in \mathcal{A}} q(a) \log p_k(a) \tag{7}$$

Since $q(a) = 0$ for $a \notin \mathcal{F}_{\text{hier}}(\sigma_k)$, the cross-entropy only accumulates terms over feasible activities. Crucially, the gradient $\partial\mathcal{L}/\partial z_a = p_k(a) - q(a)$ is non-zero for all activities: for infeasible activities where $q(a) = 0$, the gradient equals $p_k(a) > 0$, which pushes their probabilities down. This shifts probability mass toward the hierarchical feasible set rather than only re-ranking within it. Only the activity head is fine-tuned; the encoder and outcome head keep their warm-up weights.

CCOD has several properties that distinguish it from prior approaches. Unlike logic-as-loss methods that balance $\mathcal{L}_{\text{data}} + \lambda \cdot \mathcal{L}_{\text{logic}}$, CCOD has no compliance-versus-outcome tradeoff parameter. Compliance is enforced by construction, and outcome optimization happens within the compliant space. The soft target encodes a full ranking over feasible activities weighted by outcome merit. The fine-tuning is one-shot and non-iterative: feasible sets and outcome lifts are pre-computed once, making the approach both simple and efficient.

4.4 Phase 3: Hierarchical Inference

At inference time, the prescriptor scores all activities for a given prefix σ_k using Eq. 3 and selects the top-K candidates. Each candidate is checked in rank order against the full knowledge base Φ, and the first compliant candidate is recommended. If no candidate is compliant, the reasoner relaxes to Φ_{hard} and recommends the highest-scoring activity in $\mathcal{F}\text{hard}(\sigma_k)$ over the full alphabet. If $\mathcal{F}\text{hard}(\sigma_k) = \emptyset$, the prefix is irreparable and the top-ranked candidate is returned unfiltered.

4.5 Computational Cost

Each Declare constraint is compiled once into a small deterministic finite automaton (typically two to five states), so compilation is a negligible one-time cost. For a prefix σ_k, computing the feasible set evaluates every candidate activity against every constraint automaton, which is $O(|\mathcal{A}| \cdot |\Phi|)$, where each evaluation is an $O(1)$ cached state transition. In our datasets $|\mathcal{A}| \leq 36$ and $|\Phi|$ is on the order of 10 to 25, giving at most a few hundred automaton lookups per prefix. CCOD adds no online overhead: feasible sets and outcome lifts are pre-computed once before fine-tuning, and fine-tuning itself is one-shot. The resulting feasible sets are small in practice (on average 1.3 to 3.0 activities under the full knowledge base, and 8.5 to 25 under hard constraints only), so candidate scoring at inference is inexpensive. Because the knowledge base is fixed at training time, adapting to changed soft constraints at runtime requires only recompiling the affected automata and re-running this one-time pre-computation, with no change to the model architecture.

5 Experimental Design

5.1 Datasets

We evaluate NeSyPA on four real-world event logs, summarized in Table 1. BPIC 2012 [6] records a loan application process at a Dutch financial institution with 36 distinct activities and approximately 13,000 cases. BPIC 2017 [7] covers a similar loan application process with 26 activities and approximately 31,000 cases; following established practice, we evaluate on the two concept-drift subsets of cases starting before week 22 and after week 28, yielding distinct test sets. Traffic Fines [12] records an Italian road traffic fine management process with 11 activities and approximately 150,000 cases. The outcome variable is application acceptance for BPIC 2012, offer acceptance for BPIC 2017, and timely fine payment for Traffic Fines.

Data Preparation. We split each log temporally by case start time into training (60%), validation (20%), and test (20%) partitions, so that no case spans more than one partition. From every trace we extract all prefixes of length $k \in [2, n-1]$, each labeled with its ground-truth next activity, rather than sampling or fixing a single prefix length. This yields 42,587 (BPIC 2012), 86,240 (BPIC 2017 before), 113,896 (BPIC 2017 after), and 52,984 (Traffic Fines) test prefixes.

Table 1. Dataset overview

Dataset	Domain	Cases	Act.	Hard constr.	Outcome
BPIC 2012	Loan application	~13K	36	9 CF + 1 DA	Accepted
BPIC 2017	Loan application	~31K	26	8 CF	Offer accepted
Traffic Fines	Fine management	~150K	11	10 CF	Paid on time

For each dataset, the constraint knowledge base Φ is constructed from two complementary sources. Hard constraints Φ_{hard} encode process-specific business rules that the authors formulated from publicly available process descriptions and the corresponding regulations, rather than through a formal expert-elicitation study: e.g., the mandatory ordering of verification and approval steps in loan applications (BPIC 2012 and 2017), and the legally prescribed notification sequences in fine management (Traffic Fines). We treat these as reliable by assumption; validating them with domain experts is left to future work. Soft constraints Φ_{soft} are discovered from all training traces using frequency-based Declare mining [5] with support ≥ 0.3 and confidence ≥ 0.85, representing organizational process patterns. Data-aware constraints combine hard conditional rules with decision-tree-mined rules over event attributes. This hybrid construction ensures that constraints encode knowledge orthogonal to the outcome-optimizing model's training signal.

Under the full knowledge base, 5.9–35.1% of test prefixes are irreparable (no activity satisfies all constraints simultaneously), confirming that the combination of hard and soft constraints frequently leads to over-constrained states and motivating the need for hierarchical relaxation.

5.2 Baselines

We compare NeSyPA against four baselines, all using the same Transformer encoder and the same constraint knowledge base Φ for fair comparison. **B1 (Unconstrained):** The neural prescriptor with no constraint mechanism, producing a raw top-1 recommendation. **B2 (Post-hoc):** The neural prescriptor with post-hoc hierarchical filtering at inference, establishing a strong filtering baseline that isolates the value of CCOD training beyond filtering alone. **B3 (RL):** A reinforcement learning prescriptor following the approach described by Agarwal et al. [1], with constraint violations penalized in the reward function using the full knowledge base. **B4 (NeSy-PPM):** An adaptation of the neurosymbolic predictive monitoring approach by Mezini et al. [13] from the predictive to the prescriptive setting. The original method trains a next-activity predictor with a differentiable LTL$_f$ logic loss to steer predictions toward constraint-compliant activities. Our adaptation retains this logic-as-loss training (using the full knowledge base Φ) but extends the architecture with an outcome prediction head trained jointly via binary cross-entropy. At inference, candidates are ranked by a prescriptive score $\alpha \cdot p(a) + (1-\alpha) \cdot \Delta\hat{y}(a)$, combining the activity probability $p(a)$ with the estimated outcome lift $\Delta\hat{y}(a)$, and the top candidate is recommended.

5.3 Metrics

The primary metric is the *Effective Compliance Rate* (ECR), defined as the fraction of recommendations that are both outcome-improving and constraint-compliant:

$$\text{ECR} = \frac{|\{i : \text{outcome-improving}(i) \wedge \text{compliant}(i)\}|}{|\text{all recommendations}|} \tag{8}$$

ECR captures both dimensions simultaneously: a method with high compliance but poor outcomes, or vice versa, scores low. Supporting metrics include: the Outcome Improvement Rate (OIR), the fraction of recommendations that are outcome-improving regardless of compliance; and the Constraint Satisfaction Rate (CSR), the fraction satisfying all constraints regardless of outcome. A recommendation is outcome-improving if the frozen B1 outcome head assigns a strictly higher positive-outcome probability to the extended prefix. A recommendation is outcome-improving if the frozen B1 outcome head assigns a strictly higher positive-outcome probability to the extended prefix.

5.4 Experimental Protocol

Experiment 1: Main comparison. All five methods are evaluated on all four test sets. We report ECR, OIR, and CSR with means and standard deviations over 5 random seeds per method and dataset.
Experiment 2: Ablation. Four variants of NeSyPA are compared: the full method, no-CCOD (B2 with hierarchy, isolating CCOD's contribution), no-hierarchy (full KB only, isolating hierarchy's contribution), and uniform targets (uniform distribution over $\mathcal{F}_{\text{hier}}$ instead of outcome-weighted). We report ECR, OIR, and CSR over 5 seeds.
Experiment 3: Sensitivity to soft constraint quality. On BPIC 2012, we inject 10%, 20%, and 30% spurious constraints into Φ_{soft} only, since hard constraints are assumed reliable whereas soft constraints are sensitive to log quality. The model trains and infers with the noisy knowledge base; metrics are evaluated against the original clean knowledge base. We report ECR, OIR, CSR, and the noisy-KB relaxation rate.

Implementation Details. Each model is trained with early stopping on the validation split, and metrics are reported as means and standard deviations over 5 random seeds. Hyperparameters were fixed, guided by preliminary runs on the validation split, to the following values: the prescriptive score weight $\alpha = 0.3$, the outcome loss weight $\beta = 1.0$, the CCOD temperature $\tau = 0.5$, and the candidate set size $K = 5$; candidates are drawn from the $K' = 10$ most probable activities, and the outcome lift in Eq. 3 is rescaled to $[0, 1]$ as $(\Delta\hat{y}_k(a) + 1)/2$ before mixing.

6 Results and Discussion

6.1 Main Results

Table 2 presents the main comparison across all methods and test sets. NeSyPA achieves the highest ECR on all four test sets, with gains of +6.7pp (BPIC 2012), +9.7pp (BPIC 2017 after), +9.9pp (BPIC 2017 before), and +0.03pp (Traffic Fines) over the respective best baselines.

Table 2. Experiment 1: Main results (mean ± std over 5 seeds). Best ECR per dataset in **bold**

Dataset	Method	ECR	OIR	CSR
BPIC 2012	B1 Unconstrained	.476±.069	.711±.054	.646±.025
	B2 Post-hoc	.502±.047	.709±.052	.681±.001
	B3 RL	.426±.078	.673±.098	.622±.061
	B4 NeSy-PPM	.470±.034	.697±.057	.649±.023
	NeSyPA	**.569±.030**	.849±.031	.677±.004
BPIC 2017 aft.	B1 Unconstrained	.318±.043	.505±.061	.644±.000
	B2 Post-hoc	.322±.043	.506±.062	.648±.001
	B3 RL	.339±.033	.577±.044	.605±.013
	B4 NeSy-PPM	.324±.045	.516±.064	.644±.001
	NeSyPA	**.437±.058**	.718±.072	.648±.001
BPIC 2017 bef.	B1 Unconstrained	.354±.044	.547±.052	.647±.002
	B2 Post-hoc	.356±.044	.547±.052	.650±.000
	B3 RL	.316±.105	.502±.163	.623±.013
	B4 NeSy-PPM	.358±.046	.555±.059	.647±.001
	NeSyPA	**.456±.048**	.735±.062	.650±.001
Traffic Fines	B1 Unconstrained	.381±.076	.532±.132	.834±.121
	B2 Post-hoc	.384±.080	.428±.079	.940±.002
	B3 RL	.398±.080	.445±.078	.940±.001
	B4 NeSy-PPM	.325±.056	.554±.122	.753±.111
	NeSyPA	**.399±.080**	.454±.081	.941±.000

On the BPIC datasets, the ECR gains are driven by dramatically higher OIR. NeSyPA achieves OIR values of 0.849 (BPIC 2012), 0.718 (BPIC 2017 after), and 0.735 (BPIC 2017 before), compared to 0.505–0.711 for the baselines. CSR remains comparable across methods (~0.60–0.68 on BPIC datasets). On Traffic Fines, OIR differences are smaller (0.454 for NeSyPA vs. 0.428–0.554 for baselines), resulting in a negligible ECR margin of +0.03pp.

B3 (RL) and B4 (NeSy-PPM) perform at or below B1 and B2 on most datasets. B3 shows particularly high variance on BPIC 2017 before (coefficient of variation = 0.33). B4 performs poorly on Traffic Fines (ECR = 0.325), with CSR degrading to 0.753, well below the filtering baselines (0.940).

Why CSR is below 1. CSR is computed against the full knowledge base, so it can fall below 1 even though hard-constraint compliance is guaranteed by construction. When the full feasible set is empty, hierarchical relaxation deliberately drops the soft constraints and recommends an activity satisfying only the hard constraints; these recommendations, together with the small fraction of prefixes that remain irreparable, are counted as soft-constraint violations and lower CSR. This is by design, not a failure of enforcement. For the same reason, the marginal CSR difference between B2 and NeSyPA on BPIC 2012 (0.681 vs. 0.677) lies within one standard deviation (±0.001 and ±0.004): both methods apply the same hierarchical filtering at inference, so the gap reflects only the different top-K candidate sets produced after CCOD training, not a difference in how compliance is enforced.

6.2 Ablation

Figure 2 presents the ablation results across all four test sets. CCOD is the dominant contributor: the full model exceeds the no-CCOD variant (B2 with hierarchy) by +7.1pp (BPIC 2012), +11.5pp (BPIC 2017 after), +10.0pp (BPIC 2017 before), and +1.5pp (Traffic Fines). The gap is largest on BPIC datasets ($|\mathcal{F}_{\text{full}}| \approx 2$–3 over 26–36 activities) and smallest on Traffic Fines (11 activities, already-high baseline CSR).

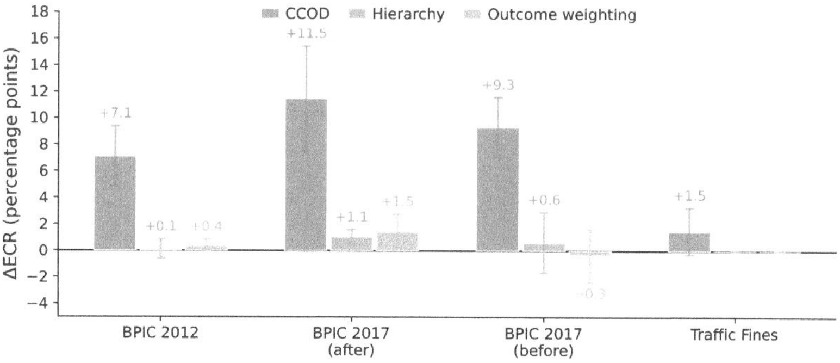

Fig. 2. ECR contribution of each component across datasets. CCOD provides the dominant improvement (+7 to +12pp on BPIC datasets), while hierarchy contributes structurally rather than incrementally

Hierarchical relaxation provides minimal incremental ECR lift: +0.1pp (BPIC 2012), +1.1pp (BPIC 2017 after), +0.6pp (BPIC 2017 before), and approximately 0pp (Traffic Fines). Without hierarchy, the irreparable rate is 31.8% (BPIC 2012) and approximately 35% (BPIC 2017), reduced to 2.3% and approximately 0% respectively; approximately 30–35% of test prefixes on BPIC datasets are hierarchically recovered.

Outcome weighting provides modest and inconsistent benefit over uniform targets: +0.4pp (BPIC 2012), +1.5pp (BPIC 2017 after), −0.3pp (BPIC 2017 before), and approximately 0pp (Traffic Fines).

6.3 Sensitivity to Soft Constraint Quality

Figure 3 presents the noise sensitivity results on BPIC 2012. ECR drops from 0.574 to approximately 0.345 (−40%) at 10% noise and plateaus. Hierarchical relaxation triggers for approximately 65% of training prefixes under noise (versus 30% at baseline), shifting CCOD targets from $\mathcal{F}_{\text{full}}$ ($\approx$ 3 activities) to $\mathcal{F}_{\text{hard}}$ ($\approx$ 22.5 activities). OIR *increases* from 0.854 to 0.918 while CSR *decreases* from 0.677 to 0.400.

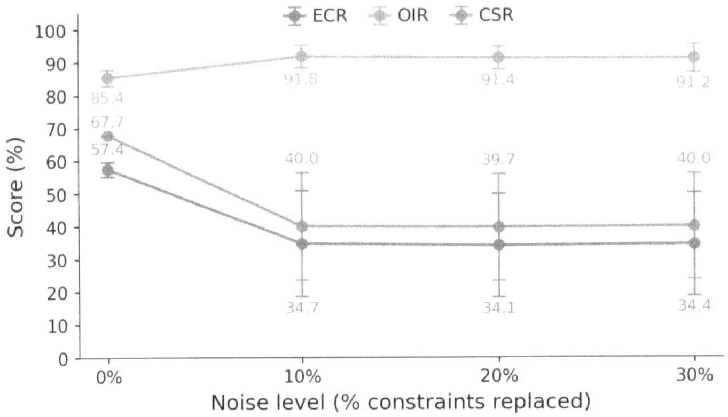

Fig. 3. Sensitivity to soft constraint noise on BPIC 2012. ECR and CSR drop sharply at 10% noise and plateau, while OIR increases as diffuse hard-only CCOD targets make the model more outcome-aggressive. Error bars show standard deviation over 5 seeds

Per-seed variation is high at noise levels above 0% (ECR std $\approx$ 0.18). The per-seed noisy relaxation rate correlates strongly with ECR degradation ($r = -0.91$). A single `not_co_existence` constraint between two common activities can render $\mathcal{F}_{\text{full}} = \emptyset$ for more than 90% of prefixes, while constraints on rare activities have minimal effect; the damage saturates at 10% noise.

6.4 Discussion

The experimental results illuminate the mechanism by which NeSyPA achieves its effectiveness and reveal important design considerations.

The primary mechanism of CCOD is outcome distillation within the feasible space. Pre-computing feasibility and constructing outcome-weighted targets teaches the model to produce candidates with high outcome lift among feasible activities. The inference-time filtering then selects the best compliant recommendation from this improved candidate set. The model does not need to internalize compliance at the raw top-1 level; it needs to produce a top-K set rich in outcome-improving feasible activities. The dramatically higher OIR for NeSyPA (0.849, 0.718, 0.735 on BPIC datasets versus 0.505–0.711 for baselines) confirms this mechanism. Meanwhile, CSR parity across methods (~0.60–0.68) confirms that CCOD improves outcome quality without sacrificing constraint satisfaction, with no compliance-versus-outcome tradeoff. The failure of B4 (NeSy-PPM) on Traffic Fines (CSR = 0.753 vs. 0.940 for filtering baselines) further illustrates the fragility of logic-as-loss approaches for strict sequential constraints, where gradient-based compliance signals are insufficient.

The absolute ECR values (roughly 0.40 to 0.57) should be read against an attainable ceiling rather than the nominal maximum of 1. A recommendation counts toward ECR only when it is both compliant and outcome-improving, where outcome-improving means that the predicted positive-outcome probability strictly increases when the recommended activity is appended. For many prefixes no feasible activity raises the predicted outcome, for instance when the case is already on a favorable trajectory or its outcome is largely determined, so an ECR of 1 is unreachable even for an oracle restricted to the feasible set. The meaningful signal is therefore the relative margin over baselines on the same ceiling, where NeSyPA improves ECR by +6.7 to +9.9pp on the BPIC datasets, driven by the higher OIR reported above.

The ablation and sensitivity experiments together clarify the distinct role of each constraint tier. CCOD's gains originate primarily from prefixes where the full feasible set is non-empty: tight feasible sets ($|\mathcal{F}_{\text{full}}| \approx 3$) produce highly concentrated soft targets that teach the model precisely which compliant activities yield the highest outcome lift, directly fulfilling the paper's core motivation. The ablation (Experiment 2) shows that hierarchy adds minimal incremental ECR (approximately 0 to +1.1pp across datasets), which might suggest it is unnecessary. However, without hierarchy, 31.8% of prefixes on BPIC 2012 would be irreparable: no recommendation can be made and no CCOD training signal can be computed for them. Hierarchy reduces this to 2.3%, enabling both inference coverage and CCOD training for those prefixes. Yet, as the ablation reveals, CCOD operates on diffuse hard-only targets ($|\mathcal{F}_{\text{hard}}| \approx 22.5$) for these recovered prefixes, yielding only modest and mixed ECR gains. Hierarchy's value is therefore structural rather than incremental: it ensures the system remains operational and trainable for cases that would otherwise be excluded entirely. The sensitivity analysis (Experiment 3) confirms that soft constraint quality is critical for CCOD effectiveness: when noisy soft constraints trigger frequent hierarchical relaxation, targets become diffuse and the training signal weakens sharply ($r = -0.91$ between relaxation rate and ECR).

Several limitations should be noted. CCOD effectiveness is sensitive to soft constraint quality: when noisy soft constraints trigger frequent hierarchical relaxation, CCOD targets become diffuse and the model fails to learn soft constraint compliance. This is inherent to the compilation approach, since the quality of compiled targets cannot exceed the quality of the input constraints. The frozen outcome head used for computing outcome lifts is trained on correlational rather than causal data, introducing potential bias in the outcome estimates. The method assumes Declare-expressible constraints and cannot handle arbitrary procedural constraints. Finally, the binary relaxation mechanism (i.e., enforcing all soft constraints or dropping all of them) is a notable limitation. When a single low-confidence soft constraint conflicts with otherwise feasible activities, the entire soft constraint set is discarded rather than selectively relaxing only the conflicting constraint. Because each soft constraint is already compiled into an individual DFA, this limitation is architecturally bounded: identifying which constraint(s) cause an empty feasible set requires at most n DFA evaluations per activity, so selective relaxation is computationally tractable. The challenge is therefore not in feasibility checking but in the selection policy, i.e., deciding which constraints to retain when not all can be simultaneously satisfied, and in how variable constraint subsets across prefixes affect the stationarity of CCOD training targets.

7 Threats to Validity

Internal Validity. Soft constraint mining quality affects all constraint-aware methods equally, as the same knowledge base Φ is used for all methods. Hyperparameters were fixed across methods to values guided by preliminary runs on the validation split, rather than tuned per method, which could understate the best achievable performance of individual baselines. To ensure fair comparison, B2 receives the same hierarchical relaxation mechanism as NeSyPA at inference time, isolating the contribution of CCOD training. All methods share the same Transformer encoder architecture and training infrastructure.

External Validity. We evaluate on four event logs, providing reasonable but not exhaustive generalizability. All datasets come from the standard BPM benchmark suite, and performance on processes with fundamentally different characteristics, such as highly unstructured creative processes or very short traces, remains to be investigated. The constraint knowledge bases were constructed by the authors, with hard constraints following published regulations and soft constraints discovered via mining; different constraint formulations could yield different results.

Construct Validity. ECR captures the dual objective of prescriptive monitoring but weights compliance and outcome improvement equally. In practice, one dimension may be more important than the other, and a weighted variant of ECR could better reflect domain-specific priorities. The frozen B1 outcome head used for both outcome evaluation and CCOD target computation may introduce systematic bias if the outcome predictor has learned spurious correlations from the training data.

8 Conclusion

We presented NeSyPA, a neurosymbolic method for prescriptive process monitoring that addresses two practical challenges: over-constrained knowledge bases that leave substantial fractions of cases without feasible recommendations, and the lack of mechanisms to compile constraint knowledge into the neural training process. Hierarchical Constraint Relaxation distinguishes hard constraints from relaxable soft constraints, reducing irreparable rates from 31.8% to 2.3% while preserving hard constraint compliance. Constraint-Compiled Outcome Distillation creates feasibility-masked, outcome-weighted targets that teach the model to produce candidates with dramatically higher outcome quality, with no iterative loop and no λ tradeoff parameter. Experiments on four real-world event logs demonstrate that NeSyPA achieves the best Effective Compliance Rate on all test sets, with gains of +6.7 to +9.9 percentage points over the strongest baselines on BPIC datasets, driven primarily by improved outcome quality within the compliant recommendation space.

Several directions for future work emerge from our findings. Granular per-constraint relaxation could selectively relax individual soft constraints based on confidence or importance scores, avoiding the coarseness of binary relaxation. Multi-step lookahead could extend feasibility checking to consider whether an activity opens or closes future feasible paths, not just immediate compliance. Online CCOD could support incremental target updates as new cases complete, rather than one-shot fine-tuning. Finally, extending the prescriptor to recommend activity-time pairs would enable meaningful temporal constraint enforcement.

References

1. Agarwal, P., Gupta, A., Sindhgatta, R., Dechu, S.: Goal-oriented next best activity recommendation using reinforcement learning. CoRR **abs/2205.03219** (2022)
2. Bozorgi, Z.D., Teinemaa, I., Dumas, M., Rosa, M.L., Polyvyanyy, A.: Process mining meets causal machine learning: Discovering causal rules from event logs. In: van Dongen, B.F., Montali, M., Wynn, M.T. (eds.) ICPM, pp. 129–136. IEEE (2020)
3. Bozorgi, Z.D., Teinemaa, I., Dumas, M., Rosa, M.L., Polyvyanyy, A.: Prescriptive process monitoring for cost-aware cycle time reduction. In: Ciccio, C.D., Francescomarino, C.D., Soffer, P. (eds.) ICPM, pp. 96–103. IEEE (2021)
4. Branchi, S., Francescomarino, C.D., Ghidini, C., Massimo, D., Ricci, F., Ronzani, M.: Learning to act: A reinforcement learning approach to recommend the best next activities. In: Ciccio, C.D., Dijkman, R.M., del-Río-Ortega, A., Rinderle-Ma, S. (eds.) BPM Forum, pp. 137–154. LNBIP, Springer (2022)
5. Ciccio, C.D., Mecella, M.: On the discovery of declarative control flows for artful processes. ACM Trans. Manag. Inf. Syst. **5**(4), 24:1–24:37 (2015)
6. van Dongen, B.: BPI challenge 2012. 4TU.ResearchData (2012)
7. van Dongen, B.: BPI challenge 2017. 4TU.ResearchData (2017)
8. Fahrenkrog-Petersen, S.A., Tax, N., Teinemaa, I., Dumas, M., Leoni, M.D., Maggi, F.M., Weidlich, M.: Fire now, fire later: alarm-based systems for prescriptive process monitoring. Knowl. Inf. Syst. **64**(2), 559–587 (2022)

9. Gaikwad, D., van der Aalst, W.M.P., Park, G.: Neuro-symbolic process anomaly detection. In: Fuentes, L., Plebani, P., Combi, C., Reijers, H. (eds.) CAiSE. Springer, LNCS (2026)
10. Giacomo, G.D., Vardi, M.Y.: Linear temporal logic and linear dynamic logic on finite traces. In: Rossi, F. (ed.) IJCAI, pp. 854–860. IJCAI/AAAI (2013)
11. Hinton, G.E., Vinyals, O., Dean, J.: Distilling the knowledge in a neural network (2015). CoRR abs/1503.02531
12. de Leoni, M., Mannhardt, F.: Road traffic fine management process. 4TU.ResearchData (2015)
13. Mezini, A., Umili, E., Donadello, I., Maggi, F.M., Mancanelli, M., Patrizi, F.: Neuro-symbolic predictive process monitoring. CoRR **abs/2509.00834** (2025)
14. Pesic, M., van der Aalst, W.M.P.: A declarative approach for flexible business processes management. In: Eder, J., Dustdar, S. (eds.) BPM Workshops, pp. 169–180. Springer, LNCS (2006)
15. Santis, F.D., Park, G., van der Aalst, W.M.P., Zanichelli, F.: Compliance-aware predictive process monitoring: A neuro-symbolic approach. In: Fuentes, L., Plebani, P., Combi, C., Reijers, H. (eds.) CAiSE. Springer, LNCS (2026)
16. Santis, F.D., Park, G., Zanichelli, F.: Neuro-symbolic artificial intelligence for real-time sepsis mortality prediction. Eng. Appl. Artif. Intell. **177**, 114920 (2026)
17. Santis, F.D., Park, G., Zanichelli, F.: Neuro-symbolic learning for predictive process monitoring via two-stage logic tensor networks with rule pruning. CoRR **abs/2603.26944** (2026)
18. Shoush, M., Dumas, M.: Prescriptive process monitoring under resource constraints: A causal inference approach. In: Munoz-Gama, J., Lu, X. (eds.) ICPM Workshops, pp. 180–193. Springer, LNBIP (2021)
19. Shoush, M., Dumas, M.: When to intervene? prescriptive process monitoring under uncertainty and resource constraints. In: Ciccio, C.D., Dijkman, R.M., del-Río-Ortega, A., Rinderle-Ma, S. (eds.) BPM Forum, pp. 207–223. LNBIP, Springer (2022)
20. Teinemaa, I., Dumas, M., Rosa, M.L., Maggi, F.M.: Outcome-oriented predictive process monitoring: Review and benchmark. ACM Trans. Knowl. Discov. Data **13**(2), (2019)
21. Vaswani, A., et al.: Attention is all you need. In: Guyon, I., et al., (eds.) NeurIPS, pp. 5998–6008 (2017)
22. Weinzierl, S., Dunzer, S., Zilker, S., Matzner, M.: Prescriptive business process monitoring for recommending next best actions. In: Fahland, D., Ghidini, C., Becker, J., Dumas, M. (eds.) BPM 2020. LNBIP, vol. 392, pp. 193–209. Springer, Cham (2020). https://doi.org/10.1007/978-3-030-58638-6_12

All Eyes on the Workflow: Automated and Efficient Event Discovery from Video Streams

Marco Pegoraro[1(✉)], Jonas Seng[2], Dustin Heller[2], Wil M. P. van der Aalst[1], and Kristian Kersting[2]

[1] Chair of Process and Data Science, RWTH Aachen University, Ahornstr. 55, 52074 Aachen, Germany
{pegoraro,wvdaalst}@pads.rwth-aachen.de

[2] Artificial Intelligence & Machine Learning Lab, Technical University of Darmstadt, Hochschulstr. 1, 64289 Darmstadt, Germany
jonas.seng@tu-darmstadt.de, kersting@cs.tu-darmstadt.de

Abstract. Disciplines such as business process management and process mining aid organizations by discovering insights about processes on the basis of recorded event data. However, an obstacle to process analysis is data multi-modality: for instance, data in video form are not directly interpretable as events. Existing approaches rely on a dictionary of activity label as input, cannot provide frame-by-frame labeling explanations, or rely on superseded computer vision techniques. In this work, we present SnapLog, an approach to extract event data from videos by converting frames to feature vectors using image embeddings and performing temporal segmentation through frame-wise similarity matrices. A generalized few-shot classification is then used to assign labels to the video segments, yielding labeled, timestamped sub-sequences of frames that are interpretable as events. Conventional process mining techniques can be used to analyze the resulting data. We show that our approach produces logs that accurately reflect the process in the videos.

Keywords: Process mining · Computer vision · Multimodal data · Uncertain event data

1 Introduction

Process mining has become a key instrument for understanding how processes are actually executed in practice. By analyzing event logs, organizations can discover process models, detect deviations, and identify opportunities for improvement. The effectiveness of these analyses, however, depends on the availability of event data in a structured log format. In many real-world settings, this assumption is increasingly violated: process-relevant evidence is collected in multimodal form,

M. Pegoraro and J. Seng—Contributed equally to this work.

F. M. Maggi et al. (Eds.): BPM 2026, LNCS 16936, pp. 521–538, 2027.
https://doi.org/10.1007/978-3-032-37777-7_31

such as videos, images, sensor streams, and audio, rather than directly as timestamped event records. For example, in an airport, process-relevant evidence may come from surveillance videos of passenger movements, barcode scans of boarding passes and baggage tags, audio announcements, and sensor data from security checkpoints or baggage handling systems. Rather than being recorded directly as event logs, these heterogeneous data sources cannot directly describe operations in processes such as check-in, security screening, boarding, or baggage transfer. As a consequence, a large and growing share of operational behavior remains inaccessible to conventional process mining methods.

Video data are an important example of this gap. Cameras are now pervasive in manufacturing, logistics, healthcare, retail, and domestic environments, and they capture rich information about the execution of physical activities. However, from the perspective of process mining, videos are difficult to use directly. They encode process behavior implicitly in high-dimensional spatio-temporal signals, rather than explicitly in event abstractions such as activity labels and timestamps. Bridging this representation mismatch is therefore a central challenge: if videos can be transformed into event logs reliably, then process mining can be applied to domains where event data are currently unavailable.

This challenge is also a practical one. Manually annotating videos into process events is labor-intensive, expensive, and difficult to scale. It requires domain expertise, careful temporal segmentation, and consistent labeling across recordings. At the same time, fully supervised video understanding approaches often require large labeled datasets, which are rarely available for process-specific activities. In process analysis scenarios, one may have only a few labeled examples of each activity, while still needing a meaningful event log for downstream mining. This creates a strong need for methods that can exploit modern computer vision embeddings while remaining effective in low-label regimes.

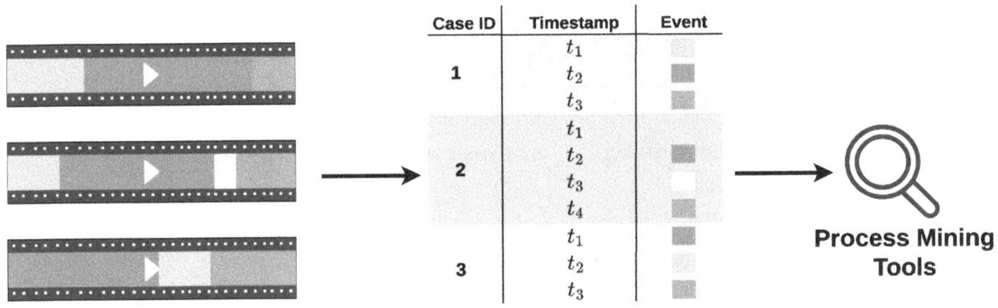

Fig. 1. SnapLog extracts logs from video data. Our method, SnapLog, constructs an event log from video data which can then be used downstream by standard process mining algorithms

In this paper, we address this need by proposing SnapLog, a pipeline that *extracts event logs from video data via temporal segmentation and few-shot classification*. The motivation is twofold. First, we seek to make process mining applicable to video-centric environments by producing standard event-log abstractions from raw visual observations. Second, we aim to do so in a way that is realistic for practice, where labeled examples are scarce and process activities may vary in appearance across recordings. Our approach combines frame-level semantic grouping with generalized few-shot video classification, yielding labeled and timestamped video segments that can be interpreted as process events. The resulting logs can then be analyzed with conventional process mining techniques, enabling process discovery and comparison across multiple recordings (Fig. 1).

An additional motivation for our approach is that video-based event extraction is inherently *uncertain*; for instance, some clips may support multiple plausible labels. Rather than viewing this uncertainty only as noise, it can be *represented explicitly and passed on to the process mining stage*. In fact, our method can output not only a deterministic event log (via the most likely label per segment), but also an uncertain event log, by retaining the probability distribution over candidate activity labels for each extracted event. This is highly relevant to recent work on uncertain and probabilistic event data, which shows that uncertainty-aware process mining can preserve ambiguity instead of forcing premature decisions [17]. Our pipeline complements these efforts by providing a computer-vision-based mechanism that can generate such uncertainty-aware logs from raw video observations. The resulting uncertain logs may then be analyzed through specialized analysis techniques available in literature, which are able to account for the probabilistic or non-deterministic information associated to event attributes (in our setting, activity labels).

Related Work. Within process mining, video data have first been discussed precisely in the context of uncertain event data by Gal and Cohen, as a natural source for uncertain logs [2]. More generally, the processing of unstructured and multimodal event data is regarded as an important contemporary challenge in the automatic analysis of processes [11]. Though research on event data abstraction from video data is in its infancy, previous attempts have been made: for instance, Lepsien et al. [13,15] described a pipeline for extracting events from video feeds in the context of pig farming. While the setting of their work include a focus on object tracking (since they abstract a process case with a single pig) and a fully known activity label dictionary, in our work we consider separate videos depicting process cases (thus, with a focus on segmentation) and a partially known dictionary of labels (motivating the few-shot classification). Other existing approaches adopt substantially different designs, including direct frame-agnostic video-to-activities extraction [3] or a general software architecture for log extraction from videos [12,26], rendering both methodologically highly distinct from our work. A recent approach to video-based process mining [8] relies on a heavy cascade of Visual Language Models (VLMs) and LLMs. While this enables zero-shot classification, the high inference costs and dependency on external API providers create significant barriers for large-scale industrial use. Furthermore, such pipelines often necessitate massive cloud-based compute clusters, complicating the handling of sensitive organizational data. In contrast, SnapLog is designed for resource-constrained, on-premise deployment. By utilizing lean models optimized for few-shot adaptation (< 20 samples), we eliminate the need for specialized high-performance computing (HPC) infrastructure. Our approach can be executed on standard local hardware with one or few GPUs, offering a cost-effective and privacy-preserving alternative that is orders of magnitude more scalable for extensive video archives. SnapLog and the logs resulting from the evaluation are openly available online[1].

The remainder of the paper is structured as follows. Section 2 introduces the basic computer vision and process mining notions needed to describe our approach. Section 3 illustrates our video-to-log pipeline. Section 4 presents the results of a quantitative evaluation of our extraction method. Lastly, Sect. 5 concludes the paper.

2 Preliminaries

The video-to-log pipeline requires a combination of concepts from the fields of machine learning, computer vision, and process mining. In this section, we present the necessary preliminary notions.

Event Data. Our target representation is the event log, a collection of records tracking the activities that take place in a process. In this work, we will only consider the control-flow, i.e., each event is abstracted through an activity label, a timestamp marking the beginning of the event, and the timestamp at completion. Formally, for a set of activity labels $\mathcal{A}$ and a totally ordered universe of timestamps $\mathcal{T}$. Then, an event is a triple $e = (a, t_{start}, t_{end}) \in (\mathcal{A} \times \mathcal{T} \times \mathcal{T})$, a trace $\sigma \in \mathcal{P}(\mathcal{A} \times \mathcal{T} \times \mathcal{T})$ is a set of events, and an event log $L \in \mathcal{B}(\mathcal{P}(\mathcal{A} \times \mathcal{T} \times \mathcal{T}))$ is a multiset of traces. Conversely, in uncertain events activity labels are probabilistic: $e_U = (p_{e_U}, t_{start}, t_{end})$, where $p_{e_U}(a)$ is a discrete probability distribution over the set of labels $\mathcal{A}$. An uncertain trace σ_U is a set of uncertain events, and an uncertain log L_U is a multiset of uncertain traces.

Our starting point for event log extraction will be a set of videos, for which we will present a processing pipeline.

Video Data and Feature Embeddings. Consider an input video $\mathbf{X} \in \mathbb{R}^{C \times T \times H \times W}$ consisting of T frames with height H, width W, and C color channels. The common practice to solve downstream tasks such as classification on video data is to first encode the input video into a dense, lower-dimensional vector representation, a so-called embedding. This representation efficiently encodes the spatio-temporal information required to solve the downstream task. Typically, neural networks are used as encoders; specifically, convolutional neural networks (CNNs) and Vision Transformers (ViT) have been shown to yield embeddings useful for many downstream tasks [14,20].

Since our pipeline for log extraction consists of first splitting a video into semantically meaningful subsequences, followed by classifying these subsequences to obtain event labels, we employ two encoding mechanisms: One that

[1] https://github.com/J0nasSeng/SnapLog.

fully preserves temporal structure (for semantic grouping of frames) and one for final classification (event label assignment).

To split the video into semantically meaningful subsequences, we rely on clustering frame-wise representation. Since the Vision Transformer (ViT) architecture has been shown to learn robust, state-of-the-art features from image data that produce well-separated and semantically grounded clusters in an Euclidean-like embedding space [19,20], we leverage a pre-trained ViT [6] in our pipeline. Given a video $\mathbf{X} \in \mathbb{R}^{T \times H \times W \times C}$ consisting of T discrete frames, each frame $\mathbf{x}_t$ is processed independently to generate a frame-level embedding. Formally, an input image $\mathbf{x} \in \mathbb{R}^{H \times W \times C}$ is reshaped into a sequence of N flattened 2D patches $\mathbf{x}_p \in \mathbb{R}^{N \times (P^2 \cdot C)}$, where P is the patch size and $N = \frac{HW}{P^2}$. These patches are mapped to a D-dimensional latent space via a linear projection $\mathbf{W}_e \in \mathbb{R}^{(P^2 \cdot C) \times D}$. To maintain spatial context, a position embedding $\mathbf{e}_{pos}$ is added to the projected patches. The resulting sequence, augmented by a learnable `[CLS]` token, serves as the input to a Transformer encoder:

$$\mathbf{Z}_{in} = [\mathbf{x}_{class}; \mathbf{x}_p^1 \mathbf{W}_e; \mathbf{x}_p^2 \mathbf{W}_e; \ldots; \mathbf{x}_p^N \mathbf{W}_e] + \mathbf{e}_{pos}$$

The encoder applies L layers of Multi-Head Attention (MHA) and Multi-Layer Perceptrons (MLP). In this work, we utilize the state of the `[CLS]` token at the final layer L as the representative embedding for the frame [6].

To obtain final event classifications, the R(2+1)D architecture is used, which encodes video data while preserving crucial spatio-temporal information, thereby enabling high classification performance [22]. To facilitate downstream log extraction, the video is encoded by a function $\mathbf{Z} = f(\mathbf{X})$, where $\mathbf{Z} = (\mathbf{z}_1, \mathbf{z}_2, \ldots, \mathbf{z}_T)$ represents a sequence of T spatiotemporal embeddings. The function f is realized through the R(2+1)D architecture, which factorizes a standard 3D convolutional filter of size $d \times k \times k$ into two separate operations. Specifically, for each convolutional block, the architecture first applies a spatial convolution with a weight tensor of size $1 \times k \times k$, immediately followed by a temporal convolution with a weight tensor of size $d \times 1 \times 1$. Formally, if $*_s$ denotes the spatial convolution and $*_t$ denotes the temporal convolution, the intermediate transformation for a hidden layer h is given by:

$$\mathbf{h}_{i+1} = \sigma(*_t(\sigma(*_s(\mathbf{h}_i))))$$

where σ denotes a non-linear activation function, such as ReLU. This specific factorization better preserves spatiotemporal information than full 3D convolutions. A final global temporal pooling layer is used to obtain a fixed-sized representation $\mathbf{z} \in \mathbb{R}^d$ that can be used for downstream few-shot classification [22].

Temporal Action Segmentation (TAS). The objective of TAS is to partition an untrimmed video into a sequence of semantically cohesive segments, representing distinct process activities or states [5,23,24]. Given the sequence of embeddings $\mathbf{Z} = (\mathbf{z}_1, \mathbf{z}_2, \ldots, \mathbf{z}_T)$ extracted in the previous steps, we employ an unsupervised approach to cluster frames that exhibit high semantic similarity. We model the TAS task as a clustering problem. Because the embeddings $\mathbf{Z}$

are obtained from a ViT encoder whose embedding space exhibits a geometry similar to the Euclidean space [19,20], we opt for the K-Means algorithm as a clustering algorithm. K-Means has been proven to work well in Euclidean spaces with hundreds of dimensions [16], while e.g., density-based clustering tends to struggle in such high-dimensional spaces [1]. Formally, for a set of embeddings $\mathbf{Z}$, K-Means seeks to partition the observations into k sets $S = \{S_1, S_2, \ldots, S_k\}$ so as to minimize the within-cluster sum of squares (WCSS):

$$\arg\min_{S} \sum_{i=1}^{k} \sum_{\mathbf{z} \in S_i} \|\mathbf{z} - \mu_i\|^2$$

where μ_i is the centroid of points in S_i. This approach provides a low-complexity method for splitting a given set of frame embeddings into semantically meaningful subsequences, enabling downstream event classification.

Few-Shot Video Classification. Few-shot video classification is a method for an existing classification model to learn new unseen classes using only a few new examples [7,25]. In our pipeline, we adopt a few-shot video classification technique inspired by the generalized approach illustrated in [27]. In this approach, it is assumed that there is a small dataset of labeled videos. These videos show examples of new classes the model should learn to recognize. The pre-trained backbone is used to obtain a video encoding $\mathbf{z}$ which is then fed into a linear classifier $\mathbf{W}_c$. The classifier $\mathbf{W}_c$ is then learned (or fine-tuned) by minimizing cross-entropy loss on the dataset showing the new classes. Therefore, adding new classes to the model to be recognized reduces to learning a small linear model that leverages the encodings of the previously pre-trained encoder.

3 Method

Our video-to-log pipeline consists of two key stages: the temporal action segmentation (Fig. 2), and the few-shot video classification (Fig. 3). We now describe the different stages of our pipeline in detail.

3.1 Temporal Action Segmentation

In our pipeline, Temporal Action Segmentation (TAS) aims to identify subsequences in a given video that represent consecutive frames of events in an unsupervised way. Treating TAS as an unsupervised learning problem enables us to detect potentially relevant events without the expense and time required for labeling. To perform TAS, our pipeline first applies a pre-trained ViT to each video frame to obtain a sequence of frame embeddings that compactly represent the content of each frame. These embeddings are then clustered using K-means, yielding semantically meaningful subsequences of the input video via unsupervised clustering. These subsequences are then passed downstream for classification.

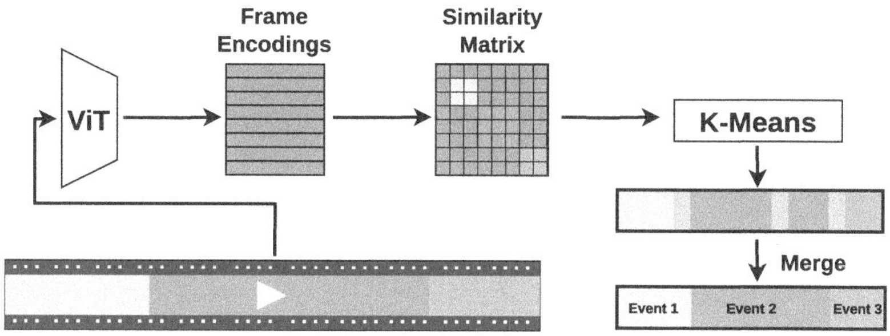

Fig. 2. Video Segmentation. From left to right: After the extraction of frame-wise embeddings using a ViT, the embeddings are used to compute a similarity matrix. By applying K-Means on the similarity matrix, atomic events are formed. These are merged into larger action segments using a greedy merging algorithm that pairs up atomic events based on their proximity in the embedding space, thus achieving a semantically grounded grouping

To find semantically meaningful subsequences in $\mathbf{X}$, we aim to identify event boundaries, e.g., when frames changes occur. Inspired by the approach in [16], K-Means clustering is used to find atomic events by clustering $\mathbf{Z}$. A simple merge algorithm combines extremely short events, reducing fragmentation in the action segmentation.

Embedding Extraction. Our approach employs a frame-wise feature encoding of an input video. A pre-trained ViT is used to compute $\mathbf{Z} = (\mathbf{z}_1, \ldots, \mathbf{z}_T)$ where each $\mathbf{z}_i = f(\mathbf{x}_i)$ is the frame embedding of the i-th image computed by the ViT f. Thus, each frame in a video $\mathbf{X} \in \mathbb{R}^{T \times H \times W \times C}$, represented as a sequence of RGB images $(\mathbf{x}_1, \ldots, \mathbf{x}_T)$, is encoded independently. Here, T is the number of frames in the video, H is the frame height, W is the frame width, and C is the number of channels (3 for RGB images). The embedding sequence $\mathbf{Z}$ is a real-valued matrix with dimension $T \times d$ where d is the dimension of the latent space learned by the ViT. The ViT was pre-trained on Imagenet, a large-scale corpus designed for image classification, often used to pre-train vision models.

Encoding the video frame-by-frame was a deliberate choice as it enables us to assign each frame to a specific cluster when identifying semantically meaningful subsequences. Since the event classification requires a small labeled dataset of subsequences showing specific events, we can directly show identified subsequences to a human user who can provide the required labels and assess the quality of identified subsequences.

The Frame-Wise Similarity Matrix. After computing frame-wise embeddings, semantically similar frames must be grouped into subsequences of the original video to identify events. To achieve this, a similarity matrix $\mathbf{D} \in \mathbb{R}^{T \times T}$ is computed. This matrix contains pairwise distances between the feature vectors for all frames of a video. The cosine similarity measure is used to compute each element of $\mathbf{D}$, i.e., $d_{ij} = 1 - \frac{\mathbf{z}_i \cdot \mathbf{z}_j}{||\mathbf{z}_i|| ||\mathbf{z}_j||}$. The clustering algorithm must consider the temporal axis since frames closer together w.r.t. the time dimension should be more likely clustered together. This allows repeated actions to be clustered separately as the clustering algorithm will separate them as two events with the same action. To achieve this, the frame index of each frame is appended as a new row of $\mathbf{D}$. Thus, we obtain a matrix $\mathbf{D}' = \mathbf{D}||\mathbf{i}$ where $||$ denotes the row-wise concatenation and $\mathbf{i} = (1, \ldots, T)$. Next, $\mathbf{D}'$ is normalized as $d'_{ij} = \frac{d'_{ij}}{\mathbf{D}'_i}$, and a softmax operation is performed, i.e., $d'_{ij} = \frac{\exp(d'_{ij})}{\sum_k \exp(d'_{ik})}$.

Note that each column of $\mathbf{D}'$ is a contextualized representation of a frame w.r.t. the entire video: Each column j represents one frame and encodes distances of j to all other frames, thereby contextualizing j. While the sequence of encoded frames $\mathbf{Z}$ could also be used to cluster frames, $\mathbf{Z}$ does not provide a contextualized representation of each frame. Adding global context has been shown to stabilize clustering [16].

Temporal Event Clustering. To cluster temporal events from the similarity matrix $\mathbf{D}'$, K-Means is applied, which assigns a cluster index $c \in \{1, \ldots, k\}$ to each frame. The clusters found by K-Means yield *atomic* events, i.e., consecutive frames with the same cluster assignment. We term these events atomic, as K-Means tends to find short events spanning too few frames. However, simply setting k to a larger value, results in too coarse clusters, often spanning frames of multiple events. Since k is the only parameter we can control in K-Means, it does not provide fine-grained control over the length of clustered events. Hence, a simple yet effective merging algorithm is used to merge shorter atomic events into adjacent events to form composite events. These events then form the subsequences that are ultimately classified in the downstream few-shot classification.

Merging atomic events is done by first defining a process-dependent minimum event length l_{min}. We then merge events that are shorter than l_{min} into a larger event using a greedy approach: For each sub-threshold event, the algorithm identifies its chronologically adjacent neighbors. If an event has only one neighbor (i.e., is at the start or end of the video), it is merged with its neighboring event. If an event is flanked by two segments, it is merged with the neighbor whose cluster centroid exhibits the smallest Euclidean distance to its own. This ensures that our merging strategy combines atomic events that semantically belong together with high probability. In cases where neighbors differ in their cluster assignment and the distances are equidistant, the algorithm prioritizes the preceding neighbor to maintain temporal consistency. If both neighbors share the same cluster assignment, the event inherits that assignment, again striving to maximize probability of combining atomic events that semantically belong together. This iterative pass ensures that the final sequence consists only of events that satisfy the minimum length requirement. Note that each step of our greedy merging strategy is computationally efficient while it aims to pair up atomic events that are too short to be plausible alone but semantically belong to the same event of the process. See Algorithm 1 for the pseudocode.

Algorithm 1: Temporal Event Merging and Refinement

Input : Sequence of atomic events E, threshold l_{min}, cluster centroids C
Output: Refined sequence of composite events E'

```
foreach event e_i in E do
    if length of e_i < l_min then
        Identify neighbors N adjacent to e_i on the temporal axis;
        if |N| == 1 then
            Merge e_i into the single neighbor;
        else
            n_prev, n_next ← neighbors in N;
            if cluster assignment of n_prev == cluster assignment of n_next then
                Merge {n_prev, e_i, n_next} into one composite event;
            else
                d_prev ← distance between centroids of e_i and n_prev;
                d_next ← distance between centroids of e_i and n_next;
                if d_prev ≤ d_next then
                    Merge e_i into n_prev;
                else
                    Merge e_i into n_next;
return E';
```

3.2 Few-Shot Video Classification

Once our pipeline has segmented the video into subsequences potentially containing frames representing events using TAS, we apply a few-shot event classification next to assign an event label to each identified segment.

Event Classification. To classify the identified video segments, we employ a flexible few-shot approach that facilitates rapid adaptation to new event types, akin to [27]. This rapid adaptation allows our approach to perform well with only a few labeled instances, enabling cheap and fast adaptation of pre-trained models. We utilize a 34-layer R(2+1)D encoder g, pre-trained on the Sports-1M dataset [9], to map an event $\mathbf{X_e} \in \mathbb{R}^{T_e \times H \times W \times C}$ from the output of Alg. 1 (i.e., $\mathbf{X_e} \in E'$) to a dense latent representation $\mathbf{z} \in \mathbb{R}^d$, where $\mathbf{z} = g(\mathbf{X})$. A linear classification head $\mathbf{W}_f \in \mathbb{R}^{d \times m}$ then produces the class distribution via $p(a|\mathbf{X_e}) = \text{softmax}(\mathbf{W}_f \cdot \mathbf{z})$ for classes $a \in \mathcal{A}$, which we interpret as activities.

Model Adaptation and Augmentation. To optimize the encoder for domain-specific process mining, we employ a two-stage adaptation strategy that separates feature refinement from task specialization. Initially, we fine-tune the encoder g using a base dataset $\mathcal{D}_b$. This stage incentivizes g to extract spatio-temporal embeddings that are semantically aligned with the target environment, effectively bridging the gap between general-purpose video features and the nuances of process-related actions. Following this refinement, the model undergoes few-shot adaptation to the specific task defined by a limited labeled dataset

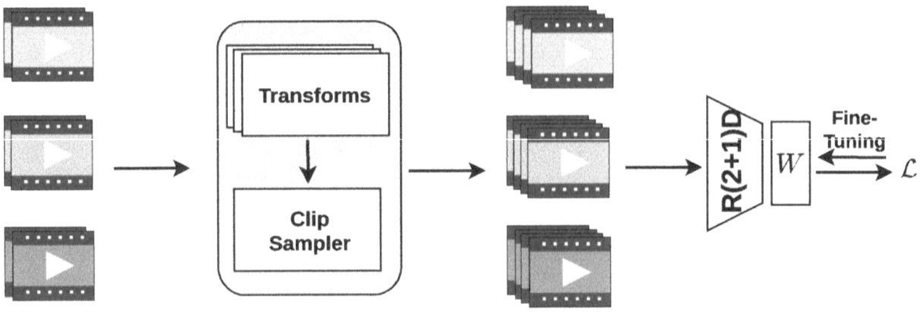

Fig. 3. Finetuning. Given a few labeled examples for novel classes, we first perform data augmentation by applying transformations (gray-scale, rotations, etc.) to increase the number of given labeled examples. Then, a new head is added to the R(2+1)D encoder and learned via gradient descent

$\mathcal{D}_c$. During this phase, the encoder g is frozen to preserve the learned domain features, while only the new classification head $\mathbf{W}_f$ is optimized via cross-entropy loss. This separation of concerns prevents the model from overfitting to the sparse samples in $\mathcal{D}_c$ while allowing it to learn new activity classes rapidly. To further bolster model robustness against the limited sample size, we apply consistent stochastic augmentations across all frames. These include geometric rotations $r \in \{90°, 180°, 270°\}$, photometric brightness scaling $b \in [0.5, 1.5]$, Gaussian noise injection, and grayscale conversion. This augmentation pipeline ensures that the learned process embeddings remain invariant to the environmental and sensor-induced variability common in real-world video streams.

Clip Sampling Strategy. Training and evaluation are performed on clips of 16 consecutive frames extracted from the event sequences. We chose 16 frames to capture enough informative frames for accurate event classification while keeping the computational load as small as possible. We compare two distinct sampling regimes: non-overlapping and overlapping clips. While non-overlapping sampling ensures each frame belongs to a unique clip, the overlapping strategy allows frames to reappear across non-identical clips. This latter approach significantly increases the volume and diversity of training data available per epoch, which is particularly beneficial in the few-shot setting.

Event Log Abstraction. Lastly, we abstract the labeled segments as events. We build an event log L by creating a trace $\sigma \in L$ for each video $\mathbf{X}$. Let E' be the list of segments corresponding to the video $\mathbf{X}$. For the segment $\mathbf{X_e}$, let $t_{start}^{\mathbf{X_e}}$ (respectively, $t_{end}^{\mathbf{X_e}}$) be the timestamp of the first (respectively, last) frame of $\mathbf{X_e}$. Then, we may abstract a traditional (certain) process trace through maximum likelihood estimation, with $\sigma = \{(\arg\max_a p(a|\mathbf{X_e}), t_{start}^{\mathbf{X_e}}, t_{end}^{\mathbf{X_e}}) \mid \mathbf{X_e} \in E'\}$. Since we know the probability distribution of all the labels, we have the option of recording it in the dataset, obtaining an uncertain event log with probabilistically uncertain activity labels: $\sigma_U = \{(p(a|\mathbf{X_e}), t_{start}^{\mathbf{X_e}}, t_{end}^{\mathbf{X_e}}) \mid \mathbf{X_e} \in E'\}$. Note that, in both the certain and uncertain event log translations, segments have unique start and end timestamps, and the resulting trace will have one distinct event per segment. This ensures that no pair of events in a trace are associated to overlapping timestamp intervals; as a consequence, extracted timestamps are certain and result in totally ordered traces. This results in an uncertain log of type $[A]_{\mathbb{W}}$ in the taxonomy of [18].

4 Evaluation

The goal of our evaluation is to quantify the fidelity of the event logs extracted from raw video and their utility for downstream process mining. We adopt a multi-stage validation strategy to isolate the performance of each pipeline component: first, we assess TAS accuracy; second, we evaluate action classification on the resulting segments. Finally, we demonstrate that the synthesized event logs successfully reconstruct a meaningful and representative process model, proving the end-to-end viability of our lean, few-shot approach for real-world video data.

4.1 Experimental Setup

Datasets. We evaluate our pipeline on unconstrained real-world environments using two datasets.

The TUM Kitchen Dataset [21] serves as our baseline for high-fidelity process extraction in a stable environment. We selected 16 videos of a brownie-baking process, characterized by a consistent background and fine-grained annotations. To bridge the gap between low-level atomic actions (e.g., "take spoon") and the higher-level activities required for meaningful process mining, we utilized a standardized relabeling step. By leveraging an LLM to aggregate these segments into 16 compound activities (e.g., "prepare batter"), we transformed the raw action recognition data into a structured format suitable for process discovery. The LLM was used to generate aggregation rules *once*, grounded in label semantics. These rules were then manually checked for correctness and applied to the entire dataset to ensure consistency of the new labels.

The Epic Kitchens-100 Dataset [4] provides a rigorous test of our pipelines robustness to environmental variability. Unlike the TUM dataset, Epic Kitchens features "in-the-wild" recordings across 45 diverse kitchen setups. We sampled 24 videos focusing on "kitchen cleaning" across 10 different environments. While most egocentric datasets suffer from extreme task variance (e.g., different recipes), cleaning behaviors exhibit the structural consistency necessary for reliable process modeling. This selection allows us to demonstrate that our few-shot approach can generalize across shifting backgrounds and heterogeneous camera angles without requiring extensive retraining for each new scene.

Model Adaptation. Our experimental evaluation utilizes 16 videos from the TUM dataset and 24 from Epic Kitchens, yielding 182 and 294 segments respectively. This distribution presents a rigorous low-data challenge, with an average of only 11 and 7 segments per class. To evaluate few-shot performance, we partition these into base, test, and novel splits; the TUM dataset comprises 114, 22, and 46 segments, while Epic Kitchens contains 196, 24, and 74. To increase robustness of our model and our evaluation, we apply stochastic augmentations–including rotation, brightness scaling, and noise–to enhance model invariance across these sparse samples. This process expands the TUM segments to 570 (base), 110 (test), and 230 (novel). Similarly, the Epic Kitchens augmented splits reach 980 (base), 120 (test), and 370 (novel).

We utilize a 34-layer R(2+1)D backbone pre-trained on Sports1M [9], processing 16-frame RGB clips at 112 × 112 resolution. The architecture is first fine-tuned on base videos (using Adam [10], $\eta = 0.001$, 10 epochs) to align the features with the target process domain. To maximize temporal coverage in training, we sample either 10 non-overlapping or 100 overlapping clips per segment.

For few-shot adaptation to novel activities, we optimize a linear classifier ($\eta = 0.01$, 10 epochs) on novel videos while keeping the backbone frozen. At inference, we aggregate predictions across multiple clips per segment to ensure a stable event classification, reporting Top-1 and Top-3 accuracies as indicators of the log's reliability for downstream process discovery. For a given k, Top-k accuracy measures the percentage of instances where the true label is among the k highest-probability predictions generated by the model.

4.2 Results

Segmentation. Our analysis of the temporal action segmentation reveals that a cluster count of $k = 7$ provides the most balanced representation of the feature space for both datasets. Silhouette score evaluations across $k \in \{3, 5, 7\}$ indicate that while lower values of k yield marginally higher scores (0.563 for TUM and 0.544 for Epic Kitchens at $k = 3$), they fail to capture the underlying cluster diversity (see Fig. 4).

At $k = 7$, silhouette scores for both TUM (0.498) and Epic Kitchens (0.487) approach the ideal 0.5 threshold. This configuration results in uniform cluster thickness and superior spatial distribution in the PCA-reduced feature space, preventing the overextension of clusters observed at lower k values (see Fig. 5).

By validating these segments against the ground truth annotations, we find that the pipeline achieves a frame-wise accuracy of 74.3% for TUM and 67.7% for Epic Kitchens. Despite the complexity of the unconstrained environments, the segmentation preserves the structural fidelity of the underlying processes. Most action boundaries align closely with the ground truth, ensuring that the essential sequence of activities remains intact for subsequent process discovery.

Classification Results. Our few-shot classification results demonstrate that high-fidelity event labels can be extracted even within the extreme constraints of a low-data regime. Despite having only a few videos available for the classification of novel activities, the R(2+1)D backbone–averaged over four random seeds–yielded stable performance with a standard deviation of only 2%–3%.

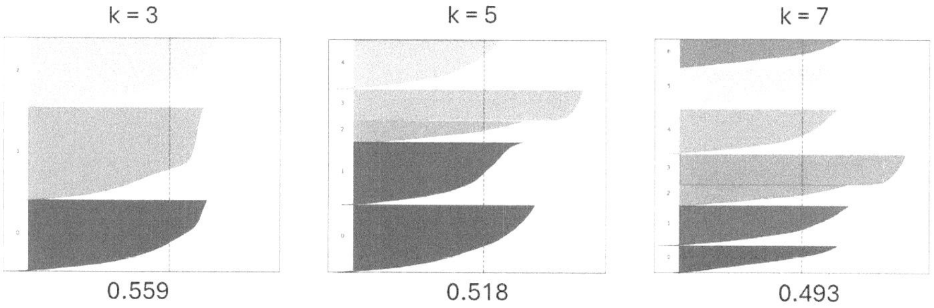

Fig. 4. Silhouette scores results. The average silhouette scores of a single video from the TUM dataset for $k = 3$, 5, and 7 clusters (average represented by the red line). The thickness of the cluster bars indicates the cluster size and the length of the cluster indicates a larger separation to other clusters. $k = 7$, although having a lower average silhouette score, provides the most balanced clustering for our purposes (Color figure online)

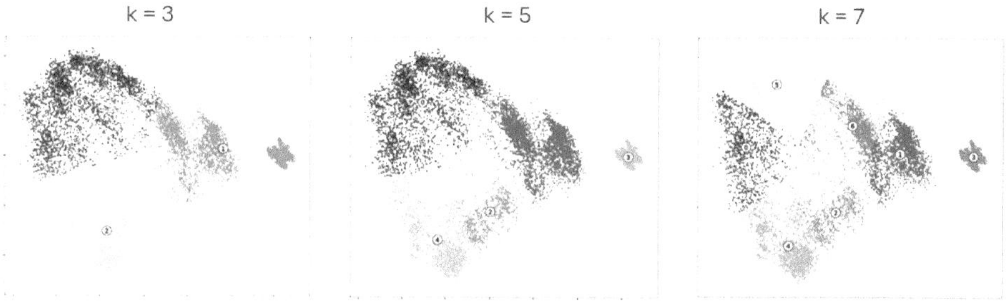

Fig. 5. Our pipeline produces semantically meaningful video segmentations. Applying K-Means on encoded frames yields semantically meaningful segmentations. It can be seen that $k = 7$ yields approximately equally sized clusters, indicating a good choice of k. Here, we show a projection of the clustered embeddings to a 2-d space using Principal Component Analysis (PCA)

Table 1. The averaged results of the few-shot classification. Four methods are evaluated varying between including augmented videos (Aug) or only using base videos (Base) and sampling 10 non-overlapping clips per epoch (10-Clips) or 100 overlapping clips (100-Clips)

Accuracy %	TUM		Epic Kitchens	
Method	Top-1	Top-3	Top-1	Top-3
Base-10-Clips	33.1 ± 2.0	63.1 ± 3.1	$28.2 \pm \mathbf{2.1}$	47.5 ± 2.9
Base-100-Clips	$50.7 \pm \mathbf{2.0}$	$73.7 \pm \mathbf{2.5}$	42.3 ± 2.8	$70.5 \pm \mathbf{2.4}$
Aug-10-Clips	59.9 ± 2.7	82.9 ± 2.8	52.2 ± 2.1	78.7 ± 2.6
Aug-100-Clips	$\mathbf{67.1} \pm 2.7$	$\mathbf{90.1} \pm 2.8$	$\mathbf{63.8} \pm 2.8$	$\mathbf{85.4} \pm 2.9$

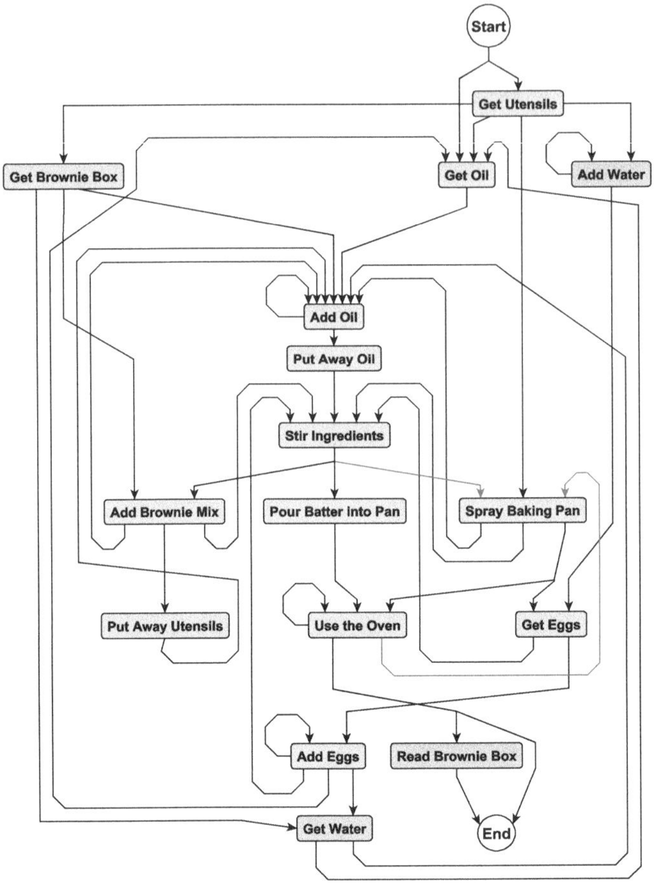

Fig. 6. A resulting process model. An example of directly-follows graph extracted from our validation in the non-augmented experiment on the TUM dataset. Red components show faults (Color figure online)

The analysis in Table 1 highlights that temporal coverage is a critical performance driver. Transitioning from 10 non-overlapping to 100 overlapping clips per segment improved Top-3 accuracy by up to 23% for the Epic Kitchens dataset. When further bolstered by spatial augmentations, Top-3 accuracies reached 90% for TUM and 85% for the more environmentally complex Epic Kitchens.

These Top-3 metrics are particularly salient given the inherent uncertainty of video-based extraction. As motivated in our introduction, certain segments support multiple plausible interpretations. Rather than forcing a potentially erroneous "hard" decision, our pipeline preserves this ambiguity by retaining the probability distribution across the top candidates. This high Top-3 accuracy confirms that our vision-based mechanism reliably captures the true activity within a narrow set of probable labels, providing a robust foundation for uncertainty-aware process mining that preserves semantic richness instead of discarding it as noise. The uncertain logs produced by SnapLog can reflect the Top-3 accuracy, by retaining the three most likely activity labels for each event. Their respective probability may be annotated as-is or, for a more refined result, may be normalized to sum to 1, rendering them immediately interpretable as discrete probability distributions.

Visual Inspection. While a full quantitative evaluation of the results of process discovery applied to the output log requires a complete case study, and is thus out of the scope of this paper, an example of a simple process model (discovered with the Process Explorer tool) of the log extracted from the TUM dataset can be seen in Fig. 6.

In general, the model shows a correct procedure to bake brownies, and most activities and connections are sensible. The logical order of cooking steps is well represented in the graph, and the process of baking brownies from a brownie mix is described with good completeness.

Some inconsistencies in ordering do appear, with *Add Water* followed by *Get Water*. The activity *Read Brownie Box* appears near the end, while it should occur near the start. There are some erroneous backward edges to *Spray Baking Pan*. Lastly, note that the activity *Use the Oven*, present in the annotations, labels both the act of preheating the oven and of inserting the pan in the oven, explaining the connection between *Spray Baking Pan* and *Use the Oven*.

In summary, despite some inaccuracies, a clear flow of the process of baking brownies can be identified through the model, which represents the overarching process throughout the videos relatively reliably.

5 Conclusion

This paper presented a pipeline for extracting process-relevant event logs from raw video through temporal segmentation and few-shot activity classification. By converting unstructured video observations into structured, timestamped events, we enable the application of process mining in domains where digital footprints are traditionally absent. Our approach is designed to be deployable on standard, non-HPC infrastructure, requiring only a small labeling effort–approximately 1520 videos–to adapt to new environments.

A key contribution of this work is the generation of uncertainty-aware event logs. By retaining class probability distributions, we preserve semantic ambiguity for downstream analysis rather than forcing premature, deterministic decisions. This is supported by our high Top-3 classification accuracies, which ensure that the true activity is captured within a narrow set of candidates even in complex kitchen environments.

Limitations and Future Work. Current limitations include the necessity for initial human labeling and a modular design that is not yet end-to-end differentiable. While our labeling requirements are low, future iterations could leverage

VLMs in-the-loop to further automate the initial annotation process while maintaining the efficiency of a lean inference pipeline. Developing a fully differentiable framework would also allow for joint optimization of the segmentation and classification stages, potentially increasing log fidelity. Besides that, our work assumes only one activity to appear in an identified video segment at a time. While not a limiting factor for applications like manufacturing lines, this assumption limits the applicability of SnapLog to more complex, diverse scenes where multiple events occur within the same video segment. Finally, at its current stage, SnapLog primarily labels activities and extracts temporal information; we lose spatial information that might be relevant for downstream process analysis and optimization. For example, having spatial information available, one could minimize transportation distances in a manufacturing line in order to increase its efficiency. As of now, such information is not extracted by SnapLog.

Future research will explore using the captured uncertainty for downstream error correction. For example, an LLM could be used as a post-processor to judge the quality of the probabilistic traces and suggest corrections based on its general knowledge of process logic. Additionally, we aim to evaluate whether these video-derived process models can be utilized in related applications, such as video reasoning and predictive machine failure, where high-level activity sequences are vital for operational insights. To mitigate the limitations of assuming that only one activity appears in each segment, SnapLog can be extended to not only identify segments relevant to activities throughout the temporal dimension, but also throughout the spatial dimension, ultimately allowing activities to co-occur in a single video segment.

Acknowledgments. The authors gratefully acknowledge the German Federal Ministry of Education and Research (BMBF) and the Ministry of Education and Research of North-Rhine Westphalia for supporting this work as part of the NHR funding.

Disclosure of Interests. The authors have no competing interests to declare.

References

1. Arcan, M.: Triples and knowledge-infused embeddings for clustering and classification of scientific documents (2026)
2. Cohen, I., Gal, A.: Uncertain process data with probabilistic knowledge: Problem characterization and challenges. In: Problems@BPM. CEUR Workshop Proceedings, vol. 2938, pp. 51–56. CEUR-WS.org (2021)
3. Corradini, F., Pettinari, S., Re, B., Rossi, L., Sampaolo, M.: Enabling process mining on multimodal robotic data. In: International Conference on Advanced Information Systems Engineering, pp. 257–269. Springer (2025)
4. Damen, D., et al.: Rescaling egocentric vision: Collection, pipeline and challenges for epic-kitchens-100. Int. J. Comput. Vision, 1–23 (2022)
5. Ding, G., Sener, F., Yao, A.: Temporal action segmentation: an analysis of modern techniques. IEEE Trans. Pattern Anal. Mach. Intell. **46**(2), 1011–1030 (2023)

6. Dosovitskiy, A., et al.: An image is worth 16x16 words: Transformers for image recognition at scale. arXiv preprint arXiv:2010.11929 (2020)
7. Ferdaus, M.M., Niles, K.N., Tom, J., Abdelguerfi, M., Ioup, E.: Few-shot learning in video and 3D object detection: a survey (2025)
8. Gavric, A., Bork, D., Proper, H.: Multimodal process mining. In: 2024 26th International Conference on Business Informatics (CBI), pp. 99–108. IEEE (2024)
9. Karpathy, A., Toderici, G., Shetty, S., Leung, T., Sukthankar, R., Fei-Fei, L.: Large-scale video classification with convolutional neural networks. In: Proceedings of the IEEE conference on Computer Vision and Pattern Recognition, pp. 1725–1732 (2014)
10. Kingma, D.P., Ba, J.: Adam: A method for stochastic optimization (2017)
11. Koschmider, A., et al.: Process mining for unstructured data: Challenges and research directions. In: Modellierung. LNI, vol. P-348, pp. 119–136. Gesellschaft für Informatik e.V. (2024)
12. Kratsch, W., König, F., Röglinger, M.: Shedding light on blind spots-developing a reference architecture to leverage video data for process mining. Decis. Support Syst. **158**, 113794 (2022)
13. Lepsien, A., Koschmider, A., Kratsch, W.: Analytics pipeline for process mining on video data. In: BPM (Forum). Lecture Notes in Business Information Processing, vol. 490, pp. 196–213. Springer (2023)
14. Madan, N., Moegelmose, A., Modi, R., Rawat, Y.S., Moeslund, T.B.: Foundation models for video understanding: a survey (2024)
15. Melfsen, A., Lepsien, A., Bosselmann, J., Koschmider, A., Hartung, E.: Describing behavior sequences of fattening pigs using process mining on video data and automated pig behavior recognition. Agriculture **13**(8) (2023)
16. Nam, J., Ahn, D., Kang, D., Ha, S.J., Choi, J.: Zero-shot natural language video localization. In: Proceedings of the IEEE/CVF International Conference on Computer Vision, pp. 1470–1479. (2021)
17. Pegoraro, M.: Probabilistic and non-deterministic event data in process mining: Embedding uncertainty in process analysis techniques. In: CAiSE (Doctoral Consortium). CEUR Proceedings, vol. 3139, pp. 37–46. CEUR-WS.org (2022)
18. Pegoraro, M., Uysal, M.S., van der Aalst, W.M.P.: Conformance checking over uncertain event data. Inf. Syst. **102**, 101810 (2021)
19. Raghu, M., Unterthiner, T., Kornblith, S., Zhang, C., Dosovitskiy, A.: Do vision transformers see like convolutional neural networks? (2022)
20. Schiappa, M.C., Rawat, Y.S., Shah, M.: Self-supervised learning for videos: a survey. ACM Comput. Surv. **55**, 1–37 (2023)
21. Tenorth, M., Bandouch, J., Beetz, M.: The TUM kitchen data set of everyday manipulation activities for motion tracking and action recognition. In: IEEE 12th International Conference on Computer Vision Workshops, ICCV Workshops, pp. 1089–1096. IEEE (2009)
22. Tran, D., Wang, H., Torresani, L., Ray, J., LeCun, Y., Paluri, M.: A closer look at spatiotemporal convolutions for action recognition. In: Proceedings of the IEEE Conference on Computer Vision and Pattern Recognition, pp. 6450–6459 (2018)
23. Wang, B., Zhao, Y., Yang, L., Long, T., Li, X.: Temporal action localization in the deep learning era: a survey. IEEE Trans. Pattern Anal. Mach. Intell. (2024)
24. Wang, P., Zeng, F., Qian, Y.: A survey on deep learning-based spatio-temporal action detection. arXiv preprint arXiv:2308.01618 (2023)
25. Wanyan, Y., Yang, X., Dong, W., Xu, C.: A comprehensive review of few-shot action recognition (2025)

26. Wördehoff, N., Egger, A., Kratsch, W., König, F., Röglinger, M.: Beyond assumptions: a reference architecture to enable unsupervised process discovery from video data. Decision Support Systems, p. 114544 (2025)
27. Xian, Y., Korbar, B., Douze, M., Torresani, L., Schiele, B., Akata, Z.: Generalized few-shot video classification with video retrieval and feature generation. IEEE Trans. Pattern Anal. Mach. Intell. **44**(12), 8949–8961 (2021)

Learning Joint Event Attribute Distributions for Realistic Event Log Generation

Nikolaos Perdikogiannis[1(✉)], Renata Medeiros de Carvalho[1], and Marwan Hassani[1,2]

[1] Department of Mathematics and Computer Science, Eindhoven University of Technology, Eindhoven, The Netherlands
{n.perdikogiannis,R.Medeiros.de.Carvalho,m.hassani}@tue.nl
[2] Department of Mathematics and Computer Science, University of Wuppertal, Wuppertal, Germany

Abstract. Generating realistic synthetic event logs remains a fundamental challenge in process mining. Existing generative approaches, such as variational autoencoders and adversarial models, often succeed in modeling individual perspectives of the process, such as control-flow, timing, or resource utilization, but fail to preserve the dependencies that connect them. To address this, we propose the TF-decoder, a novel decoder that combines a transformer backbone with a conditional normalizing flow. The transformer captures contextual dependencies along the process trace, while the flow models the joint probability of activities, timestamps, and resources, enabling the model to learn cross-perspective dependencies. The proposed model is evaluated on real-world event logs using an experimental setup that assesses marginal dimensions such as control-flow, temporal, and resource fidelity. We also evaluate dependency preservation using the Maximum Mean Discrepancy (MMD) to quantify the preservation of dependencies between event attributes. Across datasets, the TF-decoder matches the marginal performance while substantially improving the preservation of cross-attribute dependencies.

Keywords: Synthetic event log generation · Joint distribution modeling · Conditional normalizing flows · Event log simulation

1 Introduction

Process Mining (PM) techniques have been successfully applied in areas such as manufacturing, finance, and logistics; however, their use in some sectors (e.g. healthcare) remains challenging due to the sensitive nature of (patient) data. Regulations such as the GDPR restrict the sharing of identifiable data, making it difficult to exchange event logs across institutions or publish datasets for research. Particularly in outcome prediction or anomaly detection health applications, when the most interesting class is also the rarest, PM analysis can suffer

F. M. Maggi et al. (Eds.): BPM 2026, LNCS 16936, pp. 539–556, 2027.
https://doi.org/10.1007/978-3-032-37777-7_32

from insufficient training data and an inability to explore rare or hypothetical (what-if) scenarios.

Synthetic event logs provide a potential solution. By learning the statistical properties of an existing log and generating new traces that resemble real executions, generative models can support downstream analytics without exposing sensitive information. Additionally, they can ensure an increase in the training dataset size without simply copying existing traces. This harnesses the quality of prediction models by augmenting relevant data [18]. In this work, we follow this paradigm by learning from a real-world event log, and generating synthetic logs that reflect its underlying distributions.

Existing deep learning approaches for trace generation have shown promising results in reproducing control-flow and temporal behavior, but they typically restrict themselves to a few perspectives. Most methods generate activity sequences and timestamps but either ignore resource information or treat event attributes independently [8,9,13]. This is a significant limitation since real processes are inherently multiperspective: activities depend on the availability of resources, timestamps include waiting times, and some resources may perform activities faster than others. Capturing such dependencies is crucial to produce realistic logs that support process mining tasks beyond simple control-flow reproduction.

To illustrate the need for such dependencies, consider two orthopedic cases in a hospital (Fig. 1): one routine case with a meniscus tear that can be performed by a general orthopedic surgeon, and a more complex one that needs a shoulder specialist. Although both cases begin the same way, they quickly diverge after the diagnosis. The routine case proceeds quickly because each event uses readily available resources, while the complex case slows down dramatically as soon as specialist consultations and material ordering are required. The timing of each event is therefore tightly linked to its resource needs. Moreover, surgeries can be performed by various resources, and the choice of resource influences how the case unfolds. There are many ways in which the event attributes can affect each other, and therefore modeling them jointly is essential to accurately capture the interactions between them.

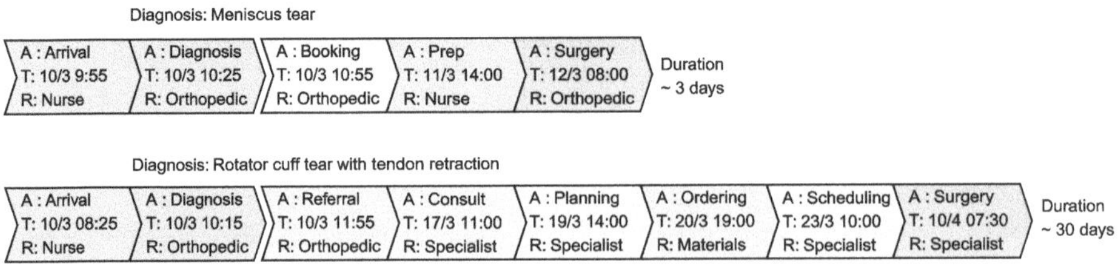

Fig. 1. Example of orthopedic cases. Meniscus tear needs a simple procedure, while rotator cuff tear is more complex

In this work, we address the challenge of capturing the dependencies between event attributes with a generative architecture. Our focus is on producing realistic synthetic logs that preserve the dependencies across the main three event attributes, activities, timestamps, and resources, ensuring that the generated traces more faithfully reflect the multiperspective nature of real processes. This enables high-fidelity synthetic data generation in privacy-sensitive domains, such as healthcare, where real event logs are scarce, hard to share, or quickly rendered obsolete due to process evolution.

Our contributions can be summarized by the following points:

(i) we propose the TF-decoder, a transformer-flow decoder that jointly models activities, resources, and timestamps for event log generation;
(ii) we introduce an evaluation protocol that combines standard marginal metrics with Maximum Mean Discrepancy (MMD) to quantify the distance between real and generated event distributions; and
(iii) we empirically demonstrate, on three real-world logs, that the TF-decoder substantially improves dependency preservation and behavioral diversity while maintaining competitive marginal fidelity.

The remainder of the paper is organized as follows. Section 2 reviews related work on synthetic event log generation. Section 3 presents the proposed TF-decoder architecture. Section 4 describes the experimental setup and the evaluation metrics, while Sect. 5 reports the empirical results and discusses the findings. Section 6 concludes and discusses ideas for future work.

2 Related Works

Previous work on synthetic event log generation can be broadly divided into simulation-based and deep-learning approaches. Simulation-based methods typically rely on process discovery to construct a simulation model, while deep learning methods aim to learn the distribution of traces directly from data.

2.1 Simulation-Based Approaches

Simulation-based approaches have long been used to generate synthetic event logs in process mining. These methods typically start by discovering a process model from an event log and enriching it with parameters such as branching probabilities, activity durations, and resource characteristics. An example is SIMOD [3], which automates the construction of simulation models from event logs by combining process discovery, parameter estimation, and simulation. While such approaches are interpretable and flexible, their quality depends on the discovered process model and the assumptions made when setting simulation parameters.

More recent work has explored hybrid settings that combine simulation and predictive models. RIMS [15] follows this direction. By incorporating machine learning predictions during runtime, it allows the simulation to adapt to the current execution context. In particular, it can account for *resource availability*, understood as the global availability of shared resources over time, and for interactions between traces that evolve concurrently. This makes it better at reproducing behavior that depends on shared resources and parallel case execution. Other simulation methods focus on optimal *resource allocation* , i.e., deciding how available resources should be assigned to competing activities or cases in order to minimize patient waiting times and hospital costs [20]. However, the effectiveness of simulation-based models is constrained by the reliance on the discovered process model and the assumptions made during its parametrization, both of which affect the behavior they reproduce.

2.2 Deep Learning-Based Approaches

Deep learning methods approach event log generation from a data-driven perspective by learning sequential patterns directly from logs. Early work mainly relied on recurrent neural networks, especially LSTMs, for next-event prediction and suffix generation [8,13]. Although effective, these methods generally rely on separate prediction heads for different attributes and therefore do not model the joint dependencies between activities, timestamps, and resources explicitly.

Generative adversarial networks have also been explored for suffix prediction and synthetic log generation. Taymouri et al. [19] proposed a GAN-LSTM architecture for suffix prediction, while ProcessGAN [12] introduced a transformer-based generator with a time-aware discriminator to generate complete traces. These methods can improve the realism of generated samples but also inherit GAN training challenges, including instability and mode collapse. Moreover, current GAN-based approaches for event logs do not support full joint modeling of activities, timestamps, and resources, and recent results suggest they may underperform compared to more stable generative models such as variational autoencoders [9], and therefore we focus our comparison only on them.

Variational Autoencoders have mainly been used in process mining for anomaly detection [11], but more recently, they have also been applied to synthetic trace generation. A first step in this direction is the conditional VAE proposed by Graziosi et al. [9], which generates sequences of activities, timestamps, and resources, together with trace attributes. Their model represents the current state of the art for multiperspective event log generation. However, its decoder is factorized into three autoregressive LSTMs, one for each perspective. As a result, dependencies across event attributes are only captured indirectly, which restricts the model's ability to learn cross attribute dependencies.

Overall, simulation-based generation offers interpretability and control, while deep learning methods allow for learning more complex behavior directly from data, but they lack interpretability. However, existing methods still have limited ability to model the joint dependencies between event attributes, which motivates the approach proposed in this work.

3 TF-Decoder: A Generative Architecture to Capture Dependencies Between Perspectives

Event logs are the primary source of data in process mining [1]. They record executions of a process in the form of **traces**, where each trace is an ordered sequence of **events** [4]. An event typically contains an activity label a_i, a timestamp t_i, and the involved resource r_i, representing the entity (e.g., a person, system or organizational unit) responsible for executing the event. Those three attributes constitute the **event attributes**:

$$x =< e_1, \ldots, e_n >, \quad e_i = (a_i, t_i, r_i).$$

Logs also include **trace-level attributes**, such as patient demographics, that remain constant for each case. An event log $\mathcal{L}$ is therefore a multiset of traces,

$$\mathcal{L} = \{x_1, \ldots, x_m\},$$

capturing multiple process perspectives: control-flow through the ordering of activities, temporal information through timestamps, and organizational behavior through resources [4]. Another useful concept is that of a variant, defined as the equivalence class of all the traces with the same control-flow [9], i.e., with the same sequence of activities. Variants highlight recurring execution patterns and distinguish common behavior from rare or exceptional cases.

We build upon the Conditional Variational Autoencoder (CVAE) of Graziosi et al. [9], the current state-of-the-art deep learning approach for synthetic event log generation. We introduce the *TF-decoder*, a new decoder that fits the CVAE model and allows for capturing the joint dependencies between the event attributes through the use of transformers and normalizing flows.

The encoder first produces a latent representation of the trace $\boldsymbol{z}$, which is combined with any trace-level conditioning variables to form a global embedding $\boldsymbol{z}_u$. At each timestep, the decoder receives the dequantized attributes of the previously generated event together with this global embedding and processes them through a transformer to obtain a contextual representation h_i. This context conditions a normalizing flow that models the joint distribution of the next event's activity, resource, and timestamp, as illustrated in Fig. 2.

3.1 Dirichlet Interpolation

Since the flows operate on the continuous domain, we relax the categorical event attributes using Dirichlet interpolation [21]. For each event $e_i = (a_i, r_i, t_i)$, given one-hot vectors for activity a_i and resource r_i, a relaxed version of each categorical variable is first obtained by sampling from a Dirichlet distribution centered around the observed category. This produces probability vectors x_a and x_r that lie within their respective simplices, preserving the geometry of the categorical domain while introducing a controlled amount of smoothness. These samples are then mapped to the Euclidean space through the **isometric log-ratio (ILR)**

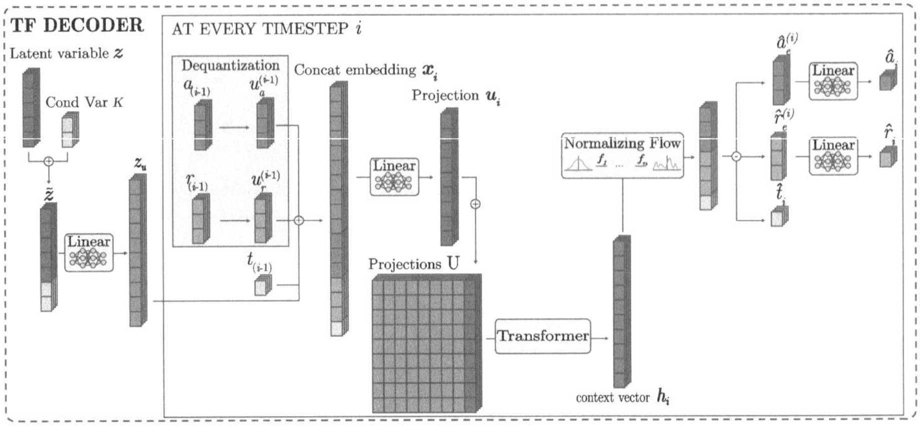

Fig. 2. Abstract view of the TF-decoder architecture. The transformer produces the context h_i, while the flow jointly models the event attributes

transformation [7], which provides an orthonormal coordinate system for the simplex. The resulting vectors, $u_a = \text{ilr}(x_a)$ and $u_r = \text{ilr}(x_r)$, serve as the continuous representations of the activity and resource attributes. The ILR transformation provides an orthonormal Euclidean representation of simplex-valued attributes.

3.2 Transformer-Based Context Encoder

The transformer backbone serves as the contextual modeling component of the decoder, responsible for capturing how past events influence future ones along the process trace. It provides the conditioning signal for the normalizing flow by producing a sequence of contextual embeddings that summarize the temporal and structural dependencies across the event history.

For instance, consider an event $e_i = (a_i, r_i, t_i)$, with the activity and resource first being dequantized into their continuous representations (u_a, u_r), while the temporal component is represented in logarithmic space as $\log t_i$. These three components are then concatenated with the global trace embedding to form the event representation:

$$x_i = \text{concat}(u_a, u_r, \log t_i, z_u).$$

Each x_i is projected through a linear layer into the transformer's embedding space u_i, as shown in Fig. 2, and a positional encoding is added to indicate its location in the sequence. This allows the model to distinguish between identical events occurring at different points in time. The projections u_i are combined into a matrix $U = [u_1 \ldots u_i]$ which is then passed through the **transformer** that models the dependencies between events through self-attention.

A causal mask is also applied to ensure that each event can only attend to previous events in the sequence, enforcing the autoregressive nature of the model.

This guarantees that the contextual embedding produced for event i, denoted as h_i, provides a temporally ordered representation of the trace, where each h_i summarizes the information available up to that point:

$$h_i = \text{Transformer}(U_{\leq i}).$$

The final contextual embedding h_i is then passed to the conditional normalizing flow, which will model the joint probability distribution of the next event's activity, resource, and timestamp.

3.3 Conditional Normalizing Flow for Joint Event Attributes

Once the transformer has encoded the process history into the context vector h_i, the conditional normalizing flow models the **joint distribution of the next event's attributes**, i.e. activity, resource, and timestamp, conditioned on the provided context h_i. The overall decoding process, with teacher-forced training and autoregressive generation, is illustrated in Fig. 3.

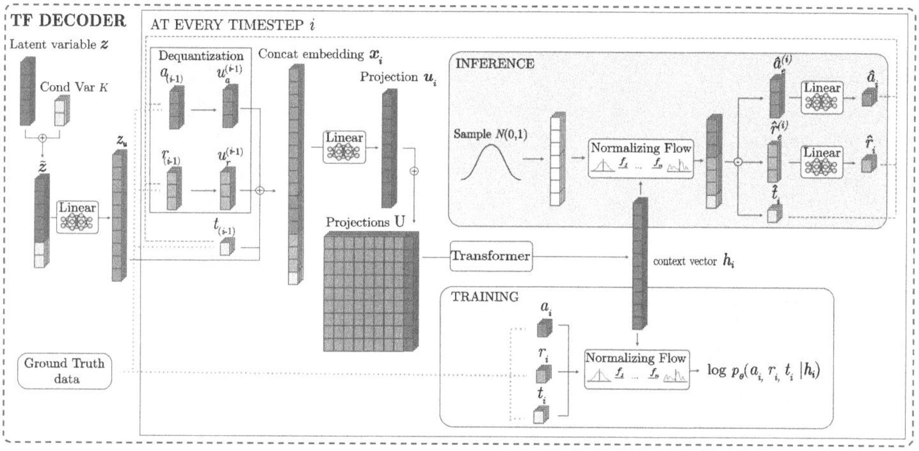

Fig. 3. TF-decoder architecture. At training time, the model uses teacher forcing, while at inference it generates autoregressively

During training, at each timestep, the dequantized event representation

$$v_i = \text{concat}(u_{a_i}, u_{r_i}, \log t_i),$$

is paired with the corresponding transformer context h_i, which conditions every coupling layer of the flow.

Normalizing flows model complex probability densities via a sequence of invertible transformations with tractable Jacobians. A coupling layer splits the input into two parts, using one part to parametrize a transformation of the other, which enables efficient likelihood computation while allowing rich, context-dependent transformations. In our implementation, the conditional density is parametrized using a rational-quadratic spline (RQS) flow consisting of 12 coupling layers, each with 24 bins and alternating binary masks, following the formulation of Durkan et al. [6]. RQS transformations define smooth, monotonic piecewise functions that are more expressive than affine transformations, making them well suited for modeling the non-linear and multi-modal temporal dynamics present in event logs. The flow receives the transformer output h_i representing the context and modulates the spline parameters at each layer.

During training, the model uses teacher forcing: the ground-truth event at each step is dequantized, transformed into ILR space, and provided to the flow. The flow f_θ learns an invertible mapping between the continuous event tuple and a standard Gaussian latent variable:

$$z_i = f_\theta^{-1}(v_i|h_i), \quad v_i = f_\theta(z_i|h_i).$$

Because the mapping is invertible, the log-likelihood of each event can be computed exactly using the change-of-variables formula:

$$\log p_\theta(v_i|h_i) = \log p_Z(z_i) + \log \left| \det J_{f_\theta^{-1}}(v_i|h_i) \right|.$$

where p_Z denotes the standard Gaussian base density.

Two additional terms are added to account for the transformations introduced during dequantization. For the *Dirichlet correction*, we subtract the log-density of the Dirichlet noise used to relax the categorical attributes, ensuring that the model does not treat the artificial noise as data. For the *ILR correction*, we subtract the log determinant of the ILR transformation that maps the simplex points to the Euclidean space. These adjustments ensure that the model assigns likelihood to the original discrete categories and not to the relaxed samples.

The total training objective is the sum of the corrected per-event log-likelihoods across the trace. The parameters of both the transformer and the flow are optimized jointly, ensuring that the contextual representations h_i produced by the transformer evolve to best support the density modeling of the flow.

At inference time, the process is reversed. A Gaussian latent variable is sampled, transformed through the conditional normalizing flow into a continuous event representation, and mapped back to discrete attributes via the inverse ILR operation. The generated event is fed back into the transformer to update the context for the next timestep, enabling full autoregressive generation. This procedure is illustrated in the bottom part of Fig. 3.

4 Experimental Setup

Datasets The evaluation is conducted on three real-world event logs that differ in size, heterogeneity, and temporal structure. The **Sepsis** [14] log contains hospital cases related to sepsis treatment, with the patient's age as the trace attribute. **BPIC2012_A** [5] describes a loan application process, with the loan amount as the trace attribute. The **Emergency ORT** log is a private orthopedic emergency dataset, where resources were constructed using domain knowledge, and the main trace attribute is patient age. Summary statistics for all datasets are reported in Table 1. Each dataset is preprocessed using the same preprocessing steps as in [9] and then split chronologically into training, validation, and test subsets with a 60%-20%-20% ratio. To ensure stable results, models generate **10 synthetic logs per dataset**, each containing the same number of traces as the test set. This repeated sampling accounts for stochasticity during both model training and autoregressive generation. The code and event logs necessary to reproduce the experiments presented in this paper are publicly available[1].

Table 1. Datasets summary statistics

Dataset	Traces	Variants	Activities	Resources #	Avg. Trace Length	Max Trace Length	Avg. trace cycle time	Total duration
Sepsis	782	619	16	25	17.16	185	38.2 d	~1.57 yrs
BPIC2012_A	4685	3222	36	63	38.9	175	18.7 d	166 d
Emergency_ORT	2421	1715	35	20	13.9	56	2.62 h	1.75 yrs

Marginal Metrics. To assess individual process perspectives, we adopt the metrics introduced by Chapela-Campa et al. [2] and used by Graziosi et al. [9]:

- **Control-Flow:** *Control-Flow Log Distance (CFLD)* and *2-Gram Distance (2GD)* quantify sequence similarity and directly-follows patterns.
- **Temporal:** *Cycle Time Distribution (CTD)* evaluates end-to-end cycle times, while *Relative Event Distribution (RED)* compares the timing of events relative to the case start.
- **Resource**: *Circadian Workforce Distribution (CWD)* captures daily and weekly resource usage patterns.
- **Conformance**: *Conformance Score (CONF)* measures the proportion of Declare constraints (mined at 90% support) satisfied by generated traces.

Maximum Mean Discrepancy (MMD). To assess the quality of the generated logs, it is important to not only compare the marginal distributions but also the full joint distribution over event attributes. The *Maximum Mean Discrepancy (MMD)* provides a kernel-based approach for this purpose [10], measuring the distance between the distributions of real and generated data without requiring exact likelihoods. Given real $\{x_i\}_{i=1}^{n} \sim R$ and generated $\{y_j\}_{j=1}^{m} \sim G$ samples, the squared MMD with respect to a positive semi-definite kernel k is

$$\text{MMD}^2(R, G, k) = \mathbb{E}_{x,x' \sim R}\left[k(x, x')\right] + \mathbb{E}_{y,y' \sim G}\left[k(y, y')\right] - 2\mathbb{E}_{x \sim R, y \sim G}\left[k(x, y)\right].$$

[1] https://doi.org/10.5281/zenodo.20644985.

In practice, the expectations are approximated by averaging the kernel values over all pairs of events within and across the two datasets. Intuitively, the MMD quantifies how similar the two datasets are by comparing how strongly the events from the same distribution relate to each other versus how strongly they relate across distributions.

A crucial component of MMD is the choice of kernel, as it determines how the similarities between events are quantified. We compare the categorical attributes, activity and resource, using an exponential Hamming distance kernel that returns 1 when the two attributes match and e^{-1} otherwise. For the continuous timestamp attribute, we use a standard RBF (Gaussian) kernel. Those base kernels are combined multiplicatively to yield the joint kernel, defined as

$$k(x, y) = k_A(A_x, A_y)\, k_R(R_x, R_y)\, k_T(T_x, T_y),$$

where k_A, k_T, and k_R are the respective kernels for activities, timestamps, and resources. The resulting kernel is characteristic, ensuring that MMD $= 0$ only when the distributions are identical.

However, the MMD as computed has one drawback. Because categorical combinations are sparse, most event pairs have low similarity, and frequent activities disproportionately influence the metric. The sparsity can therefore drive the full-joint MMD toward low values. To validate the results, we also compute the per-activity MMD by conditioning on $A = a$ and averaging across activities, i.e.,

$$\text{per-activity MMD} := \frac{1}{n_a} \sum_a \text{MMD}\big(R(T, R|A = a), G(T, R|A = a)\big)$$

where n_a denotes the number of activities. Conditioning on activity yields denser comparisons and balances the influence of frequent and rare activities.

Density and Coverage. The generative quality is further evaluated using the Diversity & Coverage (D&C) metrics of Naeem et al. [16]. These metrics are state-of-the-art for assessing deep generative models, particularly for embedding space distributions. They construct a data manifold by creating neighborhood spheres around real data points, defined as $B(X_i, \text{NND}_k(X_i))$, where $\text{NND}_k(X_i)$ is the distance from X_i to the kth nearest neighbor. Density counts how many of these neighborhoods contain each generated sample Y_j, formally defined as

$$D(X, Y) := \frac{1}{kM} \sum_{j=1}^{M} \sum_{i=1}^{N} 1\{Y_j \in B(X_i, \text{NND}_k(X_i))\}. \tag{1}$$

Density can exceed 1 because a generated sample can lie in multiple neighborhoods. Coverage measures diversity by quantifying the fraction of real samples whose neighborhoods contain at least one generated sample. Formally, coverage is defined as

$$C(X, Y) = \frac{1}{N} \sum_{i=1}^{N} 1\{\exists j \text{ s.t. } Y_j \in B(X_i, \text{NND}_k(X_i))\}. \tag{2}$$

Coverage yields a value between 0 and 1. Together, Density and Coverage provide a robust quantification of fidelity and diversity by evaluating how generated samples populate the neighborhood structure formed around real data points.

Baselines. We compare the proposed TF-decoder against the Conditional Variational Autoencoder (CVAE) of Graziosi et al. [9], which represents the state-of-the-art generative model for multiperspective event log generation. CVAE is trained using the hyperparameters reported in the original study.

5 Results

5.1 Marginal Perspectives

Across datasets, the TF-decoder achieves **comparable or better** marginal performance than the baseline in control-flow, temporal, and resource dimensions. On the **control-flow perspective**, it slightly outperforms the baseline, with the differences being more pronounced in the Sepsis and BPIC2012 logs, as seen in Fig. 4.

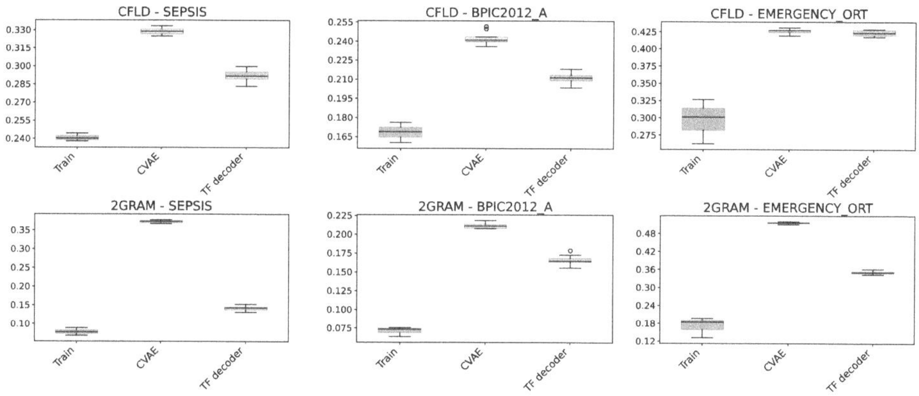

Fig. 4. Control-Flow metrics: Control-Flow Log Distance (CFLD) on top and 2-Gram Distance below. Lower is better, since metrics indicate distance between the test data and the generated logs. ('Train' shows the train-test distance.)

On **temporal fidelity**, both models reproduce timing distributions well, with the TF-decoder exhibiting higher variability, as indicated by the height of the boxplot in Fig. 5. Performance on the **resource perspective** is weaker than training for both models (see Fig. 6), reflecting the inherent difficulty of reproducing daily and weekly activity patterns using absolute temporal features.

For **conformance**, results vary per dataset (see Fig. 7). For Sepsis, the baseline achieves the highest conformance, whereas for BPIC2012 the TF-decoder performs best. For Emergency, the baseline again achieves the highest score, although this metric is sensitive to model variability: models generating more new, unseen traces might be harmed in terms of conformance.

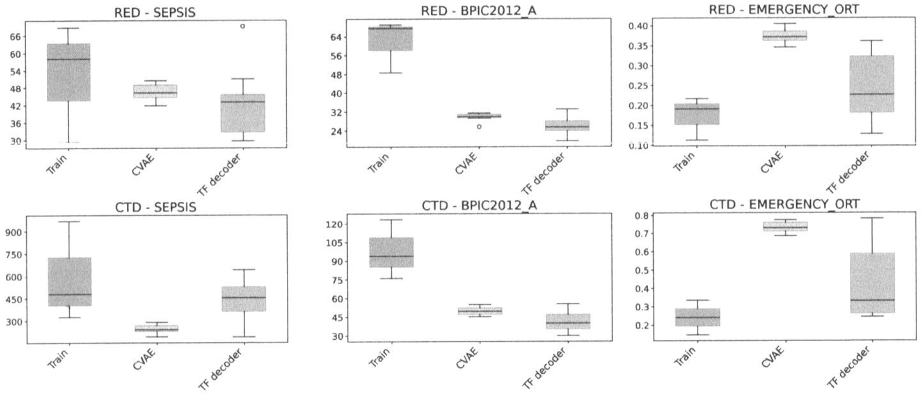

Fig. 5. Temporal metrics: Relative Event Distribution (RED) on top and Cycle Time Distribution (CTD) below. Lower is better, since metrics indicate distance between the test data and the generated logs. ('Train' shows the train-test distance).

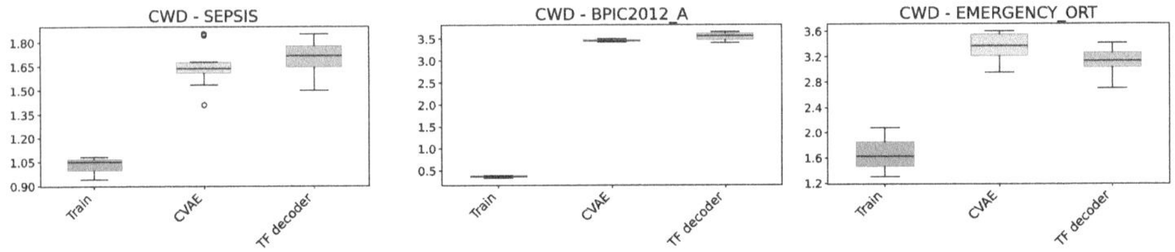

Fig. 6. Resource metric: Circadian Workforce Distribution (CWD). Lower is better, since metrics indicate distance between the test data and the generated logs. ('Train' shows the train-test distance.)

5.2 Joint Perspective (MMD & per-Activity MMD)

To evaluate the preservation of the joint distribution of activities, timestamps, and resources, we use the MMD metric to quantify the distance between the test set and the generated logs. As seen in Fig. 8, the generated logs from the TF-decoder are closer to the test data than the generated logs from CVAE. The values for MMD are overall lower, which can be partly attributed to the sparsity of the space. Despite this, the TF-decoder consistently outperforms CVAE across datasets. This observation is also verified by the per-activity MMD plots in Fig. 8. The difference between the models is most significant in the BPIC2012 dataset, which can partly be attributed to its size.

5.3 Trace Space Coverage and Variability

To evaluate whether the generated traces cover the space of the real data, we use the Density and Coverage metrics presented in Table 2. The metrics are computed by combining all 10 generated logs into a single log. We notice that the coverage of the generated logs is significantly higher for the TF-decoder,

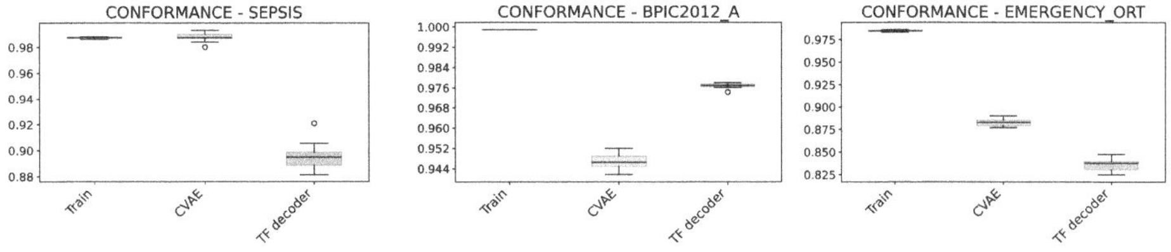

Fig. 7. Conformance metric: Conformance Score (CONF). Higher is generally better, since it indicates more traces conform to mined constraints

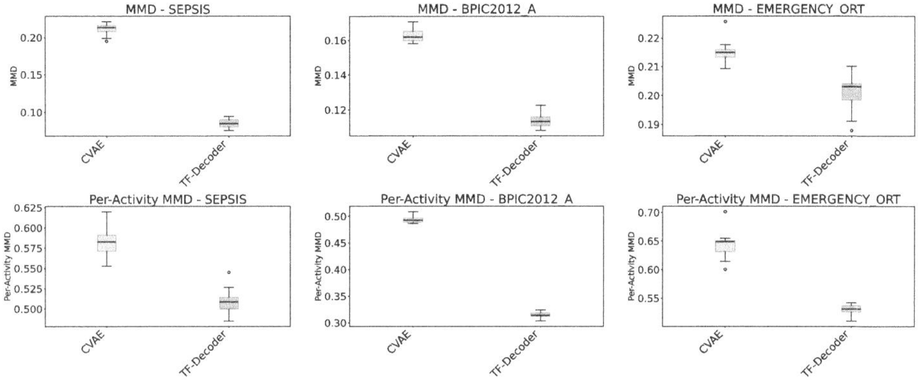

Fig. 8. Joint metric: The top row shows MMD; the bottom row shows per-activity MMD. Lower is better (Distance to real data)

consistently above 95%, while the CVAE is around 65%, except for the BPIC2012 log, for which it only captures part of the real data space. The density metric is similar for all logs, indicating that the generated data fall close to real data.

Table 2. Density and Coverage metrics introduced in [16]. Density captures the fidelity of the space while coverage captures the diversity. Higher is better

Model	Sepsis		BPIC2012_A		Emergency_ORT	
	CVAE	TF-decoder	CVAE	TF-decoder	CVAE	TF-decoder
Density	5.13	4.85	5.40	5.93	5.56	5.18
Coverage	64.82%	**98.51%**	38.99%	**98.58%**	65.77%	**97.93%**

Statistics from the generated logs further confirm this effect. The TF-decoder consistently produces more variants and recovers more low-frequency activities, while the baseline often fails to generate the least frequent events, as seen in Table 3. Unlike the baseline, the TF-decoder robustly generates the entire vocabulary of activities and resources.

Table 3. Statistics from the generated logs, compared to the training and test set. The closer the numbers are to the training data, the better

Dataset	Model	Traces	Variants	Activities	Resources #	Avg. Trace Length	Avg. trace cycle time	Total duration
Sepsis	Train	469	452	16	25	17.2	45.7 d	569 d
	Test	157	152	16	24	17.1	21 d	220 d
	CVAE	157	97	11	7	13.9 ± 0.11	15.0 ± 0.65 d	312.0 ± 21.32 d
	TF-decoder	157	156	16	25	15.5 ± 0.30	34.8 ± 3.04 d	448.9 ± 31.75 d
BPIC2012	Train	2811	2374	36	63	40.1	19.3 d	148 d
	Test	937	864	36	48	38.2	15.3 d	47 d
	CVAE	937	918	32	48	38.9 ± 0.15	13.4 ± 0.10 d	97.7 ± 2.30 d
	TF-decoder	937	901	36	63	33.1 ± 0.05	14.3 ± 0.11 d	154 ± 48.5 d
Emergency (ORT)	Train	1452	1258	35	20	14.0	2.52 h	310 d
	Test	485	451	34	20	14.2	2.76 h	223 d
	CVAE	485	347	29	19	13.4 ± 0.07	2.52 ± 0.01 h	89.5 ± 3.12 d
	TF-decoder	485	479	35	20	14.6 ± 0.09	2.68 ± 0.06 h	41.3 ± 1.55 d

5.4 Memorization and Privacy

To assess patient privacy preservation in real data, we evaluate whether the models memorize the training set. Although patient-specific information is removed, the model should not overfit and reproduce the exact training sequences. For each synthetic log, we measure how many generated traces match training activity sequences and, among those, how many also match the corresponding resource sequence. The results are shown in Table 4. Across BPIC2012 and Emergency, no generated trace matches a training trace at the full event level. For Sepsis, only one match occurs across ten generated logs. Overall, low overlap indicates that most generated traces exhibit novel activity-resource combinations.

Table 4. Variant analysis. *Train overlap* shows how many generated traces reproduce an activity sequence observed in the training set, while *Resource Alignment* shows how many also match the corresponding resource sequence (lower match is better)

			Train overlap		**Resource Alignment**	
Dataset	**Model**	**Traces**	Seen	Unseen	Match	Mismatch
Sepsis	CVAE	157	23.80 ± 1.47	133.20 ± 1.47	0.10 ± 0.10	23.70 ± 1.51
	TF-decoder	157	3.30 ± 0.73	153.70 ± 0.73	0.10 ± 0.10	3.20 ± 0.76
BPIC2012	CVAE	937	15.50 ± 1.45	921.50 ± 1.45	0.00 ± 0.00	15.50 ± 1.45
	TF-decoder	937	64.30 ± 1.97	872.70 ± 1.97	0.00 ± 0.00	64.30 ± 1.97
Emergency (ORT)	CVAE	485	1.90 ± 0.46	483.10 ± 0.46	0.00 ± 0.00	1.90 ± 0.46
	TF-decoder	485	14.10 ± 1.34	470.50 ± 1.29	0.00 ± 0.00	14.10 ± 1.34

5.5 Training and Inference Times

Given the increased size and complexity of the TF-decoder, we report in Table 5 the total training and inference times, together with the model sizes in terms of trainable parameters. Training times are from the best performing model, trained until validation loss converges. Inference times are for generating the 10 synthetic logs of the size of the test set used for evaluation. Variation in trainable parameters of the TF-decoder is related to the vocabulary sizes of the categorical variables. Our one-hot encoding approach increases model size for higher vocabulary sizes, unlike the CVAE model that uses embedding layers.

Table 5. Reported training and inference times, along with the model size in terms of trainable parameters

Model	Sepsis		BPIC2012		Emergency_ORT	
	CVAE	TF-decoder	CVAE	TF-decoder	CVAE	TF-decoder
Training Time (hh:mm)	0:45	9:05	2:46	21:48	0:34	13:43
Inference Time (mm:ss)	0:15	1:13	4:41	24:44	1:03	3:34
Model Size (M params)	1.3	36	1.3	42.3	1.3	37.4

5.6 Discussion

The results highlight several important aspects of the TF-decoder's behavior. First, although the model does not consistently outperform the baseline on all marginal metrics and all datasets, it achieves **comparable control-flow, temporal, and resource fidelity**. This indicates that replacing the conventional factorized decoder with a transformer-flow architecture does not compromise the model's ability to reproduce individual process perspectives. In particular, the TF-decoder maintains competitive performance on temporal and resource perspectives while slightly improving in terms of control-flow behavior. These findings show that the architecture remains robust on traditional dimensions despite fundamentally changing the decoding mechanism.

The most notable improvements appear in the **preservation of cross-perspective dependencies**, captured by the MMD metric. Across all datasets, the generated logs from the TF-decoder are closer to the test data than the baseline. This behavior is consistent with the architectural design: the conditional normalizing flow jointly models activity, resource, and time attributes under a shared context, allowing the model to learn structured relationships that factorized decoders struggle to represent. The MMD results therefore provide quantitative evidence of improved modeling of the joint distribution of event attributes.

Beyond dependency preservation, the TF-decoder also demonstrates improved coverage of the activity and resource spaces. The frequency analysis shows that it consistently **generates the full set of activities and resources** present in the training data, including low-frequency elements that other models tend to miss. This is also verified by the higher coverage values the TF-decoder achieves. Together, these results indicate that the model captures a broader range of process behavior while maintaining a coherent event structure.

The variability analysis further confirms that the TF-decoder does **not rely on memorization**. Across BPIC2012 and Emergency, no generated trace matches a training trace exactly when comparing activities and resources, and only one such instance occurs in Sepsis, which is understandable given the smaller traces and repetitive structure. Even when activity sequences overlap with the training data, the corresponding resource sequences almost never do, indicating that the model generates coherent but novel combinations of event attributes.

Overall, the findings indicate that joint modeling meaningfully improves dependency preservation, categorical coverage, and trace diversity without compromising on marginal behavior. The TF-decoder, therefore, represents an effective approach for generating coherent and realistic synthetic event logs.

6 Conclusions and Future Work

This work introduced the TF-decoder, a decoder that combines a transformer backbone with a conditional normalizing flow to jointly model the activity, resource, and temporal dimensions of event logs. By modeling these perspectives under a unified density, the approach addresses a longstanding limitation of existing methods, which typically treat event attributes independently. Experiments on three real-world datasets show that the TF-decoder matches marginal performance in control-flow, timing, and resource fidelity while substantially improving on joint distribution modeling, as measured by the MMD metric. Our model also demonstrates stronger behavioral diversity, generating the full spectrum of activities and resources, and producing a richer set of trace variants whose distribution simulates the real distribution without memorizing it.

Despite these strengths, some limitations remain. Resource modeling remains challenging for all evaluated models, including the TF-decoder, suggesting that additional modeling choices are needed to capture organizational dynamics. The combination of transformer and flow also introduces computational overhead, slowing training and inference, particularly as the vocabulary size increases.

Several future directions emerge from these findings. First, incorporating richer organizational context, such as shift patterns, availability constraints, or hierarchical roles helps recreate more realistic resource behavior. Second, while this work focuses on activities, timestamps, and resources, future work should systematically explore alternative or additional attributes when consistently available. Third, although normalizing flows provide expressive density estimation, their computational cost motivates exploring more efficient architectures and alternative encodings for high-cardinality categorical variables. Another direction for future work could also explore copula-based statistical approaches to explicitly model and preserve joint probability distributions, as a complement to deep learning methods, similar to the prior work of Restrepo et al. [17].

Finally, comparing generative approaches with simulation-based methods could further clarify their respective strengths in reproducing process behavior.

References

1. Van der Aalst, W., Weijters, T., Maruster, L.: Workflow mining: discovering process models from event logs. IEEE Trans. Knowl. Data Eng. **16**(9), 1128–1142 (2004)
2. Chapela-Campa, D., Benchekroun, I., Baron, O., Dumas, M., Krass, D., Senderovich, A.: Can i trust my simulation model? measuring the quality of business process simulation models. In: Di Francescomarino, C., Burattin, A., Janiesch, C., Sadiq, S. (eds.) Business Process Management, pp. 20–37. Springer Nature Switzerland, Cham (2023)
3. Chapela-Campa, D., López-Pintado, O., Suvorau, I., Dumas, M.: Simod: automated discovery of business process simulation models. SoftwareX **30**, 102157 (2025)
4. De Weerdt, J., Wynn, M.T.: Foundations of process event data. In: Process Mining Handbook, pp. 193–211. Springer International Publishing Cham (2022)
5. van Dongen, B.: BPI challenge 2012 (2012)
6. Durkan, C., Bekasov, A., Murray, I., Papamakarios, G.: Neural spline flows. Adv. Neural. Inf. Process. Syst. **32** (2019)
7. Egozcue, J.J., Pawlowsky-Glahn, V., Mateu-Figueras, G., Barcelo-Vidal, C.: Isometric logratio transformations for compositional data analysis. Math. Geol. **35**(3), 279–300 (2003)
8. Evermann, J., Rehse, J.R., Fettke, P.: Predicting process behaviour using deep learning. Decis. Support Syst. **100**, 129–140 (2017)
9. Graziosi, R., et al.: Generating multiperspective process traces using conditional variational autoencoders. Process Sci. **2**(1), 8 (2025)
10. Gretton, A., Borgwardt, K.M., Rasch, M.J., Schölkopf, B., Smola, A.: A kernel two-sample test. J. Mach. Learn. Res. **13**(1), 723–773 (2012)
11. Krajsic, P., Franczyk, B.: Variational autoencoder for anomaly detection in event data in online process mining. In: ICEIS (1), pp. 567–574 (2021)
12. Li, K., Yang, S., Sullivan, T.M., Burd, R.S., Marsic, I.: ProcessGAN: generating privacy-preserving time-aware process data with conditional generative adversarial nets. ACM Trans. Knowl. Discov. Data **18**(9), 1–31 (2024)
13. Lin, L., Wen, L., Wang, J.: MM-Pred: a deep predictive model for multi-attribute event sequence. pp. 118–126
14. Mannhardt, F.: Sepsis Cases-event Log (2016)
15. Meneghello, F., Francescomarino, C.D., Ghidini, C., Ronzani, M.: Runtime integration of machine learning and simulation for business processes: time and decision mining predictions. Inf. Syst. **128**, 102472 (2025)
16. Naeem, M.F., Oh, S.J., Uh, Y., Choi, Y., Yoo, J.: Reliable fidelity and diversity metrics for generative models. In: III, H.D., Singh, A. (eds.) Proceedings of the 37th International Conference on Machine Learning. Proceedings of Machine Learning Research, vol. 119, pp. 7176–7185. PMLR (2020)
17. Restrepo, J.P., Rivera, J.C., Laniado, H., Osorio, P., Becerra, O.A.: Nonparametric generation of synthetic data using copulas. Electronics **12**(7), 1601 (2023)
18. van Straten, S., Padella, A., Hassani, M.: Leveraging data augmentation and siamese learning for predictive process monitoring. In: International Conference on Cooperative Information Systems, pp. 70–87. Springer (2025)

19. Taymouri, F., La Rosa, M., Erfani, S.M.: A deep adversarial model for suffix and remaining time prediction of event sequences. In: Proceedings of the 2021 SIAM International Conference on Data Mining (SDM), pp. 522–530. SIAM (2021)
20. Vinci, F., Aloini, D., Benevento, E., Stefanini, A., Zen, F., de Leoni, M.: Improving organizational processes in healthcare through simulation-driven resource allocation: methodology and real-world case study. Artif. Intell. Med. 103391 (2026)
21. Williams, B., Yeom-Song, V.M., Hartmann, M., Klami, A.: Simplex-to-euclidean bijections for categorical flow matching (2025). https://arxiv.org/abs/2510.27480

Efficient Event Sequence Predictions by Leveraging Event Context Clustering

Simon Rauch[1,2,3](✉), Daniel Schuster[1,4], Gabriel Marques Tavares[1,3], and Thomas Seidl[1,3]

[1] Ludwig-Maximilians University Munich, Munich, Germany
{rauch,schuster,tavares,seidl}@dbs.ifi.lmu.de
[2] Division Supply Chain Services, Fraunhofer IIS, Nuremberg, Germany
[3] Munich Center for Machine Learning, Munich, Germany
[4] University of Mannheim, Mannheim, Germany

Abstract. Data-driven methods from the field of Predictive Process Monitoring (PPM) are able to perform different prominent forecasting and inference tasks, e.g., activity suffix prediction, remaining time prediction, and outcome prediction. Currently established deep learning frameworks in PPM offer the possibility to perform one specific or also multiple of these tasks inside a single predictive model. However, due to their model complexity, they suffer from high computational loads and lengthy training times, especially when the goal is to fit all possible downstream tasks into one holistic model. In contrast to that, we provide a time-efficient framework that does not rely on deep learning but nevertheless allows for the execution of any downstream task from the PPM domain while keeping the computational load reasonably small and, additionally, offering the potential for explainable results via rule extraction. By pairing event clustering on available intra-case context information with a suffix prediction model based on sequential patterns, we enable forecasting of any information associated with process events including predictions of temporal information and additional event attributes. Compared to deep learning techniques, we achieve similar or better results on common PPM tasks of suffix prediction and next timestamp prediction while performing the model training in a fraction of the time deep learning models need to be sufficiently trained.

Keywords: Predictive Process Monitoring · Event Clustering · Sequential Prediction

1 Introduction

In the field of Predictive Process Monitoring (PPM), various forecasting techniques were developed for different tasks including next activity prediction, suffix prediction as well as the prediction of Key Performance Indicators (KPIs) such as throughput time or other outcomes of process cases. Since the complexity of business processes and the interactions between different drivers of process execution

F. M. Maggi et al. (Eds.): BPM 2026, LNCS 16936, pp. 557–574, 2027.
https://doi.org/10.1007/978-3-032-37777-7_33

can be quite complex, the field of PPM adopted the use of powerful forecasting techniques like long short-term memory (LSTM) neural networks [8] and Transformer architectures [28]. Recently, also graph neural networks (GNN) [3,11] found application and showcased their performance benefits in predictive process monitoring tasks.

However, in terms of pure activity suffix prediction, recent work also showed that computationally less complex and, most importantly, less expensive techniques can even surpass established deep learning techniques in terms of predictive accuracy [21] while also posing the benefit of not being a black-box model.

In this work, we want to challenge whether this phenomenon also holds for more complex tasks such as the joint sequential prediction of activities and process KPIs, which are crucial for the assessment of the overall behavior to be expected of a process instance in the future. To achieve this, we modify an existing prediction method, the Bilaterally Expanding Subtrace Tree (BEST) [21], since it is already well-proven on the activity prediction task. By the integration of event context information into the subtrace patterns, we transform the approach into a fully-fledged event suffix prediction framework that is able to not only predict future activity suffixes but complete event suffixes along with the event attributes that are deemed as important for the end user. This is achieved by fusing event clustering with the tree-based sequential prediction method. Accordingly, the main contributions of this work are as follows.

- We build upon the BEST framework [21] for activity suffix prediction by including context clusters into a forecasting pipeline to enable full information next event, event suffix and attribute outcome forecasting.
- We achieve promising results on common real-life benchmark event logs for the PPM tasks of activity suffix prediction and next timestamp prediction in one holistic, interpretable, and easily tunable forecasting model.
- Our framework based on sequential patterns maintains a low computational load needed for training and allows the extraction of meaningful rules.

2 Related Work

We embed our work into existing research and start by discussing previous contributions from event clustering. Subsequently, we outline past research in the field of PPM, focusing on deep learning techniques developed in the past decade.

Event/Mixed Data Clustering. Regarding clustering performed on process event log data, there is a vivid research field concerned with clustering process traces covering methods that are able to identify similarities in whole process traces or multi-event patterns ([14,23,30]). Van Zelst et al. [31] review event abstraction techniques. Whereas the majority of approaches cover means of discovering different levels of granularity in business process event logs, Rehse & Fettke [22] perform event level clusterings of similar activities to build a hierarchical structure of similar activities in business processes.

In comparison to the quite extensive research on trace clustering, we take a different path and perform clustering of single event context data, i.e., mixed event attribute information, to represent different ranges of context information. Ahmad et al. [2] offer a survey of mixed data clustering in general, highlighting the existence of various approaches in the field. In this work, we choose to transform mixed event data into a numerical representation and use the lower complexity representation of event context from fitting clustering algorithms in the prediction task of event suffix prediction.

Suffix Prediction. Over the last decade, suffix prediction has been tackled using methods ranging from sequential pattern mining [9,29] to a variety of deep-learning architectures, including recurrent neural networks [12], LSTM networks [8,26], Transformer architectures [7,28], and even deep reinforcement learning (Deep RL) [19]. The mentioned deep learning models are mostly also capable of predicting temporal information of predicted activity suffixes as well, showing possibilities of predicting remaining runtimes of ongoing traces inside one trained model architecture. What unifies all of those approaches, however, is a high dependency on computational power due to complex model architectures and long training times needed to fit the network to the event log data.

Direct Remaining Time Prediction. Besides suffix prediction models, there are also deep learning architectures that are based on a direct, single-target remaining time prediction from prefix information. Those approaches tend to show improved accuracy in predicting the remaining runtimes of process traces. In an LSTM approach, Navarin et al. [17] are able to consistently outperform other deep learning approaches that base their remaining time prediction on a predicted event suffix [19]. More recently, graph neural networks (GNNs) showed even higher performance in forecasting remaining process durations, as Duong et al. [11] and Elyasi et al. [3] were able to show. While providing very accurate remaining duration forecasts, it is important to highlight that their GNN architectures suffer from even higher computational needs than existing LSTM and Transformer architectures, while providing only a single downstream task, i.e., remaining time prediction. Moreover, they neglect the prediction of event suffixes which can be valuable information for process owners.

We aim for tackling existing drawbacks by designing a lightweight but still holistic approach of event suffix prediction that is capable of forecasting future events accompanied by arbitrary attribute information. Due to fast training procedures and the potential for explainable insights, we offer an easily integrable and comprehensible approach for various downstream tasks.

3 Preliminaries

We introduce key concepts of process mining and predictive process monitoring, starting with basic notations for business processes [1].

Definition 1. *(Event/Trace/Event Log) An event from the universe of events $\mathcal{E}$ is defined as a tuple $\epsilon = (c, a, t, h_1, \dots, h_m)$ describing the case id c out of a set of case identifiers $\mathcal{C}$, the executed activity a from an activity set $\mathcal{A}$ and the recorded timestamp t from the temporal domain $\mathcal{T}$ along with m event attributes $h_1, \dots, h_m$ out of a set of attributes $\mathcal{H}$. A trace σ is an ordered, finite sequence of events $\sigma = \langle \epsilon_1, \dots, \epsilon_{|\sigma|} \rangle$ belonging to the same process instance, i.e., case id. An Event Log L is a multiset of traces $\{\sigma_1, \dots, \sigma_{|L|}\}$ recorded during the execution of a given process. The activity sequence S of trace $\sigma = \langle \epsilon_1, \dots, \epsilon_{|\sigma|} \rangle$ is defined as $S(\sigma) = \langle a_1, \dots, a_{|\sigma|} \rangle$. We refer to unique sequences of activities as variants.*

Definition 2. *(Information Extraction and Event Context) Regarding an event ϵ, we introduce $X(\epsilon, I)$ as a function to extract its respective event context $x \in \mathcal{X}$ as a representation of the context information given by a set of corresponding attributes $I \subseteq \mathcal{T} \cup \mathcal{H}$. We introduce $\gamma : \mathcal{E} \to \mathcal{A} \times \mathcal{X}$ as a mapping function of events ϵ_i to tuples (a_i, x_i) of the activity and corresponding context extracted from the events ϵ_i. We use γ_i as a shorthand notation of $\gamma(\epsilon_i)$.*

Definition 3. *(l-Prefix/l-Suffix). An l-prefix $^{l}_{k}$ of an event $\epsilon_k^{(i)} \in \sigma_i$ is an event subsequence of σ_i with events $\langle \epsilon_{k-l}^{(i)}, \dots, \epsilon_{k-1}^{(i)} \rangle$ that precede $\epsilon_k^{(i)}$. An l-suffix $^{l}_{k}$ of an event $\epsilon_k^{(i)} \in \sigma_i$ is an event subsequence of σ_i with events $\langle \epsilon_{k+1}^{(i)}, \dots, \epsilon_{k+l}^{(i)} \rangle$ that succeed $\epsilon_k^{(i)}$. We refer to the full event prefix $_k$ or full event suffix $_k$ of an event $\epsilon_k^{(i)}$ if $l = k - 1$ or $l = |\sigma_i| - k$, respectively. In similar fashion, we can extract l-prefixes/l-suffixes of $\gamma(\epsilon_k^{(i)})$ containing tuples of activities and event context.*

With SPEC-EP, we tackle the problem of Next Event Prediction and Event Suffix Prediction, which we define as follows:

Definition 4. *(Problem Definition) Event prediction describes the task of predicting any unknown events $\epsilon_p^{(i)} \in {}^{(i)}_{k} \subseteq \sigma_i$ of event suffix $^{(i)}_{k}$ of a running process instance given event prefix $^{(i)}_{k}$. The prediction targets are a non-empty subset of event information $I' \subseteq \{a, t, h_1, \dots, h_m\}$ of an unknown event $\epsilon_p^{(i)}$ with $p > k$. We refer to Next Event Prediction for $p = k + 1$ and Event Suffix Prediction for $p \in [k + 1, |\sigma_i|]$, predicting all events until the end of trace σ_i.*

We aim to tackle both problems with a focus on next activity ($I' = \{a\}, p = k + 1$) and activity suffix prediction ($I' = \{a\}, p \in [k + 1, |\sigma_i|]$) as well as next timestamp ($(I' = \{t\}, p = k + 1)$) and remaining time prediction ($I' = \{t\}, p = |\sigma_i|$) by combining clustered event context information together with BEST [21] into a sequential prediction model of activities and corresponding context.

4 SPEC-EP: Subtrace Patterns and Event Clustering for Event Prediction

In this section, we describe the proposed framework leveraging **S**ubtrace **P**atterns and **E**vent **C**lustering for **E**vent **P**rediction (SPEC-EP) that extends BEST [21] into being capable of full event information suffix prediction. A repository containing a reference to an implementation of the method as well as all datasets used and results gathered is available online[1] In the following, we describe the building blocks of our framework and elaborate on the needed adjustments to develop BEST into a fully-fledged event suffix prediction model that is capable of forecasting any attribute information accompanying future events of ongoing process instances. Our framework is based on two combined modules that process event information and sequentially forecast future events. A visual representation of it is presented in Fig. 1 and described in the following subsections which are, respectively, linked to the elements of the framework graphic.

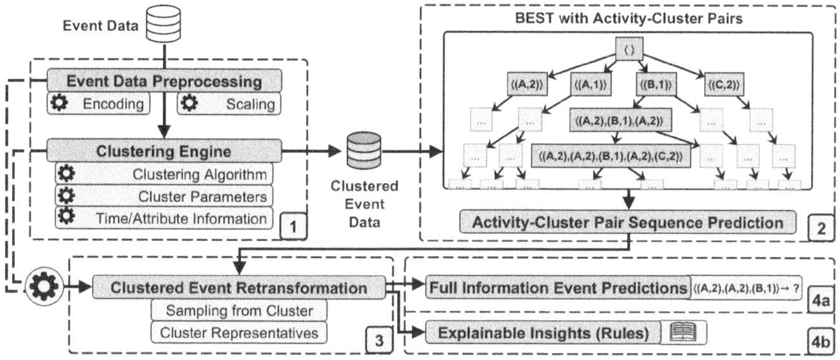

Fig. 1. SPEC-EP framework combining event context clustering (1) and predictive model training (BEST) from clustered event data (2) containing an event log's activity (sub-)sequences along with the context available (e.g. $\langle\langle A, 2\rangle, \langle B, 1\rangle, \langle A, 2\rangle\rangle$ in the process traces. Inverse transformation of preprocessing operations (3) allows for full event information suffix predictions (4a) and rule extraction (4b).

4.1 Clustering Event Context Information

Since attribute information of process events in most cases is of a mixed format, i.e., discrete/ordinal attributes like executing resource information or numerical features like time-related information, costs, or other numeric process KPIs, we first perform an encoding of the discrete/ordinal event data to arrive at a completely numerical representation (see step *Event Data Preprocessing* in

[1] https://doi.org/10.5281/zenodo.20426332.

Fig. 1). As our goal is to leave the representation of sequential transitions between events to BEST, we aim to represent single-event discrete attribute information numerically instead of encoding full process traces. Hence, we abstain from using embedding techniques like word2vec or count2vec and instead transform the discrete single-event attributes by means of one-hot encoding. This also ensures that our framework stays as simple, lightweight and interpretable as possible.

To ensure equal weighting of the different attributes in the clustering, we scale the numerical attribute data by standardizing it (zero mean, unit variance) to avoid higher emphasis on subsets of attributes due to their numerical range.

Subsequent to this preprocessing, we cluster the resulting encoded context information (see step *Clustering Engine* in Fig. 1) with a cluster algorithm of our choice. The only prerequisite of the chosen cluster algorithm is that we need to be able to retransform the respective cluster into the encoded dataspace, either by selecting a cluster representative or via sampling of a cluster component. To keep the parameter space reasonably small, we propose easily configurable cluster algorithms like KMeans++ [4] or Gaussian Mixture Models (GMM) [27], however, also other clustering algorithms fulfilling the ability to retransform cluster memberships into the encoded dataspace are valid choices. KMeans++ serves as a more static clustering candidate model leading to cluster representatives during retransformation whereas GMMs allow for resampling from cluster components, delivering more diverse context properties after retransformation. Moreover, the core function of the clustering is to provide a lower-dimensional representation of event context as additional input for the sequence prediction algorithm—a task for which both models are suitable. After performing the clustering, we receive clusters of context information for each event and by that an assigned cluster membership for each event of the analyzed event log. For applying existing sequential prediction models like BEST we need to prepare the patterns incorporating the cluster context information by creating a modified event log where we append the context cluster identified by the clustering algorithm onto the activity information (see Definition 2).

Logically, this increases the cardinality of the prediction target as we forecast tuples of activities and their respective context. In the worst case, this increase amounts to the cartesian product $\mathcal{A} \times \mathcal{X} = \{(a, x) | a \in \mathcal{A}, x \in \mathcal{X}\}$ of activities and possible context clusters. However, as not all activities happen in all possible contexts of events, the cardinality of the final prediction target is less drastically increased depending on the number of clusters computed[2].

4.2 Event Context Prediction with BEST

With the context clustering completed, we create the subtrace tree for BEST with the extended space of possible activity-context pairs to enable context cluster sequence prediction (see step 2 in Fig. 1). The key difference of SPEC-EP

[2] We provide an empirical assessment of the increased cardinality due to creating the conjoined activity-context pairs in SEction 5.

in comparison to BEST consists mainly in the integration of context information into the subtrace patterns. Therefore, we take the base mechanisms for the bilateral pattern expansion as well as the prediction mechanism from BEST and extend it by substituting the activities inside the patterns with activity-context pairs. Since the core mechanisms of BEST remain unchanged, we abstain from re-introducing the concepts and refer the reader to the original publication [21] for further details regarding the tree generation and forecasting procedure.

Definition 5. *(Activity-Context Subtrace Pattern) Given a sequence* $S = \langle \gamma_1^{(S)}, \ldots, \gamma_{|S|}^{(S)} \rangle$ *of activity-context pairs, we extract odd-length subtrace patterns* $P = \langle \gamma_i^{(S)}, \ldots, \gamma_{c(P)}^{(S)}, \ldots, \gamma_j^{(S)} \rangle \subseteq S$*, where* $c(P)$ *is the center index of* P *and* $\gamma_i^{(S)}$ *is the activity-context pair at index* i *of sequence* S*.*

As in the original contribution of BEST, we apply equally bilateral expansions with an l-prefix/l-suffix size of $l = 1$ and organize the resulting odd-length patterns in a hierarchical tree structure where the parent-child relationship preserves the central activity-context pair $\gamma_{c(P)}$. Patterns expanding over sequence length boundaries $i = 0$ and/or $j = |S|$ are padded with dummy tokens 'START' and 'END', respectively. Since the bilateral expansion distance, the conditional subtrace pattern probability as well as the prediction and tree structure generation concepts applied for SPEC-EP are identical to those introduced for BEST [21], we refer to the original paper for their definitions and implementations.

4.3 Transforming Predictions Into Original Event Log Data Space

Due to the preprocessing of event information into activity-context pairs, predictions from SPEC-EP are also in the form of activity-context pairs, making a full retransformation into the original event log data space necessary (see step 3 in Fig. 1). Starting with sampling from the forecasted context cluster of the predicted activity-context pair, we obtain context samples in the encoded numerical representation the clustering method was performed in. Samples can either be gathered by selecting the respective cluster representative, e.g., for KMeans++, or by extracting a sample from the respective cluster component for e.g., GMMs. We eventually decode each sample using the corresponding reverse mapping of the encoding function that was initially used and subsequently perform an inverse scaling of all initially scaled numerical attribute features. Eventually, we obtain attribute information from the original data space that serves as our predicted context for the future event suffixes for each process instance prefix (see step 4a in Fig. 1). Hence, SPEC-EP is capable of generating joint predictions of all event attributes that were used in the fitting of the clustering model.

5 Evaluation

To thoroughly evaluate our approach, we perform a series of experiments with real-world benchmark event log datasets from different domains. We evaluate different sets of model parameters and evaluate their resulting performance for the tasks of Next Activity Prediction (NAP), Activity Suffix Prediction (SFX), Next Timestamp Prediction (NTP) and Remaining Time Prediction (RTP).

5.1 Evaluation Setup

Datasets. For our evaluation, we use the following datasets: Helpdesk [18], variations of BPI'12 [10] that are commonly analyzed in existing PPM contributions (BPI'12 C—COMPLETE events, BPI'12 W—only workflow events, BPI'12 WC—only COMPLETE workflow events), BPI'13 (Closed Problems [24] , Incidents [25]), Env Permit [6], Sepsis [16], Nasa [15]. An overview of analyzed datasets along with their characteristics can be found in Table 1. Due to the introduction of context clusters and joining them with the activity identifiers into activity-context pairs (see Sect. 4), the cardinality of the prediction target increases compared to the base cardinality of the activity space (# activities in Table 1). Figure 2 shows the effect of this for the analyzed datasets. It is apparent, that the increased complexity is diminishing with increasing cluster count for each dataset due to the sparse activity and context cluster co-occurrence.

Table 1. Datasets used for performance evaluation.

Dataset	# events	# cases	# activities	mean—median—max trace length			# variants
Helpdesk	21348	4580	14	4.66	4.0	15	226
BPI'12 C	164506	13087	23	12.57	7.0	96	4336
BPI'12 W*	170107	9658	19	17.61	14.0	156	2921
BPI'12 WC	72413	9658	6	7.50	6.0	74	2263
BPI'13 Closed Problems	6660	1487	7	4.48	3.0	35	327
BPI'13 Incidents	65533	7554	13	8.68	6.0	123	2278
Env Permit	8577	1434	27	5.98	6.0	25	116
Sepsis	15214	1049	16	14.48	13.0	185	845
Nasa	73638	2566	47	28.70	28.0	50	2513

*: In these logs, the activity identifier is generated by a concatenation of the raw activity identifier and the activity lifecycle information (SCHEDULE, START, COMPLETE).

Model Setting. Since our framework is driven by the identified context clusters in the event data, we conduct experiments with context clusterings provided by the popular methods KMeans++ [4] and GMM [27]. This keeps the hyperparameter space regarding the clustering reasonably small as the only hyperparameter to be chosen is the number of clusters/components built from the available context

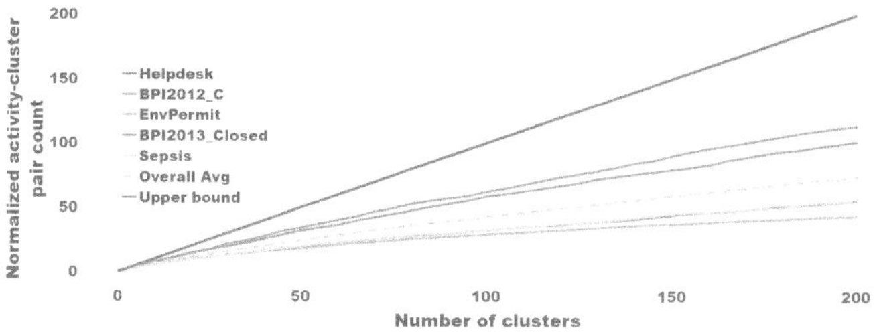

Fig. 2. Cardinality increase by coupling activities with cluster context into activity-context pairs for a subset of datasets. Activity-context pair count is normalized by the number of distinct activity identifiers per dataset and the upper bound (—$\mathcal{A} \times \mathcal{X}|$) is drawn as 45° line (red). (Color figure online)

information. We then feed the identified context clusters as additional information into BEST [21] to predict trace suffixes of ongoing process traces.

Starting at the clustering, we test different target amounts of context clusters ranging from [1, 10, 50, 100]. Regarding BEST, we train trees with pattern sizes of $[3, 5, 7, 9, 11, 15, 21]$ and the number of process stages defined by the process stage width calculated from a percentage of the overall maximum trace length of an event log. The percentage is chosen from $[0.0, 0.025, 0.1, 0.2, 0.5, 1.0]$. We refer to the original paper [21] for detailed information on those parameters. As we evaluate our approach on next temporal attribute prediction and temporal outcome prediction, we analyze predictive capabilities of different temporal attribute sets. We test different combinations of commonly used temporal features [8,26], namely, *time since midnight* (TSMN), *time since last event* (TSLE) and *time since case start* (TSCS) along with resource information and available (intra-case) case-level attributes. For RTP, we are able to calculate the overall runtime of a process either with using the cumulative sum of the predicted TSLE suffix or the last predicted TSCS attribute. In the following, we report the best value achieved from those two approaches. The best-performing attribute sets used in the evaluation are available in the results attached in the available repository, also containing results for both methods for RTP computation.

In the training of all tree models, we introduce a subtrace pattern extension cutoff that stops pattern search and matching when subtrace pattern occurrence probability drops lower than 0.01%. The training and evaluation pipeline employs a 5-fold cross validation setting where we split folds by unique case identifiers. In the resulting 80%–20% train-test-split folds, we split the contained event traces σ_i into full prefixes ${}^{(i)}_{j}$ with $j = 0, \ldots, |\sigma_i|$ where j is the length of the resulting prefix starting from an empty prefix.

With respect to runtime, it is to be noted that deep learning architectures usually rely on libraries with efficient training on GPUs. Our implementation of SPEC-EP is evaluated on CPUs. Training and inference of the evaluated deep learning architectures was performed on an NVIDIA GeForce RTX 2080 Ti GPU with 11 GB of GDDR6 VRAM, SPEC-EP was evaluated on a virtual machine with 16 virtual CPUs with base frequency of 2.10 GHz and 128 GB of RAM.

5.2 Predictive Performance

We report evaluation results of our approach for the common prediction tasks of NAP, SFX, NTP and RTP in comparison to past PPM approaches. They consist of approaches for NAP/SFX (BEST [21], Rama-Maneiro [20] (Deep RL)), models that are additionally capable of tasks NTP and RTP (Camargo [8] (LSTM), Wuyts [28] (Transformer)) and deep learning models for direct RTP (Navarin [17] (LSTM), Elyasi [3] (GNN)). See Table 2 for the complete comparison for all tasks. For SPEC-EP, we report the best results achieved across the complete parameter space for all tasks, respectively. We also report results obtained for holdout data splits (64%-16%-20%) in our supplementary repository.

Table 2. Comparison for NAP (Accuracy), SFX (Normalized Damerau-Levenshtein Similarity [5]), NTP (Mean Absolute Error (MAE) in days) RTP (MAE in days) along with respective standard deviations. The best, second best and third best approaches are shaded in green, blue and orange, respectively. For a direct comparison of our approach to approaches using minimum prefix lenghts of ≥ 2 (Elyasi et al. [3]), we report results for SPEC-EP with this minimum prefix length ($\rho_{\geq 2}$) without including it in the ranking (shaded in grey).

			BPI'12			BPI'13				
		Helpdesk	W	C	WC	Closed Problems	Incidents	Env Permit	Sepsis*	Nasa*
Camargo [8]	**NAP**	.7793 ±.0117	.4876 ±.1319	.6119 ±.0562	.4532 ±.1230	.4457 ±.0625	.6476 ±.0246	.8349 ±.0468	.4427 ±.0435	-
Wuyts [28]		.8439 ±.0069	.8161 ±.0019	.7902 ±.0083	.7149 ±.0225	.6320 ±.0025	.6485 ±.0019	.8708 ±.0135	.5566 ±.0158	.6793 ±.0158
BEST [21]		.8610 ±.0041	.8329 ±.0027	.7691 ±.0024	.6635 ±.0040	.6867 ±.0118	.7039 ±.0027	.8757 ±.0053	.5982 ±.0206	.8915 ±.0007
SPEC-EP		.8611 ±.0040	.8324 ±.0026	.7843 ±.0033	.6634 ±.0039	.6947 ±.0136	.7037 ±.0029	.8760 ±.0053	.5989 ±.0195	.8863 ±.0019
Camargo [8]	**SFX**	.8843 ±.0135	.2865 ±.0268	.3284 ±.0542	.2790 ±.0187	.6016 ±.0474	.5095 ±.0237	.8337 ±.0476	.3065 ±.0130	-
Wuyts [28]		.9191 ±.0035	.1162 ±.0058	.3016 ±.0226	.1205 ±.0097	.7321 ±.0186	.5098 ±.0263	.8705 ±.0127	.1258 ±.0247	.4404 ±.0109
Rama-Maneiro** [20]		.8989	.4000	.4711	.3829	.6933	.6230	.8369	.4285	.4981
BEST [21]		.9116 ±.0043	.3794 ±.0044	.4846 ±.0039	.3628 ±.0053	.7352 ±.0107	.5671 ±.0056	.8388 ±.0078	.4103 ±.0142	.4887 ±.0033
SPEC-EP		.9114 ±.0044	.3506 ±.0139	.4865 ±.0027	.3930 ±.0035	.7353 ±.0107	.6113 ±.0071	.8390 ±.0079	.4789 ±.0234	.4933 ±.0118
Camargo [8]	**NTP**	9.10 ±.2855	1.71 ±.0962	0.72 ±.0537	1.70 ±.0920	43.30 ±42.10	1.24 ±.7704	1.09 ±.4228	2.10 ±.2393	
Wuyts [28]		6.54 ±.1852	0.68 ±.0141	1.12 ±.0179	1.40 ±.0326	33.83 ±3.082	1.84 ±.0810	0.90 ±.3363	2.91 ±.4915	
SPEC-EP		4.89 ±.1407	0.71 ±.0307	0.71 ±.0089	1.53 ±.0201	40.40 ±2.115	1.35 ±.0847	0.99 ±.2000	2.07 ±.1414	
Camargo [8]	**RTP**	16.54 ±6.148	10.56 ±1.838	10.33 ±1.608	10.19 ±1.980	162.26 ±140.3	11.01 ±12.04	4.14 ±1.853	33.90 ±4.665	
Wuyts [28]		5.18 ±.1599	7.62 ±.1512	6.36 ±.0974	7.34 ±.3588	90.27 ±3.956	9.86 ±.8800	3.41 ±1.194	27.10 ±5.773	
Navarin [17]		34.15 ±.2732	7.83 ±.4697	7.51 ±.4891	9.61 ±1.536	129.05 ±153.9	12.35 ±12.29	4.15 ±1.856	34.22 ±4.540	
SPEC-EP		7.59 ±.3787	10.85 ±.2028	8.73 ±.2365	10.80 ±.3003	142.84 ±10.84	13.31 ±.7251	4.34 ±.8290	31.61 ±1.713	
Elyasi** [3]		4.11 ±0.04	2.70 ±0.01	2.77 ±0.02	5.07 ±0.03	37.44 ±1.49	2.23 ±0.05	2.72 ±0.08	16.48 ±0.19	
SPEC-EP($\rho_{\geq 2}$)		8.16 ±.2559	10.25 ±.3874	9.07 ±.3783	10.86 ±.3609	134.42 ±9.128	15.10 ±1.576	5.12 ±.4635	34.02 ±4.750	

*: Results for Wuyts et al. [28] for these datasets were achieved with the non-data-aware (NDA) architecture of the approach.
**: Results were taken from original publication. Differences in evaluation design are possible.

For activity prediction tasks (NAP, SFX), SPEC-EP performs very well and is able to consistently outperform competitor approaches from deep learning while being outperformed by the Deep RL approach of Rama-Maneiro et al. [20] and BEST on some of the datasets (SFX). It is to be noted that configurations with a larger amount of cluster components, in general, perform worse than configurations with fewer cluster components mined (cf. left part in Fig. 3).

Regarding the temporal attribute prediction we can see mixed results for the datasets analyzed. Performance for predicting the next timestamp of activities seems to be on par with existing approaches. SPEC-EP is able to deliver a steady and comparable performance to the LSTM and Transformer architectures where it mostly achieves either the best performance or the second best within striking distance towards the best performer per dataset (with the exception of BPI'13 Closed Problems dataset). RTP turns out to be a more difficult task for SPEC-EP, as it is mostly outperformed by existing approaches. Above all, PGTNet by Elyasi et al. [3] is the clear frontrunner while the Transformer architecture by Wuyts et al. [28] seems to be able to consistently outperform the other LSTM approach trained on direct RTP by Navarin et al. [17]. As can be seen in Table 2, performance of SPEC-EP is mostly aligned with the resulting error metrics of Camargo et al. [8] which it can outperform on some of the datasets. The analysis of total average runtimes for training and inference later in this section puts those result into perspective and—in spite of the worse results for RTP—underlines the general applicability of SPEC-EP also for the temporal attribute prediction tasks. Moreover, the approaches of Elyasi et al. [3] and Navarin et al. [17] are not capable of providing predictions of remaining trace suffixes and also have to be trained for each temporal prediction task (NTP/RTP) separately. We attribute the worse performance of SPEC-EP regarding RTP to a combination between the increase in distinct activity-context pairs, leading to a more accurate within-cluster prediction of temporal information but less accurate context cluster predictions. Vice versa, using fewer clusters increases the accuracy in predicting the correct context cluster while also increasing the overall forecasting error due to a higher within-cluster temporal variance. We can also assess the impact of integrating context information into the framework by comparing achieved metrics between SPEC-EP and BEST for NAP and SFX. Some datasets show slight improvements in prediction accuracy with larger increases for SFX on Sepsis. BPI'12 W suffers moderate negative effects regarding SFX, however, overall the results tend to increase when integrating context information.

Before delving into the runtime analysis, we provide additional analyses of our approach regarding the choice of the number of cluster components in Fig. 3. We did not find significant differences in performance or a clear overall frontrunner between the two clustering algorithms tested in our evaluation (KMeans++, GMM). We therefore abstain from reporting algorithm-specific results and rather show the overall average performance. For SFX, an increased amount of clusters introduced has an overall dereriorating effect on predicted suffix similarity. However, looking at the spread of the NDLS performance metric depicted, we can recognize a somewhat stabilizing effect of introducing context clusters into the subtrace tree. Although a decreasing effect is expected due to the increase in possible patterns found in the training procedure, SPEC-EP can still produce accurate results even with high amounts of clusters introduced. Regarding the accuracy of RTP, we can observe a clearly positive effect of introducing a larger amount of clusters. The achieved accuracy in terms of the MAE in the RTP task increases steadily as using 100 clusters in the clustering step produces the smallest achieved errors. We could not observe this behavior for all datasets analyzed as the ones with a higher spread in recorded overall times showed their minimum errors at a lower value of clusters. Those are excluded from the visualization in Fig. 3 also for better visibility of the decrease in errors for the shown datasets.

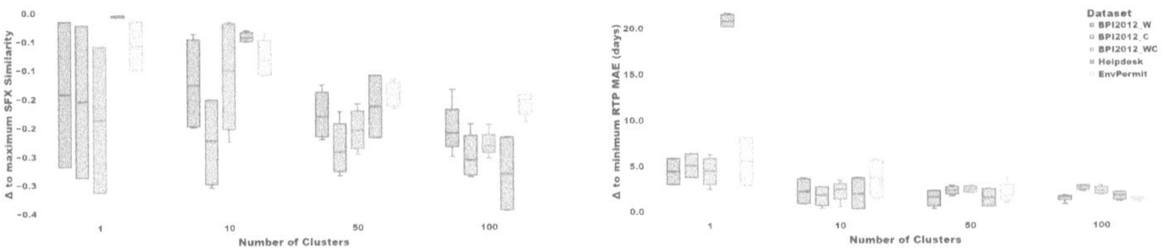

Fig. 3. Distributions of the Normalized Damerau-Levenshtein-Similarity for SFX (left) and MAE for RTP (right) in terms of distance to the minimum achieved value across all experiments for different numbers of clusters used. SFX results show a decrease in performance, whereas RTP records performance gains for higher numbers of cluster components.

Runtime Comparison. To support our claim of achieving drastically lower training times in comparison with established deep learning architectures, we provide a comparison of training and inference times. We include four competing approaches ([8,17,21,28]) trained and evaluated on identical tasks in Fig. 4. While training and inference time is higher than for BEST, SPEC-EP shows drastic improvements for training duration compared to deep learning approaches, being an order of magnitude faster on some datasets. Inference durations are still competitive compared to deep learning models, being faster for some of the datasets. While Wuyts et al. [28] can achieve very fast inference on shorter trace lengths (BPI'13 Closed, EnvPermit, Helpdesk), complex datasets show overly long inference durations due to its auto-regressive suffix prediction.

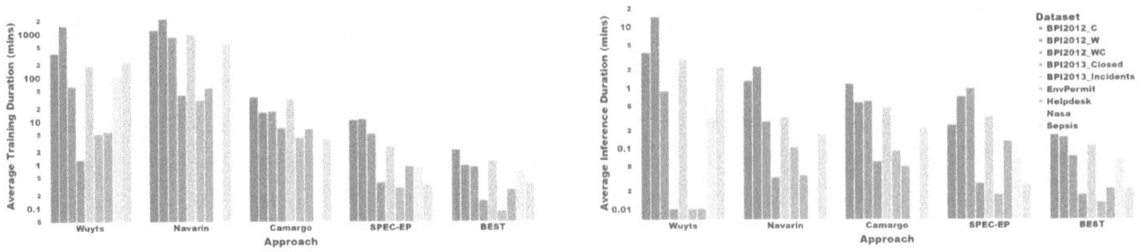

Fig. 4. Training (left) and inference (right) times (mins) averaged over all datasets for SPEC-EP and competitors (LSTM: Camargo [8], Transformer: Wuyts [28], direct RTP (LSTM): Navarin [17], BEST [21]).

5.3 Leveraging Explainability

To gain explainable insights into the conditional sequential relationships of events inside a business process, we can extract sequential rules from the mined subtrace patterns (see step 4b in Fig. 1). As one possible way to achieve that, we query probabilistic dependencies between activity-context pairs and different outcomes that those activity-context pairs are responsible for. Possible outcomes are represented by the different event attribute clusters identified during the clustering phase and, therefore, can contain any process event attribute used for the event clustering. The outcome attached to each subtrace pattern is context x extracted from the $k^{END}-1$-th activity-context pair, where k^{END} is the index of the first 'END' token reached inside the pattern. Subsequently, we scan the subtrace patterns for all possible combinations of an activity-context pair γ at the center index $c(P)$ (see Definition 5) of each pattern with its respective outcome context cluster x at index k^{END} and derive probabilistic statistics of rule importance based on Bayes' theorem [13] and the concept of *lift* [32]:

$$P(x|\gamma) = \frac{P(\gamma|x)}{P(\gamma)} \cdot P(x) \tag{1}$$

$$Lift(\gamma \Rightarrow x) = \frac{P(x|\gamma)}{P(x)} \overset{(1)}{=} \frac{P(\gamma|x)}{P(\gamma)}. \tag{2}$$

Bayes' rule specifies the belief update expressed through $\frac{P(\gamma|x)}{P(\gamma)}$ about a certain outcome x with overall occurrence (prior) probability $P(x)$ given some activity-context pair γ observed, combining it into the posterior probability $P(x|\gamma)$. From there, we can interpret the (symmetric) lift of a rule $Lift(\gamma \Rightarrow x)$ as to which extent we would update our beliefs about the likelihood of that outcome if we were to observe the activity-context pair at the center index of the pattern. $Lift(\gamma \Rightarrow x) = 1$ is interpreted as independence of x and γ and $Lift(\gamma \Rightarrow x) < 1$ and $Lift(\gamma \Rightarrow x) > 1$ represent negative and positive impact of γ on the occurrence of x, respectively.

To showcase the capabilities of SPEC-EP to supply explainable insights, we report rules extracted from the subtrace tree trained on the Helpdesk dataset with 10 context clusters in Table 3. We use TSCS as only event attribute for clustering for the sake of model parsimony and easily interpretable rules with respect to overall case duration. To be able to identify reasonably long dependency relationships and to ensure that we maximize the share of patterns that exhibit the end of a trace, we look at the tree's largest patterns.

The generated rules allow for a global assessment of important activity-context pairs with respect to temporal outcomes. For example, the occurrence of pairs *(Resolve ticket, 3)* or *(Assign seriousness, 3)* (rules 1 and 2) updates the beliefs about the final temporal outcome being situated in context cluster 5 which shows an average case duration of ≈32 days. The average event count from the start of the trace and to the end of the trace for each rule also allows for insights how early in the process the rules contribute to the decision about the

Table 3. Top-3 rules (sorted by achieved lift) identified per outcome cluster from Helpdesk dataset with respect to temporal outcomes (overall case duration). We report lift values along with the mean number of events until the end as well as from the start of a trace as a measure of farsight allowed for each specific rule. Average case durations per context cluster are listed on the right hand side.

Rule	Activity-Context Pair	⇒	Outcome Context Cluster	Avg. events from *START*	Avg. events to *END*	Lift
1	(Resolve ticket, 3)	⇒	5	2.07	3.81	2.13
2	(Assign seriousness, 3)	⇒	5	4.27	2.27	1.92
3	(Take in charge ticket, 3)	⇒	5	3.34	2.79	1.45
4	(Resolve ticket, 4)	⇒	6	2.05	4.26	3.76
5	(Take in charge ticket, 4)	⇒	6	3.17	3.67	2.82
6	(Wait, 3)	⇒	6	3.42	3.47	1.47
7	(Resolve ticket, 5)	⇒	7	2.06	4.10	2.08
8	(Wait, 4)	⇒	7	3.70	4.15	1.64
9	(Take in charge ticket, 5)	⇒	7	3.15	3.74	1.52
10	(Wait, 5)	⇒	8	3.50	3.75	3.98
11	(Resolve ticket, 6)	⇒	8	2.04	3.95	3.93
12	(Take in charge ticket, 5)	⇒	8	3.27	3.47	3.70
13	(Resolve ticket, 7)	⇒	9	2.05	4.52	10.71
14	(Take in charge ticket, 7)	⇒	9	3.24	4.00	8.94
15	(Take in charge ticket, 6)	⇒	9	3.24	3.43	6.82

Context Cluster	Avg. case duration (days)
0	0.24
1	4.85
2	10.33
3	16.57
4	24.16
5	32.14
6	38.15
7	44.31
8	50.10
9	57.59

future manifesting trace outcome. We can identify rules that allow for early information in case we observe the respective activity-context pair of over 4 events before the trace comes to an end, e.g., in rules 4, 7, 8, 13 and 14 with some of them showing high lift statistics. Considering the average trace length of the analyzed Helpdesk dataset of 4.66 (see Table 1), the knowledge about the respective activity-context pairs occurring leading to a specific temporal outcome comes available quite early. Ultimately, the probabilistic rules form a way to uncover non-linear relationships in process behavior without using complex deep learning models and allow for explainable inference regarding case outcomes.

SPEC-EP allows for flexible extraction of rules. Process owners could also uncover insights regarding other event context information besides temporal or look into short-term relationships as an alternative to focusing on the final case outcome. The latter could be achieved by extracting rules between activity-context pairs and the resulting context for the next $n << k^{END} - 1$ events.

5.4 Discussion

Considering all results and characteristics of our approach and the existing competitors, we can develop a distinguished assessment of the capabilities of SPEC-EP. While SFX and also NTP are strong suits of SPEC-EP, results for RTP (cf. Table 2) suggest that there are drawbacks in comparison to other established deep learning techniques, especially GNN-based direct RTP approaches [3]. However, we want to highlight that the direct RTP approaches [3,17] tackle a somewhat easier task. As they neglect the prediction of event suffixes, the problem complexity shrinks from a high-dimensional output to a single target value. While reducing the problem complexity drastically, they also show inflexibility when considering tasks like NTP. NTP would in principle be possible by adjusting the prediction target but would make expensive retraining of the whole model necessary. With SPEC-EP, we provide a lightweight approach that is trained orders of magnitude faster (see Fig. 4) than any of the deep learning approaches and, moreover, poses as the only candidate that has the potential of generating inherently explainable analyses due to being based on conditional probabilities of sequential patterns. We show how this could be leveraged via rule extraction for temporal outcomes in Sect. 5.3. However, predictions and rules can be extracted for different types of outcomes or certain ending activities as well. A subsequent look into the attribute value distribution of context clusters inside the rules could provide valuable insights into which activity-context patterns might be influential regarding specific questions of process owners.

We provide a coherent overview of the different model characteristics in Table 4. Eventually, we see SPEC-EP as a more than capable alternative to existing deep learning approaches for activity and attribute suffix prediction—especially for users that have a need for retraining the employed PPM model due to frequently changing process structure, i.e., process drifts, or those with no access to large computational infrastructure to efficiently make use of complex deep learning architectures that are well-tuned to a process and the task to be solved.

Table 4. Model characteristics of SPEC-EP compared with other approaches.

Approach	NAP/SFX	NTP	RTP	Inherent Explainability	Fast Training	Fast Inference
Camargo [8]	✓	✓	✓	×	×	✓
Wuyts [28]	✓	✓	✓	×	×	(✓)
Navarin [17]	×	(✓)	✓	×	×	✓
Elyasi [3]	×	(✓)	✓	×	×	✓
BEST [21]	✓	×	×	✓	✓	✓
SPEC-EP	✓	✓	✓	✓	✓	✓

5.5 Limitations

Limitations of our current framework design are showing in the case of attribute prediction for complete suffixes, i.e., at the end of process instances, as we see a performance drop with respect to the other models from NTP to RTP. We attribute this behavior of our model to the increased entropy in possible future context clusters for a given prefix, which also increases the difficulty to predict the correct context cluster that contains the respective correct range of end-of-trace attribute values. Regarding the clustering step prior to the tree generation, there is the inherent problem of process event log data of incorporating different types of process attributes and—in most cases—also mixed-type attribute information. Both aspects pose difficulties in finding a reasonable representation of the event or case attribute values due to various possible ways to weigh the different types of attribute information during encoding. One could for example put a higher emphasis on resource attributes or give a higher weight to the temporal information regarding the respective predictive task. While this is a general problem of event log data, SPEC-EP needs strict design decisions to be made in the encoding and clustering phase of the framework.

6 Conclusion

This paper presents SPEC-EP, a lightweight and explainable framework incorporating event context data clustering and sequential suffix prediction via BEST, i.e., a tree-based PPM model, for full event information suffix forecasting. The introduced event context clusters serve as representatives of different ranges of additional event attribute information and open up the opportunity of forecasting any future event attribute. We transform the existing BEST approach into a holistic forecasting framework, elevating it from a very capable forecasting approach for activity suffix prediction to a framework fit for fulfilling a range of different downstream tasks with reasonable performance while offering a fast model training and extraction of explainable insights. For future work, we aim at decreasing the hyperparameter space by structuring the search of encoding and clustering. Moreover, we want to further investigate the integration of inter-case attributes into SPEC-EP.

Disclosure of Interests. The authors have no competing interests to declare that are relevant to the content of this article.

References

1. van der Aalst, W.M.P.: Process Mining: A 360 Degree Overview, p. 3–34. Springer International Publishing (2022)
2. Ahmad, A., Khan, S.S.: Survey of state-of-the-art mixed data clustering algorithms. IEEE Access **7**, 31883–31902 (2019)
3. Amiri Elyasi, K., van der Aa, H., Stuckenschmidt, H.: PGTNet: a process graph transformer network for remaining time prediction of business process instances. p. 124–140. Springer Nature Switzerland (2024)
4. Arthur, D., Vassilvitskii, S.: k-means++: the advantages of careful seeding, Technical report Stanford (2006)
5. Boytsov, L.: Indexing methods for approximate dictionary searching: comparative analysis. ACM J. Exp. Algorithmics **16** (2011)

6. Buijs, J.: Receipt phase of an environmental permit application process (WABO), CoSeLoG project (2022). https://data.4tu.nl/articles/_/12709127/2
7. Bukhsh, Z.A., Saeed, A., Dijkman, R.M.: ProcessTransformer: predictive business process monitoring with transformer network (2021)
8. Camargo, M., Dumas, M., González-Rojas, O.: Learning accurate LSTM models of business processes. In: Hildebrandt, T., van Dongen, B.F., Röglinger, M., Mendling, J. (eds.) BPM 2019. LNCS, vol. 11675, pp. 286–302. Springer, Cham (2019). https://doi.org/10.1007/978-3-030-26619-6_19
9. Ceci, M., Lanotte, P.F., Fumarola, F., Cavallo, D.P., Malerba, D.: Completion time and next activity prediction of processes using sequential pattern mining. In: Džeroski, S., Panov, P., Kocev, D., Todorovski, L. (eds.) DS 2014. LNCS (LNAI), vol. 8777, pp. 49–61. Springer, Cham (2014). https://doi.org/10.1007/978-3-319-11812-3_5
10. van Dongen, B.: BPI Challenge 2012 (2012). https://data.4tu.nl/articles/_/12689204/1
11. Duong, L.T., Travé-Massuyès, L., Subias, A., Merle, C.: Remaining cycle time prediction with graph neural networks for predictive process monitoring. In: Proceedings of the 2023 8th International Conference on Machine Learning Technologies. p. 95–101. ICMLT 2023, ACM (2023)
12. Evermann, J., Rehse, J.R., Fettke, P.: Predicting process behaviour using deep learning. Decis. Support Syst. **100**, 129–140 (2017)
13. Gelman, A., Carlin, J.B., Stern, H.S., Dunson, D.B., Vehtari, A., Rubin, D.B.: Bayesian Data Analysis. Chapman & Hall/CRC Texts in Statistical Science, Chapman & Hall/CRC, Philadelphia, PA, 3 edn. (2013)
14. Grigore, I.M., Tavares, G.M., Silva, M.C.D., Ceravolo, P., Barbon Junior, S.: Automated trace clustering pipeline synthesis in process mining. Information **15**(4), 241 (2024)
15. Leemans, M.: NASA Crew Exploration Vehicle (CEV) Software Event Log (2017). https://data.4tu.nl/articles/_/12696995/1
16. Mannhardt, F.: Sepsis Cases - Event Log (2016). https://data.4tu.nl/articles/_/12707639/1
17. Navarin, N., Vincenzi, B., Polato, M., Sperduti, A.: LSTM networks for data-aware remaining time prediction of business process instances. In: 2017 IEEE Symposium Series on Computational Intelligence (SSCI), p. 1–7. IEEE (2017)
18. Polato, M.: Dataset belonging to the help desk log of an Italian company (2017). https://data.4tu.nl/articles/_/12675977/1
19. Rama-Maneiro, E., Vidal, J., Lama, M.: Deep learning for predictive business process monitoring: review and benchmark. IEEE Trans. Serv. Comput., 739–756 (2023)
20. Rama-Maneiro, E., Patrizi, F., Vidal, J., Lama, M.: Towards Learning the Optimal Sampling Strategy for Suffix Prediction in Predictive Monitoring, p. 215–230. Springer Nature Switzerland (2024)
21. Rauch, S., Frey, C.M.M., Maldonado, A., Seidl, T.: BEST: bilaterally expanding subtrace tree for event sequence prediction. p. 415–432. Springer Nature Switzerland (2025)
22. Rehse, J.R., Fettke, P.: Clustering Business Process Activities for Identifying Reference Model Components, p. 5–17. Springer International Publishing (2019)
23. Song, M., Günther, C.W., van der Aalst, W.M.P.: Trace Clustering in Process Mining, p. 109–120. Springer Berlin Heidelberg (2009)
24. Steeman, W.: BPI Challenge 2013, closed problems (2013). https://data.4tu.nl/articles/_/12714476/1

25. Steeman, W.: BPI Challenge 2013, incidents (2013). https://data.4tu.nl/articles/_/12693914/1
26. Tax, N., Verenich, I., La Rosa, M., Dumas, M.: Predictive business process monitoring with LSTM neural networks. In: Dubois, E., Pohl, K. (eds.) CAiSE 2017. LNCS, vol. 10253, pp. 477–492. Springer, Cham (2017). https://doi.org/10.1007/978-3-319-59536-8_30
27. Wolfe, J.H.: Pattern clustering by multivariate mixture analysis. Multivar. Behav. Res. **5**(3), 329–350 (1970)
28. Wuyts, B., Vanden Broucke, S., De Weerdt, J.: SuTraN: an encoder-decoder transformer for full-context-aware suffix prediction of business processes. In: 2024 6th International Conference on Process Mining (ICPM), p. 17–24. IEEE (2024)
29. Yap, G.-E., Li, X.-L., Yu, P.S.: Effective next-items recommendation via personalized sequential pattern mining. In: Lee, S., Peng, Z., Zhou, X., Moon, Y.-S., Unland, R., Yoo, J. (eds.) DASFAA 2012. LNCS, vol. 7239, pp. 48–64. Springer, Heidelberg (2012). https://doi.org/10.1007/978-3-642-29035-0_4
30. van Zelst, S.J., Cao, Y.: A Generic Framework for Attribute-Driven Hierarchical Trace Clustering, p. 308–320. Springer International Publishing (2020)
31. van Zelst, S.J., Mannhardt, F., de Leoni, M., Koschmider, A.: Event abstraction in process mining: literature review and taxonomy. Granular Comput. **6**(3), 719–736 (2020). https://doi.org/10.1007/s41066-020-00226-2
32. Zhang, C., Zhang, S.: Association rule mining. Lecture Notes in Computer Science, Springer, New York, NY (2006)

Linear Continuous Process Maps

Hagen Völzer(✉), Moritz Fässler, Aaron Friedrich Kurz, and Barbara Weber

Institute of Computer Science, University of St. Gallen, St. Gallen, Switzerland
{hagen.voelzer,aaron.kurz,barbara.weber}@unisg.ch

Abstract. The sequential behavior of a process captured in an event log is often represented by a simple annotated graph such as the well-known Directly-Follows-Graph (DFG). We define intuitive requirements, essentially *linearity* and *continuity*, for such graph-based process maps to capture some basic interpretability. We show that some well-known issues of interpreting the DFG are violations of those requirements, which are attributable to the tactical DFG filters as implemented in commercial tools. We study the extent to which effective visualizations can be created without these tactical filters and demonstrate that a system of linear and continuous process maps can sufficiently reduce graph complexity while accurately revealing many relevant insights. To this end, we provide a prototypical implementation of our framework based on interactive Sankey diagrams and demonstrate its usefulness on five highly used real-world event logs. Our conceptualization also opens the view for generalizations of the DFG such as the *set* abstraction, which mitigates undesired generalization commonly observed in many DFGs and allows, among other things, a precise distinction of rework and non-deterministic ordering.

Keywords: Directly-Follows-Graph · Sankey Diagrams · Exploratory Analysis · Sequential Behavior Analysis

1 Introduction

Simple annotated graphs, such as the Directly-Follows-Graph (DFG), are popular visualizations of the sequential behavior of event logs. Variants of the DFG are implemented in most commercial process mining tools, and case studies have reported their beneficial use on real-world data, e.g. [9,15]. At the same time, it is well known that DFGs often require complexity reduction, such as filtering, to become consumable for analysts. It is also known that complexity-reducing filters implemented in commercial tools can create misleading information [9,17].

Empirical studies of process mining practice highlight both the central role and the limitations of DFGs. Observational studies of analysis sessions show that DFG-based views account for a large portion of analysts' interaction time and are responsible for a significant share of discovered findings [20]. At the same time, these studies report that a substantial fraction of the observed interpretation

F. M. Maggi et al. (Eds.): BPM 2026, LNCS 16936, pp. 575–593, 2027.
https://doi.org/10.1007/978-3-032-37777-7_34

errors occurred when analysts were working with DFG-based views. Complementary interview studies also report interpretability problems when working with filtered DFGs. Analysts note that filtering produces information that does not seem consistent with the variant information, leading some analysts to state that they "do not trust the maps" when interpreting filtered DFGs [21]. Consequently, analysts frequently resort to variant visualizations, attribute statistics, case comparisons, or raw data inspection when seeking deeper insights or validating findings [19]. These observations indicate that the challenge is not only the visual complexity of DFGs, but also the lack of guarantees about how event-log behavior is represented after abstraction or filtering.

In this paper, we present a framework for graph-based visualizations of sequential behavior under complexity reduction. Our main contributions are:

- *Interpretability criteria for graph-based log visualizations.* We formalize requirements as *linear continuous maps (LCMs)* that capture a basic notion of interpretability for mappings from event logs to graphs that may reduce complexity. These criteria rule out an important source of confusion in practice: known issues of tactical DFG filters are violations of these requirements.
- *A framework for transformations.* We introduce a framework combining LCMs with other transformations. The framework includes both known and novel transformations, and we provide practical heuristics for selecting LCMs.
- *An interactive visualization implementation.* We implement the proposed framework as interactive Sankey diagrams that allow analysts to progressively adjust abstraction and information while maintaining a linear and continuous mapping between the event data and the graph structure. The implementation is publicly available [1].
- *Evaluation on real-world event logs.* We evaluate the proposed framework on five real-world event logs (*Road Fines*, *Sepsis*, loan applications from BPIC12 and BPIC17, and *Domestic Declarations* from BPIC20) and demonstrate that the proposed transformations enable sufficient complexity reduction while preserving structure in a way that visualizes meaningful sequential patterns and behavioral insights in complex event logs.

Overview. Figure 5 shows two Sankey diagrams generated from the Sepsis event log. Both diagrams visualize the same underlying process behavior, but at different abstraction levels. While the left diagram resembles a traditional DFG, the right diagram highlights different structural patterns. Both diagrams are obtained through a linear continuous map (LCM), which reduce complexity while guaranteeing that every node and edge strength precisely represents the underlying event-log behavior.

2 Requirements

2.1 Use Case: Sequential Behavior Exploration

The DFG has been used in a variety of use cases. Besides depicting the directly-follows-relation of an event log, it is also used to display performance metrics.

Moreover, it has been considered as a *predictive* process model for the event log, i.e., a model that generalizes beyond the behavior recorded in the log. In this role, the DFG can also serve as an intermediate representation for inducing more sophisticated generalizing process models such as Petri nets [4,9,17].

In this paper, we exclusively focus on the use case of exploring and visualizing the *sequential behavior* of an event log. In this setting, we consider the graph as a *descriptive* model of the log. Unlike in predictive models, generalization is a priori undesirable in a descriptive model. However, generalization is often a consequence of abstraction, which in turn is often necessary to reveal patterns in large datasets, and when we are interested in sequential patterns that may occur across multiple cases of the log. Note that an analyst can also look at the list of sequential variants of a log. Apart from the issue that the list is often too long for an analyst to inspect each variant, it neither shows the common structure of the variants nor any higher level patterns. Our perspective is to view the graph similarly to the role of analytic charts such as interactive histograms, where the transformation from data to visual representation has a well-defined and transparent semantics, and where interactive controls (e.g., the number of bins) allow the analyst to adjust the abstraction level while maintaining a clear relationship to the underlying data. To make our perspective precise, we use the following definitions.

Event Logs and Sequential Behavior. We work with *bags*, where a *bag* over a set X is a mapping $M : X \to \mathbb{N}$. $M(x)$ is called the *multiplicity* of x in M. Given a bag M, the set $\text{supp}(M) = \{x \mid M(x) > 0\}$ is called the *support* of M. M_1 is a *sub-bag* of M_2, denoted $M_1 \leq M_2$ when $M_1(x) \leq M_2(x)$ for all x. We say that M_1 is an *exact sub-bag* when, in addition, $M_1(x) = M_2(x)$ for all $x \in \text{supp}(M_1)$. Addition of bags is defined as $(M_1 + M_2)(x) = M_1(x) + M_2(x)$.

Let A be a (finite) set of symbols. An element of A is called an *event*. A finite, non-empty sequence $\sigma \in A^+$ over A is called a *trace*. An *(event) log* over A is a bag $L : A^+ \to \mathbb{N}$. We abstract from what information a symbol $a \in A$ represents, but at this point, we can think of it as usual as an activity label, in which case a trace is also often called a *(sequential) variant*.

In this paper, we are exclusively concerned with the *sequential behavior* of an event log, which, besides their supporting set of traces and the multiplicity for each trace, also consists of their *topological structure*, as explained below. For this paper, we choose the prefix structure of traces as the underlying topology (also called the *Scott topology*), which is a frequent and natural choice for sequences that represent computations. Sequences can also be equipped with other topologies: For example, sequence alignments are associated with a different topology over sequences. For two sequences $\alpha, \beta \in A^*$, we write $\alpha \sqsubseteq \beta$ when α is a *prefix* of β. Given a log L, the *bag of all prefixes* of L, denoted $L_\downarrow$ is defined as $L_\downarrow : A^* \to \mathbb{N}$ by:

$$L_\downarrow(\alpha) = \sum_{\sigma \in \text{supp}(L), \alpha \sqsubseteq \sigma} L(\sigma)$$

The prefix structure of a log can be represented as a tree with multiplicities as follows. A *multi-graph* $G = (X, N, E)$ consists of a set X and a bag of *nodes* $N : X \to \mathbb{N}$ and a bag of edges $E : X \times X \to \mathbb{N}$. Given a log L, the multi-graph $\text{Tree}(L) = (A^*, N, E)$ defined by $N = L_\downarrow$ and $E(\alpha, \alpha e) = L_\downarrow(\alpha e)$ is called the *prefix tree* of L. The prefix tree can be considered the ground truth of the sequential behavior of the log: It has all the information of the log, i.e., there is a bijection between the traces of the log and the paths of the tree that also reflect their log multiplicities. In addition, the prefix tree makes the topological structure of its traces explicit. Figure 1(a) shows an example of a prefix tree where a node represents the prefix identified as the path from the root; multiplicities of zero are omitted.

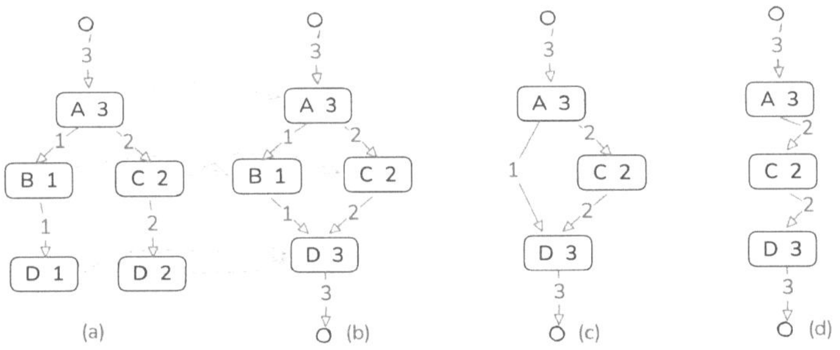

Fig. 1. *(a)* Prefix tree of the log ABD, ACD, ACD. *(b)* The corresponding DFG without filtering generated by the multi-graph morphism (dashed) from the prefix tree. *(c)* DFG of the log AD, ACD, ACD *(d)* Graph obtained in commercial tools when either using activity filter on DFG in b) or arc (a.k.a. path) filter on DFG in c).

2.2 Interactive Directly-Follows Graphs and Their Limitations

The *DFG* for an event log L over A is the multigraph (A, N, E) such that $N(a) = |\{(\sigma, i) \mid \sigma(i) = a\}|$ and $E(a, b) = |\{(\sigma, i) \mid \sigma(i) = a \text{ and } \sigma(i + 1) = b\}|$. Traditionally, the DFG for L is shown as the DFG of the log where each trace σ of L is replaced by the trace $\hat{a}\sigma\breve{a}$ where $\hat{a}$ and $\breve{a}$ are unique artificially added start and end symbols respectively. Figure 1(b) shows an example of a DFG.

Tactical Filter Issues. DFG implementations provide *tactical*, i.e., front-end implemented filters to reduce the graph complexity. It is known [9,17] that applying these filters can make it difficult for the user to interpret the resulting graph. Figure 1 shows minimal examples: (b) shows the unfiltered DFG of the log ABD, ACD, ACD; (d) shows the result of applying the *activity (threshold) filter* to that DFG as implemented in commercial tools[1]. The tactical filters as implemented in commercial tools do not have a precise public specification. A

[1] We have tested our examples against *Disco, Celonis, and Apromore*, see Supplement.

reasonable specification of the activity threshold filter is to delete activities a such that $N(a)$ is smaller than a given threshold [17], which yields the graph in Fig. 1(c). The actual graph in (d) shows a gap, i.e., *discontinuity* in the representation of AD.

The graph in Fig. 1(c) also represents the unfiltered DFG of the log AD, ACD, ACD, whereas the graph in (d) also shows the result of applying the *arc (threshold) filter* to (c), for which a reasonable specification is to delete edges (a, b) such that $E(a, b)$ is smaller than a given threshold [17]. The combination of both filters may even cause underlying graph discontinuities, e.g., nodes from which the end node cannot be reached anymore, which has been studied as *DFG unsoundness* [9]. We have created a minimal example to reproduce this condition in *Disco* (see Supplement, Supplementary Material can be found at [2]), but *Apromore* and *Celonis* seem to preserve DFG soundness, cf. also [3]. However, this does not improve matters for our use case as it makes it more complex for an analyst to understand how the visualization represents the underlying data.

Generalization. A DFG may generalize, i.e., may represent more traces than those in the log. For example, the DFG of the log $L_1 = \{AXY, BXZ\}$ also contains the traces $\{AXZ, BXY\} = L_2$. In fact, the DFGs of L_1 and L_2 are identical. The DFG is an abstraction and therefore has limited information about the log, which is not enough to reproduce the log from the DFG, but makes it easier to consume the DFG and may expose patterns that become visible only in the abstract view. Similarly, the logs $L_3 = \{AKLZ, ALKZ\}$ and $L_4 = \{AKLKZ, ALZ\}$ result in the same DFG, where L_4 contains *rework*, i.e., activity repetition, and L_3 does not. Hence the original version of a DFG does not represent rework precisely as it confuses rework with non-deterministic ordering as in this example. This issue is known. Another aspect of the generalization of the DFG is that we do not see how often activities repeat and how frequent such repetitive patterns are. More generally, we cannot reliably identify any patterns spanning more than two subsequent activities.

Some analytical charts, such as a histogram, are equipped with parameters, e.g., the number of bins, that can be chosen interactively by the analyst to adjust the abstraction level. In contrast, the minimum abstraction level of a DFG is fixed–the analyst can increase the abstraction by filtering or data pre-processing, but cannot decrease it to counter generalization when needed.

Visualization-Level Limitations. The effectiveness of conveying sequential patterns is not just a property of the mathematical object of the DFG, but also of its visualization and layout, as well as the affordances that its interactive controls provide to the analyst. For example, common DFG visualizations do not project trace multiplicities in a linear fashion, making it more difficult to see properties of the variant distribution such as a clear separation of mainstream and exceptional behavior. In general, the emerging field of visual analytics for process mining studies corresponding research questions, e.g., [6, 13]. Concretely, we know little at this point about the effectiveness of alternative visualizations of and interactions with DFGs.

2.3 Linear Continuous Process Maps

In this section, we introduce with *linear continuous maps (LCMs)* a characterization of basic interpretability for graph-based representations of event logs. Our formalization constrains how mappings from event logs to graphs may transform the underlying behavior while preserving a transparent and faithful relationship between the data and the resulting visualization. In particular, LCMs rule out the tactical filtering issues discussed in Sect. 2.2. As we demonstrate later, LCMs also enable sufficient reductions in graph complexity while still visualizing many business-relevant patterns.

Given two multi-graphs $G = (X, N, E)$ and $G' = (X', N', E')$, a *morphism* from G to G' is a mapping $\phi : X \to X'$ such that $N'(x') = \sum_{x'=\phi(x)} N(x)$ and $E'(e') = \sum_{e'=\phi(e)} E(e)$ where for $e = (x, y)$, we define $\phi(e) = (\phi(x), \phi(y))$. A *linear continuous process map (LCM)* for a log L is a morphism from the prefix tree $\text{Tree}\,(L)$ to a multi-graph G.

To construct a morphism from the prefix tree of a log, we can use any prefix abstraction, i.e., a mapping $\phi : A^* \to X$ into any set X. For example, consider the mapping $\phi(a_0, \ldots, a_n) = a_n$, which selects the last event in the prefix. The definition of LCM now lifts ϕ to map also pairs of prefixes, bags of prefixes and bags of pairs of prefixes, which together constructs the resulting multi-graph that abstracts the tree. For the example ϕ, we obtain the (unfiltered) DFG of the log. The corresponding morphism is shown in Fig. 1 from (a) to (b) through dashed arrows. We can think of a graph node as a prefix abstraction or as an equivalence class of prefixes, viz. $\alpha \equiv \beta$ iff $\phi(\alpha) = \phi(\beta)$. We will explore more useful abstractions below.

By construction, we have for two logs L_1, L_2:

$$\phi(\text{Tree}\,(L_1 + L_2)) = \phi(\text{Tree}\,(L_1)) + \phi(\text{Tree}\,(L_2)) \qquad \text{(Linearity)}$$

which means that the two logs are mapped independently and in particular, each trace of the log can be mapped independently. *Continuity* means that the graph topology is preserved by the mapping. i.e., if $\alpha \sqsubseteq \beta$ then there is a (possibly empty) path from $\phi(\alpha)$ to $\phi(\beta)$ in the graph, or more strongly: If there is a path of multiplicity at least $k > 0$ from α to β in the prefix tree, then there is a path of multiplicity at least k from $\phi(\alpha)$ to $\phi(\beta)$ in the graph.

We can imagine the graph produced by an LCM as a board where each node of the graph is a nail in the board and each trace is represented by a separate thread around different nails such that node and link sizes linearly represent the number of traces involved. The resulting multi-graph has exactly the structure of a Sankey diagram that represents the event log in a linear and continuous way. This intuition also corresponds to a formal property of the resulting graph where for each node, the inflow *essentially* equals the outflow. This is not literally true since in our setting, we may have traces that end in a given node. To keep track of such traces, we define for a given node $x \in X$ of the graph $G = (X, N, E)$:

$$\text{leakage}\,(x) = \sum_y E(y, x) - \sum_y E(x, y)$$

i.e., the *leakage* is the difference of inflow multiplicity and the outflow multiplicity. If ϕ is an LCM to G, we have that for each node $x \neq \phi(\epsilon)$ that is not the mapping of the root ϵ of the prefix tree that leakage $(x) \geq 0$, more precisely we have:

$$\text{leakage}\,(x) = \sum_{\phi(\sigma)=x} \phi(L(\sigma))$$

i.e., the leakage on the node corresponds to the multiplicity of traces σ of the original log that end in a tree node that is mapped to x. For the proof of this proposition, we can establish this property first for the prefix tree by structural induction on it and then use the definition of LCM to infer that the property is preserved by the LCM.

In the traditional setting of the DFG, where there is an unique artificial end node, it means that when mapping with an LCM, the sums of inflow multiplicities and outflow multiplicities must be precisely equal apart from the start and the end node. That means that the DFG issue examples shown in Sect. 2.2 are not compatible with an LCM.[2]

To summarize, we propose to map the traces and their multiplicities through an LCM for basic interpretability of insights on the aggregated trace level. The full prefix tree of a log does not generalize but is typically too large to be consumable whereas an LCM may map multiple prefixes onto the same graph node, thereby reducing complexity but at the cost of generalization. In the remainder of the paper, we study how to apply LCMs effectively in practice, i.e., how to reduce complexity in a way that still visualizes relevant insights.

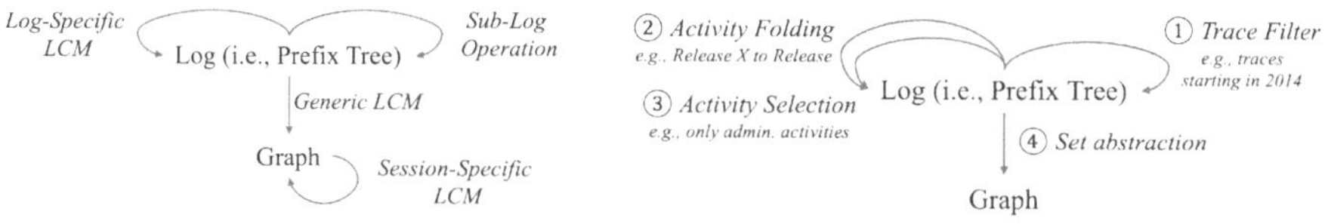

(a) Framework Overview **(b)** Framework Instance

Fig. 2. Framework overview (a) with an instantiation of a sequence of four operations for the example of the *Sepsis* log (b).

3 Framework and Prototype for Sequential Process Visualization

In our framework, the analyst can transform the analyzed process behavior at several conceptual levels in order to control and reduce complexity during exploration. We distinguish four types of transformations, each operating on a different aspect of the analysis object, cf. Figure 2a.

[2] In that traditional setting, an LCM implies *DFG-soundness* as defined by [9].

Generic LCMs. At the core of the framework are generic LCMs that construct from a given log the user-facing multi-graph that will be shown as a Sankey diagram. We consider the following example LCMs. As indicated above, in order to define an LCM, we just have to define an equivalence on the prefixes of the event log, and then the definition of the LCM will lift this to a linear continuous mapping onto a multi-graph.

In the *sequence abstraction*, two prefixes are equivalent only if they are equal. The sequence abstraction maps a log onto its prefix tree. In the *set abstraction*, two prefixes are equivalent whenever their underlying set of activities is the same. In the *bag abstraction* two prefixes are equivalent whenever their underlying bag of activities is the same. In the *last activity abstraction*, two prefixes are equivalent if their last activities in the sequence are the same. The last-activity abstraction maps a log onto its DFG. A node x in the set abstraction and the last-activity abstraction may have *self-loops*, i.e., $E(x, x) > 0$. The set abstraction and the last-activity abstractions have *self-loop-free* versions that are also linear and continuous. Intuitively, continuity is preserved by imagining that the loops are hidden inside the nodes. We will make use of that feature below. Those are canonical examples of generic maps, but many more could be imagined, e.g., we could distinguish multiplicities 1 and $k > 1$, but not between different $k > 1$. Each abstraction strikes a different balance between information maintained and complexity reduction, which we will further elaborate on in Sect. 4.

Log-specific LCMs. A log-specific LCM transforms an event log into another, more abstract event log. As indicated above, we can identify an event log with its prefix tree. As a prefix tree is a special case of a graph, an LCM can formally map from one prefix tree to another. The class of LCMs contains a wide range of potentially useful operations. Within the scope of this paper, we use the following log-specific LCMs.

An *activity selection* selects a subset $Y \subseteq X$ of activities and deletes all activities that are not in Y from each sequence. Activity selection will turn out as a powerful abstraction mechanism that effectively reduces log complexity. We will discuss in Sect. 4 how to select a useful subset of activities. Note that an activity selection may create empty sequences and the number of empty sequences could be interesting information to the analyst. In particular, it can be useful to distinguish an empty log from a log that contains only empty sequences.

An *activity folding* maps multiple activity names onto one. For example, in the Sepsis log, we can map the activities *Release A*, *Release B*, ... to the activity name *Release* in order to simplify the log. An *activity unfolding* maps a single activity name to multiple based on additional event or prefix information. For example, in the *Road Fines* log, we can map the *Payment* activity to either *Partial Payment* or *Full Payment* based on the information contained in the prefix of the event, representing whether the full amount due has been paid.

Sub-Log Operations. Like a log-specific LCM, a *sub-log* operation also transforms an event log into another event log but not by means of abstraction. Also sub-log operations can potentially reduce complexity. We distinguish *trace filters* and *sub-tracing*. Interpretability of these transformations is specific for these operations and out of scope of this paper.

Given a log L, a *trace filter* selects a sublog $L' \leq L$ by retaining only traces satisfying certain conditions. Various case filters are readily available in commercial tools. In the context of complexity reduction, there is most notably the generic *variant threshold filter* [17] which retains only the traces of the top-k sequential variants. The implementation of that filter is expected to return an exact sub-log. Less commonly available is *random case sampling*, which can also reduce complexity while retaining some properties of the variant distribution. Such a random case sampling could be designed to either deliver an exact sub-log or sub-log with random multiplicities, each of which might be useful in different contexts.

A *sub-tracing* changes the notion of a trace, i.e., the case identifier according to which events are grouped into traces. In particular, we can focus on a subprocess of the original process and hence substantially reduce complexity. For example, in the BPIC17 loan application log, each application is associated with one or more loan offers. Each loan offer has its own identifier, which allows us to zoom into the subprocess of an offer lifecycle.

Session-specific LCMs. The visualized graph can also be adjusted by the analyst through local graph operations that maintain the LCM property. This is the case whenever the local graph operation is a multi-graph morphism. A useful example operation is *merging* two or more *nodes* in the graph. The merged nodes are made equivalent changing the equivalence of the underlying LCM by coarsening but keeping linearity and continuity of the resulting map.

Together, these four transformation types provide a rich combinatorial tool set to reduce complexity while maintaining interpretable relationships between the original event log and the resulting visualization. Figure 2b illustrates a sequence of four operations on the *Sepsis* event log as an exemplary usage of the framework. Note that the sequential composition of the LCMs is again an LCM.

3.1 Design and Prototypical Implementation

Our implementation runs as a standalone web application as well as an interactive visualization inside a Jupyter notebook. An overview of the user interface (UI) is shown in Fig. 3. The UI includes with activity selection and activity folding the most important log-specific LCMs whereas the Jupyter notebook allows a data scientist to code additional LCMs on the fly using specific coding patterns (see Jupyter notebook in [1] for an example.) Three generic LCMs are implemented, which can be selected from the UI, cf. Fig. 3. The code is extensible to implement more generic LCMs.

In the Sankey diagram, links are colored by activity to see multiple occurrences of the same activity, which can happen in the set and the sequence abstractions. Nodes are colored based on their leakage: white = start node, light grey = zero leakage, dark grey = small, non-zero leakage, and black = high leakage. This feature was designed to allow the analyst to quickly see leakage, and the

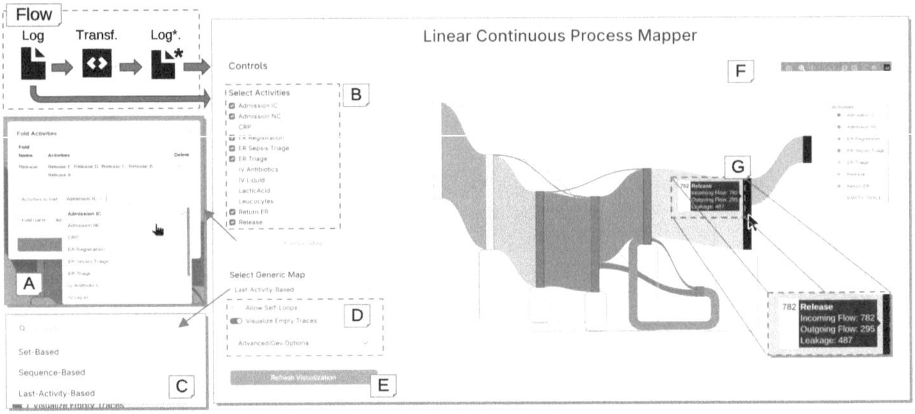

Fig. 3. Usage flow and UI of the application. Users can either provide an event log as-is or apply further transformations in a Jupyter notebook (Flow). The UI supports folding activities (A) and activity selection (B). Furthermore, users can select from the three generic LCMs: sequence-, set-, or last-activity abstraction (C) and choose whether to visualize self-loops and empty traces (D). After selecting the desired visualization controls, they can refresh the visualization (E), which refreshes the Sankey diagram (F). Here, they can freely reposition nodes to adjust the visualization to their needs. More details about edges and nodes are revealed on hover (G).

tool tip provides precise leakage numbers. The default Sankey layout is based on DFS (Depth-First-Search). Figure 4 shows the implemented generic abstractions for a simple instructive example log.

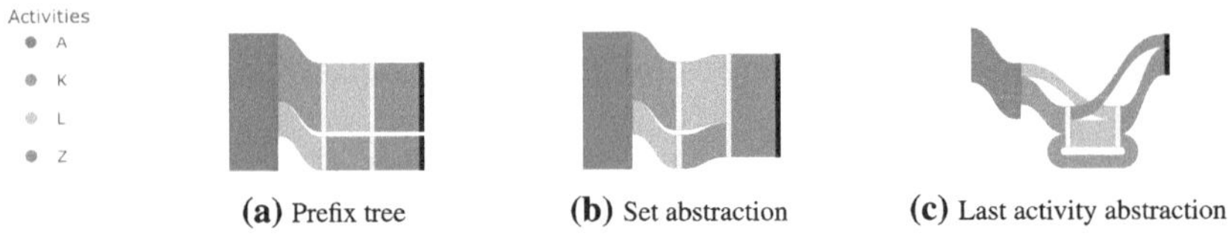

Fig. 4. Implemented generic LCMs for the log $L = A, AKLZ, AKLZ, ALKZ$

4 Evaluation

To evaluate the proposed framework, we apply it to several real-world event logs, including *Road Fines, Sepsis, BPIC12, BPIC17*, and *Domestic Declarations*. The evaluation addresses the following research question:

> *Evaluation Question:* How and to what extent does the framework allow us to reduce event log complexity in a way that visualizes relevant behavioral insights?

To evaluate this question, we took several aforementioned event logs and used specific *heuristics* to apply the framework to visualize behavioral insights. We proceed by discussing those application heuristics, then evaluate the complexity reduction achieved by their application, and finally present recurring behavioral insights across the logs that those applications produced.

LCM Application Heuristics. The primary LCM that we applied is *activity selection.* To apply activity selection effectively in practice, we exploit the heuristic that many business processes contain a main subprocess that reflects the key decisions and outcomes of each case. Such a subprocess often corresponds to the lifecycle of the primary business object, for example a contract, purchase order, loan application, or claim. Based on domain knowledge, the analyst can therefore attempt to select activities that represent the most important lifecycle states or transitions of this primary business object. We do not assume that the analyst can determine the optimal activity subset from the outset. Instead, the visualization is implemented to be reactive to changes in the selected activity set, allowing analysts to start with an initial guess and iteratively refine the selection based on the feedback provided by the visualization.

Complementing the heuristic of a primary subprocess or object lifecycle, we also present results based on secondary subprocesses as well as further simplified activity selections. These simplified views focus on specific aspects of sequential behavior, such as the overall distribution of case outcomes or the relationship between two decision points. In addition, we demonstrate how simple log-specific LCMs can further enhance the interpretability of the visualization. These abstractions allow us to systematically explore the event logs while controlling the trade-off between complexity reduction and behavioral insights.

4.1 Complexity Reduction

Several combinations of LCMs delivered sufficient complexity reduction and behavioral insights. To understand the effectiveness of each LCM, we measured the contribution of each LCM to the overall complexity reduction. Table 1 shows the complexity of each abstraction in terms of the number of *end nodes*, which we define here as a node with positive leakage. This metric takes complexity as the number of variants: in the tree, i.e., sequence abstraction, the number of end nodes equals the number of sequential variants. This is reduced in the set abstraction to the number of *set variants* where each set variant represents the equivalence class of sequential variants that share the same set of activities. In the last-activity abstraction, an end node represents the equivalence class of sequential variants that end with a specific activity. Given a suitable activity selection, the last activity often represents a process outcome.

Table 1 shows that the primary driver for complexity reduction is activity selection, in particular when the log is complex due to multiple partially independent subprocesses (there are no independent subprocesses in *Domestic Declarations*). Furthermore, we see that after the striking reduction through activity selection (column *Sequence-Subset*), the resulting number of sequential variants can still be fairly high. In those cases, set and last-activity abstraction deliver a sufficient final reduction, where last-activity reduction still reduces complexity significantly with respect to the set abstraction (with some loss of information, as we will see below.) Note that our additional first transformations change complexity only slightly (column *Sequence-Transf.*).

Table 1. Number of end nodes x (leakage $(x) > 0$) comparing full log with activity subsets under sequence-, set-, and last-activity abstraction, with potential first transformation: *Road Fines* – unfolding: *Payment* is unfolded to partial or full payment, appeal events are unfolded into *dismissals* and unsuccessful appeals. Sepsis – folding: all *"Release *"* activities are folded into a single *Release*. BPIC17 – sub-tracing: only offer object changes are selected and events are re-grouped into traces by offer id.

			Number of end nodes				
			Sequence			Set	Last Act.
Log	**Transf.**	**Subset**	**Full Log**	**Transf.**	**Subset**	**Subset**	**Subset**
Road Fines	–	Fine	231	–	48	7	4
		Prefecture Appeal		–	16	8	4
	Unfolding	Fine		315	89	20	6
Sepsis	Folding	Admin	846	842	58	9	6
Dom.Decl.	-	Declaration	63	–	62	29	4
BPIC12	–	Application	4366	–	17	12	8
		Application Sec.		-	12	12	6
BPIC17	-	Application	15930	–	102	24	6
		Offer		–	877	28	6
	Sub-tracing	Offer Sec.		12	12	12	3

4.2 Behavioral Insights

In this section, we present representative examples of behavioral insights that we were able to visualize with our framework.

Mainstream vs. Exceptional Behavior. In the Sankey diagram, the mainstream behavior is prominent as a sequence of thick links as opposed to thin links representing the exceptional behavior. All our example figures below show this clearly. For example, even though the Sepsis log exhibits a very high number of sequential variants, with no variant having a dominating number of traces, the administrative subprocess shown in Fig. 5 (a and b) reveals that the process nonetheless displays a dominant regular behavior that is clearly separated from several exceptional sequences.

Knockout Structure. Many processes exhibit a *knockout structure*, meaning that the process consists of a sequence of stages at which a knockout may occur that terminates the process. Knockout structure becomes visible in the last-activity

(a) Last activity abstraction **(b)** Set abstraction

Fig. 5. The administrative subprocess of the Sepsis log. Dominant main path with sizable rework (clearer in b) as multiple occurrences of *Admission NC*; Exceptional orderings (thin paths, clearer in b) of registration, triage and admission events; Sizable leakage after triage (patients that never get admitted to stationary care NC or IC); Also shows the proportion of first admission NC vs IC.

abstraction in the Sankey with DFS layout. For example, in a loan approval process (BPIC12, BPIC17), cf. Figure 6, a loan decline or customer cancellation ends the process, while in a Road Fine Management process, a *Payment* typically ends the process, cf. Figure 7a. In both examples, we see that the selection of a primary subprocess represents the knockout structure. The Sankey clearly shows the relative knockout contribution at each stage, e.g., the amount of cases that are paid at each position in the process, including those that get paid only after the penalty deadline. Similarly, in the BPIC17, and BPIC12 loan application objects (Fig. 6), we can see how many cases are declined or canceled late in the process. Such knockouts generate processing cost without yielding business value. The visualization therefore highlights potential opportunities for shift-left analysis (a.k.a. over-processing analysis), where earlier identification of such cases could reduce unnecessary processing effort. A similar multi-stage filtering may occur in approval processes, where a decline might not terminate the process but leads to a re-submission as in the *Domestic Declarations* log, cf. Supplement, Fig. 9.

Flow Leakage. In some processes, execution appears to stop prematurely, pointing to potential opportunities for deeper investigation. Such traces are explicitly represented as leaking nodes, colored in dark grey, or sometimes in black (in both the set abstraction and the last activity abstraction), with precise number of leaking traces available as tool tip. In the *Road Fines* log, this reveals a large number of traces that stop after the *Send Fine* activity, as can be seen in Fig. 7. In the appeal subprocess of the same log, cf. Supplement Fig. 10, leakage highlights many traces that stop after the appeal has been sent. Similarly, in the Sepsis log, a substantial number of traces stop after the *Sepsis Triage* activity, cf. Figure 5.

Alternative and Unexpected Ordering. The set abstraction directly visualizes alternative ordering of subsequent activities as multiple paths that end in the same node. For example, in the Sepsis log, cf. Figure 5b shows the main ordering

(a) BPIC12 **(b)** BPIC17

Fig. 6. Knockout structure and outcome distribution of loan application processes in last activity abstraction: (a) BPIC12 exhibiting high share of declined loans, substantial percentage of late declines and late cancellations; (b) BPIC17 exhibiting rework issues and substantial number of cases unexpectedly transiting from *Incomplete* to *Pending*.

(a) Last activity abstraction **(b)** Set abstraction

Fig. 7. The fine subprocess of the Road Fines log: Clear knockout structure (a); several exceptional, unexpected paths including *Create Fine, Payment, Send Fine* visible in (a); unexpected substantial leakage after *Send Fine*; rework as double payment, which is revealed in (b) to be more frequent in late than in early payments.

of the first three activities *ER Registration, ER Triage*, and *ER Sepsis Triage* with several exceptional paths where these three activities are ordered differently. Note that the same alternative ordering appears as back edges in the last-activity abstraction, which form cycles that cannot be distinguished from rework and hence are harder to interpret, cf. Figure 5a. Similarly, the Road Fines appeal subprocess shows several exceptional orderings that are alternative to the main flow, cf. Supplement Fig. 10. As usual, unexpected directly-follow steps show up in the DFG, i.e., last-activity abstraction. However, the corresponding set abstraction can reveal additional information due to less generalization. For example, in *Road Fines*, a *Send Fine* can exceptionally appear after a *Payment*. While the last-activity abstraction shows that, the set abstraction in addition shows that this happens only after a *Payment* directly after *Create Fine*, cf. Figure 7b.

Rework. Loops in the DFG can point to rework, but in general do not reliably indicate nor quantify rework as discussed above. In some cases however, rework is indicated already in the Sankey visualization of the DFG by help of the coloring and the DFS-layout. For example, in the BPIC17 application object, cf. Figure 6b, there is a prominent thick yellow back edge from the node representing *A_Incomplete* to the node representing *A_Validating*. Since we can see that the latter node can only be reached from the start node by going through the former, the back edge implies the repetition of *A_Validating* within the same trace, i.e., rework. To see rework precisely and always unambiguously, we can switch to the set abstraction. In the set abstraction, rework precisely manifests as a self-loop and the size of the self-loop quantifies the rework (e.g., Supplement Fig. 11 repetition of *A_Validating*, Fig. 7b repetition of *Payment*, and Fig. 5b repetition of *Admission NC*). Further detail about rework, viz. the number of traces that exhibit a specific number of repetition can be visualized by selecting the involved activities in the sequence abstraction, cf. e.g., Supplement Figs. 12 and 13.

Outcome Distribution and Predictions. If process outcomes are manifested in activity labels, which can also be lifecycle states as in the BPIC12/17 logs, then the selection of those activities directly yields a visualization of the outcome distribution of the log, e.g., cf. Supplement Fig. 14. Our graph visualization also shows whether each trace has exactly one outcome defined, i.e., we can see how many traces have no outcome (through the empty traces feature) and how many cases have more than one outcome defined. Also intermediate outcome distributions, such as whether a patient is first admitted to normal care (NC) or intensive care (IC) or the distribution of any filter stage in a knockout structure can be isolated in a visualization by selecting the corresponding activity subset.

If a process outcome is not manifested as an activity label but only implicit in the attributes of the trace prefix, a log-specific LCM might help. For example, for *Road Fines*, an activity unfolding can distinguish partial payment events from full payment events and another activity unfolding can manifest dismissals by judge or prefecture (cf. demo in Jupyter notebook provided in code repo [1]). The resulting outcome distribution is provided in Supplement Fig. 15. There, we see besides very few traces that have more than one outcome, also a large number of traces with no outcome defined.

Furthermore, if we select two decisions, i.e., two sets of outcomes manifested as activity labels, we can see how well one outcome predicts another. For example, Fig. 8 shows that the presence of *Release B* predicts the absence of *Return ER* with perfect precision and substantial recall whereas none of the other release codes predicts the presence or absence of *Return ER* with substantial precision. Supplement Fig. 16 shows another prediction example.

Transactional Behavior. Transactional behaviors occur when multiple state changes consistently occur together in a trace. Figure 8b shows a transaction with three state changes that always occur together but in different order. The detection of such a transaction is important feedback for the analyst, since they can henceforth simplify the visualization by selecting only one of the state changes in the transaction. Transactions may appear as subgraphs with no/minimal internal leakage and no/minimal flow from the inside of the subgraph to the outside of the subgraph. They can be verified by selecting only the participating activ-

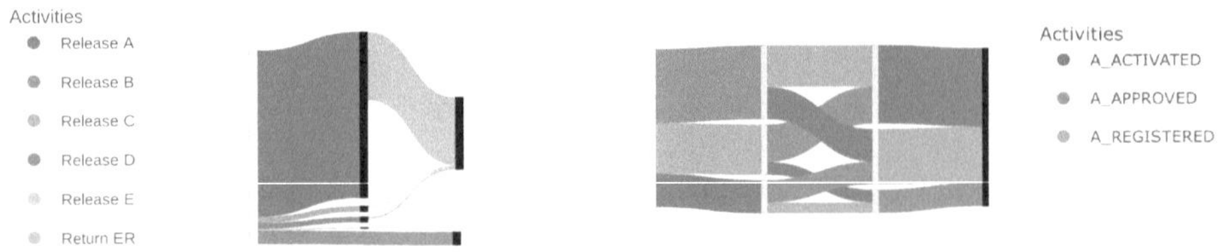

(a) Prediction in *Sepsis* **(b)** *A transaction in BPIC12*

Fig. 8. (a) Prediction of *Return ER* from the release code in the *Sepsis* log (last activity abstraction); (b) a transaction involving three state changes in the BPIC17 log (set abstraction).

ities. In fact, a transaction might not be clearly visible unless any interleaving activities that do not participate in the transaction are de-selected. Transactions may exhibit few exceptions due to noise (e.g. *CRP + Leucocytes* as well as *ER Registration + ER Triage + ER Sepsis Triage* in the *Sepsis* log). Unlike Fig. 8b, a transaction may also show consistent ordering (e.g. *A_Complete + O_Created* in BPIC17 and *Insert Fine Notification + Add Penalty* in *Road Fines*).

Sequential Variant Distribution. As discussed in Sect. 4.1, the prefix tree even of a subset of activities may be too large to be inspected closely. However, the tree gives an impression of the distribution of the sequential variants and their prefixes, e.g. whether it is compatible with a power law distribution, whether it has a dominating variant or prefix and whether the variation is created by non-deterministic ordering, leakage, or variable rework multiplicities. Supplement, Fig. 17 shows the prefix tree of the fine subprocess of the *Road Fines* log. If the number of activities is small enough, the prefix tree can also be used to resolve ambiguities that occur in the set or last-activity abstractions due to generalization, as we have exemplified this for rework above, cf. Supplement Figs. 12 and 13.

4.3 Conclusion, Limitations, and Threats to Validity

We have applied the framework to multiple real-world event logs and presented recurring behavioral patterns, which gives first evidence for the utility of our framework to find and communicate such patterns. The findings are limited by the selection of the event logs, which might not represent the full diversity of process characteristics encountered in practice. Furthermore, the evaluation identifies behavioral insights qualitatively, and the relevance of these insights was assessed by the authors rather than by independent analysts or predefined task protocols. Similarly, the evaluation does not measure analyst performance to find such behavioral patterns, which must be subject of future work. Part of that is the inter-subjective interpretability of the visualizations, which should be a dedicated evaluation goal of such controlled user studies. As the current formalization of LCM captures only necessary and not sufficient conditions for interpretability, such studies of interpretability can lead to further refinements of our framework. Furthermore, although the framework was successfully applied to large logs such as BPIC17 and Road Fines, the scalability limits of the visualization itself were not systematically evaluated.

5 Related Work

As described above, van der Aalst [17] has exemplified but not formally characterized interpretability problems of the DFG whereas Leemans et al. [9] have formalized a requirement that is weaker than LCM. Both works also discuss DFG limitations for discovery of a predictive model as well as performance mining. Also in the context of predictive discovery, Conforti et al. [4] propose a technique to filter out infrequent behavior from event logs as a preprocessing step for process discovery measuring success by increased predictive performance. Like Leemans et al. [9], Campa et al. [3] require that the filtered graph must remain a sound spanning subgraph of the original DFG, and they study how to efficiently compute a filtered DFG that maximizes the retained edge weights under this constraint. As discussed above, LCM abstraction implies such DFG soundness without the need of additional computation to enforce it.

To address limitations of DFGs, Montana et al. [13] propose to complement the DFG with a sequence alignment visualization where the alignment positions are user controlled. No additional abstractions are provided or evaluated. This is different from our approach where the alignment is always at the start (prefix-oriented) and insight is obtained from combining various abstractions.

Recent approaches extend visualization from sequences to partial orders, e.g. [14] but focus on individual instances with some extensions to local patterns such as repetitions without yet aiming at precise high-level aggregations comparable to the DFG. A broader overview over other process-mining related visualizations is provided by van den Elzen et al. [6].

Our requirements connect to established principles of visualization design. Mackinlay's expressiveness criterion requires a graphic to encode all and only the facts in the data [10]; tactical DFG filters violate this by making part of the data disappear. Relatedly, such filters are a form of design variation, against Tufte's maxim to "show data variation, not design variation" [16]. Kindlmann and Scheidegger [8] formalize the underlying intuition as *Visual-Data Correspondence*: a DFG that hides a significant change in the data for some filter setting has *jumbled* the data. Left unchecked, such behavior produces what McNutt et al. [11] call a visualization mirage: a reading that appears supported by the data but is invalidated on closer inspection. The lesson we draw, building on Elmqvist and Fekete [5], is that complexity reduction should not be mere visual removal but an aggregation that remains interpretable in visual space; our linearity and continuity requirements make this guarantee precise for process maps.

From the standpoint of the visualization community, our work falls within visual summarization techniques for event-sequence data [7]. The closest line of work is LifeFlow [18], which is essentially an icicle plot-inspired visualization of prefix trees of event sequences, the same base structure our abstractions map from. Its successor, EventFlow, added interval events and a visual query language, and subsequent work investigated sequence simplifications [12]. While our tool also allows for heuristic sequence simplifications similar to EventFlow, our framework additionally reduces complexity through a family of linear continuous abstractions that preserve, by construction, a clearly defined relationship between the visualization and the underlying log.

References

1. Online repository. https://github.com/ics-unisg/linear-continuous-process-mapper/tree/vBPM26
2. Supplementary material (part of repository). https://github.com/ics-unisg/linear-continuous-process-mapper/blob/vBPM26/paper-supplementary-material.pdf
3. Chapela-Campa, D., Dumas, M., Mucientes, M., Lama, M.: Efficient edge filtering of directly-follows graphs for process mining. Inf. Sci. **610**, 830–846 (2022). 10/gspqz4
4. Conforti, R., Rosa, M.L., Hofstede, A.H.t.: Filtering out infrequent behavior from business process event logs. IEEE Trans. Knowl. Data Eng. **29**(2), 300–314 (2017). 10/f9tv8c
5. Elmqvist, N., Fekete, J.D.: Hierarchical aggregation for information visualization: overview, techniques, and design guidelines. IEEE Trans. Visual Comput. Graphics **16**(3), 439–454 (2010). 10/db2gxw
6. van den Elzen, S., et al.: Towards multi-faceted visual process analytics. Inf. Syst. **133**, 102560 (2025). 10/qwdx
7. Guo, Y., Guo, S., Jin, Z., Kaul, S., Gotz, D., Cao, N.: Survey on visual analysis of event sequence data. IEEE Trans. Vis. Comput. Graph. **28**(12), 5091–5112 (2022). 10/gnm3xc
8. Kindlmann, G., Scheidegger, C.: An algebraic process for visualization design. IEEE Trans. Visual Comput. Graphics **20**(12), 2181–2190 (2014). 10/f6qjgj
9. Leemans, S.J.J., Poppe, E., Wynn, M.T.: Directly follows-based process mining: Exploration & a case study. In: Conference on Process Mining (ICPM 2019), pp. 25–32 (2019). 10/qvzj
10. Mackinlay, J.: Automating the design of graphical presentations of relational information. ACM Trans. Graph. **5**(2), 110–141 (1986). 10/dxdkkp
11. McNutt, A., Kindlmann, G., Correll, M.: Surfacing visualization mirages. In: Proceedings of the 2020 CHI Conference on Human Factors in Computing Systems, pp. 1–16. ACM, Honolulu HI USA (2020). 10/gnfdqp
12. Monroe, M., Lan, R., Lee, H., Plaisant, C., Shneiderman, B.: temporal event sequence simplification. IEEE Trans. Visual Comput. Graphics **19**(12), 2227–2236 (2013). 10/f5h3vn
13. Montana, L., Resinas, M., Villa-Uriol, M.C.: Addressing directly-follows graphs limitations with visualization of event sequences. In: VIPRA 2025, BPM Workshops. LNBIP (In press)
14. Siddiqui, A., Zerbato, F., Schuster, D.: Visualizing repetition in process execution variants from partially ordered event data. Inf. Syst. **138**, 102664 (2026). 10/qwdz
15. Suriadi, S., Wynn, M.T., Ouyang, C., ter Hofstede, A.H.M., van Dijk, N.J.: Understanding process behaviours in a large insurance company in australia: a case study. In: Advanced Information Systems Engineering, pp. 449–464 (2013). 10/qvzb

16. Tufte, E.R., Graves-Morris, P.R.: The visual display of quantitative information, vol. 2. Graphics press Cheshire, CT (1983)
17. van der Aalst, W.M.: A practitioner's guide to process mining: limitations of the directly-follows graph. Procedia Comput. Sci. **164**, 321–328 (2019). 10/qvzd
18. Wongsuphasawat, K., et al.: LifeFlow: visualizing an overview of event sequences. In: Proceedings of the SIGCHI Conference on Human Factors in Computing Systems, pp. 1747–1756. ACM, Vancouver BC Canada (2011). 10/d79ksg
19. Zerbato, F., Soffer, P., Weber, B.: Process mining practices: evidence from interviews. In: 20th International Conference on Business Process Management (BPM 2022), vol. 13420, pp. 268–285 (2022). 10/qvzh
20. Zerbato, F., Zimmermann, L., Vrotsou, K., Weber, B.: From analysis to findings: How do process mining analysts discover results? Inf. Syst. **135**, 102596 (2026). 10/g9z9gj
21. Zimmermann, L., Zerbato, F., Weber, B.: What makes life for process mining analysts difficult? A reflection of challenges. Softw. Syst. Model. **23**(6), 1345–1373 (2024). 10/qvzc

Continuity Matters: Improving Predictive Process Monitoring Based on Path Coherence

Maximilian Vouk[1(✉)], Thomas Grisold[1], Kate Revoredo[2], and Jan Mendling[1,2]

[1] Vienna University of Economics and Business, Vienna, Austria
maximilian.alexander.vouk@s.wu.ac.at, thomas.grisold@wu.ac.at

[2] Humboldt-Universität zu Berlin, Berlin, Germany
{kate.revoredo,jan.mendling}@hu-berlin.de

Abstract. Predictive Process Monitoring aims to forecast the future behavior of ongoing cases based on historical event log data. It is well established that predictive performance suffers when models treat cases and events as strictly isolated entities. While various approaches have attempted to integrate operational context to mitigate this limitation, the social dynamics in and around organizational processes have not been addressed in depth. We address this gap by drawing on research from the organizational science, specifically on research around organizational routines and operational friction. Specifically, we focus on the coordination friction and delays introduced when the execution context–such as the assigned resource or physical location–changes between consecutive events. This paper incorporates the concept of path coherence to explicitly quantify this situational continuity, as well as the friction that occurs when continuity breaks. We adapt this concept to event logs by engineering path coherence as an event-level feature and integrating it into a prediction pipeline based on Long Short-Term Memory models. Two variants are proposed: a resource-based coherence feature capturing responsibility handoffs and an all-attribute variant reflecting broader situational change. Using eight real-world event logs, we compare coherence-based feature sets with benchmark feature sets. The results show that coherence-based feature sets consistently reduce Mean Absolute Error, yielding statistically significant, mostly medium effect sizes, and that resource-based coherence explains the majority of the predictive performance gains.

Keywords: Predictive Process Monitoring · Remaining Time Prediction · Path Coherence · Algorithm Engineering

1 Introduction

Predictive Process Monitoring (PPM) aims to generate predictions about the future unfolding of ongoing cases in a business processes based on data captured in *event logs*. One specific type of prediction is the prediction of *remaining time*

F. M. Maggi et al. (Eds.): BPM 2026, LNCS 16936, pp. 594–610, 2027.
https://doi.org/10.1007/978-3-032-37777-7_35

until the completion of the respective case [11,26]. These predictions deliver business value by foreseeing delays in the execution of the process to proactively allocate additional resources, or to provide customers with estimates of expected waiting times.

Predictive models for remaining time conventionally encode execution history by treating the process as a sequence of structurally independent occurrences [22,27]. While this approach captures the isolated state of an event at a discrete point in time, it creates a fundamental methodological limitation: a systematic blindness to friction in coordination. By evaluating continuous execution flows as disconnected steps, current models fail to account for the inherent setup costs and operational overhead generated when execution contexts (e.g., resource assignments or physical locations) shift between consecutive events. The assumption that there are frictionless transitions obscures critical predictive signals: a sequence of tasks executed continuously by the same actor carries a fundamentally different temporal profile than a sequence delayed by constant handoffs and knowledge transfers. As Kim et al. [9] have recently demonstrated, minimizing this contextual friction through continuous execution directly impacts process resilience and speed. Consequently, there is an opportunity for PPM to overcome this isolated event representation to explicitly quantify and model the coordination friction inherent in organizational handoffs.

In this paper, we propose a theory-informed approach to feature engineering by utilizing the concept of *path coherence* [9,18] as a predictive transition signal. Path coherence describes the continuity of contextual attributes (e.g., the resource or location) between successive events. Drawing from this concept and corresponding research in the organizational sciences, allows us to leverage theory about social dynamics that are inherent in business processes [13]. Adopting a view of coordination in business processes enables us to attend to factors in business process dynamics that have not yet been considered systematically, even though they are core components of any work context involving more than one actor [18].

We operationalize path coherence as a feature that can be engineered from event log data and be employed as a predictive signal in the model's input. We empirically evaluate the efficacy of two path coherence-based feature configurations: one capturing continuity of the assigned resource, one capturing the sum of the available contextual attributes from a given event log. The evaluation is performed across eight real-world event logs using Long Short-Term Memory (LSTM) models. For each event log we compare the resulting predictive accuracy to benchmark feature sets that capture the same information from a static perspective. Our results show that the path coherence-based feature representation yield statistically significant, superior predictive accuracy to the benchmark feature representations across most event logs, as well as that the majority of the performance gain is captured by the resource-based path coherence feature.

The remainder of this paper is organized as follows. Section 2 introduces the theoretical foundation of path coherence in organizational routines and reviews related work from PPM. Section 3 outlines our proposed approach for a path coherence-based feature design. Section 4 outlines the setup for the proposal's evaluation and presents the corresponding results, which are discussed in Sect. 5. Section 6 concludes the paper.

2 Background

In this section, we present the background of our work. Firstly, we discuss the theoretical background on handoffs and path coherence from organizational science, followed by a discussion of related work with a focus on the representation of context.

2.1 Handoffs and Path Coherence in Organizational Routines

This section first establishes the theoretical relevance of these handoffs, and subsequently introduces the general definition of path coherence.

Handoffs are commonly understood as the transfer of work, responsibility, or control from one person to another in the course of an organizational process. In healthcare, for example, they are defined as "the process of transferring primary authority and responsibility for providing clinical care to a patient from one departing caregiver to one oncoming caregiver" [17]. Such person-to-person handoffs are considered critical points of vulnerability where errors, information loss, and breakdowns in continuity of care or service can occur [28]. However, handoffs are an inevitable part of organizational work: any division of labor is based on the idea that work steps are carried out across actors [21]. As work crosses functional, temporal, geographic, or technological boundaries, handoffs provide the mechanism that re-integrates distributed activities into coherent outcomes, making them central both to performance and to risk in complex sociotechnical systems [1]. Research on organizational routines–broadly understood as a concept from the organization sciences that is equivalent to business processes [13]–has emphasized the role of handoffs, precisely because routines involve "patterns of interdependent action carried out by multiple actors" [5].

Pentland et al. [18] point out that handoffs can take on different forms. They extend the notion of handoffs beyond the transition between two resources, to the similarity between two consecutive events. This perspective matters because it decouples the analysis of handoffs from a purely resource-centric view and instead treats handoffs as relations between social events characterized by multiple attributes (e.g., who, what, where, which system). In doing so, this work reveals a richer spectrum of handoffs, such as handoffs between systems, between actions, or combinations thereof, that remain invisible if one only considers person-to-person transfers, and it enables systematic assessment of what changes and what remains stable across the flow of work. Pentland et al. operationalize this notion via *(path) coherence* as a coefficient of similarity of two events (see Eq. 1), capturing the extent to which attributes are preserved across a handoff and thereby providing a quantitative handle on the continuity of organized work.

They note that path coherence might be a suitable predictor for the future state of a work system.

$$Path\ Coherence = \frac{Number\ of\ unchanged\ attributes}{Total\ attributes} \tag{1}$$

Drawing on this work, Kim et al. [9] use path coherence to predict the persistence of paths in narrative networks in the context of a disruption in an IT system of a medical facility. They assess the path coherence of all edges in the narrative network describing the routines in the facility before the disruption based on the involved resource and the location of the activity. They then show that edges with high coherence (i.e., same resource and/or location) are significantly more likely to persist after the disruption, while edges with low path coherence are more likely to dissolve. Thereby, they demonstrate the predictive power of path coherence on the level of the narrative network. In light of these results, it seems promising to integrate the idea of coherence from the organizational sciences into process prediction.

2.2 Related Work

The current state of feature encoding in PPM reveals a methodological gap regarding the representation of coordination dynamics within the process execution context. While initial approaches [26] relied solely on the control-flow perspective, modern methodologies incorporate the broader execution context. These contextual encodings can be systematically categorized into three distinct scopes: the local event context, the intra-case process context, and the global inter-case context.

At the local event level, input data to the predictive models is enriched with the event's intrinsic payload. Tax et al. [25] introduce specific representations of the timestamp. Evermann et al. [4] add an encoding of the involved resources. Navarin et al. [15] subsequently incorporate all further event attributes as categorical or numerical features.

At the intra-case level, the literature moves beyond treating resources as simple categorical features by encoding their underlying behavior within a specific process execution. Kim et al. [10] explicitly model resource experience by engineering features that reflect the individual's historical interactions with a given task.

At the global inter-case level, the input scope expands beyond the isolated process instance to encompass the broader organizational environment. Senderovich et al. [24] establish that isolated case data is insufficient, explicitly engineering features to capture interactions between concurrent cases of similar types. Pourbafrani et al. [20] further engineer features capturing inter-case dynamics, such as batching or queuing in segments with high uncertainty.

Collectively, these approaches capture varying scopes of the process execution context through specific features. However, across all these levels, existing literature obscures the social dynamics dynamics between consecutive events. After all, business process executions are relational to the extent that they include one than more actor, and frictions are an inevitable part whenever work is passed on from one actor to the next. It is surprising, then, that no approach has addressed the semantic aspect of capturing the coordination friction inherent to (resource) handoffs.

3 Proposed Approach

In this section, we present our proposed approach to operationalize path coherence as an event-level feature, enabling its integration into PPM techniques. As originally defined by Pentland et al. [18] (presented in Sect. 2.1), path coherence inherently characterizes the transition (or edge) between two consecutive events. However, established PPM approaches predominantly operate on event-level attributes. To bridge this conceptual gap, we associate the path coherence value of a transition with the second event of that transition. In a predictive setting, this mapping remains consistent, as the second event represents the exact point in time when all requisite attributes for computing path coherence become observable. For the initial event of a trace, where no prior transition exists, we establish a baseline coherence of 1.

3.1 Preliminary Definitions

To establish a formal basis for our proposed approach, this section introduces the underlying concepts of events, traces, prefixes, and their feature encodings [16,27].

Let $\mathcal{U}_{ev}$ denote the universe of events. $e \in \mathcal{U}_{ev}$ is an event, with $\#_{case}(e) \in \mathcal{U}_{case}$ the case of the event, $\#_{act}(e) \in \mathcal{U}_{act}$ its activity, $\#_{time}(e) \in \mathcal{U}_{time}$ its timestamp, and $\#_{attr}(e) : \mathcal{U}_{attr} \not\rightarrow \mathcal{U}_{aval}$ a partial function capturing an arbitrary number of additional event attributes.

Let $\mathcal{U}_{ev}^*$ denote the set of all finite sequences over the event universe $\mathcal{U}_{ev}$. A trace $\sigma = \langle e_1, \ldots, e_n \rangle$ is a sequence events such that $\forall i, j \in \{1, \ldots, n\}$, $i \leq j, \#_{time}(e_i) \leq \#_{time}(e_j)$ and $\forall i, j \in \{1, \ldots, n\}, \#_{case}(e_i) = \#_{case}(e_j)$. For any trace σ of length n, and any integer k such that $0 \leq k \leq n$, a prefix of length k is defined as the sequence $\sigma^k = \langle e_1, e_2, \ldots, e_k \rangle \in \mathcal{U}_{ev}^*$.

To apply predictive analytics to this raw sequential data, the variable-length sequences must be transformed into a structured feature format. Let $\mathcal{U}_{fval}$ denote the universe of feature values (e.g., continuous, categorical, or boolean data). A feature function $f : \mathcal{U}_{ev}^* \rightarrow \mathcal{U}_{fval}$ is a mapping that assigns a distinct, quantifiable characteristic to a trace prefix $\sigma^k \in \mathcal{U}_{ev}^*$. The resulting feature value $v = f(\sigma^k)$ encodes sequential event data to reflect the status of the process execution at $\#_{time}(e_k)$.

3.2 General Path Coherence Feature

Building upon these foundational definitions, we formalize the operationalization of path coherence as a feature. To ensure a structured integration into sequence-based models, we construct this feature sequentially across three distinct levels of abstraction: the attribute-level indicator, the event-level aggregation, and the prefix-level sequence encoding.

Attribute-Level Indicator. First, we define a binary indicator to evaluate the coherence of a single, specified attribute across a transition. Let $e_i, e_j \in \mathcal{U}_{ev}$ be two consecutive events, and let $a \in \mathcal{U}_{attr}$ be a specified event attribute. The attribute coherence indicator function $\mathbf{1}_{coh} : \mathcal{U}_{ev} \times \mathcal{U}_{ev} \times \mathcal{U}_{attr} \to \{0, 1\}$ evaluates whether the specified attribute maintains its value across the two events:

$$\mathbf{1}_{coh}(e_i, e_j, a) = \begin{cases} 1 & \#_{attr}(e_i)(a) = \#_{attr}(e_j)(a) \\ 0 & \text{otherwise} \end{cases} \tag{2}$$

Event-Level Aggregation. Subsequently, we aggregate this indicator into a *event-level* feature function. Let $A \subseteq \mathcal{U}_{attr}$ be a fixed, non-empty set of event attributes ($A \neq \emptyset$). The path coherence feature $f_{coh.A} : \mathcal{U}_{ev}^* \to [0, 1]$ (where $[0, 1] \subseteq \mathcal{U}_{fval}$) assigns a continuous feature value to any trace prefix $\sigma^k \in \mathcal{U}_{ev}^*$. This value represents the proportion of attributes within A that remain constant between the penultimate and final events of σ^k:

$$f_{coh.A}(\sigma^k) = \begin{cases} 1 & \text{if } k < 2 \\ \frac{1}{|A|} \sum_{a \in A} \mathbf{1}_{coh}(e_{k-1}, e_k, a) & \text{if } k \geq 2 \end{cases} \tag{3}$$

Prefix-Level Sequence Encoding. For a *complete prefix* σ^k, this feature is evaluated at each event by applying $f_{coh.A}$ to the ordered set of all its continuous prefixes $\{\sigma^1, \sigma^2, \ldots, \sigma^k\}$. The resulting collection of discrete feature values $\{f_{coh.A}(\sigma^1), f_{coh.A}(\sigma^2), \ldots, f_{coh.A}(\sigma^k)\}$ captures the evolving state of path coherence at every step $k \in \{1, 2, \ldots, k\}$. This collection can subsequently be encoded into sequential vectors, index-mapped matrices, or aggregated scalars depending on the architecture of the underlying predictive model [27].

3.3 Specific Feature Configurations

Following the general formulation of $f_{coh.A}$, we propose two strategies for the selection of the attribute set A, representing different focus areas of path coherence.

Resource-Based Path Coherence ($f_{coh.res}$). First, we consider only the attribute identifying the resource or person responsible for the event. This configuration is motivated the conventional understanding of handoffs as transitions of responsibility between individual resources (see Sect. 2.1). We define $A_{res} = \{a_{res}\}$, where $a_{res} \in \mathcal{U}_{attr}$ denotes the resource attribute. The resulting feature is defined as:

$$f_{coh.res}(\sigma^k) = f_{coh.A_{res}}(\sigma^k) \tag{4}$$

Holistic Path Coherence ($f_{coh.all}$). Second, we consider the comprehensive set of all available event-level attributes, consistent with the expanded notion of handoffs proposed by [18]. We define $A_{all} \subseteq \mathcal{U}_{attr}$ as the complete set of recorded event attributes, subjected to a three-step pre-processing pipeline:

1. **Redundancy Filtering:** Pairwise correlation is evaluated on the training set using Cramér's V for categorical and Pearson r for numeric features. Attributes exhibiting perfect correlation ($V = 1.0$ or $r = 1.0$) are reduced to a single representative attribute to avoid duplicate structural information.
2. **Missing-Value Handling:** If an attribute is missing in both consecutive events (e_{k-1} and e_k), it is treated as a stable state ($\mathbf{1}_{coh} = 1$); if present in only one, it denotes a disruption ($\mathbf{1}_{coh} = 0$).
3. **Variance Filtering:** Zero-variance attributes that remain constant across the entire log are removed, as they artificially inflate baseline coherence without offering a predictive signal.

The resulting holistic feature is defined as:

$$f_{coh.all}(\sigma^k) = f_{coh.A_{all}}(\sigma^k) \tag{5}$$

An exemplary representation of the two strategies is shown in Fig. 1, illustrating the transformation of an event trace into its corresponding feature representations. For each event e_k, the feature is computed based on the prefix σ^k ending at that event.

The resource-based feature $f_{coh.res}$ considers only the Resource attribute, effectively functioning as a binary indicator. As shown in the example, its value is 1.0 for the initial event (by definition) and the fourth event, where the resource remains unchanged from the preceding event. Conversely, for the remaining events where a resource transition occurs, the feature value is 0.0.

In contrast, $f_{coh.all}$ incorporates the Resource, Status, and Channel attributes, reflecting the proportion of these attributes that remain constant across transitions. For the second event, for instance, only the Status attribute remains unchanged while the other two change. This results in a feature value of $1/3 \approx 0.33$, representing the mean stability across the defined attribute set.

4 Evaluation

In this section we investigate the question: *How can remaining time prediction in PPM be improved by a path coherence-based feature?* First, we describe our methodology for evaluation, before presenting the corresponding results.

4.1 Evaluation Methodology

Our methodology draws on principles of algorithm engineering as described by [14] and mirrors the benchmarking design for features of [16].

Case	Activity	Timestamp	Resource	Status	Channel	$f_{coh.res}$	$f_{coh.all}$
C-001	Document submitted	2026-01-10 09:15:42	Applicant A	Green	Web portal	1.00	1.00
C-001	Case escalated	2026-01-10 11:03:08	Officer B	Green	Back office	0.00	0.33
C-001	Eligibility reviewed	2026-01-10 13:45:30	Manager C	Yellow	Back office	0.00	0.33
C-001	Additional info requested	2026-01-10 14:27:54	Manager C	Yellow	Email	1.00	0.66
C-001	Additional info received	2026-01-12 18:45:13	Applicant A	Green	Web portal	0.00	0.00
...						...	...

relevant for $f_{coh.res}$

relevant for $f_{coh.all}$

Fig. 1. Example event trace illustrating the computation of path coherence features using both proposed strategies.

Prediction Model and Encoding. While our proposed approach in principle is agnostic to the prediction model, we employ LSTM models to assess the impact on predictive accuracy. Prior work has consistently shown that LSTM models outperform alternative model classes across a wide range of PPM tasks [22,27]. A key advantage of LSTM models is their recurrent architecture, which enables them to naturally process event sequence prefixes of varying lengths. Accordingly, prefixes are encoded directly as tensors without the need for bucketing or padding strategies [27]. To ensure comparability across all comparative setups, we use a uniform model architecture and fixed training hyperparameters throughout the evaluation. To mitigate optimization variance and ensure the robustness of our empirical findings, all comparative setups are trained and evaluated across five distinct random seeds.

Predictive Task. The model is configured to address a regression task, namely the prediction of *remaining time*, defined as the time between the last observed event in a prefix and the completion of the corresponding process instance. Formally, for a prefix σ^k of a trace $\sigma = \langle e_1, e_2, \ldots, e_n \rangle$, the remaining time is computed as $\#_{time}(e_n) - \#_{time}(e_k)$.

Event Log Data. Each comparison is performed on eight publicly available, real-world event logs, also used in related work [16,25,27]. Within the event logs, we use the org:resource attribute to represent the resource, in accordance with the XES-specification. The characteristics of the event logs are summarized in Table 1. There, also the number of resources present in the event log, as well as the number of numeric and categorical attributes which are considered for the evaluation.

Table 1. Properties of real-world event logs.

Event log	# of activities	# of traces	# of events	# of variants	Mean case duration	# of resources	# of numeric attributes	# of categorical[1] attributes
BPI 2012 (A)[a]	10	13,087	60,849	17	8.1 days	62	1	0
BPI 2012 (O)[a]	7	5,015	31,244	168	17.4 days	60	1	0
BPI 2012 (W)[a]	7	9,658	170,107	2,643	12.7 days	69	1	1
BPI 2015 (Mun. 1)[b]	398	1,199	52,217	1,170	95.7 days	23	3	4
Helpdesk[c]	14	4,580	21,348	226	40.9 days	22	0	0
Prepaid travel cost[d]	29	2,099	18,246	202	36.8 days	2	0	1
Request for payment[e]	19	6,886	36,796	89	12.0 days	2	0	1
Sepsis[f]	16	1,050	15,214	846	28.5 days	n.a.	3	1

[1] beyond resource attribute
[a] https://doi.org/10.4121/uuid:3926db30-f712-4394-aebc-75976070e91f
[b] https://doi.org/10.4121/uuid:a0addfda-2044-4541-a450-fdcc9fe16d17
[c] https://doi.org/10.4121/uuid:0c60edf1-6f83-4e75-9367-4c63b3e9d5bb
[d] https://doi.org/10.4121/uuid:5d2fe5e1-f91f-4a3b-ad9b-9e4126870165
[e] https://doi.org/10.4121/uuid:895b26fb-6f25-46eb-9e48-0dca26fcd030
[f] https://doi.org/10.4121/uuid:895b26fb-6f25-46eb-9e48-0dca26fcd030

We apply a temporal split, training on earlier cases and evaluating on later ones. Each log is partitioned into 80% training, and 20% testing cases based on the timestamp of the last event [25, 27]. To emulate deployment, where remaining time is predicted for ongoing cases, we generate all prefixes σ^k with $2 \leq k \leq |\sigma|$ for each trace $\sigma \in L$ [25].

Comparison Groups. To rigorously evaluate the predictive value of our path coherence-based features, we systematically define multiple feature sets that serve as baselines and comparison benchmarks. We begin by establishing a foundational execution context, denoted as F_{base}. Following the approach proposed by [25], $F_{base} = \{f_{act}, f_{t1}, f_{t2}, f_{t3}\}$ comprises established event-level features: the one-hot encoded activity (f_{act}), the time since the previous event in the trace (f_{t1}), the time within the current day (f_{t2}), and the time within the current week (f_{t3}).

While F_{base} provides essential control-flow and temporal information, it natively lacks data regarding resources or other event data attributes. To isolate the specific impact of our proposed strategies, we extend F_{base} into two distinct categories: 1) coherence-augmented sets and 2) raw-attribute benchmark sets. We utilize the previously defined attribute sets A_{res} and A_{all} to construct these extensions.

For the coherence-augmented sets, we append the respective path coherence feature to F_{base}:

- $F_{coh.res} = F_{base} \cup \{f_{coh.res}\}$
- $F_{coh.all} = F_{base} \cup \{f_{coh.all}\}$

Let $F_{raw}(A)$ denote the set of directly encoded raw features for a given attribute set A, where numeric attributes are min–max normalized and categorical attributes are one-hot encoded. For the comparable raw-attribute benchmarks, we append the directly encoded attribute data to F_{base} to establish information equivalence:

- $F_{raw.res} = F_{base} \cup F_{raw}(A_{res})$
- $F_{raw.all} = F_{base} \cup F_{raw}(A_{all})$

This formulation enables four distinct comparison pairs designed to assess predictive accuracy against both minimal baselines and standard attribute encodings:

1. $(F_{coh.res}, F_{base})$: Evaluates resource coherence against the (control-flow and temporal) baseline.
2. $(F_{coh.res}, F_{raw.res})$: Evaluates resource coherence against raw resource encoding.
3. $(F_{coh.all}, F_{base})$: Evaluates holistic attribute coherence against the (control-flow and temporal) baseline.
4. $(F_{coh.all}, F_{raw.all})$: Evaluates holistic attribute coherence against the raw encoding of all available attributes.

Statistical Evaluation. The comparisons are evaluated using the *Mean Absolute Error*, following [22,27]. The MAE measures the arithmetic mean of all absolute prediction errors and is formally defined as $\text{MAE} = \frac{1}{n}\sum_{i=1}^{n}|y_i - \hat{y}_i|$, where $i \in \{1, 2, \ldots, n\}$ indexes the predictions, $y_i \in \mathbb{R}$ denotes the true value, and $\hat{y}_i \in \mathbb{R}$ the predicted value. To ensure statistical robustness and guard against dependency biases inherent to prefix-level evaluations, we aggregate our findings at the independent case level. For each unique case, we compute a case-level MAE_σ by averaging the absolute prediction errors across all its constituent prefixes and across the five random seeds, which are then averaged across all cases to yield the overall MAE. To complement this practical measure, we assess the statistical significance of observed differences using the *Wilcoxon signed-rank test* [3,8], a non-parametric alternative to the paired t-test that does not assume normality. Specifically, for each comparison, we define the paired differences in case-level absolute error as $\delta_\sigma = MAE_\sigma^{\text{benchmark}} - MAE_\sigma^{\text{coh}}$, where σ indexes the independent cases in the test set. The one-sided Wilcoxon signed-rank test is then applied to the set of differences $\{\delta_1, \ldots, \delta_m\}$ across all m cases to examine whether the errors obtained with the path coherence-based features are significantly smaller than those of the corresponding benchmark feature set. Formally, the null hypothesis H_0 states that the benchmark feature set produces case-level absolute errors no larger than those of the path coherence-based features. For each test, we report both the p-value and Rosenthal's r as a measure of effect size [6].

4.2 Evaluation Results

This section reports the results of comparing path coherencebased feature sets with their corresponding benchmark representations across eight real-world event logs. We first present the MAE-based performance comparison, followed by an analysis of the statistical significance of the observed differences. Overall, the results indicate that coherence-based encodings lead to lower prediction errors than both the base and raw feature representations in most cases.

MAE-Based Performance Comparison. Figure 2 reports the MAE obtained for all feature set comparisons across the eight evaluated event logs. For the *Helpdesk* event log, only resource-based feature sets are considered, as the data does not contain additional attributes beyond the resource, and consequently no $F_{coh.all}$ and $F_{raw.all}$ configurations are evaluated. Similarly, the *Sepsis* event log contains no resource information, hence no $F_{coh.res}$ and $F_{raw.res}$ configurations are evaluated.

Across the majority of evaluations, the coherence-based feature sets outperform both the foundational baseline (F_{base}) and the raw-attribute benchmarks ($F_{raw.res}$, $F_{raw.all}$). Coherence-based features lead to lower prediction errors with relative MAE reductions ranging from approximately 0.1% (10.44 vs. 10.46 days on the *BPI 2012 O* event log) to 22.2% (6.62 vs. 8.51 days on the *Helpdesk* event log). The mean relative difference in MAE across all comparisons is 3.9% (median 1.9%).

When comparing the two coherence strategies, a highly notable outcome is that the narrow, resource-based feature set $F_{coh.res}$ captures the vast majority of the predictive performance gains. In most event logs, extending the representation to include all remaining attributes in $F_{coh.all}$ yields only marginal additional improvements or results in comparable, sometimes higher, MAE values. This indicates a strong empirical asymmetry in process context: modeling human resource behavior through path coherence provides a highly potent abstraction for remaining time prediction, whereas aggregating additional environmental or systemic attributes introduces data redundancy that limits further model optimization.

While the coherence features generally prevail, specific exceptions exist where a coherence-based feature yields a higher error than one or both of its benchmarks:

- **BPI 2012 A:** The resource coherence feature set $F_{coh.res}$ yields a marginally higher MAE (9.02 days) compared to $F_{raw.res}$ (9.01 days).
- **BPI 2015 (Mun. 1):** The resource coherence feature set $F_{coh.res}$ yields a higher MAE (61.30 days) compared to $F_{raw.res}$ (61.05 days).
- **Sepsis:** The holistic coherence feature set $F_{coh.all}$ yields almost the same MAE (72.40 days) as F_{base} (72.40 days).

Furthermore, the results indicate that raw attribute encodings offer little benefit over the base feature set. Both $F_{raw.res}$ and $F_{raw.all}$ frequently exhibit MAE values that are similar to or higher than those of F_{base}, highlighting that simply incorporating raw attribute information is insufficient for the model to learn the transitional semantics. Structuring this information through path coherence leads to more effective and robust representations.

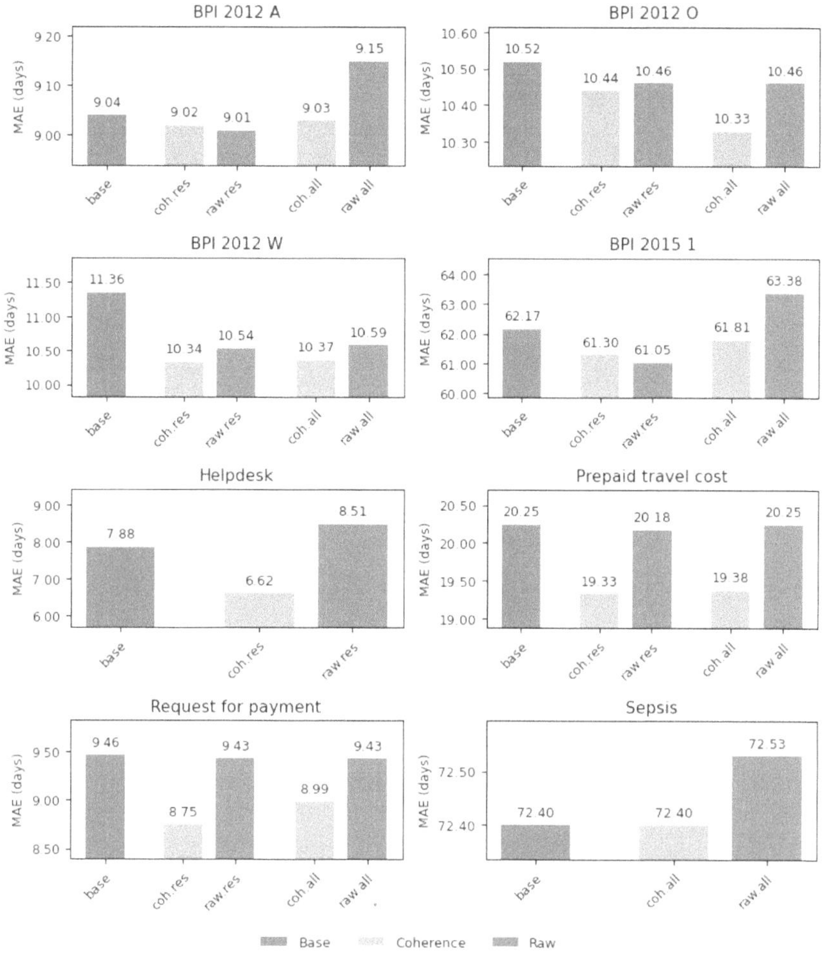

Fig. 2. MAE per event log and feature set.

Statistical Significance Analysis. Table 2 reports the p-values and effect sizes, measured using Rosenthal's r, obtained from the one-sided Wilcoxon signed-rank tests. As with the MAE analysis, no results for the all-attribute configurations are reported for the *Helpdesk* event log and no results for the resource-attribute configurations are reported for the *Sepsis* event log due to the absence of the corresponding attributes.

The analysis of statistical significance generally confirms the observations from the MAE-based comparison. For the vast majority of configurations, the Wilcoxon signed-rank test yields $p < 0.001$, providing strong evidence against the null hypothesis that the benchmark feature sets produce absolute errors no larger than those of the path coherence-based features.

Table 2. p-value and effect size from one-sided Wilcoxon signed-rank test per event log and comparison group.

Event log	$(F_{coh.res}, F_{base})$ p	r	$(F_{coh.res}, F_{raw.res})$ p	r	$(F_{coh.all}, F_{base})$ p	r	$(F_{coh.all}, F_{raw.all})$ p	r
BPI 2012 A	0.016	0.042	1.000	0.084	1.000	0.081	0.000	0.205
BPI 2012 O	0.000	0.152	0.699	0.012	0.000	0.319	0.000	0.213
BPI 2012 W	0.000	0.329	0.000	0.097	0.000	0.310	0.000	0.110
BPI 2015 (Mun. 1)	0.107	0.080	0.612	0.018	0.009	0.154	0.000	0.219
Helpdesk	0.000	0.484	0.000	0.574	n.a.	n.a.	n.a.	n.a.
Prepaid travel cost	0.000	0.399	0.000	0.401	0.000	0.386	0.000	0.386
Request for payment	0.000	0.641	0.000	0.641	0.000	0.467	0.000	0.443
Sepsis	n.a.	n.a.	n.a.	n.a.	0.375	0.022	0.174	0.065

However, mirroring the MAE results, the tests fail to establish statistical significance in a few specific configurations:

- **BPI 2012 A:** The resource coherence feature set $F_{coh.res}$ does not achieve significance against $F_{raw.res}$ ($p = 1.000$). Furthermore, the holistic feature set $F_{coh.all}$ shows no significant improvement over F_{base} ($p = 1.000$), though it is significant compared to $F_{raw.all}$ ($p < 0.001$).
- **BPI 2012 O:** The resource coherence feature set $F_{coh.res}$ does not achieve significance against $F_{raw.res}$ ($p = 0.699$).
- **BPI 2015 (Mun. 1):** The resource coherence feature set $F_{coh.res}$ does not achieve significance against F_{base} ($p = 0.107$) and $F_{raw.res}$ ($p = 0.612$).
- **Sepsis:** The holistic feature set $F_{coh.all}$ does not achieve significance against F_{base} ($p = 0.375$) and $F_{raw.all}$ ($p = 174$).

For the statistically significant comparisons, the effect sizes provide additional insight into the magnitude of the observed improvements. Following common guidelines for Rosenthal's r, values of approximately 0.1 are considered small effects, around 0.3 are medium effects, and 0.5 or above denote large effects [6]. In our evaluation, most significant comparisons fall within the medium range, indicating practically meaningful improvements. The largest effect sizes are observed for the *Request for Payment event* log (with values up to 0.641) and the *Helpdesk* event log (with r values reaching up to 0.574), both corresponding to large effects. Across all statistically significant comparisons, effect sizes range between 0.042 and 0.641, with a mean of 0.332 (median 0.329). These results indicate that, while the magnitude of improvement varies across event logs, the gains achieved by path coherence-based features are systematically non-negligible and practically relevant across varying process environments.

From a practical perspective, while these absolute numerical improvements are modest, they systematically reduce scheduling uncertainty across diverse workflows. In human-centric processes, narrowing this error variance directly stabilizes operational forecasting. By reliably quantifying situational continuity, the model provides early-warning signals that allow managers to proactively allocate resources and optimize capacity planning [12].

5 Discussion

The empirical evaluation of path coherence-based features yields several insights into the role of execution continuity in PPM. By shifting the focus from isolated events to the transitional friction between them, this study opens new avenues for understanding and anticipating process behavior.

The experimental outcomes demonstrate that integrating coherence-based features systematically reduces the MAE in remaining time predictions across the evaluated event logs. Notably, the resource-based path coherence configuration accounted for the majority of these performance gains, while the holistic all-attribute variant provided negligible additional benefit. This discrepancy suggests that responsibility handoffs are the primary drivers of transitional delays. When further non-resource attributes are aggregated into a single coherence metric, they appear to introduce informational noise that dilutes the critical signals generated by pure resource handoffs. Specifically, because the holistic variant averages all attributes, a critical drop in resource continuity is diluted if secondary environmental attributes remain constant. Blindly incorporating all attributes introduces informational noise by capturing features that lack operational relevance, lowering the overall signal-to-noise ratio. Consequently, predictive pipelines benefit most from highly specific, targeted measures of coordination friction rather than generalized metrics of situational change.

Those findings challenge the prevailing notion in PPM, where traces are predominately modeled as structurally independent events [10,15,25]. By demonstrating the predictive power of path coherence, this research substantiates the theoretical premise that organizational friction carries predictive value. This establishes a theoretical foundation for treating process executions not merely as chronological sequences of events, but interleaved execution paths [9].

From a broader perspective, our work demonstrate how process prediction can benefit from research from the organizational sciences. It has often been argued that BPM and organizational sciences offer opportunities for cross-fertilization [13], but typically, these calls have taken a one-directional angle: how research from BPM can enhance research in the organizational sciences [7,19]. Our study is a one of the first accounts to stress the benefits from the organizational research for BPM, and PPM more specifically. Any process that involves more than one actor yields social dynamics [2], and the organizational sciences provide a language (in the form of concepts and theories) to attend to those. In our case, we have been shifting our attention to handoffs, but related efforts should by no means stop here. There are many more findings from this field to inform and refine the theory behind algorithmic engineering [14].

For organizational research, the study provides evidence for the predictive relevance of handoff theory at the level of individual process instances. Extending prior work on narrative networks [9], we show that path coherence not only explains observed patterns ex post but also predicts a performance-relevant outcome (remaining time) in operational business processes.

One aspect worth highlighting is that path coherence is operationalized as a proxy for coordination friction through syntactic value-equality matching between consecutive events. This proxy preserves a sufficiently strong signal to systematically improve remaining time estimation. This result suggests that meaningful coordination patterns can be captured using lightweight representations, while also motivating future work to investigate richer operationalizations of path coherence that incorporate additional organizational and semantic information.

An important boundary is that while our formulation is generic, its predictive power remains domain-dependent. The impact of a coherence drop relies on systemic constraints: a location change implies severe friction in healthcare [9], but is negligible in software development. While resource assignment is a robust proxy, integrating explicit domain knowledge to weight specialized attributes remains critical to systematically improve predictive performance [23,24].

Building on these promising results, we identify four primary avenues for future research to further advance the path coherence concept. First, incorporating *domain knowledge* presents a significant opportunity. Future studies could explore how semantically weighting attributes–tailored to specific operational contexts–might enhance predictive accuracy. Second, expanding the *architectural and task scope* is promising, inviting the application of coherence features within transformer architectures and extending the objective from remaining time estimation to next-activity or outcome predictions. Third, extending *algorithmic horizons* offers rich potential, such as using multi-step windows or decaying moving averages to capture long-range historical continuity rather than focusing solely on immediate adjacent transitions. Fourth, enhancing *process explainability* remains a key goal. Future diagnostic tools could leverage the nature of path coherence to pinpoint specific friction-driving handoffs, providing practitioners with transparent, human-interpretable explanations.

6 Conclusion

This paper addresses a methodological gap in PPM by integrating the organizational theory of path coherence to systematically quantify the continuity of execution contexts. By engineering resource-based and holistic path coherence-based features within a remaining time prediction pipeline based on LSTM models, our evaluation across eight real-world event logs demonstrates a statistically significant reduction in prediction errors compared to benchmark feature sets. The analysis reveals that the resource-based coherence variant drives the majority of these performance gains, highlighting the impact of responsibility handoffs on process durations. Methodologically, these results establish that structured, semantic abstractions of contextual dynamics yield more suitable predictive signals than the mere ingestion of raw event attributes.

Reproducibility. The source code required to reproduce our evaluation is available under https://github.com/maalevo/Path_Coherence (https://doi.org/10.5281/zenodo.20409662)

Acknowledgments. The research was supported by the Einstein Foundation Berlin under grant EPP-2019-524; by the German Federal Ministry of Research, Technology and Space under grant 16DII1433; by Deutsche Forschungsgemeinschaft under grant 496119880; and by the EU Erasmus+ program under grant 2025-2-LI01-KA220-HED-000384433 ("Educating Business Students for Human-AI Partnerships: Training Human Competences and Critical Thinking Skills"). Kate Revoredo is funded by the Berliner Chancengleichheitsprogramm (BCP) as part of the DiGiTal Graduate Program.

References

1. Bailey, D.E., Leonardi, P.M., Chong, J.: Minding the gaps: understanding technology interdependence and coordination in knowledge work. Organ. Sci. **21**(3), 713–730 (2010). https://doi.org/10.1287/orsc.1090.0473
2. Brocke, J.V., et al.: Process science: the interdisciplinary study of socio-technical change. Process Sci. **1**(1), 1 (2024). https://doi.org/10.1007/s44311-024-00001-5
3. Demšar, J.: Statistical comparisons of classifiers over multiple data sets. J. Mach. Learn. Res. **7**, (2006)
4. Evermann, J., Rehse, J.R., Fettke, P.: A deep learning approach for predicting process behaviour at runtime. In: Business Process Management Workshops. vol. 281, pp. 327–338. Springer (2017). https://doi.org/10.1007/978-3-319-58457-7_24
5. Feldman, M.S., Pentland, B.T.: Reconceptualizing organizational routines as a source of flexibility and change. Adm. Sci. Q. **48**(1), 94–118 (2003). https://doi.org/10.2307/3556620
6. Fritz, C.O., Morris, P.E., Richler, J.J.: Effect size estimates: current use, calculations, and interpretation. J. Exp. Psychol. Gen. **141**(1), 2–18 (2012). https://doi.org/10.1037/a0024338
7. Grisold, T., Wurm, B., Mendling, J., Brocke, J.V.: Using process mining to support theorizing about change in organizations. In: Hawaii International Conference on System Sciences 2020 (HICSS-53) (2020)
8. Hennig, M.C., Schmidt, R.: Leveraging temporal graphs for enhancing transformer-based predictive process monitoring. In: Business Process Management, vol. 16044, pp. 291–307 (2026). https://doi.org/10.1007/978-3-032-02867-9_18
9. Kim, I., Pentland, B.T., Frank, K.A., Wolf, J.R.: Path coherence and disruption in routine dynamics. Organ. Sci. **36**(2), 862–880 (2025). https://doi.org/10.1287/orsc.2022.16749
10. Kim, J., Comuzzi, M., Dumas, M., Maggi, F.M., Teinemaa, I.: Encoding resource experience for predictive process monitoring. Decis. Support Syst. **153**, 113669 (2022). https://doi.org/10.1016/j.dss.2021.113669
11. Maggi, F.M., Di Francescomarino, C., Dumas, M., Ghidini, C.: Predictive monitoring of business processes. In: Advanced Information Systems Engineering, vol. 8484, pp. 457–472. Springer (2014). https://doi.org/10.1007/978-3-319-07881-6_31
12. Marquez-Chamorro, A.E., Resinas, M., Ruiz-Cortes, A.: Predictive monitoring of business processes: a survey. IEEE Trans. Serv. Comput. **11**(6), 962–977 (2018). https://doi.org/10.1109/TSC.2017.2772256

13. Mendling, J., Berente, N., Seidel, S., Grisold, T.: The philosopher's corner: pluralism and pragmatism in the information systems field: the case of research on business processes and organizational routine. ACM SIGMIS Database **52**(2), 127–140 (2021). https://doi.org/10.1145/3462766.3462773
14. Mendling, J., Leopold, H., Meyerhenke, H., Depaire, B.: Methodology of algorithm engineering. ACM Comput. Surv. **58**(4), 1–38 (2026). https://doi.org/10.1145/3769071
15. Navarin, N., Vincenzi, B., Polato, M., Sperduti, A.: LSTM networks for data-aware remaining time prediction of business process instances. In: 2017 IEEE Symposium Series on Computational Intelligence (SSCI), pp. 1–7 (2017). https://doi.org/10.1109/SSCI.2017.8285184
16. Oyamada, R.S., Tavares, G.M., Junior, S.B., Ceravolo, P.: Enhancing predictive process monitoring with time-related feature engineering. In: Advanced Information Systems Engineering, vol. 14663, pp. 71–86 (2024). https://doi.org/10.1007/978-3-031-61057-8_5
17. Patterson, E.S., Wears, R.L.: Patient handoffs: standardized and reliable measurement tools remain elusive. Joint Comm. J. Qual. Patient Saf. **36**(2), 52–61 (2010). https://doi.org/10.1016/S1553-7250(10)36011-9
18. Pentland, B.T., Recker, J., Wyner, G.: Rediscovering handoffs. Acad. Manage. Discoveries **3**(3), 284–301 (2017). https://doi.org/10.5465/amd.2016.0018
19. Pentland, B.T., Vaast, E., Wolf, J.R.: Theorizing process dynamics with directed graphs: a diachronic analysis of digital trace data. MIS Q. **45**(2), 967–984 (2021). https://doi.org/10.25300/MISQ/2021/15360
20. Pourbafrani, M., Kar, S., Kaiser, S., Van Der Aalst, W.M.P.: Remaining time prediction for processes with inter-case dynamics. In: Process Mining Workshops, vol. 433, pp. 140–153 (2022). https://doi.org/10.1007/978-3-030-98581-3_11
21. Puranam, P., Alexy, O., Reitzig, M.: What's "new" about new forms of organizing? Acad. Manag. Rev. **39**(2), 162–180 (2014)
22. Rama-Maneiro, E., Vidal, J., Lama, M.: Deep learning for predictive business process monitoring: review and benchmark. IEEE Trans. Serv. Comput. 1–1 (2022). https://doi.org/10.1109/TSC.2021.3139807
23. Schuster, D., Van Zelst, S.J., Van Der Aalst, W.M.: Utilizing domain knowledge in data-driven process discovery: a literature review. Comput. Ind. **137**, 103612 (2022). https://doi.org/10.1016/j.compind.2022.103612
24. Senderovich, A., Francescomarino, C.D., Maggi, F.M.: From knowledge-driven to data-driven inter-case feature encoding in predictive process monitoring. Inf. Syst. **84**, 255–264 (2019). https://doi.org/10.1016/j.is.2019.01.007
25. Tax, N., Verenich, I., La Rosa, M., Dumas, M.: Predictive business process monitoring with LSTM neural networks. In: Advanced Information Systems Engineering, vol. 10253, pp. 477–492 (2017). https://doi.org/10.1007/978-3-319-59536-8_30
26. Van Der Aalst, W., Schonenberg, M., Song, M.: Time prediction based on process mining. Inf. Syst. **36**(2), 450–475 (2011). https://doi.org/10.1016/j.is.2010.09.001
27. Verenich, I., Dumas, M., Rosa, M.L., Maggi, F.M., Teinemaa, I.: Survey and cross-benchmark comparison of remaining time prediction methods in business process monitoring. ACM Trans. Intell. Syst. Technol. **10**(4), 1–34 (2019). https://doi.org/10.1145/3331449
28. Wohlauer, M.V., Arora, V.M., Horwitz, L.I., Bass, E.J., Mahar, S.E., Philibert, I.: The patient handoff: a comprehensive curricular blueprint for resident education to improve continuity of care. Acad. Med. **87**(4), 411–418 (2012). https://doi.org/10.1097/ACM.0b013e318248e766

Collaborative Model Optimization in Federated Stochastic Process Discovery

Hootan Zhian(✉), Rajkumar Buyya, and Artem Polyvyanyy

The University of Melbourne, Melbourne, Australia
hzhian@student.unimelb.edu.au, {rbuyya,artem.polyvyanyy}@unimelb.edu.au

Abstract. Process discovery focuses on constructing process models that accurately represent system behavior from event logs generated by these systems. Traditional methods rely on centralized event logs, which pose significant challenges in distributed environments where organizations are reluctant to share raw data due to privacy concerns. Federated process discovery has emerged as a promising approach to enabling collaborative model generation without compromising data privacy. However, existing federated approaches often produce overly complex process models, limiting their practical utility. This paper introduces an approach that enables organizations to collaboratively construct, via iterative refinements, a global process model from local models discovered from their own data. The iterations are guided by feedback from organizations on the quality of the current global model. By formulating process model refinement as a decision-tree pruning problem, our approach systematically reduces unnecessary model complexity while ensuring high model quality. Experiments with industrial logs show that our approach often yields smaller, more accurate models than a baseline federated method, comparable in quality to models discovered from centralized logs, and has runtime comparable to that of the centralized discovery approach.

Keywords: Stochastic process mining · federated process discovery · cross-silos process discovery

1 Introduction

With the rapid digitalization of organizational operations, information systems increasingly record data in the form of event logs, which capture detailed information about operational activities. This trend has elevated the importance of process mining as a means of extracting knowledge from event logs to discover, monitor, and improve processes. In particular, process discovery aims to automatically construct descriptive process models from such logs [2,19]. These models are often represented as Directly-Follows Graphs (DFGs) [3]. A DFG is a directed graph whose nodes describe actions and whose edges capture "can-follow" ordering relations between actions. Numeric weights associated with

F. M. Maggi et al. (Eds.): BPM 2026, LNCS 16936, pp. 611–629, 2027.
https://doi.org/10.1007/978-3-032-37777-7_36

nodes and edges indicate the number of observations of the corresponding actions and subsequent executions of the corresponding actions in the event log, respectively. Numeric weights associated with nodes and edges indicate, respectively, the number of observed occurrences of the corresponding actions and the number of times one action is directly followed by another in the event log.

To evaluate the quality of a discovered model with respect to an event log, several evaluation criteria have been proposed [10]. Recall, also known as fitness, reflects how well the model reproduces the behavior recorded in the log. Precision evaluates the extent to which the model avoids allowing unsupported behavior beyond what is observed in the log. Generalization captures the ability of the model to avoid overfitting by allowing plausible yet unseen behavior. Finally, simplicity measures the structural complexity of the model, favoring compact representations without unnecessary elements.

A plethora of process discovery algorithms have been proposed to construct models with good quality characteristics. Recent studies have explored optimization-based approaches, including metaheuristic techniques for automated process discovery based on single-objective and multi-objective search [8, 9]. Yan et al. [25] propose an approach that optimizes imprecise sub-processes using frequency-based filtering. In addition, multi-objective optimization strategies have been applied to discover Pareto-optimal models that balance quality dimensions such as model size and entropic relevance [5,29]. Entropic relevance is a stochastic conformance measure that compares the likelihood of observing traces in event logs and stochastic process models, capturing how closely a model reflects the observed probability distribution of behavior [20]. Moreover, it offers a balanced assessment of model quality by jointly reflecting precision and recall [6]. However, these approaches generally assume centralized access to event logs. Cross-organizational settings pose significant challenges for process discovery because organizations are often reluctant to share event logs [4,16].

Federated process discovery (FPD) has been proposed to overcome the limitations of centralized discovery, including privacy and security concerns [13, 16,18,21], data transfer costs [16,24], and scalability and efficiency challenges [1,11,24,27]. Zhian et al. [28] introduce the *FedGASPD* approach, in which organizations, or *clients*, share models discovered locally from their own data, referred to as *local models*, rather than raw event logs. A central server then aggregates these local models to derive a *global model* that describes the overall process executed across the clients. This strategy increases the level of abstraction of the shared information, thereby improving privacy and reducing data transfer requirements. *FedGASPD* produces models of quality comparable to those obtained from centralized event logs. Because discovery is performed locally on smaller event logs and the aggregation procedure is efficient, *FedGASPD* can often construct models faster than centralized discovery. However, the models discovered by *FedGASPD* may still become unnecessarily complex, as aggregating complete local models can preserve infrequent or client-specific behavior. This limitation highlights the need for further improving the quality of federated discovery results.

This paper presents *OptFedGASPD*, in which clients collectively optimize the global process model constructed by *FedGASPD* using feedback on the global model quality with respect to clients' event logs. The approach systematically prunes the aggregated global model, reducing unnecessary structural complexity while ensuring model quality for each participating organization. Its key innovation lies in enabling collaborative optimization of the global model under the coordination of a central server by casting the pruning task as a decision-tree optimization problem. Experimental results on industrial event logs show that *OptFedGASPD* often produces smaller and more accurate models than *FedGASPD*. Although pruning introduces additional computation, the overall runtime remains comparable to that of *FedGASPD* and often still outperforms discovery from centralized event logs using *GASPD*.

Specifically, the evaluation gives positive answers to these two research questions.

RQ1. Can clients, under central coordination, systematically optimize the quality and size of the global process model constructed by *FedGASPD*?
RQ2. Is *OptFedGASPD* comparable to *FedGASPD* in terms of runtime performance and often faster than *GASPD*?

The remainder of the paper is structured as follows. Section 2 surveys related work on federated process discovery. Section 3 introduces basic concepts used in the subsequent discussions. Section 4 presents *OptFedGASPD*. Section 5 discusses the evaluation of the proposed method using an open-source implementation over publicly available industrial event logs. Finally, Sect. 6 concludes the paper by summarizing the contributions and outlining directions for future research.

2 Related Work

Decomposition and divide-and-conquer strategies have long been explored as alternatives to centralized process discovery methods. These approaches aim to address scalability and complexity challenges by breaking down event logs into smaller, more manageable components. Carmona et al. [11] propose methods to derive partial behavioral views from event logs, either by leveraging region theory to construct and combine local model components or by partitioning logs into tightly related event groups and discovering models for each partition. These models can then be assembled into a global process model or analyzed independently. Similarly, van der Aalst [1] introduces a generic framework for divide-and-conquer process discovery, which partitions actions, projects logs onto these partitions, and composes the resulting models. This framework employs strategies such as single-entry-single-exit (SESE) fragments to facilitate partitioning. Yan et al. [26] rely on this idea to propose RPSTHD, a multi-level decomposition approach that combines the Refined Process Structure Tree (RPST), heuristic mining, and Integer Linear Programming (ILP) to efficiently analyze large event logs by breaking them into smaller parts. Andersen et al. [7] propose EdgeMiner as a distributed process mining approach that avoids centralizing event logs by performing computation directly at data sources (e.g., sensor nodes). Instead of sharing raw event data, each node locally observes events, determines predecessorsuccessor relations, and constructs a partial footprint matrix (FM) of these relations. To generate a global view, a central entity collects and merges these partial FMs, forming a complete FM equivalent to one derived from a centralized event log. This global FM is then used as input to process discovery algorithms to extract the overall process model.

The concept of federated process discovery was introduced by van der Aalst [4], which aims to enable organizations to collaboratively perform process discovery without sharing raw event data. Rafiei et al. [22] introduce a privacy-aware federated conformance-checking method designed to assess the correctness of cross-organizational process models while safeguarding sensitive information. Two main strategies for sharing information during federated process discovery are sharing filtered event data and sharing abstract representations of event data. Limited works on merging abstract data representations into a global model. Rennert et al. [23] develop a method for discovering DFGs from federated event logs without requiring a trusted third party or direct data sharing. Their approach uses fully homomorphic encryption to securely compute a global DFG, ensuring that sensitive information remains protected. Rafiei and van der Aalst [21] propose a privacy-preserving federated process discovery by introducing an abstraction-based method that uses directly-follows relations to represent event log abstractions. This method employs a handover mechanism to capture trace transitions between organizations, enabling collaboration while limiting exposure to sensitive data. Nucciarelli et al. [18] introduce a federated adaptation of the Alpha discovery algorithm. The focus is on enabling privacy-preserving process discovery in healthcare. Each client computes an FM, similar to the approach by Andersen et al. [7]. A master node then combines the FMs from the clients in a way that preserves the ordering relations and ensures consistency in the aggregated FM. The aggregated FM is subsequently used to construct the final process model, which is captured as a Petri net. In other efforts, federated process discovery approaches have also been proposed to reduce communication overhead and improve scalability. For example, Khan et al. [16] design a protocol for federated process discovery using dependency graphs and a trusted central server. Rojo et al. [24] explore federated process discovery in distributed environments, such as smartphone networks, by either discovering models locally or aggregating raw traces centrally. While local discovery reduces communication costs, the quality of the resulting global model often suffers from aggregating intermediate local models.

Zhian et al. [28] introduce *FedGASPD*, a federated *stochastic process discovery* approach designed to address privacy and scalability challenges of centralized discovery. The models constructed by *FedGASPD* describe traces and the likelihoods of observing them. Unlike other approaches that focus on assembling local event logs or intermediate abstractions, *FedGASPD* merges local process models on the server node, reducing the need for raw data sharing and intermediate model aggregation. The approach builds on the *GASPD* algorithm, which applies grammatical inference techniques to extract stochastic language models from event logs [5].

3 Preliminaries

GASPD uses the *ALERGIA* grammatical inference algorithm to learn accurate stochastic representation of the traces in the input event log [5]. The language models constructed by *ALERGIA* are Deterministic Frequency Finite Automata (DFFAs) [12].

Definition 3.1. (Deterministic Frequency Finite Automata) A *Deterministic Frequency Finite Automaton* (DFFA) is a tuple $(Q, \Lambda, q_0, \mathbb{I}, \mathbb{F}, \delta)$, where:

- Q is a finite nonempty set of *states*;
- Λ is a finite set of *actions*, called the *alphabet*;
- $\mathbb{I} : Q \to \mathbb{N}$ is the *initial state frequency function*, with one *initial state* $q_0 \in Q$ for which it holds that $\mathbb{I}(q_0) > 0$ and for all $q \in Q$, $q \neq q_0$, it holds that $\mathbb{I}(q) = 0$;
- $\mathbb{F} : Q \to \mathbb{N}$ is the *final state frequency function*; and
- $\delta : Q \times \Lambda \times Q \to \mathbb{N}$ is the *transition frequency function*, such that $\forall q \in Q\, \forall a \in \Lambda\, \forall q' \in Q\, \forall q'' \in Q : ((\delta(q, a, q') > 0 \wedge \delta(q, a, q'') > 0) \implies (q' = q''))$.

┘

The notation $\delta(q, a, q') = n$ indicates that there is a transition from state q to state q' labeled a occurring n times. Figure 1 shows a DFFA with states $\{p_0, \ldots, p_3\}$, initial state p_0, $\mathbb{I}(p_0) = 20$, $\mathbb{F} = \{(p_0, 0), (p_1, 10), (p_2, 5), (p_3, 5)\}$, and $\delta(p_0, a, p_1) = 10$, $\delta(p_0, b, p_2) = 10$, $\delta(p_1, b, p_2) = 2$, $\delta(p_2, a, p_1) = 2$, and $\delta(p_2, b, p_3) = 5$.

Algorithm 1 summarizes the *FedGASPD* procedure. The algorithm constructs local models for each event log in L (lines 1–3) and then merges them into a global DFFA (line 4). The resulting DFFA can be converted into a Stochastic Directed Action Graph (SDAG), a DFG-like model that supports duplicate action nodes. Multiple local DFFAs are discovered independently (in parallel) at each client using the *GASPD* algorithm. Then, the best models are selected for each client using the *SELECTDFFA* procedure and sent to the server, which constructs the list of best local models (line 3); note that "$\circ$" is the list concatenation operation. Finally, the best models from clients are merged on the server into the resulting DFFA using the *MergeDFFAs* function on line 4.

To implement merging, *FedGASPD* relies on the unfolding of DFFAs. Figure 3 shows a prefix of the unfolding of the DFFA in Fig. 2. The construction starts from the initial state of the input automaton and proceeds iteratively. At each step, a state in the current prefix is selected and extended using one of its outgoing transitions in the input automaton. Each such extension introduces a fresh state representing the target of the chosen transition, ensuring that

Algorithm 1: *FedGASPD* [28]

Input: Initial population size p, number of generations m, number of parents k to generate offspring, and a non-empty list of event logs $L = \langle L_1, \ldots, L_n \rangle$

Output: A DFFA discovered from L

```
M ← ⟨⟩;
parallelfor i ∈ [1 .. n] do
    M ← M ∘ ⟨SELECTDFFA(GASPD(p, m, k, L_i))⟩;
return MergeDFFAs(M);
```

the unfolding has a tree structure. To guide the construction, only states with unexplored outgoing transitions are selected for expansion. If the target state of a transition has already appeared d (depth) of times among the ancestors of the current node in the prefix, the expansion is stopped; otherwise, if the input DFFA contains a cycle, the unfolding procedure would continue indefinitely. The automaton in Fig. 3 is constructed for the depth of two ($d = 2$).

The frequencies of states and transitions in the prefix automaton are computed by multiplying the corresponding frequencies in the input automaton by the probability of reaching the respective occurrence during execution. This probability is estimated from the state and transition frequencies of the input automaton.

Although *FedGASPD* effectively addresses privacy and scalability concerns, the resulting models may still exhibit unnecessary complexity. The following example illustrates this issue. Consider two event logs: $L_1 = [\langle a \rangle^8, \langle b \rangle^3, \langle a, b \rangle^2, \langle b, a \rangle^2, \langle b, b \rangle^5]$ and $L_2 = [\langle a \rangle^4, \langle a, a \rangle^4, \langle a, a, b \rangle^2, \langle b \rangle^{10}]$. Figure 1 shows A_1, the DFFA discovered from event log L_1, while Fig. 2 shows A_2, the DFFA discovered from event log L_2.

The merging process described in *FedGASPD* [28] combines these local models into an aggregated model, A^{agg}, shown in Fig. 4, by embedding a prefix unfolding of A_2, shown in Fig. 3, into A_1. After the merging process, the resulting model A^{agg} contains states that are not visited by traces from one or both event logs.

For instance, the states $\{q_3, q'_1, q'_3, q''_1\}$ are not visited with respect to L_1, while the states $\{q'_3, q''_1, p_3\}$ are not visited with respect to L_2. Moreover, q_3 represents behavior that appears only in L_2, p_3 represents behavior that appears only in L_1, and q'_3 and q''_1 represent behavior that appears in neither L_1 nor L_2. These states not only increase the structural complexity of the aggregated model but also reduce its overall quality [17]. To address this issue, our proposed approach enables clients to collaboratively prune the aggregated model, thereby obtaining a more accurate and compact representation.

Figure 5 shows an SDAG equivalent to Fig. 4. In *FedGASPD*, DFFAs are used as intermediate stochastic representations during local model discovery and model aggregation, because they provide a convenient formalism for capturing trace frequencies and supporting operations such as unfolding and merging.

However, the final process models constructed by *FedGASPD* are Stochastic Directed Action Graphs (SDAGs). An SDAG extends the DFG proposed in [5] by allowing duplicate actions, thereby enabling a more faithful representation of stochastic process behavior. The procedure for translating a DFFA into an SDAG is described in the original *GASPD* paper [5]. Thus, while DFFAs serve as the formal basis for inference and aggregation, SDAGs provide the final process representation used for interpretation and analysis.

4 Approach

FedGASPD aggregates local models into a global process model without requiring event log sharing. However, it often produces overly complex aggregated models. This complexity arises from directly merging diverse local models, which introduces unvisited states that do not meaningfully contribute to the overall behavior of the aggregated model. To address this challenge, we propose *OptFedGASPD*, an optimized federated stochastic process discovery approach that systematically refines the aggregated model by pruning unnecessary states. Unlike *FedGASPD*, which terminates once the aggregated model is constructed, *OptFedGASPD* introduces a collaborative pruning phase in which clients, under the orchestration of a central server, jointly optimize the aggregated model. The pruning decisions are based solely on model-level information and model quality measurements, thereby revealing no additional information about the event data.

Next, Sect. 4.1 describes the pruning procedure, while Sect. 4.2 embeds the pruning into the overall *OptFedGASPD* discovery routine.

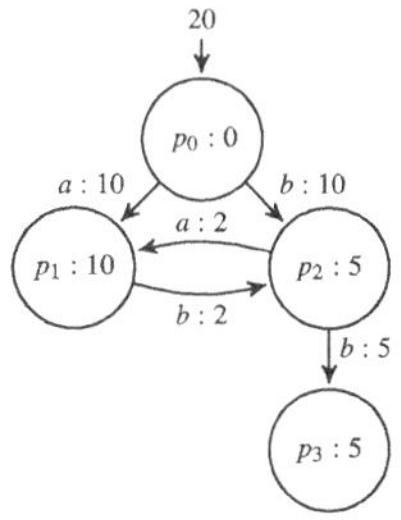

Fig. 1. DFFA A_1

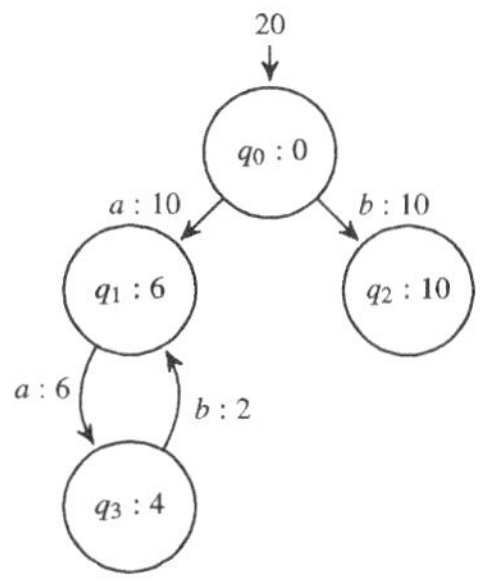

Fig. 2. DFFA A_2

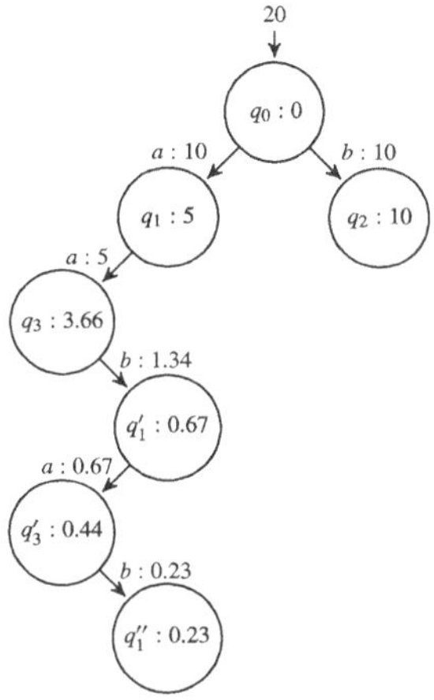

Fig. 3. DFFA A_2^u

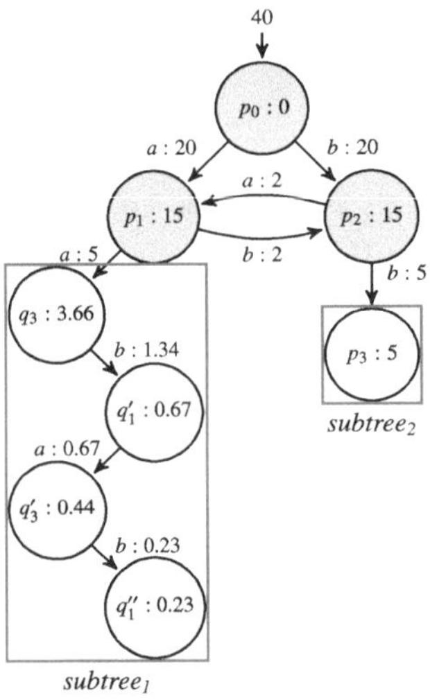

Fig. 4. DFFA A^{agg}

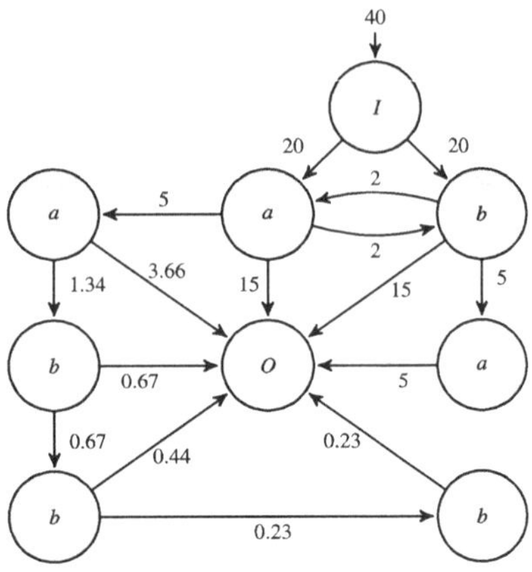

Fig. 5. SDAG A^{agg}

4.1 DFFA Pruning

The DFFA pruning problem is closely related to classical decision-tree pruning under accuracy constraints, where the goal is to remove subtrees while ensuring the resulting model does not exceed a prescribed classification error threshold. The computational complexity of related optimal pruning problems has been rigorously studied. Hyafil and Rivest [15] showed that constructing optimal binary decision trees is NP-complete, while more recent work has established that optimal decision-tree pruning under accuracy constraints is NP-hard [14]. Our problem has a similar combinatorial structure, requiring the selection of which parts of a tree-shaped or unfolded model to retain while satisfying quality constraints. However, DFFA pruning further involves global quality requirements across multiple clients.

Our DFFA pruning approach focuses on states in subtrees within the model. To identify the subtree states of a DFFA, we view the automaton as a directed graph whose nodes are the states and whose edges represent transitions with positive frequency. We then look for subgraphs that have the shape of a directed rooted tree. In such a subgraph, one state is selected as the root, every other state has exactly one parent within the subgraph, and all states in the subgraph are reachable from the root.

We only consider tree-shaped subgraphs that are attached to the rest of the DFFA through their root. The root must have one incoming edge connecting it to a state outside the tree-shaped subgraph. At the same time, no other state in the tree-shaped subgraph may have an incoming or outgoing edge to a state outside the subgraph. Thus, the root is the only interface between the tree-shaped part and the rest of the automaton.

Finally, we retain only maximal such subgraphs. That is, a tree-shaped subgraph is kept only if it cannot be extended with additional states while preserving the same properties. In this context, maximality is understood with respect to the attached branch itself: the root is the boundary state through which the tree-shaped branch is connected to the rest of the DFFA, while the predecessor of the root remains outside the subtree.

In the DFFA shown in Fig. 4, two maximal attached tree-shaped subgraphs are identified. The first one, denoted by $subtree_1$, consists of the states q_3, q_1', q_3', and q_1''. These states form a directed chain rooted at q_3. The root q_3 is connected to the rest of the DFFA through the transition from p_1 to q_3, while none of the other states in the subtree has any connection to states outside the subtree. Thus, q_3 is the only interface between $subtree_1$ and the rest of the model. The second maximal attached tree-shaped subgraph, denoted by $subtree_2$, consists only of the state p_3. This is a singleton tree whose root is p_3. The state p_3 is connected to the rest of the DFFA through the transition from p_2 to p_3. The states p_0, p_1, and p_2 are not included in these tree-shaped subgraphs. They form the surrounding non-tree part of the DFFA, in particular because p_1 and p_2 are connected by transitions in both directions.

We denote by Q_T the set of all states that belong to at least one maximal tree-shaped subgraph identified as described above. For the example automaton in Fig. 4, it holds that $Q_T = \{q_3, q_1', q_3', q_1'', p_3\}$. Let $\mathrm{Anc}(q)$ denote the set of states $p \in Q_T$ such that p and q belong to the same maximal tree-shaped subgraph and q is reachable from p by a non-empty directed path within that subgraph. In the example DFFA in Fig. 4, it holds that $\mathrm{Anc}(q_3) = \emptyset$, $\mathrm{Anc}(q_1') = \{q_3\}$, $\mathrm{Anc}(q_3') = \{q_3, q_1'\}$, $\mathrm{Anc}(q_1'') = \{q_3, q_1', q_3'\}$, and $\mathrm{Anc}(p_3) = \emptyset$. Although q_3 is reachable from p_1 and p_3 is reachable from p_2 in the full DFFA, we have $p_1 \notin \mathrm{Anc}(q_3)$ and $p_2 \notin \mathrm{Anc}(p_3)$.

The objective is to maximize the total number of pruned states in the DFFA model. The decision variable x_q, cf. Constraint C1, indicates whether state q is removed during pruning; a value of 1 means that q is removed, while a value of 0 means that q is preserved after pruning. The decision variable h_q, see Constraint C2, indicates whether the pruning effect of state q is subsumed by the pruning of one of its ancestors. A value of 1 means that an ancestor of q is pruned, and therefore the effect of pruning q is ignored. A value of 0 means that no ancestor of q is pruned. Constraint C3 states that if an ancestor state q' is selected for pruning, then all states in the subtree below q', including q, must also be selected for pruning. Constraint C4 states that if any ancestor of state q, denoted by q', is selected for pruning, then the effect on entropic relevance associated with pruning q is ignored. Constraint C5 ensures that the topmost selected state in a pruned subtree contributes its effect on the entropic relevance of the pruned model. Finally, Constraint C6 imposes quality constraints for clients, ensuring that the degradation caused by pruning remains below the threshold τ for each client. Here, ER_c denotes the entropic relevance of the model discovered by *FedGASPD* with respect to the event log of client c, before any states are pruned. $ERC_{c,q}$ denotes the change in the entropic relevance of the model with respect to the event log of client c when the subtree rooted at state q is pruned. The parameter τ is the maximum acceptable relative degradation in entropic relevance for the pruned model.

$$x_q \in \{0, 1\} \qquad q \in Q_T \qquad \text{(C1)}$$

$$h_q \in \{0, 1\} \qquad q \in Q_T \qquad \text{(C2)}$$

$$x_q \geq x_{q'} \qquad q \in Q_T, q' \in \text{Anc}(q) \qquad \text{(C3)}$$

$$h_q \geq x_{q'} \qquad q \in Q_T, q' \in \text{Anc}(q) \qquad \text{(C4)}$$

$$h_q \leq \sum_{q' \in \text{Anc}(q)} x_{q'} \qquad q \in Q_T \qquad \text{(C5)}$$

$$\tau \geq \frac{\sum_{q \in Q_T} ERC_{c,q} \cdot (x_q - h_q)}{ER_c} \qquad c \in C \qquad \text{(C6)}$$

The objective function is defined as follows. The optimization problem aims to prune as many tree states as possible while keeping the relative degradation in entropic relevance within the threshold τ for every client. In this way, the final model balances compactness with fidelity to local client behavior and the preservation of a meaningful global representation.

$$\max_{x,h} \sum_{q \in Q_T} x_q \qquad \text{(OF)}$$

To solve this MILP, we use Google OR-Tools (`com.google.ortools`), an optimization toolkit that provides solvers for integer and linear programming problems. Figure 6 shows the DFFA model obtained by pruning the A^{agg} model in Fig. 4 with $\tau = 0.1$. The corresponding SDAG model is shown in Fig. 7.

The SDAGs in Figs. 5 and 7 have sizes 25 and 19, respectively. Thus, the optimized SDAG is smaller while preserving the same entropic relevance values, namely 2.103 and 1.820, with respect to the corresponding client logs.

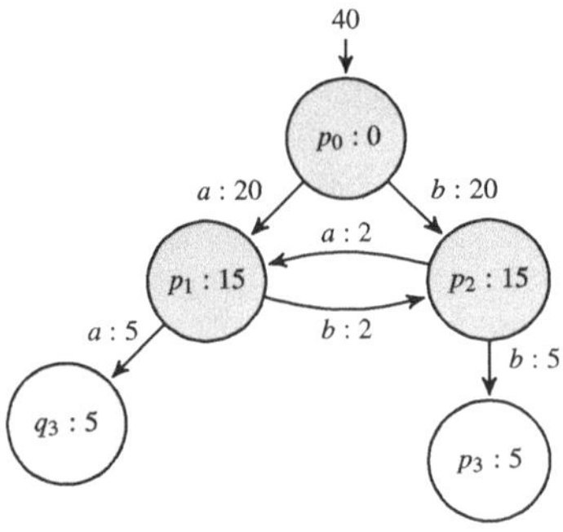

Fig. 6. DFFA A^{opt}

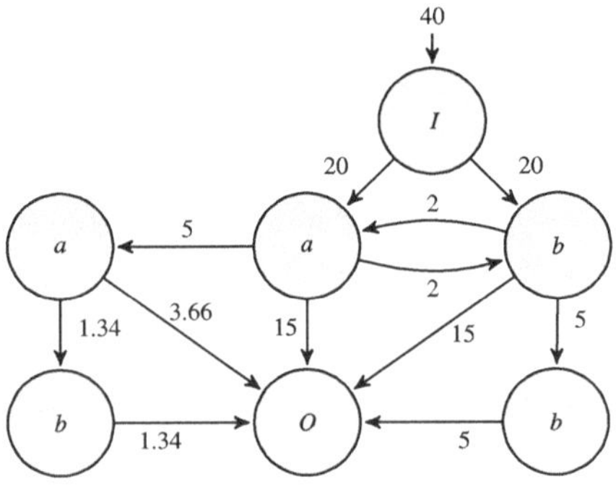

Fig. 7. SDAG A^{opt}

4.2 Optimized Federated Stochastic Process Discovery

Algorithm 2 summarizes the *OptFedGASPD* procedure. The algorithm begins by invoking *FedGASPD* (cf. Algorithm 1) in line 1 to construct the aggregated model A^{agg}. Given a set of local event logs of clients, the server requests each client to perform process discovery on its local event log. Each client applies *GASPD* to discover Pareto-optimal models. From these Pareto-optimal models, one model is selected and returned to the server. The server then merges the received local models to produce A^{agg}. Once A^{agg} has been constructed, the optimization phase of *OptFedGASPD* begins.

In line 2, the algorithm identifies the set of subtree states Q_T using the approach introduced in Sect. 4.1. For example, in Fig. 4, a merged model constructed from the event logs L_1 and L_2 introduced in Sect. 3, two subtrees are identified: one with the root state q_3 and another with the root state p_3. The subtree rooted at p_3 consists of the single state p_3. Therefore, it holds that $Q_T = \{q_3, q'_1, q'_3, q''_1, p_3\}$.

Algorithm 2: *OptFedGASPD*

Input: Initial population size p, number of generations m, number of parents k to generate offspring, entropic relevance degradation threshold τ, and a non-empty list of clients $C = \langle c_1, \ldots, c_n \rangle$

Output: A^{opt}, an optimized DFFA obtained by pruning the aggregated DFFA constructed from the local event logs of the clients in C

1 $A^{agg} \leftarrow FedGASPD(p, m, k, C.logs)$;
2 $Q_T \leftarrow GetSubtreeStates(A^{agg})$;
3 $ER \leftarrow \emptyset$;
4 **parallelfor** $c \in C$ **do**
5     $ER_c \leftarrow c.EvaluateDFFA(A^{agg})$;
6     $ERC \leftarrow \{(ER_c, \emptyset)\}$;
7   **for** $q \in Q_T$ **do**
8         $A \leftarrow PruneDFFAState(A^{agg}, q)$;
9         $ERC_{c,q} \leftarrow c.EvaluateDFFA(A) - ER_c$;
10         $ERC \leftarrow ERC \cup \{(ERC_{c,q}, q)\}$;
11     $ER \leftarrow ER \cup \{(ERC, c)\}$;
12 $A^{opt} \leftarrow PruneDFFA(A^{agg}, Q_T, ER, \tau)$;
13 **return** A^{opt};

In line 3, the algorithm initializes the container *ER*, which stores the changes in entropic relevance for each client resulting from the pruning of each state in Q_T in A^{agg}. This container is populated by each client in the for loop of lines 4–11. Each client computes the corresponding values independently, and the server collects the resulting client-specific containers into *ER*. In the iteration of this for loop for client c, first, the *EvaluateDFFA* function is invoked in line 5 to compute the entropic relevance of A^{agg} to the local event log of client c. The notation *c.EvaluateDFFA* (A^{agg}) denotes the evaluation of automaton A^{agg} with respect to the local event log stored at client c. For example, ER_c, of the A^{agg} DFFA in Fig. 4 is 2.103 for event log L_1 at client c_1 and 1.820 for L_2 at client c_2. In line 6, the container ERC is initialized, which stores the baseline entropic relevance and changes of entropic relevance values associated with pruning states in Q_T. The pair $(ER_c, \emptyset)$ represents the entropic relevance before pruning. Then, in the loop of lines 7–10, for each state in Q_T, the change in entropic relevance caused by its pruning is computed. First, the pruned automaton A is constructed in line 8. The function *PruneDFFAState* constructs the automaton obtained by pruning state q and the subtree rooted at q. Then, the change in entropic relevance $ERC_{c,q}$ is established in line 9; a positive change indicates a degradation in the model's quality. Finally, in line 10, the change value is stored in the ERC container. For example, the change values associated with pruning states in Q_T of the running example are $\{(0, q_3), (0, q_1'), (0, q_3'), (0, q_1''), (2.377, p_3)\}$ for event log L_1 at client c_1 and $\{(2.843, q_3), (2.819, q_1'), (0, q_3'), (0, q_1''), (0, p_3)\}$ for event log L_2 at client c_2. The entry $(2.377, p_3)$ for L_1 indicates that pruning state p_3 in A^{agg} increases the entropic relevance from 2.103 to $2.103 + 2.377 = 4.480$, resulting in a degradation of model quality of 2.377. In contrast, removing p_3 does not affect client c_2 as the corresponding degradation in entropic relevance is zero. After processing all states in Q_T, the container ERC of the client is added to ER in line 11.

Once all clients have computed the effects of candidate pruning operations on entropic relevance, the server solves the optimization problem introduced in Sect. 4.1 to identify the states to be pruned and then applies the pruning to construct the automaton A^{opt}, refer to line 12. The resulting automaton is the final output of *OptFedGASPD*. Considering the running example, the resulting automaton is shown in Fig. 6. Finally, A^{opt} can be converted into the SDAG representation shown in Fig. 7 [5].

To construct this result, the $\tau = 0.1$ threshold was used. When $\tau = 0.1$, the total degradation for client c_1 must not exceed $0.1 \times 2.103 \approx 0.210$, while for client c_2 it must not exceed $0.1 \times 1.820 = 0.182$. Consider the state $p_3 \in Q_T$. The degradation caused solely by pruning p_3 is 2.377, which exceeds the allowable threshold. Therefore, p_3 is not pruned. Similarly, pruning q_3 or q_1' would violate the degradation threshold. Thus, for $\tau = 0.1$, the set of pruned states is $\{q_3', q_1''\}$. These states can be pruned because their removal is not expected to increase the entropic relevance for either client.

5 Evaluation

This section presents an evaluation of *OptFedGASPD* and compares it with its predecessors, *FedGASPD* [28] and *GASPD* [5], which serve as federated and centralized baselines, respectively.

All experiments were conducted on a machine equipped with a 13th Gen Intel® Core™ i7-1355U processor running at 1.70 GHz and 16 GB of RAM. We implemented the *FedGASPD* and *OptFedGASPD* algorithms and made the source code publicly available.[1] For evaluation, we used nine real-world event logs obtained from the IEEE Task Force on Process Mining repository.[2] These logs cover a broad range of process behaviors. Table 1 summarizes their main characteristics.

Table 1. Characteristics of the event logs used in experiments.

Event log	#Actions	#Traces	#Distinct traces	Max. trace length	Avg. trace length
BPI-2012	24	13,087	4,366	175	20.03
BPI-2013	4	7,554	1,511	123	8.67
BPI-2017	26	31,509	15,930	180	38.15
BPI-2018	34	2,861	2,498	680	61.58
BPI-2019	42	251,734	11,973	990	6.33
Request for Payment	34	6,449	753	27	11.18
Prepaid Travel Cost	19	6,886	89	20	5.34
Sepsis Cases	16	1,050	846	185	14.48
Hospital Billing	18	10,000	1,020	217	4.51

Table 2 lists the parameters used in the evaluation. The parameters p and m are used by *GASPD* to discover local process models at clients, denoting the initial population size and the number of generations, respectively, to use in the genetic optimization of *GASPD*. The parameter mms limits the size of the model extracted at each client; specifically, models with more than 200 nodes and arcs are discarded. In addition, models with more than 200 nodes and arcs produced by *FedGASPD* after merging local models are rejected. The parameter $\tau = 0.2$ specifies the maximum acceptable degradation in entropic relevance (ER) for each client in the optimization problem. Finally, n denotes the number of computational nodes, or clients, used to simulate cross-organizational environments. Each event log is randomly split across up to eight nodes, ensuring that each node receives approximately the same share of the log traces.

Table 2. Experiment settings.

Parameter	Description	Value(s)
p	Population size	50
m	Number of generations	50
mms	Maximum accepted model size	200
τ	Maximum accepted ER degradation	0.2
n	Number of clients	2, 4, 8

[1] https://github.com/HzhianUnimelb/BPM2026Code.
[2] https://www.tf-pm.org/resources/logs.

Table 3. Difference in the number of models contributed to the global Pareto front.

Event log	*FedGASPD* vs *OptFedGASPD*		
	2 nodes	4 nodes	8 nodes
BPI-2012	+11	+7	+8
BPI-2013	+1	+1	+2
BPI-2017	+8	+12	+5
BPI-2018	+3	+8	+5
BPI-2019	+5	+4	+2
Request for Payment	-3	+4	+7
Prepaid Travel Cost	+7	+10	+9
Sepsis Cases	+4	+5	+9
Hospital Billing	+5	+3	+4

Table 4. Execution times of *GASPD*, *FedGASPD*, and *OptFedGASPD* algorithms.

Event log	Execution time (in seconds)						
	GASPD	*FedGASPD*			*OptFedGASPD*		
		2 nodes	4 nodes	8 nodes	2 nodes	4 nodes	8 nodes
BPI-2012	417.0	178.2	83.5	61.0	179.5	84.2	61.6
BPI-2013	84.1	35.4	26.2	11.1	36.7	27.2	11.5
BPI-2017	3,122.8	1,459.9	514.8	342.4	1,451.4	516.2	343.2
BPI-2018	2,891.0	1,325.3	165.4	76.3	1,326.8	166.3	76.9
BPI-2019	2,279.5	2,216.2	471.2	251.0	2,217.3	472.5	251.8
Request for Payment	1,402.2	685.9	332.6	229.4	686.0	342.2	236.3
Prepaid Travel Cost	172.4	94.0	57.2	46.2	95.7	58.6	46.9
Sepsis Cases	422.7	233.7	64.0	31.4	235.8	65.4	32.0
Hospital Billing	63.0	32.5	21.1	16.2	33.6	21.3	16.4

Figure 8 shows the characteristics of the Pareto-optimal models discovered from the event logs by *OptFedGASPD* and *FedGASPD* for different numbers of computational nodes (clients), while Table 3 compares the overall quality of the Pareto-optimal models. The numeric entries in Table 3 report by how many more models the number of *OptFedGASPD* models included in the Pareto front of all models, both discovered by *OptFedGASPD* and *FedGASPD* models, exceeds the number of *FedGASPD* models on this global Pareto front. For example, the value +11 for the BPI-2012 event log and 2 nodes indicates that, after jointly comparing all models and removing those dominated by others, *OptFedGASPD* contributes 11 more models to the global Pareto front than *FedGASPD*. Overall, Table 3 shows that, for most event logs used in our evaluation, a larger proportion of models discovered by *OptFedGASPD* is in the global Pareto front than models discovered by *FedGASPD*. The only exception is the Request for Payment log in the 2-node scenario.

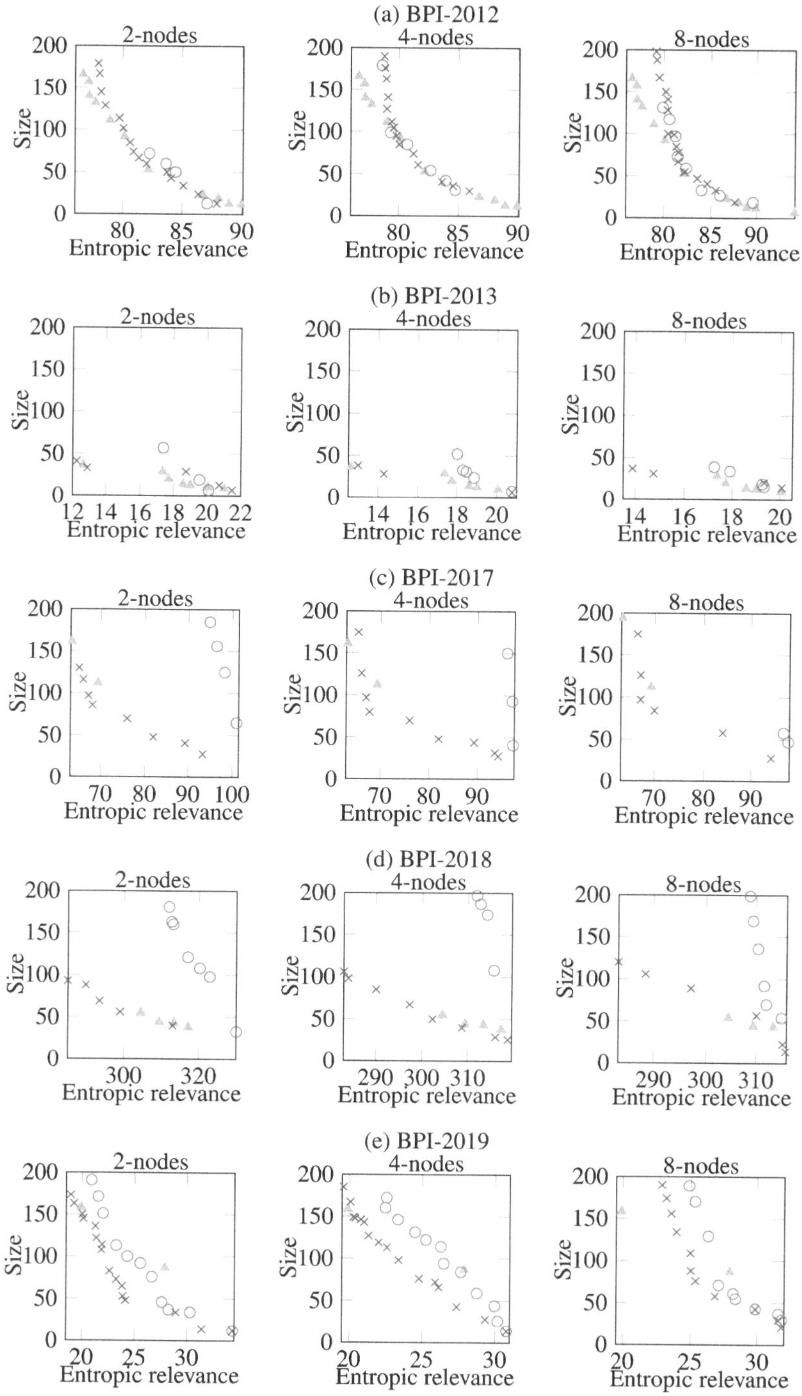

Fig. 8. Size and entropic relevance of DFFAs discovered by *GASPD*, *FedGASPD* and *OptFedGASPD*; smaller sizes and entropic relevance values signify better models.

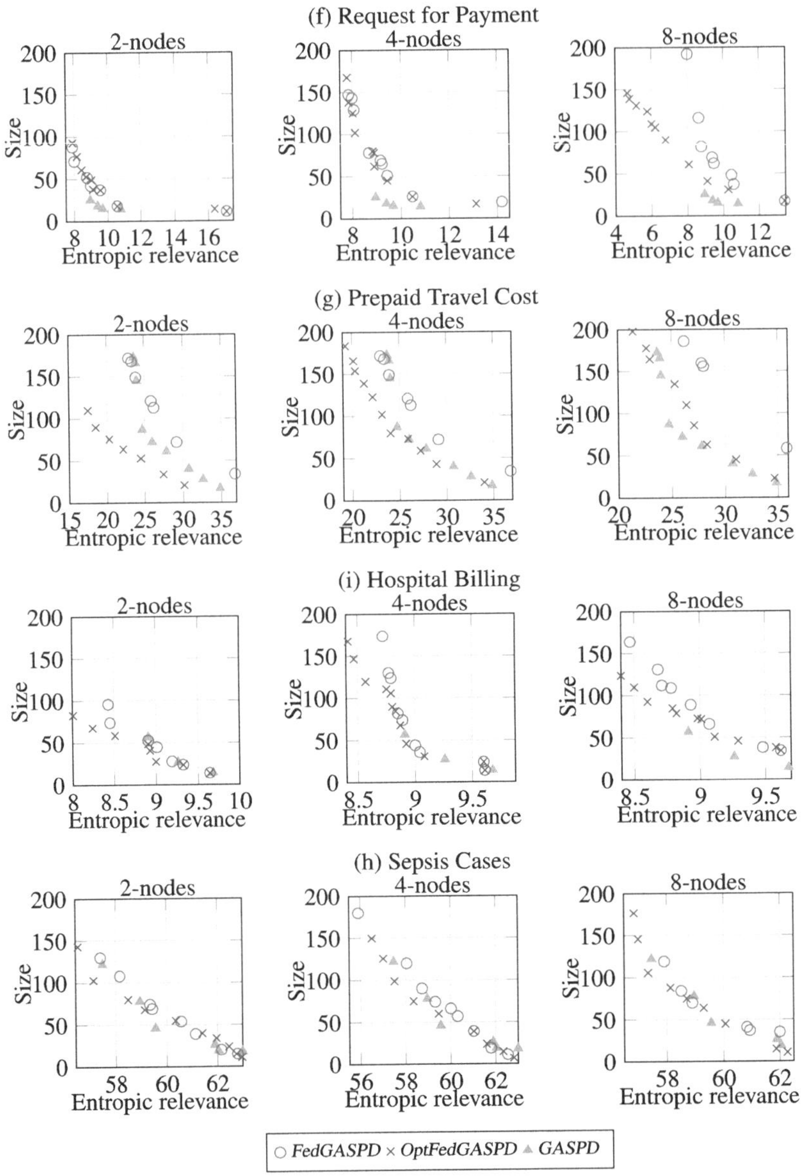

Fig. 8. *(continued)*

The results indicate that *OptFedGASPD* discovers more accurate and more compact models than *FedGASPD* more often. This improvement stems from the collaborative pruning process carried out by clients under server coordination. Specifically, the optimization removes infrequent local behavior that would otherwise add unnecessary complexity to the global model discovered. At the same time, behavior that is globally important is naturally emphasized by its recurring presence in local models contributed by multiple clients, ensuring its preservation in the optimized global model. These findings support a positive answer to RQ1: *yes*, in *OptFedGASPD*, clients collaboratively optimize the quality and size of the global process model constructed by FedGASPD.

Table 4 summarizes runtimes of *GASPD*, *FedGASPD*, and *OptFedGASPD*. Overall, the results show that the federated approaches consistently achieve substantial runtime reductions compared to the centralized *GASPD*. In addition, because the model size is capped at 200, the computational cost of the pruning phase in *OptFedGASPD* remains much lower than that of constructing the aggregated model itself. As a result, the extra runtime introduced by *OptFedGASPD* relative to *FedGASPD* is negligible. For instance, the maximum pruning time is approximately 2.8 s for the BPI-2017 event log in the 2-node scenario. These results suggest a positive answer to RQ2: *yes*, compared with centralized *GASPD*, *OptFedGASPD* is often faster, while its runtime remains comparable to that of *FedGASPD*.

6 Conclusions

This paper introduces *OptFedGASPD*, an approach to federated stochastic process discovery that addresses the limitations of existing methods in cross-organizational environments. By formulating the optimization of the resulting process model as a decision-tree pruning problem, *OptFedGASPD* systematically reduces the unnecessary complexity of the constructed model while preserving its quality for each participating organization. The proposed approach enables organizations to collaboratively optimize the constructed model through orchestration by a central server, ensuring privacy preservation and minimizing communication overhead. Experimental evaluations on real-world industrial event logs demonstrate that *OptFedGASPD* often produces smaller, more accurate models than the approach without pruning, namely *FedGASPD*. The pruning effectively eliminates unvisited states introduced during the aggregation process. These results highlight the potential of *OptFedGASPD* to improve the practicality of federated process discovery in cross-organizational settings. While the current work focuses on pruning states in subtrees disconnected from the main loops of the aggregated model, future research will aim to extend the pruning strategy to encompass all elements that contribute to the aggregated model.

References

1. van der Aalst, W.: Decomposing Petri nets for process mining: a generic approach. Distrib. Parallel Databases **31**(4), 471–507 (2013)
2. van der Aalst, W.: Process Mining—Data Science in Action. Springer, 2nd edn. (2016)
3. van der Aalst, W.: A practitioner's guide to process mining: limitations of the directly-follows graph. Proc. Comp. Sci. **164**, 321–328 (2019)

4. van der Aalst, W.: Federated process mining: exploiting event data across organizational boundaries. In: SMDS, pp. 1–7, IEEE (2021)
5. Alkhammash, H., Polyvyanyy, A., Moffat, A.: Stochastic directly-follows process discovery using grammatical inference. In: CAiSE, LNCS, vol. 14663, pp. 87–103. Springer (2024)
6. Alkhammash, H., Polyvyanyy, A., Moffat, A., García-Bañuelos, L.: Entropic relevance: a mechanism for measuring stochastic process models discovered from event data. Inf. Syst. **107**, 101922 (2022)
7. Andersen, J., Rathje, P., Imenkamp, C., Koschmider, A., Landsiedel, O.: EdgeMiner: distributed process mining at the data sources. In: SAC, pp. 705–713. ACM (2025)
8. Augusto, A., Dumas, M., La Rosa, M.: Metaheuristic optimization for automated business process discovery. In: Hildebrandt, T., van Dongen, B.F., Röglinger, M., Mendling, J. (eds.) BPM 2019. LNCS, vol. 11675, pp. 268–285. Springer, Cham (2019). https://doi.org/10.1007/978-3-030-26619-6_18
9. Augusto, A., Dumas, M., La Rosa, M., Leemans, S.J.J., vanden Broucke, S.K.L.M.: Optimization framework for DFG-based automated process discovery approaches. Softw. Syst. Model. **20**(4), 1245–1270 (2021). https://doi.org/10.1007/s10270-020-00846-x
10. Buijs, J.C.A.M., van Dongen, B.F., van der Aalst, W.: Quality dimensions in process discovery: the importance of fitness, precision, generalization and simplicity. Int. J. Cooper. Info. Syst. **23**(01), 1440001 (2014)
11. Carmona, J., Cortadella, J., Kishinevsky, M.: Divide-and-conquer strategies for process mining. In: Dayal, U., Eder, J., Koehler, J., Reijers, H.A. (eds.) BPM 2009. LNCS, vol. 5701, pp. 327–343. Springer, Heidelberg (2009). https://doi.org/10.1007/978-3-642-03848-8_22
12. Carrasco, R.C., Oncina, J.: Learning stochastic regular grammars by means of a state merging method. In: Carrasco, R.C., Oncina, J. (eds.) ICGI 1994. LNCS, vol. 862, pp. 139–152. Springer, Heidelberg (1994). https://doi.org/10.1007/3-540-58473-0_144
13. Gatta, R., et al.: On the feasibility of distributed process mining in healthcare. In: Rodrigues, J.M.F., et al. (eds.) ICCS 2019. LNCS, vol. 11540, pp. 445–452. Springer, Cham (2019). https://doi.org/10.1007/978-3-030-22750-0_36
14. Harviainen, J., Sommer, F., Sorge, M., Szeider, S.: Optimal decision tree pruning revisited: algorithms and complexity. In: ICML, Proceedings of Machine Learning Research, vol. 267 (2025) PMLR / OpenReview.net
15. Hyafil, L., Rivest, R.L.: Constructing optimal binary decision trees is NP-complete. Inf. Process. Lett. **5**(1), 15–17 (1976)
16. Khan, A., Ghose, A., Dam, H.: Cross-silo process mining with federated learning. In: Hacid, H., Kao, O., Mecella, M., Moha, N., Paik, H. (eds.) ICSOC 2021. LNCS, vol. 13121, pp. 612–626. Springer, Cham (2021). https://doi.org/10.1007/978-3-030-91431-8_38
17. Muñoz-Gama, J., Carmona, J.: A fresh look at precision in process conformance. In: Hull, R., Mendling, J., Tai, S. (eds.) BPM 2010. LNCS, vol. 6336, pp. 211–226. Springer, Heidelberg (2010). https://doi.org/10.1007/978-3-642-15618-2_16
18. Nucciarelli, L., et al.: Towards distributed process discovery in healthcare: Testing and proving the feasibility of the federated alpha+ algorithm. In: Artificial Intelligence in Medicine, LNCS, vol. 15735, pp. 294–299. Springer (2025)
19. Polyvyanyy, A.: Notes of a process scientist: process discovery. In: Mining a Scientist's Process, pp. 368–383. Springer (2026)

20. Polyvyanyy, A., Moffat, A., Garcia-Banuelos, L.: An entropic relevance measure for stochastic conformance checking in process mining. In: ICPM, pp. 97–104. IEEE (2020)
21. Rafiei, M., van der Aalst, W.: An abstraction-based approach for privacy-aware federated process mining. IEEE Access **11**, 33697–33714 (2023)
22. Rafiei, M., Pourbafrani, M., van der Aalst, W.: Federated conformance checking. Inf. Syst. **131**, 102525 (2025)
23. Rennert, C., Albers, J., Leemans, S.J.J., van der Aalst, W.: Your secret is safe with me: federated directly-follows graph discovery. In: ICPM, pp. 1–8. IEEE (2025)
24. Rojo, J., Garcia-Alonso, J., Berrocal, J., Hernández, J., Murillo, J.M., Canal, C.: SOWCompact: a federated process mining method for social workflows. Inf. Sci. **595**, 18–37 (2022)
25. Yan, J., et al.: Enhancing process discovery by optimizing imprecise sub-processes. IEEE Trans. Serv. Comput. **19**(1), 337–350 (2026)
26. Yan, Z., et al.: Decomposed and parallel process discovery: a framework and application. Futur. Gener. Comput. Syst. **98**, 392–405 (2019)
27. Zhian, H.: Federated stochastic process discovery: definition, benefits, and challenges. In: BPM Doctoral Consortium, CEUR Workshop Proceedings, vol. 4032, pp. 40–45 (2025). CEUR-WS.org
28. Zhian, H., Buyya, R., Polyvyanyy, A.: Federated stochastic process discovery using grammatical inference. In: CAiSE, LNCS, vol. 15702, pp. 76–93, Springer (2025a)
29. Zhian, H., Buyya, R., Polyvyanyy, A.: Multi-objective metaheuristics for effective and efficient stochastic process discovery. In: BPM, LNCS, vol. 16044, pp. 469–486, Springer (2025b)

Track III: Management

Heuristic Work Crafting: A BPM-Inspired Approach to Individual Process Redesign

Iris Beerepoot(✉), Inge van de Weerd, Thomas Martens, Maria Peeters, and Hajo A. Reijers

Utrecht University, Utrecht, The Netherlands
{i.m.beerepoot,i.vandeweerd,t.a.m.martens,m.peeters,h.a.reijers}@uu.nl

Abstract. Knowledge workers have considerable autonomy in how they structure their daily tasks, yet systematic guidance for improving individual work practices remains scarce. This paper introduces *CRAFT@work*, an approach that bridges job crafting theory from Work & Organisational Psychology with process redesign heuristics from the BPM field to support self-directed work improvement. We developed a catalogue of 29 individual work redesign heuristics, adapted from established process redesign heuristics, and evaluated them through an intervention study with 12 knowledge workers across diverse occupational contexts. The intervention consisted of individualised coaching sessions in which participants reflected on their current work, created a Personal Crafting Plan, and subsequently enacted and reflected on their chosen heuristics over a four-week period. Results show that the heuristics are broadly applicable across the occupational contexts included in this study, with every heuristic considered relevant by at least five participants. Almost all participants were able to enact meaningful changes to their work, with several reporting concrete improvements. The most impactful heuristics specifically target the ambiguities and information bottlenecks that slow down or complicate work, making BPM a uniquely situated discipline for addressing the friction that knowledge workers experience.

Keywords: Process redesign · Individual work redesign · Job crafting · Knowledge workers

1 Introduction

The Business Process Management field offers a rich body of knowledge for improving the way work is organised and improved on an organisational level. The well-established process redesign heuristics [20] provide practical, field-derived guidelines for making changes to existing processes to improve performance. Since their initial publication, these organisational guidelines have been evaluated, extended, and applied in a wide range of settings, including the area

F. M. Maggi et al. (Eds.): BPM 2026, LNCS 16936, pp. 633–650, 2027.
https://doi.org/10.1007/978-3-032-37777-7_37

of resource allocation (e.g., [15]). Resource allocation is concerned with clarifying which resources take care of which tasks, but it does not say how resources should organise their tasks once they have been allocated. While business processes provide high-level structures, individual workers routinely decide how to sequence their activities, which tasks to prioritise, when to delegate, and how to recover from interruptions. Knowledge workers, in particular, typically enjoy considerable latitude over how they structure and execute their daily work [10]. This latitude, though valuable, comes with a challenge: without systematic guidance, workers may unknowingly establish inefficient habits or miss opportunities to meaningfully improve their work processes.

A growing body of research has drawn attention to human-centered dimensions that complement BPM's traditionally quantitative focus. This includes the work on *benevolent process design* [9,22], *technostress* [25], and *employee well-being* [6]. Yet, despite this human-centric turn, methodological support for process redesign has remained anchored at the organisational level. Redesign is typically performed by process analysts or improvement teams to restructure inter-unit workflows, resource allocation schemes, and coordination mechanisms. Empirical studies have shown that positive process innovations, however, may also emerge in a bottom-up manner, by workers demonstrating their agency by reorganising work or deviating from procedure when the situation asks for it [21]. In the present study, we shift our focus to exactly those workers who participate in multiple processes simultaneously, have genuine agency over how they execute their daily work, and stand to benefit directly from targeted work redesign.

Work & Organisational Psychology addresses this bottom-up agency under the concept of *job crafting*: the proactive, self-initiated changes that employees make to the tasks, relationships, and cognitive framing of their work [3,23,26]. Such bottom-up work redesign has been shown to enhance engagement, reduce burnout, and improve job performance [18,19,24]. However, job crafting frameworks are motivationally rich but structurally lean: they characterise the dimensions of self-initiated redesign without providing process-oriented guidance on what to actually change. Conversely, BPM redesign heuristics offer precise, actionable interventions, but have traditionally been applied top-down by process analysts rather than by the workers themselves. Combining the two traditions addresses both gaps simultaneously: job crafting supplies the motivational and agentic foundation, while BPM heuristics supply the structural vocabulary that turns crafting intentions into concrete redesign actions.

This paper bridges the two traditions by translating the BPM process redesign heuristics into actionable individual work redesign actions. Specifically, we present a novel approach to Clarify, Reduce, Align and Focus Tasks at Work (CRAFT@work), a structured method for guiding individual knowledge workers through the systematic redesign of their own work without requiring a process analyst or manager to make changes. We evaluate the approach through a field study with 12 knowledge workers across diverse occupational contexts, each of whom applied self-selected heuristics over a four-week period and reflected on the results. In doing so, we make three contributions. First, we develop and verify a catalogue of individual work redesign heuristics, adapted from classical BPM process redesign heuristics. Second, we propose a hands-on, practical method that operationalises these heuristics and allows knowledge workers to apply them to revise their way of working. Third, we provide empirical evidence demonstrating the applicability and perceived usefulness of the framework across a range of occupational contexts. The remainder of the paper is structured as follows. Section 2 introduces the conceptual background. Section 3 describes the study's design, while Sect. 4 presents the results of the approach. Section 5 discusses the implications of our study, before concluding in Sect. 6.

2 Theoretical Background

How assigned work is executed is rarely fixed, and even in routine roles employees can exert influence on the essence of their work [3,26]. The concept of job crafting captures this bottom-up agency at the job level, describing how employees actively shape and redefine aspects of their work [26]. It extends classic job design research by emphasising that jobs evolve through employee action rather than only through managerial, top-down specification [3,26]. Contemporary organisational conditions make this self-initiated adaptation increasingly salient, since workers are expected to create resourceful work environments that fit changing needs and contexts [12]. Adopting job crafting as the core lens for individual work redesign helps explain how employees proactively shape their work at the micro level. Therefore, in this paper we ground individual work redesign in job crafting theory and operationalise it with BPM-inspired heuristics and process components. Figure 1 illustrates our conceptual framework.

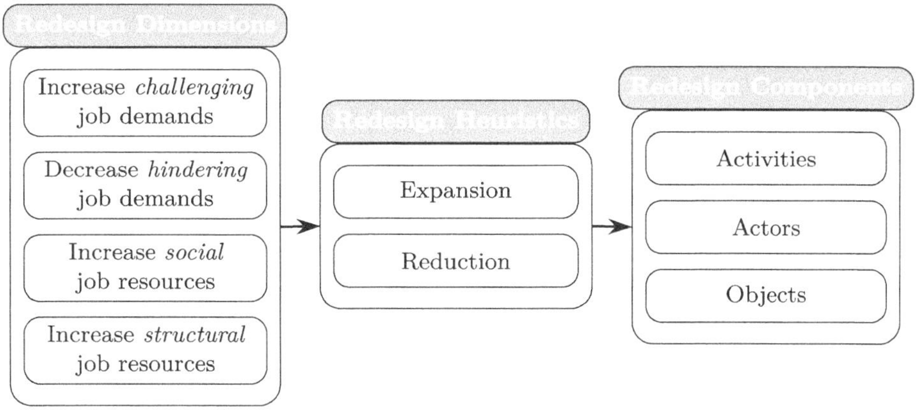

Fig. 1. Conceptual framework of individual work redesign.

2.1 Redesign Dimensions

Job crafting is about making work more meaningful, engaging, or satisfying [11]. For this reason, it is different from role innovation or related concepts, which are generally about addressing a problem or inefficiency at work [23,26]. Job crafters change aspects of their jobs in the interest of enhancing benefit for themselves [23], which may, in turn, also benefit the workplace [3]. To make such benefits more concrete, job crafting is often positioned in terms of the Job Demands and Resources (JD-R) model. In the JDR model, job demands are the physical, psychological, social, or organizational aspects of work that require sustained effort and can entail costs, whereas job resources are aspects that help achieve work goals, reduce demands and their costs, and foster growth, learning, and development [13]. Together, they influence employee well-being and performance. Job crafting is then seen as the act of balancing job demands and resources with the individual's skillset and desires. Specifically, job crafting is believed to consist of four *dimensions*: (1) increasing challenging job demands,(2) decreasing hindering job demands, (3) increasing social job resources, and (4) increasing structural job resources.

2.2 Redesign Heuristics

In terms of the crafting itself, there are two dominant streams of literature [19]. *Role-based crafting* focuses on the boundaries of the role, i.e., what the individual is doing (task crafting), with whom they are interacting (relational crafting), and how they view the work (cognitive crafting) [26]. The other stream, which is referred to as *resource-based crafting*, is directly tied to the job demands and resources [23]. Examples of resource-based crafting are searching for frequent feedback from colleagues or cutting a task to reduce workload. While the focus is different, both perspectives include a distinction between expansion or reduction of work or resources, a distinction broadly analogous to approach versus avoidance crafting, or promotion versus prevention-oriented crafting [7,18,19]. The first category generally refers to active approaches to improve some aspects of work. This category is linked to the objective of increasing challenging demands and resources. The second is the opposite, and includes reducing certain aspects of work in the interest of decreasing hindering job demands.

Whichever perspective is taken, the job crafting literature characterises what kind of change workers pursue, but does not specify what can be changed. As a result, a worker who wants to reduce coordination load may know what they want, but not where to start or what options are available to them. The process redesign heuristics from the BPM field may address this gap: they offer a structured and comprehensive vocabulary for reasoning about what could be changed, albeit at the management level. In the present study, we will explore whether the redesign heuristics established at the process level may also benefit individuals redesigning their work at their own initiative.

2.3 Redesign Components

Where the job crafting literature stops at providing examples of how employees might adapt their work, the BPM field offers valuable constructs to help employees redesign their work on a daily basis. In the interest of anchoring redesign tactics to the daily work of individuals, we adopt the core *components* of business processes: activities, actors and objects [14].

As an example, let us consider an assistant professor in search of a way to make their work more meaningful or enjoyable. Let us assume the person is interested in achieving two objectives: (1) to decrease hindering job demands, namely reducing coordination load and time pressure, while also (2) increasing social job resources, namely increasing support from colleagues. Now, consider a redesign heuristic such as *contact reduction*. The professor might benefit from consolidating student questions into a single weekly online office hour in the interest of decreasing hindering demands. Another option to decrease hindering demands might be to apply *triage*, by processing email twice daily using a simple urgent/important/defer/delegate rule, routing non-urgent items to a weekly admin block. For increasing social resources, the professor might opt for *split responsibilities*, by assigning a teaching assistant first-line ownership of routine student queries, meaning they would only need to handle exceptions. Table 1 depicts the anchoring of these heuristics to the different components and objectives. In this study, we explore whether our process redesign heuristics are applicable in the context of individual work redesign and whether they can be integrated in a systematic approach for individuals to improve their work.

Table 1. Example application of redesign heuristics to individual work.

Heuristic	Associated activities	Associated actors	Associated objects	Objective
Contact reduction	Answer questions about the course	Students	Office-hour	Decrease coordination load
Triage	Taking care of admin	Students, colleagues, external partners	Email inbox	Decrease time pressure
Split responsibilities	Answer questions about the course	Teaching assistant; students	n.a.	Increase colleague support

3 Research Method

To evaluate the approach, we adapted the process redesign heuristics to the individual worker level and conducted work redesign interventions with 12 knowledge workers and one additional participant serving as a pilot. The adaptation was carried out by two members of the author team, each with over seven years of experience teaching BPM and process redesign heuristics; one of whom is also the original author of the process redesign heuristics. This ensured that the adaptation preserved not only the surface logic of each heuristic, but also the intent and scope boundaries that underpin the original catalogue. It followed a consistent principle: for each heuristic, we preserved the underlying logic while shifting the locus of action from the organisation to the individual. Where the original heuristics address inefficiencies across resources and are typically enacted by process analysts, the individual work redesign heuristics address inefficiencies in an individual's daily work and are enacted by the worker themselves[1]. To support and document the adaptation process, each heuristic was visualised as a pair of BPMN-style diagrams contrasting the organisational and individual-level situation, serving as a discussion tool for the adaptation process. They are available in the Zenodo repository alongside the full heuristic catalogue.

Table 2 provides examples of the adaptation of heuristics from organisational to individual. For some heuristics, little adaptation was needed to fit the individual level. For example, *Activity composition* already targeted the bundling of small tasks for one resource in the interest of reducing setup times. Other heuristics required deliberate reframing. For example, *Split responsibilities*, originally aimed at eliminating shared ownership between functional units, became *Clarify ownership*. Here, the focus shifts from restructuring organisational boundaries to the individual worker proactively discussing and clarifying who is responsible for what in their day-to-day work. Another example is *Buffering*, which targets the organisational inefficiency of repeatedly requesting information from external sources by advocating for proactive subscription to data feeds. At the individual level, the underlying principle of reducing delays caused by information unavailability is preserved, but the scope is narrowed: rather than redesigning information flows between units, the individual worker takes it upon themselves to ensure that the information, templates, and notes they regularly need are readily at hand. We renamed this heuristic *Increase information readiness*, to better fit the new focus and clarify its meaning to a broad audience.

The intervention participants were invited based on their occupation as classified by the International Standard Classification of Occupations (ISCO-08)[2]. Specifically, we included occupations from the major groups *Managers*, *Professionals*, and *Armed forces occupations*, as these are the categories that comprise occupations at Skill Level 4. Skill Level 4 occupations are characterised by high levels of autonomy, making them particularly suitable for the study of job crafting. Each participant belonged to a distinct ISCO-08 minor group, ensuring heterogeneity in occupational backgrounds while maintaining a consistent skill level across the sample. An overview of the participants and their intervention

[1] Detailed information about the heuristic titles and descriptions, the adaptation of heuristics from organisational to individual, participant results, and handouts (in Dutch and English) are available on Zenodo: https://doi.org/10.5281/zenodo.20522954.

[2] For information about the ISCO standard: https://edu.nl/k64y6.

Table 2. Examples of heuristic adaptation from organisational to individual level.

Original heuristic	Adapted heuristic	Rationale
Activity composition	Combine minor activities	Core logic unchanged; title simplified for a non-analyst audience.
Resequencing	Change the order of activities	Core logic unchanged; renamed to plain language accessible to workers without BPM training.
Triage	Differentiate the approach to activities	Original splits cases across resources; adapted version shifts focus to how one individual varies their own approach depending on case characteristics.
Split responsibilities	Clarify ownership	Original targets restructuring of functional boundaries; adapted version targets the individual proactively clarifying responsibilities in day-to-day collaboration.
Buffering	Increase information readiness	Original advocates subscribing to data feeds at system level; adapted version narrows scope to the individual ensuring their own information, templates, and notes are readily at hand.
Flexible assignment	Handle backup activities separately	Original concerns resource scheduling by management; adapted version reframes as the individual maintaining a pool of tasks to switch to when circumstances change.

timelines is presented in Table 3. The first one, P0, served as a pilot for testing the study design and assess understanding of the framework. After the pilot, small changes were made to the explanation of the dimensions and descriptions of the heuristics to clarify their meaning.

Our approach, *CRAFT@work*, blends the individual job crafting exercise of Berg et al. [4] with procedural elements from the team method of Van den Heuvel et al. [16]. The former contains steps for reflecting on the current job (as-is situation) and an ideal version of the job (to-be situation), closely mirroring BPM approaches. The latter consists of a job crafting workshop, a weekly logbook, and a reflection meeting. Three of the authors of the present study each guided interventions with four participants each. To ensure consistency across facilitators, the team discussed the pilot experience collectively and agreed on what to attend to in subsequent sessions. Throughout the study, facilitators held regular check-ins to align on emerging observations, and results were recorded continuously in a shared spreadsheet accessible to all facilitators. The initial session was further standardised through the use of structured handouts, ensuring all participants received the same information (albeit customised to their occupation) and followed the same steps regardless of facilitator. The weekly logbook and closing session used identical questions across all participants.

Table 3. Overview of study participants and intervention timeline.

ID	Occupation (ISCO code)	Initial session	Start appl.	Closing session
P0	Commissioned armed forces officer (110)	15.11.25	24.11.25	9.1.26
P1	System administrator (2522)	22.12.25	5.1.26	6.3.26
P2	Landscape architect (2162)	22.12.25	5.1.26	25.2.26
P3	Assistant professor (2310)	30.12.25	5.1.26	2.3.26
P4	Marketing manager (1221)	30.12.25	5.1.26	25.2.26
P5	IT manager (1330)	1.1.26	5.1.26	3.3.26
P6	Manufacturing manager (1321)	1.1.26	5.1.26	6.3.26
P7	Services manager (1439)	13.1.26	19.1.26	2.3.26
P8	Policy administration professional (2422)	12.1.26	26.1.26	5.3.26
P9	Market researcher (2431)	14.1.26	26.1.26	4.3.26
P10	Physiotherapist (2264)	23.1.26	26.1.26	4.3.26
P11	Lawyer (2611)	25.1.26	26.1.26	6.3.26
P12	Secondary education teacher (2330)	25.1.26	26.1.26	2.3.26

For each participant, the process consisted of three phases: an initial work redesign session, a four-week enactment period, and a closing reflection session. In the **initial session**, the facilitator held a one-on-one meeting with the participant to explain the relevant constructs and jointly draft a Personal Crafting Plan (PCP), as exemplified in [16]. We used printed handouts containing three elements mirroring our conceptual framework: (1) a brief explanation of the redesign dimensions, (2) an overview of individual work redesign heuristics, and (3) a list of object types (which include different types of actors) and work activities expected to be relevant to the participant's occupation, generated using the prompts outlined in [2]. These handouts, available on Zenodo, constitute the basis for the PCP: participants select applicable dimensions, heuristics, activities, and objects by ticking the relevant items, and add a brief description of where and how they intend to enact each chosen heuristic in their own work[3]. Note that we included object types rather than object instances, as the latter cannot be generated for a participant a priori. For example, it is to be expected that an assistant professor is involved with students in their daily work, but we do not know the names of the students. We assume that object types provide sufficient inspiration for participants to contextualise redesign ideas to their own work and distinguish the to-be situation from the as-is situation. During the session, we followed the handout checklist, verifying which redesign dimensions,

[3] Researchers or practitioners wishing to apply CRAFT@work can use the interactive ISCO-08 viewer in the Zenodo repository, which lists potential object types and activities for all Skill Level 4 occupations, as a basis for creating handouts for additional occupations.

object types, activities, and heuristics applied to the participant's work. Participants were then asked to select 2–4 applicable heuristics, each linked to specific activities and objects in their work and tied to a concrete redesign objective.

During the **enactment period**, participants applied their PCP over four weeks, maintaining a brief weekly logbook coordinated through a scheduled survey sent via WhatsApp or Signal. Each survey asked participants to specify (1) which heuristics they applied that week, (2) to which work aspects they applied them, and if applicable, (3) which obstacles were in the way. In the **closing session**, we reflected on the success of the approach, asking participants (1) which heuristics they were able to apply and what effect these had; (2) which they were unable to apply and why; and (3) which single change most directly arose from the method and what effect it had on which dimension.

4 Results

This section presents the results of the work redesign interventions conducted with 12 knowledge workers across a range of occupations. We structure the results around the three phases of the *CRAFT@work* approach: the initial session, in which participants identified applicable redesign dimensions and heuristics; the enactment period, in which they applied their PCP over four weeks; and the closing session, in which they reflected on the outcomes.

4.1 Initial Sessions

Figure 2 presents the applicability of each heuristic as assessed during the initial sessions. Overall, the results suggest that the heuristics are broadly applicable across the occupational contexts in this sample: each heuristic was considered applicable by five of the participants or more, and none was universally rejected. The four most widely applicable heuristics, *Clarify ownership*, *Sort cases by type*, *Change the order of activities*, and *Increase information readiness*, were each flagged by 9 out of 12 participants, suggesting that these represent widespread opportunities for work redesign among knowledge workers.

Even heuristics that might intuitively seem difficult to apply on an individual level turned out to be applicable in some situations. *Rely on trustworthy sources*, for instance, was formulated as *Trusted party* in the original process redesign heuristic, and is concerned with delegating validation to certified external parties. At the individual level, this translates to a worker proactively seeking out the expertise of a certified party within the context of their own work. An example of this was provided by P2, a landscape architect. He noted that he might choose to consult an expert from an advocacy organisation for people with physical disabilities when this is relevant to a landscape design. Not all heuristics proved applicable, however, and P2's case also illustrates how dependencies on others can constrain individual redesign. He noted that *Handle work per case*, for example, was not feasible in practice, as his work often required

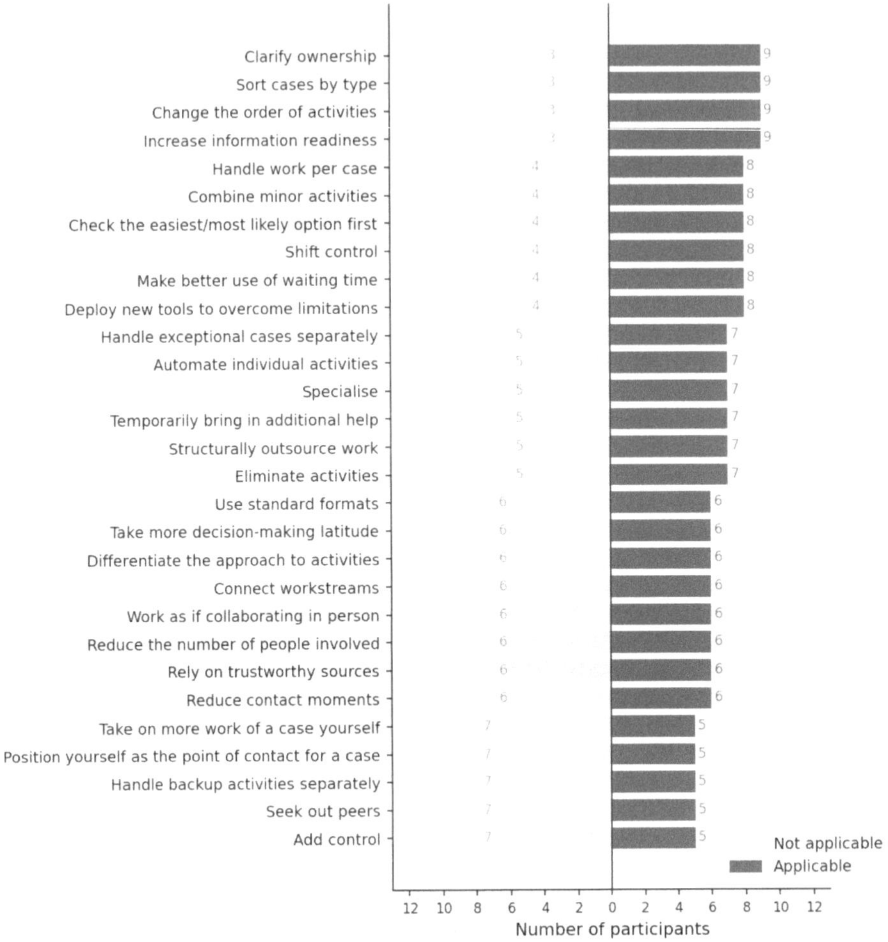

Fig. 2. Applicability of heuristics.

waiting for input from external stakeholders, making the sequential handling of cases difficult to sustain in practice.

A further pattern that emerged was that participants sometimes experienced heuristics not as isolated actions but as natural combinations. P7 noted that *Increase information readiness* and *Use standard formats* reinforced each other in practice, since having information readily at hand and having standardised ways of processing it served the same underlying goal. Similarly, P8 linked *Connect workstreams* and *Work as if collaborating in person*, finding that together they support better team coordination. Based on the applicable heuristics and participant statements, information friction appears to be a pervasive challenge in knowledge work, one that workers can meaningfully address through their own crafting efforts.

The most frequently selected dimension was *Decrease hindering job demands*, which suggests that the participants primarily saw opportunities for work redesign in reducing friction and unnecessary burden in their work. The heuristics most strongly associated with this dimension include *Automate individual activities*, *Clarify ownership*, *Shift control*, and *Increase information readiness*, each of which can be understood as ways of removing obstacles or ambiguities that slow down or complicate work. *Increase social job resources* forms the second most prominent node and is primarily associated with *Connect workstreams*. *Increase structural job resources* can count *Specialise* and *Sort cases by type* as its strongest connections, suggesting that participants seeking to build structural resources tended to focus on developing expertise and organising their work more systematically. *Increase challenging job demands* is the smallest node, with only a handful of matches: it was selected by only one participant, who linked three heuristics to this dimension. Notably, this participant was the most junior in the sample. The limited uptake of this dimension may reflect the nature of the occupations under study, i.e., highly educated professionals in demanding roles are perhaps less likely to seek out additional challenges, and more likely to focus on reducing the burden of their existing workload.

A further noteworthy pattern is that heuristics do not map one-to-one onto dimensions. Several heuristics appear connected to multiple dimensions, reflecting the fact that the same heuristic can serve different crafting goals depending on the individual and their occupational context. One participant, for instance, linked the heuristic *Take on more work of a case yourself* to decreasing hindering job demands, reasoning that handling more of a case independently would save time otherwise lost to delegation. For another participant, however, taking on more work of a case may be plausibly translated into an *increase* in hindering job demands. This is an inherent feature of individual work redesign: the meaning and purpose of a crafting action is ultimately determined by the worker enacting it. However, tying heuristics to dimensions is essential for a meaningful reflection on the outcomes of the intervention.

4.2 Enactment

During the four-week enactment period, participants applied the heuristics selected in their PCP and reported their progress through weekly logbook entries. Overall, the logbook entries suggest that the majority of participants were able to enact at least some of the planned changes, with several reporting concrete improvements to their work. At the same time, the degree of success varied considerably across participants, reflecting differences in occupational context, workload, and the nature of the heuristics selected.

Several participants demonstrated clear and systematic enactment of their plan. P5, the IT manager, applied two of the three selected heuristics over the course of the intervention: he began using Teams Premium to automate meeting notes, set up Confluence pages to centralise information for his team, and introduced inbox rules to manage notifications from his project management system. P7, a services manager, reported a similarly structured approach, bundling comparable tasks into single working sessions using standard formats, reorganising her folder structure, and proactively reaching out to colleagues in similar roles to explore collaboration. P4, a marketing manager, successfully introduced standardised process templates for three types of product launches, which were well received within her organisation and adopted for continued use. P11, a lawyer, began the enactment period by raising with her manager the need for a secretary to receive additional training, enabling her to delegate those tasks more effectively. In subsequent weeks, she pushed back on attempts to shift secretarial tasks onto the lawyers and took more initiative in her own casework.

Other participants enacted their plans partially, often due to external circumstances rather than a lack of motivation. P3, an assistant professor, found that the intervention coincided with the busiest period of the academic year, leaving little room to restructure the work. He did manage to block dedicated time for a strategic writing project, an application of *Handle exceptional cases separately*, but noted that the pressure of consecutive deadlines made sustained enactment difficult. P6, an industry manager, similarly reported a slow start due to the demands of the new year, though he did manage to delegate a number of smaller tasks to a colleague. By the end of the enactment period, he had managed to make structural changes.

P10, a physiotherapist, also faced a pressured schedule, but nonetheless managed to enact several of the planned heuristics. She began completing patient files fully before moving on to the next case, asked new patients to fill in their own questionnaires, freeing up time for the parts of the file only she could complete, and in quieter moments aimed at researching specific knee pathologies as part of her specialisation goal. A particularly creative adaptation emerged when she began scheduling patients with similar conditions simultaneously, allowing her to train them together. She noted that the changes she applied reduced the amount of work she needed to do outside of her contracted hours.

4.3 Closing Sessions

The closing sessions served primarily as a reflection on the intervention, asking participants to assess overall outcomes and name impactful changes they had made, if applicable. Among the twelve participants, ten reported partial or full enactment of their plans. The two participants who reported the least change attributed this primarily to external dependencies. Both planned on implementing a system that would streamline work, but the system could not be implemented within the intervention window.

A recurring theme across the sessions was that the most meaningful changes often entailed reshaping how responsibilities and expectations were distributed with others. P2, the landscape architect, identified *Clarify ownership* as the most meaningful change; he provided a specific example of a multi-stakeholder project where advisors from different disciplines send queries to a shared address and where the team often waited for him to distribute messages. By making explicit who picks up what type of incoming request from the email inbox, he reduced a persistent coordination bottleneck. He linked this directly to a decrease in hindering job demands for him, and more room for colleagues to grow their own responsibilities. Another example came from P10, the physiotherapist, who reported that the renewed clarity she had developed about how she wanted to use her administrative time gave her a concrete basis for a conversation with her manager, in which she successfully negotiated for those blocks to be recognised as paid working time.

The most substantial, measurable outcome across the study came from P7, the services manager, who reported in her closing session that the full event programme for her organisation had been planned through to the end of the academic year, for the first time ever at this calendar point. She believed this to be a direct result of the heuristics she applied during the enactment period, noting in particular that clustering work per case and applying standard formats across workstreams allowed her to execute all elements of a workstream consecutively rather than spreading them across the week. A more subtle but telling outcome came from P9, a market researcher, who reported a concrete instance of applying *Take more decision-making latitude*. Faced with last-minute feedback on a report due on a Friday afternoon, she decided not to incorporate the suggestions, reasoning that the report was already sound. She noted that this decision left her without stress, and described it as something she would carry forward into her future work.

P8 offered a complementary reflection on the value of the intervention itself: she noted that participation had prompted her to reflect on her work responsibilities at a depth she said she would otherwise only reach during a formal performance review. In her view, the study had made explicit what would normally remain implicit until an annual assessment moment. She also offered a broader reflection on the conditions under which the approach is most likely to succeed, noting that individual crafting efforts would be more effective if entire teams engaged in the process simultaneously, since heuristics such as *Connect workstreams* and *Work as if collaborating in person* ultimately depend on the willingness of colleagues to coordinate.

5 Discussion

The results of our field study support the core premise of this work: that process redesign heuristics, originally developed to guide organisational process improvement, can be meaningfully translated to the level of individual work. In this section we discuss the theoretical and practical implications of our approach, consider the mechanisms through which the approach drove change, reflect on the observed dominance of demand-reduction crafting, and outline directions for future work.

5.1 Extending BPM Heuristics to the Micro-Level

The central theoretical contribution of this paper is the successful extension of process redesign techniques to the individual level. The fact that all 29 heuristics were considered applicable by several participants, and that the four most widely applicable heuristics (*Clarify ownership*, *Sort cases by type*, *Change the order of activities*, and *Increase information readiness*) were found almost universally applicable, provides initial evidence that the redesign logic underlying organisational process improvement is applicable at the personal level. Our results suggest that analogous patterns may govern the micro-level execution of individual work. The four most frequently chosen heuristics in particular suggests that they may function as entry points for individual work redesign more broadly, and as strong candidates for inclusion in practical guides or training for knowledge workers. By adapting one of our BPM techniques for usage by process participants rather than management, we directly respond to the call of Björn [5] to shift from designing processes for management, towards designing them for the participants whose interests process design ultimately serves.

The empirical findings also invite a refinement of the conceptual framework. As noted in the results, the heuristics do not always map directly to either expansion or reduction. The same heuristic can serve either depending on who enacts it and in what context, or neither, i.e., in situations where the *amount of work* is not expanded or reduced, but the *order or content of the work* is changed. This suggests that the binary axis, while useful as an orienting framework, undersells the complexity of the heuristics in practice.

5.2 The Reflective Structure as a Mechanism of Change

A recurring theme in the closing sessions was that participants valued not only the specific heuristics they enacted, but also the structured reflection that the intervention invited. This points to a mechanism that operates beyond the heuristics themselves: the combination of a PCP and a weekly logbook appears to have functioned as a structured self-monitoring system. By asking participants to commit to specific heuristics, map them to concrete work activities and objects, our approach helps translate intention into action. The PCP and logbook can in this sense be understood as a soft commitment device [8]: a voluntary arrangement that helps individuals follow through on plans they would otherwise defer. Future work could demonstrate whether a self-administered version of the approach, with or without an additional commitment device, can preserve the reflective benefits that the facilitated version offered.

5.3 The Dominance of Demand Reduction

An empirical contribution is that the *Decrease hindering job demands* dimension accounted for substantially more heuristic-dimension associations than the other three dimensions, and was the primary crafting objective for the majority of participants. Our sample appears to exemplify a profile in which workers are already working at or near capacity and consequently prioritise removing friction and burden over seeking additional challenges. This is consistent with the results from a meta-analysis of Lichtenthaler and Fischbach [19], who report that workers in high-pressure conditions tend towards reduction-focused strategies. They show that burnout is actually an antecedent of reduction-focused crafting, not just an outcome of job demands and resources. This makes an approach such as CRAFT@work all the more important: by providing concrete redesign actions, the approach gives workers actionable levers for demand reduction to break the loss cycle of burnout → reduction-focused crafting → more burnout.

We also observed that crafting orientation may be moderated by career stage: earlier-career workers may lean towards expansion crafting, while more senior professionals in already saturated roles lean towards reduction. Kooij et al. [17] provide a theoretical account of this shift: they identify *accommodative crafting* as a distinct mode that becomes more salient for older workers as personal resources grow more constrained, characterised by strategies focused on regulating losses such as delegating lower-priority tasks and reducing workload rather than seeking growth or new challenges.

5.4 Limitations and Future Work

This study has a number of limitations that point to directions for future work. With 12 participants spanning 12 distinct occupational contexts, the study is exploratory in nature, and findings on heuristic applicability and dimension-heuristic associations should be understood as indicative rather than generalisable. Larger-scale and longitudinal studies tracking participants well beyond the four-week window would allow more in-depth quantitative analysis of the applicable heuristics and dimension-heuristic relations, as well as an assessment of whether enacted changes are sustained and generalised to other aspects of work. Although we took measures to mitigate facilitator bias, the involvement of three author-facilitators remains a limitation that future work could address through independent facilitation.

A particularly interesting avenue would be an integration with event data and process mining tools, which could provide more objective indicators of change in work patterns and could make such large-scale and longitudinal studies feasible. The PCP could function as the normative process model and creates a natural basis for conformance checking. The normative process model could be checked against the individual's computer behavior logs, such as Active Window Tracking (AWT) data, i.e., automatically collected logs of applications and windows that are active at any given moment [1]. Braakman et al. [6] demonstrate that AWT data can be used to detect recurring work organisation patterns. Such analysis could serve as a measurement instrument for evaluating the intervention: pre- and post-intervention AWT data could reveal whether a participant who committed to a heuristic indeed shows increases in that work pattern, and the implications for their work practices as a whole.

A further direction concerns the derivation of new heuristics. The present study took a top-down approach, adapting existing organisational heuristics to the individual level. A complementary bottom-up approach, deriving heuristics directly from observed individual needs, could enrich the catalogue and potentially surface redesign strategies that in turn inform organisational-level redesign.

6 Conclusion

This paper set out to investigate whether the process redesign heuristics can be meaningfully applied at the level of individual work, and incorporated into an approach that builds on job crafting theory to improve the work of individuals. All 29 heuristics were considered applicable by multiple participants spanning twelve distinct occupational contexts, and the majority of participants were able to enact concrete and meaningful changes to their work during the four-week intervention. We make three contributions. First, we present a verified catalogue of individual work redesign heuristics, adapted from established BPM process redesign heuristics and shown to be applicable across the knowledge work occupations included in this study. Second, we introduce *CRAFT@work*, a structured method that operationalises these heuristics by anchoring them to activities and objects. This combination transforms job crafting from a motivational framework into a methodological one, providing workers with concrete redesign tactics. Third, we provide empirical evidence from a field study providing initial evidence of the practical applicability of the approach and its value across a range of occupational contexts. Together, these contributions extend the reach of BPM redesign strategies to the micro-level of individual work, responding to the field's growing interest in human-centered process improvement. By equipping workers to systematically redesign their own work using the conceptual toolkit that BPM has long applied at the organisational level, *CRAFT@work* opens a new direction for participant-centered BPM research and practice.

Acknowledgments. We sincerely thank all participants who took part in our study, applied our approach in their work and reflected on the results with us, and our research platform Future of Work at Utrecht University for its support in this interdisciplinary collaboration.

References

1. Beerepoot, I., et al.: A window of opportunity: active window tracking for mining work practices. In: 2023 5th International Conference on Process Mining (ICPM), pp. 57–64. IEEE (2023)
2. Beerepoot, I., Dani, V.S., Lu, X.: Exoar: expert-guided object and activity recognition from textual data. arXiv preprint arXiv:2512.03790 (2025)
3. Berg, J.M., Dutton, J.E., Wrzesniewski, A.: What is job crafting and why does it matter?, Technical report, University of Michigan Ross School of Business (2008)
4. Berg, J.M., Dutton, J.E., Wrzesniewski, A.: Job crafting and meaningful work. In: Dik, B.J., Byrne, Z.S., Steger, M.F. (eds.) Purpose and Meaning in the Workplace, pp. 81–104. American Psychological Association, Washington, DC (2013)

5. Bjørn, P.: From workflow to flow of work: extending bpm research with CSCW agendas. In: Senderovich, A., Cabanillas, C., Vanderfeesten, I., Reijers, H.A. (eds.) BPM 2025. LNCS, vol. 16044, pp. 3–9. Springer, Cham (2025). https://doi.org/10.1007/978-3-032-02867-9_1
6. Braakman, M.A., et al.: Mining for well-being: the potential of process mining for evaluating employee well-being. In: Gdowska, K., Gómez-López, M.T., Rehse, J.R. (eds.) BPM 2024. LNCS, vol. 534, pp. 180–191. Springer, Cham (2024). https://doi.org/10.1007/978-3-031-78666-2_14
7. Bruning, P.F., Campion, M.A.: A role-resource approach-avoidance model of job crafting: a multimethod integration and extension of job crafting theory. Acad. Manag. J. **61**(2), 499–522 (2018)
8. Bryan, G., Karlan, D., Nelson, S.: Commitment devices. Annu. Rev. Econ. **2**(1), 671–698 (2010)
9. Chandrasiri, T., et al.: Beyond profit: the role of benevolent business processes in building long-term firm success. In: Senderovich, A., Cabanillas, C., Vanderfeesten, I., Reijers, H.A. (eds.) BPM 2025. LNCS, vol. 16044, pp. 489–504. Springer, Cham (2026). https://doi.org/10.1007/978-3-032-02867-9_29
10. Davenport, T.H.: Thinking for a Living: How to Get Better Performances and Results from Knowledge Workers. Harvard Business Press (2005)
11. Demerouti, E.: Design your own job through job crafting. Eur. Psychol. (2014)
12. Demerouti, E.: Job crafting revisited: current insights, emerging challenges, and future directions. Annu. Rev. Organ. Psych. Organ. Behav. **13** (2026)
13. Demerouti, E., Bakker, A.B., Nachreiner, F., Schaufeli, W.B.: The job demands-resources model of burnout. J. Appl. Psychol. **86**(3), 499 (2001)
14. Dumas, M., Rosa, L.M., Mendling, J., Reijers, A.H.: Fundamentals of Business Process Management. Springer, Heidelberg (2018). https://doi.org/10.1007/978-3-662-56509-4
15. Goel, K., Fehrer, T., Röglinger, M., Wynn, M.T.: Not here, but there: human resource allocation patterns. In: Di Francescomarino, C., Burattin, A., Janiesch, C., Sadiq, S. (eds.) BPM 2023. LNCS, vol. 14159, pp. 377–394. Springer, Cham (2023). https://doi.org/10.1007/978-3-031-41620-0_22
16. Van den Heuvel, M., Demerouti, E., Peeters, M.: Succesvol job craften door middel van een groepstraining. Scherp in werk **5**, 27–49 (2012)
17. Kooij, D.T., Tims, M., Kanfer, R.: Successful aging at work: the role of job crafting. In: Bal, P., Kooij, D., Rousseau, D. (eds.) Aging Workers and the Employee-Employer Relationship, pp. 145–161. Springer, Cham (2014). https://doi.org/10.1007/978-3-319-08007-9_9
18. Lazazzara, A., Tims, M., De Gennaro, D.: The process of reinventing a job: a meta-synthesis of qualitative job crafting research. J. Vocat. Behav. **116**, 103267 (2020)
19. Lichtenthaler, P.W., Fischbach, A.: A meta-analysis on promotion-and prevention-focused job crafting. Eur. J. Work Organ. Psy. **28**(1), 30–50 (2019)
20. Reijers, H.A., Mansar, S.L.: Best practices in business process redesign: an overview and qualitative evaluation of successful redesign heuristics. Omega **33**(4), 283–306 (2005)
21. Reineke, M., Löhr, B., Aßbrock, A., Bartelheimer, C., Beverungen, D.: From temporary fixes to informed decisions-design echelons for evaluating workarounds. In: Senderovich, A., Cabanillas, C., Vanderfeesten, I., Reijers, H.A. (eds.) BPM 2025. LNCS, vol. 16044, pp. 541–557. Springer, Cham (2025). https://doi.org/10.1007/978-3-032-02867-9_32

22. Rosemann, M., Ostern, N., Voss, M., Bandara, W.: Benevolent business processes-design guidelines beyond transactional value. In: Di Francescomarino, C., Burattin, A., Janiesch, C., Sadiq, S. (eds.) BPM 2023. LNCS, vol. 14159, pp. 447–464. Springer, Cham (2023). https://doi.org/10.1007/978-3-031-41620-0_26
23. Tims, M., Bakker, A.B.: Job crafting: towards a new model of individual job redesign. SA J. Ind. Psychol. **36**(2), 1–9 (2010)
24. Tims, M., Bakker, A.B., Derks, D.: The impact of job crafting on job demands, job resources, and well-being. J. Occup. Health Psychol. **18**(2), 230 (2013)
25. Wells, K., Syed, R., Bandara, W., Chukwudi, I.: Light and dark presence of technostress in RPA: systematic review of qualitative data. In: Senderovich, A., Cabanillas, C., Vanderfeesten, I., Reijers, H.A. (eds.) BPM 2025. LNCS, vol. 16044, pp. 558–580. Springer, Cham (2025). https://doi.org/10.1007/978-3-032-02867-9_33
26. Wrzesniewski, A., Dutton, J.E.: Crafting a job: revisioning employees as active crafters of their work. Acad. Manag. Rev. **26**(2), 179–201 (2001)

Guiding Process Harmonization: A Classification Scheme for Differences Between BPMN Models

Matthias Biermanns[1](✉), Adrian Rebmann[2], and Jana-Rebecca Rehse[1]

[1] University of Mannheim, Mannheim, Germany
{matthias.biermanns, rehse}@uni-mannheim.com
[2] SAP Signavio, Berlin, Germany
adrian.rebmann@sap.com

Abstract. When transforming their IT systems, organizations must harmonize their existing business processes with vendor-provided reference models. This harmonization requires a detailed comparison of underlying BPMN models to identify differences between them. However, existing research provides limited support for identifying concrete, business-interpretable process differences, which means that organizations have little guidance in their process harmonization efforts. To bridge the gap, we propose a novel classification scheme that is tailored to guide process harmonization in workshop settings. Our scheme lifts the representation of process differences from atomic, element-level mismatches to practitioner-oriented change types. It was developed through an iterative approach based on real-world model pairs and validated via expert interviews from both practice and academia. By offering a shared vocabulary and conceptual foundation, the classification scheme facilitates communication across technical and business stakeholders and provides a basis for conducting efficient and effective harmonization workshops.

Keywords: Process Harmonization · Process Model Comparison · Digital Transformation · Reference Models

1 Introduction

In the course of digital transformation, organizations increasingly replace legacy IT landscapes with standard software, for example when adopting cloud-based platforms [18]. Such software relies on reference process models that specify how business processes are implemented. Adhering to these specifications is essential for realizing the benefits of standard software, such as scalability and maintainability [32]. However, organizations usually already operate established process implementations that fit their organizational context [27]. It is hence often challenging to align existing processes with the reference processes embedded in the software [29]. Process harmonization addresses this challenge by systematically identifying and reconciling differences between existing processes and reference processes to derive a consistent and implementable process design [26].

F. M. Maggi et al. (Eds.): BPM 2026, LNCS 16936, pp. 651–669, 2027.
https://doi.org/10.1007/978-3-032-37777-7_38

Process harmonization is hence necessary for system transformation but challenging in itself [27,29,32]. It involves numerous stakeholders and requires a step-by-step comparison between the existing process and the reference process. For example, in ERP cloud transformations, business consultants, process managers and domain experts typically hold long workshops to identify, understand, and evaluate differences between the two. As both sets of processes are typically represented in the de-facto standard of Business Process Model and Notation (BPMN) [30], process modeling expertise is required, but not sufficient: Stakeholders also need to understand the relevance and implications of process differences. However, existing research offers little support for such a business-centric process comparison, such that organizations lack guidance, leaving process harmonization manual and resource-intensive.

To address this problem, this paper proposes a classification scheme that allows for a systematic classification of differences between two BPMN models. The classification scheme is grounded in real-world model pairs and validated with experts from both practice and academia. By offering a shared vocabulary and conceptual foundation, it supports consistent interpretation, facilitates communication across technical and business stakeholders, and provides a basis for conducting efficient and effective harmonization workshops.

In the remainder of this paper, Sect. 2 first presents related work, before Sect. 3 outlines our research method. Section 4 presents the developed classification scheme in detail. Section 5 describes the evaluation of the proposed scheme based on expert interviews. Finally, Sect. 6 concludes the paper with a discussion of limitations and directions for future research.

2 Related Work

This section reviews related research from business process management (BPM).

Model Analysis and Comparison. Much research is dedicated to the analysis and comparison of business process models, with a focus on structural characteristics to uncover recurring constructs and variations. For example, studies on process model patterns and anti-patterns analyze common structural deviations to improve modeling quality [15,20]. Other research incorporates broader analytical perspectives, evaluating process differences through performance indicators and contextual information [4,8]. These works capture structural and contextual characteristics of process models, but they lack a dedicated focus on systematically identifying and interpreting concrete differences between models. In particular, they primarily highlight the existence or frequency of deviations through algorithmic analysis and visualization techniques, but provide limited support for explaining the nature of differences

Similarity Measures. Another line of process comparison research focuses on measuring the similarity between two models. By employing structural or semantic matching techniques, such as aligning activities [12] or utilizing correspondence frameworks like ICoP [34], they calculate pairwise similarity scores.

These techniques have been applied in practical contexts like business process reengineering [31], and model similarity search [13], but defining and evaluating robust similarity measures is prone to inconsistencies [28]. These approaches reduce complex structural and semantic differences to numerical scores, abstracting from their concrete nature, which is needed to guide process harmonization.

Difference Detection. Another research stream focuses on detecting and analyzing concrete differences between process models. These approaches operate directly on structural elements to detect syntactic differences between processes [11] using various techniques, including node and edge analysis [21], event structures [3], refined process structure trees [14], hierarchical decomposition [5]. Behavioral comparison methods can also classify deviations into diagnostic relations such as causality, conflict, or concurrency [2]. These approaches significantly advance automated difference detection, but they primarily emphasize technical analysis. User-centric support for interpreting these differences remains scarce, with only a few exceptions targeting expert process analysts [5,11].

Process Harmonization. Another research stream builds on identified differences and harmonizes process models to derive a unified representation. These approaches aim to align process variants by balancing standardization and local adaptation. Previous studies proposed frameworks to help practitioners determine the appropriate degree of process standardization [27] and identified key factors that influence harmonization [26]. Other approaches take a methodological perspective on process harmonization, either by consolidating multiple as-is models into a single to-be model using general structured procedures [16] or by applying decomposition techniques to identify recurring process fragments [23].

Merging approaches aim to create a consolidated model that subsumes all input models. They typically rely on algorithms, for example based on function graphs [17], or by constructing configurable process models that capture multiple variants [22]. More recent work approaches consolidation by combining structural matching with feature-based techniques, such as identifying frequent subgraphs across models [19]. These approaches provide structured support for aligning process models, but they focus on consolidation and decision-making, rather than systematically identifying and interpreting the underlying model differences.

Difference Classification. Building on this line of work, several studies categorize process model differences to support human interpretation. For example, Weber et al. [33] introduce a catalog of change patterns, such as inserting or deleting process fragments, to guide controlled model evolution in process-aware information systems. However, their focus lies on enabling technical system adaptations rather than explaining differences between existing models to users. Addressing model comparison more directly, Dijkman [9,10] introduces a classification of structural change types between similar business processes. Although providing important foundations for describing model differences, this classification is restricted to a limited subset of BPMN elements and is derived deductively based on a small set of example models, thus lacking empirical insights into how practitioners actually perceive and communicate deviations.

This paper proposes an empirically derived classification scheme for BPMN model differences, designed to support harmonization workshops.

3 Research Method

This paper follows an inductive, expert-driven approach to develop and evaluate a classification scheme for differences between two BPMN models. As shown in Fig. 1, our research method consists of three phases:

(1) Model Pair Selection,
(2) Iterative Coding,
(3) Expert Interviews and Refinement, and

The refined scheme and application procedure were then the subject of a summative evaluation, described in Sect. 5. Additional material related to the development and evaluation of the classification scheme is available in our repository.[1]

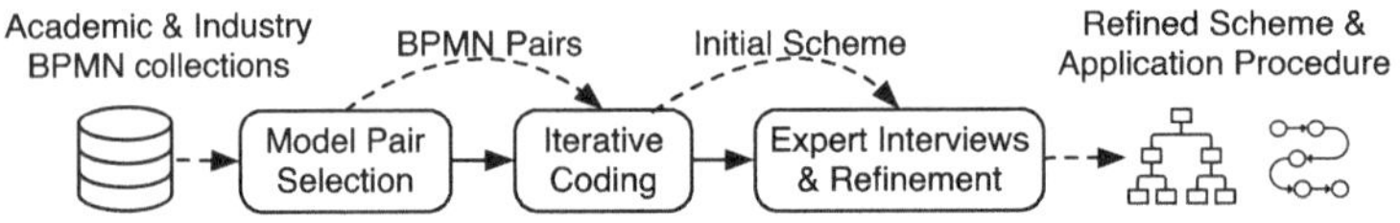

Fig. 1. Research method.

Model Pair Selection. To ensure that our classification scheme offers practical value for users, we set out to derive it inductively based on a large set of real-life models. The dataset construction comprised four steps:

(1) dataset selection,
(2) model filtering,
(3) model pairing, and
(4) purposive sampling.

The derived set of BPMN model pairs represent similar business processes, which allowed us to uncover differences in how modelers express process logic, scope, and notation. This reflects the intended harmonization setting and the methodological requirement that compared models share a common process context, ensuring that any differences identified remain meaningful and interpretable.

Dataset Selection. To obtain a heterogeneous set of models, we selected three datasets with different origin and modeling characteristics. The first one is a collection of vendor-provided reference process models, which represent standardized processes commonly used for process-oriented workshops. The second

[1] The online repository containing supplementary materials can be accessed at https://doi.org/10.6084/m9.figshare.32508516.

one is the BPMN for Similarity Metrics benchmark [1], commonly used to test similarity measures. The third one is Signavio Academic Models (SAM) [30], which contains a large corpus of BPMNs created by students and practitioners. Together, these datasets offer a variety of perspectives on modeling practices.

Filtering and Grouping (SAM Only). Unlike the Vendor Models and the Similarity Benchmark, which provide predefined groups of related models that can be directly used for pairing. The SAM dataset contains over a million independently created models from which we first had to identify suitable candidates for pair generation. We therefore started the filtering with the following criteria:

(a) modeled in BPMN 2.0,
(b) modeled in English,
(c) assigned a domain or industry category [24], and
(d) at least 10 elements.

We then normalized and screened model titles to remove non-descriptive ones that would not allow for partner identification. Following that, models were grouped through heuristic estimation of process relatedness. In three stages, they had to have at least one suitable partner not to be excluded:

(a) title similarity,
(b) activity label similarity, and
(c) structural similarity based on Graph Edit Distance (GED).

Table 1. Overview of dataset characteristics after pairing

Dataset	Vendor Reference Models	Similarity Benchmark [1]	SAM [30]
Pairing	Within Best Practice	Given in paper	Heuristic estimation
#Models	28	745	371
#Pairs	109	1490	1138

Title and label similarities were computed using sentence embeddings [25] with cosine similarity thresholds of 0.8. GED thresholds avoided including identical or incomparable models. This process resulted in groups of similar models.

Model Pairing. We then generated model pairs for each dataset. For the Vendor Models, pairing is straightforward by picking any two alternative representations of the same process. For the Similarity Benchmark, we adopted the pairings defined by the authors, which include an original model and several systematically generated structural variations. For SAM, we selected partners by combining every model with every other one within the same model group.

Purposive Sampling. The resulting dataset included 2,737 BPMN model pairs across three datasets (see Table 1). To enable manual analysis while preserving diversity in model characteristics, we employed purposive sampling. Specifically, we used a stratified approach to select 100 model pairs reflecting dataset size and practical relevance: 60 pairs from SAM, 30 from the Vendor Models, and 10 from the Similarity Benchmark. We then clustered each dataset based on the complexity metrics NOAJS, MCC, and CFC [6] to identify representative groups of model pairs and randomly selected a single pair from each cluster.

Iterative Coding of Model Differences. This phase identified and consolidated differences between the model pairs to derive a structured understanding of recurring difference types. Following a grounded theory approach [7], we derived types inductively through an iterative coding process (see accompanying repository). In the initial coding, we manually compared each model pair, inspecting and documenting differences in commonly used elements (pools, lanes, activities, subprocesses, events, gateways, and flows) [30]. The initial coding was conducted by a single researcher. We tracked the inclusion or omission of specific constructs, as well as variations in activity sequences, branching, and synchronization. When we found a difference, we compared it with the set of defined types. If it matched an existing type, it was assigned to that. If not, a new type was created. During this phase, the coding favored a granular differentiation of types, hence new instances were only assigned to existing types if they exhibited the same conceptual characteristics. Otherwise, a new type was introduced. For example, two models contained the same activities in a different order, which was neither a *change in content* nor in *branching structure*. We created the type *change of order* to describe different sequential order relations. Each observed variation was recorded as a difference instance tied to its model pair.

In the second iteration, we revisited the initial coding and consolidated recurring difference types by merging similar ones. For example, *Choice Expansion* and *Guideline Relaxation* both described adapting the branching structure for a more flexible execution, so they were merged into *Execution Flexibility Change*. Further, we grouped types based on the nature of the difference they describe (rather than the affected BPMN modeling layer), which led to a first hierarchy of the classification scheme. For example, granularity-related differences were grouped together, even when they affected different elements, such as activities or resource assignments. The outcome of this phase was an initial, hierarchically organized classification scheme with five groups (Functional Content, Execution Logic, Process Interface, Representational, and Meta Differences) and 22 types, each with a conceptual description and modeling example. Subsequent grouping of types was then based on the expert interviews described below.

Expert Interviews and Refinement. After developing the initial classification scheme, we conducted expert interviews for its formative evaluation and refinement. This phase aimed to assess the scheme's conceptual clarity and applicability. To ensure that the evaluation reflects informed perspectives on BPMN modeling and process analysis, eight experts were selected based on predefined criteria:

(1) At least one year of BPM experience (average: 5.88), to ensure familiarity with the BPM domain,

(2) familiarity with at least one BPMN modeling tool, to ensure practical knowledge of process modeling, and
(3) practical involvement in BPM-related projects or published papers within the past three years (average: 8.88), to ensure that participants are still active in the field.

The interviews consisted of three phases (see Table 2). The collected feedback was used to refine the classification scheme, clarifying type definitions and resolving conceptual overlaps. We introduced prioritization and combination rules to guide the classification of ambiguous differences and added optional subtypes to preserve conceptual distinctions within types. The scheme structure was consolidated by merging related types, which reduced the number of types from 22 to 13 and aligned groups with modeling layers, resulting in four groups.

Table 2. The three parts of our expert interviews

(1) Introduction: The classification scheme was presented to the experts. The structure of the scheme and the meaning of the individual difference types were explained to establish a shared understanding before the evaluation began.
(2) Think-aloud Application: The experts applied the classification scheme to six preselected BPMN model pairs. While identifying and classifying model differences for each model pair, they verbalized their reasoning (think-aloud), which allowed us to observe how the experts interpreted the difference types, how they navigated the scheme, and where ambiguities or uncertainties arose during application.
(3) Semi-structured Interview: The experts evaluated the scheme by answering twelve questions, which addressed aspects such as the clarity of the difference types, the differentiation between the types' concepts, and the scheme's perceived completeness. The interview format allowed the experts to elaborate on their answers and raise additional points that had emerged during the think-aloud phase.

4 Classification Scheme

In this section, we present the classification scheme (full version in accompanying repository). We first give an overview, before describing each difference type group and an application procedure.

4.1 Overview and General Considerations

We first introduce the overall structure of the classification scheme before explaining its hierarchical organization, the multi-type nature of model differences, and the principles that guide its application and interpretation.

Overall Structure. The classification scheme organizes BPMN model differences into four groups, as shown in Fig. 2. The first group focuses on content, capturing the elements present in the models. The second group addresses control-flow, describing behavioral differences in activity execution. The third group considers representational differences, including visual and syntactic modeling variations. The fourth group captures differences that cannot be meaningfully interpreted, but instead affect the validity of the comparison itself.

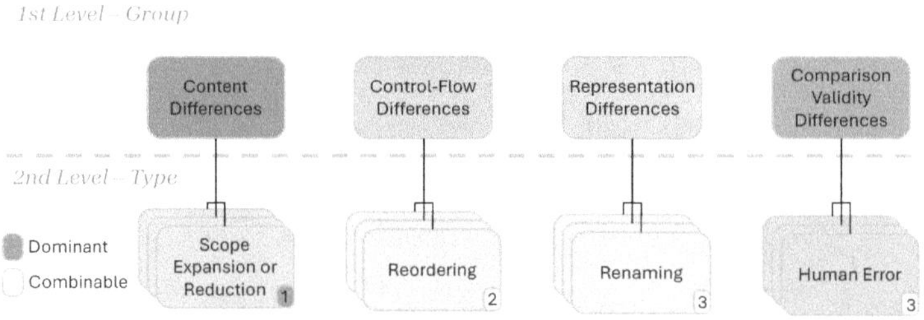

Fig. 2. Overview of the classification hierarchy.

Hierarchical Organization. The classification scheme is organized into three levels. The highest level are the groups above. The second level consists of types, which specify the concrete forms of differences within each group. On the third level, some types are divided into subtypes to further differentiate recurring forms of how the difference is expressed. This hierarchy allows for a coarse-grained understanding of differences by identifying the affected modeling layer and a fine-grained analysis in terms of different types and subtypes of differences.

Multi-type Classification. The classification of differences often requires multiple types, potentially spanning several layers. For example, sequential activities executed by the same resource in Model A can be executed in parallel by different resources in Model B. Therefore, the classification scheme supports a compositional perspective, allowing the joint application of multiple types.

Prioritization and Combination. To support consistent classifications, the scheme establishes rules for prioritization and combination. Because some difference types are more fundamental, they take precedence when multiple interpretations are possible. Some types are complementary and can be combined to capture different aspects of the same difference. Types marked as "dominant" override other applicable types of the same or lower priority, whereas "combinable" types can be assigned jointly. These rules ensure that combinations of types remain consistent and interpretable when analyzing complex model variations. For example, a change to a process start event may be interpreted both as an element class conversion and a process interface change (cf. Sect. 5.2). In such cases, prioritization rules guide analysts towards the dominant interpretation.

Direction-Agnostic Interpretation. The scheme is intended to be applied to a pair of BPMN models. In this setting, the interpretation of a difference depends on the analysts' perspective, for example whether one model is considered as-is or to-be. The same difference may hence be interpreted in opposite directions. This rationale is reflected throughout the scheme by defining types direction-agnostic.

4.2 Difference Type Groups

In the following, we introduce and characterize the individual groups and types. An overview of minimal exampled can be found in the accompanying repository.

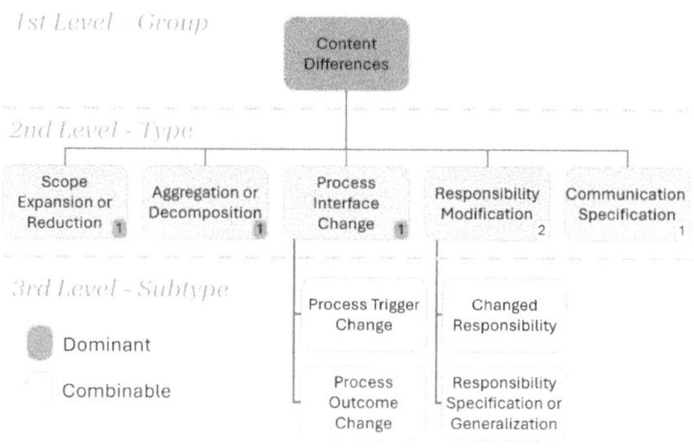

Fig. 3. Content differences.

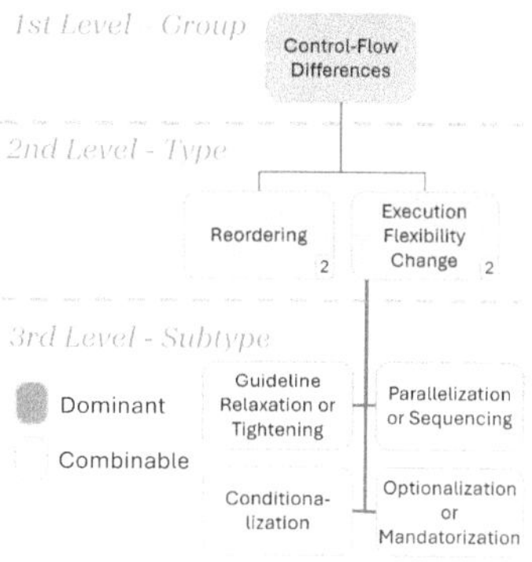

Fig. 4. Control-flow diff.

Content Differences. Differences in this group (Fig. 3) stem from the nature of the modeled elements. They address changes to the functional scope of a process, the granularity of elements, the interactions between participants, and the definition of process boundaries. This group corresponds to the content layer and contains five types, two are further refined into subtypes.

Scope Expansion or Reduction. This type captures changes to the functional scope of a model by adding or removing elements.

Aggregation or Decomposition. This type captures granularity changes of functional elements without changing their semantics. This includes dividing one element into multiple or vice versa, reflecting varying degrees of abstraction.

Process Interface Change. This type captures modifications to the entry or exit points of a process, altering how it is initiated or terminated. It has two subtypes:

- *Process Trigger Change:* Change to the initiating conditions of a process.
- *Process Outcome Change:* Change to the defined outcomes of a process.

Responsibility Modification. This type captures changes to responsibilities for activities or events, without altering the elements themselves. It has two subtypes:

- *Changed Responsibility:* Reassignment to a different organizational role.
- *Responsibility Specification or Generalization:* Changes to the granularity of responsibility modeling without altering communication patterns.

Communication Specification. This type captures changes in the interactions between process participants. This includes adding, removing, or modifying message flows, pools, or lanes, altering communication but not responsibilities.

Control-Flow Differences. Differences in this group (Fig. 4) capture changes to the control flow logic, particularly execution order, branching, and activity concurrency. This group contains two types, one is divided into subtypes.

Reordering. This type captures changes to the activity sequence along a single path without altering branching structures, concurrency, or reachability. The order of activities is modified but the overall control flow remains unchanged.

Execution Flexibility Change. This type captures control-flow modifications that change the number of execution paths independent of scope. It has four subtypes:

- *Guideline Relaxation/Tightening:* Replacing strict execution rules with more lenient ones or vice versa.
- *Conditionalization:* Executing process parts only under specific conditions.
- *Parallelization/Sequencing:* Altering sequential vs. parallel execution.
- *Optionalization/Mandatorization:* Adding/removing optional execution paths.

Representation Differences. Differences in this group (Fig. 5) stem from visual or syntactic representation, e.g., renaming or repositioning an element. They capture modeling decisions that affect labels, layout, notation, or element classes. Representation differences are typically combinable since they often occur alongside functional or execution-related changes. However, they can also appear independently when two models describe the same process behavior, but differ in their visual or notational realization. Although these differences do not necessarily affect process behavior, distinguishing them from content and control-flow differences helps analysts avoid attributing harmonization relevance to purely representational variations. This group corresponds to the representation layer and contains four types, two are further refined into subtypes.

Renaming. This type captures modifications to the label of an element without altering its semantics. This occurs when the wording of an activity, event, etc. changes, but its functional role and execution behavior remain the same.

Element Class Conversion. This delta describes the transformation of a BPMN element from one class to another without altering the scope of the process.

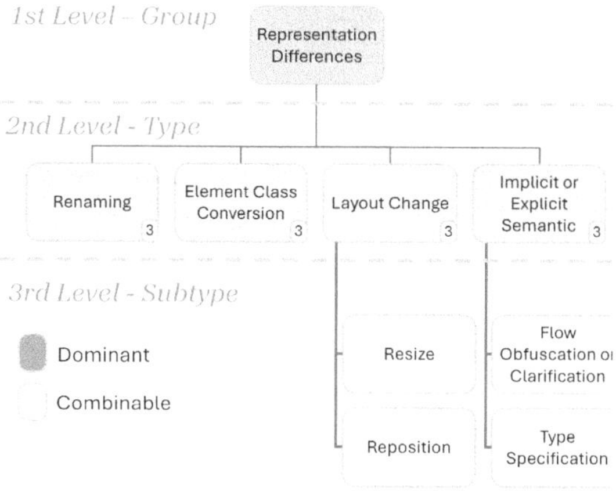

Fig. 5. Representation differences.

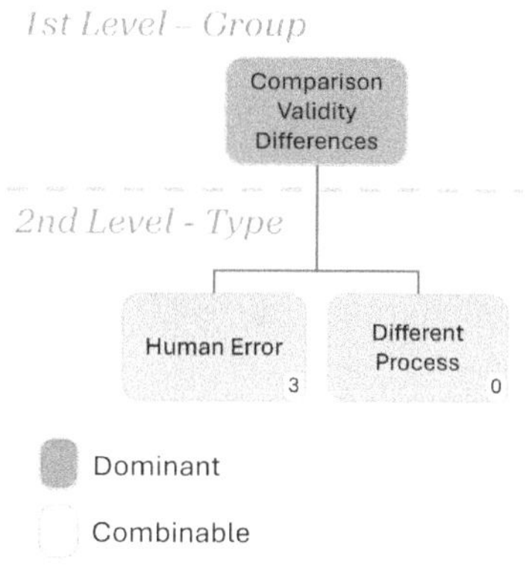

Fig. 6. Comparison validity diff.

This subcategory captures cases where the modeling construct changes while the underlying process functionality remains stable.

Layout Change. This type captures modifications to the visual arrangement of elements on the modeling canvas without affecting their semantics, relationships, or responsibilities. It is refined into two subtypes:

- *Resize:* Changes to the visual dimensions of an element.
- *Reposition:* Changes to the placement of an element on the canvas.

Implicit or Explicit Semantics. This type captures changes between implicit and explicit representations of process behavior. It reflects differences in how modeling constructs are used to express behavior, rather than changes to the behavior itself. It is refined into two subtypes:

- *Flow Obfuscation or Clarification:* Insertion or deletion of gateways that make control-flow more or less explicit without altering it.
- *Type Specification:* Changes to activity or event type within the same class.

Comparison Validity Differences. Differences in this group (Fig. 6) cannot be meaningfully attributed to the content, control flow, or representation layer. In practice, they stem from modeling mistakes or indicate that the models in question are not comparable in a meaningful way. This group hence ensures analytical completeness and avoids forced classification into layer-specific types. It contains two types, which are both combinable with other types.

Human Error. This type captures differences that stem from unintentional errors rather than deliberate modeling decisions. Such errors can be identified independent of the intended process semantics and do not reflect systematic differences between two models. This type acknowledges that empirical model comparisons may reveal inconsistencies or syntactic inaccuracies that should not be interpreted as meaningful content, execution, or representation differences.

Different Process. This type is assigned if two BPMN models depict fundamentally different processes, making a comparison of differences infeasible because the underlying process context diverges. The type serves as a boundary condition of the classification scheme, indicating that the compared models compared do not share a sufficiently common reference point for detailed analysis.

4.3 Application Procedure

To help process analysts compare BPMN models in preparation for process harmonization workshops, we complement the scheme with an application procedure, divided into two phases: (1) assessment and (2) classification.

Assessment Phase. This phase supports analysts in determining whether the classification scheme can be applied to compare two BPMN models in question. Its outcome is an informed decision on whether they are sufficiently comparable for the scheme to provide analytical support. The phase structures the initial comparison and establishes preliminary correspondences between the models:

(1) *Select the process models for comparison:* Analysts identify the two BPMN models to be analyzed and determine their respective roles within the comparison, e.g., as-is/to-be, regional variants, or alternative implementations.
(2) *Gain an overview of models:* Analysts review both models to understand their overall structure, major milestones, and interfaces. This helps establish a mental map of the processes before focusing on detailed differences.
(3) *Identify identical elements:* Analysts detect elements (tasks, events, or gateways) that appear structurally and semantically equivalent. These elements provide anchor points that facilitate the subsequent alignment of the models.
(4) *Identify match-generating types:* Analysts detect elements that likely correspond across the two models but were, e.g., renamed or specified. This further aligns the models and prepares the subsequent difference analysis.

With these steps, analysts *check the applicability* of the classification scheme. The comparison should only proceed if the models are sufficiently comparable. As a practical guideline, at least 20–50% of flow node elements (tasks, events, gateways) should correspond between the models. If this threshold is not met, the models likely represent different processes. In that case, proceeding with the classification is still possible, if the analyst deems it useful.

Classification Phase. This phase supports analysts in systematically classifying the differences between the two BPMN models by means of the classification scheme. Building on the correspondences established during assessment, the identified differences are analyzed and classified in a structured manner. Depending on the context, this can be approached either sequentially along the process flow or iteratively by revisiting elements and refining classifications:

(1) *Identify remaining differences sequentially:* Analysts traverse both models to detect differences between corresponding elements. To observed differences, they assign the appropriate category from the classification scheme.
(2) *Refine classifications where appropriate:* When beneficial, analysts can use subtype assignments to capture the nature of a difference more precisely. Such refinements should only be applied where they provide additional analytical value and do not introduce unnecessary complexity.
(3) *Resolve overlapping assignments:* In some cases, multiple difference types may be applicable to the same element. Such conflicts are resolved based on priority and behavior: Higher-priority types take precedence over lower-priority ones, dominant types override those of the same or lower priority, and combinable types may co-exist when no higher-priority dominant type applies. In rare cases where dominant types of the same priority conflict, analysts should select the more specific one based on the type definitions.

This procedure enables analysts to systematically identify and communicate differences between BPMN models. By structuring these differences using the proposed classification scheme, the approach supports collaborative interpretation of process variants in harmonization and transformation initiatives.

5 Evaluation

We evaluated the proposed classification scheme through a second round of expert interviews, to assess its applicability for practical BPMN model comparison. The evaluation covered (1) *completeness and coverage* of the scheme, (2) *conceptual soundness* of the category structure, including grouping and granularity of types, and (3) *practical applicability* during model comparison. We first present the setup and a concrete application scenario before reporting on the results.

5.1 Evaluation Setup

In contrast to the first interview round during development (see Sect. 3), this round served as a confirmatory evaluation of the revised classification scheme presented in Sect. 4. To rule out potential bias from prior involvement, we invited eight new interview partners (E9-E16; four practitioners and four academics). Participants were selected using the same criteria applied in the earlier interviews: They had more than one year of experience in BPM (average: 4.38), were familiar with at least one BPMN modeling tool and had been involved in at least one BPM-related project within the past three years (average: 9.13).

The interviews followed the same structure as in the first round (see Table 2). Because the purpose of this round was to validate the applicability and clarity of the revised scheme, the tasks and questions remained the same. Following the interview, participants additionally completed a short survey consisting of four Likert-scale questions and seven open-ended questions. The survey questions were derived from the evaluation objectives of this study and addressed perceived difficulties during the categorization process, the extent to which the framework supported the task, potential structural improvements to the scheme, and the overall applicability of the framework for model comparison. More specifically, they were designed to assess the completeness and coverage, conceptual soundness, and practical applicability of the proposed classification scheme.

5.2 Application Scenario

To illustrate the application of the classification scheme, consider the example in Fig. 7, which presents two BPMN models that describe the handling of adverse reactions after vaccination. This is one of the model pairs that experts had to examine in the second part of the interviews. Both models represent the same general procedure, but differ in structure and semantics.

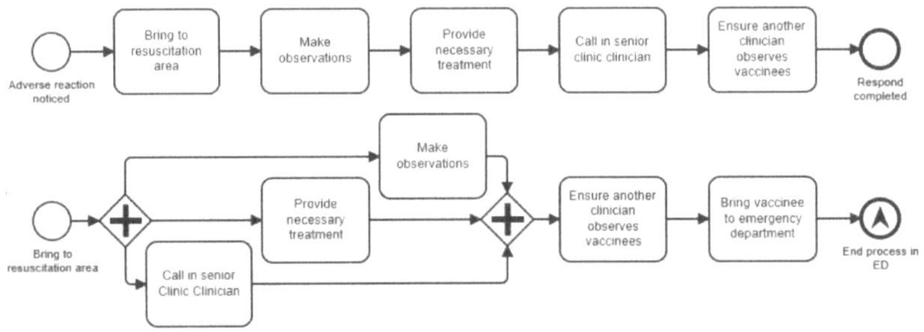

Fig. 7. Application example

All experts consistently identified the following differences:

- The additional activity *Bring vaccinee to emergency department* in the lower model was classified as Scope Expansion.
- The altered execution structure of the *observation*, *treatment*, and *consultation* activities were interpreted as a form of parallelization, corresponding to the execution flexibility change type.

Expert assessments diverged when interpreting the following differences:

- Most experts described the difference at the start as a *process interface change*. Some additional identified a *scope reduction* (E10, E12) or an *element class conversion* (E13). One described it solely as an *element class conversion* (E16).
- Six experts interpreted the end event difference as a *process interface change*, two as *implicit or explicit semantics representation* with *renaming* (E10, E13).

These cases suggest that certain model differences can be plausibly interpreted using multiple types, depending on the analytical perspective.

The example also shows that experts occasionally deviated from the scheme's prioritization and combination rules. For example, E13 assigned both *element class conversion* and *process interface change* to the end event difference, where the prioritization rules would suggest selecting one dominant interpretation. These cases were rare, but they show that the rules provide guidance rather than enforcing a strictly deterministic classification. Additionally, the think-aloud sessions allowed us to observe how experts approached the task. Some closely followed the application procedure described in Sect. 4.3, whereas others relied more on directly comparing the models visually. This also manifested in the interviews, for example, in the number and types of differences identified.

5.3 Evaluation Results

We analyzed interview and survey results, using qualitative coding to identify recurring themes across participants, aligned with the evaluation criteria described above (see Table 3). The coded feedback can be found in the accompanying repository. Only themes that appeared at least twice were considered. Overall, the feedback from participants was predominantly positive.

Table 3. Summary of feedback themes across participants.

Category	Theme	Participants
Difficulties with Task	Exhausting for large models	E9, E11, E14
	Many Categories	E9, E12, E13, E15
	Uncertainty with type combinations	E11, E12, E14
	Manual comparison fatigue	E10, E15
Supportive Aspects	Guidance & Task Overview	E9, E10, E12, E14, E15
	Clear Categorization	E9, E10, E11, E13
	Communication Support	E10, E11
	Completeness, Coverage & Discoverability	E10, E12, E13, E15
Applicability Threshold	Supporting previous Threshold	E12, E13
	Always Applicable	E10, E14
	Context Dependent	E11, E12, E14
Observations	Looking for Substitution type	E9, E10, E11, E13
	Human Errors used for modeling errors	E13, E16

Completeness and Coverage. Experts consistently indicated that the classification scheme provides sufficient coverage to capture differences between BPMN models. They emphasized that the scheme supported a systematic discovery of model differences and ensured that relevant aspects were not overlooked. One expert noted, "I was able to allocate every change" (E12); another highlighted that the scheme "helped me not to forget about a certain type of difference"

(E13). During several interviews, experts initially searched for a potential *substitution*. However, in all cases, the identified differences could be expressed using existing types. Therefore, this suggestion appeared only once in the survey responses.

Conceptual Soundness. Experts generally confirmed that the conceptual structure of the classification scheme is coherent and well-defined. In particular, they emphasized that the types are clearly separated and logically grouped. One expert stated "the categories inside the framework are well separated" (E9); another noted that the scheme "brought great clarity to the task" (E13). Feedback on potential structural changes was very limited: One expert suggested merging categories and one suggested splitting a category into additional subtypes. No expert proposed assigning any type to a different group.

Practical Applicability. Experts found the categorization task manageable, rating its difficulty as 2.5 (median) resp. 2.75 (mean) on a five-point scale. This aligns with the first interview round and indicates that the scheme provides useful guidance for the task's analytical effort. Experts emphasized that the clear separation of types and easily comprehensible definitions (median 5, mean 4.63) supported a systematic decomposition of model differences. They noted that the scheme "helped with problem deconstruction & overview" (E10) and "provided a taxonomy to follow" (E11). The groups were also perceived as highly beneficial for structuring the task (median 5, mean 4.5). However, experts also highlighted the challenges of applying the scheme to larger or more complex models. They reported fatigue from manual comparison as well as difficulties in consistently handling multiple applicable categories or overlapping type combinations. Some experts noted that the initial increase in the number of categories led to a higher perceived cognitive effort, although they found that the process felt "almost like routine" (E13) by the final task and "all a matter of learning" (E16).

Regarding the scheme's scope, the applicability threshold identified in the first interview iteration was largely confirmed. Experts indicated that the scheme provides meaningful insights when roughly 20–50% of elements can be mapped between models, with some arguing it could remain useful even outside this range. Overall, the evaluation demonstrates that the proposed classification scheme offers a structured, applicable approach for describing differences between BPMN model pairs, requiring only minor individual adjustments.

6 Conclusion

This paper presents a classification scheme to systematically identify differences between BPMN models for process harmonization. Derived from real-world model pairs and refined through expert interviews, it translates model mismatches into business-interpretable difference types. Our evaluation confirms that it is conceptually sound, featuring clearly separated categories that cover a broad range of practical comparison contexts. Experts found the scheme supportive and the categorization manageable, demonstrating its practical utility.

Limitations. Our work is subject to several limitations. First, the empirical foundation of the scheme is based on approximately 100 model pairs, so it may not capture all possible model differences. For example, variations of loop constructs are currently not explicitly considered and elements such as data artifacts and repositories are omitted. Since they occur only infrequently in practice [30], we consider this acceptable; it is doubtful whether their inclusion would justify the additional complexity of the classification scheme. Second, our evaluation relied exclusively on the perspectives of BPMN experts. This was necessary to validate the conceptual soundness of the scheme, but we have not yet evaluated its effectiveness in its intended use case, i.e., stakeholder communication in workshop settings. Finally, applying the scheme to very large process models may introduce significant cognitive complexity, which is paradoxically the exact scenario where systematic harmonization support is most critical.

Future Work. To address these limitations, future work could pursue several directions. One should be the development of supportive artifacts and a visualization of model differences to ensure minimal cognitive load. We consider this a prerequisite for improving communication with non-experts in practice. Future research could also explore how artificial intelligence might complement the proposed scheme, for example, to assist analysts in identifying correspondences or pre-structuring potential differences. This could enhance the approach's scalability while maintaining the interpretability of manual analysis. Furthermore, we aim to conduct a broader empirical validation of the scheme by applying it directly within actual process harmonization workshops. Such studies would allow us to assess whether and how the scheme can lead to more effective and efficient harmonization outcomes in practice. Conducting them would require the aforementioned tooling and visualization support to operationalize the scheme for practical use. Beyond the scenario considered here, future research could also investigate the applicability of the scheme in other settings that require systematic process comparison, such as aligning as-is and to-be models or reconciling process variants across organizational units. Finally, we plan to refine the scheme by exploring its combination with existing literature, such as change patterns [33], to provide an even more robust foundation for enterprise IT transformations.

Disclosure of Interests. The authors have no competing interests to declare that are relevant to the content of this article.

References

1. Anam, R., Muhammad, T.: A process model collection with structural variants and evaluations. IJACSA **11**(5) (2020)
2. Armas-Cervantes, A., Baldan, P., Dumas, M., Garcia-Banuelos, L.: BP-diff: a tool for behavioral comparison of business process models. BPM Demos., 1–5 (2014)
3. Armas-Cervantes, A., Baldan, P., Dumas, M., García-Bañuelos, L.: Diagnosing behavioral differences between business process models: an approach based on event structures. Inf. Syst. **56**, 304–325 (2016)

4. Ballambettu, N.P., Suresh, M.A., Bose, R.P.J.C.: Analyzing process variants to understand differences in key performance indices. In: Dubois, E., Pohl, K. (eds.) CAiSE 2017. LNCS, vol. 10253, pp. 298–313. Springer, Cham (2017). https://doi.org/10.1007/978-3-319-59536-8_19
5. Cao, B., et al.: ProDiff: a process difference detection method based on hierarchical decomposition. IEEE TSC **15**(1), 513–526 (2022)
6. Cardoso, J., Mendling, J., Neumann, G., Reijers, H.A.: A discourse on complexity of process models. In: Eder, J., Dustdar, S. (eds.) BPM 2006. LNCS, vol. 4103, pp. 117–128. Springer, Heidelberg (2006). https://doi.org/10.1007/11837862_13
7. Chun Tie, Y., Birks, M., Francis, K.: Grounded theory research: a design framework for novice researchers. SAGE Open Med. **7** (2019)
8. Cordes, C., Vogelgesang, T., Appelrath, H.-J.: A generic approach for calculating and visualizing differences between process models in multidimensional process mining. In: Fournier, F., Mendling, J. (eds.) BPM 2014. LNBIP, vol. 202, pp. 383–394. Springer, Cham (2015). https://doi.org/10.1007/978-3-319-15895-2_32
9. Dijkman, R.: A classification of differences between similar business processes. In: EDOC, pp. 37–47. IEEE (2007)
10. Dijkman, R.: Feedback on differences between business processes. BETA publicatie: Working Papers, Technische Universiteit Eindhoven (2007)
11. Dijkman, R.: Diagnosing differences between business process models. In: Dumas, M., Reichert, M., Shan, M.-C. (eds.) BPM 2008. LNCS, vol. 5240, pp. 261–277. Springer, Heidelberg (2008). https://doi.org/10.1007/978-3-540-85758-7_20
12. Dijkman, R., Dumas, M., Garcia-Banuelos, L., Kaarik, R.: Aligning business process models. In: EDOC, pp. 45–53. IEEE (2009)
13. Dijkman, R., Dumas, M., García-Bañuelos, L.: Graph matching algorithms for business process model similarity search. In: Dayal, U., Eder, J., Koehler, J., Reijers, H.A. (eds.) BPM 2009. LNCS, vol. 5701, pp. 48–63. Springer, Heidelberg (2009). https://doi.org/10.1007/978-3-642-03848-8_5
14. Fan, J., Wang, J., An, W., Cao, B., Dong, T.: Detecting difference between process models based on the refined process structure tree. Mob. Inf. Syst. pp. 1–17 (2017)
15. Fellmann, M., Koschmider, A., Laue, R., Schoknecht, A., Vetter, A.: Business process model patterns: state-of-the-art, research classification and taxonomy. BPMJ **25**(5), 972–994 (2019)
16. Fernandez, J., Bhat, J.: Addressing the complexities of global process harmonization. In: Handbook of Research on Complex Dynamic Process Management, pp. 368–385. IGI Global (2010)
17. Gottschalk, F., van der Aalst, W.M.P., Jansen-Vullers, M.H.: Merging event-driven process chains. In: Meersman, R., Tari, Z. (eds.) OTM 2008. LNCS, vol. 5331, pp. 418–426. Springer, Heidelberg (2008). https://doi.org/10.1007/978-3-540-88871-0_28
18. Hasan, M.H., Osman, M.H., Admodisastro, N.I., Muhammad, M.S.: Legacy systems to cloud migration: a review from the architectural perspective. J. Syst. Softw. **202** (2023)
19. Huang, Y., Li, W., Liang, Z., Xue, Y., Wang, X.: Efficient business process consolidation: combining topic features with structure matching. Soft. Comput. **22**(2), 645–657 (2018)
20. Koschmider, A., Laue, R., Fellmann, M.: Business process model anti-patterns: a bibliography and taxonomy of published work. In: ECIS (2019)
21. Küster, J.M., Gerth, C., Förster, A., Engels, G.: Detecting and resolving process model differences in the absence of a change log. In: Dumas, M., Reichert, M.,

Shan, M.-C. (eds.) BPM 2008. LNCS, vol. 5240, pp. 244–260. Springer, Heidelberg (2008). https://doi.org/10.1007/978-3-540-85758-7_19
22. La Rosa, M., Dumas, M., Uba, R., Dijkman, R.: Business process model merging: an approach to business process consolidation. ACM TOSEM **22**(2), 1–42 (2013)
23. Milani, F., Dumas, M., Matulevičius, R.: Decomposition driven consolidation of process models. In: Salinesi, C., Norrie, M.C., Pastor, Ó. (eds.) CAiSE 2013. LNCS, vol. 7908, pp. 193–207. Springer, Heidelberg (2013). https://doi.org/10.1007/978-3-642-38709-8_13
24. Rebmann, A., Schmidt, F.D., Glavaš, G., van der Aa, H.: On the potential of large language models to solve semantics-aware process mining tasks. Process Sci. **2**(1), 10 (2025)
25. Reimers, N., Gurevych, I.: Sentence-BERT: sentence embeddings using Siamese BERT-networks. In: IJCNLP, pp. 3982–3992. ACL (2019)
26. Romero, H., Dijkman, R., Grefen, P., Van Weele, A., de Jong, A.: Measures of process harmonization. Inf. Softw. Technol. **63**, 31–43 (2015)
27. Romero, H., Dijkman, R., Grefen, P., van Weele, A.: Harmonization of business process models. In: Daniel, F., Barkaoui, K., Dustdar, S. (eds.) BPM 2011. LNBIP, vol. 99, pp. 13–24. Springer, Heidelberg (2012). https://doi.org/10.1007/978-3-642-28108-2_2
28. Schoknecht, A., Thaler, T., Fettke, P., Oberweis, A., Laue, R.: Similarity of business process models - a state-of-the-art analysis. ACM Comput. Surv. **50**(4), 1–33 (2018)
29. Schönreiter, I.M.: Methodologies for process harmonization in the post-merger integration phase: a literature review. BPMJ **24**(2), 330–356 (2018)
30. Sola, D., Warmuth, C., Schäfer, B., Badakhshan, P., Rehse, J.R., Kampik, T.: SAP signavio academic models: a large process model dataset. In: Montali, M., Senderovich, A., Weidlich, M. (eds.) ICPM 2022. LNCS, vol. 468, pp. 453–465. Springer, Cham (2022). https://doi.org/10.1007/978-3-031-27815-0_33
31. Tka, M., Ghannouchi, S.A.: Comparison of business process models as part of BPR projects. Procedia Technol. **5**, 427–436 (2012)
32. Tregear, R.: Business process standardization. In: vom Brocke, J., Rosemann, M. (eds.) Handbook on Business Process Management 2. IHIS, pp. 421–441. Springer, Heidelberg (2015). https://doi.org/10.1007/978-3-642-45103-4_18
33. Weber, B., Reichert, M., Rinderle-Ma, S.: Change patterns and change support features – enhancing flexibility in process-aware information systems. Data Knowl. Eng. **66**(3), 438–466 (2008)
34. Weidlich, M., Dijkman, R., Mendling, J.: The ICoP framework: identification of correspondences between process models. In: Pernici, B. (ed.) CAiSE 2010. LNCS, vol. 6051, pp. 483–498. Springer, Heidelberg (2010). https://doi.org/10.1007/978-3-642-13094-6_37

Towards Agent-Based Automation beyond RPA: An Exploratory Design Science Study in the Public Sector

Christian Czarnecki[1(✉)], Philipp Münzner[2], Alina Chircu[3], and Eldar Sultanow[2]

[1] FH Aachen, Aachen, Germany
czarnecki@fh-aachen.de
[2] Capgemini, Nuremberg, Germany
{philipp.muenzner,eldar.sultanow}@capgemini.com
[3] University of New Mexico, Albuquerque, NM, USA
achircu@unm.edu

Abstract. Large organizations increasingly explore large language models (LLMs) to support knowledge-intensive process work. While Robotic Process Automation (RPA) has proven effective for rule-based and structured activities, many process improvement opportunities remain inaccessible due to unstructured natural-language artifacts that require human interpretation. This paper investigates how agentic artificial intelligence (AI) systems can support process improvement by transforming interpretative work rather than merely automating execution tasks. Following a Design Science Research (DSR) approach, we study an exploratory prototype implemented in collaboration with a large public-sector organization responsible for nationwide social benefits administration. The artifact integrates LLM-based agents into a software development business process by transforming unstructured user interface (UI) specification documents into structured JSON-based UI representations, executable code components, and automatically generated test cases. The study explores how agentic capabilities affect process structures, handover artifacts, and human roles within an existing workflow. The artifact is evaluated in a real-world setting using a mixed-method approach combining quantitative efficiency measurements and qualitative expert assessments. Results indicate significant reductions in manual transformation effort as well as improved standardization and reduced interpretative ambiguity. At the same time, the findings highlight the need for structured specification practices and human validation mechanisms. The paper provides empirically grounded design insights into how agentic AI can support business process improvement beyond traditional RPA in public-sector contexts.

Keywords: Agentic Business Process Improvement (Agentic BPI) · Business Process Management (BPM) · Generative Artificial Intelligence (GenAI) · Public Sector Modernization · Robotic Process Automation (RPA) · User Interface (UI) Specifications

F. M. Maggi et al. (Eds.): BPM 2026, LNCS 16936, pp. 670–686, 2027.
https://doi.org/10.1007/978-3-032-37777-7_39

1 Introduction

Automation has long been a central theme in Business Process Management (BPM) research and practice. Robotic Process Automation (RPA) has become an established approach for automating repetitive, high-volume tasks across organizational contexts, including the public sector [5]. By mimicking human interactions with existing information systems, RPA enables organizations to realize automation benefits without extensive modifications to legacy infrastructures.

However, the literature also emphasizes that traditional RPA is inherently limited to structured and deterministic scenarios. RPA systems rely on explicit rules, stable interfaces, and predictable execution logic. When processes involve unstructured information, contextual interpretation, or evolving requirements, RPA solutions often become brittle and costly to maintain [5]. These limitations become increasingly relevant as organizations seek to automate not only execution tasks but also upstream activities such as specification, interpretation, and quality assurance.

Recent advances in artificial intelligence (AI), and especially in generative artificial intelligence (GenAI) as well as large language models (LLMs), suggest a potential shift toward more adaptive forms of automation. Emerging studies in public administration indicate that LLM-based agents can interpret natural-language artifacts, coordinate multi-step tasks, and support knowledge-intensive activities under real-world constraints [3, 16, 24]. However, from a BPM perspective, it remains unclear how such agentic systems relate to established automation paradigms and how they can be meaningfully embedded into organizational processes.

While broader definitions of agentic AI emphasize autonomy, goal-directed behavior, and adaptation [1], this study adopts a pragmatic BPM perspective and uses the term to refer to LLM-based software agents that autonomously perform multi-step transformation tasks through reasoning, artifact generation, and interaction with external tools and systems [14].

To address this gap, this paper explores agentic business process improvement as a direction toward more adaptive, agent-based automation beyond RPA. Following a Design Science Research (DSR) approach, we present an exploratory prototype developed in collaboration with the German Federal Employment Agency (Bundesagentur für Arbeit, BA), one of the largest public-sector IT organizations in Germany responsible for nationwide employment and social benefit administration. The artifact integrates LLM-based agents into a software engineering-related business process and supports the transformation of unstructured specification documents into structured development artifacts. The guiding research question is: How can LLM-based agents be embedded into business processes to support automation beyond traditional RPA?

2 Background: from RPA to Agentic Automation

2.1 Robotic Process Automation

Automation in public-sector organizations has traditionally relied on rule-based information systems and, more recently, Robotic Process Automation (RPA). RPA mimics human interactions with existing information systems, typically through graphical user interfaces rather than deep system integration [12, 20, 21]. This approach, which allows organizations to automate operational activities without extensive changes to legacy infrastructures, has been adopted widely across both private and public-sector contexts [3, 5, 7].

From a technical perspective, RPA bots execute predefined sequences of activities governed by explicit business rules [5]. These scripts interact with applications through user interfaces (UIs) or application programming interfaces (APIs) and follow deterministic control-flow logic, including conditions, loops, and decision branches [12]. As a result, RPA is particularly effective for tasks characterized by stable input structures, clearly defined procedures, and predictable outcomes.

Despite these advantages, the literature consistently highlights structural limitations of rule-based automation [2, 5]. Because RPA uses predefined rules and stable data formats, it is sensitive to changes in interface structures, data representations, or process logic [7]. RPA struggles when process execution requires interpretation, contextual reasoning, or flexible adaptation [5]. Such situations commonly arise in administrative processes where textual documents must be interpreted and transformed into structured information [3].

In public administration, the interpretation and transformation of textual information into structured records often remains a labor-intensive activity. Prior to recent advances in GenAI, automated text processing relied largely on statistical and machine-learning-based approaches that required substantial domain-specific configuration. Recent research suggests that LLMs can support document classification, summarization, information extraction, and decision support across a wide range of domains while reducing effort and maintaining output quality [8, 25]. However, practical applications in large-scale, security-sensitive public-sector environments remain limited, highlighting the need for domain-specific solutions that adapt general-purpose LLMs to administrative contexts [16]. Taken together, these developments indicate a shift in public-sector automation from purely rule-based execution toward support for interpretation, reasoning, and transformation activities.

While RPA remains effective for highly structured and rule-based tasks, the growing availability of GenAI technologies opens new opportunities for automating activities that require interpretation, reasoning, and transformation of unstructured information.

2.2 From Cognitive to Agentic Automation

Recent advances in GenAI, and particularly in LLMs, have accelerated the transition from rule-based task automation toward systems capable of interpreting context, performing reasoning steps, and coordinating actions across heterogeneous tools and information sources [2]. In BPM research, this development is commonly discussed under the concept of cognitive automation, which extends traditional automation approaches such as RPA with AI capabilities for handling semi-structured information and supporting decision-making processes [9, 14].

One emerging realization of cognitive automation is the use of LLM-based agents. Such agents operate within iterative reasoning loops in which they analyze the current situation, select appropriate actions, and invoke external tools or systems through interfaces such as APIs or UIs. This approach allows automation logic to be implemented as tool-mediated interaction rather than as rigid execution scripts, enabling more flexible handling of varying inputs and process contexts [7].

Recent BPM research also discusses a shift toward process autonomization, where processes are guided by goals rather than detailed execution instructions and dynamically determine how tasks should be performed within defined constraints [13].

Overall, the literature points to a shift from deterministic task automation toward adaptive and agent-based process execution. While such approaches offer greater flexibility, they also introduce new challenges regarding governance, transparency, trust, exception handling, monitoring responsibilities, and human oversight [6, 17].

2.3 Generative Multi-Agent Architectures

Multi-agent systems (MAS) have long been studied within distributed AI as a paradigm for coordinating autonomous computational entities that collectively solve complex tasks [23]. In recent years, the integration of GenAI, particularly LLMs, has opened a new line of research in which agents are equipped with capabilities for natural language reasoning and content generation [3]. Emerging frameworks such as CrewAI and TaskMatrix enable the orchestration of multiple specialized agents that interact through structured workflows to accomplish complex reasoning tasks [4, 15]. Rather than executing isolated tasks, these agent-based architectures enable coordinated problem solving, including planning activities, integrating feedback, and iteratively refining intermediate results.

Within document processing scenarios, MAS architectures have been explored as a mechanism to improve the quality of automated text analysis and summarization. Agent-based approaches can incorporate iterative self-evaluation, prompt refinement, or distributed processing strategies in which individual agents analyze separate sections of a document before their outputs are consolidated [25]. In the context of large or heterogeneous documents, such architectures allow the identification of relevant sections, the allocation of summarization tasks across agents, and the aggregation of partial summaries into a coherent final result [3, 19].

Despite this growing interest, most applications of generative MAS remain largely experimental and rarely examined in regulated, complex operational environments. Thus, empirical evidence from real settings, including public-sector settings, remains limited [3]. Addressing this gap, the present study examines how generative agents can be embedded into an administrative workflow within a public-sector organization and how they can support the transformation of unstructured specification artifacts.

3 Research Method: Design Science Research

This study follows a Design Science Research (DSR) paradigm, which is well suited to contexts requiring the intentional design and systematic evaluation of artifacts that address salient organizational problems while generating generalizable knowledge contributions [10, 18].

3.1 DSR Positioning and Research Cycle

This study is an exploratory and formative DSR consistent with prior articulations of DSR as an iterative, learning-oriented process. Rather than advancing a fully specified method or reference architecture, the study focuses on examining and structuring an emergent design space at the intersection of BPM, RPA, and LLM-based agents. The research follows an iterative DSR cycle consisting of problem identification, artifact design and instantiation, demonstration and evaluation, and reflective abstraction of design-relevant insights [11, 22]. The artifact serves as the primary vehicle for inquiry, with design knowledge derived from its development, instantiation, and evaluation in the organizational setting.

3.2 Problem Identification and Relevance

The problem addressed in this study originates from software engineering processes within the information technology (IT) landscape of a large public-sector organization. The German Federal Employment Agency (Bundesagentur für Arbeit, BA) is responsible for managing employment and welfare processes for millions of citizens and represents one of the largest public-sector IT environments in Germany. With more than 40,000 internal users and a large-scale transactional workload, the organization operates highly complex administrative processes and software infrastructures. In the context of demographic change and ongoing workforce transitions, the organization faces increasing pressure to improve efficiency and maintain long-term maintainability of its application landscape while ensuring compliance, traceability, and data protection requirements. The application supports the structured capture, processing, and presentation of legally relevant information required for calculating benefit entitlements.

The existing system is implemented as a classical client-server architecture using a Java Swing client. As part of an ongoing modernization initiative, the organization is transforming the application into a web-based architecture. While the underlying business logic remains largely stable, the migration introduces a substantial transformation effort: more than 300 domain-specific dialogs must be analyzed, redesigned, and reimplemented in the new technological environment.

Each domain dialog represents a structured representation of legally defined facts, such as income situations, household compositions, or eligibility conditions. Despite differences in domain content, these dialogs follow recurring structural patterns. Typically, each subject area consists of both a list or tree overview and a corresponding detail dialog, resulting in approximately 600 UIs that need to be transformed during the migration.

Historically, UIs were implemented manually based on textual specification documents written in domain-specific language. These documents describe fields, validation rules, interaction logic, and accessibility requirements primarily in unstructured natural language. Translating these descriptions into executable UIs and associated test artifacts requires substantial interpretive effort by developers and is both time-consuming (e. g, in some cases taking up to several days per dialog depending on complexity) and error prone.

From an automation perspective, this transformation step lies largely outside the effective scope of traditional RPA because it does not have structured inputs and predefined rules [5]. Consequently, the interpretive step between specification and implementation becomes the primary bottleneck in the transformation process.

To address this challenge, the organization introduced an intermediate representation called JSON-UI, intended to formalize UI descriptions in a structured and machine-readable format. JSON-UI abstracts recurring dialog patterns such as field definitions, input types, validation constraints, and display configurations, thereby creating a bridge between domain specifications and technical implementation. However, generating JSON-UI structures from existing documentation remained largely a manual activity, which motivated the exploration of agentic LLM-based automation to support the transformation process.

3.3 Artifact Design and Instantiation

The designed artifact is an exploratory prototype that embeds LLM-based agents into the specification-to-implementation process. Rather than directly generating executable UIs from unstructured specification documents, the artifact introduces a staged transformation architecture that separates semantic interpretation from technical implementation. The core design idea is to formalize interpretive work performed by developers to translate textual interface specifications into executable UI artifacts.

Design Rationale

The artifact design was guided by four core requirements derived from the modernization context: (1) reducing manual interpretation effort when transforming unstructured GUI specifications, (2) preserving compatibility with existing frontend development workflows, (3) enabling incremental automation without requiring complete process redesign, and (4) supporting human validation and governance in ambiguity-sensitive transformation tasks.

To address these requirements, a central design decision was the introduction of JSON-UI as a structured intermediate representation between specification and implementation. The primary challenge in the modernization project was not code generation itself but the interpretation of unstructured specification documents containing domain-specific terminology, implicit assumptions, and free-text logic. By formalizing dialog structures, UI elements, validation rules, and interaction logic in a machine-readable format, JSON-UI externalizes interpretive work previously performed implicitly by developers. This creates an explicit transformation stage that supports both automated processing and human validation while enabling systematic downstream generation of implementation artifacts.

Structural Characteristics of the Artifact

The artifact operationalizes recurring structures observed across more than 300 domain dialogs in the legacy application. Although the dialogs differ in domain content, they follow common structural conventions that can be formalized and reused. These include dialog titles and operational modes, standardized form layouts, field definitions, input types, grouped composite elements, and validation constraints.

Examples include predefined input elements (such as text fields, dropdown menus, radio buttons, checkboxes, or date pickers) and composite structures representing grouped domain concepts (such as address structures or validity periods). Additional configuration elements define visibility conditions and validation rules for specific fields. By explicitly modeling these recurring structures in JSON-UI, the artifact reduces structural variability and enables systematic transformation.

Agent-Based Transformation Architecture

The artifact implements a two-stage agent architecture that transforms unstructured UI specifications into executable front-end artifacts. As shown in Fig. 1, the architecture consists of two coordinated transformation chains. The first chain converts textual specification documents into structured JSON-UI artifacts. The second chain generates executable front-end code from these structured artifacts. This allows the interpretive step between specification and implementation to be addressed explicitly.

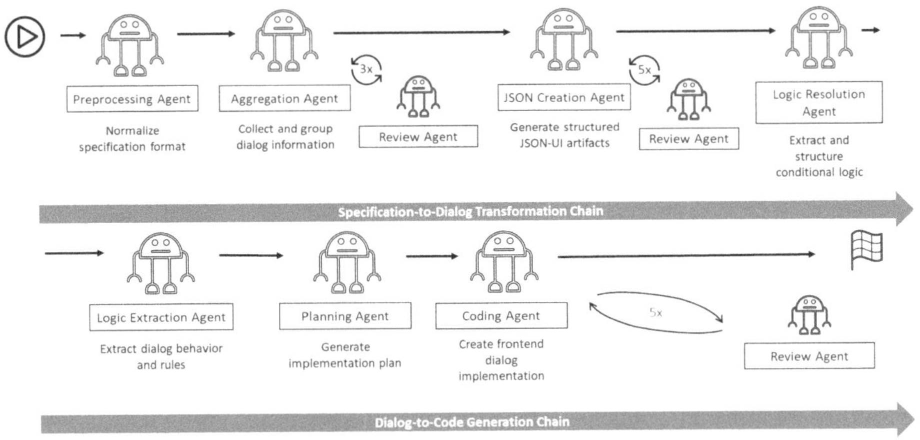

Fig. 1. Agent-based transformation architecture

Specification-to-Dialog Chain

The first transformation chain addresses the interpretive step between natural-language specifications and structured UI representations. It consists of four specialized agents operating sequentially.

The **Preprocessing Agent** performs an initial format transformation of the source documents. In the current instantiation, these documents are converted into a normalized textual representation that can be processed more reliably by downstream agents.

The **Aggregation Agent** reorganizes and groups the extracted information by collecting distributed specification elements (such as field descriptions, interaction constraints, and display attributes) and consolidating them into coherent dialog structures.

The **JSON Creation Agent** then generates structured JSON-UI artifacts by mapping the aggregated information to predefined schema elements. In this step, domain-specific textual descriptions are translated into formal UI components, configuration parameters, and dialog structures.

Finally, the **Logic Resolution Agent** processes remaining free-text logic contained in the specifications. This agent interprets and formalizes conditional logic or other implicit rules that cannot be directly derived from structured descriptions.

The output of this chain is a machine-readable JSON-UI representation of the intended dialog. This representation serves as the central intermediate artifact connecting specification and implementation.

Dialog-to-Code Generation Chain

The second transformation chain generates executable front-end artifacts from the JSON-UI representation.

The process begins with the **Logic Extraction Agent**, which analyzes the JSON-UI artifact and extracts the dialog logic relevant for implementation. This includes dynamic dialog behavior and conditional interaction rules.

The **Planning Agent** then analyzes the project context and derives a step-by-step implementation plan based on the available technology stack and the existing project structure.

Based on this plan, the **Coding Agent** generates the front-end implementation artifacts. Using predefined templates and development conventions, it creates or populates the necessary source files representing dialog components and associated configuration elements.

Finally, the **Review Agent** performs automated quality checks on the generated artifacts. If structural inconsistencies or implementation issues are detected, corrective iterations are triggered by returning the identified issues to the Coding Agent.

Together, the two chains form a staged transformation architecture that separates semantic interpretation from implementation generation while enabling automated support for both stages.

In addition to the final review step, intermediate validation mechanisms are applied throughout the transformation pipeline. These checks include schema validation, structural consistency checks, and format verification of generated artifacts. In cases where inconsistencies are detected, corrective iterations can be triggered automatically. This layered validation approach improves robustness of the transformation pipeline and reduces the propagation of structural errors to downstream processing stages.

Table 1 provides an overview of the agent components that constitute the transformation pipeline and their respective responsibilities.

Instantiation Context

The prototype was instantiated at the German Federal Employment Agency within their ongoing modernization initiative described in sect. 3.2. The instantiation followed a hybrid architecture that combines LLM-based agents with deterministic processing components. Tasks that require semantic interpretation - such as identifying UI structures or resolving implicit specification logic - are handled by LLM agents. In contrast, deterministic transformations, including document preprocessing, format normalization, and schema validation, are implemented through conventional scripts and utility functions. This separation reduces the influence of non-deterministic model behavior in technically

Table 1. Artifact components of the agentic transformation pipeline

Component	Role in the Pipeline	Output
Preprocessing Agent	Normalizes specification documents and converts heterogeneous formats into machine-readable text	Normalized specification text
Aggregation Agent	Collects and groups distributed specification elements into coherent dialog structures	Structured dialog elements
JSON Creation Agent	Maps aggregated information into the JSON-UI schema	JSON-UI artifact
Logic Resolution Agent	Extracts and formalizes rules described in free-text specifications	Structured dialog logic
Logic Extraction Agent	Identifies interaction logic and behavioral rules from JSON-UI artifacts	Implementation-relevant dialog logic
Planning Agent	Derives a generation strategy aligned with the frontend technology stack	Implementation plan
Coding Agent	Generates frontend UI components from structured JSON artifacts	Executable UI code
Review Agent	Performs automated validation and structural quality checks	Validated implementation artifacts

sensitive transformation steps while preserving the flexibility of LLM-based reasoning where interpretation is required.

The underlying language model used for the agentic pipeline is Claude 4.5 Sonnet, accessed via GitHub Copilot. Agent orchestration is implemented using GitHub Copilot, while preprocessing and technical transformations are implemented using Python.

Importantly, the artifact was designed from a human-in-the-loop perspective, intended to complement rather than replace existing development workflows. Generated JSON-UI artifacts and resulting implementation artifacts are reviewed by developers or domain experts before further use. Human work therefore shifts from manual construction of user interfaces toward validation and refinement of generated artifacts. This design reflects the inherent ambiguity of natural-language specifications and ensures controlled integration of agent-based automation within organizational processes.

4 Demonstration and Evaluation

The demonstration was conducted using authentic UI specification documents from the modernization project at the German Federal Employment Agency. During this phase, the artifact was applied to three differently structured dialogs (one of which was complex) – with the selected examples being representative of approximately half of all dialogs. The results indicate that the agentic transformation pipeline can successfully transform textual interface specifications into structured representations and executable UI artifacts, while human validation remains necessary when specifications contain ambiguities or incomplete information.

The evaluation phase followed a formative DSR logic and assessed the artifact's utility within the real transformation process at the German Federal Employment Agency. The goal was to analyze how agent-supported automation affects the efficiency, quality, and organization of the specification-to-implementation workflow. Two complementary evaluation perspectives were considered: (1) quantitative efficiency effects, and (2) qualitative assessment of output quality and process impact. This mixed evaluation approach aligns with prior DSR-based evaluations of agentic systems in public-sector environments [3]. The evaluation focused primarily on the transformation of unstructured UI specification documents into structured JSON-UI artifacts through the Specification-to-Dialog chain. The Dialog-to-Code chain was instantiated and demonstrated as part of the overall artifact, but the quantitative and qualitative evaluation mainly addressed the interpretive transformation stage, as this represented the primary organizational bottleneck and the most mature part of the implemented artifact.

The evaluation compared the traditional manual transformation workflow with the agent-supported process based on JSON-UI artifact generation. In the traditional process, developers manually interpret textual UI specifications and transform them into implementation artifacts and test cases. Changes to specifications require repeated interpretation and manual propagation into the implementation. In contrast, the agent-supported process transforms specifications into structured JSON-UI artifacts that serve as reusable implementation blueprints. This shifts the role of developers from manual artifact creation toward review and validation of generated outputs.

Within the project, unstructured interface descriptions were increasingly perceived as unsuitable for systematic automation. A structured intermediate representation based on JSON was therefore introduced to enable automated processing and reuse. In the agent-supported process, specification documents are first transformed into JSON-UI artifacts using the Specification-to-Dialog transformation chain. These artifacts capture dialog structure, UI elements, validation rules, and interaction logic in a machine-readable representation. The JSON-UI artifacts can subsequently be used to generate executable front-end components through the Dialog-to-Code Generation chain. Generated UI prototypes can already be used during the solution proposal phase, supporting earlier validation of UI concepts. During implementation, Angular-based UI components can be generated automatically from the structured artifacts and extended with individual functionality where required. In this process, the role of developers shifts from manual artifact creation toward reviewing and validating automatically generated artifacts.

4.1 Quantitative Results

To assess efficiency effects, the effort required to transform UI specifications into structured development artifacts was compared before and after introducing the agent-supported processes. Initial project experience indicates that agent-supported transformation of interface descriptions reduces the effort required to produce JSON artifacts by approximately 40% compared to manual creation. The reported reduction refers specifically to the effort required for transforming unstructured specification documents into structured JSON-UI artifacts within the Specification-to-Dialog transformation stage. In the traditional process, generating structured dialog representations from specification documents requires substantial manual interpretation effort. Project participants reported that the creation of JSON-based dialog artifacts could take several days for complex dialogs. The agent-supported pipeline significantly reduces this effort by automatically extracting dialog structures and validation rules from the textual specifications, allowing developers to focus primarily on reviewing and refining the generated artifacts.

The proof-of-concept implementation suggests overall time savings of approximately 30%, even when accounting for prompt engineering and review activities. These estimates were derived from comparative project assessments and prototype-based evaluations conducted within the modernization initiative and should therefore be interpreted as exploratory project observations rather than statistically validated performance measures. These improvements arise primarily from reducing manual interpretation and structuring activities during the early stages of the transformation process. In particular, the automated generation of JSON-UI artifacts reduces the effort required for translating textual specifications into implementation-ready structures, while also enabling partial automation of downstream tasks such as prototype generation and baseline test creation.

The quantitative evaluation was based on a comparative assessment of 16 representative transformation cases from the ongoing modernization project at the German Federal Employment Agency. The evaluated cases covered multiple dialog categories with varying complexity levels, including standard input dialogs, overview dialogs, and dialogs with extended validation logic.

Effort estimates for the manual baseline were derived from historical implementation data and project experience collected over approximately the last five years. The agent-supported approach was evaluated using prototype-based transformation scenarios and structured expert assessments conducted within the project context. To improve comparability across cases, dialogs were grouped according to recurring structural characteristics and implementation complexity. The evaluation therefore focused primarily on transformation scenarios characterized by recurring UI patterns and structurally comparable specification artifacts, which represent the dominant class of dialogs within the project.

The estimates were additionally calibrated using expert assessments from ten experienced project members, including frontend developers, technical architects, AI architects, and domain experts involved in both legacy and web-based implementation activities. The aggregated effort units represent comparative implementation estimates combining activities such as specification interpretation, JSON artifact creation, review, and transformation preparation. For example, the manual Specification-to-Dialog process included approximately 24 effort units for implementation activities and 9.5 effort units for review activities, while the agent-supported process consisted of approximately 6 effort units for specification correction and review activities and 12.45 proportional effort units for one-time agent and transformation logic development. Across the evaluated scenarios, the comparative assessment resulted in a reduction from 33.5 to 18.45 aggregated effort units for the Specification-to-Dialog transformation stage and from 103 to 63.94 aggregated effort units for the broader Specification-to-Implementation workflow. To avoid overstating precision, the reported savings are presented conservatively as approximate reductions of 40% and 30%, respectively.

Aggregated across multiple dialogs, these reductions translate into significant effort savings and improved scalability of the transformation process. Additional observations were collected for selected development activities that recur across the modernization process. Table 2 summarizes the observed effects of the agent-supported transformation across different stages of the Specification-to-Implementation pipeline. The results indicate that efficiency gains primarily occur in activities characterized by recurring structural patterns and rule-based interpretation tasks. Quality assurance activities benefit particularly from automation once structured representations are available. At the same time, highly individualized components remain difficult to automate and therefore continue to require manual implementation.

Table 2. Observed effects of agent-supported transformation across the pipeline

Pipeline Stage	Manual Process	Agentic Process	Effect
Specification interpretation	Manual interpretation of UI specifications	Spec2Dialog extraction and structuring	Reduced interpretation effort
UI artifact creation	Manual creation of JSON artifacts	Automatic JSON-UI generation	~40% effort reduction[1]
Implementation preparation	UI structure derived from text	JSON-UI as implementation blueprint	Improved consistency
Iterative updates	Reinterpretation of modified specifications	Regeneration from JSON-UI artifacts	Reduced rework
Overall transformation	Manual specification-to-implementation workflow	Agent-supported pipeline	~30% time savings[2]

[1] 33.5 vs. 18.45 aggregated effort units
[2] 103 vs. 63.94 aggregated effort units

4.2 Qualitative Insights

The qualitative findings were consolidated from multiple perspectives within the modernization project at the German Federal Employment Agency, including frontend development, software architecture, domain-specific specification design, and practical applicability within the existing system landscape. The assessment was based on structured feedback from review cycles, technical evaluations of generated artifacts, and observations collected during the application of the prototype to real and representative dialog specifications.

The evaluation focused not only on whether a dialog could be generated technically, but also on whether the resulting artifacts were suitable for the project context. Particular attention was paid to comprehensibility from a domain perspective, maintainability from a technical perspective, consistency with the existing web-client architecture, and suitability for subsequent reviews by both developers and domain experts.

The findings were mainly based on input from ten experienced project members and were consolidated by the project architect as well as analyzed along recurring observation categories, including:

- reduction of domain-specific interpretation ambiguities,
- earlier identification of incomplete or ambiguous requirements,
- improved comparability of dialogs through standardized JSON structures,
- better alignment between business specifications, UI structures, and technical implementation.

These categories represent the recurring themes identified across review activities and prototype applications. The qualitative evaluation complements the quantitative assessment, as efficiency gains alone do not adequately capture whether generated dialogs are sufficiently robust from a domain perspective, technically maintainable, and scalable within the complex application landscape of the German Federal Employment Agency. The qualitative evaluation showed that the approach is particularly effective for recurring dialog patterns. Specialized dialogs, highly individualized interaction logic, and domain-specific exceptional cases continue to require additional analysis and manual implementation effort. The qualitative findings therefore not only confirm the efficiency potential of the approach but also help define the practical boundaries within which agent-supported transformation can be applied effectively.

In sum, qualitative feedback suggests that the agent-supported approach not only accelerates transformation activities but also promotes earlier formalization of requirements. Structured artifacts therefore act as boundary objects that reduce interpretive variability and support more consistent implementation and testing processes. Consequently, development effort shifts from repetitive implementation tasks toward business and technical quality assurance.

5 Discussion: Towards Agentic Business Process Improvement

The case illustrates how LLM-based agents can extend automation beyond the scope of traditional RPA. Whereas RPA primarily automates deterministic execution steps, the artifact described in this paper demonstrates how agent-based systems can support process activities that involve interpretation, restructuring of information, and preparation of implementation artifacts. In this sense, agents do not merely replace human actions but operate as intermediaries that transform unstructured process inputs into structured representations that can be further processed automatically.

From a BPM perspective, this observation suggests a shift in how automation opportunities are conceptualized. Traditional automation approaches focus on task execution within existing process structures. In contrast, the presented artifact shows that agentic systems can intervene earlier in the process lifecycle by supporting activities such as requirement interpretation, artifact structuring, and knowledge transformation. These activities have historically remained largely manual because they require contextual understanding and flexible reasoning.

We therefore interpret the results as an early indication of what we refer to as Agentic Business Process Improvement, that is, the use of agentic AI to support and improve process activities involving interpretation, knowledge transformation, and artifact generation. Rather than fully autonomizing processes, agentic systems can augment existing organizational workflows by formalizing previously informal knowledge work and enabling more structured handover artifacts between process stages. In the studied case, this shift becomes visible in the introduction of JSON-UI artifacts that act as boundary objects between domain specification and technical implementation.

Although the artifact was instantiated within a public-sector modernization project, the underlying transformation logic is not limited to this specific context. The approach is particularly applicable to organizational environments in which recurring structured artifacts are derived from unstructured or semi-structured specifications. This can be found in typical enterprise software modernization projects, where large numbers of structurally comparable interfaces, forms, or process artifacts must be interpreted and transformed across evolving technology environments. Similar conditions can be found in domains such as insurance, healthcare, banking, and telecommunications, where complex regulatory requirements, large application landscapes, and recurring specification artifacts create comparable transformation challenges.

At the same time, the findings highlight important boundary conditions for agentic process improvements. Human oversight remains necessary to validate generated artifacts, resolve ambiguities, and ensure compliance with domain-specific requirements. Rather than eliminating human involvement, the artifact shifted work toward review, exception handling, validation, and process-level quality assurance. This suggests that agentic process improvement approaches require governance structures that combine automation capabilities with explicit validation and accountability mechanisms.

Several practical design insights emerged during the development of the artifact. First, dividing the transformation process into multiple specialized agents proved beneficial for managing complex reasoning tasks. Second, separating semantically complex interpretation tasks from deterministic technical transformations improved reliability of the pipeline. Third, constraining agent behavior through explicit intermediate representations such as JSON-UI helped reduce ambiguity and stabilize downstream processing. These observations suggest that successful integration of agentic systems into organizational processes requires not only careful orchestration of agent capabilities but also governance mechanisms that support reviewability, maintainability, and process-level quality assurance.

6 Limitations and Conclusion

This paper explored the potential of agent-based automation beyond traditional RPA using an exploratory DSR approach. Based on a prototype implemented in collaboration with a large public-sector organization responsible for nationwide social benefits administration, the study investigated how LLM-based agents can be embedded into an organizational process to support the transformation of unstructured UI specification documents into structured JSON-based UI artifacts and executable implementation components.

The results illustrate how agentic systems can complement existing automation approaches by addressing interpretive activities that have traditionally remained outside the scope of rule-based automation. In the examined case, LLM-based agents support the transformation of natural-language specifications into structured artifacts that can subsequently be used for automated UI generation and testing.

Several limitations should be considered when interpreting the results of this study:

First, the evaluation was conducted within a single public-sector modernization initiative. Although the identified transformation patterns are likely to occur in other domains, the findings may still be influenced by organization-specific processes, legacy architectures, and documentation practices. Consequently, external validity remains limited and further studies across additional organizational contexts are required.

Second, the quantitative results are based on exploratory project assessments and prototype-based evaluations rather than controlled experiments. While the reported effort reductions were mainly derived from historical project data, the results should be interpreted as indicative observations rather than statistically generalizable performance measurements.

Third, the qualitative findings are based on structured project observations and expert feedback. As the evaluation was embedded into an ongoing organizational project, participant perceptions and interpretations may have influenced the assessment of the artifact's utility and process impact.

Finally, the evaluated approach primarily addresses recurring and structurally comparable transformation tasks. Highly individualized dialogs and exceptional implementation scenarios still require substantial manual development effort and were therefore outside the primary scope of our proposed artifact.

For BPM research, the study provides empirically grounded design insights into how agentic AI can support Agentic Business Process Improvement beyond deterministic task automation. For practice, the findings demonstrate how public-sector organizations can experiment with LLM-based agents while maintaining human oversight and governance mechanisms within existing development workflows. Future research should investigate how our approach can be applied across different process types and organizational settings, and how governance mechanisms can be designed to ensure reliability, transparency, and trust in operational environments.

References

1. Acharya, D.B., Kuppan, K., Divya, B.: Agentic AI: autonomous intelligence for complex goals—A comprehensive survey. IEEE Access. **13**, 18912–18936 (2025). https://doi.org/10.1109/ACCESS.2025.3532853
2. Agostinelli, S., Marrella, A., Mecella, M.: Research challenges for intelligent robotic process automation. In: Di Francescomarino, C., Dijkman, R., Zdun, U. (eds.) Business Process Management Workshops, pp. 12–18. Springer, Cham (2019)
3. Chircu, A., Czarnecki, C., Neumüller, T., Sebrak, S., Sultanow, E., Winzer, F.: Generative agents at work: redesigning administrative processes at the German Federal Employment Agency. In: Proceedings of the Hawaii International Conference on System Sciences (HICSS), pp. 4359–4368 (2026). https://doi.org/10.24251/HICSS.2026.519
4. Collective, O. (2023) CrewAI: A framework for multi-agent coordination with LLMs. Available at: https://github.com/joaomdmoura/crewAI (Accessed: 28 May 2026).
5. Czarnecki, C., Fettke, P.: Robotic process automation: positioning, structuring, and framing the work. In: Czarnecki, C., Fettke, P. (eds.) Robotic Process Automation: Management, Technology, Applications, pp. 3–24. De Gruyter, Berlin (2021). https://doi.org/10.1515/9783110676693-001
6. Czarnecki, C., Delgado, A., González Enríquez, J., Khan, F.A., Weber, B.: RPA technology and trust. In: del Río Ortega, A., Marrella, A., Reijers, H.A., Wilde, A (eds.) Improving Trust between Humans and Software Robots in Robotic Process Automation (Dagstuhl Seminar 24292). Dagstuhl Reports 14(7), 70–75 (2024).
7. Czarnecki, C., Hong, C.G., Schmitz, M., Dietze, C.: Enabling digital transformation through cognitive robotic process automation at deutsche Telekom services Europe. In: Urbach, N., Röglinger, M., Kautz, K., Alias, R.A., Saunders, C., Wiener, M. (eds.) Digitalization Cases, vol. 2. Management for Professionals. Springer, Cham (2021). https://doi.org/10.1007/978-3-030-80003-1_7
8. Dubey, A., Sharma, R., Diwakar, M., Singh, P., Mishra, A.K., Lamba, S.: GenAI-based news article analysis chatbot. In: 4th International Conference on Advancement in Electronics & Communication Engineering (AECE), pp. 476–480. IEEE (2024)
9. Engel, C., Ebel, P., Leimeister, J.M.: Cognitive automation. Electron. Mark. **32**(1), 339–350 (2022)
10. Hevner, A.R., March, S.T., Park, J., Ram, S.: Design science in information systems research. MIS Q. **28**(1), 75–105 (2004)
11. Hevner, A.R.: A three cycle view of design science research. Scand. J. Inf. Syst. **19**(2), 87–92 (2007)
12. Hofmann, P., Samp, C., Urbach, N.: Robotic process automation. Electron. Mark. **30**(1), 99–106 (2020). https://doi.org/10.1007/s12525-019-00365-8
13. Janiesch, C., Kowalkiewicz, M., Rosemann, M.: Process autonomization: rethinking business process management. In: Senderovich, A., Cabanillas, C., Vanderfeesten, R., H. A. (eds.) Bpm 2025, LNBIP, vol. 564, pp. 361–377. Springer, Cham (2026)
14. Kaltenpoth, S., Skolik, A., Müller, O., Beverungen, D.: A step towards cognitive automation: integrating LLM agents with process rules. In: Senderovich, A., Cabanillas, C., Vanderfeesten, R., H. A. (eds.) Bpm 2025, LNCS, vol. 16044, pp. 308–324. Springer, Cham (2026)
15. Liang, Y., et al.: Taskmatrix. Ai: completing tasks by connecting foundation models with millions of apis. Intell. Comput. **3**, 0063 (2024). https://doi.org/10.34133/icomputing.0063
16. Lindgren, I., Madsen, C.Ø., Munk, A.Ø., van Veenstra, A.F.: Artificial Intelligence and the Public Sector: Transforming Public Services. Springer, Cham (2024)
17. Mirispelakotuwa, I., Syed, R., Wynn, M.T.: automation to agitation: unveiling RPA-induced technostress. In: Senderovich, A., Cabanillas, C., Vanderfeesten, R., H. A. (eds.) Bpm 2025, LNBIP, vol. 564, pp. 378–394. Springer, Cham (2026)

18. Peffers, K., Tuunanen, T., Rothenberger, M.A., Chatterjee, S.: A design science research methodology for information systems research. J. Manag. Inf. Syst. **24**(3), 45–77 (2007)
19. Shukla, N.K., Katikeri, R., Raja, M., et al.: Generative AI approach to distributed summarization of financial narratives. In: Proceedings of the 2023 IEEE International Conference on Big Data (BigData), pp. 2872–2876. IEEE (2023)
20. Syed, R., et al.: Robotic process automation: contemporary themes and challenges. Comput. Ind. **115**, 103162 (2020)
21. van der Aalst, W.M.P., Bichler, M., Heinzl, A.: Robotic process automation. Bus. Inf. Syst. Eng. **60**(4), 269–272 (2018)
22. Vaishnavi, V., Kuechler, W.: Design science research in information systems. In: Hevner, A., Chatterjee, S. (eds.) Design Research in Information Systems: Theory and Practice, pp. 19–50. Springer, Boston (2004)
23. Wooldridge, M.: An Introduction to MultiAgent Systems, 2nd edn. John Wiley & Sons, Chichester (2009)
24. Zanini, M., Gil-Garcia, J.R., Luna-Reyes, L.F.: The generative AI revolution in the public sector: opportunities, challenges, and a research agenda. Commun. ACM. **67**(5), 30–32 (2024)
25. Zhang, H., Yu, P.S., Zhang, J.: A systematic survey of text summarization: from statistical methods to large language models. ACM Comput. Surv. (2024)

Identifying Order Processing Types for Circular Production

Tim Hommen[1](✉), Michael Riesener[1,2], and Benedict Janssen[1]

[1] Laboratory for Machine Tools and Production Engineering (WZL), at RWTH Aachen University, Aachen, Germany
t.hommen@wzl.rwth-aachen.de

[2] Fraunhofer Institute for Production Technology IPT, Aachen, Germany

Abstract. The successful establishment of circular production offers significant potential for companies, including reduced material costs, increased supply chain resilience, access to new business opportunities, and compliance with increasingly stringent sustainability regulations. In circular production, material loops are closed through the take-back of used products and their value-preserving recovery, refurbishment, or remanufacturing. However, the required order processing differs substantially from the order processing logic of conventional linear value creation. These differences arise from new sub-processes such as reverse logistics and reprocessing, necessary adaptations of established order processing activities, and possible changes to the underlying business model.

To support companies in addressing these challenges, this paper identifies and characterizes order processing types for circular production. Similar to established types in linear production, such as make-to-stock or engineer-to-order, a typology of order processing for circular production can provide a structured understanding of different order processing logics. A combined research design integrating a systematic literature review of 17 typological approaches with an empirical analysis of 65 industrial case studies and twelve expert interviews is applied. The result is a set of eleven order processing types, characterized along four dimensions: return initiation, Re-X concept, Re-X order type, and replacement condition.

Keywords: Order Processing · Remanufacturing · Process Configuration · Circular Production

1 Introduction

Regulatory pressure on manufacturing companies has grown substantially in recent years, with sustainability requirements becoming increasingly binding, particularly in Europe [1]. Beyond compliance, companies face mounting economic challenges: geopolitical instabilities and fragile global supply chains expose material procurement to disruptions that were long considered unlikely [2]. In high-wage countries, these dynamics compound existing cost pressures, making it harder to produce competitively priced goods [3].

F. M. Maggi et al. (Eds.): BPM 2026, LNCS 16936, pp. 687–703, 2027.
https://doi.org/10.1007/978-3-032-37777-7_40

The dominant logic of industrial value creation has long followed a linear trajectory, in which manufacturing a product means extracting primary raw materials, processing them, and eventually discarding the result. This model ties economic output directly to primary resource consumption. Circular production challenges this assumption [1]. By taking back used products and recovering, refurbishing, or remanufacturing them, companies can retain value already embedded in materials and components rather than replacing it from scratch. Strategies such as remanufacturing and refurbishment are increasingly recognized not only as environmentally preferable but as economically attractive alternatives to conventional, linear manufacturing [4].

Broader adoption of circular production, however, remains limited. Where it has succeeded, specific product and market conditions have proven favorable, such as in the remanufacturing of industrial engines or commercial vehicle components, where high product value and established return structures make the business case tractable. Transferring these successes to other contexts is not straightforward. Order processing sits at the heart of this challenge. From a Business Process Management perspective, the transition to circular production constitutes a process redesign problem: sub-processes such as reverse logistics, disassembly, and reconditioning extend the established order processing, and a broader range of viable business model configurations expands the design space beyond the configurational range covered by existing typologies. As the process through which a company fulfills its obligations to customers, it must be reconfigured substantially when circular production is introduced. Reverse logistics, reconditioning, and disassembly are among the activities that need to be integrated into workflows originally designed for linear production [5].

To support companies in navigating this reconfiguration, this paper develops and characterizes order processing types for circular production. Analogous to the types established in linear production contexts, such as make-to-stock or engineer-to-order, such a typology provides orientation for companies seeking to understand which configuration of order processing is suited to their specific situation. Against this background, the paper addresses the following research question: *Which distinct types of order processing can be identified for circular production, and how do they differ in their constitutive characteristics?* Section 2 introduces the theoretical background on circular production, order processing, and order processing types. Section 3 develops and applies an approach for identifying types in circular production. Section 4 presents the resulting types. Section 5 concludes with a summary and directions for further research.

2 Theoretical Background

This section introduces the foundational concepts underlying the paper: circular production, order processing, and order processing types in linear manufacturing.

2.1 Circular Production

Industrial value creation has traditionally followed a linear logic, in which raw materials are processed into products that are eventually discarded at the end of their useful life. Circular economy represents a regenerative approach aimed at minimizing resource consumption and waste generation by implementing strategies to slow, reduce, and close material loops [6]. In contrast to the linear "take-make-waste" model, it follows a "reduce, reuse, recycle" philosophy, in which products are collected after use and their materials and components are systematically recovered for subsequent production cycles [7]. Among the various R-strategies available to companies, refurbishment and remanufacturing are of particular relevance for manufacturing contexts, as they allow companies to reduce the consumption of primary resources and the extent of new manufacturing required while still bringing high-quality products to market [8].

- **Refurbishment** involves the comprehensive restoration of used products to their original performance level, resulting in high-quality, fully functional products [9].
- **Remanufacturing**, in turn, combines used and new components to produce goods that meet or even exceed the functionality of newly manufactured products [9]. This includes functional upgrades and re-assembly [10].
- **Recycling**, by contrast, returns materials to industrial production at a lower value level and is therefore considered only when no higher-value recovery option is feasible [11].

This paper focuses on processes beyond recycling, emphasizing strategies that retain product value through component-level interventions. Fig. 1 shows the positioning of these strategies within the industrial value chain.

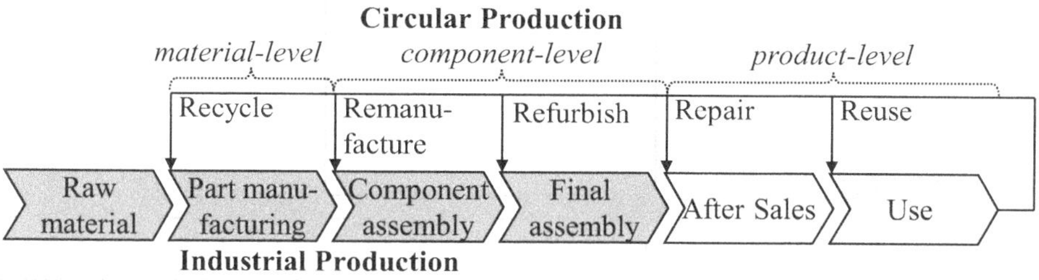

Fig. 1. Allocation of circular production in the value chain

2.2 Order Processing

Order processing describes the temporally connected sequence of processes, sub-processes, and activities through which a manufacturing company fulfills its obligations to customers [12]. It spans a broad set of functions from sales and design through procurement, parts manufacturing, and assembly to shipping and cuts across multiple organizational units [13]. This cross-functional character makes order processing both a key driver of overall company performance and a domain where uncoordinated decisions across organizational units can quickly undermine the coherence of the whole [14].

When a company transitions to circular production, the scope of order processing expands significantly. Processes with no equivalent in linear production must be introduced and coordinated with existing structures. Reverse logistics, for instance, becomes a prerequisite for obtaining used products in the first place. Disassembly, cleaning, inspection, and reconditioning must be integrated into the production workflow, and their outputs must be synchronized with procurement and assembly planning. At the same time, existing processes from linear production need to be adjusted to meet circular production requirements. A generic model of order processing in circular production that maps these additional sub-processes was developed previously [15] and is illustrated in Fig. 2.

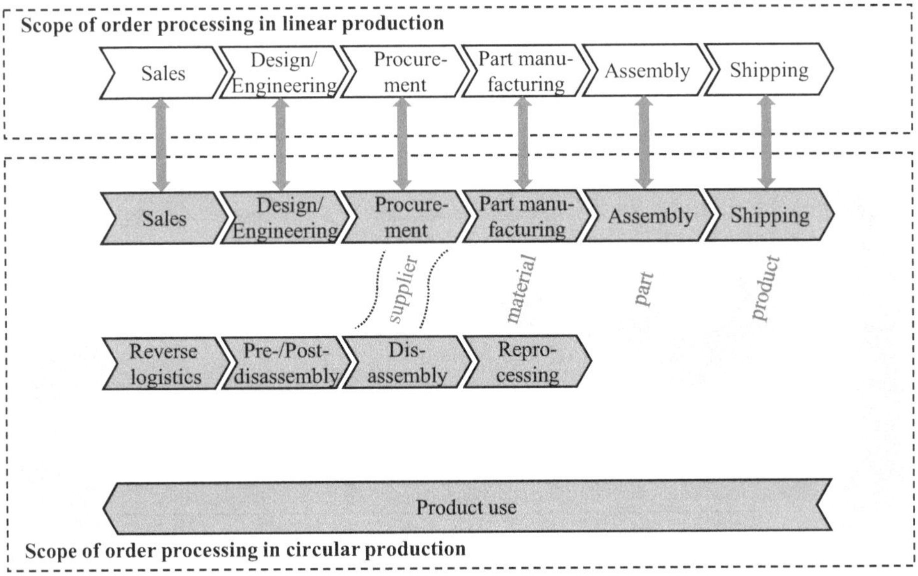

Fig. 2. Generic model of order processing in circular production

Within the field of Business Process Management (BPM), order processing represents a cross-functional business process that integrates customer-facing, planning, and execution activities [41]. The transition to circular production constitutes a process redesign challenge that affects the design and configuration phase of the BPM lifecycle: established process activities must be adapted, new activities such as reverse logistics and reprocessing must be integrated, and existing process variants must be reconfigured to accommodate the resulting design space [42]. The intersection of BPM and Circular Economy has been discussed under the label of Green BPM [43, 44], yet existing contributions focus primarily on the environmental assessment and improvement of single processes rather than on the systematic configuration of cross-functional process types for circular production. The typology developed in the present paper addresses this gap by providing a configuration artefact for the BPM design phase in circular production contexts.

2.3 Order Processing Types in Linear Production

Order processing in manufacturing companies is structured according to the prevailing order processing type, which groups companies whose order processing follows a comparable logic [16]. The constitutive characteristic of each type is the position of the order penetration point, defined as the transition from customer-anonymous to customer-specific task execution [17]. Four principal types are distinguished in linear production: Engineer-to-Order, Make-to-Order, Assemble-to-Order, and Make-to-Stock [18]. While these represent conceptually distinct archetypes, hybrid configurations are common in practice, particularly where the order penetration point varies across product groups.

To systematically describe and differentiate these types, Schomburg developed a morphological framework for linear order processing comprising eight core features, including product range, order release type, and production structure [19]. Büdenbender subsequently extended this framework with features such as assembly organization and the treatment of customer change requests [20]. These morphologies enable companies to position their order processing configuration relative to established types and assess its alignment with product and market requirements.

An equivalent typology for circular production does not yet exist. Existing contributions address selected aspects in isolation, e.g. reverse logistics structures or business model configurations, but none provides a comprehensive characterization of order processing types that spans the full scope of circular production. Identifying and characterizing such types would offer companies a structured basis for configuring their order processing accordingly. Section 3 therefore develops an approach to derive these types systematically.

The typology proposed here also has to be positioned against related frameworks in adjacent domains. Reverse supply chain typologies [29, 36] characterize the structure of return-oriented supply networks but do not address the configuration of order processing as a cross-functional business process. Circular economy business model taxonomies [45] operate at the strategic level of value proposition and value capture and abstract from operational process configuration. The present typology therefore complements rather than duplicates these contributions by addressing the operational process configuration layer in circular production specifically.

3 Research Design

The research design follows a design-oriented logic in the tradition of Design Science Research [46]. The typology is constructed as a descriptive design artefact that supports the process configuration phase of the BPM lifecycle in circular production contexts. Two methodological challenges shape the design. First, the configuration space of circular order processing is broader than in linear production because of sub-processes such as reverse logistics, disassembly, and reconditioning, and because of a wider set of viable business model configurations. Second, circular production lacks the long empirical track record of linear manufacturing, so that documented implementations are comparatively few, heterogeneous, and in many cases not yet fully mature.

The approach presented here therefore pursues two parallel identification strands, as illustrated in Fig. 3. The two strands are executed in parallel rather than sequentially in order to mitigate confirmation bias: a sequential design in which case studies are analysed only after the literature review would risk that empirical observations are forced into pre-existing literature-based categories. Parallel execution preserves the discriminatory power of the case-based analysis and allows both strands to be reconciled in a separate type finalization phase. Each strand contributes a distinct epistemic function. The literature-based strand grounds the typology in the existing body of typological knowledge and prevents the reinvention of established constructs. The case study-based strand secures empirical validity against documented industrial implementations and ensures coverage of configurations that have not yet entered the typological literature. In the *literature-based identification* strand, a systematic literature review serves as the foundation for clustering identified types, conducting a cluster-wise analysis, and ultimately combining these clusters into derived types. Concurrently, the *case study-based identification* strand collects documented industrial implementations, enriches them through supplementary expert interviews, and derives types through structured analysis and clustering. In the subsequent *type finalization* phase, the types identified through both strands are matched against each other, validated for completeness, and characterized into a coherent typology. The following subsections describe each phase in turn.

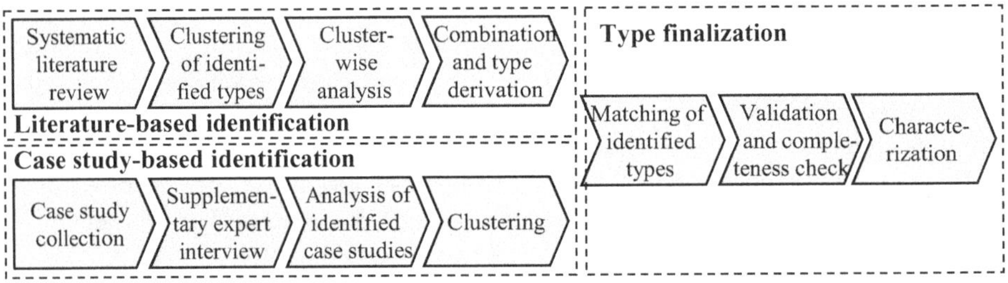

Fig. 3. Approach for identifying order process types in circular production

3.1 Literature-Based Identification

The literature-based identification strand draws on a *systematic literature review* (SLR) to surface existing typologies of order processing relevant to circular production. The SLR follows the thematic search strategy outlined by Pianos and Krüger, using structured search strings composed of Boolean operators and wildcards across the databases Scopus, Web of Science and ScienceDirect [21]. Search strings were derived from the sub-processes of order processing in circular production established in prior work [15], covering terms related to order processing types, circular production, remanufacturing, and reverse logistics. After duplicate removal, title screening, abstract screening, and full-text review, the process yielded 13 relevant typological approaches describing aspects of order processing in linear or circular production. Four relevant approaches that were already known to the authors were also included in the further analysis. Fig. 4 illustrates the screening and selection statistics. Title and abstract screening was guided by four inclusion criteria: (i) the publication addresses a typology, classification, or systematic differentiation of configurations relevant to order processing or to one of its constitutive sub-processes; (ii) the application context covers either linear manufacturing of physical goods or one or more of remanufacturing, refurbishment, or reverse logistics; (iii) the typology is described at a level of detail that allows its characterizing features and feature specifications to be reconstructed; (iv) the publication is available in English or German. Exclusion criteria included purely conceptual position papers without explicit typological structure, contributions limited to recycling at the material level, and duplicate publications of the same typology by the same authors.

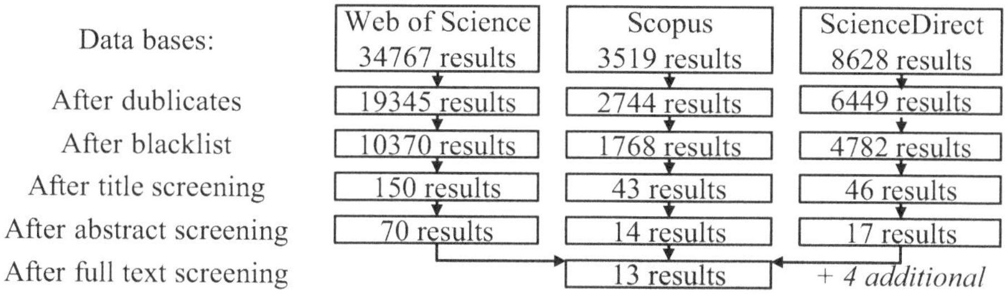

Fig. 4. Overview on the screening and selection statistics of the SLR

The 17 identified approaches do not describe order processing at the same level of completeness. Rather, each addresses a particular dimension of order processing. To make them comparable and combinable, the approaches are assigned to the four descriptive clusters in the subsequent *clustering* step: order initiation, value creation, cooperation form, and product return. Table 1 provides an overview of the 17 approaches and their respective cluster assignments.

In the *cluster-wise analysis step*, the approaches within each cluster are examined using similarity matrices that assess the degree of overlap in features and specifications used by different authors. Approaches with high similarity are progressively merged into intermediate types, while those using divergent features are used to enrich and extend these intermediates. This step reduces the 17 source typologies to a consolidated set of intermediate types per cluster.

During this analysis, the value creation cluster warrants particular attention. The approaches assigned to it differentiate types primarily by the condition of returned products and the depth of disassembly required. In practice, however, companies engaged in circular production must generally be capable of handling returned products across a range of conditions, which limits the discriminatory power of this dimension at the type level. It is therefore not used as a primary differentiating criterion in the combination step. The approaches in this cluster differentiate types by the R-strategy applied and by the depth of disassembly required. However, returned cores arrive in heterogeneous conditions, so that a company engaged in circular production must typically be capable of applying multiple R-strategies across different product structure levels simultaneously; the specific R-strategy applied to an individual core is therefore a consequence of core condition rather than a structural design decision that shapes order processing.

In the *combination and type derivation* step, intermediate types from the remaining three clusters order initiation, cooperation form, and product return, are combined into complete circular order processing types. Since each cluster covers only one aspect of order processing, a full type must integrate contributions from all three. Similarity

Table 1. Identified approaches in the four descriptive clusters

Approach		# of types	Order initiation	Value creation	Cooperation	Product return
Kritchanchai (1999)	[22]	4	•			
Schuh et al. (2012)	[23]	4	•			
Bendul et al. (2019)	[24]	4	•			
Bufka (2004)	[25]	5	•			
Novoszel (2011)	[26]	19	•	•		•
Li et al. (2013)	[27]	2		•		
Keivanpour (2013)	[28]	2		•		
Gobbi (2008)	[29]	9		•		•
Dominguez (2021)	[30]	2			•	
Schiegg (2005)	[31]	5			•	
Tian et al. (2014)	[32]	3			•	
Thierry et al. (1995)	[33]	3			•	
Zhu et al. (2025)	[34]	3			•	
Oflazgil (2017)	[35]	5				•
Östlin et al. (2008)	[36]	7				•
Zheng (2016)	[37]	5				•
Rogers (1999)	[38]	16				•

matrices assess which combinations are coherent and realistic. Types assessed as highly correlated are directly merged; those with moderate correlation are retained after intuitive validation. A subsequent redundancy check eliminates types that differ only in characteristics without operational relevance for order processing configuration, such as the underlying motivation for product return rather than its structural implications. This process yields a consolidated set of literature-based order processing types carried forward into the type finalization phase, illustrated in Fig. 5.

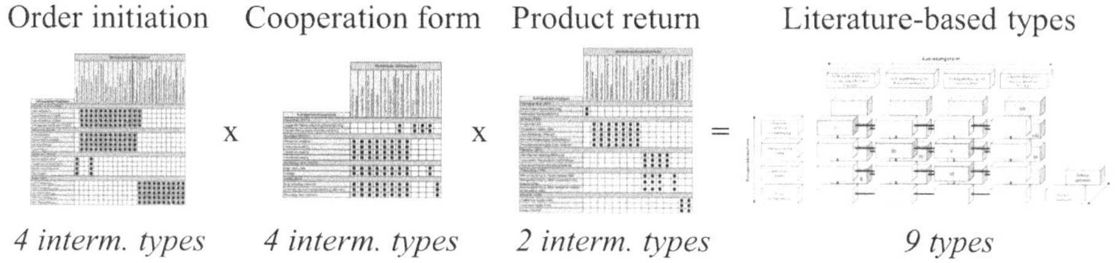

Fig. 5. Overview on approach for literature-based identification

3.2 Case-Study-Based Identification

The case study-based identification strand draws on documented industrial implementations of circular production to surface order processing types grounded in observable practice. Case studies were collected from multiple sources: a case study database maintained by the European Remanufacturing Network, a set of benchmarking studies compiled within the TRANSFORM-CE project [39], targeted searches for case studies across scientific databases including Scopus, Web of Science, and ScienceDirect, as well as cases drawn from industry experience of the Laboratory for Machine Tools and Production Engineering (WZL). To extend coverage, a structured query was submitted to ChatGPT to generate additional candidates. All candidates, regardless of source, were verified against company websites and independent online sources, and retained only where sufficient evidence was available. Each candidate was assessed against the defined scope of circular production, excluding pure recyclers and independent third-party remanufacturers. Only cases for which sufficient primary or verifiable secondary information could be confirmed were retained. This process yielded a final sample of 65 case studies spanning six industries, with a clear concentration in industrial manufacturing, mechanical engineering, and the automotive sector. The approach for the following case-study-based identification is illustrated in Fig. 6.

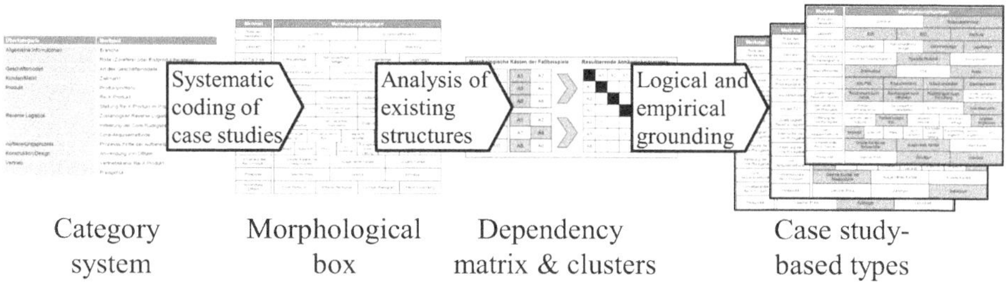

Fig. 6. Overview on approach for case study-based identification

To structure the analysis, a *category system* was used based on the generic order processing framework for circular production established in prior work [15]. The resulting categories cover key dimensions of circular order processing including business model, reverse logistics, reprocessing, sales, product scope, and network structure. Case data were systematically coded against these categories and their associated feature specifications, producing a populated *morphological box* as the basis for further analysis.

To address gaps in the case study data and deepen understanding of causal relationships within the identified configurations, the case study analysis was supplemented by structured expert interviews. Twelve interviews were conducted with representatives from companies across the sample, covering topics including order initiation logic, core acquisition methods, reprocessing organization, and network structure. Interview content was transcribed and integrated into the morphological box.

In the subsequent clustering step, a *dependency matrix* was constructed from the 65 populated morphological boxes, capturing the co-occurrence frequencies of feature specifications across cases. Relative co-occurrence probabilities were calculated and classified into three significance levels, allowing dependency clusters to be identified systematically. The three significance levels classify relative co-occurrence probabilities into high, medium, and low bands. In the cluster formation step, feature-specification pairs exhibiting high significance in at least one direction are considered as candidates; a pair is included in a cluster only if the reverse conditional probability also reaches at least medium significance. When additional specifications are added to an existing cluster, all pairwise co-occurrences within the cluster must reach at least medium significance. Eleven such *clusters* were derived, each representing a characteristic combination of feature specifications that recurs across multiple cases. These clusters served as the structural input for the *type derivation* step, in which logically coherent and empirically grounded clusters were combined and refined into eight distinct case study-based order processing types for circular production. The interview guideline, category system, similarity and dependency matrices, and an anonymized case overview are available at https://doi.org/10.5281/zenodo.20533423.

3.3 Type Finalization

The type finalization phase consolidates the outputs of both identification strands into a coherent typology. In the *matching step*, the literature-based and case study-based types are systematically compared. Types with substantial overlap in their constitutive characteristics are merged, while types that appear in only one strand but represent a plausible and distinct configuration are retained for further consideration.

The *validation and completeness check* then assesses whether the consolidated set of types covers the full range of realistic order processing configurations in circular production. This involves verifying that all types from both strands are accounted for and that no relevant configuration evident in the empirical material remains unrepresented. Types that prove redundant, differing only in characteristics without operational relevance for order processing configuration, are eliminated at this stage.

In the *characterization* step, each confirmed type is described in terms of its constitutive features in four dimensions, which are used to characterize the types after the two strands have been matched. The resulting typology is presented in Sect. 4.

4 Order Processing Types in Circular Production

Building on the type finalization described in Sect. 3, this section presents the resulting order processing types for circular production. The type finalization process revealed that the identified types differ along a manageable set of recurring characteristics that proved consistently relevant across both the literature-based and case study-based strands. These characteristics were consolidated into four dimensions that capture the essential structural differences between types without requiring the full granularity of the underlying morphological box. The dimensions and their specifications are introduced first, followed by a description of each identified type.

4.1 Dimensions of Order Processing Types in Circular Production

The four dimensions were derived from the characteristics that most consistently differentiated the types identified across both strands during the matching and characterization phase. Rather than reproducing the full set of features from the morphological analysis, they represent the level of abstraction at which types become practically distinguishable and communicable. The four dimensions and their respective specifications are described in the following.

(I) **Return Initiation** describes how used products or cores are returned to the company. *Push* places the initiative with the current user, who actively triggers the return through a deposit, incentive, or take-back system. *Contractual-Pull* refers to returns that are contractually anchored and initiated by the company, such as after a defined usage period or as part of an agreed product exchange. *Marketplace-Pull* describes procurement of cores through the market, via intermediaries or directly from end customers, without a prior contractual relationship.

(II) **Re-X Concept** describes the structural relationship between the returned product and the resulting output. *Components-as-Re-X-Product* denotes configurations in which individual components constitute the Re-X product itself. *Components-in-New-Product* refers to recovered components integrated into a newly manufactured product. *Closed-Loop-Re-X-Product* describes configurations in which the overall product structure is preserved and returned to the same use context. *Open-Loop-Re-X-Product* covers cases where components are recombined with those of other products, dissolving the original product structure.

(III) **Re-X Order Type** adopts the logic of linear order processing types to describe the production logic of the Re-X process. *Engineer-to-Order*, *Make-to-Order*, *Assemble-to-Order*, and *Make-to-Stock* carry the same meaning as in linear production and describe the degree to which the Re-X process is coupled to individual customer orders.

(IV) **Replacement Condition** captures whether the return of a used product is directly coupled with the delivery of a replacement product to the customer, as in a subscription based product-service-systems.

4.2 Description of the Identified Order Processing Types

Fig. 7 maps the eleven identified types against the four dimensions. Several types span multiple specifications within a single dimension, reflecting the configurational breadth of circular production. Together, the eleven types cover the range of configurations identified across the literature and case study analysis and confirmed during the finalization phase. Types that share their specification in one or two dimensions are not redundant; their distinctness rests on the combination of specifications across all four dimensions and on the resulting operational implications.

	I			II				III				IV
	Push	Contractual-Pull	Marketplace-Pull	Components-as-Re-X-Product	Components-in-New-Product	Closed-Loop-Re-X-Product	Open-Loop-Re-X-Product	Engineer-to-Order	Make-to-Order	Assemble-to-Order	Make-to-Stock	Replacement
Re-Make-to-Aftermarket	●		●	●							●	●
Extract-to-Stock	●	●			●			●	●	●	●	
Upgrade-to-Order	●					●				●		
Upgrade-to-Stock	●					●					●	
Re-Make-to-Order		●				●	●		●			
Re-Assemble-to-Order		●					●			●		
Re-Make-to-Stock		●				●	●				●	
Service-to-Stock		●				●					●	
Service-to-Service		●				●	●			●	●	●
Match-to-Transfer			●						●			
Buyback-to-Stock			●			●					●	

Fig. 7. Order processing types in circular production

The identified types are briefly described below:

- **Re-Make-to-Aftermarket** serves the spare parts market by supplying reprocessed components sourced from returned cores, allowing manufacturers to meet aftermarket demand without producing equivalent new parts. Returns are push-initiated or marketplace-based, often handled through intermediaries. The reprocessing process is standardized at remanufacturing or refurbishment level, and finished components are held in stock for immediate delivery.

  Example: Manufacturer of level sensors for the process industry

- **Extract-to-Stock** takes back products and disassembles them to recover components that are channeled into new product manufacturing. The focus lies on individual component value rather than product identity, and the returned product serves primarily as a source of parts rather than an object of restoration. Returns can be push- or contractual-pull-based, and recovered components enter standard production planning as incoming parts.

  Example: Manufacturer of power tools for professional use

- **Upgrade-to-Order** performs reprocessing as an on-site service at the customer's premises. While the scope of reprocessing follows defined standards rather than being fully individualized, the configuration of the service, the sales process, and decisions about upgrade content are shaped through close and continuous customer interaction. The type is particularly suited to large, complex, or permanently installed systems where transport would be disproportionately costly.

Example: Manufacturer of measuring machines

- **Upgrade-to-Stock** is characterized by products that remain in the ownership of the customer throughout their use phase. The manufacturer anticipates future functional needs and develops upgrades accordingly, which customers can purchase and implement independently. The model requires the manufacturer to produce and stock upgrade components in advance, and places particular demands on product architecture, as modular design and upgrade compatibility are prerequisites for the concept to function reliably.

  Example: Manufacturer of modular smartphones

- **Re-Make-to-Order** integrates Re-X products into existing supplier relationships with OEMs through contractually secured framework agreements. Returns are governed by contractual-pull arrangements, and production follows the delivery call-off schedules of the OEM relationship, making this type structurally the closest to established linear supplier order processing.

  Example: Transmission manufacturer (Tier 1 supplier for OEM Re-X product)

- **Re-Assemble-to-Order** produces individually configurable products from returned products, with final assembly initiated only after receipt of a customer order. Component-level stocks of recovered parts are maintained, while final configuration is deferred to the customer order, balancing inventory efficiency with product flexibility.

  Example: Refurbisher of IT server systems

- **Re-Make-to-Stock** produces standardized Re-X products from returned products for stock, fully decoupled from individual customer orders. Contractual-pull return arrangements provide the planning stability required for stock-oriented production.

  Example: Manufacturer of brake calipers for commercial vehicles

- **Service-to-Stock** takes back products from product-service systems after the contracted usage period, reprocesses them, and offers them for direct sale in a new use cycle. By selling reprocessed products rather than returning them to the PSS, the type enables simultaneous coverage of different customer groups and price segments.

  Example: Manufacturer of CT systems

- **Service-to-Service** embeds products in a usage- or outcome-oriented PSS in which return, reprocessing, and redeployment are fully integrated within the same system. Cores originate exclusively from the company's own PSS fleet. Upgrades and rapid availability of reprocessed products are operationally critical, as customers expect consistent performance throughout their service contract.

  Example: Manufacturer of forklifts

- **Match-to-Transfer** acts as an intermediary between first and second users of complex B2B products, coordinating repurchase and customer-specific reprocessing. The reprocessing scope is negotiated individually with the incoming customer. The type leverages the manufacturer's technical knowledge and reputation to create value that generic third-party reprocessors cannot replicate.

  Example: Manufacturer of machine tools

- **Buyback-to-Stock** opportunistically repurchases products from the market when economically viable reprocessing and a resale market exist. Reprocessing depth is intentionally moderate. The competitive advantage over independent reprocessors lies in the manufacturer's technical expertise, product knowledge, and brand reputation.

  Example: Manufacturer of professional audio systems

The eleven types cover the range of configurations identified across both identification strands. They differ not only in production logic but also in the underlying return structure and the relationship between the returned product and the Re-X output. The Re-X order type dimension retains the established linear terminology because the underlying logic – the coupling degree between customer order and production process – applies equally to the reprocessing flow. The conceptual extension required by circular production, in particular the reverse material flow and the dependence on returned cores, is captured by the three additional dimensions rather than by modifying the order type concept itself.

In practice, companies with diverse product portfolios may implement more than one type simultaneously for different product groups. The typology is therefore intended not as a strict classification but as an orientation for companies seeking to understand and configure their order processing in circular production.

5 Conclusion and Further Research

This paper addresses the question of which distinct order processing types can be identified for circular production and how they differ in their constitutive characteristics. Starting from an extension of linear order processing to include the sub-processes specific to circular production, a combined literature- and case study-based approach was developed and applied. The literature-based strand drew on a systematic review of 17 typological approaches, which were clustered, analyzed, and combined into candidate types. The case study-based strand collected and analyzed 65 industrial implementations, supplemented by twelve expert interviews, and derived types through dependency analysis and clustering. Following a matching, validation, and characterization phase, eleven order processing types for circular production were identified and described along four constitutive dimensions: return initiation, Re-X concept, Re-X order type, and replacement condition. The types range from stock-oriented aftermarket supply and opportunistic buyback models to fully integrated product-service systems, reflecting the breadth of configurations that circular production can take in practice.

Several directions for further research follow from this work. To provide companies with genuine operational guidance, the identified types require more detailed characterization. For this purpose, the types will be mapped against a comprehensive morphological box developed in parallel work [40], which describes order processing in circular production at the level of individual features and their specifications. This mapping will allow the characteristic configuration of each type to be expressed in operational terms. Beyond characterization, companies need a configuration logic that enables them to identify a suitable type for a given product scope and to derive the associated configuration decisions for their order processing. Developing such a procedure represents a central next step. Finally, the developed framework requires industrial application to validate its practical applicability and to identify potential refinements based on real-world feedback.

Acknowledgments. Funded by the Deutsche Forschungsgemeinschaft (DFG, German Research Foundation) under "EXC-2023 Internet of Production – 390621612".

Acknowledgement of AI Assistance. Parts of this paper were developed with the support of large language models, which assisted in the formulation and editorial refinement of the research findings. The scientific content, methodological decisions, and all results presented are the sole responsibility of the authors.

References

1. Hariyani, D., Mishra, S., Hariyani, P., Sharma, M.K.: Drivers and motives for sustainable manufacturing system. Innov. Green Dev. **2** (2023)
2. Bednarski, L., Roscoe, S., Blome, C., Schleper, M.C.: Geopolitical disruptions in global supply chains: a state-of-the-art literature review. Prod. Plan. Control, 1–27 (2023)
3. Pieterse, J.N.: Global rebalancing: crisis and the east–south turn. Dev. Chang. **42**, 22–48 (2011)
4. Bjørnbet, M.M., Skaar, C., Fet, A.M., Schulte, K.Ø.: Circular economy in manufacturing companies: a review of case study literature. J. Clean. Prod. **294**, 126268 (2021)
5. Wei, S., Tang, O., Sundin, E.: Core (product) acquisition management for remanufacturing: a review. Jnl Remanufactur. **5**, 1–27 (2015)
6. Kraft, M.H.G., Christ, O., Scherer, L.A.: Management der Kreislaufwirtschaft: Positionierung und Gestaltung zirkulärer Unternehmen. Springer, Wiesbaden (2022)
7. Korhonen, J., Nuur, C., Feldmann, A., Birkie, S.E.: Circular economy as an essentially contested concept. J. Clean. Prod. **175**, 544–552 (2018)
8. Urbinati, A., Chiaroni, D., Chiesa, V.: Towards a new taxonomy of circular economy business models. J. Clean. Prod. **168**, 487–498 (2017)
9. DIN Deutsches Institut für Normung: remanufacturing (Reman) – quality classification for circular processes (DIN SPEC 91472) (2023)
10. Schuh, G., Schmitz, S., Lukas, G., Niwar, L., Welsing, M., Calchera, R.: Framework for Circular Production Economy (2023)
11. Steinhilper, R.: Remanufacturing: the Ultimate Form of Recycling. Fraunhofer-IRB-Verl, Stuttgart (1998)
12. Otto, A.: Auftragsabwicklung. In: Klaus, P. (ed.) Gabler Lexikon Logistik, 5th edn, pp. 19–26. Springer Gabler, Wiesbaden (2012)

13. Pfohl, H.-C.: Logistic Systems, 9th edn. Springer, Berlin, Heidelberg (2018)
14. Wildemann, H.: Auftragsabwicklungsprozeß. Leitfaden für eine kundenorientierte Neuausrichtung und Kundenbindung, 24th edn. TCW (2018)
15. Hommen, T., Riesener, M., Schmitz, S., Neumann, H.: Order processing in circular production: development of a configuration framework. Procedia CIRP. **130**, 1680–1686 (2024)
16. Schuh, G., Schmidt, C.: Grundlagen des Produktionsmanagements. In: Produktionsmanagement, pp. 1–62. Springer (2014)
17. Wiendahl, H.-P.: Handbuch Fabrikplanung, 2nd edn. Hanser, München, Wien (2014)
18. Olhager, J.: The role of the customer order decoupling point in production and supply chain management. Comput. Ind. **61**, 863–868 (2010)
19. Schomburg, E.: Entwicklung eines betriebstypologischen Instrumentariums zur systematischen Ermittlung der Anforderungen an EDV-gestützte Produktionsplanungs-und -steuerungssysteme im Maschinenbau (1980)
20. Büdenbender, W.: Ganzheitliche Produktionsplanung und -steuerung. Springer, Berlin, Heidelberg (1991)
21. Pianos, T., Krüger, N.: Erfolgreich recherchieren - Wirtschaftswissenschaften. In: Erfolgreich recherchieren - Wirtschaftswissenschaften. De Gruyter Saur (2014)
22. Kritchanchai, D., MacCarthy, B.L.: Responsiveness of the order fulfilment process. Int. J. Oper. Prod. Manag. **19**, 812–833 (1999)
23. Schuh, G., Stich, V. (eds.): Produktionsplanung und -steuerung, 4th edn. VDI-Buch. Springer Vieweg, Berlin, Heidelberg (2012)
24. Bendul, J.C., Erfurth, T.: Evolution of global manufacturing networks and xKD supply chains. IFAC-PapersOnLine. **52**, 1349–1354 (2019)
25. Bufka, D.: Typologisierung von Order-to-Delivery-Prozessen in der Automobilindustrie. Fraunhofer IRB Verl, Stuttgart (2004)
26. Novoszel, T.: Gestaltung einer integrativen Reverse Supply Chain, 1st edn. Apprimus-Verlag, Aachen (2011)
27. Li, X., Li, Y., Saghafian, S.: A hybrid manufacturing/remanufacturing system with random remanufacturing yield and market-driven product acquisition. IEEE Trans. Eng. Manag. **60**, 424–437 (2013)
28. Keivanpour, S., Kadi, D.A., Mascle, C.: Economic sustainability of end-of-life vehicle recycling infrastructure under uncertainty a fuzzy logic approach. In: Proceedings of 2013 International Conference on Industrial Engineering and Systems Management (IESM), pp. 1–6 (2013)
29. Gobbi, C.: The Reverse Supply Chain: Configuration, Integration and Profitability, 1st edn. Kgs, Lyngby (2008)
30. Dominguez, R., Cannella, S., Framinan, J.M.: Remanufacturing configuration in complex supply chains. Omega. **101**, 102268 (2021)
31. Schiegg, P.: Typologie und Erklärungsansätze für Strukturen der Planung und Steuerung in Produktionsnetzwerken. Shaker, Aachen (2005)
32. Tian, G., Chu, J., Hu, H., Li, H.: Technology innovation system and its integrated structure for automotive components remanufacturing industry development in China. J. Clean. Prod. **85**, 419–432 (2014)
33. Thierry, M., Salomon, M., van Nunen, J., van Wassenhove, L.: Strategic issues in product recovery management. Calif. Manag. Rev. **37**, 114–136 (1995)
34. Zhu, H., Hu, J., Yang, Y.: Towards a circular supply chain for retired electric vehicle batteries. Int. J. Prod. Econ. **282**, 109556 (2025)
35. Oflazgil, K.: Simulationsbasierte Bewertung Alternativer Reverse Supply Chains, 1st edn. Apprimus-Verlag, Aachen (2017)

36. Östlin, J., Sundin, E., Björkman, M.: Importance of closed-loop supply chain relationships for product remanufacturing. Int. J. Prod. Econ. **115**, 336–348 (2008)
37. Zheng, X., Xu, F.: Multi-case study on business model of reverse supply chain in electronics industry under extended producer responsibility. In: ICSSSM, pp. 1–6. IEEE, Kunming, China (2016)
38. Rogers, D.S., Tibben-Lembke, R.S.: Going Backwards: Reverse Logistics Trends and Practices. Reverse Logistics Executive Council, Reno, Nev (1999)
39. The Remanufacturing Case Study Tool. https://www.remanufacturing.eu/case-study-tool.php (2026). Accessed 20 March 2026
40. Hommen, T., Schuh, G., Schmitz, S., Hermann, A.: Mapping order processing in circular production. In: Proceedings of the Conference on Production Systems and Logistics: CPSL, vol. 2025, (2025)
41. Dumas, M., La Rosa, M., Mendling, J., Reijers, H.A.: Fundamentals of Business Process Management, 2nd edn. Springer, Berlin, Heidelberg (2018)
42. Rosemann, M., vom Brocke, J.: The six Core elements of business process management. In: vom Brocke, J., Rosemann, M. (eds.) Handbook on Business Process Management 1, 2nd edn, pp. 105–122. Springer, Berlin, Heidelberg (2015)
43. Seidel, S., Recker, J., vom Brocke, J.: Green business process management. In: Green Business Process Management: Towards the Sustainable Enterprise, pp. 3–13. Springer, Berlin, Heidelberg (2012)
44. Couckuyt, D., Van Looy, A.: A systematic review of green business process management. Bus. Process. Manag. J. **26**(2), 421–446 (2020)
45. Lüdeke-Freund, F., Gold, S., Bocken, N.M.P.: A review and typology of circular economy business model patterns. J. Ind. Ecol. **23**(1), 36–61 (2019)
46. Peffers, K., Tuunanen, T., Rothenberger, M.A., Chatterjee, S.: A design science research methodology for information systems research. J. Manag. Inf. Syst. **24**(3), 45–77 (2007)

A Case Study Into Factors of Intelligent Process Automation in Audit

Dries Jarijch[1(✉)], Mieke Jans[1,2], and Jan Mendling[3,4,5]

[1] Hasselt University, Hasselt, Belgium
{dries.jarijch,mieke.jans}@uhasselt.be
[2] Maastricht University, Maastricht, The Netherlands
[3] Humboldt-Universität zu Berlin, Berlin, Germany
jan.mendling@hu.berlin.de
[4] Weizenbaum Institute, Berlin, Germany
[5] Vienna University of Economics and Business, Vienna, Austria

Abstract. Artificial Intelligence is changing the way process automation is implemented in practice. Intelligent Process Automation (IPA) integrates new AI capabilities, offering efficient mechanisms to automate tasks and overarching processes. Despite research on this topic, little is known about how organizations decide where to use which IPA capabilities and to what extent in order to mitigate potential negative effects. In this paper, we address this knowledge gap by conducting a case study with the FAA and their currently developed IPA for financial audit processes. We identify three major groups of factors and inductively derive corresponding propositions. Our research contributes empirical knowledge on design considerations organizations make to optimize the impact of AI and IPA.

Keywords: BPM · Intelligent Process Automation · Audit

1 Introduction

Recent advances in Artificial Intelligence (AI) change the way how processes can be automated. *Intelligent Process Automation (IPA)* integrates such new AI-capabilities for efficiently automating tasks and overarching processes. In this way, techniques such as large language models (LLM), optical character recognition (OCR), machine learning (ML) can be woven into automated processes. Organizations increasingly explore IPA in various use cases, but struggle to make informed design decisions about where to reply on AI-based automation and to what extent.

Prior research describes which classical process automation approaches have been found to fit specific processes, which benefits can be expected, and which risks have to be mitigated. Such research describes how these benefits and negative effects impact selection criteria for processes that are meant to be automated using workflow systems [21] or robotic process automation [1,14,30]. General

F. M. Maggi et al. (Eds.): BPM 2026, LNCS 16936, pp. 704–719, 2027.
https://doi.org/10.1007/978-3-032-37777-7_41

threats of automation in terms of skill degradation or over-reliance are discussed in engineering psychology [18] and accounting information systems [29]. What we do not know is how organizations make design decisions about where to use which IPA capability to balance beneficial and detrimental outcomes.

In this paper, we address the research question: how do organizations decide on where to use IPA and to which extent. We conduct a case study with the Flemish Audit Authority (FAA) on how they develop an IPA for financial audit processes. We inductively derive corresponding propositions. In this way, our research contributes empirical knowledge on design considerations organizations make with the aim to optimize AI and IPA impacts.

This paper is structured as follows. Section 2 summarizes the general background of process automation and human factors research. Section 3 describes our applied research method. Section 4 presents the case of FAA and its IPA project. Section 5 provides details of our findings, along with corresponding propositions. Section 6 discusses the implications of our findings before Sect. 7 concludes the paper.

2 Background

In this section, we review the background of intelligent process automation and its impact, followed by a summary of technology dominance theory and key findings on automation and human factors.

2.1 Intelligent Process Automation

Intelligent Process Automation (IPA) refers to process automation approaches that integrate AI components. The IEEE defines IPA as "a preconfigured software instance that combines business rules, experience-based context determination logic, and decision criteria to initiate and execute multiple interrelated human and automated processes in a dynamic context" [15]. In contrast to purely deterministic automation, IPA supports *AI-augmented process execution* across end-to-end workflows [8].

By combining RPA-style task automation with AI-based pattern recognition and language capabilities, IPA extends automation to tasks with unstructured inputs and less rule-bound execution [2,6,14,30]. From an implementation perspective, IPA inherits workflow and governance challenges known from RPA and adds AI-specific concerns, such as data quality, explainability, compliance constraints, and continuous monitoring [6,13,14,20,25].

2.2 Human Factors of Automation

IPA implementation challenges are often rooted in human–automation interaction. While IPA is typically justified by efficiency and quality gains, successful adoption depends on user acceptance, role reconfiguration, and the development of new skills for building, operating, and governing automation solutions [20,24,25].

Human–automation interaction research emphasizes that automation changes work rather than merely replacing it, creating new coordination demands and new risks for human performance [18]. Beyond intended performance gains and workload reduction, reported unintended consequences include overreliance on automation, loss of situation awareness, and skill degradation [10,16,19].

One way to address these challenges is to refrain from full automation. Accordingly, research distinguishes a degree of automation [18]. Tasks can also be decomposed into four stages of Human Information Processing (HIP)—information acquisition, information analysis, decision and action selection, and action implementation—each of which can be discussed separately in terms of its degree of automation [18].

Negative consequences of automation are also described by the Theory of Technology Dominance (TTD). It argues that intelligent decision aids can dominate novices' decision processes, potentially leading to biased judgments and reduced learning when the system's guidance substitutes for the user's own knowledge structures [3,4]. Decision-aid design choices, such as transparency and explanations, may mitigate or exacerbate such dominance effects [28].

These human factors are particularly salient for IPA, which can increasingly substitute human work in decision-making or supervisory control tasks. This may affect workers' ability to anticipate system behavior, diagnose failures, and maintain domain knowledge for exception handling [14,24]. Understanding how organizations decide on the degree of automation in IPA is therefore key to anticipating potential interferences with employees' skills.

3 Research Method

To answer the research question 'how do organizations decide on where to use IPA and to which extent?', we use an in-depth case study as a research design. This is grounded in an interpretive epistemology that sees the world as socially constructed [7,31]. More specifically, we aim to uncover the viewpoints and sensemaking processes of various stakeholders. The case setting is an IPA implementation at the Flemish Audit Authority, executed by a Big Four auditing company and its partners. In this way, our research strategy aligns with the epistemological stance of interpretivism in information systems research [5,17].

3.1 Data Collection

The data collection was conducted in two distinct phases. The first phase of data collection was exploratory in nature with the intent of case familiarization. The second phase was focused on gathering data relevant to answering our research question. Across both stages, we employed a multi-method approach to data collection.

At the start of the research project, we conducted exploratory interviews with key stakeholders involved in the IPA project. In total, 8 interviews were carried out between October 2024 and September 2025. The interview participants included senior and junior auditors [A1A4], software developers [D1D4], and the project manager [M1]. The interviews with the auditors were always linked to observations of the audit process. These observations took 30 to 90 min per observation instance. Similarly, interviews with the developers always resulted in observations of the IPA system as it was being developed. In parallel, during this period, we were permitted to participate in the CATE project's steering meetings. All these primary data collection techniques resulted in a rich collection of memos.

In addition to the primary sources of data, we were given access to a broad range of secondary data sources. These include training documents for junior- and mid-level auditors, development documentation, BPM diagrams of the entire audit process, and communication documents to higher management stakeholders. These secondary data sources were used throughout the research process to validate our findings.

The second phase of the data collection consisted primarily of semi-structured interviews targeted at answering our research question. In total, 5 interviews were conducted, resulting in approximately 5 h of transcribed interview data. The interview participants included auditors [A1, A5], software developers [D1, D2], and the project manager [M1]. We used an interview guide[1] that was applied in a flexible manner and iteratively refined as our understanding of the project evolved. The guide focused on how intelligent automation was implemented within the system and on participants' concerns about it. These questions reflect our emphasis on the anticipated consequences of automation and on how stakeholders' perspectives developed over time.

3.2 Data Analysis and Coding Procedure

For the analysis of the data derived from transcripts, field notes, and memos, we adopted procedures from a more interpretive variant [11] of the Straussian grounded theory methodology [27]. To ensure methodological rigor, we additionally adhered to the guidelines proposed by [23] and [9].

The researcher responsible for the field observations conducted an initial inductive coding of first-order concepts using the software ATLAS.ti. These first-order concepts were regularly discussed with a second author, on the basis of which second-order themes (see Footnote 1) were developed. To ensure data-triangulation, the second-order themes were subsequently validated and refined using memos and documentation from the first phase of data collection, additional interview data, and a further group discussion involving all members of the research team. This mixed-methods approach is applied in a manner similar to the one described in [26].

In contrast to the Straussian variant of grounded theory, the Gioia methodology does not necessitate a selective coding phase focused on constructing a

[1] The interview guideline and codes from transcripts are available in an online repository: https://tinyurl.com/48dany4j.

single, central storyline. Instead, drawing on the second-order themes, four distinct theoretical propositions on the degree of automation in the context of IPA were derived.

4 The Case at the Flemish Audit Authority

Our case study examines *Continuous Auditing based on Technological Evolution and Data Mining* (CATE), an innovation project initiated by the Flemish Audit Authority (FAA) and executed by a Big Four accounting firm. The project aims to move from periodic, document-heavy audits toward continuous, technology-supported assurance by combining workflow automation with AI components (e.g., optical character recognition and large language models).

CATE targets first- and second-level controls in audits of European Structural and Investment Funds. In this setting, Management Authorities perform first-level controls to verify eligibility and compliance, while the FAA performs second-level controls. CATE was launched to reduce costs and increase assurance by rethinking and partially automating the audit workflow.

The auditing firm differentiates control tasks into low-, medium-, and high-complexity categories. Low-complexity controls involve straightforward checks (e.g., matching invoice identifiers and dates). Medium-complexity controls require cross-document verification against regulatory requirements (e.g., validating the presence and order of documents in a procurement timeline). High-complexity controls concern procedural compliance in public procurement and require substantial domain expertise to interpret extensive documentation.

Across all control tasks, CATE targets automation along the four stages of Human Information Processing (HIP) [18]: information acquisition, information analysis, decision and action selection, and action implementation. The resulting automation profile varies systematically with task complexity. In particular, low- and medium-complexity controls are more strongly automated, whereas high-complexity controls include more human confirmation and oversight. Figure 1 summarizes the envisioned levels and types of automation for the three complexity categories.

5 Findings

This section presents the findings from our case study, grouped into three dimensions. First, we discuss how IPA differs for the phases of information processing in our case. Second, we describe how IPA is applied along different levels of task complexity. Third, we describe the challenges organizations perceive for human work in the context of IPA. For each dimension, we derive propositions based on our interview data and provide support through quotes.

5.1 Information Processing Phases

Based on our interviews, the eagerness to automate a task seems to depend on the type of task that needs to be automated. Dependent on the phase of HIP, different levels of willingness to automate surfaced. Overall, we formulate a general

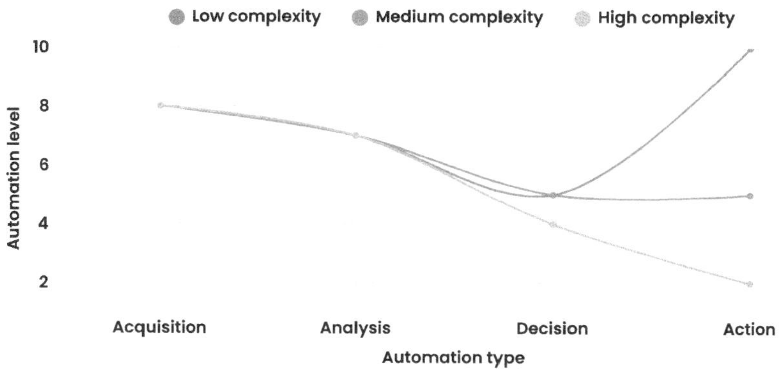

Fig. 1. The levels and types of automation (as defined in [18]), envisioned in the CATE project for low-, medium-, and high-complexity control tasks.

proposition, followed by a more detailed discussion of our findings, structured around the four automation types presented in [18]:

P1: The phase of information processing impacts the level of intelligent automation an organization is willing to allow for an activity.

Acquisition. The interview data suggests that organizations are eager to apply intelligent process automation to document gathering activities. Respondents consistently describe these tasks as both highly structured and operationally burdensome. Document requirements are often clearly defined and standardized, reducing the need for professional judgment, as reflected in:

> "[for some controls] it is always clear: if you have a certain cost for a machine, we expect an invoice. For labor costs, we expect the payslip. [...]. So we know what we are supposed to receive, we know what it should look like. And I don't like clicking a download button only to open it on my computer, so the demand for automation is very high." - [M1]

This indicates that, currently, the expected inputs are largely predictable for most controls. This was further confirmed through field observations of the auditors. This level of standardization reinforces the perception that such tasks are well-suited for automation.

Additionally, respondents emphasized the fragmented and inefficient nature of the current document retrieval process. Documents are frequently dispersed across multiple systems and formats, requiring significant manual effort to locate, extract, and organize. As illustrated by [M1]:

> "You have a transaction and there's an invoice with it, but then somewhere in another part of the interface, there's a government contract, because

that's for multiple lines. And then there's also a zip file somewhere, and that still needs to be unzipped. And there are all kinds of documents in there. That's a lot of work. And if something, like automation, can just do all that neatly, [...] Great." - [M1]

From field observations, we know the process involves navigating multiple interfaces and handling unstructured files. This fragmentation not only increases workload but also strengthens the perceived value of automation solutions, specifically agentic automation, to aggregate and structure relevant information.

Taken together, these findings suggest that the demand for intelligent automation is driven both by the routine nature of document requirements and by the inefficiencies of the existing process. From the findings, we construct the following proposition:

Proposition 1.1: Organizations are eager to apply intelligent automation to the activity of gathering documents for information processing.

Analysis. The interview data indicate that organizational attitudes toward automating document analysis are not uniformly positive or negative, but rather conditional on the nature of the task. In particular, respondents expressed eagerness for automating simple, routine aspects of analysis, such as basic information extraction from an invoice. As stated by [D2]:

> "There is no learning curve for a human being there either. Reading a date from an invoice, that is more of a frustrating task than something I really learn something from." - [D2]

Additionally, greater openness to automation was also reported for data extraction across different lengthy documents. As [M1] expressed:

> "Extracting information from [documents], that's high demands again. I was talking about that timeline; if I don't have to read 50 pages of documents to find one date – yeah, cool." - [M1]

However, this openness is accompanied by clear caution when tasks require professional judgment or critical evaluation. Respondents emphasized that analytical work often involves interpreting context, recognizing patterns, and applying experience-based reasoning. As [M1] explained:

> "Ah, those suppliers, are these really players I can expect here for this government contract? After a while you get to know quite a few suppliers. Not all of course, but you do see some recurring suspects popping up in all those government contracts. [...]. Does that mean AI can't do that? No. Does that mean you actually still need to be able to be critical of the AI? Yes." - [M1]

This suggests that while intelligent automation is feasible, automated outputs are not always viewed as self-sufficient, but also as inputs that require human validation.

These findings point to
Proposition 1.2: Organizations are eager to apply intelligent automation to the activity of analyzing documents, but want to keep a human in the loop, where professional judgment is required.

Decision. The interview data indicates reluctance to automate decision selection, due to concerns of external consequences and accountability. Respondents emphasized that delegating approval or rejection decisions to automated systems carries risk, as responsibility for outcomes cannot be fully transferred. [D1] highlighted that organizations fear customer complaints arising from automated decisions, noting that:

> "They [organizations] want to maintain that collaboration [with their customers] well. And so, a fear among organizations today is: if we start automatically approving and rejecting tomorrow, with little human review, what if we receive complaints from customers. [...]. Those will actually be complaints about things we didn't do ourselves, but that our machine did." - [D1]

This illustrates a perceived accountability gap, where organizations remain responsible for decisions despite not directly making them.

In response to these concerns, video material made for upper management stakeholders revealed that investments are made in mechanisms such as explainability to make automated decisions more transparent. As [D2] also explained:

> "That is why we put so much effort into explainability. How exactly do we arrive at that decision? Something that might seem strange to an auditor at first glance, but when you see the reasoning and what the decision is based on, it might actually be okay." - [D2]

However, while such measures may increase acceptance, they do not eliminate the underlying concern regarding responsibility for outcomes. As [M1] highlights, while reluctance is partially driven by technical limitations, there is a persistent discomfort with delegating a final judgment to automation.

> "At the moment, it is agreed upon to always keep the auditor in the loop. This means that if an audit does not work, the auditor does not have to redo 100% of that work. He naturally already receives the entire analysis for this. And that LLM will also say, "I think this is okay," or "I think it isn't." We certainly allow that. But we do leave the final decision to the individual, especially for very difficult controls." - [M1]

The reluctance to apply intelligent automation to decision selection, seems primarily driven by concerns about accountability, even when decision-making processes are technologically supported. Based on findings, we propose
Proposition 1.3: Organizations are reluctant to apply intelligent automation to the activity of decision selection.

Action. All of the interviewees indicate a favorable attitude toward automating action implementation. The implementation of intelligent process automation is characterized as a procedural and low-risk activity, involving the application of predefined rules or the communication of already established outcomes. As noted by [M1]:

> "I have performed all checks and there is a thousand euros on the invoice. You reported ten thousand. And that is the only error. Nothing is stopping us from automating that right now and sending an email to that project partner. Dear client, this is the error. Please resubmit. [...] I am not even afraid of the little texts the LLM might generate. That it might hallucinate something. No, because it is only two sentences. You can have it generated super clearly." - [M1]

This suggests that once a decision is selected for simple control tasks, there is minimal concern for the automatic execution when assisted by an LLM. Auditors [A1] and [A5] also report feeling less inclined to check the correctness of simple controls, as they report having a high level of trust in the system. However, as control task complexity increases, the willingness to automate decreases due to technological limitations.

In addition to being perceived as low-risk for low-complexity controls, action implementation tasks were also described as operationally burdensome in their current form. Respondents pointed to inefficiencies in communication and follow-up processes, which often rely on fragmented email exchanges and long wait times. [D1] expanded on this:

> "What happens now is really opening an invoice and saying, 'Ah, it's illegible,' or whatever. And that causes frustration; it requires communication; it involves a lot of project management, which today often takes place via email, for example. In good cases, we have case management systems, but often things aren't followed up via email, it gets lost, and then there is frustration for both the grant provider and the grant recipient. We can really eliminate that through automation." - [D1]

These findings suggest that organizations are receptive to automating action implementation for low-complexity controls because such tasks are viewed as procedural and standardized. Organizations still feel reluctance for higher complexity controls, due to technological limitations. As such, we propose
Proposition 1.4: Organizations are eager to apply intelligent automation to the activity of action implementation, given sufficient technological capabilities.

5.2 Task Complexity

When examining the interview data through the lens of task complexity level, the interviews revealed that task complexity influences the profile of the intelligent process automation (the profile refers to the combined type and level of automation). This is further supported by the insights generated by *proposition*

1.4. For low-complexity controls, full automation is often feasible and considered low-risk. As explained by [M1]:

> "This isn't a "from now on, everything is like this" [fully automatic] either. I just think, regarding those low-complexity controls, if they are deemed wrong, the text is automatically generated, that decision is automatic, and in that area, we can go fully automatic. With the more difficult controls, it will take a bit longer. And part of that is because, technologically speaking, there are currently still too many hallucinations in the system because the models don't know the legislation well enough and have therefore essentially made up laws. [...]. But even that is changing quickly.' - [M1]

This illustrates that straightforward tasks can be fully automated, whereas more complex tasks still require additional guardrails during implementation. High-complexity or borderline cases are approached more cautiously. Development documentation reveals automation for complex tasks currently involves more human oversight. Auditor [A5] mentions how the profile of the automation is such that it supports professional judgment rather than replace it when tasks involve ambiguity, risk, or interpretation. As further elaborated by [D2]:

> "That kind of thing has also been our major focus during the recent development phase of those more complex public procurement audits. Truly providing added value to the auditor. Not being afraid to push audits to a manual level if it is a borderline case, and providing the best possible foundation for the auditor to rely on." - [D2]

From these findings, we put forward
Proposition 2: The complexity of a control task has an impact on the profile of the intelligent automation, with higher complexity leading to more human oversight.

5.3 Human Impact

The human impact factor addresses how the role of a human auditor will change as process automation is increasingly applied. [M1] highlights in his interview how the knowledge and expertise of an auditor remain relevant, even as the working environment alters:

> "Prompt checking is always going to be linked to that skill [of auditing]. If you are an expert in public procurement at an organization, and you say 'We are going to use a prompt here to prepare the award reports, [...] and I have a very good prompt, because I am an expert'. Yes, that expertise remains necessary there. It is just that the way of working has changed completely. But in itself, that same skill always remains. [...]. And so that knowledge is retained in a way." - [M1]

Additionally, interviews with the auditors even revealed the potential for skill development. Based on their limited experience with IPA, the system seems to provide cognitive support rather than limiting it. As Auditor [A5] expresses:

> "If I assume that CATE is working optimally on all the documents and controls implemented within it, then I think that can indeed provide cognitive support as well, because it might already start giving a bit more of a hint of... Pay attention to this or pay attention to that." - [A5]

Auditor [A1] expressed similar views regarding the cognitive support provided by the IPA. The data do not indicate a strong concern about the potential negative effects of this transition on skills. Instead, respondents generally framed these changes as a natural evolution of the role, implying that skill adaptation is both expected and manageable.

The shift away from routine work and towards a supervisory role implies that auditors will be increasingly exposed to more difficult cases. Currently, the work of an auditor consists mostly of simple controls, with an occasional complex control. As automation relieves the auditors of simple tasks and controls, the mental workload placed on an auditor is likely to increase. According to [M1] this implies a more quick uptake of expertise:

> "Those juniors [starting auditors] are just going to learn a lot more very quickly." - [M1]

However, as [M1] further elaborates:

> "I can imagine that if you only get very difficult problems, that is going to become very tiring. [...]. Information overload, fatigue, that is certainly a thing." - [M1]

This comment highlights that organizations are more concerned for information overload as a result of intelligent automation, rather than a loss of skills. Especially when the auditor is increasingly relieved of low-complexity tasks. Auditor [A1] expressed the following sentiment after working with the IPA for two weeks:

> "I view it more from a time efficiency perspective. So essentially, I only need to focus on the complex audits and not the simpler ones, which actually makes it feel less burdensome. But I must say, there isn't an auditor with years of experience with CATE who could truly say: Okay, performing only complex audits for two years now, that is quite burdensome." - [A1]

This implies that short-term usage can reduce the cognitive workload experienced by auditors, but we cannot make any statements on potential long-term effects.

Taken together, two prominent aspects of human impact come forward. We formulate the following two propositions.

Proposition 3: Organizations recognize that intelligent automation alters required skill sets, resulting in a shift of competencies. This required shift does not impact eagerness to apply intelligent automation.

Proposition 4: Organizations are eager to apply intelligent automation, despite the concern for information overload for system users.

6 Discussion

In this section, we discuss our findings. First, we provide an integrated discussion on how the propositions relate to each other. Second, we examine how benefits and challenges known from RPA and workflow systems remain relevant for IPA and how they manifest in our case. Third, we discuss how concerns from human factors research are reflected in, and addressed by, concrete IPA design decisions. Fourth, we discuss study limitations and derive directions for future research.

6.1 Integrated View on Propositions

Our findings indicate that there are two primary types of dimensions that impact the organizations decisions on when to use IPA and to which extent. First, we discuss task-related dimensions. Second, we discuss the dimensions related to human factors.

The task-related dimensions refer to the phase of information processing that is involved and the overall complexity of the task. These dimensions are touched upon in both proposition 1 and proposition 2. The interviews indicate that three principles are at the core of what determines the level of automation of a task. These principles are explainability, accountability, and technical limitations. As such, the degree of automation for a task seems to vary, depending on the combination of task dimensions that are related to these the three principles.

Besides the activity-related dimensions, there are also dimensions related to human factors. These include the concern for skill loss and the concern for information overload, as described by proposition 3 and proposition 4. A core factor that connects these dimensions is cognitive load. The propositions taken together suggest that IPA systems should be designed in a way that provides sufficient cognitive load to develop newly needed competencies, while not causing information overload to the users.

In summary, organizations make implementation and design decisions for IPA based on explainability, accountability, technical limitations, and the amount of cognitive load experienced by the users.

6.2 General Process Automation Challenges

Our findings suggest that IPA retains the classical value proposition of workflow systems and RPA: automating routine work and reducing avoidable cognitive load. At the same time, IPA inherits well-known implementation challenges, notably the need for end-to-end coordination of process redesign, governance, and operations [20,25,30]. IPA also amplifies these challenges through AI-specific risks, particularly around reliability and explainability [6,14]. Particularly in an audit context, this plays an important role, as financial auditors are personally liable for the opinions they issue [12].

In our case, these risks are managed through a differentiated automation profile driven by task complexity (*proposition 2*). Low-complexity controls are considered comparatively low-risk and therefore see a higher degree of automation (e.g., agentic document retrieval, OCR-based extraction). For high-complexity controls, concerns about explainability and non-deterministic failure modes (e.g., hallucinations in LLM-based reasoning) translate into tighter human-in-the-loop control points and a more supportive (rather than substitutive) role of automation.

This points to a broader implication for IPA governance: the decision of "how much" to automate is not only a question of technical feasibility, but also of risk exposure and accountability. Future research should test the generalizability of this complexity-based logic and investigate whether it mitigates or, conversely, fosters automation bias and inappropriate reliance that has been observed for other automated decision aids [16].

6.3 Human Factors Challenges for IPA

Our case also confirms key insights from human-automation interaction research: automation changes work, reconfigures roles, and shifts skill requirements. In particular, IPA design choices in our case support auditors in the phases of information acquisition and low-stakes action implementation, while maintaining human accountability for judgments and decisions (*proposition 1*). This design logic aligns with recommendations to avoid undesirable technology dominance and to calibrate reliance through transparency and explainability [29].

Consistent with *proposition 3*, respondents did not frame automation as "de-skilling" per se, but as a reallocation of competencies. This aligns with the concept of "re-skilling" [22,24]. The observed design emphasis is on preventing overreliance and maintaining core auditing skills by reserving judgment-intensive controls for the auditor and by using automation outputs as inspectable inputs [22,29].

At the same time, *proposition 4* highlights a human risk that remains insufficiently addressed in the current project: as low-complexity work is automated, auditors are exposed to a higher share of complex and exception cases. However, this may increase sustained cognitive load and, in turn, risk fatigue and burnout, particularly for novices. This contrasts the risk of knowledge loss, as anticipated in literature [14,24]. Future work should examine workload trajectories and learning effects longitudinally after IPA deployment, and evaluate concrete interventions (e.g., adaptive task allocation, training, and escalation policies) that balance efficiency gains with sustainable work design.

6.4 Limitations and Future Research

This study is exploratory and subject to several limitations. First, our propositions are derived from a single in-depth case in a specific public-sector audit context. Within the field of auditing, personal accountability and explainability is often considered an essential part of the job, unlike other workfields, which may limit direct generalizability to other industries, regulatory regimes, or organizational cultures. Additionally, due to the smaller setting of the case, only one manager was allocated to this project, limiting the diversity of insights derived from a management perspective. Despite this limitation, the paper still provides a theoretical and analytical contribution by illuminating propositions that can be used for future theory development, refinement, or extension. Second, CATE was still under development during most of the data collection, which means our analysis primarily captures design rationales and anticipated impacts rather than measured outcomes in routine operation. Third, our evidence base relies largely on qualitative accounts (interviews and observations), which can be affected by retrospective bias and by the perspectives of the stakeholders accessible to the research team.

These limitations motivate a research agenda that (i) empirically tests and validates the propositions across multiple organizations and IPA settings, (ii) complements qualitative insight with evaluative evidence (e.g., field experiments, usage logs, performance and error metrics), and (iii) explicitly measures human outcomes such as appropriate reliance, skill development, and well-being over time.

7 Conclusion

In this paper, we investigated how organizations make design decisions about where to use IPA capabilities and to what extent. Based on a case study in the context of an IPA initiative for financial audit processes, conducted by a Big Four accounting company, we inductively derived propositions on how information-processing phases, task complexity, and human impact considerations shape automation design. Our findings provide empirically grounded guidance on how organizations trade off efficiency gains against reliability, explainability, and human factors concerns, and they lay a foundation for future deductive and evaluative research on IPA design and governance.

References

1. van der Aalst, W.M.P., Bichler, M., Heinzl, A.: Robotic process automation. Bus. Inf. Syst. Eng. **60**(4), 269–272 (2018). https://doi.org/10.1007/s12599-018-0542-4
2. Afrin, S., Roksana, S., Akram, R.: AI-enhanced robotic process automation: a review of intelligent automation innovations. IEEE Access **13**, 173–197 (2025). https://doi.org/10.1109/ACCESS.2024.3513279, https://ieeexplore.ieee.org/abstract/document/10781408
3. Arnold, V., Collier, P.A., Leech, S.A., Sutton, S.G.: Impact of intelligent decision aids on expert and novice decision-makers' judgments. Account. Finan. **44**(1), 1–26 (2004)https://doi.org/10.1111/j.1467-629x.2004.00099.x, https://onlinelibrary.wiley.com/doi/abs/10.1111/j.1467-629x.2004.00099.x
4. Arnold, V., Sutton, S.: The theory of technology dominance: understanding the impact of intelligent decision aids on decision makers' judgments. Adv. Acc. Behav. Res. **1**, 175–194 (1998)
5. Berente, N., Yoo, Y.: Institutional contradictions and loose coupling: postimplementation of NASA's enterprise information system. Inf. Syst. Res. **23**(2), 376–396 (2012). https://doi.org/10.1287/isre.1110.0373, https://pubsonline.informs.org/doi/abs/10.1287/isre.1110.0373

6. Chakraborti, T., et al.: From robotic process automation to intelligent process automation. In: Asatiani, A., et al. (eds.) BPM 2020. LNBIP, vol. 393, pp. 215–228. Springer, Cham (2020). https://doi.org/10.1007/978-3-030-58779-6_15
7. Diaz Andrade, A.: Interpretive research aiming at theory building: adopting and adapting the case study design. Qual. Rep. **14**(1), 42–60 (2000). https://doi.org/10.46743/2160-3715/2009.1392, https://nsuworks.nova.edu/tqr/vol14/iss1/3
8. Dumas, M., et al.: AI-augmented business process management systems: a research manifesto. ACM Trans. Manage. Inf. Syst. **14**(1), 11:1–11:19 (2023). https://doi.org/10.1145/3576047, https://dl.acm.org/doi/10.1145/3576047
9. Eisenhardt, K.M., Graebner, M.E.: Theory building from cases: opportunities and challenges. Acad. Manag. J. **50**(1), 25–32 (2007). https://doi.org/10.5465/AMJ.2007.24160888
10. D Endsley, M.R., Kiris, E.O.: The out-of-the-loop performance problem and level of control in automation. Hum. Fact. **37**(2), 381–394 (1995). https://doi.org/10.1518/001872095779064555
11. Gioia, D.A., Corley, K.G., Hamilton, A.L.: Seeking qualitative rigor in inductive research: notes on the Gioia methodology. Organ. Res. Meth. **16**(1), 15–31 (2013). https://doi.org/10.1177/1094428112452151
12. Gu, H., Duan, H.K., Vasarhelyi, M.A.: Challenges for leveraging explainable artificial intelligence in audit procedures. J. Emerg. Technol. Account. **23**(1), 59–70 (2026). https://doi.org/10.2308/JETA-2023-044
13. Herm, L.V., Janiesch, C., Helm, A., Imgrund, F., Hofmann, A., Winkelmann, A.: A framework for implementing robotic process automation projects. IseB **21**(1), 1–35 (2023). https://doi.org/10.1007/s10257-022-00553-8
14. Herm, L.-V., Janiesch, C., Reijers, H.A., Seubert, F.: From symbolic RPA to intelligent RPA: challenges for developing and operating intelligent software robots. In: Polyvyanyy, A., Wynn, M.T., Van Looy, A., Reichert, M. (eds.) BPM 2021. LNCS, vol. 12875, pp. 289–305. Springer, Cham (2021). https://doi.org/10.1007/978-3-030-85469-0_19
15. IEEE: IEEE guide for terms and concepts in intelligent process automation. IEEE Std 2755-2017, pp. 1–16, September 2017. https://doi.org/10.1109/IEEESTD.2017.8070671
16. Manzey, D., Reichenbach, J., Onnasch, L.: Human performance consequences of automated decision aids: the impact of degree of automation and system experience. J. Cogn. Eng. Decis. Making **6**(1), 57–87 (2012). https://doi.org/10.1177/1555343411433844
17. Mueller, S.K., Mendling, J., Bernroider, E.W.: The roles of social identity and dynamic salient group formations for ERP program management success in a post-merger context. Inf. Syst. J. **29**(3), 609–640 (2019). https://doi.org/10.1111/isj.12223
18. Parasuraman, R., Sheridan, T., Wickens, C.: A model for types and levels of human interaction with automation. IEEE Trans. Syst. Man Cybern. Part A Syst. Hum. **30**(3), 286–297 (2000). https://doi.org/10.1109/3468.844354, https://ieeexplore.ieee.org/document/844354
19. Parasuraman, R., Manzey, D.H.: Complacency and bias in human use of automation: an attentional integration. Hum. Fact. **52**(3), 381–410 (2010). https://doi.org/10.1177/0018720810376055
20. Plattfaut, R., Borghoff, V., Godefroid, M., Koch, J., Trampler, M., Coners, A.: The critical success factors for robotic process automation. Comput. Ind. **138**, 103646 (2022). https://doi.org/10.1016/j.compind.2022.103646, https://www.sciencedirect.com/science/article/pii/S0166361522000434

21. Reijers, H.A., Vanderfeesten, I., van der Aalst, W.M.: The effectiveness of workflow management systems: a longitudinal study. Int. J. Inf. Manage. **36**(1), 126–141 (2016)
22. Rinta-Kahila, T., Penttinen, E., Salovaara, A., Soliman, W., Ruissalo, J.: The vicious circles of skill erosion: a case study of cognitive automation. J. Assoc. Inf. Syst. **24**(5), 1378–1412 (2023). https://doi.org/10.17705/1jais.00829, https://aisel.aisnet.org/jais/vol24/iss5/2/
23. Runeson, P., Höst, M.: Guidelines for conducting and reporting case study research in software engineering. Empir. Softw. Eng. **14**(2), 131–164 (2009). https://doi.org/10.1007/s10664-008-9102-8
24. Schlegel, D., Kraus, P.: Skills and competencies for digital transformation – a critical analysis in the context of robotic process automation. Int. J. Organ. Anal. **31**(3), 804–822 (2023). https://doi.org/10.1108/IJOA-04-2021-2707, https://www.proquest.com/docview/2796212383/abstract/C9A97C554DC64974PQ/1
25. Schlegel, D., Rosenberg, B., Fundanovic, O., Kraus, P.: How to conduct successful business process automation projects? An analysis of key factors in the context of robotic process automation. Bus. Process. Manag. J. **30**(8), 99–119 (2024). https://doi.org/10.1108/BPMJ-06-2023-0465
26. Sergeeva, A.V., Faraj, S., Huysman, M.: Losing touch: an embodiment perspective on coordination in robotic surgery. Organ. Sci. **31**(5), 1248–1271 (2020). https://doi.org/10.1287/orsc.2019.1343, https://pubsonline.informs.org/doi/10.1287/orsc.2019.1343
27. Strauss, A., Corbin, J.: Basics of Qualitative Research: Techniques and Procedures for Developing Grounded Theory, 2nd edn. Sage Publications Inc., Thousand Oaks, CA, US (1998)
28. Sutton, S.G., Arnold, V., Holt, M.: How much automation is too much? Keeping the human relevant in knowledge work. J. Emerg. Technol. Account. **15**(2), 15–25 (2018). https://doi.org/10.2308/jeta-52311, https://publications.aaahq.org/jeta/article/15/2/15/9279/How-Much-Automation-Is-Too-Much-Keeping-the-Human
29. Sutton, S.G., Arnold, V., Holt, M.: An extension of the theory of technology dominance: capturing the underlying causal complexity (2022). https://doi.org/10.2139/ssrn.4101835, https://papers.ssrn.com/abstract=4101835
30. Syed, R., et al.: Robotic process automation: contemporary themes and challenges. Comput. Ind. **115** (2020). https://doi.org/10.1016/j.compind.2019.103162, https://www.sciencedirect.com/science/article/pii/S0166361519304609
31. Walsham, G.: Interpretive case studies in IS research: nature and method. Eur. J. Inf. Syst. **4**(2), 74–81 (1995). https://doi.org/10.1057/ejis.1995.9

Semi-automated vs. Manual Process Model Generation: A Comparative Study of the Process of Process Modeling

Maximilian Möller(✉), Luca Franziska Hörner, and Manfred Reichert

Ulm University, James-Franck-Ring 1, 89081 Ulm, Germany
maximilian.moeller@uni-ulm.de

Abstract. Due to recent advances in large language models (LLMs), semi-automated approaches to business process modeling have emerged. These fundamentally shift the modeling task from manual model creation to prompt-based model generation and refinement. Prior research has focused on the capability of LLMs to generate (BPMN 2.0) process models, but little evidence exists on the way the involvement of an LLM affects the process of process modeling itself. This paper investigates how semi-automated process modeling with the assistance of an LLM-based chatbot differs from manual process modeling (i.e., with the use of a common modeling tool) with respect to efficiency, cognitive load, and process model quality. A controlled eye-tracking study was conducted with 24 participants, with each participant creating process models manually as well as semi-automatically with an LLM-based chatbot. *Efficiency* was measured through task completion time, *cognitive load* was assessed through eye-tracking metrics and NASA-TLX scores, and *process model quality* was evaluated with respect to syntactic and semantic correctness. Our results show that semi-automated model creation significantly increases efficiency and reduces both objective and subjective cognitive load. Notably, these advantages do not come at the expense of process model quality. The LLM-generated process models exhibited only moderate syntactic errors and achieved very high semantic accuracy. Overall, the findings highlight the potential of LLM-based semi-automated process modeling to improve the process of process modeling.

Keywords: LLM-based Process Modeling · BPMN 2.0 · Eye-Tracking · Cognitive Load · Process Model Quality

1 Introduction

Business process modeling constitutes a central task in process management, as process models form the basis for process analysis, improvement and automation, as well as organizational communication [6]. Usually, process models are created manually by domain experts, business analysts or other stakeholders using graphical modeling tools (e.g., using BPMN 2.0 as a modeling notation).

F. M. Maggi et al. (Eds.): BPM 2026, LNCS 16936, pp. 720–736, 2027.
https://doi.org/10.1007/978-3-032-37777-7_42

This approach requires humans to translate textual process descriptions into semi-formal representations by mapping natural language to modeling elements, and correctly linking these elements according to syntactic rules [16].

Recent advances in large language models (LLMs) have led to the emergence of a different approach. Instead of creating process models element by element, users can generate process models semi-automatically from textual process descriptions by interacting with an LLM-based chatbot and iteratively refining the LLM-generated process models through conversational exchange. Accordingly, the process of process modeling shifts from manually creating process models with a graphical process modeling tool to formulating prompts that describe the intended process and iteratively refining them until the models reflect the desired process [5,12].

The growing interest in LLM-assisted process modeling is driven by its potential benefits. LLM-based approaches range from fully automatic process model generation (e.g., Zero-Shot prompting) to semi-automatic approaches that allow for the iterative refinement of process models through conversations between user and chatbot [9,27]. Corresponding approaches can increase the efficiency of creating process models, and lower the technical barriers associated with BPMN 2.0 process modeling [14]. In particular, individuals without extensive expertise in process modeling are enabled to create appropriate process models [8]. In Zero-Shot prompting scenarios, users can obtain an initial process model immediately and then refine it iteratively, instead of laboriously creating the process models step by step starting with an empty modeling canvas.

Despite increasing interest in LLM-based process modeling, research has primarily focused on the capabilities of LLMs to generate process models [13,20,26]. Comparatively little attention has been paid to the process of process modeling in the context of LLM-assisted process modeling. In particular, there exists limited empirical evidence on how the modeling process itself changes when process modelers interact with an LLM-based chatbot instead of manually creating process models from scratch with a process modeling tool. More specifically, little is known about how these changes affect cognitive load. Although prior research has extensively examined cognitive load in manual process modeling [31], it remains unclear how semi-automatic modeling affects the modeling process itself [17]. Consequently, it has not yet been well understood how the interaction with an LLM-based chatbot influences cognitive load, efficiency, and process model quality compared to manual modeling [18]. Against this background, we formulate the following research question.

RQ: How does semi-automatic process modeling with an LLM-based chatbot differ from manual process modeling in terms of efficiency, cognitive load, and resulting model quality?

To address this research question, we conducted a controlled eye-tracking study employing a within-subject design. Participants created process models manually as well as semi-automatically using the LLM-based chatbot BPMN-Gen (https://bpmngen.de) [9]. By combining objective and subjective measures of cognitive load with performance indicators such as efficiency and process model quality, the study provides a comprehensive perspective on the differences between semi-automated and manual process modeling.

The remainder of the paper is structured as follows. Section 2 presents the theoretical background, and Sect. 3 discusses related work. Section 4 details the study design, materials, participants, and evaluation metrics. The empirical results are presented and discussed in Sect. 5, while Sect. 6 addresses limitations and threats to validity. Finally, Sect. 7 concludes the paper and outlines directions for future research.

2 Backgrounds

This section provides theoretical backgrounds, covering semi-automatic process modeling and the assessment of cognitive load during process modeling tasks.

2.1 Semi-automatic Process Modeling

Business process modeling refers to the creation of semi-formal representations of business processes to support process analysis, communication, and automation. In practice, BPMN 2.0 is widely used as language and notation to model business processes [21]. Manual BPMN 2.0 modeling requires domain experts, business analysts, or other stakeholders to translate informal process descriptions into syntactically correct and semantically meaningful process models – a task that can be a cognitively demanding and time-consuming [33].

LLM-assisted process modeling shifts the task from manually creating process models to specifying process requirements in natural language and refining the generated process models. Users interact with the LLM-based chatbot through textual instructions, while the LLM generates BPMN 2.0 process models based on the given process descriptions [15]. In this study, we employ BPMNGen [9], an LLM-based chatbot we developed for the semi-automated creation of process models from textual process descriptions (cf. Fig. 1). BPMNGen can be connected to a variety of LLMs (e.g., Claude Opus 4.5, GPT-5.4), accepting prompts that range from complete process descriptions to incremental instructions. As output, BPMNGen provides BPMN 2.0 process models fitting the given process description. It also supports various prompting strategies, ranging from Zero-Shot to more elaborated approaches (e.g., Chain-of-Thought and Tree-of-Thought), allowing users to guide model generation through the way instructions and contextual information are formulated. Through conversational refinement, BPMNGen further allows users to iteratively improve the generated process model by using follow-up prompts. Routine modeling tasks are handled by the LLM, while the role of the user shifts toward reviewing, validating, and guiding the process model development [9].

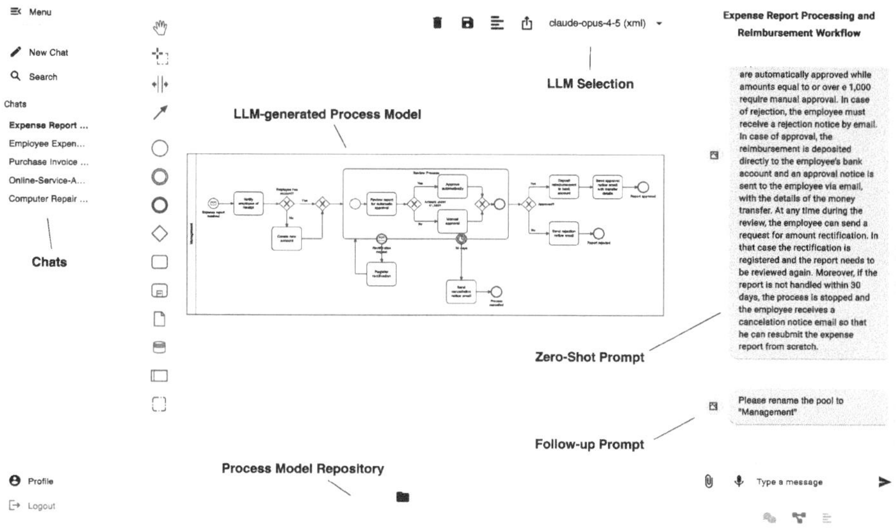

Fig. 1. User Interface of BPMNGen.

2.2 Cognitive Load in Process Modeling Tasks

Business process modeling imposes substantial cognitive load, as modelers need to interpret process semantics and map the natural language process description to semi-formal modeling constructs, while ensuring syntactic and semantic correctness of the specified model. Accordingly, cognitive load constitutes a key factor that influences modeling performance and user experience [19,29].

According to Cognitive Load Theory [28], human working memory has a limited capacity for processing and storing information. When the cognitive demands of a task exceed these limits, individuals struggle with processing information efficiently, leading to increased mental effort, slower task completion, and higher error likelihood. Consequently, performance tends to decline when tasks require excessive demands on cognitive resources. Therefore, maintaining a low cognitive load is important for modeling tasks.

Cognitive load can be assessed with complementary objective and subjective measures. Eye-tracking provides objective indicators of visual-cognitive processing during modeling tasks. In particular, the *number of fixations* and the *mean fixation duration* are commonly used to approximate visual search behavior and processing effort. Though eye-tracking does not directly measure cognitive load, these metrics are established proxies for cognitive processing demands in complex interaction tasks [22,31].

Subjective cognitive load can be captured with standardized questionnaires, like the NASA-TLX questionnaire [7]. The mental demand dimension is particularly relevant for modeling tasks, as it reflects the perceived cognitive load required by humans. Combining eye-tracking metrics with subjective cognitive load assessments enables a more comprehensive and robust assessment of the cognitive load in process modeling [30].

3 Related Work

Prior research on manual process modeling has extensively investigated modeling behavior, modeling strategies, cognitive load, and process model quality. Wolf & Soffer examined the use of routing patterns to support BPMN model creation and experimentally evaluated their impact on process model quality and modeling time [32]. Building on the cognitive perspective of modeling, Pinggera et al. introduced Modeling Phase Diagrams as a formal method to analyze modeling behavior based on logged interaction data. Their work evaluated how modeling phases relate to the quality of the resulting process models [25]. Complementing this analytical perspective, Claes et al. introduced PPMChart as a visualization technique to explore modeling activities [4]. The evaluation demonstrated how different modeling patterns and strategies can be identified across user sessions. In a related study, Pinggera et al. identified different modeling styles and analyzed their influence on efficiency and model quality [24].

Research on cognitive aspects of process modeling further extended this line of work by integrating eye-tracking and cognitive load measurements. Pinggera et al. combined modeling phase diagrams with eye-tracking data to analyze how fixation patterns correspond to both comprehension and modeling tasks [23]. Similarly, Burattin et al. integrated eye-tracking data to refine phase detection, providing a more fine-grained understanding of modeling and comprehension phases [2]. Advancing this line of work, Burattin et al. proposed an automated machine learning approach to detect five distinct modeling phases based on interaction and eye-tracking data, demonstrating high classification accuracy [3]. From a cognitive psychology perspective, Neurauter et al. investigated the influence of cognitive abilities and cognitive load on process model creation [19]. The provided study evaluated the relationship between working memory, cognitive load, and process model quality. Finally, Ivanchikj et al. presented an NLP-based approach to generate BPMN process models semi-automatically from textual descriptions, evaluating syntactic correctness and tool applicability [10].

Together, these studies provided a comprehensive understanding of manual process modeling, modeling strategies, cognitive load, and efficiency. However, they did not investigate LLM-assisted process modeling.

With the emergence of LLMs, research has increasingly investigated their application to process modeling. Existing work primarily focuses on process model generation quality, conversational modeling approaches, usability, and efficiency aspects of LLM-supported systems. Kourani et al. proposed an LLM-based framework for the automated generation and iterative refinement of process models, integrating prompting strategies and quality guarantees within a system-oriented architecture. Their evaluation focused on syntactic correctness and exportability to standard notations rather than on conversational interaction patterns [14]. Moving toward conversational settings, Safan & Köpke introduced a framework for LLM-based conversational modeling that combines generative AI with model checking to ensure syntactic correctness [27]. The extended version BPMN-Chatbot++ further supports collaboration diagrams and data flows, evaluating process model quality and user acceptance in a chatbot-based interaction [26]. Addressing efficiency aspects, Köpke & Safan analyzed token usage and cost-efficiency in conversational LLM-based process modeling, demonstrating how optimized prompting strategies reduce computational overhead while maintaining process model correctness [13].

From a user-centered perspective, Klievtsova et al. explored conversational process modeling in which domain experts iteratively created and redesigned process models by interacting with a chatbot [12]. In a subsequent study, the same authors compared LLM-generated and human-created process models, evaluating user preferences and perceived process model quality through a survey [11]. Focusing on quality and usability, Hörner et al. evaluated their chatbot for process modeling with respect to usability, cognitive load, and semantic quality [8]. In a more comprehensive assessment, they analyzed syntactic, semantic, and pragmatic quality dimensions as well as process model comprehensibility [9].

Additional work investigated technical and comparative aspects of LLM-based process modeling systems. Licardo et al. introduced BPMN Assistant, an LLM-based system for creating BPMN process models using a structured JSON intermediate representation and evaluated editing performance, latency, and success rates [15]. Beyond individual systems, Drakopoulos et al. compared multiple LLM-powered BPMN generation tools with a structured quality framework assessing clarity, correctness, and completeness [5]. Shifting the focus to comprehension, Möller et al. conducted an eye-tracking study comparing LLM-generated and human-created process models, evaluating cognitive load and comprehension performance during the reading and comprehension of already generated process models [18].

Overall, the presented studies demonstrated that LLMs – either embedded in chatbot-based conversational systems or integrated into system-oriented frameworks – are capable of generating and refining BPMN process models. The primary focus of these studies lies on the quality, comprehensibility, usability, or efficiency of the generated process models, whereas the LLM-based modeling process itself has not yet been systematically analyzed.

4 Study Context and Settings

This section provides an overview of the study design, the materials[1] used, the participants, and the evaluation metrics.

4.1 Study Design and Materials

The study followed a controlled within-subject design in which participants completed two process modeling tasks under different modeling approaches. Two

[1] All study materials can be found and downloaded online at: https://doi.org/10.5281/zenodo.20354756.

process descriptions (P_A and P_B) of comparable complexity were used as study materials (P_A: 106 words, 784 characters; P_B: 102 words, 778 characters). Each participant modeled one process description manually with the modeling tool BPMN.io [1] and the other semi-automatically with BPMNGen. Note that BPMNGen also relies on BPMN.io for the visualization of the generated process models, ensuring comparable graphical representations across both modeling approaches. A counterbalanced four-group design was used (see Table 1) to control for potential ordering and learning effects, ensuring that both modeling approaches and the process descriptions P_A and P_B were evenly distributed across the two tasks.

Table 1. Task Assignment

Group	Task 1	Task 2
1	Manual modeling of P_A	Semi-automatic modeling of P_B
2	Manual modeling of P_B	Semi-automatic modeling of P_A
3	Semi-automatic modeling of P_A	Manual modeling of P_B
4	Semi-automatic modeling of P_B	Manual modeling of P_A

First, participants received an introduction to the study procedure and both modeling approaches, i.e., manual modeling with BPMN.io and semi-automatic modeling with BPMNGen. Prior to the study tasks, participants were further equipped with the *Pupil Labs Neon* eye-tracker to record eye movements throughout the modeling sessions. For each task, one process description (P_A or P_B) was presented to the participants. Moreover, the task completion time was recorded from the moment the description was provided until the participant indicated that his or her process model was complete. In the semi-automatic approach, participants were instructed to first enter the full process description as a Zero-Shot prompt in BPMNGen to generate an initial process model and then to refine it iteratively. BPMNGen can be connected to different LLMs to generate and refine process models. For this study, Gemini 2.5 Pro was selected as the underlying LLM to produce and adapt the process models, as preliminary internal studies indicated that this LLM currently achieves the strongest performance in process model generation. In the manual modeling approach, process models were created directly using BPMN.io. After each task, participants completed the NASA-TLX questionnaire to assess subjective cognitive load. This procedure was repeated for both process descriptions. After completing both tasks, for all process models their syntactic and semantic quality was evaluated.

4.2 Study Participants

The study involved 24 participants recruited from various university BPM courses. To ensure a sufficiently high and homogeneous level of modeling competence, participants were required to complete a short BPMN 2.0 knowledge quiz prior to the study. The quiz assessed syntactic and semantic aspects of BPMN 2.0, and only students who achieved a pre-specified performance threshold were admitted to the study. This procedure ensured that all participants possessed solid conceptual knowledge and practical familiarity with BPMN 2.0, thereby reducing potential confounding effects caused by insufficient modeling expertise. An overview of the demographic details of the participants is provided in Table 2.

Table 2. Demographic Details of the Study Participants ($N = 24$)

Variable	Category	N (%)
Gender	Female	9 (37.5)
	Male	15 (62.5)
Age	<25 years	13 (54.2)
	≥25 years	11 (45.8)
Education	≤Bachelor Degree	22 (91.7)
	≥Master Degree	2 (8.3)

4.3 Study Metrics

We applied the following metrics to assess the modeling efficiency and cognitive load of the study participants, as well as the process model quality.

Efficiency

- **Task Completion Time (TCT).** The time in seconds from the start of the modeling task until the time the participants indicated that their process model is completed.

Cognitive Load was measured with the following metrics:

- **Number of Fixations (NF).** The total number of eye fixations recorded during the task, indicating visual attention and processing effort.
- **Mean Fixation Duration (MFD).** The average duration of all fixations (in ms), reflecting how much cognitive effort is spent on each process element.
- **NASA-TLX Overall Score (O_{NASA}).** A subjective measure of cognitive load collected via the NASA-TLX questionnaire, which includes six dimensions: Mental Demand (M_{NASA}), Physical Demand (Ph_{NASA}), Temporal Demand (T_{NASA}), Performance (P_{NASA}), Effort (E_{NASA}), and Frustration (F_{NASA}). Each dimension is rated on a scale from 0 to 100, and aggregated into an overall score, where higher values indicate greater cognitive load.

Process Model Quality was assessed along the following two dimensions:

- **Syntactic Quality.** Operationalized by counting the number of syntactic errors (SE) of each process model, with fewer errors indicating higher syntactic quality.
- **Semantic Quality.** Assessed with a predefined scoring scheme[2] we derived from the respective process description. For each process description, relevant process elements were assigned points, resulting in a maximum of 10 achievable semantic points (SP) per process model. Elements considered particularly important for the correct behavioral representation of the process received higher points than simpler structural elements. Higher scores indicate a stronger alignment between the process model and the process description.

5 Results and Discussion

This section presents the study results. Descriptive statistics concerning the efficiency and cognitive load during the modeling tasks can be found in Table 3, the corresponding inferential statistics in Table 4, and descriptive as well as inferential results for process model quality in Table 5. The normality was assessed using the Shapiro-Wilk test. As some NASA-TLX dimensions were not normally distributed (Ph_{NASA}, E_{NASA}, and O_{NASA}), we analyzed them using Wilcoxon signed-rank tests – all remaining variables were examined using paired-samples t-tests.

Efficiency. Regarding the descriptive results, task completion time (TCT) was substantially lower when using the semi-automatic approach (M = 744.06 s, MDN = 740.32) compared to manual modeling (M = 1063.55 s, MDN = 1071.10), indicating a considerable reduction in modeling time.

Inferential analysis using a paired-samples t-test confirmed the statistical significance of this difference, $t(23) = -2.398$, $p = .025$, with a medium effect size ($d = 0.49$). Consequently, semi-automatic modeling significantly increased efficiency. The magnitude of the effect not only indicates statistical but also practical relevance. Delegating modeling steps to an LLM reduces the need to create process models element by element, enabling the participants to focus on reviewing and refining the generated LLM outputs.

Two of the 24 study participants considered the process model the LLM generated based on the initial Zero-Shot prompt being sufficient, i.e., they did not perform any further modifications. Across all participants, the median number of additional prompts used to refine the process model was two, indicating that only few iterations were required. Note that this small number of required interactions emphasizes the efficiency advantage gained with the semi-automated approach.

[2] All study materials can be found and downloaded online at: https://doi.org/10.5281/zenodo.20354756.

Table 3. Descriptive Statistics for Efficiency and Cognitive Load

Variable	Semi-automatic Modeling			Manual Modeling		
	M	MDN	SD	M	MDN	SD
Efficiency						
TCT	744.06	740.32	423.78	1063.55	1071.10	314.33
Cog. Load						
Eye-Tracking						
NF	1465.04	1521.00	752.16	1867.88	1817.50	608.35
MFD	353.82	356.16	72.16	442.23	453.36	97.26
NASA-TLX						
M_{NASA}	37.92	35.00	24.93	57.29	62.50	21.92
Ph_{NASA}	7.71	5.00	9.67	10.00	5.00	13.99
T_{NASA}	18.96	10.00	20.59	29.79	22.50	25.85
P_{NASA}	29.17	20.00	26.73	31.67	25.00	22.78
E_{NASA}	29.58	20.00	24.84	57.71	65.00	23.17
F_{NASA}	32.71	27.50	29.15	33.54	25.00	27.44
O_{NASA}	26.01	21.66	17.45	36.67	37.08	15.90

Note. M = mean; MDN = median; SD = standard deviation.

Table 4. Inferential Statistics for Efficiency and Cognitive Load

Variable	t-Test				Wilcoxon Sign.-Rank Test		
	t	df	d	p	W	r	p
Efficiency							
TCT	−2.398	23	0.49	**.025***	–	–	–
Cog. Load							
Eye-Tracking							
NF	−1.633	23	0.33	.116	–	–	–
MFD	−7.873	23	1.16	$<$ **.001*****	–	–	–
NASA-TLX							
M_{NASA}	−2.677	23	0.55	**.013***	–	–	–
Ph_{NASA}	–	–	–	–	29.5	0.135	.475
T_{NASA}	−1.770	23	0.36	.090	–	–	–
P_{NASA}	−0.353	23	0.07	.728	–	–	–
E_{NASA}	–	–	–	–	26.5	0.617	**.002****
F_{NASA}	−0.087	23	0.02	.932	–	–	–
O_{NASA}	–	–	–	–	81.0	0.402	.050

Note. t = t-statistic of the paired-samples t-test; df = degrees of freedom; d = Cohen's d (effect size). W = Wilcoxon signed-rank test statistic; r = effect size. $^{*}p < .05$; $^{**}p < .01$; $^{***}p < .001$.

Cognitive Load - Eye-Tracking. The descriptive results reveal that both eye-tracking indicators were lower when using the semi-automatic approach. The number of fixations (NF) decreased from 1867.88 to 1465.04, and the mean fixation duration (MFD) decreased from 442.23 ms to 353.82 ms.

Inferential analysis revealed that the difference for the NF was not statistically significant ($t(23) = -1.633$, $p = .116$). This indicates that the participants had similar visual interactions. In contrast, MFD was significantly lower when using the semi-automatic approach $t(23) = -7.873$, $p < .001$, with a large effect size ($d = 1.16$). This substantial effect indicates reduced processing effort per visual element when using the LLM-based approach.

Although the overall NF did not differ significantly, the shorter MFD indicates that the participants required less cognitive processing time per fixation during semi-automatic modeling. Although participants visually engaged with the interface to a similar extent, the cognitive load required to process individual elements was lower.

Cognitive Load - NASA-TLX. Regarding the descriptive results, several subjective workload dimensions were lower when using the semi-automatic approach. Mental demand (M_{NASA}) decreased from 57.29 to 37.92, and effort (E_{NASA}) decreased from 57.71 to 29.58.

For normally distributed dimensions, paired-samples t-tests were conducted. M_{NASA} was significantly lower for the semi-automatic approach, $t(23) = -2.677$, $p = .013$, with a medium effect size ($d = 0.55$). In contrast, temporal demand (T_{NASA}), performance (P_{NASA}), and frustration (F_{NASA}) did not show statistically significant differences (all $p > .05$), i.e., effect sizes remained small.

For non-normally distributed variables, Wilcoxon Signed-Rank Tests were applied. E_{NASA} was significantly lower for the semi-automatic approach, $W = 26.5$, $p = .002$, with a medium-to-large effect size ($r = 0.617$). Physical demand (Ph_{NASA}) and overall workload (O_{NASA}) did not reach statistical significance.

Overall, no significant differences were found for Ph_{NASA}, T_{NASA}, P_{NASA}, or F_{NASA}. Regarding T_{NASA}, the participants were not subject to strict time constraints, as they were instructed to complete the task until being satisfied with their process model. Hence, the perceived time pressure was expected to be relatively low and comparable between both modeling approaches. Similarly, no differences in Ph_{NASA} were expected, as both approaches required participants to sit at a desk interacting with a computer-based modeling environment (i.e. BPMNGen vs. BPMN.io). The physical interaction effort was therefore comparable across the two modeling approaches. The absence of significant differences in perceived performance can be explained with the task structure: Participants were free to finalize their process models until being satisfied, likely leading to comparable levels of self-reported performance in both modeling approaches. Finally, the frustration levels were relatively low and similar in both approaches. This indicates that none of the two modeling approaches introduced substantial emotional strain. In general, our findings indicate that the LLM-based approach significantly reduces cognitive load without negatively affecting participants' perceived time pressure, physical demands, performance, or emotional experience.

Overall Interpretation. Taken together, the results demonstrate that semi-automatic process modeling with an LLM-based chatbot increases modeling efficiency on the one hand and reduces key dimensions of cognitive load on the other. Objective measures (MFD) and subjective workload assessments (M_{NASA} and E_{NASA}) converge indicating reduced cognitive processing effort. At the same time, dimensions such as Ph_{NASA}, T_{NASA}, and F_{NASA} remain unchanged, indicating that the semi-automatic approach specifically alleviates mental processing demands associated with syntactic rule application and structural composition, rather than fundamentally changing all dimensions of the task.

The large effect observed for MFD ($d = 1.16$) and the moderate effects for TCT ($d = 0.49$) and M_{NASA} ($d = 0.55$) indicate that the benefits of LLM-supported modeling are not merely marginal improvements but reflect meaningful reductions in cognitive load. By externalizing syntactic rule application and structural composition to the LLM, semi-automatic modeling shifts the task from manual process model creation toward process model review and refinement, thereby reducing cognitive load while maintaining perceived performance.

Process Model Quality. While the results indicate that semi-automatic modeling reduces cognitive load and increases modeling efficiency, these advantages would be of low practical value if they came at the expense of lower process model quality. Therefore, generated process models – either manually or semi-automatically – were systematically evaluated with respect to their syntactic and semantic quality.

Syntactic quality was assessed by counting the number of syntactic errors (SE), whereas semantic quality was measured using a predefined scoring scheme, deriving the semantic points (SP) from the respective process descriptions. The results are presented in Table 5. As both variables (SE and SP) met the assumptions for parametric testing, paired-samples t-tests were conducted.

Table 5. Descriptive and Inferential Statistics for Process Model Quality

Variable	Semi-Auto. Modeling		Manual Modeling		t	df	d	p
	M	SD	M	SD				
Quality								
SE	3.21	1.91	4.88	4.37	-1.641	23	0.33	.114
SP	9.41	1.28	6.82	2.37	4.336	23	0.89	$<$ **.001*****

Note. $^{***}p < .001$.

Regarding the descriptive results, semi-automatically generated models with the use of the LLM-based chatbot showed fewer SE (M = 3.21, SD = 1.91)

compared to manually created process models (M = 4.88, SD = 4.37). However, this difference did not reach statistical significance, $t(23) = -1.641$, $p = .114$, with a small effect size ($d = 0.33$). This indicates that both approaches produced comparable levels of syntactic correctness.

In contrast, a clear difference emerged for semantic quality. Semi-automatically generated process models achieved substantially higher SP (M = 9.41, SD = 1.28) than manually created process models (M = 6.82, SD = 2.37). This difference was statistically significant, $t(23) = 4.336$, $p < .001$, with a large effect size ($d = 0.89$). The magnitude of this effect indicates a practically meaningful advantage of the semi-automatic approach in terms of semantic alignment.

Taken together, these findings indicate that the increased efficiency and reduced cognitive load associated with semi-automatic modeling are not reached at the expense of process model quality. Syntactic quality remains comparable between both approaches, whereas semantic quality is significantly higher for the semi-automatic approach. This indicates that delegating structural composition to the LLM does not compromise correctness but may even enhance the faithful translation of textual process descriptions into BPMN 2.0 process models.

Answer to RQ. Semi-automatic process modeling, i.e., the use of an LLM-based chatbot the human modeler interacts with to initially create and iteratively refine business process models, significantly increases modeling efficiency and reduces both objective and subjective cognitive load compared to manual modeling. Note that these advantages do not come at the expense of process model quality, which we consider as a fundamental prerequisite for any LLM-supported conceptual modeling approach. The semi-automatically generated process models exhibit only a moderate number of syntactic errors and achieve a very high semantic accuracy, with an average of 9.41 out of 10 points. Overall, the results indicate that LLM-based modeling not only improves efficiency and reduces cognitive load, but is capable of producing high-quality process models as well.

6 Limitations and Threats to Validity

The findings of our study need to be interpreted in light of several limitations. One limitation concerns the experimental design. Although a controlled within-subject design with counterbalancing was applied to reduce inter-individual variance and learning effects, strategy adaptation effects cannot be fully excluded. Participants completed two modeling approaches requiring fundamentally different types of interaction. While manual modeling involves constructing process models step by step using a graphical modeling tool, LLM-based modeling requires formulating prompts and evaluating generated process models. Consequently, the observed effects cannot be attributed solely to the LLM itself, but may also result from differences in modeling workflow, interface design, or the availability of an initially generated process model after Zero-Shot prompting. These differences in interaction style may have influenced cognitive load and task performance beyond the intended comparison. At the same time, this distinction reflects realistic usage and therefore increases the practical relevance of our study.

Another limitation relates to the operationalization of cognitive load. Cognitive load was measured using eye-tracking metrics (number of fixations and mean fixation duration) in combination with NASA-TLX scores. These complementary measures provide both objective and subjective perspectives on workload. However, eye-tracking offers indirect indicators of cognitive processing demands and may reflect visual search behavior or interface characteristics as well. Likewise, subjective workload assessments capture perceived rather than directly observable cognitive load. Thus, the applied measures approximate, but do not directly observe the underlying cognitive processes involved in process modeling.

A further limitation concerns the assessment of process model quality, which was evaluated using syntactic error counts and a predefined semantic scoring scheme. Although this ensured a structured and transparent evaluation, semantic assessment involves interpretative elements despite predefined criteria.

Another limitation arises from the controlled study setting. The experiment relied on two process descriptions of comparable and moderate complexity to ensure experimental control. Although this strengthens internal consistency, real-world modeling scenarios may involve higher complexity, ambiguity, changing requirements, or collaborative modeling settings not represented in this study.

Finally, the characteristics of the participant sample and the investigated tool need to be considered. The participants were university students with verified BPMN 2.0 knowledge. The qualifying quiz ensured a homogeneous and sufficiently high competence level, thereby reducing expertise-related confounding effects. Nevertheless, students may differ from experienced practitioners in modeling strategies and domain expertise, which might limit generalizability. In addition, the relatively small sample size further limits the generalizability of the findings, as some observed effects may be influenced by sample-specific characteristics.

In addition, the results are specific to the examined LLM-based chatbot, BPMNGen, and the underlying LLM used in this study, Gemini 2.5 Pro. This LLM was selected based on results from a pre-study in which it demonstrated particularly strong performance for the task of process model generation. Although BPMNGen can be connected to different LLMs, this study intentionally focused on a single LLM to ensure a controlled experimental setting. Future studies are currently conducted to investigate the effects of alternative LLMs, including open-source LLMs, on the process of process modeling.

7 Conclusion and Future Work

This paper investigated how semi-automatic process modeling with an LLM-based chatbot differs from manual BPMN 2.0 modeling with a graphical modeling tool regarding efficiency, cognitive load, and process model quality. Based on a controlled within-subject eye-tracking study, the findings consistently indicate that semi-automatic modeling leads to substantial reductions in both objective and subjective cognitive load while at the same time improving modeling efficiency. Participants required significantly less time to complete modeling tasks when using and interacting with the LLM-based chatbot, and both eye-tracking metrics and NASA-TLX scores revealed lower cognitive processing demands.

Importantly, these advantages did not come at the expense of process model quality. The semi-automatically generated process models exhibited only a moderate number of syntactic errors, indicating solid syntactic quality, and achieved a very high level of semantic quality. These results demonstrate that delegating modeling tasks to an LLM with which the user then may interact can yield process models that can not only be efficiently produced but also show a high syntactic and semantic quality as well.

Overall, the results indicate that semi-automatic process modeling represents a promising approach that alleviates cognitive demands and increases efficiency without compromising, and potentially even enhancing, semantic correctness. By shifting the modeling process from element-wise construction toward prompt-based generation and refinement, LLM-based tools may fundamentally change the way process models are created. Despite the growing body of research on LLM-based process model generation, little empirical evidence exists on how such tools affect the process of process modeling itself. This study contributes to addressing this gap by providing empirical insights into how LLM-based assistance influences efficiency, cognitive load, and process model quality.

In future research we will further examine the findings of this study in alternative study designs to minimize potential carryover effects and strengthen the validity of the results. In addition, combining eye-tracking and questionnaire data with additional cognitive or behavioral measures will provide a more detailed understanding of how the support of LLMs influences the modeling process. Investigating more complex, realistic, and collaborative modeling scenarios will further help to better reflect real-world conditions. Finally, studies involving expert modelers and different LLM-based tools are needed to assess how well the results generalize beyond the present setting.

Disclosure of Interests. The authors have no competing interests to declare that are relevant to the content of this article.

References

1. bpmn.io. https://bpmn.io/. Accessed 09 Mar 2026
2. Burattin, A., Kaiser, M., Neurauter, M., Weber, B.: Eye tracking meets the process of process modeling: a visual analytic approach. In: Dumas, M., Fantinato, M. (eds.) BPM 2016. LNBIP, vol. 281, pp. 461–473. Springer, Cham (2017). https://doi.org/10.1007/978-3-319-58457-7_34
3. Burattin, A., Kaiser, M., Neurauter, M., Weber, B.: Learning process modeling phases from modeling interactions and eye tracking data. Data Knowl. Eng. **121**, 1–17 (2019)

4. Claes, J., Vanderfeesten, I., Pinggera, J., Reijers, H.A., Weber, B., Poels, G.: A visual analysis of the process of process modeling. IseB **13**(1), 147–190 (2015)
5. Drakopoulos, P., Malousoudis, P., Nousias, N., Tsakalidis, G., Vergidis, K.: Do LLMs speak BPMN? An evaluation of their process modeling capabilities based on quality measures. Computation **14**(1), 10 (2026)
6. Dumas, M., La Rosa, M., Mendling, J., Reijers, H.A.: Fundamentals of Business Process Management, vol. 1. Springer, Heidelberg (2013). https://doi.org/10.1007/978-3-662-56509-4
7. Hart, S.G., Staveland, L.E.: Development of NASA-TLX: results of empirical and theoretical research. Adv. Psychol. **52**, 139–183 (1988)
8. Hörner, L.F., Möller, M., Reichert, M.: LLM-based process modeling: evaluation of usability and semantic quality. In: International Conference on LLMs, Springer Nature (2025, accepted for publication)
9. Hörner, L.F., Möller, M., Reichert, M.: Automatically generating BPMN 2.0 process models from natural language process descriptions: challenges, framework, quality assessment. Bus. Inf. Syst. Eng., 1–25 (2026)
10. Ivanchikj, A., Serbout, S., Pautasso, C.: From text to visual BPMN process models: design and evaluation. In: Proceedings of the 23rd ACM/IEEE International Conference on Model Driven Engineering Languages and Systems, pp. 229–239 (2020)
11. Klievtsova, N., Benzin, J.V., Mangler, J., Kampik, T., Rinderle-Ma, S.: Process Modeler vs. Chatbot: is generative AI taking over process modeling? In: In: Delgado, A., Slaats, T. (eds.) International Conference on Process Mining, pp. 637–649. Springer, Cham (2025). https://doi.org/10.1007/978-3-031-82225-4_47
12. Klievtsova, N., Kampik, T., Mangler, J., Rinderle-Ma, S.: Conversationally actionable process model creation. In: Comuzzi, M., Grigori, D., Sellami, M., Zhou, Z. (eds.) Cooperative Information Systems, CoopIS 2024. LNCS, vol. 15506, pp. 39–55. Springer, Cham (2025). https://doi.org/10.1007/978-3-031-81375-7_3
13. Köpke, J., Safan, A.: Efficient LLM-based conversational process modeling. In: Gdowska, K., Gómez-López, M.T., Rehse, J.R. (eds.) Business Process Management Workshops, BPM 2024. LNBIP, vol 534, pp. 259–270. Springer, Cham (2025). https://doi.org/10.1007/978-3-031-78666-2_20
14. Kourani, H., Berti, A., Schuster, D., van der Aalst, W.M.: Process modeling with large language models. In: van der Aa, H., Bork, D., Schmidt, R., Sturm, A. (eds.) Enterprise, Business-Process and Information Systems Modeling, BPMDS EMMSAD 2024. LNBIP, vol. 511, pp. 229–244. Springer, Cham (2024). https://doi.org/10.1007/978-3-031-61007-3_18
15. Licardo, J.T., Tankovic, N., Etinger, D.: BPMN assistant: an LLM-based approach to business process modeling. arXiv arXiv:2509.24592 (2025)
16. Mendling, J.: Metrics for Process Models: Empirical Foundations of Verification, Error Prediction, and Guidelines for Correctness, vol. 6. Springer, Heidelberg (2008). https://doi.org/10.1007/978-3-540-89224-3
17. Möller, M.: Towards the comprehensibility of manually, automatically, and semi-automatically created process models: a conceptual framework. In: Pufahl, L., Rosenthal, K., España, S., Nurcan, S. (eds.) Intelligent Information Systems, CAiSE 2025. LNBIP, vol. 557, pp. 294–301. Springer, Cham (2025). https://doi.org/10.1007/978-3-031-94590-8_35
18. Möller, M., Hörner, L.F., Reichert, M.: Comprehending LLM- vs. human-generated business process models: An eye-tracking study. In: International Conference on LLMs, Springer Nature (2025, accepted for publication)

19. Neurauter, M., et al.: The influence of cognitive abilities and cognitive load on business process models and their creation. In: Davis, F.D., Riedl, R., vom Brocke, J., Léger, P.-M., Randolph, A.B. (eds.) Information Systems and Neuroscience. LNISO, vol. 10, pp. 107–115. Springer, Cham (2015). https://doi.org/10.1007/978-3-319-18702-0_14
20. Nour Eldin, A., Assy, N., Anesini, O., Dalmas, B., Gaaloul, W.: Nala2BPMN: automating BPMN model generation with large language models. In: Comuzzi, M., Grigori, D., Sellami, M., Zhou, Z. (eds.) Cooperative Information Systems, CoopIS 2024. LNCS, vol. 15506, pp. 398–404. Springer, Cham (2025). https://doi.org/10.1007/978-3-031-81375-7_27
21. Object Management Group: BPMN 2.0: Business process model and notation (2011). https://www.omg.org/spec/BPMN/2.0/. Accessed 20 Jan 2025
22. Petrusel, R., Mendling, J., Reijers, H.A.: How visual cognition influences process model comprehension. Decis. Support Syst. **96**, 1–16 (2017)
23. Pinggera, J., et al.: Investigating the process of process modeling with eye movement analysis. In: La Rosa, M., Soffer, P. (eds.) BPM 2012. LNBIP, vol. 132, pp. 438–450. Springer, Heidelberg (2013). https://doi.org/10.1007/978-3-642-36285-9_46
24. Pinggera, J., et al.: Styles in business process modeling: an exploration and a model. Softw. Syst. Model. **14**(3), 1055–1080 (2015)
25. Pinggera, J., et al.: Tracing the process of process modeling with modeling phase diagrams. In: Daniel, F., Barkaoui, K., Dustdar, S. (eds.) BPM 2011. LNBIP, vol. 99, pp. 370–382. Springer, Heidelberg (2012). https://doi.org/10.1007/978-3-642-28108-2_36
26. Safan, A., Köpke, J.: BPMN-Chatbot++: LLM-based modeling of collaboration diagrams with data. In: Proceedings of the BPM (2025)
27. Safan, A., Köpke, J.: A framework for LLM-based conceptual modeling: application to BPMN collaboration diagrams. In: International Conference on Conceptual Modeling (2025)
28. Sweller, J.: Cognitive load theory. Psychol. Learn. Motiv. **55**, 37–76 (2011)
29. Weber, B., et al.: Measuring cognitive load during process model creation. In: Davis, F.D., Riedl, R., vom Brocke, J., Léger, P.-M., Randolph, A.B. (eds.) Information Systems and Neuroscience. LNISO, vol. 10, pp. 129–136. Springer, Cham (2015). https://doi.org/10.1007/978-3-319-18702-0_17
30. Winter, M., Neumann, H., Pryss, R., Probst, T., Reichert, M.: Defining gaze patterns for process model literacy-exploring visual routines in process models with diverse mappings. Exp. Syst. Appl. **213**, 119217 (2023)
31. Winter, M., Pryss, R., Probst, T., Baß, J., Reichert, M.: Measuring the cognitive complexity in the comprehension of modular process models. IEEE Trans. Cogn. Dev. Syst. **14**(1), 164–180 (2020)
32. Wolf, I., Soffer, P.: Supporting BPMN model creation with routing patterns. In: Iliadis, L., Papazoglou, M., Pohl, K. (eds.) CAiSE 2014. LNBIP, vol. 178, pp. 171–181. Springer, Cham (2014). https://doi.org/10.1007/978-3-319-07869-4_15
33. Zimoch, M., Pryss, R., Probst, T., Schlee, W., Reichert, M.: The repercussions of business process modeling notations on mental load and mental effort. In: Daniel, F., Sheng, Q.Z., Motahari, H. (eds.) BPM 2018. LNBIP, vol. 342, pp. 133–145. Springer, Cham (2019). https://doi.org/10.1007/978-3-030-11641-5_11

The Process Canvas

Michael Rosemann[1](✉), Oktay Turetken[2], and Christoph Buck[3]

[1] Centre for Future Enterprise, Queensland University of Technology, Brisbane QLD,, Australia
m.rosemann@qut.edu.au

[2] Industrial Engineering and Innovation Sciences, Eindhoven University of Technology, Eindhoven, Netherlands
o.turetken@tue.nl

[3] Computer Science, Technical University Augsburg, Augsburg, Germany
christoph.buck@hs-augsburg.de

Abstract. Business processes as organisational assets need to be evaluated from integrated viewpoints that go far beyond the common operational view of being an executable, efficient process. Established BPM artefacts, however, do not support a comprehensive, multi-dimensional process description on a strategic level. This disconnect motivated the development of the Process Canvas, which allows describing the feasibility, viability, desirability and responsibility of a business process. Developed using Design Science Research, the Process Canvas integrates these four views in a new artefact and, with this, facilitates a previously not supported comprehensive assessment and contextualisation of a process. Its application in three semi-naturalistic and two naturalistic cases exemplifies its applicability and utility, and demonstrates its analytical and explanatory capability. This is further confirmed by a survey with 32 process experts.

Keywords: Process Canvas · Responsibility · Context-aware Process · Business Model Canvas

1 Introduction

Driven by the goal of seeking ways to improve the cost, time, and quality of a process, the last century of BPM research has been invested in developing methods to identify and address pain points along business processes. This has made the flawless execution of a process the common northstar of BPM. As a result, there is an advanced body of knowledge, methods, and tools available to address the elements immediate to a process – its events, activities, data, and resources [1].

However, such a control flow-centric reduction of a process falls short when it comes to supporting an understanding of what a process enables, as it does not sufficiently capture the process's attractiveness to the external market, its key capabilities and resources it leverages, its economic viability, or the societal and ethical responsibilities it must fulfil. The same is true for requests for responsible business processes [2].

Researchers argue that business processes should prioritise flexibility and adaptability over rigid specifications [3–5]. Studies on digital innovation emphasise the need for

F. M. Maggi et al. (Eds.): BPM 2026, LNCS 16936, pp. 737–754, 2027.
https://doi.org/10.1007/978-3-032-37777-7_43

process design but advocate a lightweight approach [5, 6]. Recent work conceptualises process "framing" as specifying the goals and constraints that define the permissible space for process enactment (i.e., autonomy within boundaries), rather than prescribing full execution scripts [7]. However, despite these insights, there is no consensus on how organisations can effectively operationalise light process design.

This motivated the research presented in this paper. Our intention was to design an artefact that helps elevate a business process from its operational focus (execution, efficiency) to become an asset that is comprehensively positioned against essential dimensions that matter to today's firms. Therefore, as a first design choice, we looked into domains such as business modelling and human-centric design where such broader considerations are common [8]. Our design goal was to develop and validate a new artefact that facilitates a rich, integrated understanding and assessment of business processes, incorporating strategic, customer-centric, viability-oriented, and responsible dimensions.

Therefore, our research objective was to develop a Process Canvas that captures process-related information across the dimensions of desirability, feasibility, viability, and responsibility. The Canvas focuses on the strategic layer of a process - why the process exists, what value it creates, and how it aligns with organisational goals along these four dimensions. It aims to facilitate stakeholder-informed and adaptable process (re)design in support of organisational transformation. The distinctive value of the Canvas lies in treating a business process not only as an executable flow of activities, but as a strategic, service-oriented, resource-dependent, value-generating, and responsibility-bearing organisational asset. Its contribution is therefore not the addition of isolated elements to existing process representations, but the provision of a unified, consolidating process-level artefact that enables stakeholders to reason more comprehensively about trade-offs among desirability, feasibility, viability, and responsibility [8, 9]

To address our research objective, we followed Design Science Research (DSR) [10] and developed the Process Canvas as a new artefact for process management. To demonstrate its applicability, we applied it to three semi-naturalistic processes and two business cases. We also conducted an online survey with 32 process experts.

This paper is structured as follows. In Sect. 2, we cover related work that aims to extend the focus of business processes. Section 3 presents our Design Science Research approach. The key contribution of this paper is discussed in Sect. 4, where we present the Process Canvas, which is applied in Sect. 5. Section 6 presents the evaluation of our artefact with BPM experts. Section 7 provides a discussion of our paper, followed by our conclusion outlining limitations and future research opportunities.

2 Related Work

The concept of a canvas-based approach has gained significant traction following the introduction of the Business Model Canvas (BMC) by Osterwalder and Pigneur [8]. The BMC offers a structured yet flexible framework for visualising and refining business models, helping organisations map out key elements while considering desirability, feasibility, and viability aspects [8]. Despite its widespread adoption in business, the BMC (and its derivatives) has yet to establish a strong foothold within the BPM domain

[11]. Available research has focused so far only on the link between business models, represented through BMC, and processes [12].

The core trifecta of the Business Model Canvas, feasibility, desirability, and viability, was first presented by the design and consulting firm IDEO as the three criteria critical for a successful product. Feasibility refers to the technical, as well as the legal possibility of creating a product or process. Viability represents the sustainability of a product or process. Desirability captures the extent to which a product or process is, in fact, wanted by its customers and whether it meets their needs and desires. In the field of design thinking, [12] argue that the classical view of the trifecta also needs to embed responsibility as an overarching criterion.

Koutsopoulos et al. [13] developed the "business process canvas" as a structured approach for designing and analysing processes, drawing inspiration from the Business Model Canvas (BMC) [8]. Their canvas captures key elements of the process, including goals, participants, and resources. By emphasising tasks, activities, exceptions, and outcomes, it primarily focuses on the feasibility dimension and does not adequately address desirability, viability, and responsibility. Similarly, the "process profile" approach [1] supports defining the key process components, such as vision (objective), customer, triggers, outcomes, and resources, but shares the same limitations in adequately incorporating other critical dimensions.

Missikoff [14] proposes the "business process analysis canvas", a framework structured into eight knowledge sections designed to support professionals in conducting business analysis. While the canvas provides basic descriptions of a business process - including triggers, inputs, outputs, key actors, and objects - it is focused on process analysis within the early phases of software development projects.

Another approach in this context is the "BPM Billboard" proposed by vom Brocke et al. [15]. The BPM Billboard is designed to plan, scope, and integrate BPM projects and programs with the objective of enhancing alignment between BPM initiatives, organisational goals, contextual factors, and capabilities. It is structured around five key dimensions: BPM's strategic alignment with the organisation, its relation to organisational context, BPM-related capabilities, BPM-related projects, and the results achieved from these initiatives. While the BPM Billboard can serve BPM governance and strategic decision-making, its unit of analysis is not a single business process.

In BPM practice, several approaches are commonly used to contextualise a business process. One such approach is SIPOC (Suppliers, Inputs, Processes, Outputs, Customers), a high-level process-mapping tool used in Six Sigma, Lean, and process improvement initiatives [16]. Similarly, IGOE (Inputs, Guides, Outputs, Enablers), developed by the Process Renewal Group, extends beyond narrowly defined process documents by incorporating additional dimensions [17].

Taken together, the existing artefacts in BPM provide valuable but only partial support for broadening the understanding of business processes. BPMN and related modelling approaches focus on process executability; SIPOC, IGOE, process profiles, and earlier process canvases [1, 14, 16, 17] support process scoping and operational contextualisation; the BMC and business model–process alignment methods [8, 11] focus on business-model-level value logic; service blueprinting and customer journey approaches [18, 19] emphasise customer experience and service interaction; and digital responsibility [20] and responsible innovation frameworks [21] articulate ethical and societal design principles. However, these perspectives remain distributed across separate artefacts, levels of analysis, and disciplinary traditions. Extending or layering existing artefacts requires process stakeholders to move across different units of analysis, vocabularies, and modelling purposes. In contrast, the Process Canvas proposed here contributes a unified process-level artefact that brings these perspectives together and enables stakeholders to analyse and design a business process as a feasible, desirable, viable, and responsible organisational asset.

3 Research Design

To develop the Process Canvas as a new artefact, we chose Design Science Research (DSR) [10] for two reasons: First, DSR is a well-recognised artefact development method in information systems and BPM research [22]. Second, DSR seeks to design and develop innovative artefacts, concepts, and frameworks with robust applications addressing a problem of high practical relevance [23]. In designing and developing the Process Canvas as a new artefact for process management, we followed the DSR process proposed by Peffers et al. [24]. Fig. 1 presents our research design.

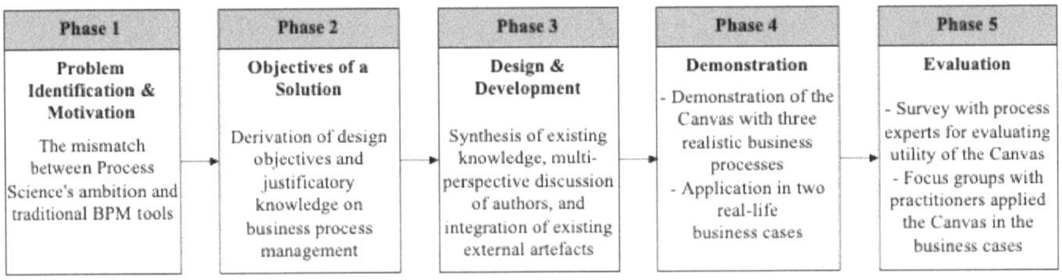

Fig. 1. Research design

The motivation of our research rested on the observation that BPM professionals eager to contextualise a business process in a strategic, market-exposed setting, also addressing emerging new requirements (e.g., AI-triggered responsibility demands), did not have sufficient access to supportive artefacts. As discussed above, available solutions were either too operational, not process-specific or not comprehensive. Scanning the literature beyond the BPM discipline for guidance, we identified business modelling and human-centric design as a source of inspiration for the *solution objectives* in the form of requirements. As a result, we made the design choice to ground the *development* of our artefact in the established feasibility, desirability and viability trifecta, extended with contemporary demands for responsible processes [9]. A design objective is to create an artefact that enables (re) designing processes under consideration of these four perspectives, addressing multiple stakeholders [25].

To *demonstrate* its use and validate its applicability, we applied the Process Canvas to three semi-naturalistic processes: the traveller journey in a shared e-bike service, loan processing, and Amazon's fulfilment and delivery process following a formative semi-naturalistic approach [23]. We also applied it in two real-life cases (Sect. 5) for a naturalistic *evaluation*. In addition, we conducted an online survey with 32 process experts as target users to evaluate the utility of the Process Canvas (Sect. 6).

4 The Process Canvas

The Process Canvas combines the three dimensions of successful products – feasibility, viability, and desirability - with a new dimension – responsibility. Our design choice to include the latter has been motivated by calls for responsible Business Process Management, which also manifested in the thematic focus of BPM 2024 and an entirely new responsible BPM workshop at BPM 2025 [2]. This led to the four constitutional dimensions of feasibility, desirability, viability, and responsibility. Fig. 2 presents the Process Canvas, which we discuss in detail in the following.

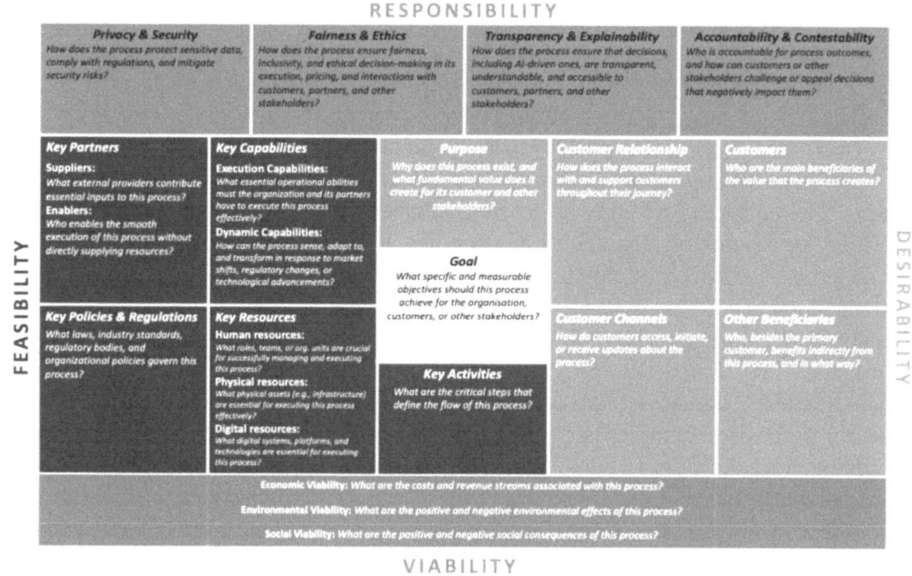

Fig. 2. Process Canvas

4.1 The Desired Process

Desirability captures the extent to which there is a market for the process expressed by defined cohorts of customers and other stakeholders, relationships, and channels via which they can interact with the process. The call to integrate value contribution into process design has been a key focus in value-based process management [26]. Thus, the following elements are considered relevant for the desirability of processes, i.e., by whom, but also how the process is desired.

Process Purpose defines the overarching reason for the existence of the process. It is the long-lasting description of the net value of the process and is often defined as the social license to operate. The process purpose aligns with the organisation's purpose, strategic goals, and business models [27, 28]. It emphasises the value the process creates not just for its customers but, in particular, other stakeholders [11, 29].

Process Goal defines the objectives the process aims to achieve and can be split into internal and external goals. Internal goals are common to BPM and include, for example, minimising cost/time-to-serve or ensuring compliance. External goals are customer-facing and defined from their perspective (e.g., call resolution in one call, real-time response, maximum self-guidance). The process goals need to be specified in a measurable way [1]. They operationalise the process purpose with clear objectives and are essential in directing the improvement and execution of process activities by prioritising tasks, allocating resources, and making informed decisions [30]. Thus, purpose captures the enduring rationale of the process, while goals operationalise that rationale into specific objectives for design, execution, and improvement. For example, the purpose of a shared e-bike traveller journey process may be to enable accessible and sustainable urban mobility, whereas its goals may include maintaining a target level of bike availability, reducing booking time, or increasing successful trip completion.

Customer represents the primary beneficiary who derives value from the process. As the target of the process [1], the customer plays a crucial role not only as a recipient but often also as a co-creator of value [31]. Clearly identifying the distinct customer cohorts and personas served by the process and understanding their perspectives is essential to ensuring that the process effectively creates, enhances, and sustains value.

Other Beneficiaries represent the secondary stakeholders who benefit indirectly from the process. Clearly defining these stakeholders ensures a holistic approach to value creation, aligning the process with broader interests beyond the immediate customer [32]. This can enhance stakeholder alignment, inform responsible process design and execution, identify shared value opportunities [33], and help mitigate risks by considering regulatory, ethical, and ecosystem-wide impacts [34]. While customers directly receive or use the process outcome, other beneficiaries benefit indirectly or have a legitimate stake in the process outcome.

Customer Relationships define the nature of engagement between the process and its customer. They can be, for example, transactional, personalised, self-assisted, or specific to the extent to which the engagement occurs with digital versus human agents. Customer relationships indicate how the process interacts with and supports customers throughout their journey.

Customer Channels are the different media through which customers interact with the process for accessing, initiating, or receiving updates about the process. These touchpoints influence customer experience and satisfaction, making the management of these interactions important to the overall customer journey [18].

4.2 The Feasible Process

A common process lifecycle includes the definition of a newly or re-designed business process. However, the extent to which this process is actually technically or legally feasible is only an implicit, and not an explicit concern of process design. This is the reason for the feasibility cluster in the Process Canvas.

Key Process Activities are the essential tasks or actions that drive the process from initiation to completion. In BPM, process activities are considered one of the main components of a process definition and make up, together with events, the control flow of a business process [1]. The activities selected must be executable and accessible.

Key Partners comprise external stakeholders providing critical resources or capabilities for process execution (**Suppliers**) and those that facilitate execution but are not direct suppliers (**Enablers**). Research highlights that effectively managing supplier relationships enhances supply chain performance, underscoring the need to integrate key partners into business processes [35]. Understanding the roles of key partners and their alignment with business processes is important for value co-creation and operational efficiency, as defining these partners enables the organisation to ensure capability alignment, resource availability, and process resilience.

Key Execution Capabilities are the process-specific organisational abilities that the organisation and its partners require to execute the process and use the embedded resources effectively. Organisations with explicitly defined capabilities can better manage resource allocation across their processes and refine competencies over time [36]. In general, well-defined capabilities contribute to better coordination and integration across processes, fostering resilience [37].

Key Dynamic Capabilities refer to an organisation's ability to sense changes in its environment, adapt its processes in response, and transform them to maintain alignment with evolving market conditions, regulations, or emerging technologies [38]. Firms with strong dynamic capabilities are better able to sense market shifts, quickly reconfigure processes, integrate new technologies effectively, and adapt to regulatory changes and evolving customer demands [38, 39]. While execution capabilities ensure the reliable performance of the current process, dynamic capabilities support the adaptation and reconfiguration of the process under changing conditions.

Key Resources encompass the essential assets required for effective process execution. This includes **Human Resources**, referring to the critical roles, teams, or organisational units assigned to the process; **Physical Resources**, referring to tangible assets such as infrastructure, equipment, and facilities; and **Digital Resources**, which include software, algorithms, data, platforms, and digital tools that support and enable the execution of the process.

Key Policies and Regulations represent the organisational policies, regulations, and the associated regulatory bodies or authorities that govern the process, ensuring legal compliance, industry standards, and organisational governance [40]. Incorporating regulations and relevant directives into process management can help organisations better adapt to changing legal landscapes. Additionally, defined policies help organisations improve governance and align processes with strategic objectives [41].

4.3 The Viable Process

Processes are typically designed with a strong cost bias as minimising the costs-to-serve has been a long-lasting goal of BPM. Economic viability, however, goes beyond a one-dimensional cost metric and also considers the revenue impact of a process. Moreover, following the Triple Bottom Line framework, viability also involves environmental and societal aspects [42]. While every firm has environmental and social impacts, these aspects have only recently gained significant attention in both practice and research, primarily among sustainability-focused scholars [43].

Economic Viability refers to the financial aspects of the process, including the costs involved in its execution and the revenue generated. Though concepts such as activity-based costing have facilitated an identification of cost drivers and total costs of execution, the revenue impact of a business process has only recently emerged as an area of explicit BPM interest [44]. Understanding the financial implications of a process enables organisations to evaluate the return on process investment.

Environmental Viability refers to the ecological impact of the process, including resource consumption and sustainability efforts [45]. Incorporating environmental considerations into process management helps organisations to align with sustainable development strategies, to minimise negative environmental effects, and to respond to societal pressures for environmental responsibility and 'green processes' [46].

Social Viability relates to the societal impact, including inclusivity, well-being, and potential unintended consequences. Explicit consideration of social viability supports organisations in assessing the impact of a process and mitigating negative ones to help contribute positively to social welfare [47].

4.4 The Responsible Process

Organisations are moving beyond profit maximisation to prioritise sustainability, stakeholder values, and ethical considerations [9, 48]. They also use AI agencies within their processes, which raises questions about the bias, fairness, security, etc., of those processes [49]. These challenges have led to the emergence of the concept of "digital responsibility" [20]. In response, BPM research is evolving beyond its traditional focus on executability and the dimensions outlined above, and considers *responsibility* as well [2, 50]. As data-driven systems become more widespread, integrating ethics into process management is recognised as essential to ensure that business processes reflect ethical values, such as transparency, accountability, privacy, and fairness [51].

Privacy and Security represents data protection, compliance, and fraud prevention. It defines how the process safeguards sensitive information, mitigates security risks, and maintains transparency in responsible automation. Security and privacy are among the most mentioned values in the process management research area [48].

Fairness and Ethics refers to the measures ensuring that the process operates in a non-discriminatory, ethical, and inclusive manner. It defines how the process guarantees equal access, fair pricing, moral transparency, and consistent service quality while upholding ethical standards in data usage, decision-making, and interactions with customers, partners, and other stakeholders [52].

Transparency and Explainability define how clearly process decisions, logic, and outcomes are communicated to customers, partners, and other stakeholders. This includes making AI-driven and automated decisions understandable and ensuring stakeholders can access relevant information.

Accountability and Contestability define the mechanisms that ensure responsibility for process outcomes [53] and provide stakeholders with the means to challenge, review, and rectify unfair or incorrect decisions [54].

Some level of overlap between Key Policies and Regulations and the responsibility elements is intentional, because the same concern may appear both as a formal feasibility constraint and as a process-level safeguard. While the former captures the legal, policy, standard-setting, and regulatory conditions under which the process can be executed, the latter captures how these and broader ethical expectations are enacted through safeguards for privacy, security, fairness, transparency, explainability, accountability, and contestability.

5 Exemplary Assessment of the Process Canvas

5.1 Semi-Natural Cases

We applied the Process Canvas to three semi-natural processes using secondary data - the traveller journey in a shared e-bike service, general loan processing, and Amazon's fulfilment and delivery process - to demonstrate its use and provide evidence of its applicability and potential to be used for its intended purpose of use.[1] We chose these processes as they encapsulate key aspects critical for effective process analysis and improvement from different viewpoints and sectors. This allowed a first assessment of the broad applicability of the Process Canvas. All three cases share characteristics such as multiple stakeholders, regulatory compliance, and considerations of various viability dimensions, while integrating technology and aligning operational tasks with strategic goals. In the following and in Fig. 3, we report only on the simplest and most intuitive of these three cases to showcase the utility of the Process Canvas, i.e. the shared e-bike service. This populated Process Canvas provides further clarity on how to interpret each of these elements in the Canvas.

The *traveller journey* process provides seamless green mobility for city commuters. It aims to balance e-bike availability, accessibility, and reliability and includes tasks such as bike reservation, navigation support, issue handling, and payment. Multiple stakeholders - such as the service operator, maintenance provider, municipality, and travellers - are involved, with a focus on economic viability as well as environmental and social impact. The process is designed to adapt to demand shifts, integrate AI for predictive maintenance, and comply with privacy, safety, and sustainability regulations.

5.2 Applications in Real-Life Business Cases

Beyond these three exemplary, semi-naturalistic cases, we wanted to better understand the application of the Process Canvas and its net-benefit in naturalistic BPM settings. Thus, we were eager to study the use of the Process Canvas in two cases by experienced BPM professionals. The first case was conducted within Powerlink Queensland, a *provider of electricity transmission services* that owns, develops, operates, and maintains the high-voltage electricity transmission network in the Australian state of Queensland. It employs more than 1,500 people. Aligned with the resource-intensive nature of the organisation, important areas for the application of BPM include asset management and network operations management. The second case was conducted at a *global asset*

[1] All three cases can be found on the web page www.processcanvas.org.

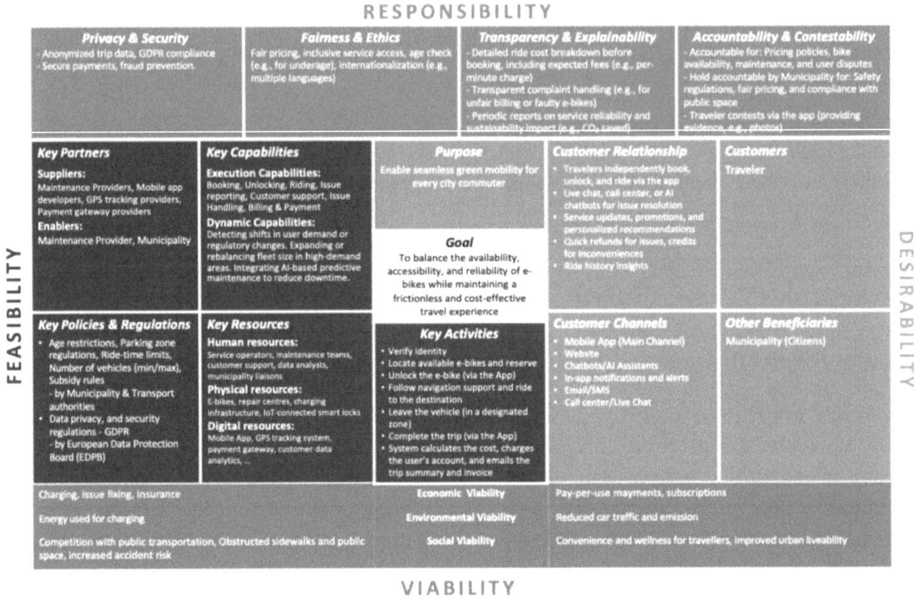

Fig. 3. Process Canvas of the Traveller Journey Process in a Shared E-bike Service

finance company headquartered in the Netherlands with around 6,000 employees that helps businesses access equipment, technology, and software through financing across various industries. Both cases were conducted in October 2025, and no researcher was involved in these deployments to ensure a self-guided, non-biased application. In the Powerlink case, the canvas was deployed by a Senior Business Architect and an Enterprise Business Analyst, whereas in the second case, an Enterprise Architect and a Process Architect deployed it. The applications were followed by focus group meetings to gather feedback from the practitioners who applied the Canvas.

In the Powerlink case, the selected process (Identify Network Asset Need) was a part of the asset management capability and was dedicated to evaluating the requirement for asset renewal based on an asset health index. The process was not facing any external customers and sought to minimise asset-related expenses while securing well-conditioned assets. This process was on level 2 within Powerlink's four-level hierarchy, i.e., on a higher level of granularity. The team started to populate purpose, goal, and activity first, as they were seen as the central point of reference for all subsequent assessments. Using an online version of Canvas, they then worked through it, starting with desirability and then viability, feasibility, and responsibility.

In the second case, the Process Canvas was applied to a high-level process within the IT Management capability, specifically focusing on the value stream that describes the lifecycle of a digital product from portfolio management to operations. This process was chosen because it represented a crucial end-to-end flow within the organisation and involved multiple stakeholders. The application sequence mirrored that of the first case. The practitioners recognised that, depending on the nature of a process and the focus of its desired transformation, different clusters might become more relevant. For example, if AI were to play a significant role, feasibility and responsibility might become crucial.

We structured the primary data gained from these two cases according to the Canvas evaluation criteria proposed by Fatima et al. [55]. In both cases, the Process Canvas was assessed as complete and even overly comprehensive for the processes studied. For example, for Powerlink, the absence of external customers meant that parts of the desirability cluster were less relevant. At the same time, it was unclear if external regulators are seen as key partners or, in fact, as customers. However, the inability to complete all sections did not affect Canvas's utility. There were no obvious specific elements that the users wanted included in the Canvas.

The practitioners found Canvas helpful for clarifying the broader context of the process. There was a good overall fit between Canvas and the *real-world requirements* for capturing additional contextual process information. However, these requirements are a function of the audience addressed by this Canvas, but the Canvas's audience is not explicitly stated as part of it. It was proposed to consider positioning the use of the Canvas within the process lifecycle so that process designers would know when to use it (e.g., before, during, or after modelling a business process). Both elements, audience and positioning the Canvas within established, commonly used methods, were regarded as important meta-information that should be added to the Canvas.

Regarding the Canvas's internal *consistency*, some issues arose. The difference between the process's purpose and its goal was seen as not clear enough. A few elements (e.g., privacy and security) were considered too vaguely defined. In light of these comments, we refined the definitions of the elements to better clarify their differences. One challenge in both cases was finding the right place in Canvas to capture regulatory aspects. For instance, in Powerlink's industry setting, regulations and relationships with regulatory bodies are important factors in the design and compliance of a process. The users considered 'accountability and contestability', 'privacy and security', as well as 'key partners' and 'key policies and regulations' (feasibility) to capture regulatory information. It was decided that regulation can indeed be modelled in multiple, complementary parts of the Canvas, a fact that was seen as requiring further clarification.

The perception was that the Canvas's *level of detail* and the overall relevance of the four dimensions depended on the process's level within an organisational process hierarchy. The higher the level is, the more designers can have an actual impact on these dimensions. The further a process is positioned lower in the process hierarchy, the more the elements in these four dimensions are perceived as defined and decided. Instead, the aims of transparency (the process as a guide for conducting work) and executability (the process as a model for a systems-supported workflow) become important.

In sum, the application was overall seen as a validation of the Canvas's *robustness*. The practitioners indicated that they used the Canvas to create *an insightful holistic description* of a process. Compared with traditional process and enterprise architecture models, which provide structural insights but may be less accessible to non-modelling experts, the Canvas helped connect strategic intent with operational design elements. Furthermore, practitioners state that the canvas helped them to find a common language, which helped them work more efficiently and with higher strategic alignment. This suggests that the Canvas complements existing BPM approaches by providing a process-level framing of strategic value that considers the demands of external stakeholders and emerging responsibility requirements.

6 Evaluation Using a Survey

To further evaluate the Process Canvas, we examined its *utility* [23], specifically, how useful and easy to use it is perceived by its target users, namely, BPM professionals. To this end, we conducted an online survey with 32 process experts. The majority of experts (22) had over 10 years of experience, primarily in BPM, process architecture, and digital transformation. They rated themselves as either very familiar or extremely familiar with process analysis and design. The experts came from diverse industries and countries, including 11 from Australia, 8 from Germany, and 5 from the Netherlands, with additional experts from Switzerland, Brazil, Belgium, the Philippines, New Zealand, South Africa, the USA, and Canada.

For the first part of the survey questionnaire, we used the core constructs of the Technology Acceptance Model (TAM), which is frequently used in IS research and DSR [56], to account for the *perceived usefulness* and *perceived ease of use* of novel technologies and design artefacts, as well as users' *intention to use* them [57]. Fig. 4 presents the TAM-based survey items and experts' responses.[2] The results indicate a general positive perception across the measured constructs. The Process Canvas is seen as effective (Q1), useful (Q4), and easy to use (Q5), with skill development anticipated (Q6). Though clarity opinions are divided (Q5), future preference is high (Q9), signalling motivation for adoption. The Canvas's multi-dimensional approach and concise overview are valued. Positive intention for use is evident (Q10).

In addition to demographic and TAM-based questions, the survey included open-ended questions, enabling us to capture expert evaluations of the Process Canvas's strengths and weaknesses. We discuss these findings below.

The Process Canvas receives considerable acclaim for its ability to facilitate a multifaceted, holistic understanding of processes, enabling stakeholders to consider crucial dimensions often overlooked in conventional process analyses (e.g., E5). Experts particularly value the structured framework, which resonates with the widely adopted Business Model Canvas, thus promoting ease of assimilation and application (E7, E11). The visual clarity and organised presentation of intricate process dynamics are consistently emphasised as instrumental in fostering effective communication and synergistic collaboration amongst stakeholders (E3, E14). Its propensity to inspire a broad, contextualised perspective can support expansive process discovery, facilitating the identification of optimisation opportunities and mitigating potential oversights (E3, E6). Its alignment with contemporary corporate imperatives, encompassing ethical, social, and environmental considerations, is deemed invaluable for responsible process design and management (E5, E14). Furthermore, integrating elements from scoping diagrams (SIPOC, IGOE) and the Business Model Canvas positions the Process Canvas as an effective tool for

[2] The survey questionnaire and experts' responses are available at: *https://sites.google.com/view/processcanvas*

structuring exploratory work and strategic deliberations, delivering a concise, synthesised overview of process attributes (E6, E10). The Canvas was considered to have the potential to foster a more conscientious and comprehensive approach to process engineering and re-engineering initiatives (E2, E9).

A primary concern articulated by experts pertains to the complexity and potential for cognitive overload inherent in the Process Canvas (e.g., E1). This is evidenced by suggestions for a simplified iteration to enhance usability (E1). Apprehension regarding the Canvas's accessibility was noted, prompting recommendations for more articulate guidance for its use (E5, E22). Challenges in practical implementation were identified, with requests for illustrative, populated exemplars to facilitate application (E6). Several participants observed a deficit in explicit emphasis on critical elements such as inputs, outputs, key performance indicators, and process interfaces, which are deemed essential for rigorous process analysis (E10, E13, E22). The limited guidance available for populating Canvas emerged as a weakness (E6). Finally, some respondents suggested specific responsibility methodologies to be included within the framework (E2).

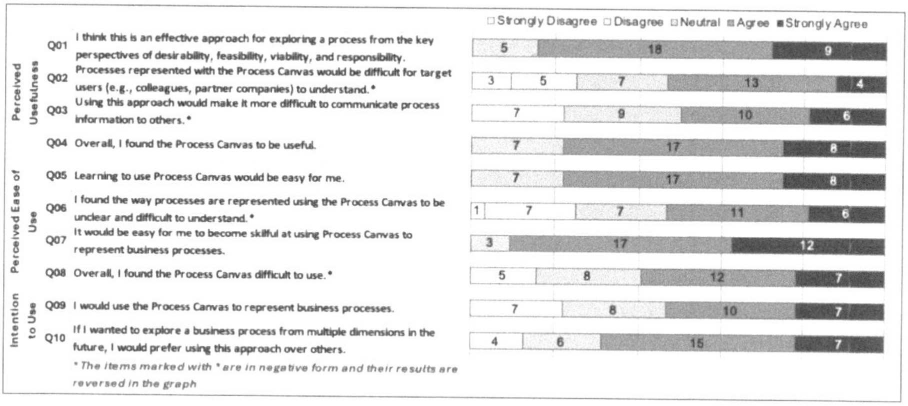

Fig. 4. Survey results on the perceived usefulness, ease of use, and intention to use the Canvas

Experts frequently drew parallels between the Process Canvas and the Business Model Canvas (e.g., E8, E14), noting its familiar structure. Several experts suggested its utility in framing discovery workshops and strategic discussions (E11, E14). Integration with existing methodologies, such as scoping diagrams (E7) and BPMN (E11), was also proposed. Furthermore, the need for rigour in the application (E10) and for clear communication of Canvas's purpose (E3) was emphasised. Some experts expressed interest in aligning the Canvas with enterprise architecture deliverables (E11), while others requested a more practical approach (E3, E22). The dimension of responsibility was seen as a very forward-looking improvement (E7).

7 Discussion

As there is more to a business process than its compliant execution and continuous improvements in time, cost, and quality, BPM requires approaches that integrate strategic, market-facing considerations into the design and management of business processes. One such quasi-standard framework is the desirability, feasibility, and viability assessment as developed by IDEO, which is widely used in product and service development. This is why we selected these three dimensions and combined them with a fourth dimension that considers new requirements resulting from the uptake of AI and data-intensive solutions – responsibility. Together, they form the Process Canvas.

Depending on the nature of the process and its strategic narrative, the Process Canvas will differ substantially and either focus on resourcing or marketing positioning. For example, *feasibility-focused* business processes are grounded in unique, often hard-to-replicate resources or dynamic capabilities of the firm. Examples are Walmart's advanced logistics, the Mumbai Dabbawalas's nearly flawless execution of food delivery, or, more recently, AI-empowered business processes in the financial industry [58]. In contrast, *desirability-focused* business processes are characterised by well-defined and well-served customer cohorts. Examples are Zara's rapid replenishment process with its ability to respond to the fashion trends quickly, and Amazon's processes that are adapted to customer preferences based on A/B testing.

Viability-focused business processes are less centred on internal feasibility or external desirability but stand out because of their ability to balance process input and output in economic, environmental, or societal terms. Uber demonstrates how a revenue lens on processes creates an economically unique Process Canvas as the organisation continuously generates new revenue channels by scaling a largely existing process capability. Patagonia is a showcase for a business process with an environmentally compelling viability, as multiple process mechanisms ensure a low carbon footprint.

Finally, we see the rise of business processes with a focus on their responsibility. These processes might not be the most attractive in terms of classical process performance criteria (time, cost, quality), but have their unique strength. A compelling example is Ben & Jerry's, which emphasises integrating social and environmental responsibility into its operations. The company's processes prioritise Fair Trade sourcing, carbon reduction, and social justice advocacy, aiming to embed ethical considerations into their production and reduce their ecological footprint [59].

8 Conclusion

This paper proposes the Process Canvas as a new artefact that adds four fundamental new perspectives on business processes: desirability, feasibility, viability, and responsibility. This allows a more structured consideration of enquiries into business processes that previously have not received sufficient support. For example:

- what are the customer cohorts attracted to this process, and what type of channels and relationships do we maintain with them?
- what are the requirements and prerequisites (e.g., partners, activities, resources) to ensure the desired business process is actually feasible?

- how viable is the process in terms of its economic, environmental, and societal net position?
- is the business process responsible as per important criteria of responsibility (e.g., fairness, security, accountability)?

This paper is a first step towards a new, high-impact process artefact. Within the constraints of a conference, it has to come with *limitations*. First, there is no underlying (semi)formalisation of the Process Canvas (e.g., via a meta-model). This means that the interrelationships between its elements are currently underspecified. Second, the empirical validation of the Process Canvas in terms of its utility and actual impact is still in its infancy, though this paper provides first insights in terms of its applicability via three cases and initial utility perceptions by global BPM experts. Third, we applied it to represent high-impact business processes, but its usefulness and scalability for smaller, routine processes remain uncertain. Fourth, the Process Canvas has not yet been embedded in existing BPM methods and tools (e.g., process architecture, BPM maturity model, process lifecycle), meaning it remains a stand-alone artefact.

The Process Canvas does not aim for completeness in terms of capturing every possible element of a business process; rather, it includes those elements that are repeatedly highlighted in the extant literature as critical for understanding and shaping processes from a strategic, stakeholder-informed, and innovation-oriented perspective. This design choice reflects our focus on practical relevance and conceptual clarity rather than exhaustive representation, aligning with the purpose of the Canvas as a facilitative, rather than prescriptive, artefact.

Extending established process considerations via this Process Canvas catalyses a variety of future research opportunities, including but not limited to research questions such as (1) What constitutes successful configurations across the elements of the four clusters of the Process Canvas? Are there even normative configurations? What would be a blue ocean Process Canvas? (2) What are the trade-offs within the various elements of the Process Canvas? (3) How can available process information (e.g., log files, digital traces) be used to monitor elements within the Process Canvas? (4) What differentiates back-office processes (behind the line of visibility) from customer-facing processes in terms of their respective Process Canvas?

These questions demonstrate the fertile nature of the Process Canvas for upcoming research. However, as the empirical evidence demonstrates, the Canvas is already in its current form a valuable addition to the existing landscape of BPM artefacts.

References

1. Dumas, M., La Rosa, M., Mendling, J., Reijers, H.A.: Fundamentals of Business Process Management, 2nd. edn. Springer Verlag (2018)
2. Gal A, Santoro F, Grisold T, ER M (2025) Call for Responsible BPM Forum - BPM-2025. https://www.bpm2025seville.org/calls/responsible-forum/. Accessed 1 June 2026
3. Distel, B., Plattfaut, R., Kregel, I.: How business process management culture supports digital innovation: a quantitative assessment. Bus. Process. Manag. J. **29**, 1352–1385 (2023)
4. Baiyere, A., Salmela, H., Tapanainen, T.: Digital transformation and the new logics of business process management. Eur. J. Inf. Syst. **29**, 238–259 (2020)

5. Mendling, J., Pentland, B.T., Recker, J.: Building a complementary agenda for business process management and digital innovation. Eur. J. Inf. Syst. **29**, 208–219 (2020)
6. Benner, M.J.: Dynamic or static capabilities? Process management practices and response to technological change. J. Prod. Innov. Manag. **26**, 473–486 (2009)
7. Dumas, M., Fournier, F., Limonad, L., et al.: AI-augmented business process management systems: a research manifesto. ACM trans Manag. Inf. Syst. **14** (2023)
8. Osterwalder, A., Pigneur, I.: Business Model Generation: a Handbook for Visionaries, Game Changers and Challengers. Wiley, New Jersey, US (2010)
9. Baldassarre, B., Calabretta, G., Karpen, I.O., et al.: Responsible design thinking for sustainable development: critical literature review, new conceptual framework, and research agenda. J. Bus. Ethics. **195**, 25–46 (2024). https://doi.org/10.1007/S10551-023-05600-Z
10. Hevner, A.R., March, S., Park, J., Ram, S.: Design science in information systems research. MIS Q. **28**, 75–105 (2004)
11. Lara Machado, P., van de Ven, M., Aysolmaz, B., et al.: Methods that bridge business models and business processes: a synthesis of the literature. Bus. Process. Manag. J. **29**, 48–74 (2023). https://doi.org/10.1108/BPMJ-08-2022-0396
12. Lara Machado, P., Van De Ven, M., Aysolmaz, B., et al.: Navigating business model redesign: the compass method for identifying changes to the operating model. Bus. Inf. Syst. Eng. **66**, 607–638 (2024). https://doi.org/10.1007/S12599-024-00892-5
13. Koutsopoulos, G., Bider, I.: Business process canvas as a process model in a nutshell. In: BPMDS 2018, LNBIP 318, pp. 49–63. Springer Verlag (2018)
14. Missikoff M (2022) A Knowledge-driven Business Process Analysis Canvas. Arxiv: 2201.06860
15. J, v.B., Mendling, J., Rosemann, M.: Planning and scoping business process management with the BPM billboard. In: Business Process Management Cases, vol. 2. Springer (2021)
16. ASQ (2016) SIPOC Diagram. https://tinyurl.com/4u7zk8hb. Accessed 1 June 2026
17. BPTrends (2013) IGEO Diagram (Scope Diagram). https://tinyurl.com/yw9s8xcr. Ac20-03-2026
18. Lemon, K.N., Verhoef, P.C.: Understanding customer experience throughout the customer journey. J. Mark. **80**, 69–96 (2016). https://doi.org/10.1509/JM.15.0420/
19. Bitner, M.J., Ostrom, A.L., Morgan, F.N.: Service blueprinting: a practical technique for service innovation. Calif. Manag. Rev. **50**, 66–94 (2008). https://doi.org/10.2307/41166446/
20. Trier, M., Kundisch, D., Beverungen, D., et al.: Digital responsibility: a multilevel framework for responsible digitalization. Bus. Inf. Syst. Eng. **65**, 463–474 (2023). https://doi.org/10.1007/S12599-023-00822-X/
21. Stilgoe, J., Owen, R., Macnaghten, P.: Developing a framework for responsible innovation. Res. Policy. **42**, 1568–1580 (2013). https://doi.org/10.1016/J.RESPOL.2013.05.008
22. vom Brocke, J., Baier, M.S., Schmiedel, T., et al.: Context-aware business process management: method assessment and selection. Bus. Inf. Syst. Eng. **63**, 533–550 (2021). https://doi.org/10.1007/S12599-021-00685-0
23. Venable, J., Pries-Heje, J., Baskerville, R.: FEDS: a framework for evaluation in design science research. Eur. J. Inf. Syst. **25**, 77–89 (2016)
24. Peffers, K., Tuunanen, T., Rothenberger, M.A., Chatterjee, S.: A design science research methodology for information systems research. J. Manag. Inf. Syst. **24**, 45–77 (2014). https://doi.org/10.2753/MIS0742-1222240302
25. Moder, L., Fehrer, T., Röglinger, M.: Design principles for process improvement and innovation systems. Bus. Process. Manag. J. (2025). https://doi.org/10.1108/BPMJ-10-2024-0975
26. Buhl, H.U., Röglinger, M., Stöckl, S., Braunwarth, K.S.: Value orientation in process management: research gap and contribution to economically well-founded decisions in process management. Bus. Inf. Syst. Eng. **3**, 163–172 (2011)

27. Al-Debei, M.M., Avison, D.: Developing a unified framework of the business model concept. Eur. J. Inf. Syst. **19**, 359–376 (2010)
28. Rosemann M, vom Brocke J (2015) The six Core elements of business process management. In: Brocke J vom, Rosemann M (eds) Handbook on Business Process Management 1. Springer.
29. Globocnik, D., Faullant, R., Parastuty, Z.: Bridging strategic planning and business model management – a formal control framework to manage business model portfolios and dynamics. Eur. Manag. J. **38**, 231–243 (2020). https://doi.org/10.1016/j.emj.2019.08.005
30. Hammer, M.: What is business process management? In: vom, B.J., Rosemann, M. (eds.) Handbook on Business Process Management 1, pp. 3–16. Springer (2015)
31. Vargo, S.L., Lusch, R.F.: Service-dominant logic: continuing the evolution. J. Acad. Mark. Sci. **36**, 1–10 (2008)
32. Turetken, O., Grefen, P., Gilsing, R., Adali, O.E.: Service-dominant business model Design for Digital Innovation in smart mobility. Bus. Inf. Syst. Eng. **61**, 9–29 (2019)
33. Porter, M.E., Kramer, M.R.: Creating Shared Value. Harward Business Review Jan-Feb (2011)
34. Tantalo, C., Priem, R.L.: Value creation through stakeholder synergy. Strateg. Manag. J. **37**, 314–329 (2016). https://doi.org/10.1002/SMJ.2337
35. Teller, C., Kotzab, H., Grant, D.B., Holweg, C.: The importance of key supplier relationship management in supply chains. Int. J. Retail Distrib. Manag. **44**, 109–123 (2016). https://doi.org/10.1108/IJRDM-05-2015-0072
36. Eisenhardt, K.M., Martin, J.A.: Dynamic capabilities: what are they? Strateg. Manag. J. **21**, 1105–1121 (2000). https://doi.org/10.1002/1097-0266(200010/11)21:10/11
37. Grant, R.M.: Toward a knowledge-based theory of the firm. Strateg. Manag. J. **17**, 109–122 (1996). https://doi.org/10.1002/SMJ.4250171110
38. Teece, D.J., Pisano, G., Shuen, A.: Dynamic capabilities and strategic management. Strateg. Manag. J. **18**(7), 509–533 (1997)
39. Wilden, R., Gudergan, S.P., Nielsen, B.B., Lings, I.: Dynamic capabilities and performance: strategy, structure and environment. Long Range Plan. **46**, 72–96 (2013)
40. Elgammal, A., Turetken, O., van den Heuvel, W., Papazoglou, M.: Formalizing and applying compliance patterns for business process compliance. Softw. Syst. Model. **15**, 119–146 (2016)
41. Weill, P., J., W.R.: IT Governance: how Top Performers Manage IT Decision Rights for Superior Results. Harvard Business Review Press (2013)
42. Elkington, J.: Accounting for the triple bottom line. Meas. Bus. Excell. **2**, 18–22 (1998). https://doi.org/10.1108/EB025539
43. Szopinski, D., Massa, L., John, T., et al.: Modeling business models: a cross-disciplinary analysis of business model Modeling languages and directions for future research. Commun. Assoc. Inf. Syst. **51**, 39 (2022)
44. Rosemann, M.: Explorative process design patterns. In: BPM 2020 LNCS, vol. 12168, pp. 349–367. Springer Science and Business Media Deutschland GmbH (2020)
45. vom Brocke, J., Seidel, S., Recker, J.: Green Business Process Management: Towards the Sustainable Enterprise. Springer (2012)
46. Sheth, J.N., Sethia, N.K., Srinivas, S.: Mindful consumption: a customer-centric approach to sustainability. J. Acad. Mark. Sci. **39**, 21–39 (2011). https://doi.org/10.1007/S11747-010-0216-3
47. Alomoto, W., Niñerola, A., Pié, L.: Social impact assessment: a systematic review of literature. Soc. Indic. Res. **161**, 225–250 (2022). https://doi.org/10.1007/S11205-021-02809-1
48. Kern, C.J., Poss, L., Kroenung, J., Schönig, S.: Navigating the moral maze: a literature review of ethical values in business process management. Bus. Process. Manag. J. **30**, 343–370 (2024). https://doi.org/10.1108/BPMJ-06-2024-0504/
49. Human, S., Neumann, G., Alt, R.: Human-centricity in a sustainable digital economy. In: Proceedings of the 54th HICSS (2021)

50. Chandrasiri, T., Croes, N., Bandara, W., et al.: Beyond profit: the role of benevolent business processes in building long-term firm success. In: BPM 2025 LNCS, vol. 16044. Springer (2026)
51. Mendling, J., Decker, G., Reijers, H.A., et al.: How do machine learning, robotic process automation, and Blockchains affect the Human factor in business process management? CAIS. **43**, 19 (2018). https://doi.org/10.17705/1CAIS.04319
52. Santoro, F.: Ethics at the Core: the future of BPM. In: Marrella, A., Resinas, M., Jans, M., Rosemann, M. (eds.) 22nd Int. Conference on BPM. Springer (2024)
53. Friedman, B., Kahn, P.H.: Human agency and responsible computing: implications for computer system design. J. Syst. Softw. **17**, 7–14 (1992). https://doi.org/10.1016/0164-1212(92)90075-U
54. Winkler, T.: Human values as the basis for sustainable software development. In: International Symposium on Technology and Society, pp. 37–42. IEEE (2018)
55. Fatima, S., Desouza, K.C., Buck, C., Fielt, E.: Public AI canvas for AI-enabled public value: a design science approach. Gov. Inf. Q. **39**, 101722 (2022). https://doi.org/10.1016/J.GIQ.2022.101722
56. Moody, D.L.: The method evaluation model: a theoretical model for validating information systems design methods. In: ECIS, vol. 2003, p. 79 (2003)
57. Davis, F.D.: Perceived usefulness, perceived ease of use, and user acceptance of information technology. MIS Q. **13**, 319–340 (1989). https://doi.org/10.2307/249008
58. Davenport TH, Redman TC How to Marry Process Management and AI. Harward Business Review Jan-Feb (2025)
59. Activism | Ben & Jerry's. https://www.benjerry.com/values. Accessed 1 June 2026

Agent Behavior Mining: Generative AI Agent Governance in Business Processes

Hoang Vu[1(✉)], Maximilian Körner[1], Adrian Rebmann[1], Gabriel Kevorkian[1], Michael Perscheid[1], Gregor Berg[1], and Timotheus Kampik[1,2]

[1] SAP, Berlin, Germany
{h.vu,maximilian.koerner,adrian.rebmann,gabriel.kevorkian, michael.perscheid,gregor.berg,timotheus.kampik}@sap.com
[2] Umeå University, Umeå, Sweden

Abstract. As organizations increasingly deploy generative AI agents to automate business processes, they face a governance dilemma: although these agents can increase operational flexibility, their non-deterministic nature challenges the control and standardization that Business Process Management seeks to enforce. This paper addresses this *invisible autonomy risk* by introducing *Agent Behavior Mining*, a governance capability that enables the application of process mining techniques to render generative AI agent decision-making observable and traceable. We (1) improve the understanding of generative AI agent behavior through an event data model that translates granular agent activities–including reasoning traces, tool usage, and token costs–into standardized process logs; (2) instantiate the data model in a multi-agent order-to-cash implementation, demonstrating how process managers can leverage agent logs to detect policy deviations and quantify operational variability; and (3) evaluate the perceived practical utility of the approach in an exploratory study with 18 industry practitioners. The results indicate that practitioners view behavioral transparency as a prerequisite for trust and consider the ability to examine agent reasoning as an important governance requirement for the next generation of AI-driven business processes.

Keywords: AI Governance · Generative AI Agents · Business Process Management · Process Mining · Agent Behavior Mining

1 Introduction

As organizations increasingly deploy generative AI (GenAI) agents in business processes, they face an *autonomy-control paradox* [16]: granting agents autonomy can increase operational flexibility, yet Business Process Management (BPM) governance depends on control, standardization, and auditable compliance [22]. This paradox exposes organizations to an *invisible autonomy risk*: unlike deterministic, rule-based task automation, GenAI agents driven by large language

F. M. Maggi et al. (Eds.): BPM 2026, LNCS 16936, pp. 755–772, 2027.
https://doi.org/10.1007/978-3-032-37777-7_44

models (LLMs) operate through opaque, non-deterministic reasoning. This opacity creates risks, including undetected policy violations and compliance failures, as well as unmanaged resource consumption [28].

Process owners face a *black box challenge*: they observe outputs but lack visibility into *why* decisions were made and *how* resources were used. Existing BPM governance mechanisms, built for repeatable execution, lack constructs to analyze probabilistic behavior of GenAI agents, rendering governance difficult.

To address these challenges, we propose *Agent Behavior Mining* (ABM) as a technology-enabled BPM governance capability that operationalizes visibility for organizational control. ABM establishes (1) *observability* of agent decisions, (2) *accountability* for agent outcomes, and (3) *controllability* by enabling targeted interventions such as policy guardrails and prompt updates. This supports the detection of compliance drift, containment of variable costs, and the development of empirical trust needed to scale agent adoption. We make three contributions:

1. **Event Data Model for Agent Behavior:** We introduce an event data model that provides the basis for turning agent activities—including reasoning traces and token costs—into standardized event logs, extending *eXtensible Event Stream* (XES) standards for GenAI agent governance.
2. **Operational Risk Detection:** We instantiate the data model by implementing a multi-agent Order-to-Cash (O2C) scenario, demonstrating how process mining methods (discovery, conformance, variant analysis) quantify compliance deviations, cost anomalies, and behavioral variability.
3. **Practical Utility Assessment:** We conduct an exploratory study with 18 industry practitioners assessing ABM's practical utility and how behavioral transparency influences trust and adoption decisions in critical workflows.

Our findings indicate that ABM is feasible and perceived as potentially useful, with behavioral variability emerging as a core governance concern. Practitioners expect transparency gains with ABM and view reasoning-level traceability as a prerequisite for trust. Overall, we argue that ABM can provide a practical foundation for moving from pilots to managed, accountable agent deployments.

The remainder of this paper is structured as follows. Section 2 reviews related work. Section 3 introduces a multi-agent system (MAS) scenario to illustrate the invisible autonomy risk. Section 4 derives requirements and introduces the ABM event data model. Section 5 instantiates the model, applies process mining techniques, and validates its utility through a practitioner survey. Finally, Sect. 6 discusses implications and limitations, followed by the conclusion in Sect. 7.

2 Background

Our work relates to governance in BPM, algorithmic accountability and transparency, process mining, and tracing GenAI agents.

Governance in BPM. Governance in BPM provides the structural mechanisms, decision rights, accountability frameworks, and performance metrics necessary to align process execution with organizational strategy. This aligns with the six core elements of BPM, which position governance as a distinct capability alongside strategic alignment and people [22]. Seminal work by Weill and Ross [27] defines IT governance as the specification of decision rights to encourage desirable behavior. In traditional BPM, this governance relies on bounded, deterministic workflows where deviations are treated as manageable exceptions. However, the deployment of GenAI agents creates an *autonomy-control paradox* [16]: organizations want agents to be flexible, yet this flexibility undermines the standardization that governance requires. Unlike earlier forms of task automation, such as Robotic Process Automation [2] or classical MAS [14], which followed strict rules and protocols, GenAI agents operate through probabilistic reasoning. This introduces risks of unobservable compliance drift that traditional governance mechanisms struggle to detect.

Algorithmic Accountability and Transparency. As algorithms take over operational tasks, they create what Kellogg et al. call a new *contested terrain of control* [15], widening an *accountability gap* [20] where the inability to monitor run-time execution prevents responsible AI operationalization. In the context of business processes, algorithmic accountability requires that an automated system provides a justifiable, auditable record of its decision-making—not just of *what* it decided, but of *why*. As AI systems become embedded in core business operations, this form of transparency becomes a prerequisite for effective governance. Prior work on human trust in AI shows that opaque decision logic can undermine trust, as process owners cannot readily understand why a system acted as it did [11]. Recent regulatory frameworks, such as the EU AI Act or the NIST AI Risk Management Framework, codify the need for explainability and auditability in high-risk AI systems. However, these frameworks often remain high-level principles without specifying operational mechanisms for business processes. We argue that ABM addresses this gap by repositioning agent governance from a strategic liability into an organizational capability, enabling the behavioral transparency necessary for accountable GenAI agent deployment.

Process Mining. Process mining offers a data-driven approach for process analysis by reconstructing execution flows from event logs, revealing deviations between designed processes and execution reality [1]. Standardized event data models like XES [12] provide the technical foundation but treat resources as passive executors. Recent advances in *Object-Centric Event Data* (OCED) [9] address complex, many-to-many interactions between business objects (e.g., orders, items, shipments) by storing objects and their attributes separately and linking them to events. However, while OCED captures the *structural* reality of interconnected processes, it also does not consider agents as first-class citizens. To address this, *Agent System Mining* (ASM) [25] centers the analysis on the agents themselves, advancing the field by discovering agent systems [26] and formalizing their interactions through *Agent System Event Data* (ASED) [24]. Extending the XES meta-model, ASED introduces a separate entity for agents, which captures the roles agents play and the organizations to which they belong [24].

Neither OCED nor ASED addresses the unique characteristics of GenAI agents, such as reasoning capabilities or token-based costs. More importantly, because both approaches extend the XES meta-model, they require adapted or entirely new process mining algorithms to leverage their data. In contrast, our event data model strictly adheres to the XES standard. By using tailored event types and attributes rather than altering the meta-model, our approach remains fully compatible with off-the-shelf process mining tools, significantly lowering the entry barrier for GenAI agent governance. Crucially, our proposed model does not conflict with existing frameworks. It can integrate with ASED, which allows agent metadata (e.g., roles) to be stored in separate entities and linked to events, and with OCED to explicitly model the objects agents interact with. In both cases, users can continue to leverage the GenAI-specific elements introduced in our framework.

Tracing GenAI Agents. Emerging *AgentOps* platforms, such as LangSmith, LangFuse, and Phoenix, already provide token-level cost tracking, latency monitoring, error detection, and reasoning traceability for LLM-based agents [6], often with OpenTelemetry compatibility [18]. However, a critical gap remains: designed primarily for machine learning and DevOps engineers, these platforms surface individual execution traces rather than structurally formatted event data [17]. Because they lack standard event log semantics, where low-level executions must be categorized into standardized, non-unique activity types, these traces cannot be readily analyzed using process discovery, conformance checking, or variant analysis.

To bridge this gap, ABM provides the structural foundation required to perform process mining directly on these agent traces. It structures the event data to capture essential information for process improvement on *accountability*, *decision causality*, *operational performance*, and *financial impact*. In contrast with other approaches, ABM treats *tool usage*, *LLM model configuration*, *token cost*, and *reasoning traces* as first-class process attributes. This capability allows organizations to move beyond monitoring *what* agents do, toward greater visibility into *how* they reason, *what* they cost, and *how well* their execution aligns with business policies. We argue that process mining can evolve from a retrospective analysis tool into an active enabler of *trustworthy AI*, providing the fact-based visibility that can help operationalize high-level governance principles.

3 The Invisible Autonomy Risk

Consider a MAS that manages an O2C process for a coffee shop (Fig. 1). Four autonomous GenAI agents coordinate the work: the *Order Agent* interprets customer intent, the *Inventory Agent* manages stock, the *Barista Agent* executes production, and the *Customer Service Agent* adjudicates refunds.

Traditional workflows embed business logic explicitly in process models, making decisions traceable to their triggering rules. GenAI agents internalize logic probabilistically, creating a governance blind spot: when the *Order Agent* handles *"large caramel latte, oat milk, refund if delayed,"* it interprets ambiguous

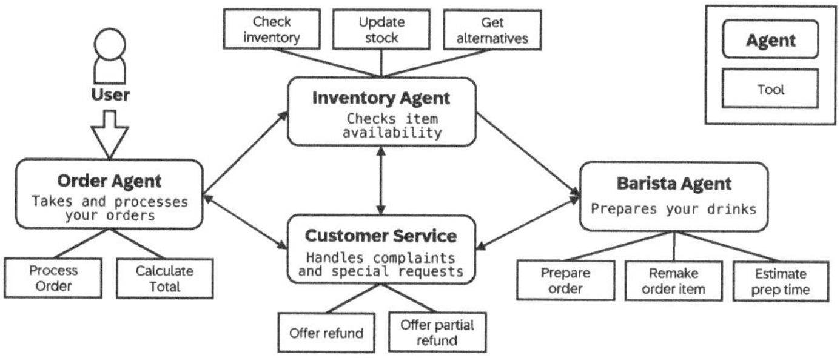

Fig. 1. Order-to-Cash multi-agent system with four GenAI agents.

terms without exposing its reasoning—leaving managers unable to verify whether a policy was applied or a new rule was hallucinated. This opacity compounds into systemic risk: unnecessary restocking by the *Inventory Agent* becomes an unmonitored financial leak, and adversarial prompts can silently induce the *Customer Service Agent* to grant undeserved refunds—both undetectable without access to reasoning chains.

Existing observability tools do not close this gap: agent frameworks produce verbose logs optimized for debugging, not governance. Without a standardized event data model translating these traces into process-level data, organizations cannot govern their digital workforce at scale—leaving them exposed to the very autonomy they sought to leverage.

4 Event Data Model for Agent Behavior

In this section, we present an event data model for GenAI agent behavior by operationalizing the ASED meta-model [24] and extending XES [12] with agent-specific attributes (Contribution 1). While XES provides the foundation for event representation, it lacks constructs for the specific characteristics of GenAI agents. We therefore derive requirements from operational goals and industry standards (Sect. 4.1) to instantiate our model (Sect. 4.2).

4.1 Model Requirements

We derive requirements from two perspectives: *goal-oriented*, addressing the operational objectives of GenAI agent deployment, and *system-driven*, stemming from agent architectures and observability standards like OpenTelemetry [18].

Goal-Oriented Requirements. Adopting the *Devil's Quadrangle* [8], we view GenAI agent deployment as process redesign targeting four performance dimensions: cost, time, quality, and flexibility.

R1 (Cost). Unlike traditional software, GenAI agents incur direct per-action costs via token usage. The model must correlate individual steps with their exact financial impact.

R2 (Time). Activity durations range from near-instantaneous tool calls to multi-step LLM reasoning with substantial latency. The model must capture this variance to assess time-efficiency relative to task complexity.

R3 (Quality). Non-deterministic execution requires tracing failures to their origin in user intent. The model must expose the reasoning chain from prompt interpretation to agent decision.

R4 (Flexibility). Agents autonomously adapt execution paths based on context. The model must capture distinct sequences to make behavioral variability observable.

System-Driven Requirements. These requirements stem from GenAI agent framework architectures, AgentOps observability standards, and the structural requirements of process mining platforms.

R5 (Behavior). The model must capture atomic events—LLM calls, tool executions, and agent delegation—to isolate logic errors from API or handoff failures and enable precise root cause analysis.

R6 (Semantics). Heterogeneous agent frameworks produce traces with varying structures and granularity. The model must align with OpenTelemetry GenAI semantic conventions [18] to ensure expressiveness and prevent information loss.

R7 (Interoperability). Proprietary formats create vendor lock-in. The model must produce standardized output aligned with the XES standard [12], accessible to any process mining platform without costly migration.

4.2 Event Data Model

Based on the above requirements, we propose the event data model visualized in Fig. 2. The model extends XES and can complement previous event data models—specifically ASED [24] and OCED [9]—by adding GenAI-specific event types and attributes that are absent from existing event data models (Sect. 2).

The event data model is organized as a shallow hierarchy where all events inherit standard attributes from the XES `concept`, `org`, and `time` extensions. The `prompt` event type captures user inputs within single or multi-turn conversations, establishing the context for subsequent system actions. In multi-turn settings, a case corresponds to a complete conversation session: each user turn is recorded as a `prompt` event sharing the same `case_id`, so the full dialogue context is preserved as a single traceable unit. All remaining event types are agent-centric, derived from an abstract `AgentEvent` type that factors out common attributes. Specific event types include `agent_start` and `agent_finish`, which track the overall execution time of agent tasks (R2). Agent operations are captured through `call_llm` and `execute_tool`, while `transfer_to_agent` represents delegation, a fundamental interaction pattern in MAS (R5). Standard attributes are adapted to support business-oriented analysis. `concept:name` and

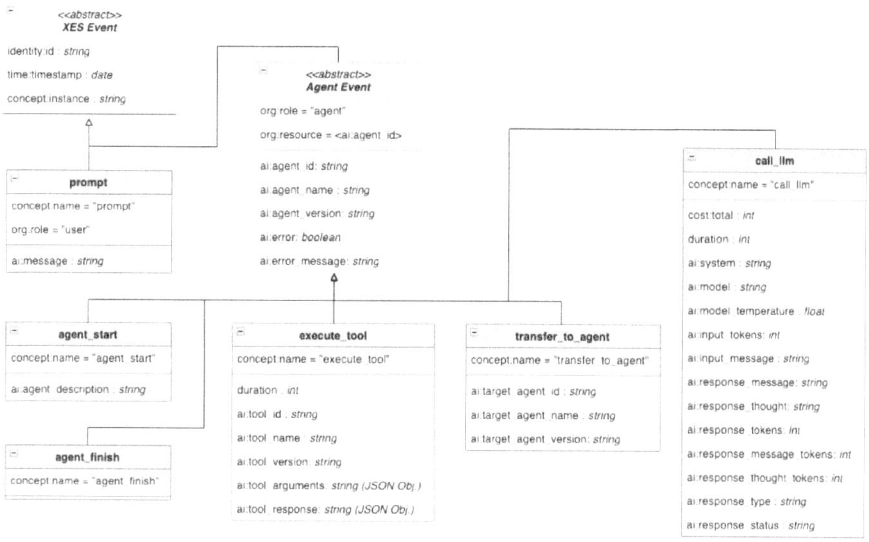

Fig. 2. Agent Event Concept Hierarchy

`concept:instance` capture the technical event type and a human-readable activity name, respectively (R4); grouping events by these names enhances understandability in process visualizations (e.g., DFGs). To assess costs (R1), various `*_tokens` attributes track LLM consumption. For long-running tasks, we introduce a `duration` attribute to complement the standard `time:timestamp` (R2). To ensure quality and transparency (R3), internal reasoning is tracked via `ai:response_thought`, while `ai:error*` attributes detail execution failures. Table 1 provides a traceability matrix summarizing how each requirement is satisfied by the attributes in the event data model. Crucially, most `ai:*` attributes align with OpenTelemetry GenAI semantic conventions (R6), ensuring broad compatibility and semantic coverage. Although not formally defined as an XES extension, the model builds upon the XES standard [12] and corresponding event logs are readily usable by process mining tools supporting the standard (R7). By satisfying these requirements (R1–R7), our model provides a practical foundation for capturing the nuances of GenAI agent behavior, enabling the practical application of existing process mining techniques, as demonstrated in the following section.

Figure 3 illustrates the model through a concrete O2C execution trace. The `prompt` event establishes case context via `ai_message` ("Can I please have an espresso for Max"), while `agent_start` attributes subsequent actions to the "Order Agent" via `ai_agent_name`. The `call_llm` event captures model identity (`gpt-4.1`), token counts (372 input, 27 output), and execution duration; `execute_tool` logs the action payload via `ai_tool_args` (`{"order":[...],...}`) to enable decision reconstruction. Together, these

Table 1. Requirement–attribute traceability

Req.	Dim.	Satisfying Attributes/Event Types
R1	Cost	`cost:total`, `ai:*_tokens` (`call_llm`)
R2	Time	`duration`, `time:timestamp`
R3	Quality	`ai:{error, error_message, response_thought}`
R4	Flexib.	`concept:name`, `concept:instance`
R5	Behavior	`agent_{start,finish}`, `call_llm`, `execute_tool`, `transfer_to_agent`
R6	Semant.	`ai:*` (OpenTelemetry-aligned [18])
R7	Interop.	XES `concept`, `org`, `time` [12]


```
- case_id: "2cd03d3f-..."
  events:
  - concept_name: "prompt"
    time: "2025-11-03T09:28:38.479"
    ai_message: "Can I please have
      an espresso for Max"
  - concept_name: "agent_start"
    time: "2025-11-03T09:28:38.482"
    ai_agent_name: "Order Agent"
    concept_instance: "[Order Agent] starts"
  - concept_name: "call_llm"
    time: "2025-11-03T09:28:38.483"
    duration: 691ms
    ai_model: "gpt-4.1-2025-04-14"
    ai_input_tokens: 372
    ai_response_message_tokens: 27
  - concept_name: "execute_tool"
    time: "2025-11-03T09:28:39.178"
    duration: 3.7s
    ai_tool_name: "process_order"
    ai_tool_args: '{"order":[...],"customer
         ":"Max"}'
    ...
```


Fig. 3. Agent event log fragment.

attributes allow organizations to distinguish cost sources, trace reasoning back to user intent, and analyze operational efficiency across execution variants.

5 Agent Behavior Mining in Practice

This section demonstrates how ABM can address governance challenges in GenAI agents. We implemented the multi-agent O2C scenario[1] to generate event data based on our proposed event data model; the repository includes the implementation, generated event logs, survey instruments, and evaluation dashboards to support conceptual reproducibility. Subsequently, we conducted a two-phase study where we first applied ABM to surface governance-supporting insights (Contribution 2). We then validated the perceived utility of these insights in a study with 18 industry practitioners (Contribution 3).

5.1 Study Design

We followed Design Science Research [13] and conducted an exploratory study to assess artifact utility in a realistic organizational context. We pursued *analytical generalization* [30], using the findings to refine theoretical propositions

[1] Repository contains the multi-agent O2C implementation, event log with 371 cases, survey instruments, and evaluation dashboards. https://github.com/A-rebmann/ABM/.

about GenAI agent governance rather than to estimate population-level effects. We recruited 18 practitioners via purposive sampling to ensure adequate diversity of roles: **strategic** (4 Enterprise Architects, 1 Product Owner, 1 IT Manager), **operational** (6 Consultants, 2 Process Managers, 1 Business Analyst), and **control** (1 Security Expert, 1 Data Scientist, 1 Developer). Participants' AI experience ranged from none (2) to beginners (9), intermediate (6) and expert (1), supporting assessment across skill levels.

Data Collection Protocol. We used a four-phase protocol to reduce common method bias [19]: (1) **Context setting** (30 min) with a standardized briefing on invisible autonomy risks and the O2C architecture; (2) **Simulation** (60 min) in which participants interacted with four GenAI agents (Order, Inventory, Barista, Service), generating 371 execution traces, including induced failures (e.g., hallucinations, loop errors), which we converted to our event data model via a framework-agnostic, extensible approach compatible with major agent frameworks; (3) **Insight elicitation** (30 min) where participants explored four dashboard views—process discovery, conformance checking, performance analysis, and variant analysis—and were asked to identify governance-relevant observations, including compliance violations, cost anomalies, and behavioral variants; and (4) a **Feedback phase** (20 min) comprising a structured survey [10] covering *Perceived Usefulness* (TAM-based [5]), *Transparency*, and *Adoption Barriers*, with Likert items and open-ended responses analyzed through thematic analysis [4]: responses were inductively coded by one researcher, grouped into recurring themes, and reviewed by a second researcher to ensure consistency. Validity was strengthened through anonymized responses, pilot-tested instruments, and immediate post-exercise administration. All agents utilized the `gpt-4.1` model; implications of this choice are discussed in Sect. 6.3.

5.2 Operational Risk Detection

Leveraging the 371 execution cases of the O2C coffee shop example, we instantiate our event data model to validate its utility for real-world governance.

Process Discovery. The directly-follows graph (Fig. 4) captures agent behavior and interactions (R5). The *Order Agent* orchestrates 1.2k instances via `process order` (348) and delegates to *Inventory Agent* (191) or *Customer Service Agent* (66); *Inventory Agent* executes `check inventory` (249) before handing off to *Barista Agent* (180). For process owners, this transforms the *black box* of agents' thought processes into a tangible process map.

Conformance Checking. Conformance checking supports failure analysis (R3) by comparing traces against specifications (Fig. 5): *Order Agent* omits `calculate_total`; *Barista Agent* skips `estimate_prep_time` and then executes `remake_order_item` (insertion violation). Model alignment with tracing standards (R6) and tool interoperability (R7) enables compliance rate assessment and violation categorization. This quantifies gaps between intended versus actual

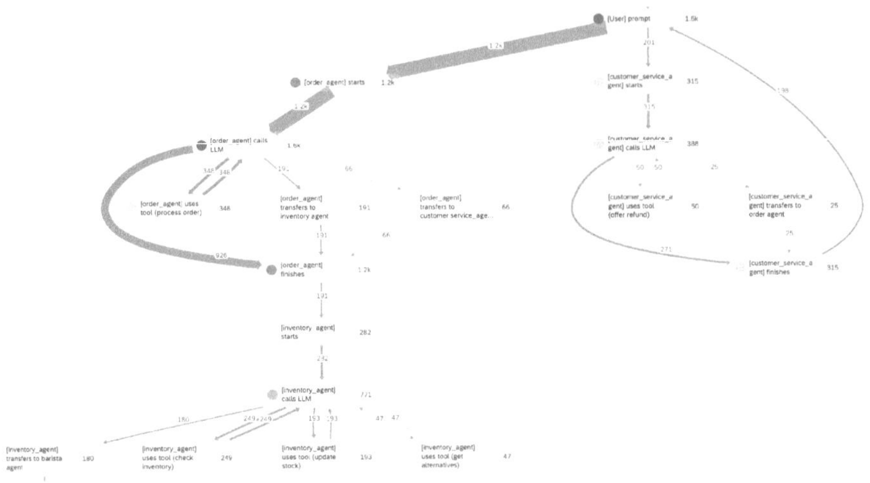

Fig. 4. Process Discovery: Visibility of Execution Patterns.

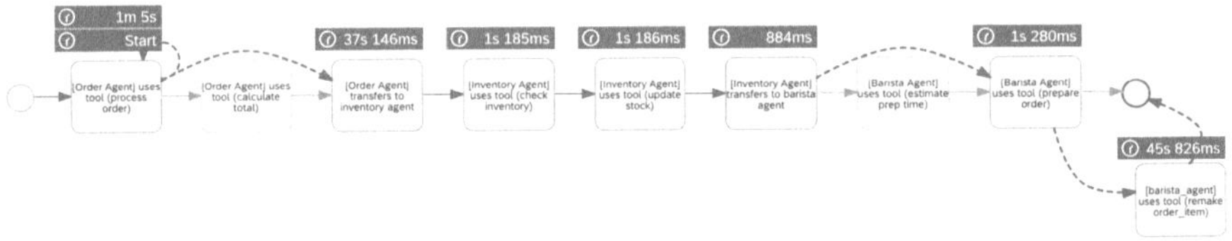

Fig. 5. Conformance Checking: Specification Alignment Assessment.

execution, supporting quality assurance for non-deterministic agents and revealing when agents prioritize customer satisfaction over compliance–providing risk officers with evidence for guardrails and retraining before systemic risks escalate.

Performance Analysis. The performance dashboard (Fig. 6) operationalizes cost and time metrics for agent behavior (R1R2). For 371 cases using `gpt-4.1-2025-04-14`, we observed 1 m 49 s average prompt-to-result time, 1.24 s average LLM-call latency, 4,294,572 total tokens (4,088,632 input, 205,940 output) at $9.82 ($8.18 input, $1.65 output), and 2.3 agents per case. Token attribution shows *barista*, *inventory*, and *order* agents each consuming 1M tokens, versus 604k for *customer service*. This enables process managers to benchmark deployments and target costly outliers.

Variant Analysis. Variant analysis captures GenAI agent variability (R4) (Fig. 7). For semantically similar orders, Variant A (13 cases): *Barista Agent* invokes `estimate_prep_time` before `prepare_order`; Variant B (4 cases): estimation skipped. Both complete via different reasoning paths. The model enables measuring variant frequency and comparing operational characteristics. We argue that ABM can make behavioral drift observable, potentially enabling orga-

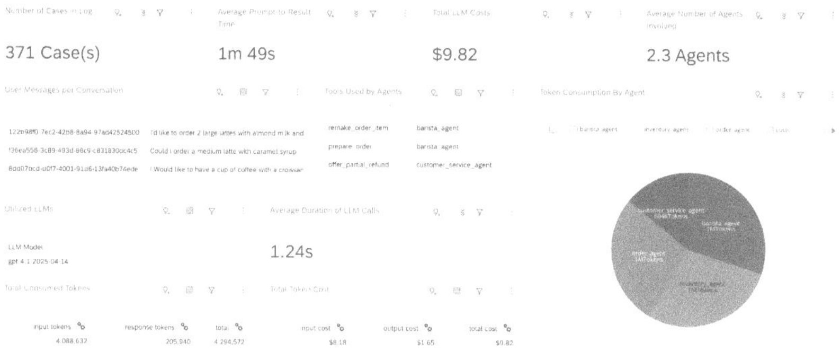

Fig. 6. Performance Analysis: Quantifiable Metrics.

Fig. 7. Variant Analysis: Observable Execution Differences.

nizations to standardize agent behavior and ensure consistent service quality despite probabilistic variance.

5.3 Practical Utility Assessment

We complemented the quantitative log analysis with practitioner feedback to assess the artifact's perceived value for decision-making (Fig. 8). Note that this phase measures TAM-based perceived usefulness rather than demonstrated governance outcomes: no objective measures of task accuracy or decision quality were collected, and participants reviewed dashboards under laboratory conditions without making real governance decisions.

Primacy of Variance Control. Behavioral inconsistency emerged as the primary governance concern: 78% of respondents selected *Process Variants* as the most valuable insight, exceeding *Agent Performance* (50%) and *Compliance* (44%) (Fig. 8a). This pattern held across roles (Process Managers to Security Experts), suggesting that, for these practitioners, visualizing and controlling

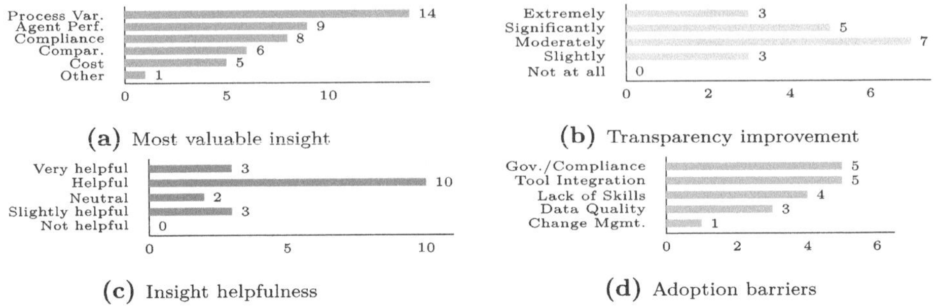

Fig. 8. Practitioner evaluation results (n = 18): perceived value, transparency improvement, insight helpfulness, and adoption barriers.

non-deterministic drift may matter more than traditional speed or cost metrics in probabilistic systems.

Transparency and Trust. ABM appeared to make opaque agent behavior more visible to participants: 83% reported "Moderate" to "Extreme" transparency improvement (Fig. 8b). This perception was consistent across experience levels; a Security Expert noted that granular logs were essential for *"seeing the stability of the process technically end-to-end,"* indicating that ABM can bridge trust barriers for risk-averse stakeholders.

Decision Support. 72% rated the insights "Helpful" or "Very Helpful" (Fig. 8c). Higher prior GenAI agent experience was associated with greater confidence in using ABM for governance, suggesting that novices may benefit from immediate transparency while experienced practitioners may be better able to translate granular traces into control actions. An Enterprise Architect requested *"a generated paragraph explaining... why the final result appeared,"* validating the capture of reasoning traces while motivating summarization layers in future iterations.

Adoption Barriers. Despite high utility, participants identified barriers to production deployment (Fig. 8d): *Governance/Compliance* (25%) and *Tool Integration* (25%) dominated, followed by *Lack of Skills* (20%). This distribution suggests a governance triad for adoption: alignment with compliance frameworks, seamless integration into established toolchains, and capability building. Qualitative feedback from a Consultant reinforced this, stressing that *"automating the integration of agent traces into [standard tools] will be key for adoption,"* implying that embedding agent governance into existing BPM suites may be preferable to standalone operational silos.

6 Discussion

As organizations move from GenAI agent pilots to productive deployments in their business processes, governance shifts from technical feasibility to operational control. This section examines the implications of ABM, outlines considerations for BPM research, and discusses validity threats and limitations.

6.1 Implications for AI Governance

Based on the observability and transparency gains evidenced above, we identify the following areas where ABM could support the governance of GenAI agents as an organizational capability, though empirical validation in production settings remains a direction for future work.

Operationalized Compliance. Conformance checking detected non-deterministic failures, such as the *Order Agent* omitting `calculate_total` (Fig. 5). This suggests ABM can facilitate the verification of rules and regulations through traceable events, potentially enabling compliance officers to pinpoint specific execution traces where policy guardrails were breached.

Strategic Resource Allocation. Performance analysis revealed token consumption disparities—a *Barista Agent* consuming nearly 1M tokens versus a *Customer Service Agent*'s 604k (Fig. 6). This granular attribution may enable targeted placement of *human-in-the-loop* controls at high-cost or high-risk points while leaving routine cases automated.

Data-Driven Continuous Improvement. By visualizing execution from user prompt to agent output (Fig. 4), ABM can enable an integration of GenAI agents into established improvement cycles. This can support a more systematic and disciplined approach to prompt engineering [23], shifting from ad-hoc changes toward metric-guided refinement.

Trust and Change Management. 83% of practitioners reported improved transparency (Fig. 8b), with security experts valuing the ability to inspect process stability. Prior research on trust in AI highlights transparency as an important antecedent to trust [11]. ABM dashboards may help stakeholders evaluate agent behavior and support change management in risk-averse settings, potentially supporting more collaborative human-agent working arrangements.

Organizational Readiness. Adoption barriers clustered evenly: tool integration (25%), governance/compliance (25%), and skills (20%) (Fig. 8d). This suggests that scaling ABM may require cross-functional teams blending process analysis with AI engineering, integration into existing BPM suites to avoid silos, and data minimization policies to address Personally Identifiable Information (PII) risks.

6.2 Implications for the BPM Discipline

Building on Dumas et al.'s agenda for explainable and adaptable AI-augmented process management [7], we argue that ABM suggests several disciplinary adaptations for extending process mining to autonomous agents.

Resources as Probabilistic Agents. Traditional BPM treats resources as executors of control flows [8]. GenAI agents embed decision logic within resources, blurring the model-executor distinction. Building on the agenda outlined above [7], this suggests BPM should complement control-flow modeling with explicit agent goals and constraints.

Prompts as Process Artifacts. Process logic has traditionally resided in BPMN models and business rules [8]. In agentic processes, prompts increasingly function as de-facto process definitions, shifting the focus of behavioral governance away from explicit models. This motivates treating prompts as governed artifacts—specified, versioned, and managed with rigor comparable to process models—while ABM links prompt configurations to execution outcomes.

From Efficiency to Safety. BPM optimization has historically prioritized time, cost, and quality [8]. Autonomous agents introduce risks (hallucinations, opaque reasoning) that elevate safety as a first-class governance concern. ABM provides event-level data to detect unsafe patterns and assess whether behavior remains within acceptable boundaries.

Variability as Intrinsic. BPM literature views variability as deviation to minimize [21]. For GenAI agents, behavioral variability is intrinsic—agents legitimately follow different reasoning paths for similar cases. ABM can make variants explicit, enabling distinction between adaptive problem-solving and non-compliant drift, and suggests the value of developing normative theories of *acceptable variance* for probabilistic executors.

From Episodic Redesign to Continuous Adaptation. Canonical BPM lifecycles assume episodic design-execution-redesign progression [3]. Agent behavior evolves with model or prompt updates even when process models remain stable. This suggests the value of lifecycle models with tighter feedback loops between mining insights and governed interventions.

6.3 Threats to Evaluation Validity

While the preceding implications highlight ABM's broad potential, our empirical findings must be interpreted with caution due to the exploratory nature and specific setup of our case study.

Construct Validity. We measured utility through perceived helpfulness. While perception precedes adoption [5], it does not equate to objective outcomes; participants may have rated potential value rather than immediate efficacy. The insight elicitation used guided dashboard exploration rather than structured tasks with measurable outcomes, and no objective measures of decision accuracy or response quality were collected, which limits the strength of this assessment.

Internal Validity. The workshop setting may have introduced novelty or desirability bias. We mitigated this through realistic scenarios, genuine failure modes, and anonymized responses; however, the short duration of the study limited the assessment of long-term learning curves.

LLM Choice. All evaluated agents ran on `gpt-4.1`. Cost, latency, behavioral variability, and reasoning trace availability are heavily model-dependent: models with stronger instruction-following may exhibit fewer conformance violations, while those without extended thinking yield no `ai:response_thought` data. The specific governance insights derived in this study may therefore differ across model families or versions.

External Validity. The 371 execution cases were generated during the workshop under controlled conditions with induced failures, rather than drawn from a production environment; the trace distribution may not reflect the frequency or variety of real-world agent behavior. Our evaluation relies on a single researcher-constructed multi-agent scenario and a limited sample of 18 practitioners, precluding statistical generalization. Furthermore, our findings reflect sequential ReAct agents [29]; alternative architectures may present distinct governance challenges. Replication across domains with different process characteristics, such as document-intensive, long-running, or regulatory-heavy processes, is needed to assess broader generalizability. Thus, our current results demonstrate analytical feasibility rather than universal applicability.

6.4 Limitations of the Proposed Framework

The proposed framework and its data model present several inherent limitations that point toward future research directions.

Reasoning Availability. ABM's failure analysis capability (R3) depends on the availability of `ai:response_thought`. However, some LLMs do not perform internal reasoning, and for models that do generate chain-of-thought content, such as OpenAI o-series or Anthropic Claude, availability may further depend on invocation parameters. Where reasoning summaries are absent, analysts can detect *that* an agent deviated but not *why*, as the agent's internal rationale remains hidden. In these cases, fields such as `ai:error_message`, `ai:tool_arguments`, and `ai:response_message` may help infer the underlying causes.

Event Granularity. The model captures events at the framework interaction level (`call_llm`, `execute_tool`, `transfer_to_agent`) but treats intra-LLM processing as atomic. Potential sub-steps, such as multi-step reasoning, within a single `call_llm` invocation remain unobservable as discrete events. Extracting finer-grained events from `ai:response_thought` content is a promising direction but requires NLP techniques and risks noise from non-deterministic generation.

Session-Level Case Notion. While each case can capture a multi-turn conversation, our framework assumes that all messages relate to the same instance or intent established by the user's first message, i.e., we assume that the user does not start a new topic or distinct request within the same conversation, but that all exchanged messages remain related to the initial prompt. Should topics change mid-session, the framework currently cannot dynamically split or join cases based on this shift. Furthermore, concurrent or asynchronous interactions may require richer case correlation beyond a shared session identifier.

Data Model Constraints. We based our event data model on XES, as it is formally standardized and widely supported in process mining tools. However, we acknowledge that once standardized, the OCED model may offer advantages for complex scenarios involving numerous concurrent agents operating on diverse, interacting business objects, which we aim to investigate in the future.

7 Conclusion

Generative AI agents promise a fundamental shift in operational efficiency, yet they introduce a critical risk: the invisible autonomy of non-deterministic decision-making. As organizations increasingly delegate core business tasks to autonomous systems, the inability to observe and audit why an agent acted becomes a strategic vulnerability. This paper bridges the gap between technical execution and organizational control by introducing Agent Behavior Mining (ABM), a governance capability that transforms opaque agent behavior into auditable, manageable process data.

We have contributed to the management of AI-driven processes by providing: (1) an event data model that captures GenAI agent behavior, enabling rigorous monitoring; (2) an implemented demonstration for detecting operational risks, such as policy deviations and costly variance; and (3) preliminary empirical evidence that ABM can enhance the transparency required for trusted adoption.

While our study represents an initial step within specific architectural boundaries, it charts a clear path for future work. Future work should (1) integrate ABM event logs with object-centric process mining [9] to reconstruct end-to-end case choreographies spanning human tasks and agent decisions in hybrid workflows; (2) design closed-loop governance mechanisms that trigger prompt updates or guardrail adjustments when ABM detects recurring conformance violations above a defined threshold; (3) conduct longitudinal field studies in real-world organizational deployments to quantify the return on investment of ABM and establish the break-even point for human-in-the-loop oversight; (4) develop methods to extract structured information from `ai:response_thought` content to enable reasoning-level conformance checking; (5) consider concurrent or asynchronous agent interactions, e.g., when multiple agents are operating in parallel. Also, while ABM is domain-agnostic, (6) applying it across process domains will establish boundary conditions for its governance capabilities.

Ultimately, ABM aims to empower organizations to move beyond blind reliance on autonomous systems. By establishing a foundation for verifiable governance, we aim to enable enterprises to harness the speed of GenAI without sacrificing the control necessary for sustainable operations.

Acknowledgments. The authors thank their SAP colleagues from Signavio and Research & Innovation, especially Thomas Kowark, Martin Sprengel, Ralf Teusner, and Robert Witt for their valuable contributions to this project.

References

1. van der Aalst, W.M.P.: Process Mining - Data Science in Action, 2 edn. Springer, Heidelberg (2016). https://doi.org/10.1007/978-3-662-49851-4
2. van der Aalst, W.M.P., Bichler, M., Heinzl, A.: Robotic process automation. Bus. Inf. Syst. Eng. **60**(4), 269–272 (2018)
3. van der Aalst, W.M.P., Netjes, M., Reijers, H.A.: Supporting the full BPM life-cycle using process mining and intelligent redesign. In: Contemporary Issues in Database Design and Information Systems Development, pp. 100–132. IGI Global Scientific Publishing (2007)
4. Braun, V., Clarke, V.: Using thematic analysis in psychology. Qual. Res. Psychol. **3**(2), 77–101 (2006)
5. Davis, F.D.: Perceived usefulness, perceived ease of use, and user acceptance of information technology. MIS Q. **13**(3), 319–340 (1989)
6. Dong, L., Lu, Q., Zhu, L.: AgentOps: enabling observability of LLM agents. arXiv preprint arXiv:2411.05285 (2024)
7. Dumas, M., et al.: AI-augmented business process management systems: a research manifesto. ACM Trans. Manage. Inf. Syst. **14**(1), 1–19 (2023)
8. Dumas, M., Rosa, L.M., Mendling, J., Reijers, A.H.: Fundamentals of Business Process Management. Springer, Heidelberg (2018). https://doi.org/10.1007/978-3-662-56509-4
9. Fahland, D., et al.: Towards a simple and extensible standard for object-centric event data (OCED)–core model, design space, and lessons learned. arXiv preprint arXiv:2410.14495 (2024)
10. Fowler Jr., F.J.: Survey Research Methods. Sage Publications (2013)
11. Glikson, E., Woolley, A.W.: Human trust in artificial intelligence: review of empirical research. Acad. Manag. Ann. **14**(2), 627–660 (2020)
12. Gunther, C., Verbeek, H.: XES - standard definition. BPM reports, BPMcenter.org (2014)
13. Hevner, A., Chatterjee, S.: Design science research in information systems. In: Design Research in Information Systems: Theory and Practice, pp. 9–22. Springer, Boston (2010). https://doi.org/10.1007/978-1-4419-5653-8_2
14. Jennings, N.R., Norman, T.J., Faratin, P., O'Brien, P.D., Odgers, B.: Autonomous agents for business process management. Appl. Artif. Intell. **14**(2), 145–189 (2000)
15. Kellogg, K.C., Valentine, M.A., Christin, A.: Algorithms at work: the new contested terrain of control. Acad. Manag. Ann. **14**(1), 366–410 (2020)
16. Mohlmann, M., Zalmanson, L.: Hands on the wheel: navigating algorithmic management and uber drivers' autonomy. In: Proceedings of the International Conference on Information Systems (ICIS) 2017. AIS (2017). https://aisel.aisnet.org/icis2017/DigitalPlatforms/Presentations/3, paper 3
17. Moshkovich, D., Mulian, H., Zeltyn, S., Eder, N., Skarbovsky, I., Abitbol, R.: Beyond black-box benchmarking: observability, analytics, and optimization of agentic systems. arXiv preprint arXiv:2503.06745 (2025)
18. OpenTelemetry Authors: Semantic conventions for generative AI systems (2024). https://opentelemetry.io/docs/specs/semconv/gen-ai/, version 1.36.0
19. Podsakoff, P.M., MacKenzie, S.B., Lee, J.Y., Podsakoff, N.P.: Common method biases in behavioral research: a critical review of the literature and recommended remedies. J. Appl. Psychol. **88**(5), 879 (2003)
20. Raji, I.D., et al.: Closing the AI accountability gap: defining an end-to-end framework for internal algorithmic auditing. In: Proceedings of the 2020 Conference on Fairness, Accountability, and Transparency, pp. 33–44 (2020)

21. Reijers, H.A.: Business process management: the evolution of a discipline. Comput. Ind. **126**, 103404 (2021)
22. Rosemann, M., vom Brocke, J.: The six core elements of business process management. In: vom Brocke, J., Rosemann, M. (eds.) Handbook on Business Process Management 1. IHIS, pp. 105–122. Springer, Heidelberg (2015). https://doi.org/10.1007/978-3-642-45100-3_5
23. Schmidt, D.C., Spencer-Smith, J., Fu, Q., White, J.: Towards a catalog of prompt patterns to enhance the discipline of prompt engineering. ACM SIGAda Ada Lett. **43**(2), 43–51 (2024)
24. Shen, Q., Polyvyanyy, A., Lipovetzky, N., Kampik, T.: Agent system event data: concepts, dimensions, applications. In: Conceptual Modeling, pp. 56–72 (2024)
25. Tour, A., Polyvyanyy, A., Kalenkova, A.A.: Agent system mining: vision, benefits, and challenges. IEEE Access **9**, 99480–99494 (2021)
26. Tour, A., Polyvyanyy, A., Kalenkova, A.A., Senderovich, A.: Agent miner: an algorithm for discovering agent systems from event data. In: Business Process Management, pp. 284–302 (2023)
27. Weill, P., Ross, J.W.: IT Governance: How Top Performers Manage IT Decision Rights for Superior Results. Harvard Business Press (2004)
28. Xi, Z., et al.: The rise and potential of large language model based agents: a survey. Sci. China Inf. Sci. **68**(2), 121101 (2025)
29. Yao, S., et al.: ReAct: synergizing reasoning and acting in language models. In: ICLR. OpenReview.net (2023). https://openreview.net/forum?id=WE_vluYUL-X
30. Yin, R.K.: Case Study Research: Design and Methods, vol. 5. Sage (2009)

Developing RPA UI Interaction Patterns to Guide Bot Design

Zeping Wang[1,2(✉)], Roy Jing Yang[1,2], Chun Ouyang[1,2], and Rehan Syed[1]

[1] School of Information Systems, Queensland University of Technology, Brisbane, Australia
{roy.j.yang,c.ouyang,r.syed}@qut.edu.au
[2] Centre for Data Science, Queensland University of Technology, Brisbane, Australia
zeping.wang@hdr.qut.edu.au

Abstract. Robotic Process Automation (RPA) automates task execution by mimicking user interactions between humans and computer systems. Nonetheless, the design and development of RPA bots often depend on the designers' personal expertise and preferences, resulting in solutions that are challenging to maintain, scale, and reuse. To address this challenge, we propose a collection of vendor-agnostic RPA design patterns that support systematic and consistent bot design, focusing on the UI interaction. Informed by the Object-Action-Interface Model, we derive and consolidate these patterns from literature in UI design, UI interaction design, and software engineering, and from research defining RPA concepts and terminology through an ontology and its extensions. These patterns are further substantiated through an analysis of three leading commercial RPA platforms to ensure their empirical relevance. To demonstrate the applicability of the identified patterns, we propose a pattern-guided approach for recommending suitable automation methods to support RPA bot design, and validate the approach using a publicly accessible RPA use case.

Keywords: Robotic Process Automation · Design Pattern · User Interface Interaction

1 Introduction

Robotic Process Automation (RPA) has attracted significant interest from both industry and academia [7,22], given its potential to deliver benefits such as reduced processing times, cost savings, and standardisation of task execution [8]. An RPA bot is a software agent [20] that automates process tasks by mimicking human interactions with computer user interfaces (UI) in a non-intrusive manner [6]. Conventional RPA bot design begins with identifying a target task and documenting human task executions in a Process Description Document (PDD). Human designers then translate PDDs into machine-executable actions specified

F. M. Maggi et al. (Eds.): BPM 2026, LNCS 16936, pp. 773–792, 2027.
https://doi.org/10.1007/978-3-032-37777-7_45

in a Solution Design Document (SDD) and implement them using an RPA tool. While vendors provide tool-specific PDD and SDD templates, these templates vary considerably in structure, terminology, and documentation requirements. Alternatively, data-driven approaches, commonly referred to as Robotic Process Mining (RPM), apply process mining techniques to UI interaction logs to extract and analyse fine-grained UI events generated during human task execution, thereby automatically discovering bot workflows [5]. However, in both conventional and data-driven approaches, systematic guidance for RPA bot design remains limited [11]. Whether designers manually construct bot workflows or intervene to refine automatically discovered ones, design decisions largely depend on individual expertise and preferences, often resulting in solutions that are difficult to maintain, scale, and reuse.

This research adopts a pattern-based design paradigm, which has been shown to be effective in addressing recurring challenges in software design [9]. Accordingly, we propose a set of vendor-agnostic RPA design patterns that focus on interactions between bots and user interfaces—a defining characteristic that differentiates RPA from other automation technologies, as our primary contribution. These patterns aim to provide RPA designers with structured guidance for consistently and effectively selecting suitable automation methods.

In developing these patterns, we introduce the Action-Object-Method-Context (AOMC) framework to guide their conceptualisation. The AOMC framework is derived from the Object-Action-Interface Model [18], capturing how bots interact with user interfaces. The patterns are grounded in literature on UI interactions, RPA ontology and taxonomy, and RPM, as well as industry standards for web and UI automation, and are further informed by an analysis of three leading commercial RPA platforms (UiPath, Automation Anywhere, and Power Automate) to ensure empirical relevance.

To demonstrate the applicability of the identified patterns, we propose a pattern-guided approach for recommending suitable automation methods based on RPA UI interaction patterns. The approach supports selecting automation techniques that address specific design requirements derived from UI logs. We illustrate this approach using a publicly available RPA use case, known as the *RPA Challenge—Fill Form* task, to demonstrate the pattern-based recommendation process and the applicability of the proposed RPA UI interaction patterns. The AOMC framework and the pattern-guided recommendation approach constitute additional contributions of this work.

The rest of this paper is organised as follows. Section 2 reviews related work. Section 3 outlines the research methodology, design principles, and conceptual foundation for RPA UI interaction patterns. Section 4 presents representative patterns. Section 5 introduces a pattern-guided approach for recommending suitable automation methods for bot design. Section 6 concludes and discusses future work.

2 Background and Related Work

RPA Ontology and Taxonomy. Völker and Weske [22] propose an ontology of RPA operations (ORPAO) to define RPA concepts and their relationships, later extending it to control-flow [23] and data perspectives [24]. They introduce an abstraction method based on ORPAO that generates compact yet expressive bot models, emphasising automation goals while abstracting from operational details [25]. While ORPAO contributes to the standardisation of terminology and modelling constructs, it offers limited guidance on bot design. Hohenadl *et al.* [10] present a taxonomy of low-level actions derived from UI logs, organising recordable user interactions into hierarchy actions, UI element actions, physical system actions and empty actions. This taxonomy provides a structured and comprehensive view of observable user behaviour captured in UI logs, though its focus remains primarily at the log-analysis level. We adopt this taxonomy as the basis for the Action building block in our AOMC framework (Sect. 3.4).

Bot Design Approaches. Conventional bot design follows a PDD-to-SDD pipeline that relies on tool-specific templates and designer expertise. In contrast, RPM applies process mining techniques on UI logs to automatically discover user interaction workflows, enabling a data-driven pipeline for bot design [3,12]. Agostinelli [3] proposes an approach that generates executable scripts directly from structured UI logs, replicating human actions without manual intervention. While this approach effectively captures observed behaviour, it largely reproduces recorded UI interactions and makes limited use of automation methods such as DOM parsing, script injection, or accessibility-layer interactions. Consequently, neither bot design approach provides systematic guidance for selecting among the automation techniques available in modern RPA platforms. Our pattern-guided approach addresses this limitation by abstracting from recorded interactions to automation objectives, recommending context-suitable automation methods, rather than replicating recorded interactions (Sect. 5).

Pattern-Based Design. Design of RPA bots, as software agents [20], can draw on established software engineering design patterns that capture reusable solutions to recurring design problems [9]. In the context of business processes, workflow patterns extend this paradigm by standardising modelling constructs and enabling reuse [17]. However, their focus is on providing guidance at the process level rather than on task execution at the UI level. Given that RPA bots operate through UI interactions, UI design directly influences their behaviour. Human-Computer Interaction research provides a strong foundation, particularly through established work on usability and interaction design. Van Welie [26] introduces UI design patterns that contextualise solutions by specifying when, how, and why they should be applied [27], and interaction patterns that address recurring UI interaction techniques. These interaction patterns encapsulate reusable knowledge at the interface layer, making them directly relevant to RPA, where bots operate through concrete UI interactions. However, these patterns are designed to support human user experience and interface usability. We adapt the pattern-based paradigm to bot-UI interaction, and employ a structured pattern representation consistent with established pattern literature (Sect. 4).

3 Designing RPA UI Interaction Patterns

3.1 Methodology

This research adopts the Design Science Research Methodology (DSRM) proposed by Peffers *et al.* [16]. It consists of six activities. **Activity 1 (Problem identification and motivation)** involved a comprehensive literature review that revealed a lack of systematic guidance for RPA bot design (Sect. 2). **Activity 2 (Definition of objectives)** established the objectives of proposing RPA UI interaction patterns by formulating three design principles grounded in the pattern-based design literature (Sect. 3.2). **Activity 3 (Design and development)** derived a conceptual foundation from the Object-Action-Interface (OAI) model [18], resulting in the Action-Object-Method-Context (AOMC) framework (Sect. 3.3), and four building blocks were populated using insights from scientific literature, industry standards, and an empirical analysis of three commercial RPA platforms (Sect. 3.4). Based on this foundation, the design patterns were developed and specified using a standardised pattern representation (Sect. 4). **Activity 4 (Demonstration)** introduces the pattern-guided recommendation approach and demonstrates its application using the *RPA Challenge—Fill Form* task as an illustrative use case (Sect. 5). **Activity 5 (Evaluation)** assesses the per-pattern tool support across analysed RPA platforms (Sect. 4) and involves the use case demonstration (Sect. 5), while identifying user studies and expert review as future work (Sect. 6). **Activity 6 (Communication)** disseminates the RPA UI interaction patterns, their conceptual foundation, the pattern-guided recommendation approach, contributing to the knowledge base on pattern-based RPA design.

3.2 Design Principles

The development of RPA UI interaction patterns is guided by three design principles drawn from established pattern-based design literature [1,9]. Accordingly, the proposed RPA UI interaction patterns should be:

- **Conceptual**: defined at an abstract level, being vendor-agnostic and domain-independent.
- **Comprehensible**: understandable without requiring extensive prior knowledge of programming.
- **Applicable**: readily applicable in practical RPA design, enabling designers to effectively apply the patterns when modelling, implementing, or analysing RPA bots.

3.3 Conceptual Foundation

To guide the conceptualisation of RPA UI interaction patterns, we adopt the Object-Action-Interface (OAI) model [18], which provides the structural vocabulary for specifying bot-UI interactions. It describes UI interaction using: **Object** (the entity manipulated), **Action** (the operation performed), and **Interface** (the medium through which the action is executed). In RPA, the same structure applies, with the bot rather than a human as the actor: the bot performs Actions on Objects through the rendered Interface. To distinguish the goal of an interaction from its target, we use **objective** for the interaction goal and reserve **object** for the target UI element.

Building on this foundation, we introduce the **Action-Object-Method-Context** (AOMC) framework to conceptualise the design of RPA UI interaction patterns. OAI is extended with two dimensions for RPA: the same action on the same object can be realised by different techniques (e.g., a click as a DOM event, a UI Automation invoke, or a simulated mouse click); and the available techniques depend on the execution environment (web, accessible desktop, or rendered-image-only application). AOMC therefore consists of four components: **Action** (what the bot does), **Object** (the target UI element), **Method** (how the action is performed), and **Context** (the execution environment). In this study, RPA UI interaction patterns are defined based on Actions, representing recurring interaction objectives. Object, Method, and Context components collectively specify alternative realisations of the same action across different scenarios.

3.4 Building Blocks

The RPA UI interaction patterns comprise four building blocks informed by the above AOMC framework. These building blocks are grounded in scientific literature and industrial standards, and were refined through an analysis of three leading RPA tools (UiPath, Automation Anywhere, and Power Automate) [21]. Table 1 summarises the conceptual and empirical sources of the building blocks.

Actions represent automation goals performed by bots based on the interaction taxonomies proposed by Hohenadl *et al.*. [10]. User interactions captured in UI logs can be categorised into: hierarchy actions, UI element actions, physical system actions, and empty actions. For bot design, physical system actions (e.g., attaching a USB device) fall outside the scope of what software bots can directly perform, and empty actions (e.g., background clicks) do not contribute to task execution. Hierarchy actions and UI element actions can be contextualised into: *extraction* (obtaining information from the interface), *modification* (changing UI element content or state), and *control* (changing the interaction context, such as opening an application, activating a control, or switching windows). The resulting Actions are summarised in Table 2.

Table 1. Conceptual and empirical sources for AOMC building blocks

AOMC component	Conceptual source	Empirical source
Action	Hohenadl et al. taxonomy of low-level actions [10]	Automation actions supported in tool design interfaces and documentation
Object	Abb & Rehse UI log data model [2]	UI element properties exposed by tools
Method	ECMA-262 [4], Component Object Model [14], Microsoft UI Automation Framework [15]	Tool-supported automation methods
Context	Völker & Weske ORPAO software/data context [23]	Tool-supported execution environments

Table 2. Actions in AOMC

Category	Action	Definition
Extraction	Find/Identify	determine the existence or location of a UI element
	Read/Extract	extract the value of a property or attribute from a UI element
	Observe	continuously monitor for specific events or changes
Modification	Write/Input/Update/Modify	input or update data into a field or element, including properties or attributes
	Delete/Remove	remove data or element
	Disable/Hide	modify an element's availability or visibility without altering its data content
Control	Open/Run	create an application instance or open a file
	Activate	trigger an element's primary action (e.g., click)
	Hover	initiate hover-based UI behaviour
	Switch context	change active window, tab, or view
	Scroll	adjust viewport to reveal hidden content
	Focus	focus on an element, usually used for preparing for input
	Refresh	reload a window or webpage

Objects correspond to interface elements upon which actions are performed. According to the user interaction log data model [2], UI elements form a tree-shaped hierarchy tied to an application and OS environment. AOMC distinguishes three object categories by how interfaces are exposed: HTML elements (web, via the DOM), UI elements (desktop or web exposing an accessibility hierarchy), and visual elements (rendered content via image recognition). The RPA platform analysis identified their commonly exposed properties (Table 3).

Methods describe the automation techniques used to execute actions on target objects. Application interfaces expose programmatic access through documented APIs [4,14,15]. When structured access is unavailable or unreliable, bots can operate through unstructured techniques, such as visual recognition and OS-level input event generation. From these foundations, four method families

Table 3. Objects and common properties in AOMC

Category	Identification properties	Content properties	State properties	Visual/Style properties	Event/Interaction properties
HTML Elements	id, class, name, type	InnerText, value	checked, disabled, hidden	width, color, font (external CSS may not be accessible)	Control patterns (Invoke, Selection, etc.)
Desktop UI Elements	Name, AutomationId/ID, Class, Control Type	Value, Text	Enabled/Disabled, Checked/Unchecked, Selected	Position, Size, Colour, Font	onclick, onchange
Visual Elements	Detected pattern, template, bounding box coordinates	OCR-extracted text	Presence/visibility on screen	Position, Size, Color, Font style	Clickable regions, hover-sensitive areas

are identified. Structured methods include *DOM parsing/manipulation*, which support content-level interaction with web pages, and *UI Automation tree parsing/manipulation*, which support accessibility-level interaction with structured UI elements. Unstructured methods include *visual recognition*, which extracts information from rendered screen content, and *hardware simulation*, which provides input through OS-level hardware events. AOMC organises method families in Table 4 by automation purpose: *extraction* (reading information from the interface) and *modification* (changing UI element content or state).

Context defines the execution environment that constrains available objects and methods. This building block is derived from ORPAO [23], which categorises RPA operation context into *software context* and *data context*. In AOMC, software context constrains available objects and methods, and data context is captured within Object properties. Based on how different software expose interfaces, three context categories are identified: *web environments*, *desktop environments*, and *screen environments*. These categories are summarised in Table 5.

4 RPA UI Interaction Patterns

Each pattern operationalises an Action from Table 2, capturing a recurring interaction objective in RPA bot design. The Object, Method, and Context components collectively specify alternative realisations (variants) across different scenarios. The resulting pattern collection comprises 13 patterns in three categories that mirror the Action categories. Table 6 summarises the full pattern collection, which is subject to extension as additional RPA interaction behaviours and automation scenarios are identified. To ensure consistency and reusability, all patterns use a standardised representation derived from established design literature [1,9], documents what the pattern enables (Description), its design intent (Motivation), its variants (Pattern variants), operational conditions (Applicability), how it is operationalised (Operation), an illustrative use (Example), and

Table 4. Methods in AOMC

Category	Method	Description
Extraction	HTML DOM parsing	Identifies HTML elements within the DOM and retrieves attributes or content directly from webpage elements.
	UI Automation tree parsing	Retrieves properties and control patterns of UI elements exposed through operating system automation or accessibility frameworks.
	Visual recognition	Detects UI elements rendered on screen and extracts information through pixel analysis, computer vision, or OCR techniques.
	Hardware simulation	Emulates human interaction to indirectly obtain information from the interface.
Modification	HTML DOM manipulation	Manipulates webpage content by executing scripts that directly modify DOM elements or their attributes.
	UI Automation tree manipulation	Programmatically modifies UI element properties or invokes interaction patterns exposed by automation frameworks.
	Hardware simulation	Simulates keyboard or mouse input by generating virtual hardware events or operating system messages to perform interactions.

tool support (Tool evaluation). Each pattern's tool evaluation indicates whether a tool directly supports (+), indirectly supports (i.e., requires a workaround or modification, ±), or does not support it (−). Three representative patterns are discussed in detail to exemplify the proposed pattern specification. The full pattern collection is documented in a supplementary file[1].

4.1 Extraction Patterns

Pattern 1: Find element
Description: The Find element pattern enables a bot to establish a reference to a target UI element. Within the AOMC model, Find element represents a UI action that connects a bot to an interface object through context-dependent identification methods.

[1] https://github.com/chester713/RPA-UI-interaction-patterns.

Table 5. Contexts in AOMC

Category	Context	Automation affordance
Web environment	Modern browser, Legacy browser, CEF-based application	HTML elements are accessible through DOM interfaces
Desktop environment	Applications exposing structured accessibility hierarchy	UI elements are accessible via system accessibility frameworks
Screen environment	Rendered GUI (virtual or physical desktop)	Visual elements are accessible via image recognition

Table 6. RPA UI interaction pattern catalogue

Category	No	Pattern	Category	No	Pattern
Extraction	1	Find element	Control	7	Open
	2	Read element		8	Activate
	3	Observe		9	Hover
Modification	4	Write element		10	Switch context
	5	Delete element		11	Scroll
	6	Disable element		12	Focus
				13	Refresh

Motivation: Automation tasks rarely operate directly on applications as a whole; instead, they interact with specific elements such as buttons, text fields, and images. The Find element pattern works as a prerequisite step by separating element identification from subsequent actions, such as input and click. Successful bot operation relies on accurate object identification.
Pattern Variants: Find element manifests with three variants, each corresponding to a distinct interaction strategy based on the interface abstraction level, as summarised in Table 7.

Table 7. Variants of Find element pattern

Variant	Action	Object	Method	Context
1	Find	HTML element	DOM parsing	Web environment
2		HTML element, UI element	UI Automation tree parsing	Web environment, Desktop environment
3		HTML element, UI element, Visual element	Visual recognition	Web environment, Desktop environment, Screen environment

Applicability: This pattern applies when the bot is capable of executing JavaScript or other scripts (e.g., IE COM), invoking accessibility patterns to traverse the UI hierarchy (e.g., the UI Automation Framework), and performing image recognition (either via an external service or a built-in function). Additionally, when the target object is an HTML element, the webpage must be fully loaded so that the bot can accurately capture all elements meeting the criteria.
Operation: The bot accesses the DOM, UI hierarchy, or the image rendered on computer screen, based on the user-defined criteria. It identifies the elements that match the criteria. The bot can be configured to identify the required number of elements (the first matching instance, all matching instances, or a specific number of instances). If no element matches the specified criteria, an appropriate error handling mechanism should be triggered.
Example: A bot automates data entry in a desktop accounting application. Before interacting with the "Submit" button, the bot applies selector criteria within the UI Automation hierarchy to locate the button element based on its accessible name and control type. Once identified, the resolved element reference is used by a subsequent interaction pattern to activate the button.
Tool Evaluation: See Table 8.

4.2 Modification Patterns

Pattern 4: Write element
Description: This pattern enables a bot to write, update, or insert data into a target UI element exposed through the application interface. It operationalises content modification within the user interface by allowing bots to change element values, attributes, or states during task execution.
Motivation: Data entry is one of the most frequent and fundamental requirements in RPA. Many business processes require bots to populate forms, submit credentials, update records, or transfer information between applications. These tasks involve inserting or updating values within user interface elements during task execution. The Write Element pattern addresses this recurring automation requirement by providing a generalised mechanism for modifying interface content regardless of how the target application exposes its user interface.
Pattern Variants: Write element pattern has three variants depending on the context, as summarised in Table 9.
Applicability: This pattern applies when a writable interface element exists, the element is successfully identified and activated for input, and at least one input method is feasible.
Operation: After locating the target element, the bot uses an appropriate writing strategy defined according to available interaction layers. The bot may directly assign the desired value through DOM or UI Automation manipulation when structured UI is accessible. If structured interaction is unavailable, the bot can resort to simulating virtual keystrokes. Either way, when input is completed, the bot removes the focus from the element to register the updated input.
Example: A bot automates entering product data into an ERP system. The bot first extracts the product name from a procurement file, then searches for

Table 8. Tool evaluation of Find element

Tool	Version	Variant	Score	Justification
UiPath	Studio 2025.0.161	Find HTML element	+	Directly supported through "Find Element"
		Find UI element	+	Directly supported through "Find Element"
		Find visual element	+	Directly supported through "Find Element"
Automation Anywhere	Community Editionbuild 29082	Find HTML element	±	Integrated into "Recorder: Capture"
		Find UI element	±	Integrated into "Recorder: Capture"
		Find visual element	±	Directly supported through "Image Recognition: Find image in window"
Power Automate	2.53.180.25037	Find HTML element	+	Directly supported through "If web page contains"
		Find UI element	+	Directly supported through "If window contains"
		Find visual element	+	Directly supported through "If image"

the product name input field in the ERP system. Once the input field is located, the bot invokes the UIA pattern `ValuePattern.SetValue(''G.Skill Trident Z5 64GB DDR5'')` to programmatically assign the value.
Tool evaluation: See Table 10.

4.3 Control Patterns

Pattern 8: Activate
Description: This pattern enables a bot to trigger or invoke the functional behaviour of a UI element. Activation typically corresponds to executing an action associated with an interactive control, such as clicking a button, selecting a menu item, opening a link, or confirming an operation. Rather than modifying content, this pattern operationalises control execution within the UI.
Motivation: Activation represents one of the most fundamental interaction objectives in automation, as many application workflows progress through user-triggered controls. Tasks such as submitting forms, navigating interfaces, confirming operations, or initiating system functions require explicit activation of

Table 9. Variants of Write element pattern

Variant	Action	Object	Method	Context
1	Write	HTML element	DOM manipulation	Web environment
2		HTML element,UI element	UI Automation manipulation	Web environment, Desktop environment
3		HTML element, UI element, Visual element	Hardware simulation	Web environment, Desktop environment, Screen environment

UI elements. This pattern addresses the frequent automation requirement of reproducing user-triggered commands.
Pattern Variants: Activate pattern has three variants depending on the context, as summarised in Table 11.
Applicability: This pattern applies when the target element represents an actionable control (e.g., button, hyperlink, menu item, checkbox), the element is successfully identified, and the automation environment allows interaction through exposed interface structures or input simulation mechanisms.
Operation: After locating the target element, the bot invokes the activation behaviour using an appropriate interaction method. For content-level activation, the bot triggers DOM events (e.g., `click()` or dispatched mouse events) through script execution. For accessibility-level activation, the bot invokes automation interfaces exposed by the UI Automation tree, such as the `InvokePattern.Invoke()` or equivalent control patterns, allowing direct execution without relying on visual rendering. When exposed interfaces are unavailable, activation can be achieved through hardware simulation, where the bot generates mouse clicks or keyboard events at the element's screen location, emulating physical user interaction.
Example: A bot processes purchase orders in an ERP system. After locating the *"Approve"* button through the UI Automation tree, the bot invokes the element using `InvokePattern.Invoke()`, triggering the approval action. The application then advances to the next processing state automatically.
Tool Evaluation: See Table 12.

5 Pattern-Guided Automation Method Recommendation

We propose a pattern-guided approach to support RPA designers in selecting suitable automation methods for specific automation scenarios identified from UI logs. Unlike RPM approaches (e.g., [3]), which generate scripts replicating recorded interactions, the approach provides guidance to designers about how an automation objective should be implemented and recommends context-suitable methods that may differ from those recorded from humans.

Table 10. Tool evaluation of Write element

Tool	Version	Variant	Score	Justification
UiPath	Studio 2025.0.161	Write HTML element	+	Directly supported through "Inject Js Script", "Invoke VBScript", "Set Text", and "Type Into".
		Write UI element	+	Directly supported through "Browser: Run JavaScript", "Browser: Call a JavaScript function", and "Recorder: Capture" when setting to "Set Text".
		Write through Hardware simulation	+	Directly supported through "Run JavaScript function on web page" and "Populate text field on web page"while input mode set to "Hardware Events".
Automation Anywhere	Community Edition build 29082	Write HTML element	+	Directly supported through "Set Text" and "Type Into" with input mode set to "Simulate", while both allow explicit overwrite/append control.
		Write UI element	±	Supported through "Recorder: Capture" when "Action to take on object" is set to "Set text".
		Write through Hardware simulation	+	Directly supported through "Populate text field in window" while "Simulate" is enabled. This function supports explicit overwrite/append control.
Power Automate	2.53.180.25037	Write HTML element	+	Directly supported through "Set Text", and "Type Into" when mode set to "Hardware Events".
		Write UI element	+	Directly supported through "Simulate keystrokes".
		Write through Hardware simulation	+	Directly supported through "Send keys" with enable "Send Text as hardware keys"

5.1 Approach Description

The proposed recommendation approach consists of two major stages: (1) Task interpretation, and (2) Method recommendation, as illustrated in Fig. 1.

Stage 1: Task Interpretation

Following the interaction log model [2], each recorded **event** in UI logs implies a corresponding **activity**, which contains an **action** and its target **object**. For instance, a `ClickButton` event yields **Action** *Click* and **Object** *Button*. Extracted Actions are mapped to the predefined action vocabulary (Sect. 3.4) for semantic consistency. *Click* is mapped to **Action** *Activate*, as it triggers the button's primary function. This mapping also applies to Objects, with *Button* resolving to an *HTML element*, a *UI element*, or a *visual element* depending on context.

Whilst automation methods can be recommended at the activity level, we suggest abstracting these activities into higher-level tasks, reflecting the automa-

Table 11. Variants of Activate pattern

Variant	Action	Object	Method	Context
1	Activate	HTML element	HTML DOM manipulation	Web environment
2		HTML element, UI element	UI Automation manipulation	Web environment, Desktop environment
3		HTML element, UI element, Visual element	Hardware simulation	Web environment, Desktop environment, Screen environment

Table 12. Tool evaluation of Activate pattern

Tool	Version	Variant	Score	Justification
UiPath	Studio 2025.0.161	Activate HTML element	+	Directly supported through "Click" while set input mode to "Chromium API".
		Activate UI element	+	Directly supported through "Click" while set input mode to "Simulate".
		Activate through Hardware simulation	+	Directly supported through "Click" while set input mode to "Hardware events".
Automation Anywhere	Community Edition build 29082	Activate HTML element	+	Directly supported through "Recorder: Capture" when "Action to take on object" is set to "Click", "Left click", "Right click", and "Double click".
		Activate UI element	+	Directly supported through "Recorder: Capture" when "Action to take on object" is set to "Click", "Left click", "Right click", and "Double click".
		Activate through Hardware simulation	+	Directly supported through "Mouse: Click".
Power Automate	2.53.180.25037	Activate HTML element	+	Directly supported through "Click link on webpage", "Click download link on web page".
		Activate UI element	+	Directly supported through "Click UI element in window".
		Activate through Hardware simulation	+	Directly supported through "Send mouse click".

tion objectives that bots should be designed to address. Recommending at this level aligns the patterns with what the bot is intended to achieve rather than with the specific interactions recorded, thereby making the underlying objectives explicit. To support this abstraction, we adopt the aggregation mechanism proposed by Völker and Weske [25] and introduce the *interaction segment*: consecutive actions that share a common execution environment (e.g., repeated value in `Application` column). These are grouped by a software context (the active application instance) and refined by data context (webpage URLs, UI identifiers, or DOM attributes), helping designers recognise coherent interaction segments.

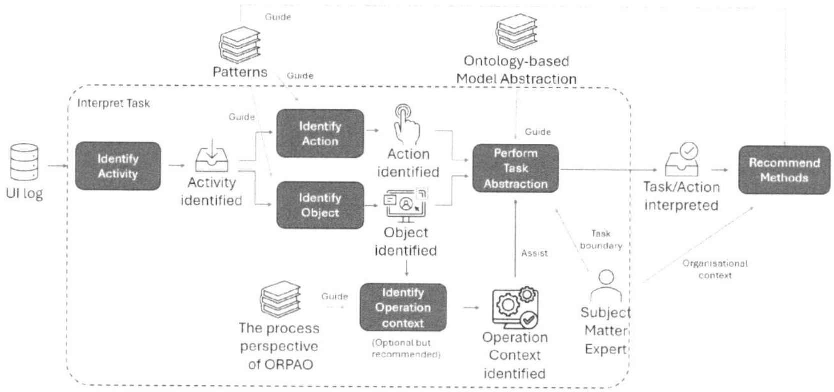

Fig. 1. Overview of pattern-guided automation method recommendation approach

Subject Matter Experts (SMEs) provide important input to task abstraction. They group the interaction segments and name the groups by the corresponding business tasks based on domain knowledge, i.e., further refining task boundaries.

Stage 2: Method Recommendation

Task interpretation yields either an identified task or a set of low-level UI activities; suitable automation methods can be recommended for either, and multiple methods may exist for the same interaction. From task interpretation, the action, object, and context can be identified; as such, the relevant pattern and its applicable methods can be matched.

When multiple methods are available, they are recommended in a priority order: (1) content-level > (2) accessibility-level > (3) visual recognition and hardware simulation. This ordering reflects differences in reliability and efficiency: content-level methods act directly on elements with the highest data fidelity; accessibility-level methods use the structured UI hierarchy and are generally more reliable in dynamic environments [19]; while visual recognition and hardware simulation methods operate on rendered images or simulated inputs, typically requiring more actions and resources for the same task [13].

For example, a webpage text field can be targeted via DOM parsing methods (HTML Tag or XPath), or through UI Automation tree parsing methods, since HTML elements are also exposed in the UI hierarchy and accessible via accessibility interfaces. In scenarios where neither is available, the bot may click at screen coordinates.

The priority is a general guideline rather than a strict hierarchy: it reflects the trade-offs among the underlying methods, and the most suitable method is context-dependent. For example, in single-page applications with auto-generated or volatile DOM identifiers, DOM-based selectors can become fragile across page renderings, and accessibility-level or visual methods may provide more stable identification. RPA designers can override the approach's recommendation based on their knowledge of the bot design target. This decision-making is also shaped by SMEs' input of organisational context information, such as compliance, governance, stakeholder roles, and responsibility allocation. Such context information is not recorded in UI logs but can be relevant to the bot design.

5.2 Use Case Demonstration

To demonstrate the applicability of the proposed pattern-guided recommendation approach, a benchmark automation scenario was conducted using the publicly available RPA Challenge—Fill Form task[2]. We completed it manually and recorded the execution with SmartRPA ActionLogger[3] The resulting UI log captures actions, interaction objects, and execution contexts, and serves as input to the approach.

A consecutive sequence of events from the log serves as an example snippet for analysis (Table 13). Only columns relevant to UI interactions are retained, and each event is assigned a unique identifier for reference.

Table 13. A snippet in UI log of RPA challenge—Fill Form task

Event ID	Application	Event	Workbook	Worksheet	Cell range	Element ID	URL
1	Microsoft Excel	Get Cell	challenge.xlsx	Sheet 1	A2		
2	Edge	clickTextField				lPupZ	https://rpachallenge.com/
3	Edge	Paste					https://rpachallenge.com/
4	Edge	changeField				lPupZ	https://rpachallenge.com/
5	Microsoft Excel	Get Cell	challenge.xlsx	Sheet 1	B2		

The UI log records events generated by the operating system or application interfaces in response to user interactions, serving as observable evidence of automation activities. A single event may correspond to one activity or implicitly represent several, which SME interpretation makes explicit. In the snippet, the **Event** column names each recorded event. Event 1 (`Get Cell`) records the retrieval of a spreadsheet cell from Microsoft Excel. Informed by the Event, Application, and Cell range columns (`Get Cell`, `Microsoft Excel`, `A2`), it is interpreted as an "*extract data from a cell in Microsoft Excel*" **activity**, mapping to **action** *Read* on **object** *data in cell.* The remaining context columns situate this *Read* within a desktop environment, so the **Read element** pattern applies with the **UI Automation tree parsing** method recommended.

Applying the **Read element** pattern requires the target object to be identified first, so an implicit "*find the target cell*" **activity** precedes the *Read*, giving **action** *Find* on **object** *cell.* Sharing the same context, it invokes the **Find element** pattern with the **UI Automation tree parsing** method.

Event 2 (`clickTextField`) occurs when the user clicks the input field (id `lPupZ`) before typing. This event corresponds to a "*focus on a text field on a*

[2] https://rpachallenge.com/.

[3] GitHub repository at https://github.com/bpm-diag/smartRPA.

web page": **activity**: **action** *Focus* and **object** *text field* element. The context marks it as an HTML element in a web environment. So the **Focus element** pattern applies with the **HTML DOM manipulation** method.

Unlike other events, no object can be directly identified from the log for Event 3. However, it is sandwiched between Events 2 and 4, which share identical context, so it is assumed to operate on the same object. Event 3 (`Paste`) thus pastes the extracted data into the text field focused by Event 2, giving **action** *Write* on **object** *Text field*; the **Write element** pattern applies with the **HTML DOM manipulation** method.

Event 4 (`changeField`) occurs when the text field is de-focused, thereby registering the change made by the *Write* action of Event 3. The corresponding activity is "*de-focus the text field*". According to the operation of the **Write element** pattern, the de-focus action is considered as an integrated step of the *Write* action, rather than a separate action.

Event 5 (`Get Cell`) retrieves cell `B2` from `Microsoft Excel` and is another instance of Event 1, so it receives the same recommendation, again preceded by an implicit *Find Cell* activity.

The context columns also reveal transitions between Excel and Edge (between Events 1–2 and 4–5). Although not explicitly logged, these implicit transitions must be handled for subsequent actions to target the correct application. From an operating system perspective, the browser itself is a desktop application, and therefore the **Switch context** pattern with the **UI Automation tree manipulation** method is recommended.

The recommendation outcome is illustrated in Fig. 2. For each recorded interaction, the figure depicts the corresponding action, object, context, applicable pattern, and the recommended automation method. The example demonstrates how the proposed patterns support designers in utilising automation methods beyond simply reproducing recorded UI actions.

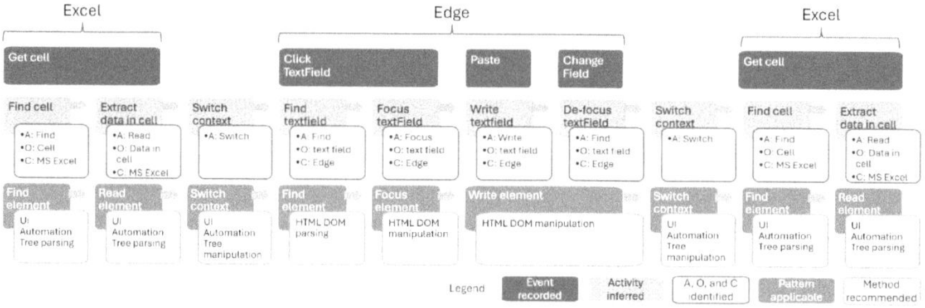

Fig. 2. Method recommendation outcome for the snippet of *RPA Challenge—Fill Form*

6 Conclusion

Despite the growing adoption of RPA, systematic guidance for bot design remains limited. This study addresses this challenge by introducing a set of RPA UI interaction patterns that provide structured guidance for selecting suitable automation methods when designing bots. To demonstrate the applicability of the patterns, we present a pattern-guided approach for recommending automation methods from UI logs, illustrated on an RPA Challenge use case. In contrast to existing RPM approaches that replicate recorded interactions, our approach abstracts from observed interactions to automation objectives and recommends context-suitable methods that may differ from the specific interactions recorded. The patterns are grounded in established conceptual frameworks and reinforced by an analysis of leading RPA platforms, making them vendor-agnostic yet practically relevant.

This work has several limitations. First, the patterns are derived from an analysis of a limited set of RPA platforms, and may not capture all variations in UI interaction techniques. Second, this work focuses specifically on UI interaction patterns in RPA bot design, excluding other aspects such as control flow, data handling, and system integration. Third, while the current evaluation comprises a per-pattern assessment of tool support across three commercial platforms and a demonstration using a publicly available use case, it does not fully assess usefulness or generalisability across diverse organisational contexts. Finally, the patterns represent an initial, non-exhaustive set of commonly observed solutions that may be extended as additional automation contexts and tools are examined.

Future work will extend the evaluation and refine the pattern collection: user studies and practitioner feedback will be used to assess the pattern's comprehensibility, usefulness, and potential for extension. Broader demonstration across heterogeneous contexts, including web, desktop, legacy, and virtualised application environments, will further evaluate the approach's applicability in diverse organisational settings. Failure modes for each pattern will be characterised systematically, drawing on practitioner experience to identify recurring breakdown scenarios.

Acknowledgments. We acknowledge contributions made by Arthur ter Hofstede to the conceptualisation of RPA UI Interaction Patterns.

References

1. van der Aalst, W.M.P., ter Hofstede, A.H.M., Kiepuszewski, B., Barros, A.P.: Parallel Databases. Workflow Patterns. Distrib. **14**, 5–51 (2003)
2. Abb, L., Rehse, J.R.: Process-related user interaction logs: State of the art, reference model, and object-centric implementation. Inf. Syst. **124**, 102386 (2024)
3. Agostinelli, S.: Generating Executable Robotic Process Automation Scripts from Unsegmented User Interface Logs, Springer (2024)
4. ECMA International: ECMA-262 ECMAScript® 2023 language specification. Tech. rep., ECMA International (2023). https://tc39.es/ecma262/

5. El-Gharib, N.M., Amyot, D.: Robotic process automation using process mining - A systematic literature review. Data Knowl. Eng. **148**, (2023)
6. Enríquez, J.G., Jiménez-Ramírez, A., Domínguez-Mayo, F.J., García-García, J.A.: Robotic process automation: a scientific and industrial systematic mapping study. IEEE Access **8**, 39113–39129 (2020)
7. Fernandez, D., Dastane, O., Omar Zaki, H., Aman, A.: Robotic process automation: bibliometric reflection and future opportunities. Eur. J. Innov. Manag. **27**(2), 692–712 (2024)
8. Flechsig, C.: The Impact of Intelligent Process Automation on Purchasing and Supply Management – Initial Insights from a Multiple Case Study. In: Logistics Management, 2021, Dresden, Germany, pp. 67–89. Springer (2021)
9. Gamma, E., Helm, R., Johnson, R., Vlissides, J.: Design patterns. Elements of reusable object-oriented software (1995)
10. Hohenadl, T., Völker, M., Stummeyer, C., Weske, M.: Democratizing robotic process mining: a conceptual framework for user actions, tasks, and RPA Bots. In: BPM2024: Blockchain, Robotic Process Automation, Central and Eastern European, Educators and Industry Forum, pp. 185–199. Springer (2024)
11. Jimenez-Ramirez, A., Reijers, H.A., Barba, I., Del Valle, C.: A Method to Improve the Early Stages of the Robotic Process Automation Lifecycle. In: Giorgini, P., Weber, B. (eds.) CAiSE 2019. LNCS, vol. 11483, pp. 446–461. Springer, Cham (2019). https://doi.org/10.1007/978-3-030-21290-2_28
12. Leno, V.: Robotic Process Mining: Accelerating the Adoption of Robotic Process Automation. UNIVERSITY OF MELBOURNE, Thesis (2022)
13. Leotta, M., Clerissi, D., Ricca, F., Tonella, P.: Visual vs. DOM-based web locators: An empirical study. In: ICWE, pp. 322–340. Springer (2014)
14. Microsoft: Component Object Model (COM). Tech. rep. (2024). https://learn.microsoft.com/en-us/windows/win32/com/component-object-model--com--portal
15. Microsoft: Microsoft UI Automation Framework. Tech. rep. (2024). https://learn.microsoft.com/en-us/windows/win32/winauto/entry-uiauto-win32
16. Peffers, K., Tuunanen, T., Rothenberger, M.A., Chatterjee, S.: A design science research methodology for information systems research. JMIS **24**(3), 45–77 (2007)
17. Russell, N., van der Aalst, W.M., ter Hofstede, A.H.M.: Workflow Patterns: The Definitive Guide. The MIT Press (2016)
18. Shneiderman, B., Plaisant, C., Cohen, M., Jacobs, S.M., Elmqvist, N.: Designing the User Interface: Strategies for Effective Human-Computer Interaction, 6th edn. Pearson Education Limited, Harlow, Essex, England (2018)
19. Spinak, J.: Model-based GUI automation. In: SoSyM (2025)
20. Syed, R.: Robotic process automation: contemporary themes and challenges. Comput. Ind. **115**, 103162 (2020)
21. Villa, A., Ray, S., Alexander, M., Joshi, S., Helsel, M.: Magic Quadrant for Robotic Process Automation. Report G00800237, Gartner (07 August 2024). https://www.gartner.com/document/code/800237
22. Völker, M., Weske, M.: Conceptualizing bots in robotic process automation. In: Ghose, A., Horkoff, J., Silva Souza, V.E., Parsons, J., Evermann, J. (eds.) ER 2021. LNCS, vol. 13011, pp. 3–13. Springer, Cham (2021). https://doi.org/10.1007/978-3-030-89022-3_1
23. Völker, M., Weske, M.: Ontology-supported modeling of bots in robotic process automation. In: ER 2022, pp. 239–254. Springer (2022)
24. Völker, M., Weske, M.: On the nature of data in RPA Bots. In: ICSOC 2023. LNCS, vol. 14420, pp. 29–37. Springer (2023)

25. Völker, M., Weske, M.: Ontology-based abstraction of bot models in robotic process automation. In: ER 2023, pp. 239–256. Springer Nature Switzerland (2023)
26. van Welie, M.: Task-based user interface design. Vrije Universiteit Amsterdam, Thesis (2001)
27. van Welie, M., van der Veer, G.C., Eliëns, A.: Patterns as tools for user interface design. In: TFWWG, pp. 313–324. Springer London (2001)

Cognitive Effects of Using LLMs for Process Modeling: An EEG Study

Neal Wessling[1,2](✉), Carl Corea[2], Elisabeth L'Orange[2], Mohammad Hossein Yamani[1], and Patrick Delfmann[2]

[1] Deloitte Neuroscience Institute, Hamburg, Germany
wessling.neal@gmail.com, mohammad.yamani@gmail.com
[2] University of Koblenz, Koblenz, Germany
{ccorea,delfmann}@uni-koblenz.de, elisabeth.lorange@gmail.com

Abstract. We investigate the cognitive effects of using large language models (LLMs) for BPMN process modeling. To this aim, we conducted electroencephalographic (EEG) experiments to measure brain activity during AI-assisted vs. non-assisted modeling. Participants were divided into two groups – a group with AI assistance ("LLM group") and a group that had to model manually ("brain-only group") – and were then asked to create a BPMN process model in the respective setting. Our results indicate significant differences in brain activities for the two groups, in particular, LLM users exhibited strongly reduced active cognitive engagement across multiple EEG measures such as theta power, alpha power and frontal alpha asymmetry. Furthermore, a post-experiment survey one week after the laboratory session also revealed that participants who used the LLM showed significantly lower memory and ownership of the model they had created. Combined with the EEG results, our findings raise concerns regarding the use of LLMs for process modeling, in particular, in regard to reduced cognitive engagement, weaker recall, and reduced ownership.

Keywords: Large Language Models (LLMs) · Process Modeling · EEG

1 Introduction

Recent advances in LLMs have been closely investigated for an application in Business Process Management (BPM). Here, the ability of LLMs to process large amounts of information and natural language make them a candidate technology for enhancing various subfields of BPM, such as process modeling [7,18,36], analysis [17,32], or process mining [2]. In this work, we focus on process *modeling*.

Using LLM for process modeling seems widely recognized as a straightforward use-case of LLMs, where the LLM can (semi-)automate process models based on input prompts and other context information [10,18]. The benefits of this approach, such as improved modeling efficiency and accessibility, have been

F. M. Maggi et al. (Eds.): BPM 2026, LNCS 16936, pp. 793–810, 2027.
https://doi.org/10.1007/978-3-032-37777-7_46

acknowledged in both academia and industry, with a rising number of works on this topic [10,13,18,36] and leading tool vendors such as *Camunda* or *Signavio* implementing AI-assisted modeling capabilities as new key features of their modeling suites (cf. [3,4]). However, despite these potential advantages, the adoption of LLMs for modeling may also introduce new *cognitive costs*, which we investigate in this work.

Research on using LLMs for *other* generative tasks (such as essay-writing) suggests that AI-usage may impair core cognitive functions such as critical engagement [8,16]. For example, participants in [16] increasingly shifted into "autopilot" strategies (copy–paste) while using LLMs rather than engaging in active knowledge construction. Furthermore, related research also suggests that LLM usage could have long-term cognitive effects, such as reduced neural engagement over time and weaker memory [16,28].

These findings raise strong concerns regarding the application of LLMs in process modeling, as they directly affect how humans can understand, evaluate, use, or even *think* about AI-generated process models in general. Orthogonal to efficiency gains, LLM-assisted modeling might introduce novel cognitive blind spots, with potential consequences for memory, error detection, and compliance. While LLM-based modeling systems are generally envisioned as human-in-the-loop approaches, this assumption might be difficult in practice, as human modelers might be ill-equipped to critically evaluate AI-generated models due to reduced cognitive involvement or automation bias [5,23]. Here, the long-term cognitive effects of LLM-usage (e.g., on memory) might also negatively affect how humans can recall and sustain process models, and might also erode model ownership over time.

To address these concerns, we present an EEG study investigating the *cognitive effects* associated with LLM-based process modeling (vs. manual modeling). Specifically, this project aims to quantify differences in brain activity, attention, memory, and error monitoring between AI-assisted vs. manual BPMN modeling. Here, our **research questions** are as follows:

1. How do participants' *brain activities* differ between manual modeling and LLM-assisted modeling?
2. How does the use of LLMs affect long-term *memory* of the modeled process?
3. How does AI involvement influence *subjective ownership* of the process model?

We continue with Preliminaries (Sect. 2), our Study Design (Sect. 3), Results (Sect. 4), Discussion (Sect. 5) and Conclusion (Sect. 6). *Supplementary material* including result data can be found at https://zenodo.org/records/20677120.

2 Preliminaries

2.1 (Problems of) LLMs for Process Modeling

Generative AI and LLMs are generally seen as a transformative technology for BPM, with many different applications within the BPM lifecycle [10,32]. One particular application is that of *process modeling*, where LLMs can assist in the generation or improvement of process models [10,13,15,18]. The actual modeling through the use of LLMs can be fully automated or semi-automated/iterative [11]. We will refer to both forms of modeling as "LLM-based modeling".

LLM-based modeling is seen as promising in both academia [9,10,12,14] and industry (e.g., [3,4]), and many recent works have shown that LLM-based modeling can be performed efficiently [13] and with high quality [6,11,14,19].

A common assumption in the literature is that LLM-based modeling systems are generally envisioned as human-in-the-loop approaches in which AI-generated models are reviewed by domain experts. For instance, the authors of [14] state that *"The [...] nature of LLMs necessitates that domain experts check the results."* (p. 647). However, based on recent research [16], the assumption that humans in the loop reliably act as rational gatekeepers may be idealizing real-world operators. In practice, when construction effort is outsourced to AI, users may be ill-equipped to critically evaluate AI-generated models, which may lead to uncritical acceptance of flawed models in LLM-based modeling. An important aspect here is that of *automation bias*, which describes the tendency of humans to overtrust automated recommendations as during LLM-use [23]. Automation bias during LLM-use has been shown in various professional domains such as healthcare [24] or law [5], and may even be increased by the typical anthropomorphism of LLMs – a core part of (conversational) LLM-based modeling [5,11].

The above considerations call for an investigation of the cognitive effects of AI-assisted modeling, as will be conducted in this work.

2.2 Cognitive Load

Process modeling requires various cognitive (sub)tasks such as processing information and problem-solving. Here, *cognitive load theory* [29] can be used to characterize the mental effort required for these tasks by humans. In general, it can be expected that the use of LLMs can lower cognitive load across several dimensions (intrinsic, extraneous, germane load) [16]. For example, participants in [27] experienced significantly less cognitive effort in an information-finding problem when using LLMs. However, these benefits can also be expected to come with (negative) trade-offs w.r.t. other dimensions, as LLM users might engage less deeply with the materials or problems. For example, in the study above [27] and [16], while participants using LLMs had less cognitive load during information finding, they also performed far worse when actually having to answer questions about the viewed materials or what they had submitted.

In this study, we measure *continuous* cognitive load during modeling (cf. [33]). To do so, we will measure physiological activity that is well-known in NeuroIS to indicate cognitive load [26], in particular, brain activity (EEG).

2.3 Electroencephalography (EEG)

EEG is a method for measuring electrical brain activity [1]. For this, electrical activity is recorded as waves by placing electrodes on the human scalp. To measure electrical waves, electrodes can be placed along different positions on the scalp, which follow an internationally standard naming system, with a letter from the area of the brain (pre-frontal (Fp), frontal (F), temporal (T), parietal (P), occipital (O), and central (C)) and an ID, e.g., "P4". Recorded waves can be classified into frequency bands. Among these frequencies, alpha (8–13 Hz) and theta (4–8 Hz) are especially central to neuroscience [1,16]: *(1) Alpha* waves are typically associated with relaxed wakefulness and decrease with attentional engagement and cognitive processing. Reductions in alpha (alpha suppression) reflect increased attentional engagement; *(2) Theta* waves are typically associated with mental effort, working memory engagement, and executive control. Increases in theta are often interpreted as markers of greater cognitive load and the recruitment of control-related neural processes during demanding tasks.

Regional patterns further refine cognitive interpretations. Frontal alpha reflects prefrontal processes such as inhibition and decision-making, and reductions in frontal alpha power accompany greater cognitive engagement [34]. Similarly, parietal theta is tightly linked to working-memory load and retrieval processes [1,16]. Furthermore, frontal alpha asymmetry (FAA) refers to the difference in alpha power between the left and right frontal cortical regions and is widely used as an index of relative frontal brain activity (cf. [25]). Greater left-frontal activation is typically associated with approach-related motivation, such as goal pursuit or positive affect. In contrast, greater right-frontal activation is commonly associated with withdrawal-related motivation such as avoidance.

2.4 Related Works

This work is mainly related to two streams of research, namely the evaluation of LLM application (in BPM) and cognitive effects research.

Regarding the first (*LLM application in BPM*), related works have predominantly focused on the evaluation of LLM application from a *technical* perspective, in particular, feasibility and the quality of the resulting models [2,6,7,11,15,19]. For example, those works typically assess the quality of LLM-created models (w.r.t. corresponding process descriptions) by means of objective model quality KPIs [6,11,15]. Other (technical) evaluation aspects also include the costs of usage, bias, or privacy concerns such as data leakage [7,12]. There are some works that evaluate cognition-related aspects of LLMs in BPM settings [11,14], however, these mostly focus on how humans perceive AI-generated output *post-hoc*, e.g., the perceived usefulness of AI-generated process models. To the best of our knowledge, this work is the first to investigate cognitive effects *during* the modeling process (in LLM-based modeling), in particular, with EEG measurement.

Regarding the second (*cognitive research*), while prior work has shown that LLM usage can alter cognitive processes in tasks such as essay-writing [16], it remains unclear whether these findings can be generalized to the domain of process modeling. The assumption that LLM-based modeling systems are generally envisioned as human-in-the-loop approaches (with human verifiers) seems especially problematic here – based on the recent research, we see evidence that the verification task in itself may introduce new forms of cognitive costs that could undermine precisely the envisaged verification role of the human modeler. In turn, a dedicated study on BPM modeling is needed to capture the specific cognitive implications of LLM-based modeling, which we present in the following.

3 Study Design

In this section, we outline the research objectives and the overall study design.

3.1 Hypotheses

Based on the research questions and findings from related works, we derive the following hypotheses:

Prior research shows that automation can reduce users' active cognitive processing [8,16]. In process modeling, LLM-generated suggestions may shift users from actively constructing models to (more passively) reviewing generated structures. We therefore hypothesize that LLM-supported modeling leads to lower neurocognitive engagement.

Hypothesis 1: *LLM-based process modeling is associated with lower overall neurocognitive engagement than manual process modeling.*

Furthermore, in terms of long-term effects, active knowledge construction typically strengthens memory, whereas more passive review can reduce processing depth and hinder internalization [16]. Thus, we hypothesize that LLM-based process modeling leads to lower long-term recall than manual modeling.

Hypothesis 2: *LLM-based process modeling is associated with reduced long-term recall of the modeled process compared to manual process modeling.*

Regarding ownership, when artifacts are partly generated by automated systems, users may feel less personal ownership (as also shown in related studies on AI-assisted creation [16]). Consequently, we hypothesize that LLM-based process modeling is associated with moderate to low subjective ownership.

Hypothesis 3: *LLM-based process modeling is associated with moderate to low subjective ownership of the resulting process model.*

Finally, regarding downstream effects of AI exposure on subsequent tasks, automation bias suggests that reliance on automated systems may reduce critical evaluation over time [5,22,24]. Similarly, a recent work shows that AI exposure in general may have negative carry-over to other tasks, e.g., exposure to AI support was shown to reduce perseverance when AI support is withdrawn [21]. In process modeling, this may impair users' ability to detect errors in models after LLM exposure. We therefore hypothesize that prior exposure to LLM-based modeling reduces error detection performance.

Hypothesis 4: *Prior exposure to LLM-based process modeling is associated with reduced error detection performance, consistent with automation bias.*

To test our hypothesis, we conduct an experiment, which we introduce next.

3.2 Experiment Procedure

Design and Tasks. The experiment follows a controlled between-subjects design to investigate differences of LLM-based vs. manual BPMN modeling. The independent variable is the modeling mode (LLM-assisted vs. manual). Dependent variables capture complementary cognitive, behavioral, and subjective outcomes, including neurocognitive engagement, cognitive load, memory recall, subjective ownership, and error detection performance.

The experiment structure is shown in Fig. 1. Participants were randomly assigned to one of two groups: a manual modeling group (*"brain-only group"*) and an LLM-assisted modeling group (*"LLM group"*). Participants completed two tasks under their condition, followed by a delayed post-experiment assessment.

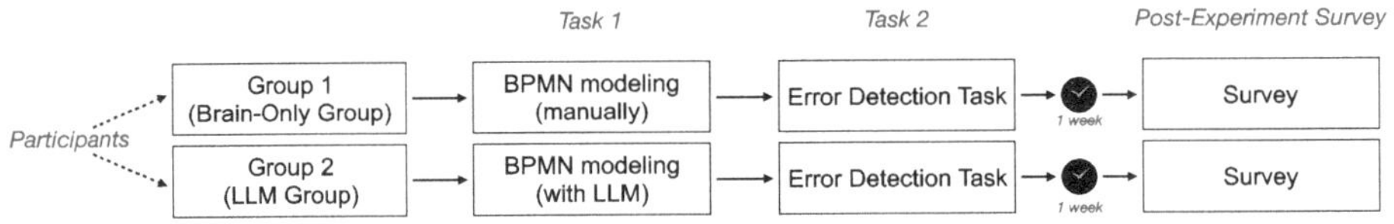

Fig. 1. Overview of Experiment.

Task 1 operationalizes the cognitive processes under different modeling modes (LLM-assisted vs. manual). In this task, participants were asked to create a BPMN process model based on a textual description. All participants were presented with the same description of a bank process on a dedicated screen. The process description was taken from the *Camunda for Research* repository[1], which offers process descriptions and solutions inspired by real-life processes. The concrete process description for Task 1 can be found in the supplementary material.

Participants created the BPMN model using a modeling environment displayed on a second screen. The brain-only group used the *bpmn.io* tool[2]. For the LLM Group, we used the open-source *BPMN Assistant* tool[3] [20], which enables LLM-based generation and modification of BPMN models. For the applied LLM, GPT 4.1 was used, as suggested by the developers of the BPMN Assistant tool.

Task 2 operationalizes error detection, enabling the investigation of potential automation bias following prior exposure to LLM-assisted modeling. In this task, all participants were shown a description of a billing process, and an *already finished* BPMN model on a second screen. The process was taken from the Camunda repository (cf. supplementary material). However, the researchers took the gold standard solution and purposefully added three mistakes (see Sect. 4.5). Participants were then asked to verify and approve the process model (however,

[1] https://github.com/camunda/bpmn-for-research.
[2] http://bpmn.io.
[3] https://github.com/jtlicardo/bpmn-assistant.

they were not told that mistakes were added purposefully). Participants were also instructed to fix errors (if applicable).

To assess memory retention, participants completed a **post-experiment survey** seven days after the laboratory session. Participants were not informed beforehand that we would send such a follow-up. The survey included structured questions about the experiment, e.g., to test whether they could remember the process descriptions. The survey can be found in the supplementary material.

Concrete Conducting of the Experiment. [4] Participants first provided informed consent and were then prepared for combined eye tracking and multichannel EEG acquisition. Participants wore a mobile eye tracking headset (*Tobii Pro Glasses*[5]). Eyetracking was used in our experiment to calibrate the EEG data, i.e., to segment which Task a participant was currently working on. Because the task involved two separate displays (one stimulus/prompt screen and one solution generation screen), the mobile tracker was calibrated to both monitors to ensure correct gaze angle mapping. EEG data was recorded using a 7 channel system headset (*Wearable Sensing DSI*[6]), where each of the seven primary electrode clusters consisted of 16 Ag/AgCl micro-electrodes averaged into a single channel-level signal. Electrode montage included *F3/F4 (frontal), C3/C4 (central), P3/P4 (parietal)*. A reference electrode (*Pz*) was positioned above the pelvic bone between *P3* and *P4*, with mastoid references at both earlobes.

Two resting-state baselines were collected: (1) Eyes-open fixation for 45 s, and (2) Eyes-closed rest for 45 s. This baseline served as subject-level calibration.

3.3 Measurements

Neurocognitive engagement was measured with EEG activity. Here, cognitive workload was characterized by evaluating both the *suppression* of alpha activity and the *enhancement* of theta activity relative to a resting baseline (alpha band power reflects the degree of cortical idling; theta band power indicates the involvement of control and working memory systems). Furthermore, frontal alpha asymmetry was computed to measure relative activity of the left versus right frontal cortex in order to quantify *approach motivation* vs. *avoidance*.

To measure memory recall, we asked participants a set of post-experiment questions about the process they had previously modeled (one week after the laboratory session). Responses were provided using numerical scales and open-text answers. A quantitative recall score was obtained through a coding procedure

[4] It is important context of this research that one of the lead authors (N. Wessling) is a licensed neuropsychologist. All experimental procedures, including the setup and application of EEG equipment, were conducted under the supervision of Mr. Wessling.

[5] https://www.tobii.com/products/eye-trackers/wearables/tobii-pro-glasses-3.

[6] https://wearablesensing.com/dsi-7/.

applied to the open responses. Details of the coding scheme and resulting scores are reported in the Results section.

To measure subjective ownership of the resulting process model, participants were asked to indicate their perceived level of ownership on a rating scale.

To measure error detection performance, participants were asked to evaluate a process model containing several embedded errors. Performance was operationalized as the number of errors correctly identified by each participant.

3.4 Participants

The study included 26 participants, of whom 17 had an academic background and 9 were professional experts. Importantly, the academic participants were primarily PhD students who have also been involved in teaching BPMN. Thus, all participants can be characterized to be novice to expert users with strong knowledge of BPMN (via a self-assessment, the average experience with BPMN of participants was around 3.5 years).

With 26 participants, the present study is within the typical sample size reported in EEG-based NeuroIS research. Two independent surveys [1,31] indicate that most EEG studies in this domain include between 10 and 24 participants, placing this study in the upper range of related EEG-based NeuroIS research.

Participants were recruited from three universities and through an open call on LinkedIn targeting BPMN professionals. Participation was voluntary and without incentive. The average age of participants was approximately 36 years.

4 Results

4.1 Preprocessing

EEG results were computed from preprocessed signals after filtering, artifact correction, and re-referencing. Each channel was segmented and transformed to the frequency domain using a discrete Fourier transformation, and power spectral densities were averaged. Alpha band power was computed by integrating the spectral energy between eight and twelve cycles per second. Theta band power was defined analogously for frequencies between five and eight cycles per second. Both metrics describe how much of the brain's electrical activity is concentrated in the respective rhythmic band. To normalize these values and make them comparable across subjects, each band specific power value was divided by the total spectral power of the channel, yielding relative alpha power and relative theta power. These measures express the proportion of cortical activity that is dominated by either alpha or theta rhythms.

Regarding measurements, alpha suppression and theta increase were quantified relative to resting state, while frontal alpha asymmetry (FAA) was computed as the log-transformed difference between right and left frontal electrodes. All metrics were computed at the electrode level and then averaged across predefined regions (frontal, central, parietal, and all signals over all electrodes) to obtain spatially aggregated indicators of cognitive load, attentional engagement, and prefrontal asymmetry.

Post-experiment survey responses were analyzed numerically, with group comparisons performed as described below.

4.2 General Results

As a general observation, the models produced by the LLM group were far less complex than those created by participants in the brain-only group. For objective comparison, we calculated a size-based complexity metric ('total number of activities and control-flow elements'; cf. [35]). The models of the brain-only group had an average complexity score of 33.3 (SD 9.1), and the models of the LLM group had an average complexity score of 22.4 (SD 3.3). Interestingly, despite producing more simplistic models, participants in the LLM group reported high satisfaction with the AI's initial suggestions. When explicitly asked to rate their satisfaction with the initial suggestion from 1–10, the LLM group reported an average score of 8.0 ($SD = 1.27$), indicating a positive perception of the AI-generated models. The changes made by LLM users to the initial AI suggestions were also mostly cosmetic (e.g., adding task types) rather than structural.

In many cases the AI tool did not include pools and lanes in its initially suggested model, and all participants but one in the LLM group did not care to add them to the final model. In contrast, virtually all participants in the brain-only group started their modeling by defining pools. This suggests that the modeling process itself may differ when supported by an LLM, with users potentially relying more on the AI's initial structure rather than actively constructing the model.

4.3 Neurocognitive Effects

Figure 2 shows EEG heatmaps[7] for the aggregated neural activity of all participants for the theta and alpha band.

In Task 1, participants in the brain-only group exhibited elevated frontal theta power. By contrast, participants in the LLM group showed substantially reduced theta power. Because frontal–midline theta is a well established neural marker of cognitive control and mental effort [30], the higher theta amplitude observed in the brain-only group indicates substantially greater exertion of executive resources. The lower theta values in the LLM group are consistent with under-engagement, reflecting shallower processing, or a cognitive "idling" during modeling [30]. Behaviorally, these neural patterns align with the observation that more than **30% of LLM group participants showed almost no increase (theta) above their resting baseline during Task 1**.

[7] The heatmaps show a top-view of the average spectral power (alpha, theta) of participants projected onto a 2-dimensional scalp surface (nose indicates direction of view). The color encodes the magnitude of oscillatory power at those scalp locations.

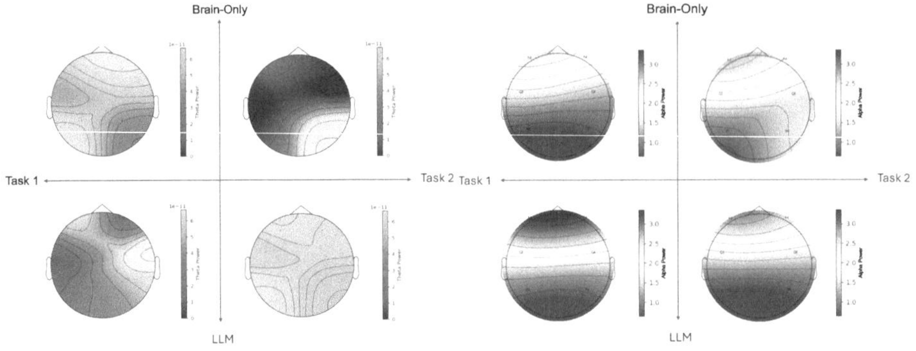

(a) Theta band (4-8 Hz) (b) Alpha band (8-12 Hz)

Fig. 2. Aggregated neural activity of all participants.

In Task 2, when both groups completed the task without LLM support, a divergence emerged in the direction of theta modulation. The brain-only group showed reduced theta relative to Task 1, indicating a practice related decrease in cognitive effort after already engaging deeply with the modeling task. In contrast, the LLM group exhibited an increase in theta power during Task 2 compared to their Task 1 baseline, demonstrating that participants had to exert considerably more mental effort once LLM support was removed. This dissociation indicates that prior reliance on AI did not ease subsequent independent task performance. Instead, the neural cost increased, consistent with evidence that sustained tool use can produce deficits when AI support is withdrawn [16].

Alpha band power provided an additional measure of cognitive workload. In Task 1, the brain-only group exhibited parietal alpha suppression (i.e., lower alpha power), which is characteristic of increased attentional engagement [25]. In contrast, the LLM group showed elevated alpha power during the same task, reflecting weaker working memory engagement [34]. In Task 2, alpha dynamics mirrored the theta findings. The brain-only group showed stable low alpha and continued suppression, reflecting practiced efficiency and flow. By contrast, the LLM group demonstrated fluctuating alpha (not visible in the static heatmap, see videos in the supplementary material), indicating unstable engagement. This could be caused by participants not having internalized task relevant strategies during Task 1.

In summary, theta and alpha indicate that LLM group participants engaged less deeply with the modeling task.

Conclusion 1. *LLM-based process modeling was associated with reduced active cognitive engagement compared to manual modeling.*

We furthermore computed frontal alpha asymmetry (FAA) (calculated as log(F4)-log(F3)) for additional insight into motivational orientation.

Figure 3 shows the FAA aggregated over all participants. During Task 1, the brain-only group showed stable positive FAA values, indicating relatively greater

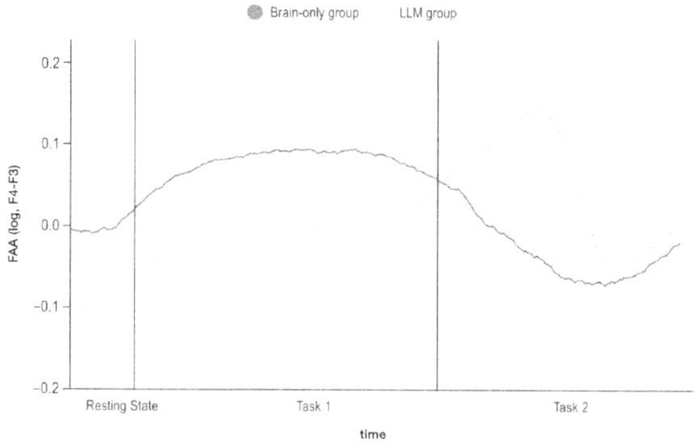

Fig. 3. FAA over both tasks.

left frontal activity associated with approach motivated engagement. At the onset of Task 2, FAA values of brain-only participants decreased to slightly negative levels. This decline may reflect reduced motivation, potentially attributable to the prior completion of a cognitively demanding task in Task 1. Consequently, the comparatively lower complexity of Task 2 may have required less attentional engagement.

The LLM group started with negative FAA values that gradually approached neutrality, indicating greater right frontal dominance typically associated with withdrawal motivation, reduced engagement, or avoidance. At the transition between tasks, FAA values in the LLM group first increased to positive levels, which may be interpreted as an initial elevation in motivational engagement in response to the new task. However, this increase was only short-lived, as FAA values subsequently shifted sharply toward negative levels. This reversal may reflect a (quick) rise in withdrawal tendencies when participants were required to work without AI assistance, potentially indicating task-related overwhelm.

Throughout Task 1, many brain-only participants showed FAA >0 for 50–70% of the time, indicating sustained approach-motivated engagement, whereas for many LLM users, FAA reached >0 for less than 40% of the time, reflecting more right-frontal dominance and reduced approach orientation.

Using the EEG data, we generated **videos** (see supplementary material) showing neural activity over time. The brain-only group displays a structured oscillatory pattern: during Task 1, frontal alpha suppression alternates with parietal theta bursts, reflecting efficient switching between control and memory. By Task 2, they enter a flow-like state with dominant parietal theta and stable alpha suppression. In contrast, the LLM group shows unstable dynamics: during Task 2, alpha “flickers” recur (instead of remaining suppressed), and theta and alpha are broadly distributed without coordinated frontal-parietal shifts. This pattern suggests inefficient neural recruitment once LLM support is removed.

4.4 Memory Recall and Ownership

To test memory recall, we conducted a post-experiment survey 7 days after the laboratory session. 11/13 participants of the brain-only group, and 10/13 participants of the LLM group responded. Participants described Task 1 in their own words, and responses were coded from −2 (no answer) to +2 (correct and detailed). Ratings were compared between groups using an independent-samples t-test; assumptions of normality and equal variance were met. The full responses and coded ratings are provided in the supplementary material.

Table 1. Memory recall comparison between groups (coded rating from −2 to 2).

	Mean	SD	t	p (2-tailed)
Group 1 (Brain-Only group)	0.45	1.51	2.3303	**0.0310**
Group 2 (LLM group)	−0.90	1.10		

Table 1 shows that the LLM group exhibited **significantly lower memory recall** compared to the brain-only group. Notably, 50% of participants in the LLM group indicated they had no recall (or provided no answer at all), whereas only one participant in the brain-only group provided no answer at all.

In addition, we also asked participants the question whether they could name the *roles* involved in Task 1 and Task 2. Responses were open-ended and were coded from −2 (no or completely incorrect answer) to 2 (fully correct answer). Partial points were assigned if some of the roles were correct. We then performed independent-samples t-tests with the coded ratings as dependent variable (data was normally distributed and met assumption of equal variance). Full responses and coding are provided in the supplementary material.

Table 2. Role recall comparison between groups (coded rating from −2 to 2).

		Mean	SD	t	p (2-tailed)
Task 1	Group 1 (Brain-Only group)	0.82	1.25	2.1821	**0.0419**
	Group 2 (LLM group)	−0.60	1.71		
Task 2	Group 1 (Brain-Only group)	1.09	1.14	2.2217	**0.0387**
	Group 2 (LLM group)	−0.30	1.70		

Table 2 shows that – for both tasks – participants of the LLM group had **significantly lower recall of the roles** than participants of the brain-only group. For Task 1, over 50% of LLM users could not recall *any* roles (only one brain-only participant failed to recall any roles). For Task 2, 40% of LLM users could not recall any roles. These results suggest that even basic structural information was not reliably retained by participants who relied on LLMs, highlighting clear differences in task comprehension and memory between groups.

The analyzed data suggests that participants of the LLM group had significantly lower memory recall and poorer retention of basic structural information like task roles. We also asked participants to give a subjective rating on how well they think they remembered Task 1. Though not reported in extensive detail due to space constraints, the data indicated that LLM users had significantly less trust in their recall.

Conclusion 2. *LLM-assisted modeling was associated with significantly lower memory recall.*

For the LLM group, we also asked participants to rate from 1–10 how much they thought the final process model from Task 1 was "their own result". LLM group participants reported a low sense of ownership, with a mean of 3.3/10 (SD = 2.45). This suggests that when process models are partially generated by automated systems, users may perceive them as less personally authored, even when they have the opportunity to edit the models. From a culture perspective, this reduced ownership could impact employees' identification with and responsibility for process artifacts, potentially affecting engagement, accountability, and the willingness to suggest improvements or maintain the models over time.

Conclusion 3. *LLM-assisted modeling was associated with low perceived ownership of the resulting process models.*

4.5 Error Detection Performance

As described in Sect. 3.2 ('procedure'), in Task 2 the researchers manually added 3 mistakes to the gold standard, shown in Fig. 4: (1) the timer event label did not match the description, (2) the order of two activities was switched, and (3) an outgoing message flow was attached to a wrong (non-sending) activity.

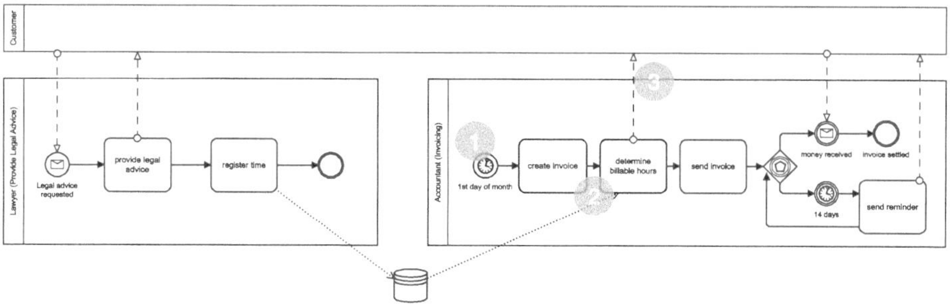

Fig. 4. Overview of the considered BPMN model for Task 2 (billing process), with mistakes (1)–(3) added by the researchers.

The number of mistakes found by the participants was then assessed by means of an independent sample t-test between the groups, with the number of corrected mistakes as the dependent variable. The numbers of corrected mistakes were normally distributed and the data met the assumption of equal variance. All models submitted by participants can be found in the supplementary material.

Table 3. Comparison of the number of detected errors (Task 2) between groups.

	Mean	SD	t	p (2-tailed)
Group 1 (Brain-Only group)	2.00	0.58	3.0907	**0.0050**
Group 2 (LLM group)	1.15	0.80		

Table 3 shows that participants in the LLM group could **detect significantly fewer errors** than participants in the brain-only group. These results suggest that prior exposure to LLM-based process modeling may reduce participants' ability to critically evaluate process models, consistent with effects commonly associated with automation bias.

Conclusion 4. *Prior exposure to LLM-based process modeling was associated with reduced error detection performance.*

5 Discussion

Our findings demonstrate systematic (neurocognitive) differences between participants who completed modeling tasks manually and those using LLM support. Across all EEG measures – theta, alpha, and FAA – the results indicate that LLM assistance reduces active cognitive engagement during modeling.

The reduced theta response observed in the LLM group in Task 1 suggests that participants engaged less deeply with the modeling task when AI provided support. The parietal alpha suppression observed in the brain-only group reflects heightened attentional allocation and sustained cognitive processing. In contrast, the relatively elevated alpha power in the LLM group during Task 1 suggests reduced attentional engagement, consistent with studies showing that alpha suppression indexes enhanced attentional focus and cognitive workload [34].

The sharp FAA reversal observed in Task 2 further illustrates the consequences of this offloading behavior. Having invested substantial effort in Task 1, the brain-only group showed decreased theta and increased alpha in Task 2, reflecting a classical practice effect. The LLM group showed the opposite pattern. These results suggest that participants relying on LLM support did not acquire durable task strategies and had to recruit executive and attentional resources more strongly once the external assistance was removed. Motivational dynamics, as indexed by FAA, reveal an additional consequence of LLM-support. The consistently positive FAA values in the brain-only group, especially in Task 1, indicate sustained approach motivation. In contrast, the LLM group's negative FAA trajectory – particularly in the second half of Task 2 – suggests a shift to withdrawal tendencies when participants had to work without AI assistance.

The post-survey aligns with this neurocognitive pattern. LLM users recalled less about their models one week after completion, which could stem from the reduced cognitive effort during modeling. Here, reduced recall could affect employees' ability or motivation for maintaining process models in the long-term.

In summary, the study identifies a consistent cost of LLM assistance. While LLM support may reduce immediate mental effort, it may also diminish attention, motivation, and long-term recall. Manual modeling promoted deeper cognitive involvement, more motivational orientation, better transfer to subsequent tasks, and stronger memory retention. These findings highlight the importance of designing AI tools that encourage active rather than passive collaboration.

As per design, our study has limitations. First, comparability of cognitive processes across conditions is inherently constrained. The concrete modeling for LLM and brain-only participants differed in cognitive demands, making direct neural comparisons difficult. Task 2 provides a more valid comparison as neither group used an LLM; however, effects observed in the LLM group cannot be attributed solely to a carry-over effect from prior AI assistance, as general trust in AI and automation bias may also have influenced their behavior. Second, the sample size of 26 participants introduces potential variance due to individual differences such as motivation, cognitive ability, familiarity with LLMs, and prior experience. These factors may interact with both task demands and neural responses. Nevertheless, the sample size is comparable to similar EEG-based NeuroIS studies [1,31] and included BPMN professionals. Finally, the study was conducted in a field-laboratory setting using dry EEG rather than a traditional gel-based laboratory system. While ecologically valid, it also increases susceptibility to movement artifacts, lower signal-to-noise ratios, and environmental variability. Capturing realistic workplace conditions necessarily limits the extent to which all confounds and noise sources can be fully controlled.

While we acknowledge that some reduction in cognitive effort (during LLM use) may in fact reflect a desirable efficiency gain due to cognitive offloading, our results do show a negative carry-over on long-term recall. Consequently, while the observed differences in neural activity should not be interpreted as evidence that one mode of work is universally superior to the other, our results show that LLM-based modeling was associated with a consistent cognitive cost. On this note, we fully acknowledge that alternative explanations to our findings cannot be fully excluded. For example, the lower model complexity of the LLM-generated process models could mean that those models required less effort to review in general, which could have impacted recall. Likewise, due to the experiment setup, participants in the brain-only condition may have benefited from a stronger cognitive "warm-up" effect before completing the subsequent error-detection task. Still, our evidence shows clear costs of LLM assistance for downstream tasks and long term recall, which underlines the presented interpretation. This is also in line with recent works showing that prior AI assistance introduces negative effects such as reduced task perseverance when the assistance is removed [21].

6 Conclusion

This study investigated the cognitive effects of LLM-based modeling. Our findings revealed systematic differences in cognitive engagement, memory retention, and analytical performance in LLM-based- vs. manual modeling. In particular, the results suggest that LLM support may reduce users' cognitive involvement and internalization of the modeled process. Two key findings here are as follows:

- LLM-based modeling promoted lower cognitive involvement, worsened transfer to subsequent tasks, and significantly weaker memory retention.
- 50% of participants in the LLM group indicated they had no recall of what they had done only 7 days after the experiment.

Overall, our findings raise some concerns regarding the use of LLMs for process modeling. While LLM-based tools can increase efficiency of modeling, these benefits may come with cognitive trade-offs that affect how users understand, evaluate, and remember the resulting models. Future work should therefore explore interaction designs that preserve the efficiency benefits of LLMs while encouraging active cognitive involvement instead of passive consumption.

Acknowledgments. We thank Stefan Wittenberg (HTW Berlin) and Matthias Weske (Hasso-Plattner-Institute) for generously allowing us to use their premises for parts of the experiments. We greatly thank the reviewers of this work for their time and feedback. Further, we thank the authors of [20] for their guidance in setting up the 'BPMN Assistant' tool used for the experiments. The applied data-collection methods (EEG) where non-invasive.

References

1. Bellos, C., Stefanou, K., Tzallas, A., Stergios, G., Tsipouras, M.: Methods and approaches for user engagement and user experience analysis based on electroencephalography recordings: a systematic review. Electronics **14**(2), 251 (2025)
2. Berti, A., Kourani, H., van der Aalst, W.M.: PM-LLM-Benchmark: evaluating large language models on process mining tasks. In: Delgado, A., Slaats, T. (eds.) ICPM 2024. Springer, Cham (2024). https://doi.org/10.1007/978-3-031-82225-4_45
3. de Boer, L.: SAP Signavio launches AI-assisted process modeler, text to process capability (2025). https://bit.ly/4lDmAgj
4. Camunda: BPMN copilot. https://docs.camunda.io/docs/components/early-access/alpha/bpmn-copilot/, camunda 8 documentation
5. Carnat, I.: Human, all too human: accounting for automation bias in generative large language models. Int. Data Priv. Law **14**(4), 299–314 (2024)
6. Drakopoulos, P., Malousoudis, P., Nousias, N., Tsakalidis, G., Vergidis, K.: Do LLMs speak BPMN? An evaluation of their process modeling capabilities based on quality measures. Computation **14**(1), 10 (2026)
7. Fettke, P., Houy, C.: Evaluating the process modeling abilities of large language models - preliminary foundations and results. arXiv preprint (2025)
8. Gerlich, M.: AI tools in society: impacts on cognitive offloading and the future of critical thinking. Societies **15**(1), 6 (2025)

9. Grohs, M., Abb, L., Elsayed, N., Rehse, J.R.: Large language models can accomplish business process management tasks. In: De Weerdt, J., Pufahl, L. (eds.) BPM, pp. 453–465. Springer, Cham (2023). https://doi.org/10.1007/978-3-031-50974-2_34
10. Hafner, A., Wittges, H., Rinderle-Ma, S.: Gen AI in business process management: a systematic review of the current state. In: AMCIS (2025)
11. Hörner, L.F., Möller, M., Reichert, M.: Automatically generating bpmn 2.0 process models from natural language process descriptions: challenges, framework, quality assessment. Bus. Inf. Syst. Eng., 1–25 (2026)
12. Kampik, T., et al.: Large process models: business process management in the age of Generative AI. arXiv preprint (2023)
13. Klievtsova, N., Benzin, J.V., Kampik, T., Mangler, J., Rinderle-Ma, S.: Conversational process modelling: state of the art, applications, and implications in practice. In: Di Francescomarino, C., Burattin, A., Janiesch, C., Sadiq, S. (eds.) International Conference on Business Process Management, pp. 319–336. Springer, Cham (2023). https://doi.org/10.1007/978-3-031-41623-1_19
14. Klievtsova, N., Benzin, J.V., Mangler, J., Kampik, T., Rinderle-Ma, S.: Process Modeler vs. Chatbot: is generative ai taking over process modeling? In: Delgado, A., Slaats, T. (eds.) International Conference on Process Mining, pp. 637–649. Springer, Cham (2024). https://doi.org/10.1007/978-3-031-82225-4_47
15. Klievtsova, N., Mangler, J., Kampik, T., Benzin, J., Rinderle-Ma, S.: How can Gen. AI empower domain experts in creating process models? In: WI. AISeL, (2024)
16. Kosmyna, N., et al.: Your brain on ChatGPT: accumulation of cognitive debt when using an ai assistant for essay writing task. arXiv preprint (2025)
17. Kourani, H., et al.: Leveraging large language models for enhanced process model comprehension. Decis. Support Syst., 114563 (2025)
18. Kourani, H., Berti, A., Schuster, D., van der Aalst, W.M.: Process modeling with large language models. In: van der Aa, H., Bork, D., Schmidt, R., Sturm, A. (eds.) International Conference on Business Process Modeling, Development and Support, pp. 229–244. Springer, Cham (2024). https://doi.org/10.1007/978-3-031-61007-3_18
19. Kourani, H., Berti, A., Schuster, D., van der Aalst, W.M.: Evaluating large language models on business process modeling: framework, benchmark, and self-improvement analysis. Softw. Syst. Model., 1–36 (2025)
20. Licardo, J.T., Tankovic, N., Etinger, D.: BPMN assistant: an LLM-based approach to business process modeling. arXiv preprint (2025)
21. Liu, G., Christian, B., Dumbalska, T., Bakker, M.A., Dubey, R.: AI assistance reduces persistence and hurts independent performance. arXiv preprint arXiv:2604.04721 (2026)
22. Lyell, D., Coiera, E.: Automation bias and verification complexity: a systematic review. J. Am. Med. Inform. Assoc. **24**(2), 423–431 (2017)
23. Mosier, K.L., Skitka, L.J.: Human decision makers and automated decision aids: made for each other? In: Automation and Human Performance, pp. 201–220 (2018)
24. Qazi, I.A., Ali, A., Khawaja, A.U., Akhtar, M.J., Sheikh, A.Z.: Automation bias in LLM assisted diagnostic reasoning among AI-trained physicians. medRxiv (2025)
25. Raufi, B., Longo, L.: An evaluation of the EEG alpha-to-theta and theta-to-alpha band ratios as indexes of mental workload. Neuroinformatics **16** (2022)
26. Riedl, R., Davis, F.D., Hevner, A.R.: Towards a NeuroIS research methodology: intensifying the discussion on methods, tools, and measurement. J. Assoc. Inf. Syst. **15**(10), 4 (2014)

27. Schmidhuber, J., Schlögl, S., Ploder, C.: Cognitive load and productivity implications in human-chatbot interaction. In: 2021 IEEE 2nd International Conference on Human-Machine Systems (ICHMS), pp. 1–6. IEEE (2021)
28. Shanmugasundaram, M., Tamilarasu, A.: The impact of digital technology, social media, and AI on cognitive functions. Front. Cogn. **2**, 1203077 (2023)
29. Sweller, J.: Cognitive load theory. Psychol. Learn. Motiv. **55**, 37–76 (2011)
30. Tan, E., et al.: Theta activity and cognitive functioning: integrating evidence from resting-state and task-related EEG research. Cogn. Neurosci. **67** (2024)
31. Torabi, M.: The application of EEG in web search research: a systematic review on sample size, active authors and results. JLIS **23**(2), 1–45 (2025)
32. Vidgof, M., Bachhofner, S., Mendling, J.: Large language models for bpm: Opportunities and challenges. In: Di Francescomarino, C., Burattin, A., Janiesch, C., Sadiq, S. (eds.) BPM, pp. 107–123. Springer, Cham (2023). https://doi.org/10.1007/978-3-031-41623-1_7
33. Weber, B., Neurauter, M., Pinggera, J., Zugal, S., Furtner, M., Martini, M., Sachse, P.: Measuring cognitive load during process model creation. In: Davis, F.D., Riedl, R., vom Brocke, J., Léger, P.-M., Randolph, A.B. (eds.) Information Systems and Neuroscience. LNISO, vol. 10, pp. 129–136. Springer, Cham (2015). https://doi.org/10.1007/978-3-319-18702-0_17
34. Woodman, G.F., Wang, S., Sutterer, D.W., Reinhart, R.M., Fukuda, K.: Alpha suppression indexes a spotlight of visual-spatial attention that can shine on both perceptual and memory representations. Psychon. Bull. **29**(3) (2022)
35. Zhou, C., Zhang, D., Chen, D., Liu, C.: Business process complexity measurement: a systematic literature review. IEEE Access **11**, 47940–47955 (2023)
36. Ziche, C., Apruzzese, G.: LLM4PM: a case study on using large language models for process modeling in enterprise organizations. In: Di Ciccio, C., et al. (eds.) Business Process Management: Blockchain, Robotic Process Automation, Central and Eastern European, Educators and Industry Forum, BPM 2024. LNBIP, vol 527. Springer, Cham (2024). https://doi.org/10.1007/978-3-031-70445-1_35

Author Index

F. M. Maggi et al. (Eds.): BPM 2026, LNCS 16936, pp. 811–812, 2027.
https://doi.org/10.1007/978-3-032-37777-7